10	IB 11	IIB 12	IIIA 13	IVA 14	VA 15	VIA 16	VIIA 17	
								He 4.002602
			5 B 10.811	6 C 12.011	7 N 14.0067	8 O 15.9994	9 F 18.998403	10 Ne 20.179
			13 Al 26.98154	14 Si 28.0855	15 P 30.97376	16 S 32.066	17 Cl 35.453	18 Ar 39.948
28 Ni 58.69	29 Cu 63.546	30 Zn 65.38	31 Ga 69.723	32 Ge 72.59	33 As 74.9216	34 Se 78.96	35 Br 79.904	36 Kr 83.80
46 Pd 106.42	47 Ag 107.8682	48 Cd 112.41	49 In 114.82	50 Sn 118.710	51 Sb 121.75	52 Te 127.60	53 I 126.9045	54 Xe 131.29
78 Pt 195.08	79 Au 196.9665	80 Hg 200.59	81 Tl 204.383	82 Pb 207.2	83 Bi 208.9804	84 Po (209)	85 At (210)	86 Rn (222)

63 Eu 151.96	64 Gd 157.25	65 Tb 158.9254	66 Dy 162.50	67 Ho 164.9304	68 Er 167.26	69 Tm 168.9342	70 Yb 173.04	71 Lu 174.967
95 Am (243)	96 Cm (247)	97 Bk (247)	98 Cf (251)	99 Es (252)	100 Fm (257)	101 Md (258)	102 No (259)	103 Lr (260)

Sponsoring Editor: Richard Stratton
Development Editors: Anne Wightman, June Goldstein
Project Editor: Susan Lee-Belhocine
Assistant Design Manager: Karen Rappaport
Production Coordinator: Frances Sharperson
Manufacturing Coordinator: Sharon Pearson
Marketing Manager: Michael Ginley

To Robert and Michael and their grandparents

Cover image illustrates mercury(II) oxide, when heated, decomposing to mercury and oxygen gas. This is a classic example of a decomposition reaction, which is further discussed on page 211.

Cover photograph by James Scherer.

Printed in the U.S.A.

Library of Congress Catalog Card Number: 91-71959

ISBN: 0-395-47205-9

ABCDEFGHIJ-R-954321

Basic Concepts of Chemistry

Fifth Edition

Alan Sherman
Middlesex County College

Sharon J. Sherman
Rutgers University

Leonard Russikoff
Middlesex County College

Houghton Mifflin Company Boston Toronto
Dallas Geneva, Illinois Palo Alto Princeton, New Jersey

Contents Overview

Contents

Preface

Basic Concepts of Chemistry, Fifth Edition, is intended for use in the one-quarter, one-semester, or two-quarter introductory or preparatory chemistry course. For more than a decade and a half, users have continued to comment positively on the student-oriented approach that has been our trademark in the previous four editions. We are often told by professors that their students enjoy reading and learning from our text. For that reason, we have retained the conversational writing style from the earlier editions and strengthened the pedagogical aids even more by adding several new features.

New to This Edition

Topical Coverage

New chapters have been added on nomenclature (Chapter 8) and the mole and chemical calculations (Chapter 9). In addition, Chapters 4, 5, and 6 that cover atomic theory and the periodic table have been rewritten and reorganized to present historical events in greater depth. This offers students a greater insight to the exciting process of discovery and inquiry. Chapter 7 has been recast to focus entirely on chemical bonding. We continue to emphasize environmental chemistry, particularly in Chapter 18.

Chemistry Applications in Today's World

The use of exciting and interesting examples of chemistry in our everyday lives has always been a hallmark of our book. In the fifth edition, this theme has been strengthened in three ways. First, a new series of boxed inserts, *Chemical Frontiers,* has been added to present areas of recent research that hold fascinating implications for the future. Topics covered in this series include ceramic engines, naturally occurring anticarcinogens, and chemical innovations that reduce energy consumption. Second, the boxed essays entitled *Career Sketch,* which describe occupations that draw upon the skills and knowledge learned in chemistry, have been completely updated. And third, topics that provide the chemical basis for understanding important natural processes, such as osmosis and buffers, have been added to the text.

Polya's Problem-Solving Framework

In addition to the strong program of problem-solving aids that have been part of previous editions, we provide in the fifth edition the general problem-solving framework of the late George Polya of Stanford University. Polya's four-step method is introduced in Chapter 2. We then go on to discuss estimation skills, stressing the importance of examining a solution so the student is sure the answer makes sense. In most chapters, we solve at least one problem using Polya's four-step method. Students can then apply this method when problem solving.

Cooperative Learning Opportunities

More than ninety years of research has shown that students working in small cooperative groups do improve their academic performance. They can solve problems together, and as part of that process, they discuss the material being studied. Areas that need clarification are identified as students teach one another.

Opportunities for cooperative problem solving are new to the fifth edition. Several exercises at the end of each chapter have been marked with a blue triangle (▲), indicating that they may be used as conventional problems or treated as cooperative problem-solving opportunities. This may be done either in class under the direction of the instructor or independently in small study groups. Additional suggestions on implementing cooperative learning are included in the *Instructor's Resource Manual*.

Features of the Book

Readability

We have maintained the readability of our text by using short paragraphs, concise and direct sentences, and a conversational tone. We also have used applications of chemistry to everyday life to provide students with concrete examples. In addition, line drawings, diagrams, and photographs emphasize important points.

Problem Solving

In each edition of *Basic Concepts of Chemistry,* we have sought to address the difficulties students encounter in solving problems. In this edition, we have maintained the same high number of worked-out examples, with solutions provided. Each example is followed by a practice exercise that offers an immediate opportunity to apply the skill just demonstrated. Answers to all practice exercises are provided at the end of the book.

End-of-chapter problems are divided into two groups: self-test exercises and extra exercises. More difficult problems are marked with an asterisk. Self-test exercises are keyed to the specific learning goals of each chapter and appear in matched pairs, with the odd-numbered problems, in most cases, answered in the back of the book. These problems help reinforce the learning of individual skills. Extra exercises are not keyed to learning goals; these exercises help students review skills in a format resembling a quiz or an exam. Answers to most of these problems appear at the end of the book.

Cumulative review problems appear after every two or three chapters. They are designed to test the student's knowledge of each set of chapters, and to reinforce material from earlier in the course. These exercises are also useful in studying for tests, midterms, and final examinations.

Study Aids

Our text has been planned to suit the backgrounds and needs of a wide variety of students. Each chapter begins with a series of learning goals. Throughout each chapter the learning goals appear in the margin next to the discussion of the corresponding material. Key terms appear in boldface and are accompanied by concise definitions. These definitions reappear in the Glossary at the end of the book. A list of key terms appears at the end of each chapter, along with a summary that reviews the most important points discussed in the chapter.

As in previous editions, a unit on mathematics (Supplement A) is included at the end of the book. It can be treated as a reference or used at the beginning of a course as a review of skills.

The fifth edition also includes a supplement that discusses study skills for chemistry students (Supplement C). Concrete suggestions are offered to help students study more efficiently. Tips for learning the material are provided, and there is a discussion of college life in general.

Complete Instructional Package

Basic Concepts of Chemistry, Fifth Edition, is part of a complete instructional package for introductory or preparatory chemistry courses. Our **Laboratory Experiments for Basic Chemistry** contains 25 experiments and two laboratory exercises. The fifth edition of the laboratory manual also contains the important safety and chemical hazards guide, which discusses potential hazards and stresses safety in the laboratory. New to the fifth edition of the laboratory manual is a *chemical disposal guide* with suggestions for disposing of various chemicals used in the experiments. Once again *Laboratory Experiments* includes the copyrighted game CHEM-DECK, which teaches the student how to write and name chemical compounds while playing card games that are a variation of gin rummy and poker.

The **Instructor's Resource Manual** accompanies both the book and the laboratory manual. It includes a test bank containing more than 900 test questions and answers, including math tests of final examination items. Each lab experiment section includes a discussion of the experiment's purpose, possible answers to questions, and a list of the quantities of chemicals needed to conduct the experiment for a class of 24 students.

The **Solutions Manual** to the fifth edition contains step-by-step solutions to the more than 2,000 problems in the text.

This edition includes a set of 52 two-color **Transparencies** of figures and tables in the text.

As before, the fifth edition of the text is accompanied by a **Study Guide** by James Braun of Clayton State College in Morrow, Georgia which offers extra explanations and problems for students who need more practice. Each chapter in the *Study Guide* corresponds to a chapter in the textbook, and contains additional worked-out examples and extra exercises for the student to solve. The opportunity for further study is offered in special review exercises that cover groups of chapters.

Acknowledgments

We thank the following people who thoroughly reviewed and commented on portions of the manuscript so that we could continue to improve its quality. Without their time, effort, and helpful suggestions, we could not have forged ahead so easily. They are

Helen Place, Washington State University
Jerry P. Suits, Brazosport College
Tom Lubin, Cypress College
Laurel E. Heyman, Owens Technical College
Charles D. Keilin, Laney College
Barbara C. O'Brien, Texas A & M University
Jerome Maas, Oakton Community College
David Taylor, Normandale Community College
Fred Redmore, Highland Community College
Joseph Wijnen, Hunter College
Carl Burmaster, Olympic College
Renee Gittler, Pennsylvania State University
Elsa C. Santos, Colorado State University
Karen Sanchez, Florida Community College
Tamar Susskind, Oakland Community College
David Winters, Tidewater Community College
Catherine Keenan, Chaffey Community College
Robert Heyer, Kirkwood Community College
John Goodenow, Lawrence Technological University

Once again, we thank those who have supported us, inspired our thoughts, and given us ideas that improved the fifth edition: Professors Dominic Macchia, John Murray, Linda Christopher, Barbara Drescher, and Booker Alexander, all of Middlesex County College; Professors Rebecca L. Lubetkin, Marylin Hulme, Aleta You Mastny, Sami Kahn, and Arlene Chasek, of the Rutgers University Consortium for Educational Equity; and Professor James Bliss of the Rutgers University Graduate School of Education. Thanks, also to Dr. Ethan Glickman, Dr. Jack Kurlansik, and the staff at the Park Dental Group. More thanks go to Mr. Max Sapirstein, Ms. Gertrude Sapirstein, Ms.

Paula Russikoff, Ms. Marlene Heller, and Dr. Stanley Kokowicz. Special thanks to Ms. Pamela D. Kance, Dr. Maryann F. Fralic, Professor Charles Oxman, and Dr. Gregg S. Teitel for their invaluable assistance with the content of the fifth edition.

Finally, we'd like to thank our sons, Robert and Michael, for their continued cooperation and assistance. The time has come when they are both old enough to study chemistry, and we hope that they, too, will marvel at its wonders.

To the Student

Chemistry is a field that grows constantly. Every day new scientific discoveries are made, and new applications of these discoveries are developed. Many of these discoveries affect our lives. There are new cures for diseases, new ways to save energy, new methods of growing foods, and better processes for keeping foods fresher for longer periods of time. There are synthetic fabrics for our clothing, new drugs to help improve our health, and new frontiers to explore throughout the universe. The list of ways in which the knowledge of chemistry is applied to make life better for all of us goes on and on.

This basic chemistry course will help you on the path to greater scientific knowledge. Perhaps it will lead you to a career in chemistry or a related field. We have taken care in writing this textbook so that it provides you with complete topical coverage presented clearly and simply. We present chemistry to you with a historical perspective so that you can experience the excitement of discovery and inquiry known by scientists. We include *Chemical Frontiers* so that you can keep abreast of many new discoveries in the field.

Problem solving is an area that receives much attention in chemistry because it helps to develop the critical thinking skills that are so important in all areas of science. We have looked at strategies that improve problem-solving ability and have included a four-step method of problem solving in the text that will provide you with a general framework to follow. By studying the numerous worked-out examples in each chapter and solving the end-of-chapter exercises, you will get a lot of practice in solving problems.

Knowing how to study is another important aspect of achieving success in chemistry. At the back of the text you will find a supplement on study skills. Spend some time reviewing this supplement. We believe that you will find useful information that can serve as a guide in your studies.

Another technique for success that we encourage is cooperative problem solving. Forming a study group is something you should consider. Meeting regularly with four or five other students to discuss material from the book and to go over problems can improve your understanding of the subject matter.

When you begin your study of chemistry you'll open the door to many exciting career opportunities in such fields as basic research, biochemistry, medicine, dentistry, nursing, pharmaceutical chemistry, engineering, and environmental chemistry, to name just a few. Throughout the text we describe science-related careers, including high-tech fields of employment, in a feature called *Career Sketch*.

As you study basic chemistry keep an open mind toward the subject. Then you'll acquire a feeling for the discipline and for the people who devote their lives to finding new knowledge about the nature of the universe. We hope our textbook will help you achieve your career goals and guide you as you grow intellectually.

Alan Sherman

Sharon Sherman

Leonard Russikoff

1 The Origins of Chemistry: Where It All Began

Learning Goals

After you've studied this chapter, you should be able to:

1. Discuss the beginnings of scientific thought.
2. Name the substances that Greek philosophers thought were basic to the composition of the earth.
3. Discuss the role of the Egyptians in the development of chemistry.
4. Describe the major goals of the alchemists.
5. Explain how Robert Boyle's book laid the foundations of modern chemistry.
6. Name the two main branches of chemistry.
7. Discuss some of the positive and some of the negative aspects of modern chemistry.

Introduction

Human beings are curious. They have always wondered about what they are and why they are here. The desire to understand their environment has made people search for explanations, principles, and laws, and modern chemistry is one result of that search.

Chemistry is the science that deals with matter and the changes it undergoes. Chemistry also focuses on energy and how it changes when matter is transformed. Chemists seek to understand how all matter behaves and to discover the principles governing this behavior.

Chemists ask fundamental questions and look for solutions in their attempt to unlock the secrets of nature. Many of their questions have been answered, and this information has built the body of knowledge that we call science.

Many of the basic concepts of chemistry have been incorporated into different fields of study. Physicists, biologists, engineers, medical and dental professionals, and nutritionists are just some of the professionals whose fields require chemical knowledge.

In this chapter we first look at the beginnings of chemistry and then briefly touch upon some areas of modern chemical research.

1.1 History

Learning Goal 1
Beginnings of scientific thought

The earliest attempts to explain natural phenomena led to fanciful inventions—to myths and fantasies—but not to understanding. Around 600 B.C., a group of Greek philosophers became dissatisfied with these myths, which explained little. Stimulated by social and cultural changes as well as curiosity, they began to ask questions about the world around them. They answered these questions by constructing lists of logical possibilities. Thus Greek philosophy was an attempt to discover the basic truths of nature by thinking things through, rather than by running laboratory experiments. The Greek philosophers did this so thoroughly and so brilliantly that the years between 600 and 400 B.C. are called the "golden age of philosophy."

Learning Goal 2
Influence of Greek philosophy on chemistry

Some of the Greek philosophers believed they could find a single substance from which everything else was made. A philosopher named Thales believed that this substance was water, but another named Anaximenes thought it was air. Some earlier philosophers believed the universe was composed of four elements—earth, air, fire, and water. The Greek philosopher Empedocles took this belief a step further, maintaining that these four elements combine in different proportions to make up all the objects in the universe.

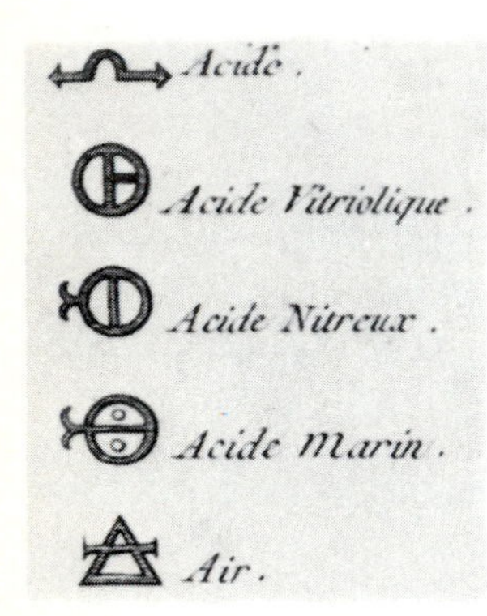

Some early alchemical symbols
(Courtesy Fisher Collection, Fisher Scientific Co., Pittsburgh, PA)

During this period, the Greek philosophers laid the foundation for one of our main ideas about the universe. Leucippus (about 440 B.C.) and Democritus (about 420 B.C.) were trying to determine whether there was such a thing as a smallest particle of matter. In doing so, they established the idea of the atom, a particle so tiny that it could not be seen. At that time there was no way to test whether atoms really existed, and more than 2,000 years passed before scientists proved that they do exist.

Learning Goal 3
Egyptians' role in the development of chemistry

While the Greeks were studying philosophy and mathematics, the Egyptians were practicing the art of chemistry. They were mining and purifying the metals gold, silver, and copper. They were making embalming fluids and dyes. They called this art *khemia,* and it flourished until the seventh century A.D., when it was taken over by the Arabs. The Egyptian word *khemia* became the Arabic word *alkhemia* and then the English word *alchemy*. Today our version of the word is used to mean everything that happened in chemistry between A.D. 300 and A.D. 1600.

Learning Goal 4
Alchemists' major goals

A major goal of the alchemists was to transmute (convert) "base metals" into gold. That is, they wanted to transform less desirable elements such as lead and iron into the element gold. The ancient Arabic emperors employed many alchemists for this purpose, which, of course, was never accomplished.

The alchemists also tried to find the "philosopher's stone" (a supposed cure for all diseases) and the "elixir of life" (which would prolong life indefinitely). Unfortunately they failed in both attempts, but they did have some lucky accidents. They discovered acetic acid, nitric acid, and ethyl alcohol, as well as many other substances used by chemists today.

The modern age of chemistry dawned in 1661 with the publication of the book *The Sceptical Chymist,* written by Robert Boyle, an English chemist, physicist, and theologian. Boyle was "skeptical" because he was not willing

Learning Goal 5
Boyle's influence on the development of modern chemistry

to take the word of the ancient Greeks and alchemists as truth, especially about the elements that make up the world. Instead Boyle believed that scientists must start from basic principles, and he realized that every theory had to be proved by experiment. His new and innovative scientific approach was to change the whole course of chemistry.

Table 1.1 chronologically outlines some of the major contributions in chemistry throughout history.

1.2 Chemistry Is a Diverse Field

Learning Goal 6
Branches of chemistry

During the 1700s and early 1800s, most chemists believed that there were two main branches of chemistry: organic and inorganic. Organic substances were thought to have been derived from living or once-living organisms; inorganic substances were said to have originated from nonliving materials. Sugars, fats, and oils were classified as organic chemicals. Salt and iron were classified as inorganic chemicals.

These two branches of chemistry still exist today, but the rules governing their classification have changed. Chemists now classify organic substances as those containing the element carbon. Inorganic substances are generally those that are composed of elements other than carbon. (Some inorganic substances are exceptions to this rule and do contain carbon.) Sugars, fats, and oils contain the element carbon and are indeed classified as organic substances. Salt and iron do not contain carbon, and both fall into the category of inorganic substances.

Other branches of chemistry include nuclear chemistry, biochemistry, pharmaceutical chemistry, analytical chemistry, environmental chemistry, and physical chemistry.

Robert Boyle
(Reprinted with permission from *Torchbearers of Chemistry* by Henry Monmouth Smith, © Academic Press, Inc., 1949)

1.3 How Chemistry Affects Our World

Learning Goal 7
Positive and negative aspects of chemistry

Drawing upon such natural resources as the air, forests, oceans, mines, wells, and farms around us, the chemical industry produces more than 50,000 different chemicals. These chemicals are used to manufacture a variety of products and to provide raw materials for other industries to use in producing their goods. It is by this process that the items required to sustain our basic human needs are made available to us.

Chemistry has played an important role in the processing of foods. Foods are processed so that they remain fresh and free of harmful toxins for a longer period of time. Chemists are hard at work seeking and researching ways to alleviate the world food shortage. Thousands of drugs to help us treat disease have become available through the application of medical knowledge in the chemical and pharmaceutical industries. Just a few of the chemically based products we use are plastics, cleansing agents, paper products, textiles, hardware, machinery, building materials, dyes and inks, fertilizers, and paints.

Table 1.1
Time Line of Chemistry*

Date	Person	Event
600 B.C.	Thales	Water is the main form of matter
546 B.C.	Anaximenes	Air is the main form of matter
450 B.C.	Empedocles	Idea that the four elements combine in different proportions
420 B.C.	Democritus	Idea of the atom
400 A.D.	Ko Hung	Attempts to find the elixir of life
1000	Avicenna	*Book of the Remedy*
1330	Bonus	*Introduction to the Arts of Alchemy*
1500		*Little Book of Distillation*
1620	Van Helmont	Foundations of chemical physiology
1625	Glauber	Contributions to practical chemistry
1661	Boyle	*The Sceptical Chymist*
1766	Cavendish	Discovery of hydrogen
1775	Lavoisier	Discovery of the composition of air
1787	Lavoisier and Berthollet	A system of naming chemicals
1800	Proust	Law of Definite Composition
1800	Dalton	Proposes an atomic theory
1820	Berzelius	Modern symbols for elements
1829	Döbereiner	Law of Triads
1860	Bunsen and Kirchhoff	Spectroscopic analysis
1869	Mendeleev	Periodic Law
1874	Zeidler	Discovery of DDT
1874	Van't Hoff and Le Bel	Foundations of stereochemistry
1886	Goldstein	Naming of cathode rays
1897	Thomson	Proposes a structure of the atom
1905	Einstein	Matter–energy relationship: $E = mc^2$
1908	Gelmo	Discovers sulfanilamide
1911	Rutherford	Proposes the nuclear atom
1913	Bohr	Proposes energy levels in atoms
1922	Banting, Best, and Macleod	Discover insulin
1928	Fleming	Discovers penicillin
1932	Urey	Discovers deuterium
1942	Fermi	First atomic pile
1945		First atomic bomb; nuclear fission
1950	Pauling	Helical shape of polypeptides

Table 1.1
Time Line of Chemistry *(continued)*

Date	Person	Event
1952		First hydrogen bomb; nuclear fusion
1953	Watson and Crick	Structure of DNA
1958	Townes and Schawlow	Develop laser beam
1969		Discovery of first complex organic interstellar molecule—formaldehyde
1970	Ghiorso	Synthesis of element number 105
1973	Cohen, Chang, Boyer, and Helling	Initiation of recombinant DNA studies
1974	Seaborg and Ghiorso	Synthesis of element 106
1979	A. Crewe	First color motion pictures of individual atoms
1982	Munzenberg and Armbruster	Synthesis of element 109
1984	Gallico	Development of test-tube skin
1985		Genetically engineered human insulin and human growth hormone become commercially available
1985	Diana	Development of compounds that kill cold viruses
1986		Nearly 800 genes mapped using genetic engineering techniques
1987		New class of high-temperature superconducting oxides with major implications for use in electronics discovered by researchers at the University of Houston and at AT&T Bell Laboratories in Murray Hill, NJ
1988	Zewail	First snapshots of chemical reactions
1989	Felix	Synthesis of human growth hormone
1990	Filiskoin	Development of chemical muscles

* Taken in part from the *Mallinckrodt Outline of the History of Chemistry,* Mallinckrodt Chemical Works, St. Louis, 1961.

At the same time, many thorny environmental problems have been created by the chemical industry. Chemists, engineers, and environmental scientists are now working together to devise safe and efficient means of destroying the toxic products that are generated along with the desired products of industrial activity. As long as funds are made available and the industry behaves responsibly, this important endeavor can progress.

Summary

Scientific thought had its beginnings in the contributions of the Greek philosophers of 600 to 400 B.C. The earliest chemistry was practiced at about that same time by the Egyptians. Later the art of chemistry—or alchemy—was taken over and extended by the Arabs. "Alchemy" now includes all of chemistry from the fourth through the sixteenth centuries A.D. The alchemists

Chemists work to produce products that improve the quality of life.
(Richard T. Nowitz/Phototake)

made several important discoveries, although they were primarily interested in transmuting base metals into gold. Modern chemistry began with the work of Robert Boyle in the mid seventeenth century. Boyle believed that theories must be founded on basic principles and must be supported by experiment.

Modern chemistry and its applications affect us in a variety of ways as we eat or sleep, work or play. Present chemical research is directed toward the solution of problems in physiology, in medicine, and in the environment, among many other areas.

CAREER SKETCHES*

Chemistry not only touches our daily lives but also is a part of many occupations. At the end of each chapter, we have included brief descriptions—career sketches—of some of these occupations. We hope you will enjoy reading about the various jobs that require an understanding of chemistry. You

* The career sketch material was adapted from two U.S. Department of Labor publications: Occupational Outlook Handbook and Dictionary of Occupational Titles (1990–1991). In addition, information about careers was obtained from the *Encyclopedia of Careers,* Eighth Edition (Chicago, Ill.: J.G. Ferguson Publishing Co., (1990).

might even find that you want to pursue one of these occupations as your educational goal. To find out more about any of the careers described, ask at your college counseling office.

CAREER SKETCH

NURSING OCCUPATIONS*

More than half of all health care workers are employed in the nursing occupations, which include registered nurses (RNs), licensed practical nurses (LPNs), nursing aides, orderlies, and attendants. In recent years a great deal of attention has been paid to the quality of nursing work life. Specialization, backed by intensive opportunities for nurses to develop expertise, and the recognition and sharing of expertise in the medical community have increased job satisfaction.

RNs: There are over one million RNs working in the United States. With the increase in educational opportunities, the growth of the profession and concern for job enrichment, nursing provides many expanded roles. These include nurse educator, nurse administrator, nurse specialist, nurse practitioner, certified nurse midwife, clinical care manager, nurse anesthetist, nurse researcher, primary nurse, and community health nurse. These expanded opportunities and roles include increased emphasis on the health, maintenance, illness prevention, and concern for the client as a whole person. Nursing has evolved into a profession and is no longer just a collection of specific skills. RNs can enroll in a two-year associates degree program, a two- or three-year diploma program, or a four-year baccalaureate program. There are also master's and doctoral programs available in nursing. Nurses are licensed in the state in which they practice.

LPNs: Licensed practical nurses function as associate nurses under the supervision of a registered nurse. LPNs also provide input to the primary nurse for development of the nursing care plan. LPNs can work for physicians and dentists. They observe patients, record information, take temperature and blood pressure, measure respiration, assist with personal hygiene, give injections and medication, and change dressings. They work in many areas of medicine and in locations including operating rooms, recovery rooms, day care facilities, physician's offices, private homes, schools, camps, health clubs, and industry. They generally take a 12- to 18-month course of study in a vocational school and must pass a written licensure examination.

Nurses, aides, orderlies, and attendants: These health care providers make beds, carry meals, assist in patient transport, give baths, and set up equipment. These positions require a high school education, on-the-job training, and some classroom training.

* *Quality of Nurse Life: Partners in Innovation.* Alberta Hospital Association Education Services, 1990.

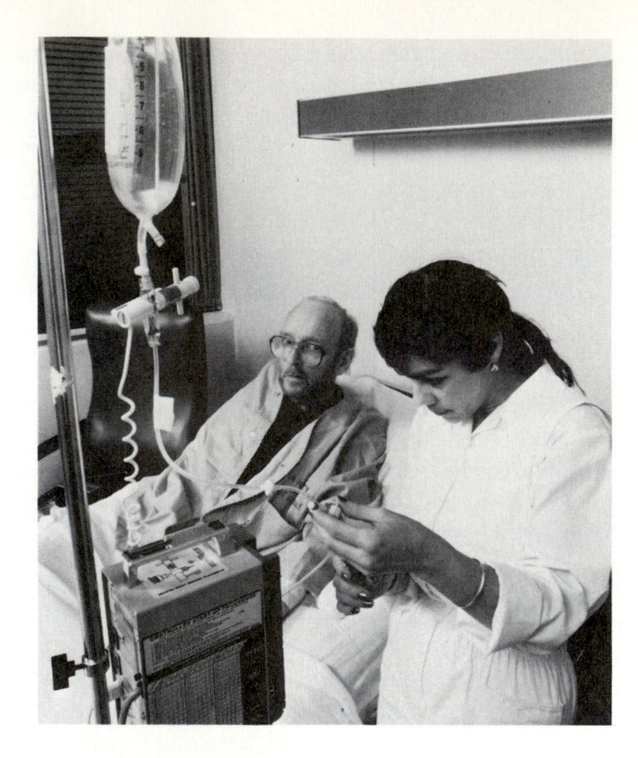

Nursing occupations provide many opportunities for professional development.
(Yoav Levy/Phototake)

CHEMICAL FRONTIERS

Chemistry is a dynamic field. Discoveries about the nature of matter are made every day by chemists all over the world. Researchers may report a newly synthesized compound that exhibits some unusual property or present a new theory that better explains a property of matter. To familiarize you with some of the new findings, we have added Chemical Frontiers, a new section of each chapter. By reading about intriguing breakthroughs made in chemistry today, you will see the impact chemistry has on our daily lives and gain a glimpse of the exciting future that lies ahead.

CHEMICAL FRONTIERS

THE USE OF BUPRENORPHINE IN FIGHTING DRUG ADDICTION*

One way to fight drug abuse is through pharmacotherapy, which uses legal drugs to help addicts overcome their dependence on illegal drugs. Many

*Adapted from "New Weapon in the War Against Drug Abuse," *What's Happening in Chemistry* (American Chemical Society, 1990).

heroin addicts, for example, take methadone as part of their treatment program. Researchers are now focusing on a new drug called *buprenorphine,* which may prove to be a potent weapon against both cocaine and heroin addiction. This dual addiction develops when users habitually smoke a mixture of crack cocaine and heroin, a form of drug abuse that is becoming more widespread.

The effects of heroin and cocaine are quite different: Cocaine is a stimulant, whereas heroin acts as a depressant, providing relaxation, drowsiness, and euphoria. Users require higher and higher dosages of these drugs to achieve the same pleasurable effects. Research indicates that there are common neural mechanisms at work in both cocaine and heroin addiction.

Buprenorphine, used as a painkiller in the hospital setting, has the ability to act against both drugs. It is believed that one part of the buprenorphine molecule attaches to the same cell receptors in the brain as does the illicit drug. This causes a pleasant sensation for the user. But another part of the molecule blocks receptors in the brain for the illicit drug, thereby blocking the euphoria associated with that drug. The blocking action increases at increased dosages, which means that an addict cannot experience euphoria as the dosage of the illicit drug rises. In addition, patients do not experience prolonged withdrawal symptoms when buprenorphine therapy stops.

At the Alcohol and Drug Abuse Research Center, Harvard Medical School–McLean Hospital, researcher Nancy K. Mello and a group of coworkers evaluated the effects of buprenorphine in cocaine-addicted experimental animals. After a single dose of the drug, most of the animals experienced a 50 percent decrease in cocaine use. Cocaine self-administration ended about two weeks after the final dose of buprenorphine was administered. The results of this research represent an important breakthrough for scientists engaged in the rehabilitation of individuals addicted to illicit drugs.

Self-Test Exercises

All exercises whose numbers are in color are answered in the back of the book. Self-test exercises are arranged in matched pairs, one below the other. Difficult problems are marked with an asterisk. Problems marked with a triangle (▲) may be used for cooperative problem solving.

Learning Goal 1: Beginnings of Scientific Thought

▲ **1.** Discuss the evolution of science from the golden age of philosophy to the birth of modern chemistry.

2. What years are called "the golden age of philosophy"? Why are those years considered the beginning of scientific thought?

Learning Goal 2: Influence of Greek Philosophy on Chemistry

3. List the four elements that Empedocles thought made up the world.

4. What did Thales think the world was composed of?

▲ **5.** What contribution did the Greek philosopher Democritus make to modern-day society?

6. Who established the concept of the atom?

7. If Empedocles had been right and the world were indeed made up of four elements, what two elements would compose paper?

▲ **8.** If the world were made up of Empedocles' four elements, what elements would compose vegetation?

Learning Goal 3: Egyptians' Role in the Development of Chemistry

9. List some of the contributions of the Egyptians to the development of chemistry.
10. The Egyptians mined and purified the metals gold, silver, and antimony. True or false?

11. While the Greeks were studying philosophy and mathematics, the Egyptians were already practicing the art of chemistry, which they called ________ .
12. Until what time in history did the art of *khemia* flourish?

Learning Goal 4: Alchemists' Major Goals

▲ **13.** What was the aim of chemistry from A.D. 300 to A.D. 1600?
14. Name the metals that the alchemists tried to turn into gold.

15. The people who tried to turn lead into gold were known as ________ .
16. What is meant by the term *elixir of life?*

Learning Goal 5: Boyle's Influence on the Development of Modern Chemistry

17. What was the message behind Robert Boyle's book *The Sceptical Chymist?*
▲ **18.** How do you think Boyle might have set about to prove that water is not an element and that it can be broken down into simpler substances?

Learning Goal 6: Branches of Chemistry

19. Name an organic chemical and an inorganic chemical.
▲ **20.** Explain by what criterion chemicals are now classified as organic or inorganic.

Learning Goal 7: Positive and Negative Aspects of Chemistry

▲ **21.** Choose one application of chemistry that is positive and one that is negative, and discuss the impact of each on society today.
22. Weigh the benefits that chemical research and applications yield to society against the problems caused by the chemical industry. What legislative measures have been taken (or are pending) to solve some of these problems?

Extra Exercises

23. Make a list of the industries that are based on the science of chemistry.
24. From the "Time Line of Chemistry" (Table 1.1), choose those events in the history of chemistry that you believe have had the greatest social importance. Give reasons for your choices.

25. Explain the benefits of buprenorphine over traditional pharmaceuticals that help fight drug abuse.

▲ **26.** If you could extend the "Time Line of Chemistry" 50 years into the future, what would you like to see happen? What advances in research do you believe will take place?

27. Can an educated public, better equipped with scientific knowledge, more easily assess the effects of science on society? Explain your answer.

28. With what is the science of chemistry concerned?

29. According to Empedocles, how would each of the following be classified? *(a)* copper *(b)* rain *(c)* lightning *(d)* nitrogen gas

▲ **30.** Explain in a few sentences how you believe chemistry has affected life in the context of one topic, such as food, clothing, or medicine.

▲ **31.** Discuss the contributions of alchemy to the science of chemistry.

32. List four ways in which chemistry has had a positive, healthy effect on life in your home. Explain each.

2 Systems of Measurement

Learning Goals

After you've studied this chapter, you should be able to:

1. Find the number of significant figures in a measurement.
2. Do calculations using the rules for significant figures.
3. Write numbers in scientific notation.
4. Find the area of any square, rectangle, circle, or triangle.
5. Find the volume of any cube, other rectangular solid, cylinder, or sphere.
6. Convert units of mass, length, and volume within the metric system using the factor-unit method.
7. Convert from a metric unit to the corresponding English unit using the factor-unit method.
8. Convert from an English unit to the corresponding metric unit using the factor-unit method.
9. Distinguish between the mass of an object and its weight.
10. Calculate the density, mass, or volume of an object when you are given the other two.
11. Convert temperatures from the Celsius to the Fahrenheit scale and vice versa.

Introduction

Chemists, as well as other scientists, use measurements when they do basic research. Measurements represent the dimensions, quantity, or capacity of things. Researchers ask questions and perform experiments. The ability to make accurate measurements of quantities such as mass, volume, temperature, or time enables them to gather the information that leads to the compilation of reliable scientific data. They analyze the data, often with the use of statistics, to look for patterns or regularities in nature. Sometimes the pattern or regularity is basic and can be stated simply to describe some natural phenomenon. The scientist may then devise a hypothesis, or speculative guess, that can explain the phenomenon. The problem-solving process continues as the scientist tests the hypothesis through further experimentation, which leads to the compilation and analysis of more measured data. Sometimes experimental results prove a hypothesis because it successfully explains a regularity or law of nature. The hypothesis then becomes a theory, or detailed explanation, which helps us describe and organize scientific knowledge.

In this chapter we look at systems of measurement so that you will have an understanding of how scientists record and manipulate quantitative data. We will begin with a look at problem solving so that you will be able to process the data you are given. Familiarity with problem-solving techniques will help you feel more comfortable as you work with units of measurement and tackle problems in all areas of chemistry.

2.1 Problem Solving

Humans have been given the gift of intelligence. We use this intelligence by solving problems. Each day we solve a variety of different problems as we deal with many types of obstacles. When a direct course of action is not clear, we devise ways to reach our goals, regardless of what it takes. Think of a typical day in your life and you will see that you deal with a whole range of problem types. You solve personal problems, social problems, scheduling problems, and financial problems, to name just a few. In your basic chemistry course, you will solve a variety of scientific problems. Like the problems we face in life, the problems you will encounter in basic chemistry require you to use judgment, originality, creativity, and independent thinking.

A great deal of research has gone into determining what makes a successful problem solver. George Polya, who is known as the "father of problem solving," tells us that when we solve problems we must search for some action to attain a goal. In other words, solving a problem does not mean just coming up with an answer. It means coming up with a strategy, or plan, to derive the answer. We will use Polya's four-step model as our general strategy. We will examine each step of Polya's model and show you how to proceed.

In preparation for taking an exam, you may have memorized many different formulas and definitions. Even though you know all the formulas and all the definitions, you may be unsure of where and when to use them. The general problem-solving framework will help you organize your thoughts and allow you to retrieve the relevant background information from your memory. It will help you sort relevant information from irrelevant information and will allow you to apply your knowledge.

One of the most important aspects of the problem-solving process is learning to ask yourself the correct questions. Questions such as "What is the unknown?" and "What information have I been given?" are a good starting point. Another important aspect is your own motivation to become adept at problem solving.

2.2 Polya's Method: The Four-Step General Model

Polya's four-step general model for problem solving will help you organize your thoughts and give you a general framework for solving any type of problem, whether it be a real life situation or a textbook problem.

Step 1: Understand the Problem

This is a basic step in problem solving. It consists of asking yourself a variety of questions to diagnose the situation. Pinpoint what you are trying to determine. Is there unknown information? What is it? Is any information assumed? Is there any additional information needed? Can you restate the problem in your own words?

Step 2: Devise a Plan

In this step you come up with a plan to solve the problem. You connect what you know with what is unknown to you. Do you know of a problem that is similar to this one? Is there a similar simpler problem that you can solve? Can you use a formula or write an equation? Can you make a table or draw a diagram to help you solve the problem?

Step 3: Carry Out the Plan

In this step you actually solve the problem. If you will be using a formula or an equation, you will set it up, plug in the numbers, and solve.

Step 4: Look Back

In this step you will look back at your work and be sure that it is correct. Did you check to be sure that the conditions of the problem were satisfied by the solution? Were all relevant data used? Is your answer reasonable? Compare your solution to your estimate. How do they compare?

When we solve the textbook examples, we will work out at least one solution per chapter using Polya's four-step method. So that you can practice using the process on your own, examine the solutions that do not include Polya's method and try to follow the steps on your own.

2.3 Essential Estimation Skills

Estimation is a close cousin to problem solving. It is the process of producing an answer that is sufficiently close so that decisions can be made. Estimation is an essential skill because it helps make problem solving more meaningful. Instead of coming up with the exact answer to a problem, estimation enables us to determine a ballpark figure for the solution. It encourages us to look at the problem and decide what a sensible answer would be, which gives us a check on our final results. Sometimes when we solve equations we plug numbers into the calculator rather thoughtlessly. We are not always sure when an unreasonable answer has resulted because we don't routinely stop and think of what would be a sensible result and what would be a nonsensical result. Estimation can produce reasonableness about computation. It can encourage us to have a greater appreciation of number size and provide a complement to our routine use of the calculator. Overall, estimation can bring more meaning to the problem-solving process.

The language of estimation includes words such as *about, close to, just about, a little less than,* and *between*. There is no one correct estimate. When you solve the problem you will find the correct answer, which will be exact.

In chemistry, we look for exact answers to problems. To be sure that the answers we come up with are reasonable and sensible, we can use estimation. When the problem is solved and an exact solution is computed, the estimate can be compared with the exact solution. This enables the problem solver to have an additional system of checking the calculation, and adds more meaning to the problem-solving process.

2.4 Significant Figures

Suppose that, in an experiment, you are measuring the temperature of a liquid. The thermometer you are using is calibrated, or marked off, only in whole degrees. Imagine that the mercury in the thermometer is halfway between 34 and 35 degrees (Figure 2.1a). You can then estimate that the temperature of the liquid is 34.5 degrees.

Now you pick up another thermometer, which is calibrated in tenths of degrees (Figure 2.1b). When you read this thermometer, it looks as though the temperature of the liquid is 34.55 degrees. It seems that the two thermometers differ in sensitivity, but which measurement is correct?

Learning Goal 1
Significant figures

To answer this question, we consider the number of significant figures in each measurement. **Significant figures** are *digits that express information that is reasonably reliable. The number of significant figures equals the number of digits written, including the last digit even though its value is uncertain*. For example, the temperature 34.5 degrees given by the first thermometer is known to be correct to three significant figures. These significant figures are the 3, the 4, and the 5. (Expressing the measurement as 34.50 degrees would mean that the temperature is known to four significant figures—the zero on the right is significant.)

The temperature 34.55 degrees given by the second thermometer shows a measurement with confidence to four significant figures. It may not be more accurate than the first measurement, but it is more *precise*. In general, the more significant figures there are, the more precise the measurement is.

Actually, neither of the two temperature measurements may be accurate. (They certainly wouldn't be accurate if the temperature of the liquid were actually, say, 47.51 degrees.) Thus **accuracy** involves *closeness to the actual dimension*. **Precision** is related to the *detail with which a measurement is known (expressed by the number of significant figures.)*

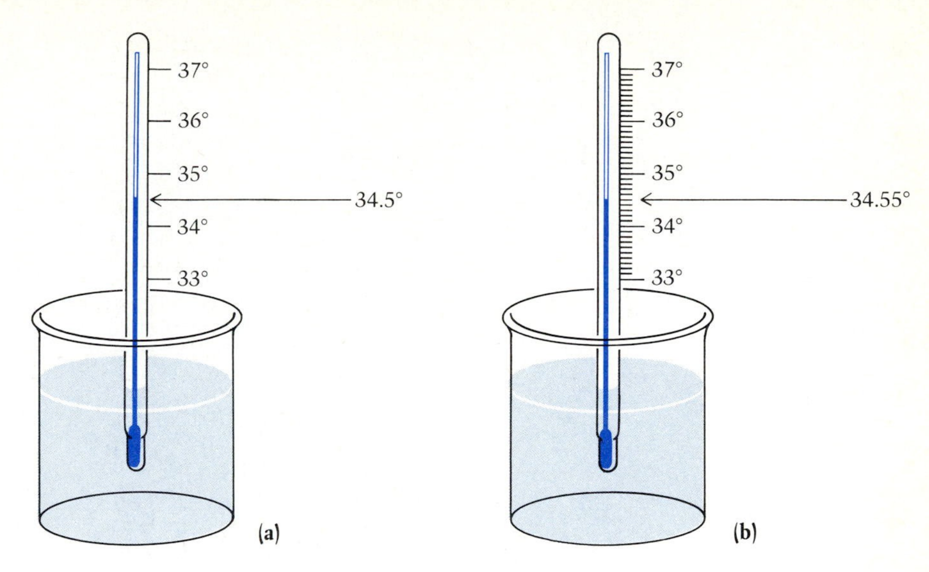

Figure 2.1
(a) We can measure the temperature of a liquid using a thermometer calibrated only in degrees. (b) We can also measure the temperature of a liquid using a thermometer calibrated in tenths of degrees.

A measuring *instrument* is considered accurate if it consistently provides measurements that are close to the actual amount. The instrument is considered to have good precision if it consistently provides the same measurement when it is used to measure the same amount. To see the difference, consider several darts being thrown at a target. If the darts land very close together, the results exhibit precision. If all the darts land on the bull's eye, they exhibit accuracy as well.

Learning Goal 2
Calculations with significant figures

Another experiment will show how significant figures play an important role in measurement. Suppose that we have to measure a certain piece of glass to find its perimeter (we do this by measuring the four edges of the glass and adding the lengths). We have two rulers, one calibrated in centimeters and the other calibrated in millimeters. We set out to measure each edge of the glass with the more precise ruler (the one calibrated in millimeters). However, when we get to the fourth side of the glass, we absent-mindedly pick up the less precise ruler (the one calibrated in centimeters). Looking at the glass, shown in Figure 2.2, we can see that three edges are measured to two decimal places and that the fourth is measured to one decimal place. What can we say about the perimeter of the glass? Is it 67.73 cm or 67.7 cm? The answer is 67.7 cm. We can report only what we know for sure—and in this case that is the answer to one decimal place. *Our least precise measurement determines the number of significant figures in our result*.

Whenever we *add or subtract* measured quantities, we must report the results in terms of the least precise measurement. We do this by **rounding off**

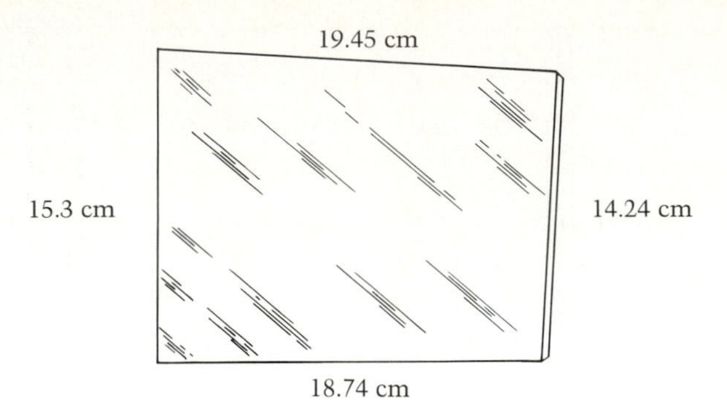

Figure 2.2
Measurement of a piece of glass

to the least number of decimal places. Our result must have no more decimal places than our least precise quantity.

To round off a number, we drop one or more digits at the right end of the number and, if necessary, adjust the rightmost digit that we keep. In some cases, dropped digits must be replaced with zeros. The rules we shall use in rounding numbers are as follows.

Rule 1 If the leftmost dropped digit is smaller than 5, simply drop the digits. Replace dropped digits with zeros as necessary to maintain the magnitude of the rounded number. Thus 7.431 rounded to two significant figures is 7.4. And 7,431 rounded to two significant figures is 7,400.

Rule 2 If the leftmost dropped digit is 5 or greater, increase the last retained digit by 1. Replace dropped digits with zeros as necessary to maintain the magnitude of the rounded number. Thus 93.56 rounded to three significant figures is 93.6. And 9,356 rounded to three significant figures is 9,360.

Here is a rounding rule that is useful in *multiplying or dividing* measured quantities.

Step 1
Count the number of significant figures in each of the quantities to be multiplied or divided.

Step 2
Report the result to the least number of significant figures determined in step 1. Round as required.

Example 2.1

Add 18.7444 and 13, and report the result to the appropriate number of significant figures.

Solution

Understand the Problem

We will add the numbers. Since 13 is the less precise measurement, we must round the sum to a whole number.

Devise a Plan

Use Rule 2, which says if the leftmost dropped digit is 5 or greater, increase the last digit by 1. In this case the leftmost dropped digit is 7, so the 1 in 31 increases to 2, and we have 32.

Carry Out the Plan

$$\begin{array}{r} 18.7444 \\ +13 \\ \hline 31.7444 \end{array} \quad \text{Round off to 32 (whole number).}$$

Look Back

The answer makes sense since 18.7444 is close to 19, and 19 + 13 = 32.

Practice Exercise 2.1 Add 12.4432 and 15. Provide the appropriate number of significant figures in your answer.

Example 2.2

Subtract 0.12 from 48.743, and use the appropriate number of significant figures in your answer.

Solution

Understand the Problem

Because we are aware of only two decimal places, 0.12 is the less precise number. Therefore we subtract and round off to two decimal places.

Devise a Plan

Use Rule 1, which says if the leftmost dropped digit is smaller than 5, simply drop the digits. In this case the leftmost dropped digit is the 3 in 48.623.

Carry Out the Plan

$$\begin{array}{r} 48.743 \\ -0.12 \\ \hline 48.623 \end{array} \quad \text{Round off to 48.62 (two decimal places).}$$

Look Back
Use mental arithmetic, and you will see that the answer makes sense.

Practice Exercise 2.2 Subtract 1.23 from 54.667, giving the appropriate number of significant figures in your answer.

Example 2.3

What is the area of a square whose side is measured as 1.5 cm?

Solution The area of a square is equal to the length of its side squared, or

$$A = s \times s = 1.5 \text{ cm} \times 1.5 \text{ cm} = 2.25 \text{ cm}^2$$

Round off to 2.3 cm^2. (We report to only two significant figures, because the measurement has only two significant figures.)

Practice Exercise 2.3 Calculate the area of a square whose side is measured as 2.5 cm.

Example 2.4

Divide 20.8 by 4, and give the result to the proper number of significant figures.

Solution The result must be rounded to one significant figure because of the one-digit divisor 4.

$$\frac{20.8}{4} = 5.2 \qquad \text{Round off to 5.}$$

Practice Exercise 2.4 Divide 48.2 by 4, giving the result to the proper number of significant figures.

Example 2.5

Multiply 20.8 by 4.1, and report the result to the proper number of significant figures.

Solution The result must be rounded to two significant figures.

$$20.8 \times 4.1 = 85.28 \qquad \text{Round off to 85.}$$

Practice Exercise 2.5 Multiply 25.5 by 3.2, giving the result to the proper number of significant figures.

Another problem arises when we use significant figures: What do we do about zeros? Are they significant figures or not? Here are some helpful rules. Read them carefully, work through Example 2.6, and then do Practice Exercise 2.6.

Rule 1 Zeros *between* nonzero digits are significant:

4.004 has four significant figures

Rule 2 Zeros to the *left* of nonzero digits are *not* significant, because these zeros show only the position of the decimal point:

0.00254 has three significant figures

0.0146 has three significant figures

0.06 has one significant figure

Rule 3 Zeros that fall at the *end* of a number are not significant unless they are marked as significant. If a zero does indicate the number's precision, we can mark it as significant by *placing a line over it*. Zeros to the right of the decimal place are always significant.

84,000 has two significant figures

$84,\bar{0}00$ has three significant figures

84,000.0 has six significant figures

Rule 4 Exactly defined quantities have an unlimited number of significant figures

4 qt = 1 gal

1 m = 100 cm

Example 2.6

In this example, we tell you how many significant figures there are in different numbers. In Practice Exercise 2.6, *you* are asked to tell *us*.

Solution

(a) 0.00087 has two significant figures.

(b) 1.004 has four significant figures.
(c) 873.005 has six significant figures.
(d) 9.00000 has six significant figures.
(e) 320,000 has two significant figures.
(f) $18\overline{0,0}00$ has four significant figures.
(g) 180,000.0 has seven significant figures.
(h) $2,\overline{000}$ has four significant figures.

Practice Exercise 2.6 Find the number of significant figures in each of the following numbers: *(a)* 0.0023 *(b)* 5.025 *(c)* 123.456 *(d)* $5,\overline{000}$ *(e)* $5,\overline{0}00$ *(f)* 12.000

2.5 Scientific Notation: Powers of 10

Learning Goal 3
Scientific notation

In science, we often deal with numbers that are *very* large or *very* small. Numbers like 100,000,000,000 (a hundred billion) or 0.0000008 (eight ten-millionths) arise frequently, and they are troublesome to work with in calculations. There is a shorthand method of writing such numbers, based on powers of 10. The number 100,000,000,000 can be written as 1×10^{11}, and 0.0000008 as 8×10^{-7}. In the first example, we moved the decimal point eleven places to the left:

$$1\,0\,0,\ 0\,0\,0,\ 0\,0\,0,\ 0\,0\,0. = 1 \times 10^{11}$$

Moving the decimal point to the left is compensated for by multiplying by a positive power of 10. In the second example, we moved the decimal point seven places to the right.

$$0.\ 0\,0\,0\,0\,0\,0\,8 = 8 \times 10^{-7}$$

Moving the decimal point to the right corresponds to multiplying by a negative power of 10. In both cases, the power that 10 is raised to is called the **exponent** of the **base number** 10. (If this causes you to hesitate, turn to Section A.3 in Supplement A and see that explanation, plus Examples A.1 through A.4.) We have

$$10^{11}$$

Exponent (the 11); Base number (the 10)

Here's how to use the shorthand method: To express a number in **scientific notation,** write the number with only one significant figure to the left of the decimal point. Multiply it by 10 raised to the number of places you moved the decimal—positive if it was moved to the left, and negative if it was moved to the right.

$$3\,800 = 3.8 \times 10^3$$

$$0.00625 = 6.25 \times 10^{-3}$$

$$100{,}000{,}000 = 1 \times 10^8$$

$$0.0000001 = 1 \times 10^{-7}$$

The number of significant figures is made clear by using scientific notation. For example, if we know the number 500 to three significant figures, we write 5.00×10^2. If we know the number 500 to only one significant figure, we write 5×10^2. Writing 5.00×10^2 is the same as writing $5\bar{0}0$ (three significant figures). Writing 5×10^2 is the same as writing 500 (one significant figure).

With a little practice, you'll soon be using scientific notation as easily as you use the more standard notation, but do not deny yourself the practice you need!

Example 2.7

Write the following numbers in scientific notation, showing the number of significant figures requested.

(a) $5{,}\bar{0}00$ (two significant figures)
(b) $48{,}\bar{0}00$ (three significant figures)
(c) 4,090,000 (three significant figures)
(d) 0.000087 (two significant figures)

Solution

(a) $5{,}\bar{0}00. = 5.0 \times 10^3$ (two significant figures)

(b) $48{,}\bar{0}00. = 4.80 \times 10^4$ (three significant figures)

(c) $4{,}090{,}000. = 4.09 \times 10^6$ (three significant figures)

(d) $0.000087 = 8.7 \times 10^{-5}$ (two significant figures)

Practice Exercise 2.7 Write the following numbers in scientific notation, showing the number of significant figures requested.

(a) $4{,}2\bar{0}0$ (two significant figures)
(b) $56{,}\bar{0}00$ (three significant figures)
(c) 6,023,000 (four significant figures)
(d) 0.00123 (three significant figures)

2.6 Area and Volume

Learning Goal 4
Areas of geometric figures

Learning Goal 5
Volumes of geometric figures

Suppose you want to know the area of a tennis court. A flat rectangular surface like a tennis court has just *two* dimensions: length and width, abbreviated l and w. As Table 2.1 shows, the area of a rectangle is equal to its length l times its width w. So the area of the tennis court would be $l \times w$.

Now suppose you want to find the volume of a cereal carton. A geometric figure like a carton (a rectangular solid) has three dimensions: length l, width w, and height h. As Table 2.1 and Figure 2.3 show, the volume of your carton is $l \times w \times h$.

Area is a *measure of the extent of a surface*. It is a two-dimensional measure that is always stated in squared units such as square feet (ft^2). **Volume** is a *measure of the capacity of an object*. It is a three-dimensional measure that is given in cubed units such as cubic feet (ft^3).

Table 2.1 lists formulas for the areas and volumes of several geometric figures. Example 2.8, which follows, shows you how to use them. Then, in Practice Exercise 2.8, you are asked to apply a volume formula on your own. You should take the time to work each practice exercise as you come to it. It will help you learn and apply what you have just read, and it will serve as a quick check of your understanding of each topic. Answers to practice exercises are provided in the back of the book.

Example 2.8

The soup can shown in Figure 2.3 has a height of 4.00 inches and a radius of 1.00 inch. What is the volume of the can?

Solution

Understand the Problem

Ask yourself what knowledge is required to arrive at a solution. A soup can has the shape of a cylinder, and therefore you need to know how to calculate the volume of a cylinder.

Devise a Plan

Looking at Table 2.1, you find that the formula for the volume of a cylinder is $V = \pi r^2 h$.

Given: $\pi = 3.14$, $r = 1.00$ inch, $h = 4.00$ inches

Carry Out the Plan

$$V = \pi r^2 h$$
$$= (3.14)(1.00 \text{ in.})^2 (4.00 \text{ in.})$$
$$= 12.56 \text{ in}^3 \quad \text{or} \quad 12.6 \text{ in}^3 \quad \text{(Three significant figures)}$$

The can holds 12.6 cubic inches of soup.

Look Back
See if the solution makes sense. Estimation can be used to give an approximate answer, telling us if our calculation is in the ballpark. We can say, "π is roughly 3, so 3 times 1 inch times 1 inch times 4 inches is about 12 cubic inches." Our estimate says that the answer should be about 12 cubic inches. Our exact calculation tells us that the answer is 12.6 cubic inches. We are definitely in the ballpark.

Practice Exercise 2.8 A cylinder has a height of 6.00 inches and a radius of 4.00 inches. What is its volume?

2.7 The English System of Measurement

The English system of units, still used in the United States today, has a built-in problem. It is not an orderly system and is therefore difficult to use. We will look at three units in the English system and then discuss this inherent difficulty.

In the English system, the *foot* is the unit of length and is divided into 12 smaller units called *inches*. The inch can be used to measure short distances, and the foot can be used to measure longer distances. To measure still longer distances there is the *yard* (which equals 3 feet) and the *mile* (which equals 5,280 feet).

The unit of weight in the English system is the *pound,* and the pound is divided into 16 smaller units called *ounces*. To measure larger weights, the *ton* may be used. The ton is equal to 2,000 pounds.

In the English system as used in the United States today, the unit of liquid volume is the *quart*. To measure smaller quantities, the *fluid ounce* is used

Volume = length × width × height
$V = l \times w \times h$

Volume = π × (radius)^2 × height
$V = \pi \times r^2 \times h$
$\pi = 3.14$

Figure 2.3
Formulas for calculating the volume of a cereal carton and a soup can.

Table 2.1
Areas and Volumes of Various Geometric Figures

Areas (Note that areas are always given in squared units: in^2, ft^2, and so on.)

1. Square	Area = side × side $A = s \times s$	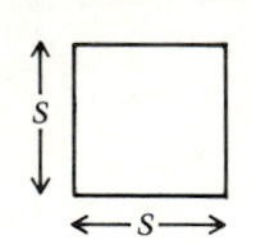
2. Rectangle	Area = length × width $A = l \times w$	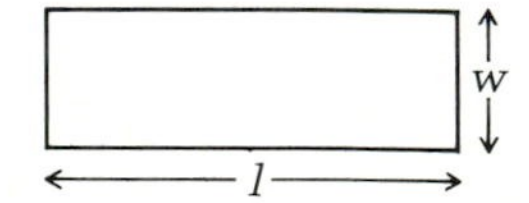
3. Circle	Area = π × (radius)^2 $A = \pi \times r^2$ $\pi = 3.14$	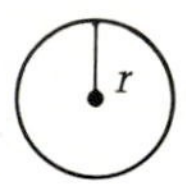
4. Triangle	Area = $\frac{1}{2}$ × base × height $A = \frac{1}{2} \times b \times h$	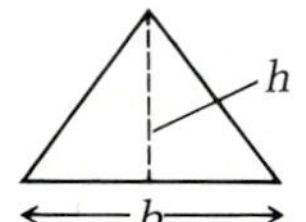

Volumes (Note that volumes are always given in cubed units: in^3, ft^3, and so on.)

5. Cube	$V = s \times s \times s$	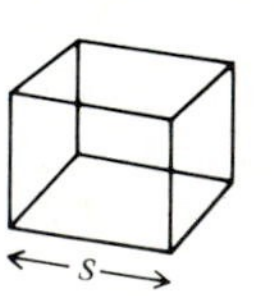
6. Carton (rectangular solid)	$V = l \times w \times h$	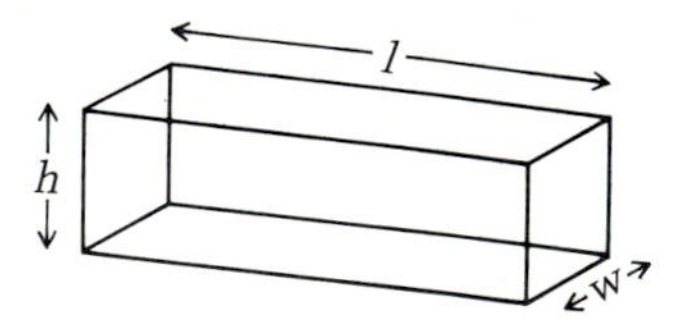
7. Cylinder	$V = \pi \times r^2 \times h$	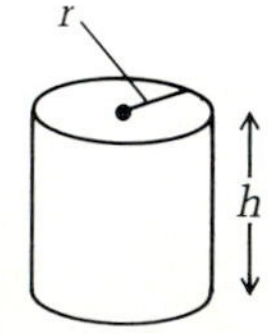
8. Sphere	$V = \frac{4}{3}\pi \times r^3$	

(32 fluid ounces equal 1 quart). Other units of liquid volume are the *pint* (2 pints equal 1 quart) and the gallon (4 quarts equal 1 gallon).

Table 2.2 lists these units. If you examine that table closely, the problem will be evident: there is no systematic relationship among units used to measure the same property. Consider the length unit. To convert inches to yards, it is necessary to know that 12 inches equal 1 foot and that 3 feet equal 1 yard. You cannot easily move from one unit to the other. The units for weight and volume are just as inconvenient to convert. Such problems led to the development of the metric system of measurement.

2.8 The Metric System of Measurement

In the late 1700s, the French decided to change their own system of measurement. To replace it they developed a logical and orderly system called the **metric system.** The advantages of this system led to its adoption in most countries of the world and in all branches of science. The British held out until 1965 and then began a changeover to the metric system.

In countries using the metric system, almost everything is measured in metric units—distances between cities, the weight of a loaf of bread, the size of a sheet of plywood. Nearly all countries have adopted the metric system. In 1975 the United States Congress passed a bill establishing a policy of voluntary conversion to the metric system and creating the U.S. Metric Board. A program of gradual conversion was the goal of this board, whose funding ended in 1982. The Office of Metric Programs, established by the U.S. Department of Commerce, now has the job of promoting the increased use of metric units of measurement by business and industry.

The metric system consists of (1) a set of standard units of measurement for distance, weight, volume, and so on, and (2) a set of prefixes that are used to express larger or smaller multiples of these units. The prefixes represent multiples of 10. This makes the metric system a decimal system of measurement. In 1960 an international group of scientists modified the metric system by adopting a system of units called the *Système International d'Unités* (International System of Units), or the SI system.

In the SI system there are seven basic units. Table 2.3 lists them. Other units, called *derived units,* are composed of combinations of basic units. (For example, the unit for speed, meters per second, is derived from the basic

Table 2.2
The English System of Measurement (as Used in the United States)

Length	Weight	Volume
12 inches = 1 foot	16 ounces = 1 pound	16 fluid ounces = 1 pint
3 feet = 1 yard	2,000 pounds = 1 ton	2 pints = 1 quart
5,280 feet = 1 mile		4 quarts = 1 gallon

Table 2.3
Basic SI (Metric) Units

Quantity	Unit	Symbol
Length	meter	m
Mass	kilogram	kg
Time	second	s
Electric current	ampere	A
Temperature	kelvin	K
Light intensity	candela	cd
Amount of substance	mole	mol

units *meter* and *second*.) The basic units were all carefully defined. Occasionally one of the definitions is modified to make it more precise or more useful. For example, the meter was first defined as one ten-millionth of the distance between the North Pole and the equator along a meridian of the earth. Later, to ensure that 1 meter meant the same thing everywhere, it was redefined as the distance between two scratches on a platinum bar that is kept at exactly the freezing point of water (zero degrees on the Celsius temperature scale) in a vault outside Paris. Still later, as more exacting measuring instruments were developed, the meter was defined as 1,650,763.73 wavelengths in vacuum of the orange-red line of the spectrum of krypton-86. You will learn more about this in Chapter 5.

Learning Goal 6
Conversion of units within the metric system

Table 2.4 lists the more commonly used metric prefixes. To use one, simply "tack it on" to the front of a metric unit. That gives a related unit that is a multiple of the original unit. For example, the prefix *kilo* has a multiplier of 1,000, so a kilometer is equal to 1,000 meters. The prefix *centi* has a multiplier 0.01, so a centigram is equal to 0.01 gram. Conversely, there are 100 centigrams in a gram.

To make converting from one unit to another simpler, we will use a method called the factor-unit method, or dimensional analysis. With this approach, any problem that requires conversion from one unit to another can be set up and solved in a similar manner. We can say that:

$$\text{Quantity wanted} = \text{quantity given} \times \text{factor unit}$$

When you multiply the quantity given by the proper factor unit, some of the units cancel to give the desired quantity. Let's see how this works. We'll convert meters to centimeters as we begin to understand this method.

$$\text{Quantity wanted} = \text{quantity given} \times \text{factor unit}$$

$$\text{Centimeters} = \cancel{\text{meters}} \times \frac{\text{centimeters}}{\cancel{\text{meter}}}$$

Table 2.4
Metric Prefixes

Prefix	Symbol	Multiplier
nano	n	0.000000001
micro	μ (Greek mu)	0.000001
milli	m	0.001
centi	c	0.01
deci	d	0.1
deka	da	10
hecto	h	100
kilo	k	1,000
mega	M	1,000,000

Note that the factor unit expresses a relationship between the quantity wanted and the quantity given. The factor unit is written in such a way that the given units cancel when you multiply the quantity given times the factor unit. Then you're left with the quantity wanted:

$$\text{Centimeters} = \cancel{\text{meters}} \times \frac{\text{centimeters}}{\cancel{\text{meter}}}$$

$$\text{Centimeters} = \text{centimeters}$$

Of course any factor unit can be expressed in two ways. For example, the relationship 1 meter = 100 centimeters can be expressed as

$$\frac{100 \text{ centimeters}}{1 \text{ meter}} \quad \text{or} \quad \frac{1 \text{ meter}}{100 \text{ centimeters}}$$

You choose the factor unit that will make the proper terms cancel.

We'll now look at a variety of examples in which the factor-unit method is helpful. As you proceed through basic chemistry, you'll find this method useful in many cases. (Supplement A, especially Examples A.5 through A.9, provides many more examples that will help you understand this strategy.)

Example 2.9

Change 40 meters to centimeters.

Solution

Understand the Problem

This example requires conversion from one unit to another. We ask "How many centimeters are there in one meter?" Table 2.4 shows that *centi* means 0.01, so there are 100 centimeters in 1 meter.

Devise a Plan

Use the factor-unit method to convert.

Carry Out the Plan

Remember that there are 100 centimeters in 1 meter. This can be written as

$$\frac{100 \text{ centimeters}}{1 \text{ meter}}$$

which reads "100 centimeters per meter"; the division line reads "per." Next we put in the numbers.

$$? \text{ centimeters} = 40 \cancel{\text{meters}} \times \frac{100 \text{ centimeters}}{\cancel{\text{meter}}} = 4{,}000 \text{ cm}$$

There are 4,000 centimeters in the 40 meters.

Look Back

We want to see if this solution makes sense. We determined that one meter equals 100 centimeters. Does it make sense that 40 meters would be 40 times larger, or 4,000 centimeters? Yes, this does make sense.

Practice Exercise 2.9 Change 35 meters to centimeters.

Example 2.10

Change $43\overline{0}$ milligrams to grams.

Solution

Understand the Problem

You will see that it is similar to Example 2.9. Therefore we can use the same strategy to devise a plan.

Devise a Plan

Again we use the factor-unit mehtod. (This time we will use abbreviations.) We ask, "How many milligrams are there in 1 g?" Table 2.4 shows that *milli* means 0.001, so 1 g equals 1,000 mg. This means there is

$$\frac{1 \text{ g}}{1{,}000 \text{ mg}} \quad \text{(which reads 1 g per 1,000 mg)}$$

Carry Out the Plan

We say that

$$g = \cancel{mg} \times \frac{g}{\cancel{mg}}$$

$$?\ g = 43\bar{0}\ \cancel{mg} \times \frac{1\ g}{1{,}000\ \cancel{mg}} = 0.430\ g$$

There is 0.430 g in $43\bar{0}$ mg.

Look Back

We ask if this solution makes sense. If 1,000 mg equals 1 g, how much of a gram would $43\bar{0}$ mg equal? Using estimation we can say that $43\bar{0}$ mg is a little less than half a gram, therefore our answer 0.430 g, which is a little less than half a gram is in the ballpark, and makes sense.

Practice Exercise 2.10 Change $35\bar{0}$ mg to grams.

Example 2.11

How many centimeters are there in 8 meters?

Solution

$$cm = \cancel{m} \times \frac{cm}{\cancel{m}}$$

$$?\ cm = 8\ \cancel{m} \times \frac{100\ cm}{1\ \cancel{m}} = 800\ cm$$

There are 800 cm in 8 m.

Practice Exercise 2.11 How many centimeters are there in 18 meters?

Example 2.12

Convert 580 millimeters to centimeters.

Solution

$$cm = \cancel{mm} \times \frac{cm}{\cancel{mm}}$$

$$?\ cm = 580\ \cancel{mm} \times \frac{1\ cm}{10\ \cancel{mm}} = 58\ cm$$

There are 58 cm in 580 mm.

If you wonder where the term

$$\frac{1\ \text{cm}}{10\ \text{mm}}$$

came from, look at a metric ruler, and you'll see that there are 10 mm in 1 cm.

Practice Exercise 2.12 Convert 660 mm to centimeters.

Example 2.13

Convert 75 millimeters to its corresponding length in *(a)* centimeters, *(b)* meters, *(c)* kilometers.

Solution

(a) $? \text{ cm} = 75 \cancel{\text{mm}} \times \frac{1\ \text{cm}}{10\ \cancel{\text{mm}}} = 7.5\ \text{cm}$

(b) $? \text{ m} = 7.5 \cancel{\text{cm}} \times \frac{1\ \text{m}}{100\ \cancel{\text{cm}}} = 0.075\ \text{m}$

(c) $? \text{ km} = 0.075 \cancel{\text{m}} \times \frac{1\ \text{km}}{1000\ \cancel{\text{m}}} = 0.000075\ \text{km}$

Practice Exercise 2.13 Convert 55 millimeters to the corresponding length in *(a)* centimeters, *(b)* meters, *(c)* kilometers.

Example 2.14

Convert 2.3 kilograms to the corresponding mass in *(a)* grams, *(b)* decigrams, *(c)* centigrams, *(d)* milligrams.

Solution

(a) $? \text{ g} = 2.3 \cancel{\text{kg}} \times \frac{1000\ \text{g}}{1\ \cancel{\text{kg}}} = 2{,}300\ \text{g}$

(b) $? \text{ dg} = 2{,}300 \cancel{\text{g}} \times \frac{10\ \text{dg}}{1\ \cancel{\text{g}}} = 23{,}000\ \text{dg}$

(c) $? \text{ cg} = 23{,}000 \cancel{\text{dg}} \times \frac{10\ \text{cg}}{1\ \cancel{\text{dg}}} = 230{,}000\ \text{cg}$

(d) $? \text{ mg} = 230{,}000 \cancel{\text{cg}} \times \frac{10\ \text{mg}}{1\ \cancel{\text{cg}}} = 2{,}300{,}000\ \text{mg}$

Practice Exercise 2.14 Convert 4.4 kilograms to the corresponding mass in *(a)* grams, *(b)* decigrams, *(c)* centigrams, *(d)* milligrams.

Table 2.5
Conversion of Units

To convert	Into	Multiply by
Length		
inches	centimeters	2.54 cm/in.
centimeters	inches	0.39 in./cm
feet	meters	0.30 m/ft
meters	feet	3.28 ft/m
Weight (mass; see Section 2.9)		
ounces	grams	28.35 g/oz
grams	ounces	0.035 oz/g
pounds	grams	454 g/lb
grams	pounds	0.0022 lb/g
Volume		
liters	quarts	1.06 qt/liter
quarts	liters	0.946 liter/qt

You should now realize that conversion between units of the metric system is not difficult. If you want to convert metric units into English units, consult Table 2.5. (More precise conversion factors can be found in Supplement B, Table B.3.)

Learning Goal 7
Conversion of metric to English units

Although it will always be convenient to be able to convert between systems, it is even more important to be able to associate metric measurements with commonly used terms. Start now to "think metric," for only in this way will the metric system become meaningful to you. The following example offers a starting point.

Example 2.15

A new car is described in an advertisement as having an overall length of 5.50 meters and an overall width of 1.50 meters. Find the dimensions of the car in feet.

Solution Consulting Table 2.5, we see that to convert meters to feet it is necessary to multiply by 3.28. In other words, 1 meter = 3.28 feet. Therefore the length and width can be determined as follows:

$$\text{Length:} \quad ?\ \text{ft} = 4.40\ \cancel{\text{m}} \times \frac{3.28\ \text{ft}}{1\ \cancel{\text{m}}} = 18.0\ \text{ft}$$

$$\text{Width:} \quad ?\ \text{ft} = 1.50\ \cancel{\text{m}} \times \frac{3.28\ \text{ft}}{1\ \cancel{\text{m}}} = 4.92\ \text{ft}$$

Practice Exercise 2.15 The dimensions of a new American car are as follows: overall length, 4.20 meters; overall width, 1.25 meters. Is this a compact car? Find its dimensions in feet.

Learning Goal 8
Conversion of English to metric units

Before we work the next example, we need to discuss the SI unit of volume. This unit, a derived unit, has the dimension (length)^3—for example, m^3 or cm^3. However, before the SI system was adopted, the metric unit of volume was the liter. One liter (1 L) is equal in volume to 1,000 cm^3, and 1 mL = 1 cm^3. For some measurements the liter is simpler to use than the SI units, and many people continue to use it in such cases. In this book, we shall use whichever unit best fits a particular situation. In the next example we work with liters, which are used in selling gasoline in all countries using the metric system.

Example 2.16

An American family visiting Mexico wants to fill its car with 15 gallons of gasoline. Mexico uses the metric system. How many liters of gasoline should the visitors ask the attendant to put in their tank?

Solution We know that 15 gallons of gasoline are the same as $6\bar{0}$ quarts (4 quarts = 1 gallon). Consulting Table 2.5, we see that to convert quarts to liters we multiply by 0.946 liter/quart. Therefore we may solve the problem as follows:

$$?\ \text{liters} = 6\bar{0}\ \cancel{\text{quarts}} \times \frac{0.946\ \text{liter}}{1\ \cancel{\text{quart}}} = 57\ \text{liters}$$

Practice Exercise 2.16 To conserve water, the Simon family installed a reducing valve and a water-metering device in their shower. If Mr. Simon uses $2\bar{0}$ gallons of water in his shower, and Mrs. Simon uses 18 gallons in hers, how many liters of water do the Simons use?

It is valuable to know that a liter is just a little more than a quart. A meter is slightly longer than a yard (it is 1 yard and 3.375 inches). When you are dealing with weights, it is helpful to remember that a nickel weighs about 5 g. If you can keep these three approximations in mind, you'll find it easier to work in the metric system.

2.9 Mass and Weight

Learning Goal 9
Mass and weight

The mass of an object and its weight are often thought of as being the same, so the words *mass* and *weight* are frequently used interchangeably. This is incorrect because, by definition, the two terms have different meanings. **Mass** is *the quantity of matter in an object*, whereas **weight** is *the gravitational force that attracts an object*.

The definition of mass implies that your body's *mass* is constant no matter where you are. Your body's *weight*, on the other hand, varies from planet to planet and even varies slightly at different places on the earth. An astronaut who weighs 180 pounds on earth weighs only 30 pounds on the moon (which has only one-sixth of the gravitational pull of the earth) and has no weight at all in outer space. But *the astronaut's mass is the same in all places*. We can define weight mathematically as

$$\text{Weight} = \text{mass} \times \text{gravity} \qquad \text{or} \qquad W = m \times g$$

The force of gravity varies from planet to planet, so the weight of an object must vary too.

2.10 Density

Learning Goal 10
Calculation of density, mass, and volume

Which is heavier: glass, iron, or wood from an oak tree? Naturally it depends on the size of each piece. However, what if all three were the same size? In other words, what if we had cubes of glass, iron, and oak wood, each with a volume of 1 cm^3? (See Figure 2.4.) Suppose we weighed each cube to determine which was the heaviest. We would find that 1 cm^3 of iron weighs 7.9 g, 1 cm^3 of glass weighs 2.4 g, and 1 cm^3 of oak wood weighs 0.6 g. We would conclude that, *for a particular volume*, iron has the greatest mass.

The concept of density enables us to express this relationship conveniently. **Density** can be defined as *the mass per unit volume of a substance or object*. The density of a substance can be determined by using the formula

$$D = \frac{m}{V}$$

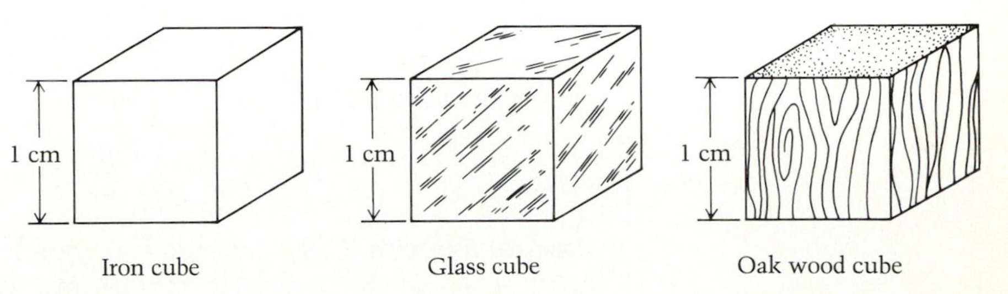

Figure 2.4
Three cubes of equal volume

where D is the substance's density, m is its mass, and V is its volume. If the mass is measured in grams and the volume in cubic centimeters, then the unit of density is

$$\frac{\text{grams}}{\text{cubic centimeters}} \quad \text{or} \quad \frac{\text{g}}{\text{cm}^3}$$

(See Supplement A for a review of solving algebraic and word equations.)

Example 2.17

A block of iron that is 5.0 cm long, 3.0 cm high, and 4.0 cm wide weighs 474 g. What is the density of iron?

Solution We first calculate the volume of the block.

$$\text{Volume} = \text{length} \times \text{width} \times \text{height}$$

$$V = 5.0\ \text{cm} \times 4.0\ \text{cm} \times 3.0\ \text{cm} = 6\bar{0}\ \text{cm}^3$$

We now know that $V = 6\bar{0}\ \text{cm}^3$ and $m = 474$ g, so all we have to do is solve the density formula for D.

$$D = \frac{m}{V} = \frac{474\ \text{g}}{6\bar{0}\ \text{cm}^3} = 7.9\ \frac{\text{g}}{\text{cm}^3}$$

Practice Exercise 2.17 A block of aluminum is 2.0 cm long, 3.0 cm high, and 5.0 cm wide, and it weighs 81.0 g. What is the density of the aluminum?

Example 2.18

Suppose you are told that 400 g of alcohol occupy a volume of 500 mL. What is the density of alcohol?

Solution In this example we are given the mass and the volume of the substance whose density we are asked to find. Note that the volume of the alcohol is given in milliliters (mL). But *1 milliliter is equal to 1 cubic centimeter*. That is, 1 mL = 0.001 L = 1 cm^3. So it is acceptable to interchange the units mL and cm^3. Now let's solve this problem.

$$D = \frac{m}{V} = \frac{400\ \text{g}}{500\ \text{cm}^3} = 0.8\ \frac{\text{g}}{\text{cm}^3}$$

Practice Exercise 2.18 Suppose you are told that 880 g of a clear, colorless liquid occupy a volume of 110 mL. What is the density of this liquid?

Example 2.19

Determine which is more dense, carbon tetrachloride or chloroform, from the following information:
(a) 16 g of carbon tetrachloride occupy a volume of $1\bar{0}$ mL.
(b) $3\bar{0}$ g of chloroform occupy a volume of $2\bar{0}$ mL.

Solution We simply calculate the density of each liquid and see which is greater:

$$D(\text{carbon tetrachloride}) = \frac{m}{V} = \frac{16 \text{ g}}{1\bar{0} \text{ cm}^3} = 1.6 \frac{\text{g}}{\text{cm}^3}$$

$$D(\text{chloroform}) = \frac{m}{V} = \frac{3\bar{0} \text{ g}}{2\bar{0} \text{ cm}^3} = 1.5 \frac{\text{g}}{\text{cm}^3}$$

The carbon tetrachloride is the more dense liquid.

Practice Exercise 2.19 Determine which is the more dense, liquid A or liquid B, given the following information:
(a) 55.0 g of liquid A occupy a volume of 10.0 mL.
(b) 25.0 g of liquid B occupy a volume of 12.5 mL.

The preceding calculations are straightforward uses of the density formula. But what if you were given the density of a material and its mass? Could you calculate its volume? Here's how to do it.

Example 2.20

The density of alcohol is 0.8 g/cm^3. Calculate the volume of 1.6 kg of alcohol.

Solution First we have to solve the density formula

$$D = \frac{m}{V}$$

for V. To do so, we multiply both sides of the equation by V. This gives us $D \times V = m$. Now we divide both sides of the equation by D. This gives us the formula we want:

$$V = \frac{m}{D}$$

We are given that $D = 0.8$ g/cm^3 and $m = 1.6$ kg, or 1,600 g. We substitute these numbers into the formula and solve for V.

$$V = \frac{m}{D} = \frac{1{,}600 \text{ g}}{0.8 \text{ g/cm}^3} = 2{,}000 \text{ cm}^3 \text{ (or 2,000 mL)}$$

Practice Exercise 2.20 The density of chloroform is 1.5 g/cm^3. Calculate the volume in cm^3 of 2.0 kg of chloroform.

One important fact about density is that most substances *expand when heated*. Therefore, when a certain mass of a substance is hot, it occupies a larger volume than it does when it is cool. This means that the density of the substance decreases as it is warmed. Think about this for a moment, and make sure you understand why it is true. Most densities reported in chemical references, such as the *Handbook of Chemistry and Physics,* are the densities at 20°C, which is about room temperature.

2.11 · Temperature Scales and Heat

When we heat a substance, we add a quantity of heat to that substance. We can then use a thermometer to measure the **temperature** of the substance. The thermometer measures the *intensity* of the heat; it tells us nothing about the quantity of heat that has entered the substance. (We shall discuss heat quantity in Chapter 12.)

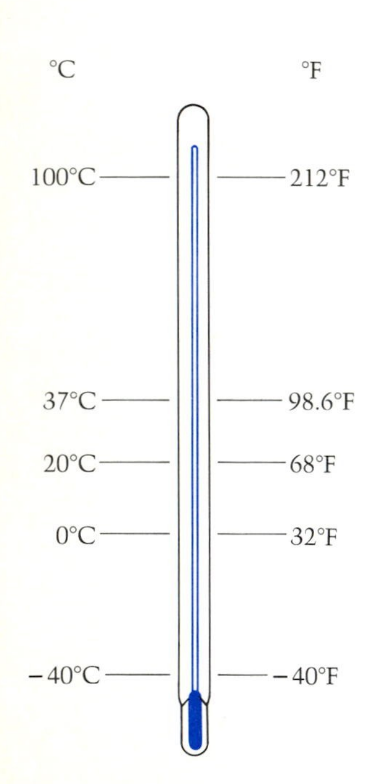

Figure 2.5
A comparison of Fahrenheit and Celsius scales

Three different temperature scales are commonly used in measuring heat intensity. Two of these—the Fahrenheit and Celsius scales—are in general use. The third, the Kelvin scale, is used mainly by scientists.

The **Fahrenheit temperature scale** was devised by Gabriel Daniel Fahrenheit, a German scientist, in 1724. On this scale (see Figure 2.5), the freezing point of pure water is at 32 degrees (32°F), and the boiling point of water is at 212 degrees (212°F). There are thus 180 Fahrenheit degrees between the freezing point and the boiling point of water.

The **Celsius temperature scale** was devised in 1742 by Anders Celsius, a Swedish astronomer. His objective was to develop an easier-to-use temperature scale; he did so by assigning a nice, round 100 Celsius degrees between the freezing and boiling points of pure water. On the Celsius scale (Figure 2.5), the freezing point of water is at zero degrees (0°C), and the boiling point of water is at 100 degrees (100°C). (The Celsius scale is also sometimes referred to as the centigrade scale.)

The **Kelvin temperature scale** is an *absolute* temperature scale. That is, its zero point (0 K) is at absolute zero, the lowest possible temperature theoretically attainable. The divisions of the Kelvin scale are the same size as Celsius degrees, but they are called kelvins (abbreviated K) rather than degrees. Chapter 13 discusses the Kelvin scale in more detail.

2.12 Converting Celsius and Fahrenheit Degrees

Learning Goal 11
Celsius and Fahrenheit temperature conversions

On the Celsius scale, there are 100 divisions between the freezing point and the boiling point of water. On the Fahrenheit scale, there are 180 divisions between these two points. Therefore 100 Celsius degrees cover the same range as 180 Fahrenheit degrees, so that 1 Celsius degree = 1.8 Fahrenheit degrees. Moreover, 0°C is equivalent to 32°F. Formulas for converting from a temperature on one scale to a temperature on the other are based on these facts. The formulas are

$$°F = (1.8 \times °C) + 32 \qquad \text{and} \qquad °C = \frac{°F - 32}{1.8}$$

Let us use these formulas to convert from Fahrenheit to Celsius degrees and from Celsius to Fahrenheit degrees. When using significant figures remember that exactly defined quantities have an unlimited number of significant figures.

Example 2.21

Convert 122°F to degrees Celsius.

Solution Substitute 122°F into the conversion formula for changing °F to °C.

$$°C = \frac{°F - 32.0}{1.8} = \frac{122 - 32.0}{1.8} = \frac{90.0}{1.8} = 50.0$$

$$122°F = 50.0°C$$

Practice Exercise 2.21 Convert 244°F to degrees Celsius.

Example 2.22

Convert $1\overline{00}$°C to degrees Fahrenheit.

Solution Substitute $1\overline{00}$°C into the conversion formula for changing °C to °F.

$$°F = (1.8 \times °C) + 32.0 = (1.8 \times 1\overline{00}) + 32.0 = 18\overline{0} + 32.0 = 212$$

$$1\overline{00}°C = 212°F$$

Practice Exercise 2.22 Convert $15\overline{0}$°C to degrees Fahrenheit.

Summary

Measurements represent the dimensions, quantity, or capacity of things. Scientists record and analyze measured data as they do research to attain knowledge while engaging in the problem-solving process. Chemistry students, too, engage in problem solving. George Polya's four-step method facilitates this process. The process includes understanding the problem, devising a plan for solution, carrying out the plan, and looking back at the results. Estimation techniques can help students assess the reasonableness of a solution.

In working with measurements, significant figures are used to convey the precision of information. The results of measurements and computations should always be rounded to the appropriate number of significant figures. Scientific notation, in powers of ten, may be used to indicate significant figures; it also simplifies computations involving very large or very small numbers.

The metric (or SI) system is a system of measurement used in most parts of the world and in all scientific work. It consists of seven basic units and a number of derived units, along with a set of prefixes that indicate multiples of these units. The unit of length is the meter, the unit of mass is the kilogram, and the unit of volume is a derived unit (cubic centimeters or, sometimes, liters). The corresponding units in the English system, as used in the United States, are the foot (length), pound (weight), and quart (volume). The factor-unit method may be used to convert from unit to unit within either system or from system to system. Mass is a measure of the quantity of matter, and weight is a measure of the gravitational force on an object. The density of an object is its mass divided by its volume. Temperatures, which indicate heat intensity, are measured on the Fahrenheit, Celsius, or Kelvin temperature scale and may be converted from one scale to another.

Key Terms

We have listed the major terms that have been defined in this chapter. Be sure that you are familiar with each of these terms. If you need to refresh your memory, refer to the indicated section or use the glossary at the back of the book.

accuracy **(2.4)**
area **(2.6)**
base number **(2.5)**
Celsius temperature scale **(2.11)**
density **(2.10)**
exponent **(2.5)**
Fahrenheit temperature scale **(2.11)**
Kelvin scale **(2.11)**
mass **(2.9)**
metric system **(2.8)**
precision **(2.4)**
rounding off **(2.4)**
scientific notation **(2.5)**
significant figures **(2.4)**
temperature **(2.11)**
volume **(2.6)**
weight **(2.9)**

CHEMICAL FRONTIERS

ENERGY FROM HOT ROCKS

As our energy needs continue to increase, a researcher from the Massachusetts Institute of Technology is grappling with a possible solution to the problem. For about 17 years, Jefferson W. Tester, director of MIT's Energy Laboratory has studied the concept of extracting energy from the granite beneath the earth's surface. Unlike other forms of geothermal energy, this form, known as hot dry rock (HDR), is not emitted at the earth's surface. To extract the energy, engineers would drill pairs of wells into the hot rock layer. Water pressure would be used to create fractures in the rock that would serve as artificial reservoirs to hold the heat. Cool water would then be sent through the first well where it would heat in the fractures. It would return to earth's surface through the second well at between 475° and 575°F.

Unlike oil and coal, the energy produced by HDR would produce no carbon dioxide or acid rain precursors to damage the environment. However, there would be small-scale environmental disruption in the mining area. Small-scale seismic risk could occur if some of the circulating water is lost in the rock, creating pressure in the surrounding area.

Old Faithful geyser, Yellowstone National Park—a source of geothermal energy. (National Park Service photograph)

According to a recent Los Alamos National Laboratory report written by Tester, within a decade HDR sources could produce energy at a cost of five to six cents per kilowatt-hour which is competitive with the price of oil. Researchers estimate that HDR resources could prove to be "orders of magnitude larger than ... all fossil and fissionable resources."

Self-Test Exercises

Learning Goal 1: Significant Figures

▲ **1.** How many significant figures are there in each of the following numbers? *(a)* 4,000.0 *(b)* 4,000 *(c)* 0.808 *(d)* 35.000 *(e)* 101,010.0 *(f)* 2×10^2 *(g)* 1.000×10^3 *(h)* 48.1×10^5 *(i)* $4,\overline{00}0$ *(j)* $4,\overline{000}$

▲ **2.** How many significant figures are there are in each of the following numbers? *(a)* 4250.0 *(b)* 0.509 *(c)* 3×10^2 *(d)* 5,000 *(e)* 32.1×10^3 *(f)* 0.00034 *(g)* $402,\overline{000}$ *(h)* 90

3. Determine the number of significant figures in each of the following numbers: *(a)* $3,\overline{000,000}$ *(b)* $3,\bar{0}00,000$ *(c)* 3×10^6 *(d)* 3.00×10^6 *(e)* 3.000000×10^6 *(f)* 0.0000305 *(g)* 0.100054 *(h)* 6.720×10^{-8} *(i)* 305,075 *(j)* 35.00

4. Determine the number of significant figures in each of the following numbers: *(a)* $4,\overline{000,000}$ *(b)* $4,\overline{00}0,000$ *(c)* 2×10^5 *(d)* 0.00123 *(e)* 4.02×10^{-5} *(f)* 23.00 *(g)* 2×10^{-1} *(h)* 750

Learning Goal 2: Calculations with Significant Figures

5. Find the perimeter of a triangle whose sides are 4.23 cm, 4.198 cm, and 4 cm. How many significant figures are there in your answer?

6. Find the perimeter of a triangle whose sides are 3.23 cm, 5.006 cm, and 3 cm. How many significant figures are there in your answer?

7. Find the area of a square whose sides are 1.3 cm. How many significant figures do you have in your answer?

8. Find the area of a square whose sides are 4.2 cm. How many significant figures are there in your answer?

▲ **9.** Find the perimeter of a four-sided object whose sides are 7.382 cm, 3.95 cm, 5.4342 cm, and 3.83 cm. (Use the proper number of significant figures.)

10. Find the area of a rectangle whose sides are 10.12 cm and 10.25 cm. (Use the proper number of significant figures.)

11. Find the area of a rectangle whose sides are 20.62 cm and 10.4 cm. How many significant figures are there in your answer?

▲ **12.** Find the area of a square whose sides are 2.1 cm each. (Use the proper number of significant figures.)

13. State the number of significant figures in each number given in Exercise 19.

14. Find the perimeter of a square whose sides are 2.1 cm each. (Use the proper number of significant figures.)

Learning Goal 3: Scientific Notation

15. Express the following numbers in scientific notation: *(a)* 8,000,000 *(b)* 0.002 *(c)* 0.00004 *(d)* 905,000 *(e)* 0.00207 *(f)* 305,000,000

16. Express the following numbers in scientific notation: *(a)* 900,000 *(b)* 45,000 *(c)* 2,970 *(d)* 2,546,000 *(e)* 0.00006 *(f)* 0.0122 *(g)* 0.056 *(h)* 20.0040

17. Express the following numbers in scientific notation: *(a)* 10,581 *(b)* 0.00205 *(c)* 1,000,000 *(d)* 802

▲ **18.** Express the following numbers in scientific notation: *(a)* 45,000,000 *(b)* $4\bar{0}0$ *(c)* $4\overline{00}$ *(d)* 400.0 *(e)* $425,\overline{000}$

▲ **19.** Express the following numbers in scientific notation: *(a)* 850,000,000 *(b)* 0.00000607 *(c)* 6,308,000 *(d)* 0.06005 *(e)* 500 *(f)* $5\overline{0}0$ *(g)* $5\overline{00}$ *(h)* 500.0 *(i)* $23,\overline{000},000$ *(j)* 0.0000000930

20. Express the following numbers in scientific notation: *(a)* 0.123 *(b)* 0.006 *(c)* 0.00601

Learning Goal 4: Areas of Geometric Figures

▲ **21.** Find the area of a rectangular room that measures 6.0 m by 3.0 m. Use the proper number of significant figures in reporting your answer.

22. Find the area of a rectangular room that measures 5.0 m by 4.0 m. Use the correct number of significant figures in reporting your answer.

23. Determine the area of this page in *(a)* square centimeters, *(b)* square inches.

24. Find the area of this page in *(a)* square meters, *(b)* square feet.

25. Find the area occupied by a rectangular house that measures $3\overline{0}$ m by $2\overline{0}$ m.

26. Find the area of a table that measures 1.5 meters by 2.0 meters.

Learning Goal 5: Volumes of Geometric Figures

27. Find the volume of a milk carton that measures 5.00 cm long, 6.00 cm wide, and 15.0 cm high.

28. Find the volume of a box that measures 5.0 m long, 4.0 m wide, and 2.0 m high.

29. Determine the volume of a cylinder-shaped can that measures 60.0 cm high and has a radius of 15.0 cm.

30. Determine the volume of a metal cylinder that is 25 cm high and has a radius of 3.0 cm.

▲ **31.** A rectangular solid—a gasoline can—has the following dimensions: length, 18.0 cm; width, 10.0 cm; and height, 25.0 cm. Determine the volume of this gasoline can.

32. There are 100 cm in 1 m. How many cubic centimeters (cm^3) are there in 2 cubic meters (2 m^3)?

33. Determine the volume of a cylinder-shaped soup can that measures 15.0 cm high and has a radius of 6.00 cm.

▲ **34.** Find the volume of a cylinder-shaped jar of applesauce that measures $1\overline{0}$ cm high and has a radius of 5.0 cm.

35. There are 12 inches in one foot. From this information, calculate the number of cubic inches in one cubic foot.

36. A cookie jar has the dimensions $4\overline{0}$ cm by 25 cm by 15 cm. Calculate its volume.

Learning Goal 6: Conversion of Units Within the Metric System

37. Convert 0.15 kg to *(a)* grams, *(b)* decigrams, *(c)* milligrams.

38. Convert 0.84 kg to *(a)* grams, *(b)* decigrams, *(c)* milligrams.

39. Convert 3.1 m to *(a)* decimeters, *(b)* centimeters, *(c)* millimeters.

40. Convert 35,000 mm to *(a)* centimeters, *(b)* decimeters, *(c)* meters, *(d)* kilometers.

41. Convert 149 mm to *(a)* centimeters, *(b)* meters, *(c)* kilometers.

42. Convert 5.5 m to *(a)* decimeters, *(b)* centimeters, *(c)* millimeters.

▲ **43.** Convert 7.850 m to *(a)* millimeters, *(b)* centimeters, *(c)* decimeters, *(d)* kilometers.

44. Convert 125 mm to *(a)* centimeters, *(b)* meters, *(c)* kilometers.

45. Convert 2.5 km to *(a)* meters, *(b)* decimeters, *(c)* centimeters.

46. Convert 1.234 m to *(a)* millimeters, *(b)* centimeters, *(c)* decimeters, *(d)* kilometers.

47. Convert $56,78\overline{0}$ mg to *(a)* centigrams, *(b)* decigrams, *(c)* grams, *(d)* kilograms.

48. Convert 7.5 km to *(a)* meters, *(b)* decimeters, *(c)* centimeters.

49. Convert $3,5\overline{00}$ mL to *(a)* liters, *(b)* deciliters.

50. Convert $15\overline{00}$ mL to *(a)* liters, *(b)* deciliters.

51. There are $1\overline{00}$ cm in one meter. From this information, calculate the number of cubic centimeters (cm^3) in one cubic meter (m^3).

▲ **52.** Convert $25,55\overline{0}$ mg to *(a)* centigrams, *(b)* decigrams, *(c)* grams, *(d)* kilograms.

Learning Goal 7: Conversion of Metric to English Units

53. Express 25.0 m in (*a*) feet, (*b*) inches.
54. Express 100.0 m in (*a*) feet, (*b*) inches.

55. Express 1.2 L in (*a*) gallons, (*b*) quarts.
56. Express 10.0 L in (*a*) gallons, (*b*) quarts.

57. Express 5.00 g in (*a*) pounds, (*b*) ounces.
58. Express 100.0 kg in (*a*) pounds, (*b*) tons.

▲ **59.** Without looking at a conversion table, determine the number of cubic feet in a cubic meter.
▲ **60.** Express $25\bar{0}$ m in (*a*) yards, (*b*) feet.

61. Express $1\bar{0}\bar{0}$ m in (*a*) yards, (*b*) feet.
Express 2.50 g in pounds.

63. There are 3.28 ft in one meter (1.00 m). From this information, calculate the number of square feet (ft^2) in one square meter (m^2).
64. Express 50.0 L in gallons.

Learning Goal 8: Conversion of English to Metric Units

65. Convert your height from feet and inches to meters.
66. Express $2\bar{0}$ gal in (*a*) liters, (*b*) milliliters.

67. Convert your weight from pounds to kilograms.
68. Express 100.0 lb in (*a*) kilograms, (*b*) grams.

▲ **69.** Express $1\bar{0}$ ft in (*a*) meters, (*b*) centimeters, (*c*) millimeters.
70. Express 5.0 ft in (*a*) meters, (*b*) centimeters.

71. Express 6.00 ft in (*a*) meters, (*b*) centimeters.
72. Express 25 ft in (*a*) meters, (*b*) centimeters.

73. Express $1\bar{0}\bar{0}$ yd in (*a*) meters, (*b*) centimeters.
74. Express 15 gal in (*a*) liters, (*b*) milliliters.

75. Express 5.00 gal in (*a*) liters, (*b*) milliliters.
▲ **76.** Express $22\bar{0}$ lb in (*a*) kilograms, (*b*) grams.

Learning Goals 9 and 10: Calculation of Density, Mass, and Volume

77. Determine the density of this textbook. You'll have to weigh it and determine its volume. Report your answer in grams per cubic centimeter.
78. A cube is 4.00 cm on each side and has a mass of 250.0 g. What is its density?

79. A cube is 5.00 cm on each side and has a mass of 600.0 g. What is its density?
80. A block of aluminum with a density of 2.70 g/cm^3 weighs 274.5 g. What is the volume of the block?

81. A block of aluminum with a density of 2.7 g/cm^3 and a mass of 549 g. What is the volume of the block?
▲ **82.** The element barium has a density of 3.5 g/cm^3. What would be the mass of a rectangular block of barium with the dimensions 2.0 cm × 3.0 cm × 4.0 cm?

▲ **83.** The element barium, which is a soft, silvery-white metal, has a density of 3.5 g/cm^3. What would be the mass of a rectangular block of barium with the dimensions 1.0 cm × 3.0 cm × 5.0 cm?
84. Calculate the mass of a cube that has a density of 2.50 g/cm^3 and is 1.5 cm on each side. Report your result to the proper number of significant figures.

***85.** A spherical balloon is filled with helium, a gaseous element. Helium has a density of 0.177 g/liter. The balloon has a radius of 3.0 cm. What is the mass of the helium in the balloon? (*Hint:* Volume of a sphere = $\frac{4}{3}\pi r^3$.)
86. A cube has a density of 0.250 g/cm^3 and a mass of 15.0 g. What is the volume of this cube?

87. A 1.00-liter container holds a block of cesium that is 1.00 cm × 2.00 cm × 3.00 cm and whose density is 1.90 g/cm^3; 14.0 g of iron, whose density is 7.86 g/cm^3; and 0.500 liter of mercury, whose density is 13.6 g/cm^3. The rest of the container is filled with air, whose density is 1.18×10^{-3} g/cm^3. Calculate the *average* density of the contents of the container. *Hint:* To calculate average density, use the formula

$$\text{Average density} = \frac{m_a + m_b + \cdots}{V_a + V_b + \cdots}$$

88. A small metal sphere has a mass of 75.0 g. The sphere is placed in a graduated cylinder containing water. The water level in the cylinder changes from 10.0 cm^3 to 20.0 cm^3 when the sphere is submerged. What is the density of the sphere?

***89.** A fish tank whose dimensions are 1.50 ft × 1.00 ft × 0.500 ft contains six fish. Fish one and two weighs 2.00 g each, fish three weighs 2.50 g, and the combined weight of fish four, five, and six is 9.80 g. The bottom of the fish tank is covered with gravel, which occupies one-eighth of the volume of the tank. The density of this gravel is 3.00 g/cm^3. The rest of the fish tank is filled with water (density 1.00 g/cm^3). Assuming that the volume occupied by the fish is negligible, calculate the average density of the contents of the fish tank.

▲ **90.** Determine the density of a rectangular solid that has the dimensions 8.0 cm × 2.0 cm × 3.0 cm and a mass of 192.0 g.

▲ **91.** Calculate the mass of a cube that has a density of 1.61 g/cm^3 and is 4.1 cm on each side. Express your answer to the proper number of significant figures.

92. A balloon contains a gaseous compound. The radius of the balloon is 1.06 cm, and the mass of the gaseous compound in the balloon is 9.80 g. What is the density of the gaseous element?

***93.** Suppose you are drinking a chocolate milk shake. Its average density is 2.00 g/cm^3. The volume of the container that holds it is $25\overline{0}$ cm^3. The milk shake consists of 90.0% milk products and 10.0% chocolate powder by volume. Assuming that the density of the milk products is 1.50 g/cm^3, calculate the amount of chocolate powder that was used to make the drink.

94. Determine the length of a metallic cube that has a density of 10.5 g/cm^3 and a mass of 672 g.

95. A cube measures 3.50 cm on each side and has a mass of $4\overline{0}0$ g. What is its density?

96. The element copper has a density of 8.92 g/cm^3. What is the mass of a cylinder-shaped piece of copper that has a radius of 2.00 cm and a height of 10.00 cm?

97. Gold has a density of 19.3 g/cm^3. A certain block of gold is a rectangular solid that measures 30.0 cm by 5.00 cm by 10.0 cm. What is its mass?

98. An empty graduated cylinder has a mass of 80.00 g. The empty cylinder is filled with exactly 30.0 cm^3 of a liquid and weighed again. The cylinder with the liquid in it has a mass of 488 g. What is the density of the liquid? If you are told that this liquid is an element, can you determine what element it is?

99. The density of lead is 11.3 g/cm^3. What is the volume of a chunk of lead that has a mass of $25\overline{0}$ g?

100. A small stone that appears to be an irregularly shaped diamond is weighed and found to have a mass of 21.06 g (about 105 carats if it is indeed a diamond). The stone is placed in a graduated cylinder that initially has 10.0 mL of water in it. When the stone is immersed in the water, the water level rises to 16.0 cm^3. What is the density of the stone? Could the stone be a diamond? (*Hint:* Check the *Handbook of Chemistry and Physics.*)

101. A small rock has a mass of 55.0 g. The rock is placed in a graduated cylinder containing water. The water level in the cylinder changes from 25.0 cm^3 to 40.0 cm^3 when the rock is submerged. What is the density of the rock?

102. A piece of copper that has a mass of 120.0 g is melted and poured into some liquid silver that has a mass of 200.0 g. The two metals are mixed and allowed to solidify. What is the average density of this metal alloy? [*Hint:* D(copper) = 8.92 g/cm^3 and D(silver) = 10.5 g/cm^3.]

***103.** An object in the shape of a sphere—for example, a hollow globe—is filled with ethyl alcohol. The sphere has a radius of 18.0 cm, and ethyl alcohol has a density of 0.800 g/cm^3. What is the mass of ethyl alcohol in the sphere? (*Hint:* Volume of a sphere = $\frac{4}{3}\pi r^3$.)

104. Determine the mass in grams of one gallon of gasoline if the density of gasoline at 25°C is 0.56 g/cm^3.

105. A cube has a density of 0.500 g/cm^3 and a mass of 25.0 g. What is the volume of this cube?

106. What mass of lead [D(lead) = 11.3 g/cm^3] occupies the same volume as $1\overline{0}0$ g of aluminum [D(aluminum) = 2.70 g/cm^3]?

Learning Goal 11: Celsius and Fahrenheit Temperature Conversions

107. Convert 28°F to degrees Celsius.

108. Convert 54.0°F to degrees Celsius.

109. Convert −40°F to degrees Celsius. What is unusual about this temperature?

▲ **110.** Convert −20.0°F to degrees Celsius.

111. Convert 20°C to degrees Fahrenheit.

112. Convert 23.0°C to degrees Fahrenheit.

▲ **113.** Change each of the following temperatures from degrees Fahrenheit to degrees Celsius. *(a)* 50°F *(b)* −94°F *(c)* 419°F *(d)* −130°F

114. Convert −20.0°C to degrees Fahrenheit.

115. Change each of the following temperatures from degrees Celsius to degrees Fahrenheit. *(a)* 95.0°C *(b)* −80.0°C *(c)* 80.0°C *(d)* 210°C

116. Convert 15.0°C to degrees Fahrenheit.

Extra Exercises

117. Would a carefully weighed object weigh the same in Death Valley as it would on the top of Mount McKinley?

▲ **118.** In filling out an application form for a company that uses the metric system, a job applicant recorded his height as 4.00 m and his weight as 200.0 kg. Did he fill the application out correctly?

119. An art dealer needs to know the volume of a piece of irregularly shaped sculpture. A chemist friend tells the dealer to submerge the object in water and measure the amount of water displaced by it. The dealer finds that 48 g of water are displaced. If the density of the water is 1.0 g/mL, what is the volume of the art object?

120. A chemistry student decides to cook a pizza in the lab. The instructions call for cooking the pizza for 10 minutes at 425°F. The oven dial, however, is set in degrees Celsius. At what temperature should the dial be set so that the pizza will be done in 10 minutes?

121. Choose a room where you live and measure its dimensions in English units; then convert each dimension to metric units.

122. Express the following relationships in scientific notation:

(a) ________ g = 1 kg
(b) ________ kg = 1 g
(c) ________ μg = 1 g
(d) ________ mg = 1 g

▲ **123.** Assuming that the following numbers are from experimental measurements, calculate the answer, and state it to the proper number of significant figures.

$$\frac{(3.12 \times 10^{6})(8.123 \times 10^{-4})}{3.1}$$

124. A temperature is measured as 16.0°F. What is this temperature in degrees Celsius?

125. Change the following temperatures to °F:
(a) 16°C *(b)* $2\overline{0}0$°C

126. Change the following temperatures to °C:
(a) $3\overline{0}0$°F *(b)* $-15\overline{0}$°F

127. The 1976 world's record for the 100-meter dash was 9.90 seconds. What would this speed be in miles per hour?

3 Matter and Energy, Atoms and Molecules

Learning Goals

After you've studied this chapter, you should be able to:

1. Explain what is meant by the scientific method.
2. Explain the Law of Conservation of Mass and Energy.
3. Explain the difference between physical and chemical properties.
4. Describe the difference between homogeneous and heterogeneous matter, between mixtures and compounds, and between compounds and elements.
5. Describe the difference between an atom and a molecule.
6. Explain the Law of Definite Composition (or Definite Proportions).
7. Explain the terms *atomic mass*, *formula mass*, and *molecular mass*.
8. Determine the formula or molecular mass of a compound when you are given the formula for the compound.

Introduction

This chapter is really the beginning of your study of chemistry. We start with discussions of the most elementary concepts, those of matter and energy. Then, in this chapter and later chapters, we build on and extend these concepts. As we do this, we shall be discussing the results of centuries of scientific research—the theories and laws of modern chemistry. These theories and laws are sometimes presented to students as though each resulted from a quick flash of insight on the part of some scientist. Actually they are the fruit of years—and sometimes decades or centuries—of hard work by many people.

3.1 The Scientific Method

Chemistry is an experimental science that is concerned with the behavior of matter. Much of the body of chemical knowledge consists of abstract concepts and ideas. Without application, these concepts and ideas would have little impact on society. Chemical principles are applied for the benefit of society through technology. Useful products are developed by the union of basic science and applied technology.

Learning Goal 1
The scientific method

Over the past 200 years, science and technology have moved forward at a rapid pace. Ideas and applications of these ideas are developed through carefully planned experimentation, in which researchers adhere to what is called the **scientific method.** The scientific method is composed of a series of logical steps that allow researchers to approach a problem and try to come up with solutions in the most effective way possible. It is generally thought of as having four parts:

1. *Observation* and classification. Scientists begin their research by carefully observing natural phenomena. They carry out experiments, which are observations of natural events in a controlled setting. This allows results to be duplicated and rational conclusions to be reached. The data the scientists collect are analyzed, and the facts that emerge are classified.

2. *Generalization*. Once observations are made and experiments carried out, the researcher seeks regularities or patterns in the results that can lead to a generalization. If this generalization is basic and can be communicated in a concise statement or a mathematical equation, the statement or equation is called a *law*.

3. *Hypothesis*. Researchers try to find reasons and explanations for the generalizations, patterns, and regularities they discover. A hypothesis expresses a tentative explanation of a generalization that has been stated. Further experiments then test the validity of the hypothesis.

4. *Theory*. The new experiments are carried out to test the hypothesis. If they support it without exception, the hypothesis becomes a theory. A theory is a tested model that explains some basic phenomenon of nature. It cannot be proven to be absolutely correct. As further research is performed to test the theory, it may be modified or a better theory may be developed.

The scientific method represents a systematic means of doing research. There are times when discoveries are made by accident, but most knowledge has been gained via careful, planned experimentation. In your study of chemistry you will examine the knowledge and understanding that researchers using the scientific method have uncovered.

3.2 Matter and Energy

We begin with the two things that describe the entire universe: *matter* and *energy*. **Matter** is *anything that occupies space and has mass*. That includes trees, clothing, water, air, people, minerals, and many other things. Matter shows up in a wide variety of forms.

Energy is *the ability to perform work*. Like matter, energy is found in a number of forms. Heat is one form of energy, and light is another. There are also chemical, electrical, and mechanical forms of energy. And energy can change from one form to another. In fact, matter can also change form or change into energy, and energy can change into matter, but not easily.

3.3 Law of Conservation of Mass and Energy

Learning Goal 2
Law of conservation of mass and energy

The **Law of Conservation of Mass** tells us that when a chemical change takes place, no detectable difference in the mass of the substances is observed. In other words, mass is neither created nor destroyed in an ordinary chemical reaction. This law has been tested by extensive experimentation in the laboratory, and the work of the brilliant French chemist-physicist Antoine Lavoisier provides evidence for this conclusion. Lavoisier performed many experiments involving matter. In one instance he heated a measured amount of tin and found that part of it changed to a powder. He also found that the *product* (powder plus tin) weighed *more* than the original piece of tin. To find out more about the added weight, he heated metals in sealed jars, which, of course, contained air. He measured the mass of his starting materials (*reactants*) and when the reaction concluded and the metal no longer changed to powder, he measured the mass of the products. In every such reaction, the mass of the reactants (oxygen from the air in the jar plus the original metal) equaled the mass of the products (the remaining metal plus the powder). Today we know that the reaction actually stopped when all of the oxygen in the sealed jar combined with the metal to form the powder. Lavoisier concluded that when a chemical change occurs, *matter is neither created nor destroyed, it just changes from one form to another* (Figure 3.1), which is a statement of the Law of Conservation of Mass.

Whenever a chemical change occurs, it is accompanied by an energy transformation. In the 1840s, more than half a century after Lavoisier, three scientists—the Englishman James Joule and the Germans Julius von Mayer and Hermann von Helmholtz—performed a number of experiments in which energy transformations were studied. They provided experimental evidence

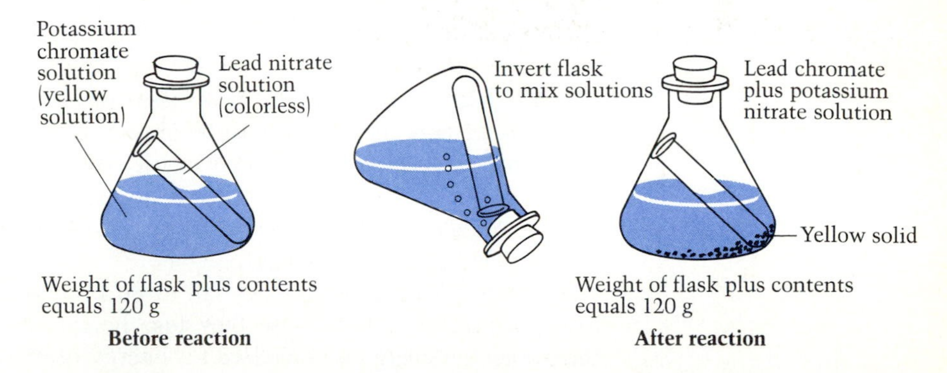

Figure 3.1
An experiment like Lavoisier's. An experimenter puts a test tube containing a lead nitrate solution into a flask containing a potassium chromate solution. The experimenter weighs the flask and contents, then turns the flask upside down to mix the two solutions. A chemical reaction takes place, producing a yellow solid. The experimenter weighs the flask and contents again and finds no change in mass.

that led to the discovery of the **Law of Conservation of Energy.** The law tells us that *in any chemical or physical change, energy is neither created nor destroyed, it is simply converted from one form to another*.

An auto engine provides a good example of how one form of energy is converted to a different form. *Electrical* energy from the battery generates a spark that contains *heat* energy. The heat ignites the gasoline–air mixture, which explodes, transforming chemical energy into heat and *mechanical* energy. The mechanical energy causes the pistons to rise and fall, rotating the engine crankshaft and moving the car.

At the same time, in the same engine, matter is changing from one form to another. When the gasoline explodes and burns, it combines with oxygen in the cylinders to form carbon dioxide and water vapor. (Unfortunately, carbon monoxide and other dangerous gases may also be formed. This is one of the major causes of air pollution, as you will see in Chapter 18.)

To appreciate the significance of these facts, think of the universe as a giant chemical reactor or system. At any given time there are certain amounts of matter and energy present, and the matter has a certain mass. Matter is always changing from one form to another, and so is energy. Besides that, matter is changing to energy and energy to matter. But *the sum of all the matter (or mass) and energy in the universe always remains the same*. This repeated observation is called the **Law of Conservation of Mass and Energy.**

3.4 Potential Energy and Kinetic Energy

Which do you think has more energy, a metal cylinder held 1 foot above the ground or an identical cylinder held 5 feet above the ground? If you dropped them on your foot, you would know immediately that the cylinder with more energy was the one that was 5 feet above the ground. But where does this energy come from?

Work had to be done to raise the two cylinders to their respective heights—to draw them up against the pull of gravity. And energy was needed to do that work. The energy used to lift each cylinder was "stored" in each cylinder. The higher the cylinder was lifted, the more energy was stored in it—due to its position. *Energy that is stored in an object by virtue of its position* is called **potential energy.**

If we drop the cylinders, they fall toward the ground. As they do so, they lose potential energy because they lose height. But now they are moving; their potential energy is converted to "energy of motion." The more potential energy they lose, the more energy of motion they acquire. *The energy that an object possesses by virtue of its motion* is called **kinetic energy.** The conversion of potential energy to kinetic energy is a very common phenomenon. It is observed in a wide variety of processes, from downhill skiing to the generation of hydroelectric power.

3.5 The States of Matter

Matter may exist in any of the three physical states: solid, liquid, and gas.

A **solid** has a definite shape and volume that it tends to maintain under normal conditions. The particles composing a solid stick rigidly to one another. Solids most commonly occur in the **crystalline** form, which means they have a fixed, regularly repeating, symmetrical internal structure. Diamonds, salt, and quartz are examples of crystalline solids. A few solids, such as glass and paraffin, do not have a well-defined crystalline structure, although they do have a definite shape and volume. Such solids are called **amorphous solids,** which means they have no definite internal structure or form.

A **liquid** does not have its own shape but takes the shape of the container in which it is placed. Its particles cohere firmly, but not rigidly, so the particles of a liquid have a great deal of mobility while maintaining close contact with one another.

A **gas** has no fixed shape or volume and eventually spreads out to fill its container. As the gas particles move about they collide with the walls of their container causing *pressure*, which is a force exerted over an area. Gas particles move independently of one another. Compared with those of a liquid or solid, gas particles are quite far apart. Unlike solids and liquids, which cannot be compressed very much at all, gases can both be compressed and expanded.

Often referred to as the fourth state of matter, **plasma** is *a form of matter composed of electrically charged atomic particles*. Many objects found in the earth's outer atmosphere, as well as many celestial bodies found in space (such as the sun and stars), consist of plasma. A plasma can be created by heating a gas to extremely high temperatures or by passing a current through it. A plasma responds to a magnetic field and conducts electricity well.

3.6 Physical and Chemical Properties

Learning Goal 3
Physical and chemical properties

Matter—whether it is solid, liquid, or gas—possesses two kinds of properties: physical and chemical. These unique properties separate one substance from another and ensure that no two substances are alike in every way. The **physical properties** are those that can be observed or measured without changing the chemical composition of the substance. These properties include state, color, odor, taste, hardness, boiling point, and melting point. A **physical change** is one that alters at least one of the physical properties of the substance without changing its chemical composition. Some examples of physical change are (1) altering the physical state of matter, such as what occurs when an ice cube is melted; (2) dissolving or mixing substances together, such as what happens when we make coffee or hot cocoa; and (3) altering the size or shape of matter, such as what happens when we grind or chop something.

Chemical properties stem from the ability of a substance to react or change to a new substance that has different properties. This often occurs in the presence of another substance. For example, iron reacts with oxygen to produce iron(III) oxide (rust). This is an example of a **chemical change.** The chemical properties can be *observed* or *measured* when a substance undergoes chemical change. The rusting of iron is an example of a chemical property of iron. When we pass an electric current through water, it decomposes to form hydrogen gas and oxygen gas. This reaction is an example of a chemical property of water.

Sometimes it is difficult to differentiate a chemical change from a physical change. In fact, physical changes almost always accompany chemical changes. Some of the signs of physical change that tell us that a chemical change has occurred include the presence of a large amount of heat or light, the presence of a flame, the formation of gas bubbles, a change in color or odor, or the formation of a solid material that settles out of a solution.

3.7 Mixtures and Pure Substances

Since matter consists of all the material things that compose the universe, many distinctly different types of matter are known. Matter that has a definite and *fixed composition* is called a **pure substance,** which is a substance that cannot be separated into any other form of matter by physical change. Some of the pure substances that you are familiar with are helium, oxygen, table salt, water, gold, and silver. Two or more pure substances can be combined to form a **mixture** whose *composition can be varied*. The substances in a mixture can be separated by physical means; we can separate the substances without chemical change.

Learning Goal 4
Types of matter

Matter can also be classified as heterogeneous or homogeneous (Figure 3.2). **Homogeneous matter** has the same parts with the same properties throughout, and **heterogeneous matter** is made up of different parts with different properties. A combination of salt and pepper is an example of heterogeneous matter, whereas a teaspoonful of sugar is an example of homogeneous matter. Another example of homogeneous matter is a teaspoonful of salt dissolved in a glass of water. We call this a **homogeneous mixture** because it is a uniform blend of two or more substances, and its proportion can be varied. In a homogeneous mixture every part is exactly like every other part. The salt can be separated from the water by physical means. Seawater and air are also examples of homogeneous mixtures.

We know that there are two types of homogeneous matter: pure substances and homogeneous mixtures. According to this classification scheme, matter can be broken down even further. Let's look at both homogeneous mixtures and heterogeneous mixtures. In a **heterogeneous mixture** different parts have different properties. A salt-sand mixture is heterogeneous because it is composed of two substances, *each of which retains its own unique properties*. It does not have the same composition or properties throughout, and its composition can be varied. A salt-sand mixture can be separated by physical

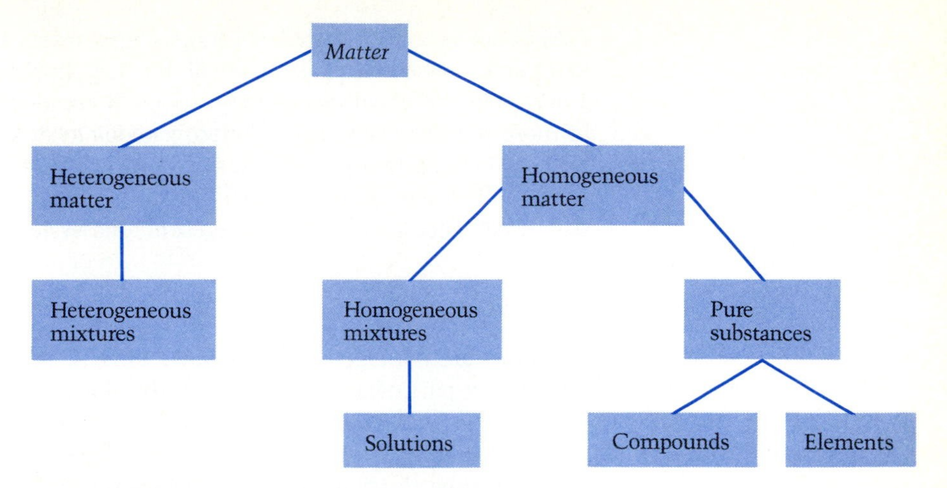

Figure 3.2
Classification of matter

means. If the mixture is placed in water, the salt will dissolve. The sand can be filtered out, and the salt can be recovered by heating the saltwater until the water evaporates (Figure 3.3).

We call any part of a system with uniform composition and properties a **phase.** A system is the body of matter being studied. A heterogeneous mixture is composed of two or more phases separated by physical boundaries.

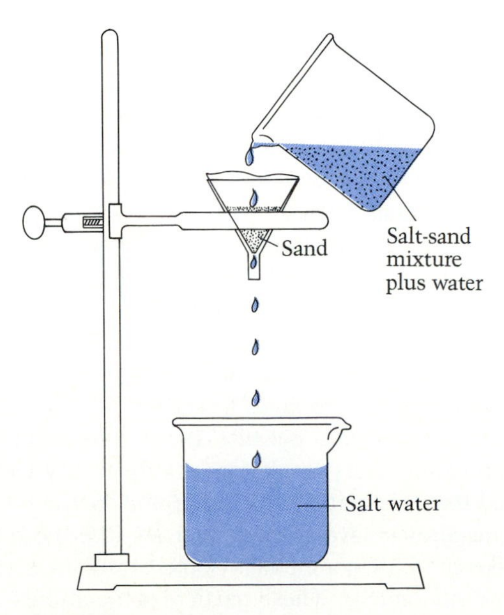

Figure 3.3
Separating sand from salt water

Additional examples of heterogeneous mixtures are oil and water (two liquids) and a tossed salad (several solids). It is important to note that although a pure substance is always homogeneous in composition, it may actually exist in more than one phase in a heterogeneous **system.** Think of a glass of ice water. This is a two-phase system composed of water in the solid phase and water in the liquid phase. In each phase the water is homogeneous in composition, but since two phases are present, the system is heterogeneous.

3.8 Solutions

Solutions are homogeneous mixtures. That is, they are uniform in composition. Every part of a solution is exactly like every other part. The salt-and-water mixture described in Section 3.7 is a solution. Even if we add more water to the solution, it will still be homogeneous, because the salt particles will continue to be distributed evenly throughout the solution. We would get the same result if we added more salt to the solution. However, we couldn't do this indefinitely. Homogeneity would end when the solution reached *saturation* (the point at which no more salt could dissolve in the limited amount of water). We will discuss this further in Chapter 15.

3.9 Elements

An **element** is *a pure substance that cannot be broken down into simpler substances, with different properties, by physical or chemical means*. The elements are the basic building blocks of all matter. There are now 109 known elements. Each has its own unique set of physical and chemical properties. (The elements are tabulated on the inside front cover of this book, along with their chemical symbols—a shorthand notation for their names.) The elements can be classified into three types: **metals, nonmetals,** and **metalloids.**

Examples of metallic elements are sodium (which has the symbol Na), calcium (Ca), iron (Fe), cobalt (Co), and silver (Ag). These elements are all classified as metals because they have certain properties in common. They have luster (in other words, they are shiny), they conduct electricity well, they conduct heat well, and they are malleable (can be pounded into sheets) and ductile (can be drawn into wires).

Some example of nonmetals are chlorine, which has the symbol Cl (note that the second letter of this symbol is a lower-case "el" and not the figure *one*), oxygen (O), carbon (C), and sulfur (S). Again, these elements are classified as nonmetals because they have certain properties in common. They don't shine, they don't conduct electricity well, they don't conduct heat well, and they are neither malleable nor ductile.

The metalloids have some properties like those of metals and other properties like those of nonmetals. Some examples are arsenic (As), germanium (Ge), and silicon (Si). These particular metalloids are used in manufacturing transistors and other *semiconductor devices*.

3.10 Atoms

Learning Goal 5
Difference between atom and molecule

Suppose we had a chunk of some element, say gold, and were able to divide it again and again, into smaller and smaller chunks. Eventually we could get a particle that could not be divided any further without losing its identity. This particle would be an atom of gold. An **atom** is *the smallest particle of an element that enters into chemical reactions.*

The atom is the ultimate particle that makes up the elements. Gold is composed of gold atoms, iron of iron atoms, and neon of neon atoms. These atoms are so small that billions of them are needed to make a speck large enough to be seen with a microscope. In 1970, Albert Crewe and his staff at the University of Chicago's Enrico Fermi Institute took the first black-and-white pictures of single atoms, using a special type of electron microscope. In late 1978, Crewe and his staff took the first time-lapse moving pictures of individual uranium atoms.

3.11 Compounds

Learning Goal 6
Law of Definite Composition

A **compound** is *a pure substance that is made up of two or more elements, chemically combined in a definite proportion by mass.* Unlike mixtures, compounds have a definite composition. Water, for instance, is made up of hydrogen and oxygen in the ratio of 11.1% hydrogen to 88.9% oxygen by mass. No matter what the source of the water, it is always composed of hydrogen and oxygen in this ratio. This idea, that *every compound is composed of elements in a certain fixed proportion,* is called the **Law of Definite Composition** (or the Law of Definite Proportions). It was first proposed by the French chemist Joseph Proust in about 1800.

The properties of a compound need not be similar to the properties of the elements that compose it. For example, water is a liquid, whereas hydrogen and oxygen are both gases. When two or more elements form a compound, they truly form a new substance.

Compounds can be broken apart into elements only by chemical means—unlike mixtures, which can be separated by physical means. More than 10 million compounds have been reported to date, and millions more may be discovered. Some compounds we are all familiar with are sodium chloride (table salt), which is composed of the elements sodium and chlorine, and sucrose (cane sugar), which is composed of the elements carbon, hydrogen, and oxygen.

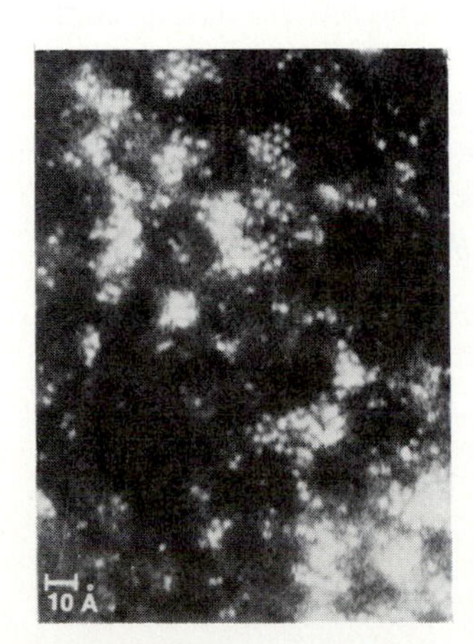

Uranium single atoms and microcrystals obtained from a solution of uranyl acetate
(Courtesy Albert V. Crewe)

3.12 Molecules

We have discussed what happens when a chunk of an element is continually divided: We eventually get down to a single atom. What happens when we keep dividing a chunk of a compound? Suppose we do so with the compound sugar. As we continue to divide a sugar grain, we eventually reach a small particle that can't be divided any further without losing the physical and chemical properties of sugar. This ultimate particle of a compound, *the smallest particle that retains the properties of the compound,* is called a **molecule.** Like atoms, molecules are extremely small, but with the aid of an electron microscope we can observe some of the very large and more complex molecules. Molecules are uncharged particles. That is, they carry neither a positive nor a negative electrical charge.

Molecules, as you might have guessed, are made up of two or more atoms. They may be composed of different kinds of atoms (for instance, water contains hydrogen and oxygen) or the same kinds of atoms (for instance, a molecule of chlorine gas contains two atoms of chlorine). (See Figure 3.4.) *The Law of Definite Composition states that the atoms in a compound are combined in definite proportions by mass*. We can see in Figure 3.4 that they are also combined in definite proportions by number. For example, water molecules always contain two hydrogen atoms for every oxygen atom.

3.13 Molecular versus Ionic Compounds

As we just learned, a molecule is the smallest *uncharged* part of a compound formed by the chemical combination of two or more atoms. Such compounds are known as *molecular compounds*, and we usually say that such compounds are composed of molecules. Water is a molecular compound composed of water molecules. Many molecular compounds are composed of atoms of nonmetallic elements that are chemically combined.

However, there are many compounds composed of oppositely charged *ions*. An ion is *a positively or negatively charged atom or group of atoms*. Compounds composed of ions are known as *ionic compounds*. These compounds are held together by attractive forces between the positive and negative ions that compose the compound. Ordinary table salt, sodium chloride, is such a compound. For these compounds, it is more proper to talk about a **formula unit** of the compound, rather than a molecule, as being *the smallest part of the compound that retains the properties of the compound*. Ionic com-

Figure 3.4
Molecules of water and chlorine

pounds are composed of metallic and nonmetallic elements. (We'll have much more to say about ions and ionic compounds when we discuss chemical bonding in Chapter 7.)

3.14 Symbols and Formulas of Elements and Compounds

We have already mentioned the chemical symbols—the shorthand for the names of the elements. In some cases, as in the symbol O (capital "oh") for oxygen, the chemical symbol is the first letter of the element's name, capitalized. Often, though, the symbol for an element contains two letters. In these cases only the first letter is capitalized; the second letter is never capitalized. For instance, the symbol for neon is Ne and the symbol for cobalt is Co. (Be careful of this: CO does not represent cobalt, but a combination of the elements carbon and oxygen, which is a compound.) Some symbols come from the Latin names of the elements: iron is Fe, from the Latin *ferrum,* and lead is Pb, from the Latin *plumbum*.

Because compounds are composed of elements, we can use the chemical symbols as a shorthand for compounds too. We use the symbols to write the **chemical formula,** which shows the elements that compose the compound. For example, sodium chloride (table salt), an ionic compound, contains one atom of sodium (Na) and one atom of chlorine (Cl). The formula unit for sodium chloride is NaCl. Water, a molecular compound, is another example: The water molecule contains two atoms of hydrogen (H) and one atom of oxygen, so we write it as H_2O. The number 2 in the formula indicates that there are two atoms of hydrogen in the molecule; note that it is written as a *subscript*. A molecule of ethyl alcohol contains two atoms of carbon, six of hydrogen, and one of oxygen. It is written as follows:

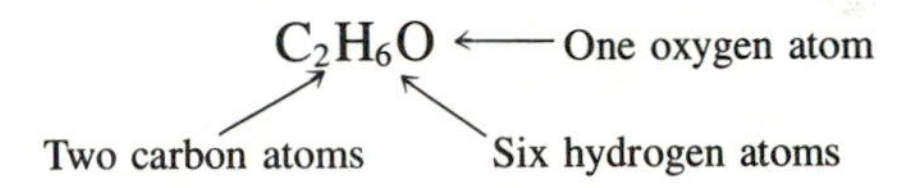

Example 3.1

State the number of atoms of each element in a molecule or formula unit of the following compounds: *(a)* $C_6H_{12}O_6$ *(b)* $Ca(OH)_2$ *(c)* $C_3H_6O_2$ *(d)* $Al_2(SO_4)_3$

Solution

Understand the Problem

We ask "What does the subscript mean?" Subscripts tell us the number of atoms in a molecule or formula unit of the substance.

Devise a Plan

Our plan will be to look at the subscripts noted in each case. We can then determine the number of atoms of each element in a molecule or formula unit of the compound.

Carry Out the Plan

(a) There are 6 atoms of C, 12 atoms of H, and 6 atoms of O.
(b) There is 1 atom of Ca, 2 atoms of O, and 2 atoms of H. (In this case, the subscript 2 means multiply everything inside the parentheses by 2.)
(c) There are 3 atoms of C, 6 atoms of H, and 2 atoms of O.
(d) There are 2 atoms of Al, 3 atoms of S, and 12 atoms of O. (In this case, the subscript 3 means multiply everything inside the parentheses by 3.)

Look Back

Recheck to be sure that you have followed the steps correctly.

Practice Exercise 3.1 State the number of atoms of each element in a molecule or formula unit of the following compounds: *(a)* C_2H_7N *(b)* $(NH_4)_2SO_4$

3.15 Atomic Mass

Learning Goal 7
Atomic mass, formula mass, and molecular mass

Suppose we want to find the relative masses of the atoms of the various elements. Suppose also that we have a double-pan balance that can weigh a single atom. We begin by assigning an arbitrary mass to one of the elements. Let's say we decide to assign a mass of 1 unit to hydrogen. (This is what chemists did originally, because hydrogen was known to be the lightest element even before the relative atomic masses were determined.)

Then, to find the relative mass of, say, carbon, we place an atom of carbon on one pan. We place hydrogen atoms on the other pan, one by one, until the pans are exactly balanced. We find that it takes 12 hydrogen atoms to balance 1 carbon atom, so we assign a relative atomic mass of 12 to carbon. In the same way for an oxygen atom, we find that it takes 16 hydrogen atoms to balance 1 oxygen atom. So we assign a relative mass of 16 to oxygen. By doing this experiment for all the other elements, we can determine all the masses relative to hydrogen (which is assigned mass 1).

Unfortunately this kind of balance has never existed, and more elaborate means had to be developed to find the relative atomic masses of the elements. But the logic we used here is the same as that used by the many scientists who determined the relative masses. (The atomic mass scale has been revised. The present scale is based on a particular type of carbon atom, called carbon-12. This carbon is assigned the value of 12 atomic mass units, amu.) The periodic table (inside the front cover) lists the relative atomic masses of the elements below their symbols. From now on, instead of referring to "relative atomic mass," we will use the simpler term **atomic mass.**

Example 3.2

Using the periodic table (inside the front cover), look up the atomic masses of the following elements. (For this exercise, round all atomic masses to one decimal place.) *(a)* I *(b)* Ba *(c)* As *(d)* S

Solution Look up the atomic mass of each element in the periodic table. Remember, the atomic mass is the number below the symbol of the element.
(a) I is 126.9.
(b) Ba is 137.3.
(c) As is 74.9.
(d) S is 32.1.

Practice Exercise 3.2 Using the periodic table (inside the front cover), look up the atomic masses of the following elements. (For this exercise, round all atomic masses to one decimal place.) *(a)* La *(b)* Fe *(c)* Ar *(d)* Sn

3.16 Formula Mass and Molecular Mass

Learning Goal 8
Formula or molecular mass of a compound from the formula

The **molecular mass** of a compound is *the sum of the atomic masses of all the atoms that make up a molecule of the compound*. The term *molecular mass* is applied to compounds that exist as molecules. For example, the molecule P_2O_5 has two phosphorus atoms and five oxygen atoms. The atomic mass of each P (to one decimal place) is 31.0, and the atomic mass of each O (to one decimal place) is 16.0. Therefore the molecular mass of the compound is

$$(2 \times 31.0) + (5 \times 16.0) = 62.0 + 80.0 = 142.0$$

(Check this calculation and see whether you get the same answer. If not, you may have forgotten that in an equation like this, multiplication and division are done *before* addition and subtraction.)

The **formula mass** of a compound is *the sum of the atomic masses of all the ions that make up a formula unit of the compound*. The term *formula mass* is applied to compounds that are written as formula units and exist mostly as ions (charged atoms or groups of atoms). For example, the formula unit of aluminum oxide is Al_2O_3. This means that a formula unit of aluminum oxide has two aluminum atoms (actually aluminum ions) and three oxygen atoms (actually oxide ions). The atomic mass of aluminum (to one decimal place) is 27.0, and the atomic mass of oxygen (to one decimal place) is 16.0. Therefore the formula mass of Al_2O_3 is

$$(2 \times 27.0) + (3 \times 16.0) = 54.0 + 48.0 = 102.0$$

Now let's try to determine the formula and molecular masses of some additional compounds.

An EPA cleanup crew wears self-contained breathing apparatus and protective clothing while cleaning up hazardous waste. Many hazardous wastes are compounds that have high molecular masses.
(Steve Delaney/EPA)

Example 3.3

Find the molecular or formula masses (to one decimal place) of the following compounds: *(a)* H_2O *(b)* NaCl *(c)* $Ca(OH)_2$ *(d)* $Zn_3(PO_4)_2$
Note: In a chemical formula, parentheses followed by a subscript mean that everything inside the parentheses is multiplied by the subscript. For example, in one formula unit of $Ca(OH)_2$, there is one Ca atom plus two O atoms and two H atoms.

Solution We must find the atomic mass of each element in the periodic table and then add the masses of all the atoms in each compound.
(a) The atomic mass of H is 1.0, and the atomic mass of O is 16.0.

$$\text{Molecular mass of } H_2O = (2 \times 1.0) + (1 \times 16.0)$$
$$= 2.0 + 16.0 = 18.0$$

(b) The atomic mass of Na is 23.0, and the atomic mass of Cl is 35.5.

$$\text{Formula mass of NaCl} = (1 \times 23.0) + (1 \times 35.5)$$
$$= 23.0 + 35.5 = 58.5$$

(c) The atomic mass of Ca is 40.1, the atomic mass of O is 16.0, and the atomic mass of H is 1.0.

$$\text{Formula mass of } Ca(OH)_2 = (1 \times 40.1) + (2 \times 16.0) + (2 \times 1.0)$$
$$= 40.1 + 32.0 + 2.0 = 74.1$$

(d) The atomic mass of Zn is 65.4, the atomic mass of P is 31.0, and the atomic mass of O is 16.0.

$$\text{Formula mass of } Zn_3(PO_4)_2 = (3 \times 65.4) + (2 \times 31.0) + (8 \times 16.0)$$
$$= 196.2 + 62.0 + 128.0 = 386.2$$

Practice Exercise 3.3 Find the molecular or formula masses (to one decimal place) of the following compounds: *(a)* Na_2CO_3 *(b)* $CoCl_2$ *(c)* Cl_2O *(d)* N_2O_4

Summary

The body of knowledge called science has been developed through the scientific method: observation and classification of data, generalization of observations, and testing of generalizations. Of importance in all the sciences is the idea that the universe is made up of only matter and energy. Matter and energy can be neither created nor destroyed, but they can be changed to other forms. And matter can be transformed into energy, and vice versa. Potential energy is energy that is stored in a body because of its position. Kinetic energy is energy that is due to motion.

Matter may exist in any of three states—solid, liquid, or gas—and may be either heterogeneous (nonuniform) or homogeneous (uniform). Mixtures are combinations of two or more kinds of matter, each retaining its own chemical and physical properties. Mixtures too may be either homogeneous or heterogeneous, and solutions are homogeneous mixtures. Elements are the basic building blocks of matter. The 109 known elements are, in turn, made up of atoms. A compound is a substance that is made up of two or more elements chemically combined in definite proportions by mass. Molecules are the smallest particles that retain the properties of a compound, and atoms are the smallest particles that enter into chemical reactions.

Each element has its own symbol, and every compound has its own chemical formula. Each element has a unique atomic mass. The atomic mass of an element is found in the periodic table. The formula mass or molecular mass of a compound is the sum of the atomic masses in a formula unit or molecule of the compound.

Key Terms

amorphous solid **(3.5)**
atom **(3.10)**
atomic mass **(3.15)**
chemical change **(3.6)**
chemical formula **(3.14)**
chemical property **(3.6)**
compound **(3.11)**
crystalline **(3.5)**
element **(3.9)**
energy **(3.2)**
formula mass **(3.16)**
formula unit **(3.13)**

gas **(3.5)**
heterogeneous matter **(3.7)**
heterogeneous mixture **(3.7)**
homogeneous matter **(3.7)**
homogeneous mixture **(3.7)**
kinetic energy **(3.4)**
Law of Conservation of Energy **(3.3)**
Law of Conservation of Mass **(3.3)**
Law of Conservation of Mass and Energy **(3.3)**
Law of Definite Composition **(3.11)**
liquid **(3.5)**
matter **(3.2)**
metalloid **(3.9)**
metal **(3.9)**
mixture **(3.7)**
molecular mass **(3.16)**
molecule **(3.12)**
nonmetal **(3.9)**
phase **(3.7)**
physical change **(3.6)**
physical property **(3.6)**
plasma **(3.5)**
potential energy **(3.4)**
pure substance **(3.7)**
scientific method **(3.1)**
solid **(3.5)**
solution **(3.8)**
system **(3.7)**

CHEMICAL FRONTIERS

NEW CERAMIC MATERIAL FOR AUTOMOBILE ENGINES OF THE FUTURE

When people think of ceramic materials, typically they think of dinnerware and pottery. Conventional ceramic materials are brittle, and they crack and break easily. Recently, however, Fumihiro Wakai and his coworkers at the Government Industrial Research Institute in Nagoya, Japan, announced that they had developed a new type of stretchable ceramic. This new ceramic was made from the compounds silicon nitride, silicon carbide, zirconium oxide, and yttrium oxide.

The scientists prepared strips of these materials and heated them to a temperature of 1,600°C. During the heating step, the strips were pulled in opposite directions. The strips stretched like taffy to more than 2.5 times their original length. Wakai and his colleagues think that this behavior is due to the formation of liquid regions within the crystalline structure of the ceramic.

This new ceramic material has the potential to revolutionize the design of automobile engines. Automobile manufacturers would like to make certain engine parts out of ceramics rather than metals, since ceramics can handle heat stress better than metals. In addition, ceramic engines could be run at higher temperatures than metal engines, which means better fuel efficiency—an important consideration for automobiles of the future. The properties of conventional ceramics do not allow them to be molded into shapes precise enough for engine parts, but this new ceramic material can be molded into shapes with little or no machining. Perhaps your next new car will have a ceramic engine.

Self-Test Exercises

Learning Goal 1: The Scientific Method

1. Explain the difference between a theory and a scientific law.

▲ **2.** Suppose you are a researcher and you believe you have found a drug that can cure the common cold. How would you use the scientific method to determine whether the drug is effective?

Learning Goal 2: Law of Conservation of Mass and Energy

▲ **3.** Explain what the Law of Conservation of Mass and Energy means in terms of the workings of our universe.

4. What is the significance of the Law of Conservation of Mass and Energy in terms of your study of chemistry?

Learning Goal 3: Types of Matter

5. Match each word on the left with its definition on the right.

(a) Homogeneous	(1) The basic building block of matter
(b) Heterogeneous	(2) The word used to describe matter that is uniform throughout
(c) Mixture	(3) A type of matter in which each part retains its own properties
(d) Compound	(4) A chemical combination of two or more elements
(e) Element	(5) The word used to describe matter that is not uniform throughout

▲ **6.** *(a)* Distinguish among an element, a compound, and a mixture.
(b) What type of matter is uniform throughout?
(c) What type of matter is not uniform throughout?

Learning Goal 4: Physical and Chemical Properties

▲ **7.** State whether each of the following processes involves chemical or physical properties:
(a) Paper burning
(b) Glass breaking
(c) An egg being fried
(d) Oil burning
(e) Wood being chopped
(f) Food being digested

8. Determine whether each of the following processes involves chemical or physical properties:
(a) Ice melts.
(b) Sugar dissolves in water.
(c) Milk sours.
(d) Eggs become rotten.
(e) Water boils.
(f) An egg is hard-cooked.

Learning Goal 5: Difference Between Atom and Molecule

9. Describe the difference between an atom and a molecule.

▲ **10.** *(a)* What is the smallest particle of matter that can enter into a chemical combination?
(b) What is the smallest uncharged individual unit of a compound that is composed of two or more atoms?

11. Classify each of the following elements as metal, metalloid, or nonmetal: *(a)* Ba *(b)* Si *(c)* O *(d)* Hg *(e)* Ge *(f)* In *(g)* U

12. Classify each of the following elements as metal, metalloid, or nonmetal: *(a)* Mn *(b)* Nd *(c)* Al *(d)* At *(e)* Pt *(f)* Cl *(g)* Ra

13. Explain the difference between Co and CO.

14. Explain the difference between Si and SI.

Learning Goal 6: Law of Definite Composition

15. State the Law of Definite Composition (Proportions).

16. Give an example of the Law of Definite Composition (Proportions).

17. State the number of atoms of each element in a molecule or formula unit of the following compounds:
(a) $C_{12}H_{22}O_{11}$ *(b)* K_2CrO_4 *(c)* $H_8N_2O_3S_2$
(d) $Zn(NO_3)_2$

▲ **18.** State the number of atoms of each element in a molecule or formula unit of the following compounds:
(a) H_2SeO_4 *(b)* $C_{21}H_{27}FO_6$ *(c)* $(NH_4)_3PO_4$
(d) $Fe_3(AsO_4)_2$

Learning Goal 7: Atomic Mass, Formula Mass, and Molecular Mass

19. What is the molecular mass of a compound?
20. What is the formula mass of a compound?

▲***21.** If in the periodic table oxygen were assigned an atomic mass of 1, what would be the atomic mass of sulfur?
***22.** If in the periodic table neon were assigned an atomic mass of 1, what would be the atomic mass of bromine?

23. Using the periodic table (inside front cover), look up the atomic masses of the following elements. (For this exercise, round all atomic masses to one decimal place.) *(a)* Rb *(b)* Cr *(c)* U *(d)* Se *(e)* As
24. Using the periodic table (inside front cover), look up the atomic masses of the following elements. (For this exercise, round all atomic masses to one decimal place.) *(a)* S *(b)* N *(c)* Li *(d)* Cs *(e)* Au

Learning Goal 8: Formula or Molecular Mass of a Compound from the Formula

▲ **25.** Determine the molecular or formula mass of each of the following compounds. (For this exercise, round all atomic masses to one decimal place.) *(a)* Al_2O_3 *(b)* $CuBr_2$ *(c)* H_3PO_4 *(d)* LiCl *(e)* $Ca(OH)_2$ *(f)* $Fe_2(SO_4)_3$ *(g)* $(NH_4)_2S$ *(h)* $CoCl_2$
26. Determine the molecular or formula mass of each of the following compounds. (For this exercise, round all atomic masses to one decimal place.) *(a)* H_2O *(b)* H_2SO_4 *(c)* NaCl *(d)* $Ca_3(PO_4)_2$ *(e)* P_2O_5 *(f)* $SrSO_4$ *(g)* C_2H_6O *(h)* SO_2

27. Determine the molecular or formula mass of each of the following compounds. (For this exercise, round all atomic masses to one decimal place.) *(a)* SiO_2 *(b)* H_2SO_3 *(c)* $Sr(OH)_2$ *(d)* RbF *(e)* $Cu(NO_3)_2$ *(f)* $CoBr_2$ *(g)* $(NH_4)_3PO_4$ *(h)* $HC_2H_3O_2$
▲ **28.** Determine the molecular or formula mass of each of the following compounds. (For this exercise, round all atomic masses to one decimal place.) *(a)* LiOH *(b)* Na_2CO_3 *(c)* $CoCl_2$ *(d)* NaBr *(e)* SO_3 *(f)* C_2H_6 *(g)* OF_2 *(h)* $(NH_4)_2SO_3$

Extra Exercises

29. Explain in detail what the chemical formula H_2O means.

▲ **30.** Make a list of heterogeneous mixtures and homogeneous mixtures that you encounter in everyday life. Do the same for elements and compounds.

31. Write the names and symbols for the 14 elements that have a one-letter symbol.

32. Write the names and symbols of all the metalloids.

▲ **33.** How many metals are there in the periodic table? How many nonmetals? How many metalloids?

34. Write the names of the 11 elements whose symbols are not derived from their English names.

35. Name the elements present in each of the following compounds: *(a)* $MgCl_2$ *(b)* N_2O *(c)* $(NH_4)_2SO_4$ *(d)* H_3PO_4

36. Write the chemical formula of each of the following, given the number of atoms in a molecule or formula unit of the compound:
(a) one nitrogen atom, two oxygen atoms (nitrogen dioxide)
(b) two sodium atoms, one sulfur atom (sodium sulfide)
(c) three potassium atoms, one arsenic atom, four oxygen atoms (potassium arsenate)
(d) two phosphorus atoms, five oxygen atoms (diphosphorus pentoxide)

37. Classify each of the following as an element, compound, or mixture:
(a) gold
(b) air
(c) carbon dioxide
(d) wine
(e) table salt

38. State whether each of the following involves a physical or chemical change:
(a) toasting bread
(b) water freezing
(c) tearing paper
(d) burning wood

39. Determine the molecular or formula mass of each of the following compounds. (For this exercise, round all atomic masses to one decimal place.) *(a)* OsO_4 *(b)* HNO_3 *(c)* $Fe(OH)_2$ *(d)* $Ba_3(PO_4)_2$

40. Explain the difference between Hf and HF.

Cumulative Review/Chapters 1–3*

Indicate whether each of the following statements is true or false.

1. Chemistry is the science that deals with matter and the changes it undergoes.

2. Organic chemicals are those chemicals that have been derived from living or once-living organisms.

3. Sugar and salt are examples of inorganic chemicals.

4. The art of *khemia* flourished until A.D. 1600.

5. Alchemists were able to transmute lead and iron into gold.

6. The idea that every theory must be proved by experiment was advanced by Robert Boyle.

7. In 420 B.C., Greek philosophers were able to prove that atoms existed.

8. Some carbon-containing compounds are classified as inorganic chemicals.

9. The two main branches of chemistry that existed during the 1700s and 1800s still exist today.

10. Nuclear chemistry, biochemistry, and analytical chemistry are three subdivisions into which chemistry can be divided.

Answer the following questions to help sharpen your test-taking skills.

▲ **11.** Find the area of a rectangular room that measures 12.5 m by 10.2 m. Use the proper number of significant figures in reporting your answer.

12. Calculate the volume of a cardboard box that measures 15.25 cm long, 12.00 cm wide, and 24.85 cm high.

13. Determine the volume of a cylindrical solid whose radius is 14.50 cm and whose height is 25.05 cm.

14. Convert 0.28 kg to (*a*) grams, (*b*) decigrams, (*c*) milligrams.

15. Convert 6.8 m to (*a*) decimeters, (*b*) centimeters, (*c*) millimeters.

16. Convert 125 mm to (*a*) centimeters, (*b*) meters, (*c*) kilometers.

*Answers to cumulative-review questions are given in the back of the book.

17. Convert 25,595 mL to (*a*) liters, (*b*) deciliters.

18. Express 50.0 m in (*a*) feet, (*b*) inches.

19. Express 25.55 g in (*a*) pounds, (*b*) ounces.

20. Determine the number of cubic centimeters in a cubic inch without using a conversion table.

21. Express 165.00 g in pounds.

22. Express 12.0 gallons in (*a*) liters, (*b*) ounces.

23. Determine the density of a cube that has a mass of 500.0 g and measures 12.0 cm on each side.

24. A spherical balloon with a volume of 113.0 mL is filled with an unknown gas weighing 20.0 g. Will the balloon float on air?

▲ **25.** A small metal sphere weighs 90.0 g. The sphere is placed in a graduated cylinder containing 15.0 mL of water. Once the sphere is submerged, the water in the cylinder measures 30.0 mL. What is the density of the sphere?

26. A small metal sphere with a density of 3.50 g/cm^3 has a mass of 937.83 g. Calculate the radius of the sphere.

27. Solve the following problems using the proper number of significant figures:

(*a*) $28.64 + 3.2$
(*b*) $125.4 \div 13.5$
(*c*) 6.55×12.1
(*d*) $98.4 - 0.12$

▲ **28.** Express each of the following numbers in scientific notation: (*a*) 5,000 (*b*) 0.0005 (*c*) 602,300,000 (*d*) 35,000,000

29. How many significant figures are there in each of the following numbers? (*a*) 5,500.0 (*b*) 0.5123 (*c*) 12.000 (*d*) 3,500

30. Convert each of the following temperatures from °F to °C: (*a*) 45.0°F (*b*) −10.0°F (*c*) 450°F (*c*) −100.0°F

31. Convert each of the following temperatures from °C to °F: (*a*) 88.0°C (*b*) −12.5°C (*c*) 65.6°C (*d*) 400°C

▲ **32.** Distinguish between an amorphous solid and a crystalline solid.

33. Determine whether each of the following processes involves chemical or physical changes:
(a) A match burns.
(b) Glucose dissolves in water.
(c) Bread becomes moldy.
(d) A piece of wood is sawed.

34. Distinguish between heterogeneous and homogeneous matter.

35. State the number of atoms of each element in a molecule or formula unit of the following compounds: *(a)* $Zn(C_2H_3O_2)_2$ *(b)* $(NH_4)_2CrO_4$

36. Why do we use the term *molecular mass* for some compounds and *formula mass* for other compounds?

37. If in the periodic table calcium were assigned an atomic mass of 1, what would be the atomic mass of mercury?

38. Using the periodic table (inside front cover), look up the atomic masses of the following elements. (For this exercise, round all atomic masses to one decimal place.) *(a)* Yb *(b)* At *(c)* P *(d)* Ag

39. Determine the molecular or formula mass of each of the following compounds. (For this exercise, round all atomic masses to one decimal place.)
(a) $Fe(C_2H_3O_2)_3$ *(b)* $(NH_4)_2SO_4$ *(c)* $C_9H_8O_4$
(d) $Co(SO_4)_2$

40. Explain in detail what the chemical formula $Al_2(SO_4)_3$ means.

41. Determine whether each of the following is an element, compound, or mixture:
(a) air
(b) arsenic
(c) carbon dioxide
(d) water
(e) gold
(f) root beer soda
(g) gasoline

42. Determine whether each of the following is an example of a heterogeneous or homogeneous mixture:
(a) beach sand
(b) ethyl alcohol and water
(c) tossed salad
(d) soda water (club soda)

43. Using the periodic table (inside front cover), write the names and symbols for all elements whose symbol begins with the letter *A* (*Hint:* There are eight.)

44. Write the names and symbols of all the nonmetallic elements.

45. Name the elements present in each of the following compounds: *(a)* K_2S *(b)* Ag_2CrO_4 *(c)* $KMnO_4$
(d) $Hg_3(PO_4)_2$

46. Write the chemical formula of each of the following, given the number of atoms in a molecule or formula unit of the compound:
(a) two nitrogen atoms, one oxygen atom (dinitrogen monoxide)
(b) two potassium atoms, one chromium atom, four oxygen atoms (potassium chromate)
(c) one nitrogen atom, three hydrogen atoms (ammonia)
(d) one strontium atom, one sulfur atom, four oxygen atoms (strontium sulfate)

47. State whether each of the following involves a physical or chemical change:
(a) a candle burning
(b) table sugar dissolving in water
(c) tooth decaying
(d) snow melting

48. Explain the difference between No and NO.

▲ **49.** A piece of paper is weighed and is then burned. The resulting ash weighs less than the paper. Is this a violation of the Law of Conservation of Mass? Explain.

50. You are given a mixture of sodium chloride (table salt) and sand. Explain how you would separate this mixture.

4 Atomic Theory, Part 1: What's in an Atom?

Learning Goals

After you've studied this chapter, you should be able to:

1. Describe Dalton's atomic theory, Thomson's model of the atom, and Rutherford's model of the atom.
2. Give the charge and mass of the electron, proton, and neutron.
3. Find the number of protons, electrons, and neutrons in an atom of an element.
4. Explain what isotopes are and give examples.
5. Use the standard isotopic notation for mass number and atomic number.
6. Calculate the average atomic mass of an element when you know the relative abundances of its isotopes.
7. Calculate the percentage abundance of two isotopes when you know the atomic mass of each isotope and the average atomic mass.
8. Use the table of relative abundances of isotopes to calculate the number of atoms or grams of a particular isotope present in a given sample.

Introduction

Around 420 B.C., Democritus developed the idea that the atom—indestructible and indivisible—was the smallest particle of matter. Twenty-two hundred years later, John Dalton reinforced this idea in his atomic theory. Yet less than a century after Dalton presented his theory, scientists had adopted a different model, in which the atom was made up of several even smaller particles. The development of that model is the subject of this chapter.

4.1 What Is a Model?

Learning Goal 1
Atomic theory and models

Before we try to discuss the atom, we should recall that until very recently no one had ever seen one. How, then, did we know what an atom was like? We really didn't, exactly. But scientists have put together observations, experimental results, and a good deal of reasoning and have come up with a

John Dalton
(Reprinted with permission from *Torchbearers of Chemistry* by Henry Monmouth Smith, © Academic Press, Inc., 1949)

model. This model of the atom is just a representation of what scientists believe the atom is like. Models can have different degrees of accuracy. Take a model of a car, for instance. It could be carved from a solid block of wood or plastic and represent just the shape of the car. Or a more detailed model could show the interior of the car as well. A still more detailed model might have a gasoline-powered motor and move just like a real car. This model would represent the car most accurately. The car model would be a physical model, whereas models of the atom are *intellectual* or *thought* models; however, they too have become more accurate as scientists have gathered more information about atoms.

4.2 Dalton's Atomic Theory

John Dalton (1766–1844) was a British schoolteacher who studied physics and chemistry. On the basis of facts and experimental evidence, he proposed an atomic theory. This theory was an attempt to explain all the different forms of matter. It has turned out to be one of the greatest contributions to chemistry since the time of the ancient Greeks. Dalton theorized that:

1. All elements are composed of tiny, indivisible particles called atoms. This idea was similar to Democritus' idea. Both men believed there was an ultimate particle—that all matter was composed of tiny indestructible and indivisible spheres.
2. All matter is composed of combinations of these atoms. In Chapter 3 we discussed how the atoms of different elements combine to produce formula units that make up compounds. For example, two atoms of hydrogen and one atom of oxygen combine to form a molecule of water.
3. Atoms of different elements are different. Dalton believed gold was different from silver because somehow the atoms of gold are different from the atoms of silver. For example, he thought they differed in mass.
4. Atoms of the same element have the same size, mass, and form. Dalton believed all gold atoms, for example, are the same as all other gold atoms in every respect.

Dalton's atomic theory laid the foundation for what is believed today. But his basic idea about the indivisibility of the atom was eventually shown to be incorrect, as was his idea that all atoms of the same element have the same size, mass, and form. Revisions of his theory have led to our current model of the atom. Let us review how our present-day model of the atom was developed.

4.3 The Discovery of the Electron

In the mid-1800s, scientists began to question whether the atom was really indivisible. They also wanted to know why atoms of different elements had different properties. Some of the answers came with the invention of the Crookes tube, or *cathode-ray tube* (Figure 4.1).

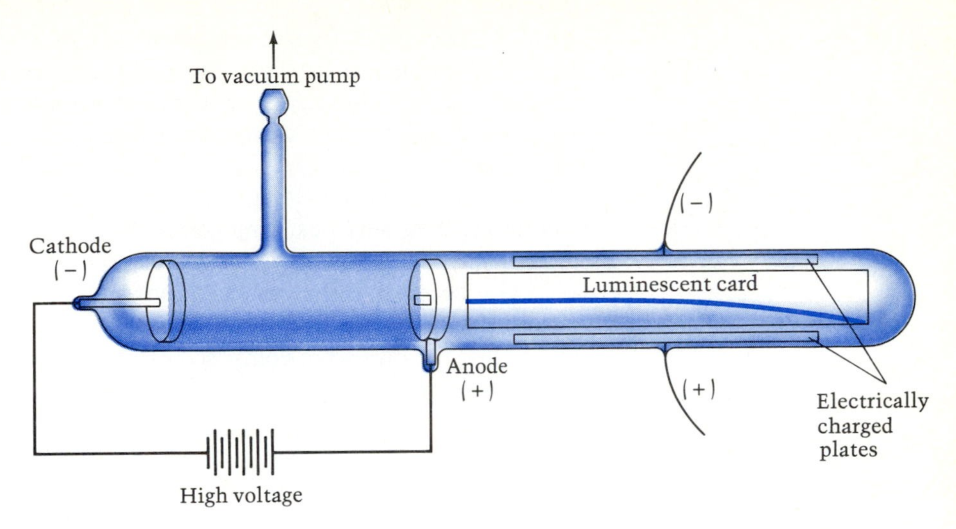

Figure 4.1
A Crookes or cathode-ray tube

The scientists of the time knew that some substances conduct electric current (that is, they are conductors), whereas other substances do not. And, with enough electrical power, a current can be driven through any substance—solid, liquid, or gas. Electric current results from the accumulation of electric charge. Electric charge may be transferred from one object to another. We find that electric charge is either positive or negative. Experiments tell us that unlike charges attract, and like charges repel. In the cathode-ray tube, a high-voltage electric current is driven through a nearly empty tube, a near **vacuum.** The tube contains two pieces of metal called *electrodes*. Each electrode is attached by a wire to the source of an electric current. The source has two *terminals,* positive and negative. *The electrode attached to the positive electric terminal* is called the **anode**; *the electrode attached to the negative terminal* is called the **cathode.** Crookes showed that when the current was turned on, a beam moved from the cathode to the anode; in other words, the beam moved from the negative to the positive terminal. Because the beam originated at the *negative* terminal, and because it was attracted to the *positive* terminal, the beam had to be negative in nature.

Learning Goal 2
Mass and charge of electrons, protons, and neutrons

What was this beam? Was it made of particles or waves? Did it come from electricity or from the metal electrodes? Physicists in Crookes's time were not sure about the answers to these questions, but they did make guesses. Whatever this beam was, it traveled in straight lines (they knew that because it cast sharp shadows). For lack of a better name, the German physicist Eugen Goldstein called the beams *cathode rays,* because they came from the cathode.

The German physicists in Crookes's time favored the *wave theory* of cathode rays because the beam traveled in straight lines, like ocean waves.

But the English physicists favored the *particle theory*. They said that the beam was composed of tiny particles that moved very quickly—so quickly that they were hardly influenced by gravity. That was why the particles moved in a straight path. (Note how a single experimental observation led to two different theories.)

Crookes proposed a method to solve the dilemma. If the beam was composed of negative particles, a magnet would deflect them (Figure 4.2). But if the beam was a wave, a magnet would cause almost no deflection. Particles would also be more easily deflected by an electric field. In 1897 the English physicist J. J. Thomson used both these techniques—magnetic and electric—to show that the rays were composed of particles (Figure 4.3). Today we call these particles **electrons.** Moreover, we understand that an electric current is actually a flow of electrons.

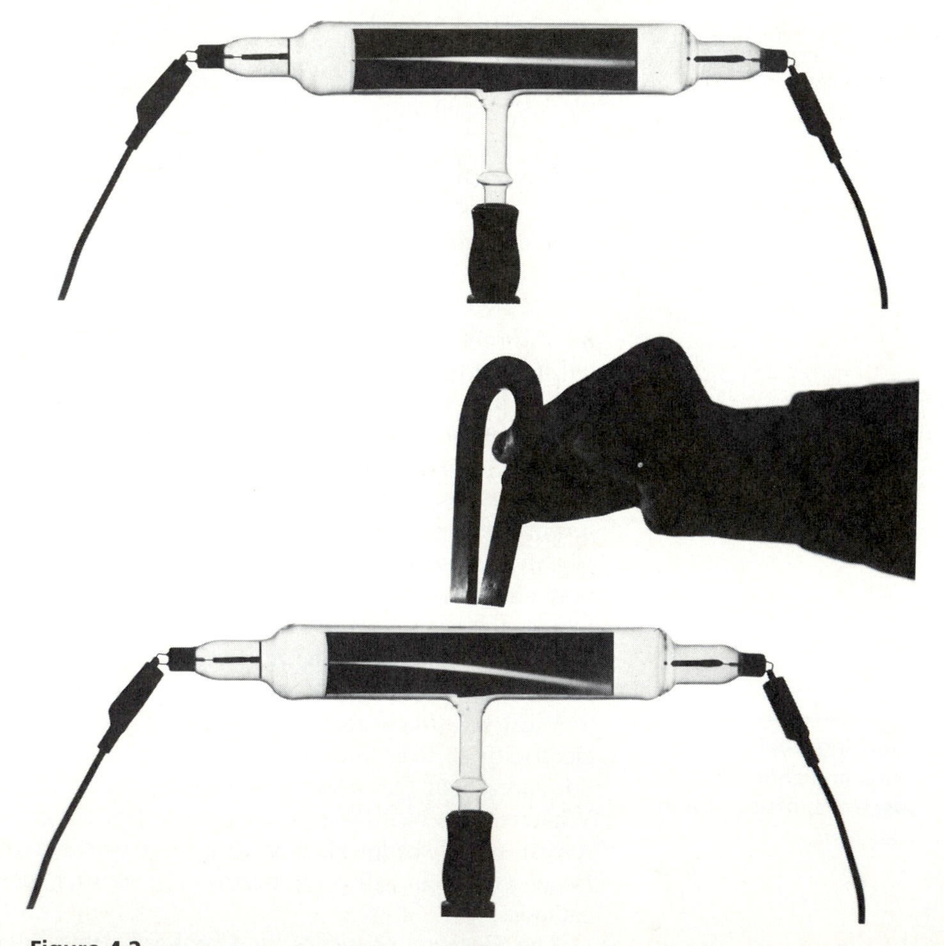

Figure 4.2
A cathode-ray tube in operation. Compare these pictures with Figure 4.3.
(Fundamental Photographs, New York)

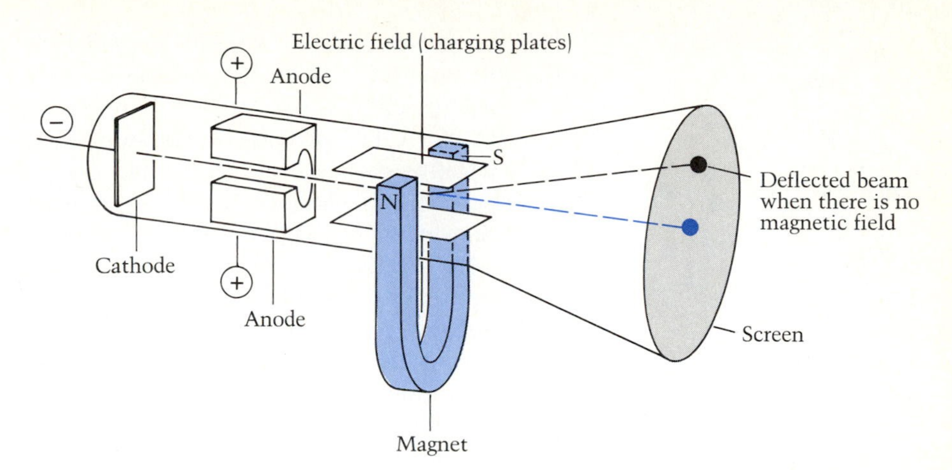

Figure 4.3
The Thomson experiment: (1) A beam of electrons (dashed line) moves from cathode toward anodes. (2) Some electrons pass between anodes. (3) Electric field causes beam of electrons to bend. This is visible on the screen. (4) But by adding a magnet to counteract deflection caused by electric field, one can make the beam follow a straight path. (5) One can then measure the strengths of the electric and magnetic fields and calculate the charge-to-mass ratio (e/m) of the electron.

Were the electrons in the beam coming from the metal electrodes or from the current source? Unless it was the metal that gave off these electrons, the atom could still be the smallest particle of matter. Proof that metals do give off electrons came from the laboratories of Phillipp Lenard, a German physicist. In 1902 he showed that ultraviolet light directed onto a metal makes it send out, or emit, electrons. This effect, known as the *photoelectric effect,* indicated that the metal atoms—and the atoms of other elements—contain electrons.

In 1911 a young American physicist named Robert Millikan calculated the mass of the electron: 9.11×10^{-28} grams. (To get an idea of how very small this is, consider the fact that it would take about 16,000,000,000,000,000,000,000,000,000, or 1.6×10^{28}, electrons to equal the weight of a standard half-ounce chocolate bar.)

4.4 The Proton

The discovery of the electron as part of the atom and as a *negative* particle of electricity was very important, but it raised many questions. Because electrons are part of all atoms, and because atoms in their normal state are electrically *neutral,* scientists reasoned, there must be something else in the atom that balances the negative electron. Otherwise, if matter had a negative charge, you would get a shock every time you touched anything.

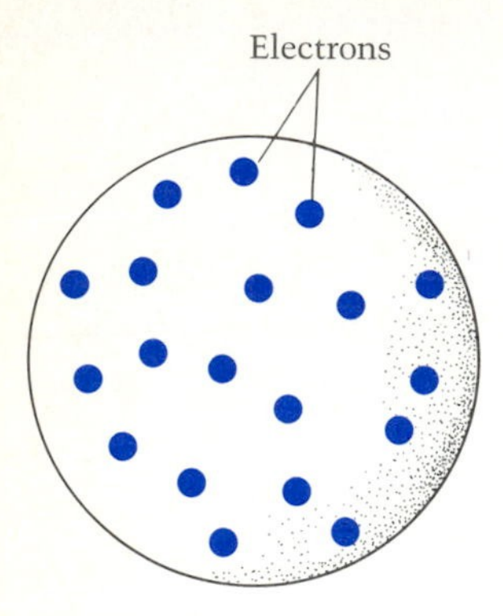

Figure 4.4
The Thomson model of the atom

This "something else," it seemed, would have to be some other sort of particle that has the same amount of charge as the electron but is positive in nature. Such reasoning led scientists to suppose that there were positive particles, which they called **protons,** in the atom. But then why could atoms give off only negative particles (electrons) and not positive ones (protons)? The British physicist J. J. Thomson suggested an answer. He thought of the atom as a sphere made up of positive electricity in which electrons were embedded (Figure 4.4). The English called this the *plum-pudding theory,* likening the electrons to the raisins in plum pudding. As Americans, we can find a more familiar analogy in a scoop of chocolate-chip ice cream. The chocolate chips represent the electrons. The ice cream represents a sea of positive electricity (composed of protons). Each element is different from the others because each has a different number of electrons and protons arranged in a different way—like different scoops of ice cream with different numbers of chocolate chips. For example, hydrogen atoms have one electron and one proton. Helium atoms have two electrons and two protons. Thomson seemed to have given scientists an idea they could agree on, but his theory still needed experimental proof.

Lord Rutherford, a former student and colleague of Thomson, attempted to test Thomson's theory. He had been working with positive particles called **alpha particles** (α particles), which are actually helium atoms with their electrons removed. He reasoned that if he shot a stream of these positive particles through a gold foil, the neutral and symmetrical "plum-pudding" gold atoms would not greatly affect the path of the alpha particles. Most of the particles would move straight through the foil (Figure 4.5). Rutherford used gold foil in his experiment because it could be hammered into very thin sheets, making the thickness of the foil resemble a single layer of atoms. However, when Rutherford actually performed the experiment, he found that about one alpha particle in 20,000 ricocheted, or bounced back out of the foil! As Rutherford said, "It was almost as incredible as if you fired a fifteen-inch shell at a piece of tissue paper and it came back and hit you."

Lord Rutherford
(Reprinted with permission from *Torchbearers of Chemistry* by Henry Monmouth Smith, © Academic Press, Inc., 1949)

To explain these peculiar results, Rutherford devised a new model for the atom, called the nuclear model. We subscribe to this model today. Rutherford said that an atom must have a *center of positive charge,* where all the protons must be located. He called this center of positive charge the **nucleus** of the atom. Rutherford felt that the nucleus of the atom must be very small compared to the overall size of the atom. He also said that the nucleus must be where the mass of the atom is concentrated. Rutherford's reasoning was that some alpha particles were tremendously deflected (Figure 4.6) because they were repelled by a high concentration of positive charge with an immovable mass. He also decided that the electrons were located around the nucleus instead of in it and that the mass of the electrons was very small compared to the mass of the protons. His reasoning here was that for so many of the particles to go through the foil undeflected, the electrons had to be located in a relatively large space outside the nucleus but still within the atom.

Later experiments similar to Rutherford's showed that the diameter of the nucleus of an atom is approximately 10^{-13} cm, whereas the diameter of the

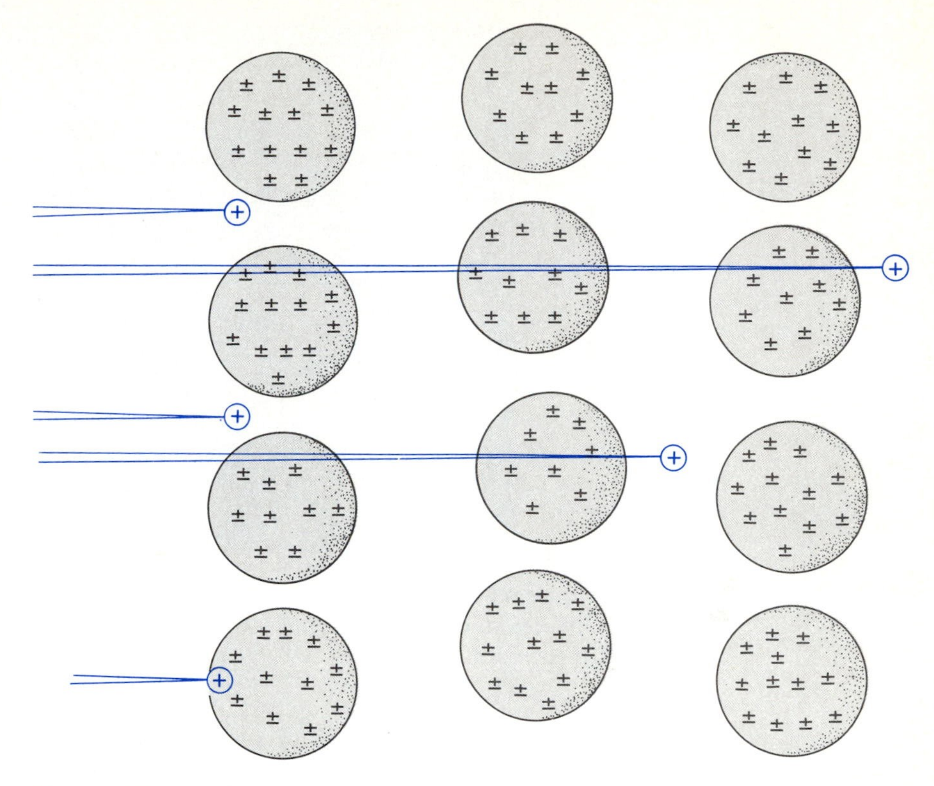

Figure 4.5
According to the Thomson model of the atom, positive particles ought to pass through a metal foil without being deflected.

whole atom is up to 100,000 times as great. Suppose we could expand an atom so that the nucleus was 1 millimeter in diameter (about the size of this dot: ·). Then the nearest electron could be as far as 50 meters (half the length of a football field) away from the nucleus! And between the nucleus and the electrons is only vast empty space.

Rutherford's model of the atom, based as it was on new experimental evidence, made the Thomson model of the atom obsolete.

4.5 The Neutron

Thus, before 1920, scientists had experimentally confirmed the existence of two basic *subatomic* particles (particles that are fundamental constituents of an atom): electrons, which are negatively charged and have a very small mass, and protons, which are positively charged and contain most of the atom's mass. But a problem came up. Scientists had defined a unit of mass called the **atomic mass unit** (abbreviated amu). They knew that the proton had a mass of 1.0072766 amu and that the electron had a much smaller mass of 0.0005486 amu. These numbers all seemed fine at first. The mass of hydrogen—which is made up of one proton and one electron—was very close

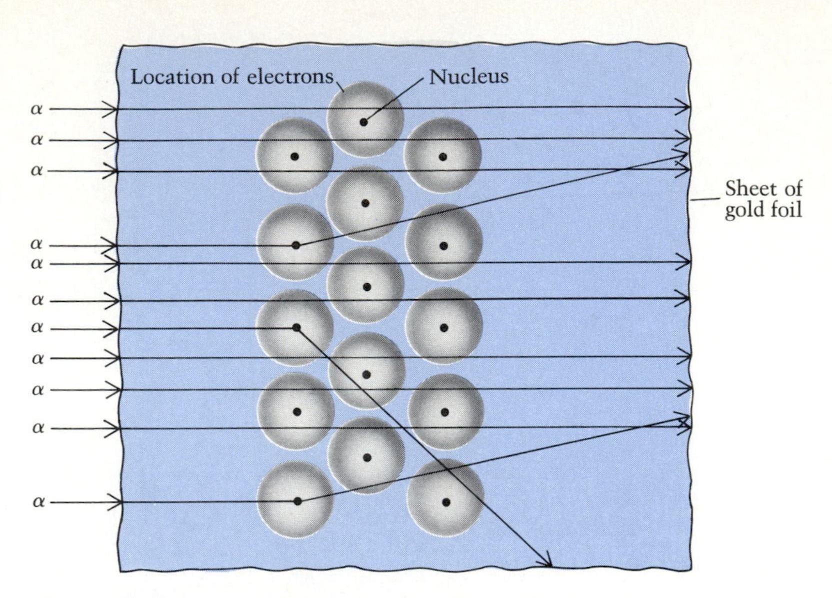

Figure 4.6
The actual way that positive particles pass through metal foil, as shown by Rutherford.

to the sum of the masses of a proton and an electron. (Remember that protons account for just about all of the atom's mass; when we are calculating the mass of an atom we can usually ignore the mass of the electrons.) The problem arose when scientists considered helium. The helium atom has two electrons and two protons, so the scientists thought it should have a mass of about 2 amu. But its mass is 4 amu. And when they looked at other elements, they were even more puzzled. Carbon has six electrons and six protons but a mass of 12 amu. Where was the missing mass?

To account for the extra mass in the atom, scientists proposed another particle, one that had no charge but had a mass equal to the proton's. In 1932, the British physicist James Chadwick (1891–1974) was able to prove the existence of a new subatomic particle called the **neutron,** which has a mass of 1.008665 amu. If this particle were also present in the atom, it would add to the mass but would not affect the charge (see Table 4.1). Scientists could now picture helium as being composed of two electrons, two protons, and two neutrons, so they could understand helium's mass of 4 amu.

4.6 Atomic Number

Learning Goal 3
Number of protons, electrons, and neutrons

The factor that makes one element different from another element is that atoms of different elements have different numbers of electrons and protons. *The number of protons in an atom of an element* is called the **atomic number** of that element and is *symbolized Z.*

The atomic number of an element is also equal to the positive charge on the nucleus and to the number of electrons in the neutral atom of that ele-

Table 4.1
The Particles in the Atom

Particle	Symbol	Charge	Mass
Proton	p	+1	1.0073 amu
Electron	e	−1	0.0005 amu
Neutron	n	0	1.0087 amu

ment. On any periodic table there is a whole number identified with each element. This is the atomic number Z. In the periodic table inside the front cover, the atomic number is given above the symbol for each element. An atom of sodium (Na), which has the atomic number 11, has 11 electrons and 11 protons (Figure 4.7). An atom of magnesium (Mg), which has the atomic number 12, has 12 electrons and 12 protons; and an atom of uranium (U), which has the atomic number 92, has 92 electrons and 92 protons. Note that the elements are listed in the periodic table in the order of their atomic numbers.

4.7 Isotopes

If you look at the periodic table, you will see that the atomic masses of most elements are not whole numbers. For example, sodium (Na) has a mass of 22.990 amu, titanium (Ti) a mass of 47.90 amu, and rubidium (Rb) a mass of 85.468 amu. Why is this so? If each proton and each neutron gives a mass of almost exactly 1 amu to the atom, then the atomic masses of the elements certainly ought to be very close to whole numbers.

Learning Goal 4
Definition of an isotope

Investigation of this question led to the discovery that *each element can exist in more than one form*. When John Dalton said that the atoms of an element must be identical in size, mass, and shape, he did not know that *isotopes,* or different forms of an element, could exist. **Isotopes** are *forms of an element that have the same number of electrons and protons but different numbers of neutrons*.

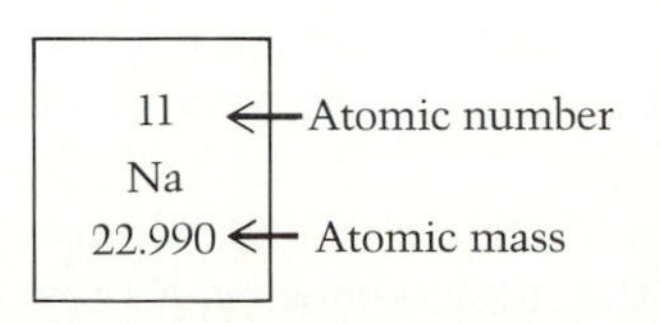

Figure 4.7
In the periodic table used in this text, an element's atomic number appears above its symbol.

Although the numbers of negative and positive charges of isotopes are identical, isotopes differ in mass. Hydrogen, for example, exists in three isotopic forms (Figure 4.8). The common form of hydrogen (called protium) contains one electron and one proton. Another form of hydrogen (called deuterium) contains one electron, one proton, and one neutron. A third form of hydrogen (called tritium) contains one electron, one proton, and two neutrons.

Note that all isotopes of an element have the same number of protons and thus have the same atomic number. However, it is possible to distinguish among isotopes by means of mass numbers. The **mass number** of an isotope is simply *the sum of the number of protons and the number of neutrons*.

Figure 4.8
Isotopes of hydrogen

Learning Goal 5
Standard notation for mass number and atomic number

A standard notation has been developed for giving the mass number and atomic number of an isotope. It consists of writing the symbol of the element with its mass number as a *superscript* (above the line) and its atomic number as a subscript (below the line). Both numbers are usually written to the left of the symbol for the element.

$$\begin{matrix}\text{Mass number} \longrightarrow \\ \text{Atomic number} \longrightarrow\end{matrix} \quad {}^{16}_{8}\text{O} \quad {}^{12}_{6}\text{C} \quad {}^{35}_{17}\text{Cl}$$

With this notation we can immediately tell the composition of the isotope in terms of protons, electrons, and neutrons. The number of protons (or, in the neutral atom, the number of electrons) is simply the atomic number of the element. The number of neutrons is the difference between the mass number and the atomic number. This notation allows us to distinguish among the isotopes of an element.

Example 4.1

Give the number of electrons, protons, and neutrons in each of the following neutral isotopes: *(a)* ${}^{238}_{92}\text{U}$ *(b)* ${}^{235}_{92}\text{U}$ *(c)* ${}^{84}_{36}\text{Kr}$

Solution The atomic number (subscript) gives the number of electrons in the neutral atom as well as the number of protons. Subtracting the atomic number from the mass number (superscript) gives the number of neutrons.

(a) Because the atomic number is 92, there are 92 electrons and 92 protons. And subtracting the atomic number from the mass number gives the number of neutrons: $238 - 92 = 146$ neutrons. So there are 92 electrons, 92 protons, and 146 neutrons in an atom of ${}^{238}_{92}\text{U}$.

(b) Again there are 92 electrons and 92 protons. And $235 - 92 = 143$ neutrons. So there are 92 electrons, 92 protons, and 143 neutrons in an atom of ${}^{235}_{92}\text{U}$. Note that ${}^{238}_{92}\text{U}$ and ${}^{235}_{92}\text{U}$ are an *isotope pair*.

(c) Here we have 36 electrons and 36 protons. And 84 − 36 = 48 neutrons. So there are 36 electrons, 36 protons, and 48 neutrons in an atom of $^{84}_{36}Kr$.

Practice Exercise 4.1 Give the number of electrons, protons, and neutrons in each of the following isotopes: *(a)* $^{28}_{14}Si$ *(b)* $^{31}_{15}P$ *(c)* $^{6}_{3}Li$

The following notations can be used to distinguish among the three isotopes of hydrogen:

$^{1}_{1}H$	$^{2}_{1}H$	$^{3}_{1}H$
Protium	Deuterium	Tritium

Learning Goal 6
Average atomic mass of an element

Learning Goal 7
Percentage abundance of isotopes

Learning Goal 8
Table of relative abundances of isotopes

The deuterium and tritium forms of hydrogen are not so common as the protium form. If a random sample of all hydrogen atoms found on the earth were taken, it would be found to consist of 99.985% protium atoms, 0.015% deuterium atoms, and a vanishingly small amount of tritium atoms. The atomic mass given in the periodic table is the *weighted average of the atomic masses* of all three isotopes. *When calculating the average atomic mass for any element, we must take into account the relative abundance of each isotope of that element and its exact isotopic mass.* For example, the exact isotopic mass of $^{1}_{1}H$ is 1.0078 amu, and the exact isotopic mass of $^{2}_{1}H$ is 2.0141. It is for this reason that the atomic mass of hydrogen shown in the periodic table is 1.008 amu. This is nearly the same as the mass of protium, the most common isotope of hydrogen. Table 4.2 gives a breakdown of the isotopes of the first 15 elements and their isotopic masses. This averaging process also explains why the atomic masses of the elements in the periodic table are not whole numbers.

If we know the percentage abundance of each isotope of an element, we can calculate the average atomic mass of the element. The following example shows how to do this. In the example we use the exact isotopic mass.

Example 4.2

Using Table 4.2, calculate the atomic mass of boron.

Solution From Table 4.2, the percentage abundance of ^{10}B is 19.6%, and its exact isotopic mass is 10.013 amu. That of ^{11}B is 80.4%, and its exact isotopic mass is 11.009 amu. Therefore the atomic mass of

$$B = (10.013\ \text{amu})(0.196) + (11.009\ \text{amu})(0.804)$$
$$= 1.96\ \text{amu} + 8.85\ \text{amu} = 10.81\ \text{amu}$$

For a review of changing percentages to decimal equivalents, see Example A.19 in in Supplement A.

Practice Exercise 4.2 Using Table 4.2, calculate the atomic mass of lithium.

Table 4.2
Naturally Occurring Isotopes of the First 15 Elements

Name	Symbol	Atomic number	Mass number	Isotopic mass	Percentage natural abundance
Hydrogen-1	$^{1}_{1}H$	1	1	1.007825	99.985
Hydrogen-2	$^{2}_{1}H$	1	2	2.01410	0.015
Hydrogen-3	$^{3}_{1}H$	1	3	3.01605	Negligible
Helium-3	$^{3}_{2}He$	2	3	3.016	0.00013
Helium-4	$^{4}_{2}He$	2	4	4.003	99.99987
Lithium-6	$^{6}_{3}Li$	3	6	6.015	7.42
Lithium-7	$^{7}_{3}Li$	3	7	7.016	92.58
Beryllium-9	$^{9}_{4}Be$	4	9	9.012	100
Boron-10	$^{10}_{5}B$	5	10	10.013	19.6
Boron-11	$^{11}_{5}B$	5	11	11.009	80.4
Carbon-12	$^{12}_{6}C$	6	12	12.0000	98.89
Carbon-13	$^{13}_{6}C$	6	13	13.003	1.11
Nitrogen-14	$^{14}_{7}N$	7	14	14.003	99.63
Nitrogen-15	$^{15}_{7}N$	7	15	15.000	0.37
Oxygen-16	$^{16}_{8}O$	8	16	15.995	99.759
Oxygen-17	$^{17}_{8}O$	8	17	16.999	0.037
Oxygen-18	$^{18}_{8}O$	8	18	17.999	0.204
Fluorine-19	$^{19}_{9}F$	9	19	18.998	100
Neon-20	$^{20}_{10}Ne$	10	20	19.992	90.92
Neon-21	$^{21}_{10}Ne$	10	21	20.994	0.257
Neon-22	$^{22}_{10}Ne$	10	22	21.991	8.82
Sodium-23	$^{23}_{11}Na$	11	23	22.9898	100
Magnesium-24	$^{24}_{12}Mg$	12	24	23.9850	78.70
Magnesium-25	$^{25}_{12}Mg$	12	25	24.9858	10.13
Magnesium-26	$^{26}_{12}Mg$	12	26	25.9826	11.17
Aluminum-27	$^{27}_{13}Al$	13	27	26.9815	100
Silicon-28	$^{28}_{14}Si$	14	28	27.9769	92.21
Silicon-29	$^{29}_{14}Si$	14	29	28.9765	4.70
Silicon-30	$^{30}_{14}Si$	14	30	29.9738	3.09
Phosphorus-31	$^{31}_{15}P$	15	31	30.9738	100

The concept of isotopes with different abundances allows us to solve two other kinds of quantitative problems. These are illustrated by the next two examples.

Example 4.3

A hypothetical element Q exists in two isotopic forms, ^{270}Q and ^{280}Q. The atomic mass of Q is 276 amu. What is the percentage abundance of each isotope? (Assume that the exact isotopic mass is the same as the mass number for each isotope.)

Solution

Understand the Problem

If we know the percentage abundance of each isotope of an element, we can calculate the average atomic mass of the element. This problem requires that we determine the percentage abundance of each element when given the atomic mass.

Devise a Plan

We will use algebra and the ability to solve an algebraic equation. We will let x equal the fraction of ^{270}Q and $(1 - x)$ equal the fraction of ^{280}Q. Then we have

$$270x + 280(1 - x) = 276$$

Carry Out the Plan

$$x = 0.40$$

$$1 - x = 0.60$$

In other words, $4\bar{0}\%$ of the element is ^{270}Q and $6\bar{0}\%$ of the element is ^{280}Q.

Look Back

We ask if the solution makes sense. Using estimation we can reason that if 50% of the element were ^{270}Q and 50% were ^{280}Q, then the atomic mass of the element should be 275. Since it is more than 275, there should be a slightly greater amount of ^{280}Q present, and that is the case with this solution.

Practice Exercise 4.3 A hypothetical element Y exists in two isotopic forms, ^{100}Y and ^{110}Y. The atomic mass of Y is 108. What is the percentage abundance of each isotope? (Assume that the exact isotopic mass is the same as the mass number for each isotope.)

Example 4.4

Using Table 4.2, find how many ^{17}O atoms there are in a sample containing 1,000,000 oxygen atoms.

Solution From Table 4.2, we find that the natural abundance of ^{17}O is 0.037%. Therefore the number of ^{17}O atoms in 1 million O atoms equals

$$(0.00037)(1{,}000{,}000) = 370 \text{ atoms of } ^{17}O$$

Practice Exercise 4.4 Using Table 4.2, find how many ^{20}Ne atoms there are in a sample containing 1,000,000 neon atoms.

4.8 What Next for the Atom?

We have traced the development of a model for the atom from the time of Dalton, who believed that the atom was an indivisible unit of matter, to the time of Rutherford, who suggested that the atom has a nucleus composed of protons and neutrons and an outer part for the electrons. But many questions were still left unanswered. How were the electrons arranged in an atom? Where were the electrons? How were they moving? How fast were they going? Could they account for the formation of compounds? Why didn't these negative electrons "fall into" the positive nucleus of the atom? After all, opposite charges attract! We will answer these questions in Chapter 5.

Summary

John Dalton's atomic theory was an attempt to explain all the different forms of matter. He theorized that (1) the atom is indivisible; (2) all atoms of the same element have the same mass, size and form; (3) all matter is composed of combinations of atoms; and (4) atoms of different elements are different. Although the first two of these ideas were later shown to be incorrect, most of Dalton's theory is the basis for what is still believed today. Dalton's model of the atom was modified as various subatomic particles—protons, electrons, and neutrons—were discovered. Thomson's theory of the atom followed his "discovery" of electrons. This British physicist theorized that the atom is a sphere made up of positive electricity in which negatively charged electrons are embedded. Lord Rutherford, a colleague of Thomson, tested Thomson's theory with an experiment in which a stream of positive (alpha) particles was supposed to pass straight through the atoms in a gold foil with only a few particles being slightly deflected. Instead, some of the alpha particles bounced back. To explain this result, Rutherford theorized that an atom has a relatively very small center where all the positively charged protons and most of the atomic mass are located. Electrons, he postulated, are located outside the nucleus, perhaps relatively far from it.

There was one difficulty with the two-particle atomic model. For most atoms, the masses of all the protons and electrons did not account for the total mass of the atom. In 1932 James Chadwick was able to prove the existence of a new subatomic particle called the neutron. The neutron has about the same mass as a proton but carries no charge. With neutrons, scientists were able to account for the total mass of atoms.

The number of protons in the atom of an element is called the atomic number of that element. The sum of the number of protons and the number of neutrons is called the mass number of the element. All the atoms of any element have the same atomic number, but they may have different mass numbers. Two or more atoms with the same atomic number but different mass numbers are called isotopes.

Key Terms

alpha particle **(4.4)**
anode **(4.3)**
atomic mass unit **(4.5)**
atomic number **(4.6)**
cathode **(4.3)**
electron **(4.3)**
isotope **(4.7)**
mass number **(4.7)**
neutron **(4.5)**
nucleus **(4.4)**
proton **(4.4)**
vacuum **(4.3)**

Career Sketch

Biochemist

As a biochemist, you will study the composition and behavior of living things. You will conduct research on such topics as reproduction, growth, and heredity. You will study the effects of foods, hormones, and drugs on various organisms and make important contributions to the sciences of medicine, nutrition, and agriculture.

You may also investigate the causes and cures for disease, identify the nutrients necessary to maintain good health, and develop chemical compounds for the control of pests. You will use a variety of instruments, such as electron microscopes, ultra-high-speed centrifuges, and sophisticated radiation-detecting equipment for isotope studies.

Some of your work may be in the area of basic research and some in applied research. For example, you may perform a study that shows how an organism produces a hormone. You may then use that knowledge to synthesize and manufacture that hormone on a mass scale using recombinant DNA technology.

To become a biochemist, you will need to take courses in biology, chemistry, physics, and mathematics. You will need a minimum of a baccalaureate degree and preferably a master's degree or doctorate in biochemistry. Biochemists are employed by pharmaceutical companies, chemical firms, hospitals, government research facilities, and universities.

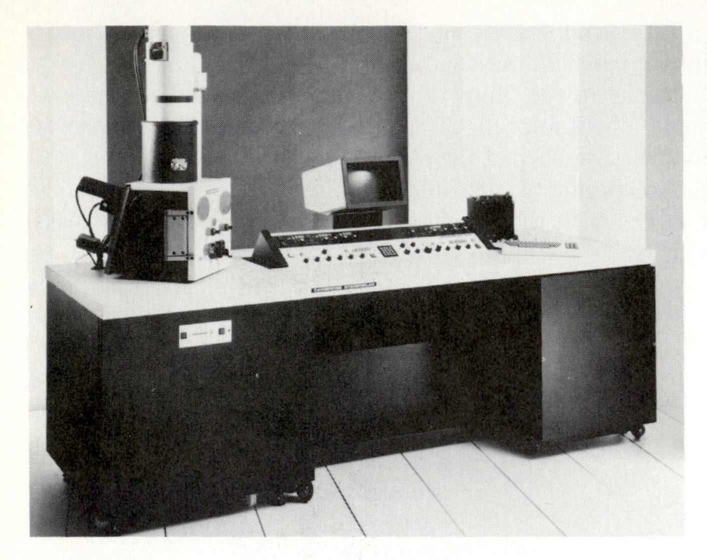

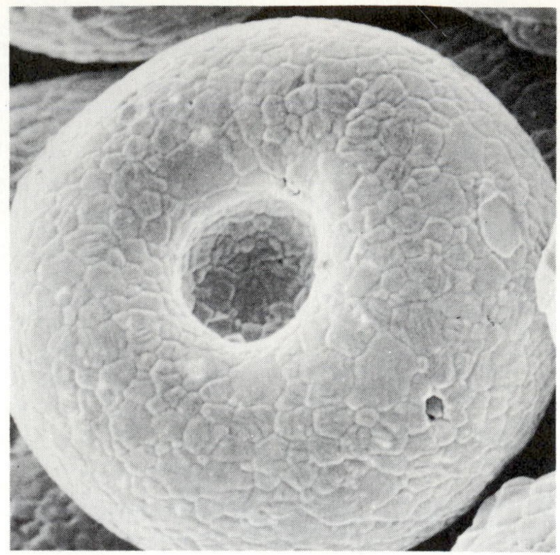

Biochemists must be familiar with many research techniques and laboratory instruments, such as the scanning electron microscope (SEM) shown on the left. Biochemists use the SEM to photograph body components. This red blood cell (right) has been magnified 10,000 times.
(Left: Cambridge Stereoscan 200 SEM courtesy Leica, Inc. Right: SEM courtesy John B. Vander Sande, Massachusetts Institute of Technology)

Self-Test Exercises

Learning Goal 1: Atomic Theory and Models

1. Explain the idea of a model. Use an example of your own.
2. Give an example of a physical model and an example of an intellectual model.

3. List the points in Dalton's atomic theory. Explain which ones had to be corrected after later experiments were performed.
▲ **4.** In what ways is Dalton's atomic theory similar to the ideas of Democritus, set forth in 420 B.C.?

5. Contrast the Thomson model and the Rutherford model of the atom.
6. Give a general outline of the construction of a Crookes tube.

▲ **7.** Explain how a television set is similar to a Crookes tube. (You may want to check an outside reference source for help in answering this question.)

8. Discuss the experiment that was conducted to test the wave and particle theories of cathode rays.

Learning Goal 2: Mass and Charge of Electrons, Protons, and Neutrons

9. How was the electron discovered?
10. Which of the three subatomic particles has the greatest mass? Which has the least mass?

11. Name the three major subatomic particles. Give their charges and approximate atomic masses (in amu).
▲ **12.** Compare the charges of the three subatomic particles and briefly explain how each charge was determined.

Learning Goal 3: Number of Protons, Electrons, and Neutrons

13. Find the number of protons (p), electrons (e), and neutrons (n) in neutral atoms of the following:
(a) $^{23}_{11}Na$ *(b)* $^{223}_{87}Fr$ *(c)* $^{238}_{92}U$

14. Determine the number of protons (p), electrons (e), and neutrons (n) in neutral atoms of the following: *(a)* $^{14}_{7}N$ *(b)* $^{15}_{7}N$

15. Find the number of protons, electrons, and neutrons in neutral atoms of the following: *(a)* $^{16}_{8}O$ *(b)* $^{17}_{8}O$ *(c)* $^{18}_{8}O$ *(d)* $^{20}_{10}Ne$ *(e)* $^{21}_{10}Ne$ *(f)* $^{22}_{10}Ne$

16. Find the number of protons, electrons, and neutrons in neutral atoms of the following: *(a)* $^{31}_{15}P$ *(b)* $^{22}_{10}Ne$ *(c)* $^{24}_{12}Mg$

17. From the information that follows, select the element that is not an isotope of the others. *(a)* 8p, 8e, 8n *(b)* 7p, 7e, 8n *(c)* 8p, 8e, 9n

▲ **18.** Which of the following describes an atom that is not an isotope of the other two? *(a)* 12p, 12e, 14n *(b)* 12p, 12e, 12n *(c)* 13p, 13e, 12n

19. Find the number of protons, electrons, and neutrons in neutral atoms of the following: *(a)* $^{57}_{25}Mn$ *(b)* $^{60}_{27}Co$ *(c)* $^{80}_{36}Kr$ *(d)* $^{128}_{52}Te$

20. Determine the number of protons, electrons, and neutrons in neutral atoms of the following hypothetical isotopes: *(a)* $^{35}_{16}X$ *(b)* $^{300}_{90}Y$

▲ **21.** The alchemists wanted to turn lead into gold. Suppose that you had the isotope of lead $^{208}_{82}Pb$, and you wanted to turn it into the isotope of gold $^{197}_{79}Au$. How many protons, electrons, and neutrons would you have to remove from the lead atom to turn it into the gold atom?

22. Which of the following describes an atom that is not an isotope of the other two? *(a)* 14p, 14e, 15n *(b)* 15p, 15e, 14n *(c)* 15p, 15e, 15n

▲ **23.** Fill in the blanks in the following table. (Use the periodic table when necessary.)

Symbol	Protons	Electrons	Neutrons	Mass number	Atomic number
$^{174}_{70}Yb$					
	59			141	
			60		44
			24	45	
		22	28		
$^{?}_{12}Mg$			13		

24. State the number of protons, electrons, and neutrons in $^{42}_{20}Ca$.

Learning Goal 4: Definition of an Isotope

▲ **25.** Define atomic number, mass number, and isotope.

26. Atoms of an element that have the same number of electrons and protons but different numbers of neutrons are called ________ .

27. Isotopes are atoms with the same number of ________ and ________ but different numbers of ________. (Place the words *protons, electrons*, and *neutrons* in the proper places.

28. *(a)* The number of protons in an atom is called the ________ ________ of that element.
(b) The sum of the numbers of protons and neutrons in an isotope is called the ________ ________ of that isotope.

Learning Goal 5: Standard Notation for Mass Number and Atomic Number

29. Write the standard isotopic notation for the following elements: *(a)* 48p, 48e, 55n *(b)* 55p, 55e, 68n *(c)* 87p, 87e, 136n *(d)* 17p, 17e, 15n

▲ **30.** Write the standard isotopic notation for the following elements: *(a)* 3p, 3e, 3n *(b)* 9p, 9e, 10n *(c)* 14p, 14e, 16n

Learning Goal 6: Average Atomic Mass of an Element

▲ **31.** The element gallium (Ga) exists in two isotopic forms with the following abundances: 60.16% ^{69}Ga and 39.84% ^{71}Ga. Calculate the atomic mass of gallium. (The exact isotopic mass of ^{69}Ga is 68.9257. The exact isotopic mass of ^{71}Ga is 70.9249.)

32. The element carbon (C) exists in two isotopic forms with the following abundances: 98.89% ^{12}C and 1.11% ^{13}C. Calculate the atomic mass of carbon.

33. The element chlorine (Cl) exists naturally in two isotopic forms with the following abundances: 75.53% ^{35}Cl and 24.47% ^{37}Cl. Calculate the atomic mass of chlorine. (The exact isotopic mass of ^{35}Cl is 34.9689. The exact isotopic mass of ^{37}Cl is 36.9659.)

34. The element magnesium (Mg) exists in three isotopic forms with the following abundances: 78.70% ^{24}Mg, 10.13% ^{25}Mg, and 11.17% ^{26}Mg. Calculate the atomic mass of magnesium.

35. The element chromium (Cr) exists naturally in four isotopic forms with the following abundances: 4.31% ^{50}Cr, 83.76% ^{52}Cr, 9.55% ^{53}Cr, and 2.38% ^{54}Cr. Calculate the atomic mass of chromium. (The exact isotopic mass of ^{50}Cr is 49.9461. The exact isotopic mass of ^{52}Cr is 51.9405. The exact isotopic mass of ^{53}Cr is 52.9407. The exact isotopic mass of ^{51}Cr is 53.9389.)

36. The element copper exists in two isotopic forms with the following abundances: 69.09% ^{63}Cu and 30.91% ^{65}Cu. Calculate the atomic mass of copper. (The exact isotopic mass of ^{63}Cu is 62.9296. The exact isotopic mass of ^{65}Cu is 64.9278.)

Learning Goal 7: Percentage Abundance of Isotopes

***37.** Bromine exists in two isotopic forms, ^{79}Br and ^{81}Br. The atomic mass of bromine is 79.90 amu. Calculate the percentage abundance of each isotope. (The exact isotopic mass of ^{79}Br is 78.9183. The exact isotopic mass of ^{81}Br is 80.9163.)

38. Boron (B) exists in two isotopic forms, ^{11}B and ^{10}B. The atomic mass of boron is 10.81 amu. Calculate the percentage abundance of each isotope.

***39.** The element antimony occurs naturally in two isotopic forms, ^{121}Sb and ^{123}Sb. The atomic mass of antimony is 121.75 amu. Calculate the percentage abundance of each isotope. (The exact isotopic mass of ^{121}Sb is 120.9038. The exact isotopic mass of ^{123}Sb is 122.9041.)

▲ **40.** Chlorine exists in two isotopic forms, ^{35}Cl and ^{37}Cl. The atomic mass of chlorine is 35.453 amu. Calculate the percentage abundance of each isotope. (The exact isotopic mass of ^{35}Cl is 34.9689 and that of ^{37}Cl is 36.9659.)

Learning Goal 8: Table of Relative Abundances of Isotopes

41. Using Table 4.2, determine how many ^{2}H atoms you will find in 10,000,000 hydrogen atoms.

42. Using Table 4.2, determine how many ^{2}H atoms you would find in 100,000 atoms of hydrogen.

43. Using Table 4.2, determine how many atoms of oxygen you would need to get 7,400 atoms of ^{17}O.

44. Using Table 4.2, determine how many atoms of oxygen you would need to get 102 atoms of the isotope ^{18}O.

45. Using Table 4.2, determine how many ^{6}Li atoms you would find in 100,000,000 lithium atoms.

46. Using Table 4.2, determine how many ^{7}Li atoms you would find in 1,000 lithium atoms.

▲ **47.** Using Table 4.2, determine how many ^{26}Mg atoms you would find in 5.00×10^6 Mg atoms.

48. Using Table 4.2, determine how many ^{14}N atoms you would find in 1.000×10^{14} nitrogen atoms.

49. Using Table 4.2, determine how many ^{13}C atoms you would find in 2.00×10^4 C atoms.

▲ **50.** Using Table 4.2, determine how many atoms of ^{22}Ne you would find in a sample of neon that contains 1×10^6 neon atoms.

51. Using Table 4.2, determine how many ^{15}N atoms you would find in 1.0×10^{15} N atoms.

52. Using Table 4.2, determine how many ^{6}Li atoms you would find in 1,000 atoms of Li.

53. Using Table 4.2, determine how many atoms of hydrogen you would need to get 150,000 atoms of ^{2}H.

54. Using Table 4.2, determine how many ^{14}N atoms you would find in exactly 100,000,000 atoms of N.

Extra Exercises

55. You have learned that atoms are composed of fundamental particles (electrons, protons, and neutrons), some of which carry an electrical charge. Explain how atoms of an element can be neutral if they consist of charged particles.

▲ **56.** Consider the following unknown atoms.

$^{200}_{80}A$ $\quad$ $^{208}_{82}B$ $\quad$ $^{222}_{86}C$ $\quad$ $^{184}_{74}D$

(a) Which atom has the most electrons in the neutral state?
(b) Which atom has the fewest neutrons?
(c) What is the mass number of element C?
(d) Which atom has the fewest protons?
(e) Identify element A, using a periodic table.

57. What is the relationship between the atomic number and the number of electrons in a neutral atom?

58. A student produces a beam of rays and can't decide whether these rays are cathode rays or alpha particles. Explain a simple experiment that the student might perform to determine what these rays are.

59. What experiment gave the proof that atoms of all elements contain electrons?

60. Indicate the number of protons, electrons, and neutrons that each of the following atoms contains:
(a) $^{13}_{6}C$ *(b)* $^{10}_{5}B$ *(c)* $^{14}_{7}N$

61. What was the problem with Dalton's statement that all the atoms of a particular element are identical in every respect?

62. How many neon-21 atoms are there in 1 million neon atoms?

63. How many electrons are needed to equal the mass of one proton?

▲ **64.** Can the atomic number of an element exceed its mass number? Explain.

65. Make a list of some of our modern conveniences that would not function if atoms were indeed indivisible.

66. Referring to Table 4.2, determine all the possible molecular masses for H_2O, using the various isotopes of hydrogen and oxygen.

5 Atomic Theory, Part 2: Energy Levels and the Bohr Atom

Learning Goals

After you've studied this chapter, you should be able to:

1. Discuss the difference between a line spectrum and a continuous spectrum.
2. Describe the Bohr model of the atom.
3. Explain the significance of shells and energy levels.
4. Differentiate between the ground state and an excited state of an atom.
5. Describe the quantum mechanical model of the atom.
6. Explain the significance of energy sublevels.
7. Define and discuss electron orbitals.
8. Write the electron configurations for various elements using the filling order diagram.
9. Explain the importance of electron configuration.

Introduction

By the early 1900s, scientists had developed a model of the atom with protons and neutrons in the center, or nucleus, and electrons outside the nucleus. But where, exactly, were the electrons? And why weren't the negative electrons attracted into the nucleus by the positive protons? In 1913 the Danish physicist Niels Bohr proposed a model to answer these and other questions. His model has since been modified in the light of later observations and research, but it still provides an excellent basis for understanding atomic structure. In this chapter, we discuss the observations that led Bohr to propose his model, as well as the model itself.

5.1 Spectra

In the mid-1600s the English scientist, astronomer, and mathematician Sir Isaac Newton allowed a beam of sunlight to pass through a prism. He found

Figure 5.1
Sunlight passing through a prism

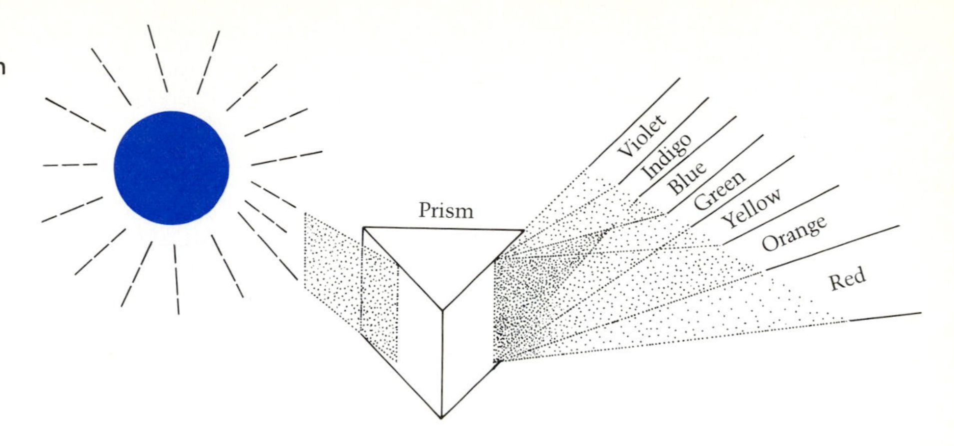

Learning Goal 1
Line spectra and continuous spectra

that the prism separated the light into a series of different colors, which we call the visible spectrum (Figure 5.1). His results were published in *Opticks* in 1704. Later experiments showed that a specific amount of energy is associated with each color of light, because light is a form of energy.

In the 1850s the German physicist Gustav Robert Kirchhoff and the German chemist Robert Bunsen developed an instrument called the spectroscope. Its main parts were a heat source (the Bunsen burner), a prism, and a telescope (Figure 5.2). Kirchhoff and Bunsen heated samples of various elements. In each case, as an element became hotter and hotter and began to glow, it produced light of its own characteristic color. When the light from the heated element was allowed to pass through the prism, it separated into a series of bright, distinct lines of various colors.

The spectrum produced by sunlight is called a **continuous spectrum**, because one color merges into the next without any gaps or missing colors. The spectrum that Bunsen and Kirchhoff found is called a **line spectrum** (Figure 5.3), because it is a series of bright lines separated by dark bands. Each of the elements produces its own unique line spectrum.

Scientists find these line spectra very useful for identifying elements in unknown samples. Analysis of line spectra has enabled scientists to identify the elements that compose the sun, the stars, and other extraterrestrial bodies. This technique is also used in chemical laboratories to identify the elements present in unknown samples.

Robert Bunsen
(Reprinted with permission from *Torchbearers of Chemistry* by Henry Monmouth Smith, ©Academic Press, Inc., 1949)

The visible spectrum produced by sunlight is only a small part of what is called the **electromagnetic spectrum** (Figure 5.4). The existence of electromagnetic waves was predicted by the British physicist James Clerk Maxwell in 1864. He assumed that electric and magnetic fields acting together produced radiant energy. He believed that radiant energy took the form of electromagnetic waves and that visible light was only one form of electromagnetic wave (Figure 5.5). We distinguish one form of electromagnetic radiation from another in terms of *wavelength*. The wavelength is the distance from one point on a wave to the corresponding point on the next wave. For electromagnetic waves, the shorter the wavelength, the higher the energy.

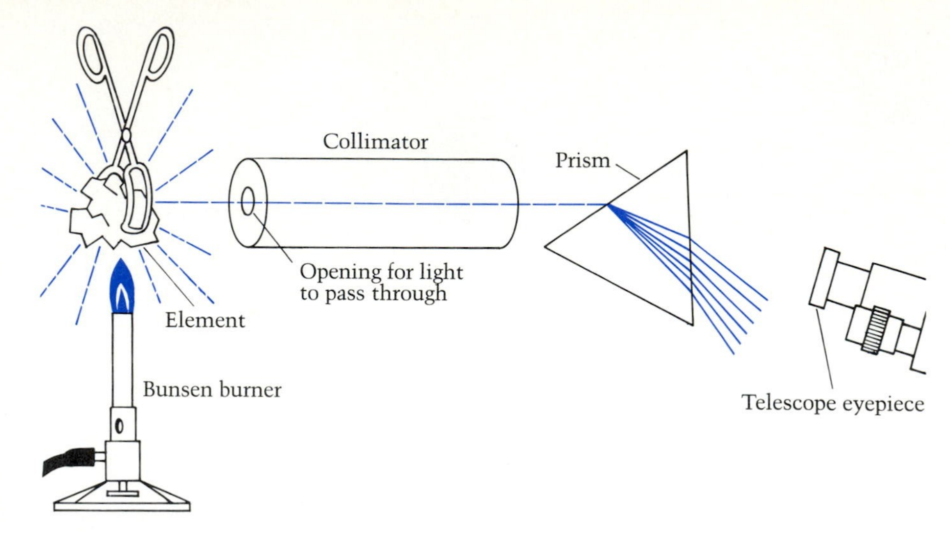

Figure 5.2
A spectroscope works as follows: An element is heated in flame of Bunsen burner. Some light from heated element passes through collimator, which is simply a brass tube with a narrow slit at both ends. The collimator directs a fine beam of light onto the prism. When light passes through the prism, it is broken into its different parts—a spectrum—and observers can see the spectrum through a telescope.

Included in the electromagnetic spectrum were both visible waves and invisible waves, each associated with a different amount of energy. For example, x rays are invisible waves with enough energy to pass through skin but not bones. Microwaves are also invisible waves, but they have enough energy to heat food. Electromagnetic waves can travel through the vacuum of space, as well as through air and water.

Maxwell's predictions were verified by the German physicist Heinrich R. Hertz in 1887. By varying an electric charge, he produced electromagnetic waves that were longer than visible light waves. This discovery eventually led to the development of radio and television.

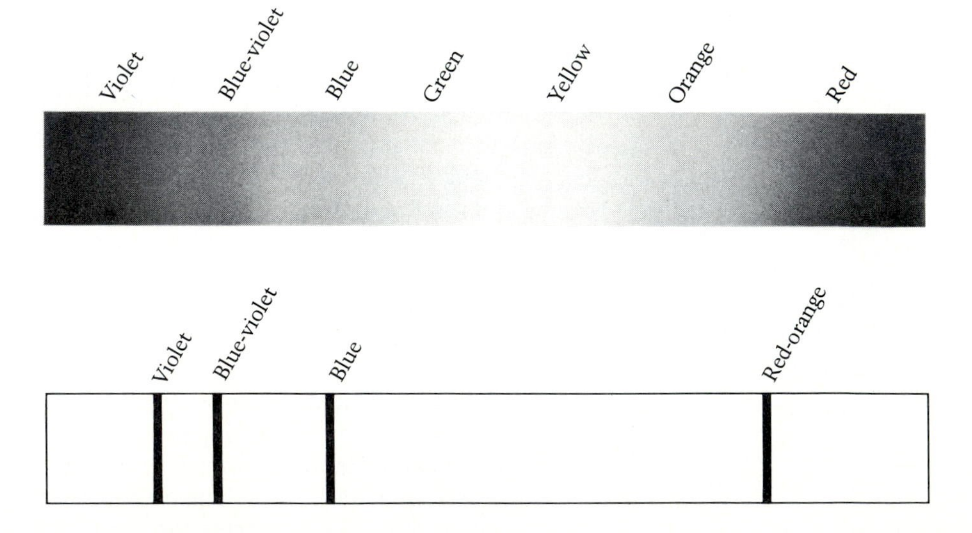

Figure 5.3
(Above) A continuous spectrum produced by sunlight. (Below) A line spectrum produced by hydrogen.

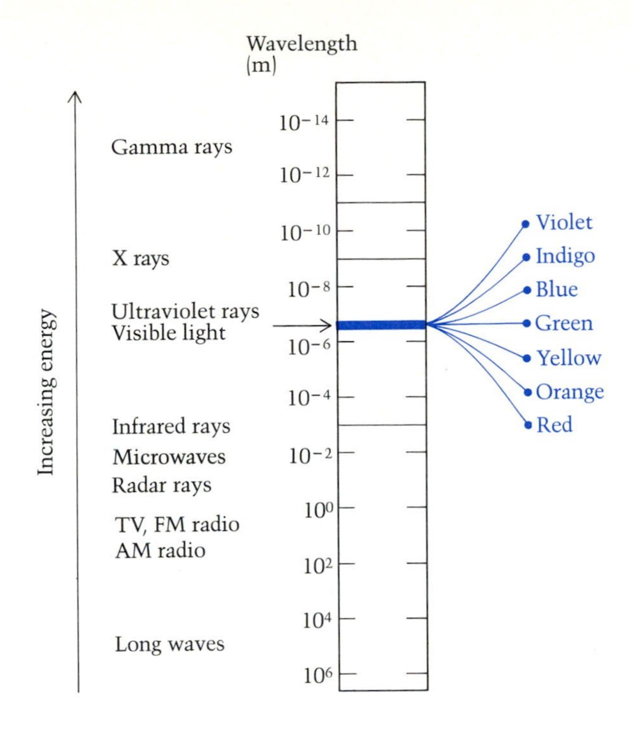

Figure 5.4
The electromagnetic spectrum

The full impact of these experiments with light was not felt until 1913, when it was discovered that the existence of line spectra held the key to the structure of the atom.

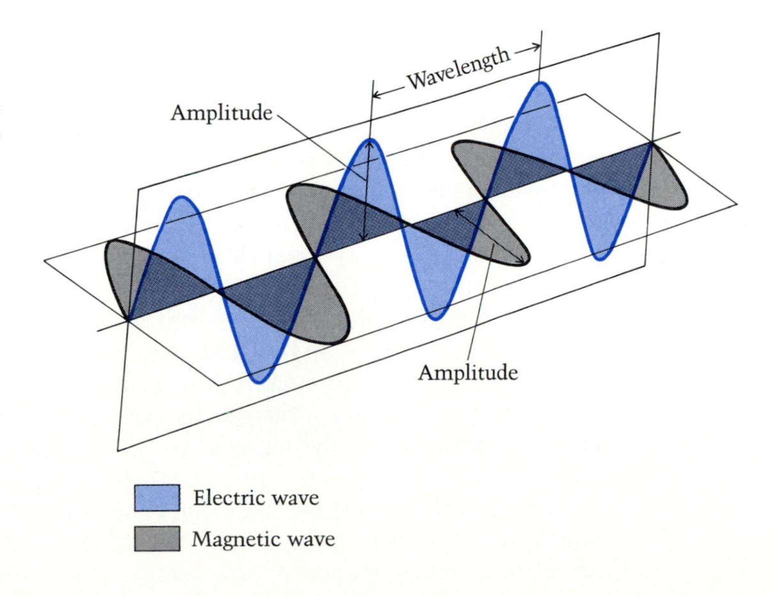

Figure 5.5
Light waves are one form of electromagnetic wave. They consist of an electric field and a magnetic field vibrating at right angles to each other.

5.2 Light as Energy

We have noted that light is a form of energy and that energy can be converted from one form to another. For example, a photoelectric cell converts light energy to electrical energy (in the form of electric current). Energy may also be absorbed and then re-emitted. For example, the luminous dial on a watch absorbs energy when it is exposed to a strong light. When it is removed from the light source, the dial re-emits this energy in the form of light: It glows.

In the experiments of Kirchhoff and Bunsen, each element was made to absorb energy in the form of heat. When the element got hot enough, it emitted this energy as light. Scientists still wondered why the elements emitted line spectra rather than continuous spectra, and they wondered what subatomic particle caused the line spectra.

In 1900, the German physicist Max Planck suggested that although light seems continuous, it is not. He said that electromagnetic radiation can be thought of as a stream of minute packets of energy called **quanta** (singular, *quantum*). The word quanta comes from the Latin, *quantus,* meaning "how much." According to Planck's quantum hypothesis, these bundles of energy (also referred to as *photons*) are the smallest packets of energy associated with a particular form of electromagnetic radiation, and each contains a specific, fixed amount of energy and has its own wavelength. Niels Bohr used the concept of quanta to explain line spectra of elements.

5.3 The Bohr Atom

Learning Goal 2
Bohr model of the atom

In 1913 Bohr suggested that electrons are responsible for line spectra. He proposed the idea that electrons travel around the nucleus of the atom in **shells,** which were described as imaginary spherical surfaces roughly concentric with the nucleus. Bohr also proposed that each shell is associated with a particular **energy level** and that shells farther away from the nucleus are associated with higher energy levels. According to this Bohr model, electrons whirl around the nucleus in specific shells, just as planets travel in specific orbits around the sun. But electrons, unlike planets, are able to jump from one shell (at one energy level) to another (at a higher energy level) when they absorb enough energy from an outside source to do so (Figure 5.6). They can also fall back to their original shells by emitting this energy.

Learning Goal 3
Shells and energy levels

Learning Goal 4
Difference between ground state and excited state of an atom

Look at the hydrogen atom in Figure 5.7. The single electron is moving around the nucleus in the shell closest to the nucleus—the shell associated with the lowest energy. In this situation, the electron is said to be in its *lowest energy level,* and the atom is in its **ground state.** As the electron moves, it neither gains nor loses energy. Its energy of motion exactly counterbalances the attraction of the nucleus. For this reason, it is not pulled into the nucleus.

Now suppose we heat our hydrogen atom. According to Bohr, the electron will jump to a high-energy shell, further from the nucleus, when it has absorbed a certain amount of heat energy. It must absorb the entire amount

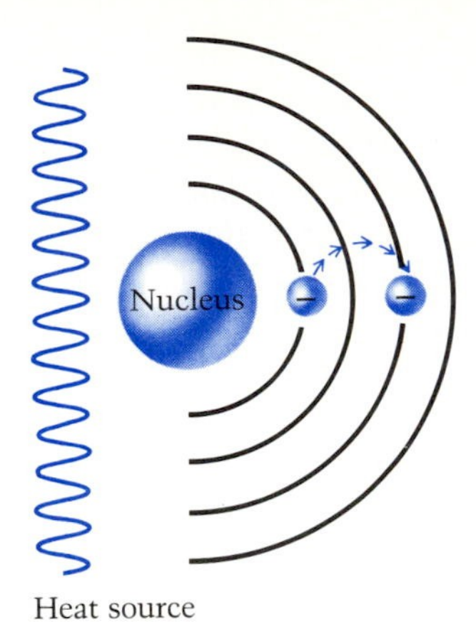

Figure 5.6
In the Bohr model of the atom, electrons "jump" from one shell to another when an outside source gives them enough energy to do so.

before it can leave the shell it occupies. It *cannot* absorb some energy, move out toward the next shell, absorb more energy and move closer to that shell, and so on until it reaches the next shell. It must absorb all the energy it needs and then instantaneously jump to the next shell. It may then absorb enough energy at that level to jump to another level (Figure 5.8). Or it may absorb enough energy to jump two or three levels at a time. But in each case it will absorb the energy and jump instantaneously. When an electron is at an energy level above its lowest level, the atom is said to be in an **excited state.** (An electron can also absorb so much energy that it escapes completely from the atom. This process, called *ionization,* is discussed in Chapter 6.)

Eventually, the electron emits some or all of the energy that is absorbed, and it falls back to a lower energy level. The energy is emitted in quanta (singular, *quantum*), the difference in energy between the higher energy shell where the electron was and the lower energy shell where it lands. If the electron spiraled down without a level-to-level requirement, it would emit a continuous stream of energy. But it moves only from one level to another and emits its energy in specific bursts that may be seen as light energy. It is this light, being emitted by electrons falling back to lower energy levels, that produces the unique line spectrum of hydrogen (or of any other element).

It is difficult to visualize electrons moving instantaneously from one energy level to another. Picture a marble rolling down a flight of stairs. The marble can rest on each step (shell), but it can never stop part of the way between steps. And, as the marble falls from step to step, it also loses (emits) energy in bursts.

5.4 Quantum Mechanical Model

Learning Goal 5
Quantum mechanical model

The Bohr model of the atom contributed a great deal to our knowledge of atomic structure. The concept of energy levels for electrons, as well as the relationship between the observed spectral lines and the shifting of electrons from one energy level to another for the hydrogen atom were important to the development of atomic theory. Unfortunately, Bohr's calculations did not hold true for heavier atoms. Scientists continued to search for information to understand atomic structure.

In the decades that followed, revolutionary discoveries were made in physics. These discoveries had an impact on chemical knowledge. Experimental evidence showed that the laws of physics that applied to large objects did not apply to very small objects. In 1924, the French physicist Louis de Broglie explained that electrons in motion had properties of waves and also had mass. Two years later the Austrian physicist Erwin Schrödinger introduced a complicated mathematical equation that described electrons as having dual characteristics. He showed that in some ways the properties of electrons are best described in terms of waves, and in other ways their properties are most similar to those of particles having mass. Physicists formalized laws that could be used to characterize the motion of electrons in what we call **quantum mechanics.**

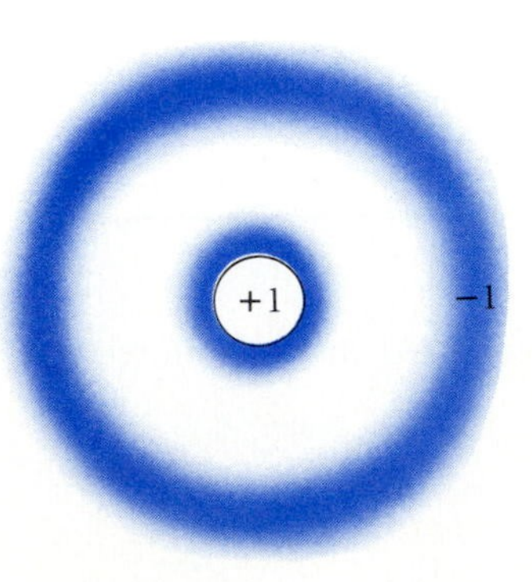

Figure 5.7
A hydrogen atom in its ground state

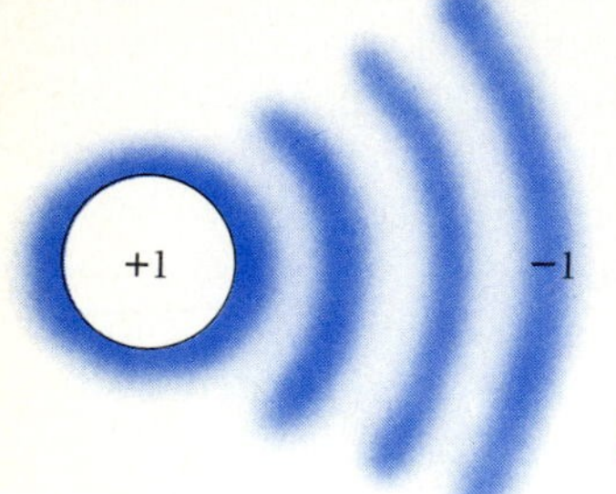

Figure 5.8
An excited hydrogen atom. Note that the electron is in energy level 3, and thus the atom is in an excited state.

Quantum mechanical concepts have replaced the ideas Bohr presented in his model, where electrons are thought to revolve about the nucleus in planetary orbits. In this model of the atom, we think of an electron *not* as being in a carefully defined orbit, but instead as occupying a volume of space. This volume of space is called an **orbital.** The orbital represents a region of space in which the electron can be found with a 95% probability.

5.5 Energy Levels of Electrons

As we learned from both the Bohr model and the quantum mechanical model, we are most likely to find an electron at a certain specified distance from the nucleus, called an **energy level** or electron **shell.** The electrons in each shell have *approximately* the same amount of energy. We will use the terms shell and energy level interchangeably.

There are two ways to identify energy levels. One method uses the letters of the alphabet beginning with the letter *K* and continuing sequentially (Figure 5.9). The farther away a level is from the nucleus, the higher the energy of an electron at that level. The *K* energy level is associated with electrons at the lowest level, whereas the *Q* energy level holds the electrons of the highest level known. The newer method of identifying electron shells uses the *principal quantum number, n*, to designate the energy levels allowed for electrons. The shell having the lowest energy level is designated $n = 1$, then the next higher level is $n = 2$, and so on, until we reach $n = 7$. We know of no atom in its ground state having electrons in an energy level higher than the seventh. Notice that the *K* energy level and the $n = 1$ energy level are exactly the same, as are the *L* energy level and the $n = 2$ energy level, and so on.

Each energy level can hold only a certain number of electrons at any one time. Table 5.1 gives the *theoretical maximum number* of electrons for each level. The lowest energy level ($n = 1$) holds a maximum of 2 electrons, while the second, third, fourth, and fifth energy levels hold 8, 18, 32, and 50 electrons, respectively. The maximum number of electrons in each level can also be determined using a simple mathematical calculation.

$$\textit{Maximum number of electrons for an energy level} = 2n^2$$

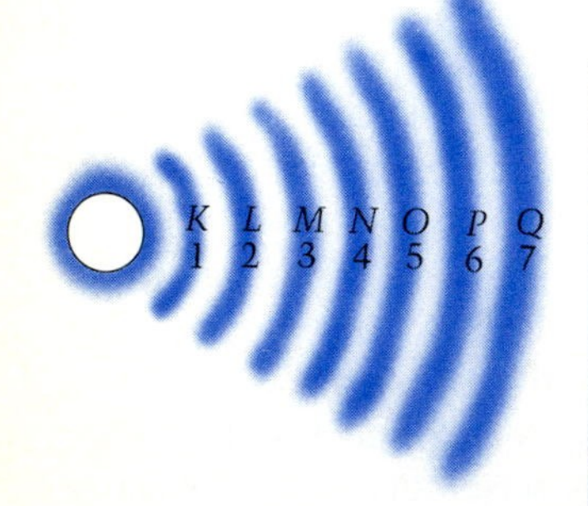

Figure 5.9
Names of the energy levels in atoms

Example 5.1

Calculate the maximum number of electrons permitted in the following:
(a) *M* shell ($n = 3$) *(b)* *P* shell ($n = 6$)

Solution

Understand the Problem

Using the number and letter designations, calculate the electron capacity for each shell.

Devise a Plan

Use the mathematical equation $2n^2$

Carry Out the Plan

(a) For $n = 3$, $2(3)^2 = 2(9) = 18$

(b) For $n = 6$, $2(6)^2 = 2(36) = 72$

Look Back

Check the mathematics and determine if these solutions make sense.

Practice Exercise 5.1 Calculate the maximum number of electrons permitted in each of the following: *(a)* N shell ($n = 4$) *(b)* Q shell ($n = 7$)

5.6 Electron Subshells

Learning Goal 6
Energy sublevels

Within each energy level we find electrons whose energies are very close in magnitude, but not exactly the same. The range in energies can be explained by the existence of **energy subshells** or **energy sublevels.** Electrons within each sublevel have exactly the same amount of energy.

To visualize the concept of energy sublevels, think of a tall office building with the unusual architectural feature depicted in Figure 5.10. The "building" is upside down. Each floor of the building represents an energy level [$K (n = 1)$, $L (n = 2)$, and so on]. Offices on each floor represent the energy sublevels. Each floor is a different size, so each has a different number of offices. In the same way, each energy level has a different number of sublevels. The sublevels are represented by the letters s, p, d, f, g, h, and i. The K energy level ($n = 1$), which is on the ground floor, has only one sublevel: the s sublevel; the L energy level ($n = 2$) has two: the s and p sublevels; and so on.

Table 5.1
Theoretical Maximum Number of Electrons for Each Energy Level

Energy level		
Letter designation	*Number*	*Maximum number of electrons*
K	1	2
L	2	8
M	3	18
N	4	32
O	5	50
P	6	72
Q	7	98

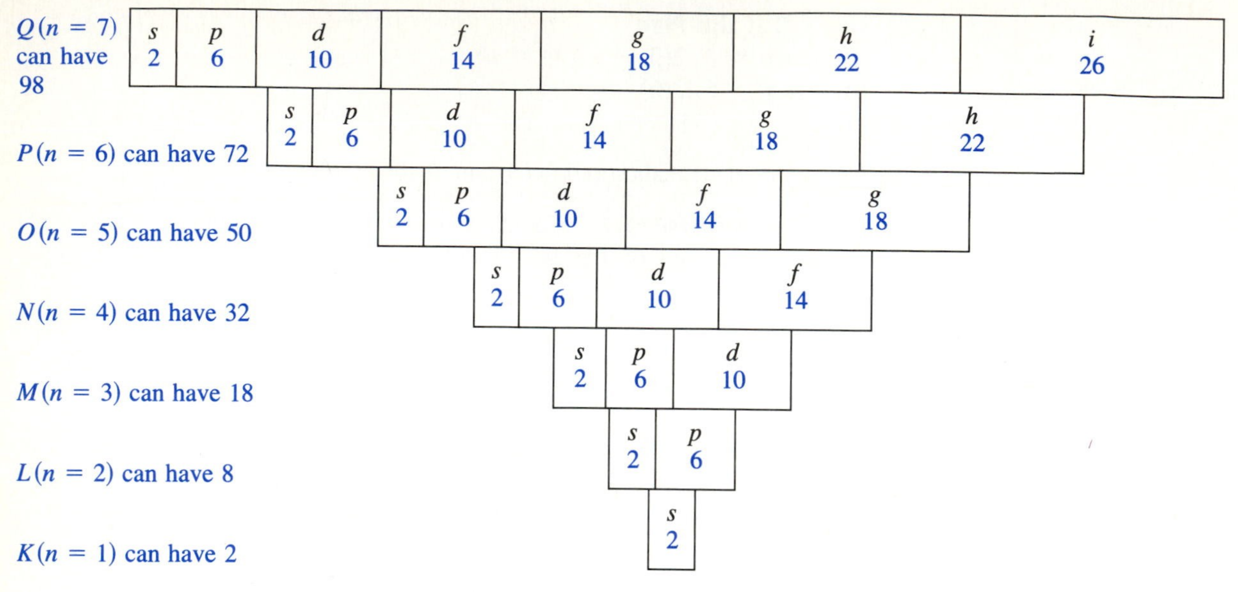

Figure 5.10
Energy levels and sublevels, showing maximum number of electrons in each

Each office of our "energy level building" is a different size and can hold only a certain number of desks. That is, each sublevel is a different size and can hold only a certain number of electrons. The *s* sublevel can hold a maximum of two electrons, and the *p* sublevel a maximum of six. Figure 5.10 shows the maximum number of electrons that each sublevel can hold. We discussed the maximum number of electrons for each energy level (2, 8, 18, 32, 50, 72, 98), and now we see that the total number of electrons that a main energy level can hold is simply the sum of the numbers of electrons that all of its *sublevels* can hold.

5.7 Electron Orbitals

Learning Goal 7
Electron orbitals

We have already looked at energy levels and sublevels. We will now look at the final concept that will help us understand how electrons are arranged about the nucleus of an atom. Electron *orbitals* are the regions of space around the nucleus of an atom where we are most likely to find an electron with a specific amount of energy. Each type of orbital has a specific shape and can hold a maximum of *two* electrons (Figure 5.11). For example, all *s* orbitals are spherical. However, the different levels of *s* orbitals occupy different regions of space around the nucleus. Figure 5.12 shows how the 1*s*, 2*s*, and 3*s* orbitals are related to each other.

The *p* orbitals are dumbbell shaped, as Figure 5.11 shows. There are three *p* orbitals in each major energy level except the first. They are called p_x, p_y,

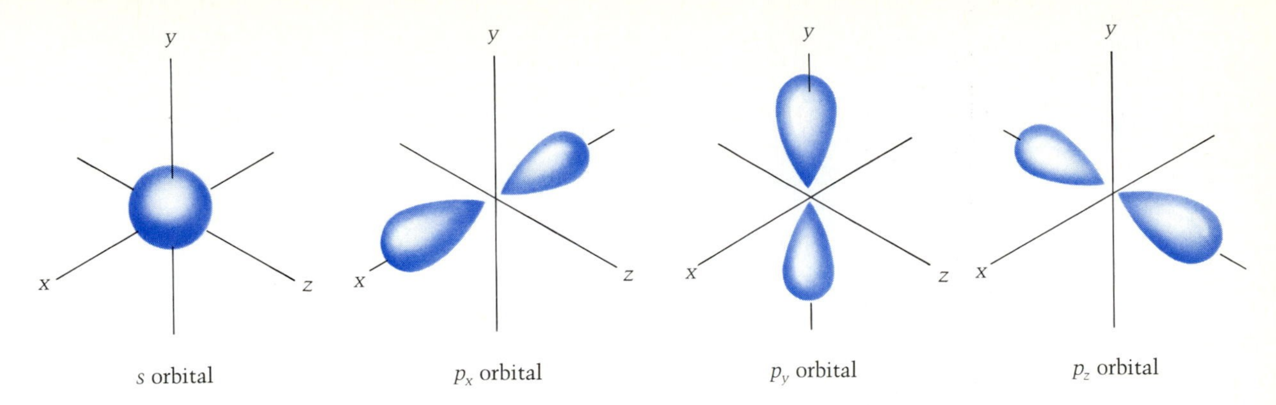

Figure 5.11
The shapes of the s, p_x, p_y, and p_z orbitals. The shape of the orbital defines the region of space in which an electron may be found with a probability of 95%. Darker areas are regions of higher probability.

and p_z, and they correspond to three axes in space. Each of the p orbitals can hold a maximum of two electrons, so the three p orbitals combined can hold a maximum of six electrons. (This total corresponds to the maximum number of electrons that a p sublevel can hold.) The p orbitals that are associated with the different energy levels occupy different regions of space around the nucleus. Figure 5.13 shows the relationship between the $2p_x$ and $3p_x$ orbitals, and Figure 5.14 shows the relationship among the $1s$, $2s$, and $2p_x$ orbitals.

Orbitals of the d and f type can hold a maximum of 10, and 14 electrons, respectively. There are five orbitals for a d sublevel and seven orbitals for an f sublevel.

Figure 5.12
The relationship of the $1s$, $2s$, and $3s$ orbitals to each other

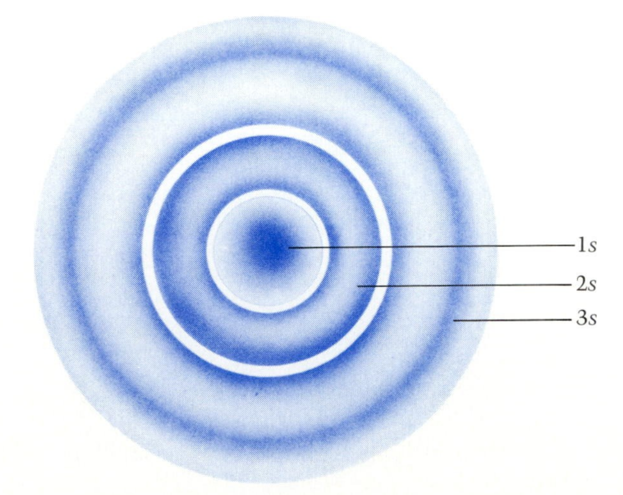

Figure 5.13
The spatial relationship of the $2p_x$ and $3p_x$ orbitals to each other

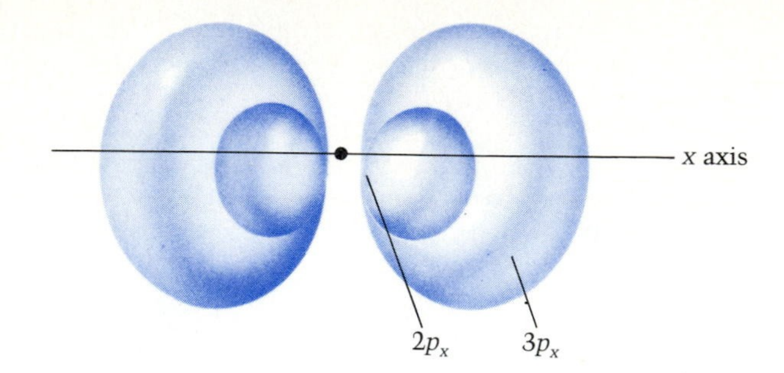

To clarify the idea of an electron being found in a specific orbital with a probability of 95%, consider the following analogy. George, a bank teller, works at the First National Bank between 10:00 A.M. and 3:00 P.M. The probability of finding George anywhere in the bank between these hours on a given day is extremely high, let us say 95%. (It is not 100% because there is always the chance that he may be ill or on vacation on a given day.)

Now suppose we make a map of the inside of the bank. We mark George's position every ten minutes during the course of the workday. For most of the day we would find him at window three. This is where we would find most of our marks on the map. However, there would also be other marks on the map where he might have walked.

This map would be very much like the orbital pictures shown in Figures 5.11 to 5.14. Where there are many marks or darker shading, we have a better chance of finding George and the electron—for example, at window three or at the edges of the orbitals in Figure 5.13. Where there are fewer marks or lighter shading, the probability of finding George or an electron is low. And the probability of finding an electron any place within the orbital (or George any place within the bank) is 95%. Just as George can take a day off, an electron can move out of its orbital and then—just as easily—move back in.

Figure 5.14
The relationship among the $1s$, $2s$, and $2p_x$ orbitals

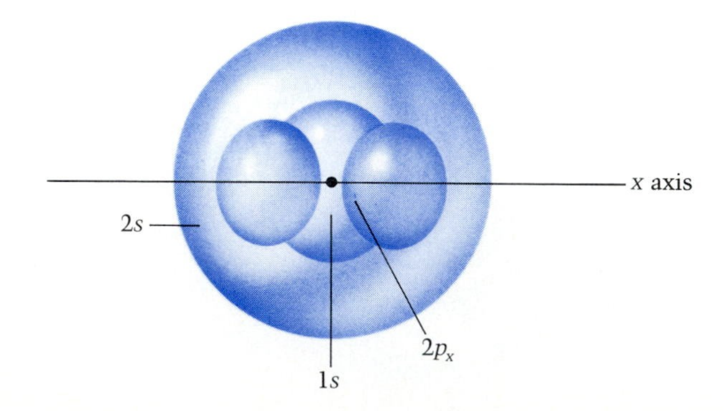

Example 5.2

For the third energy level determine *(a)* the maximum number of electrons the level could hold, *(b)* the number of sublevels, *(c)* the maximum number of electrons each sublevel could hold, and *(d)* the number of orbitals within each sublevel.

Solution

(a) For the third energy level, $n = 3$, therefore $2(n)^2 = 2(3)^2 = 2(9) = 18$.
(b) For the third level, the sublevels are $3s$, $3p$, and $3d$.
(c) The $3s$ sublevel can hold a maximum of two electrons, $3p$ can hold a maximum of six electrons, and $3d$ can hold a maximum of ten electrons.
(d) For $3s$ there is one orbital, three orbitals for $3p$ and five orbitals for $3d$.

Practice Exercise 5.2 Determine all of the above for the fourth energy level.

5.8 Writing Electron Configurations

Learning Goal 8
Writing electron configurations

The term **electron configuration** refers to the way the electrons fill the various energy levels of the atom and is what determines how an atom behaves chemically. As shown in Figure 5.7, unexcited hydrogen has its one electron in the first energy level. This seems reasonable, since the first level is associated with the lowest energy. Both of helium's electrons are also in the first energy level (Figure 5.15). Lithium, however, has three electrons. Two of them go into the first energy level, but one must go in the second level, the next-highest level. This is consistent with the fact that the first level can hold only two electrons. Atoms with four to ten electrons have two electrons in the first level and the remainder in the second level. Neon, with ten electrons, fills both the first and second levels (Table 5.2). Sodium, with eleven electrons, has full first and second levels and one electron in the third level.

Table 5.2 shows that (up to element 18) the electrons fill the energy levels in the order that we might predict. The lowest numbered level is filled before

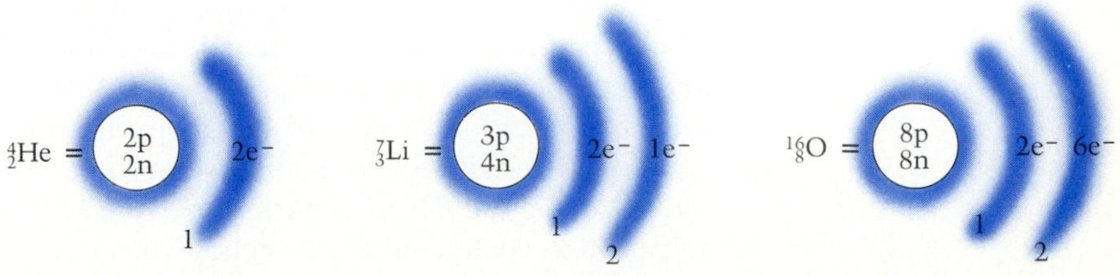

$^{4}_{2}He$ = 2p 2n 2e⁻ 1

$^{7}_{3}Li$ = 3p 4n 2e⁻ 1e⁻ 1 2

$^{16}_{8}O$ = 8p 8n 2e⁻ 6e⁻ 1 2

Figure 5.15
Some ground-state electron configurations

Table 5.2
Ground-State Electron Configurations of the First 21 Elements

Element	Atomic number	Electron configuration			
		$K(1)$	$L(2)$	$M(3)$	$N(4)$
Hydrogen	1	1			
Helium	2	2			
Lithium	3	2	1		
Beryllium	4	2	2		
Boron	5	2	3		
Carbon	6	2	4		
Nitrogen	7	2	5		
Oxygen	8	2	6		
Fluorine	9	2	7		
Neon	10	2	8		
Sodium	11	2	8	1	
Magnesium	12	2	8	2	
Aluminum	13	2	8	3	
Silicon	14	2	8	4	
Phosphorus	15	2	8	5	
Sulfur	16	2	8	6	
Chlorine	17	2	8	7	
Argon	18	2	8	8	
Potassium	19	2	8	8	1
Calcium	20	2	8	8	2
Scandium	21	2	8	9	2

any electrons are positioned in the next higher numbered level. But when we write the electron configuration for $_{19}K$, the pattern changes. We would expect the ground state configuration for potassium to include two electrons in the first level, eight electrons in the second level, and nine electrons in the third level. Instead we find that potassium includes two electrons in the first level, eight electrons in the second level, eight electrons in the third level, and one electron in the fourth energy level. Why should this last electron go into the fourth level when the third level hasn't been filled?

To understand electron configuration more fully we must look further into energy sublevels. These sublevels group electrons according to energy, and the electron configuration shows how many electrons an atom has at the various energies. Since electrons do not occupy orbitals in a random fashion, we must know in exactly what order they do occupy the orbitals. Up to element number 18, we can write electron configurations by filling the energy levels closest to the nucleus first. For elements above number 18, this method does not produce the correct configuration. We can then use the **Aufbau principle** to guide us in determining electron configuration. Aufbau comes from the German word *aufbauen,* which means to build. We must consider the overlapping of energy levels, shown in Figure 5.16. For example, the 4*s* sublevel has a lower energy than the 3*d* sublevel. Therefore the 4*s* sublevel fills with electrons before the 3*d* sublevel does, even though the 4*s* sublevel is farther away from the nucleus than the 3*d* sublevel. And the 5*s* sublevel fills with electrons before the 4*d* and 4*f* sublevels.

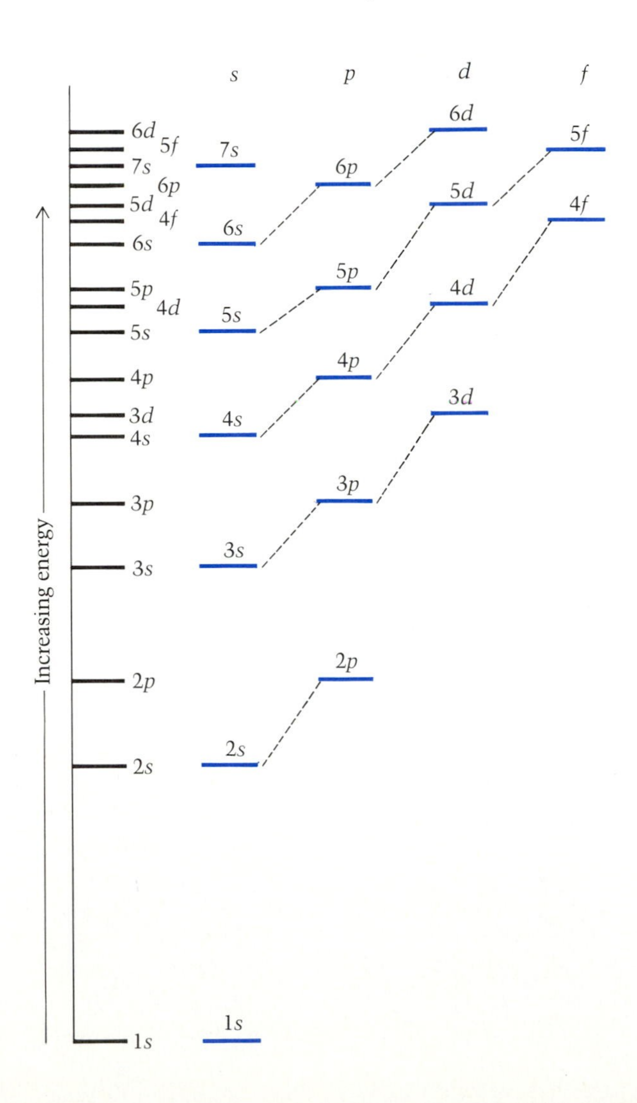

Figure 5.16
How energy levels overlap

Now you can write the electron configurations for any element with any atomic number—by simply filling the lowest energy *sublevel* first. The Aufbau diagram in Figure 5.17 will help you do this. All you have to do is follow the diagonal arrows, starting at the upper left. That is, the 1*s* sublevel is filled first, then the 2*s* sublevel, followed by 2*p*, 3*s*, 3*p*, 4*s*, 3*d*, and 4*p* sublevels, and so on. Unfortunately, even using the Aufbau diagram you will not be able to predict the electron configurations for *all* elements. There are a few exceptions.

When writing electron configurations we do not use words, we use a system involving shorthand. We represent the *K* energy level by the number 1, the *L* energy level by the number 2, the *M* energy level by the number 3, and so on. Then, for example, a 2*p* electron is an electron in the *p* sublevel of level 2.

Here are two examples of the notation we will use to write electron configurations:

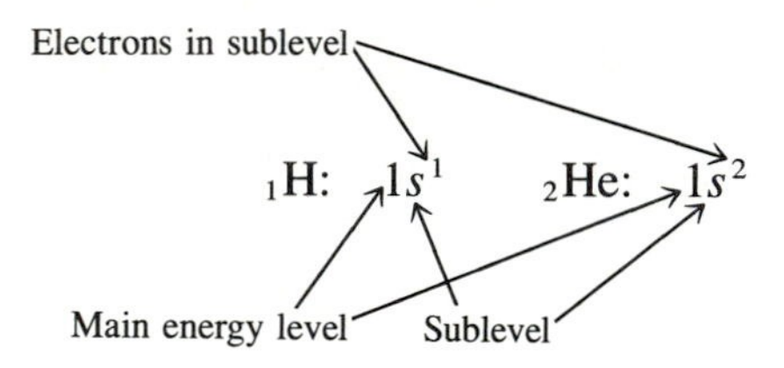

Note that the superscript to the right of the sublevel letter tells us the number of electrons in that sublevel.

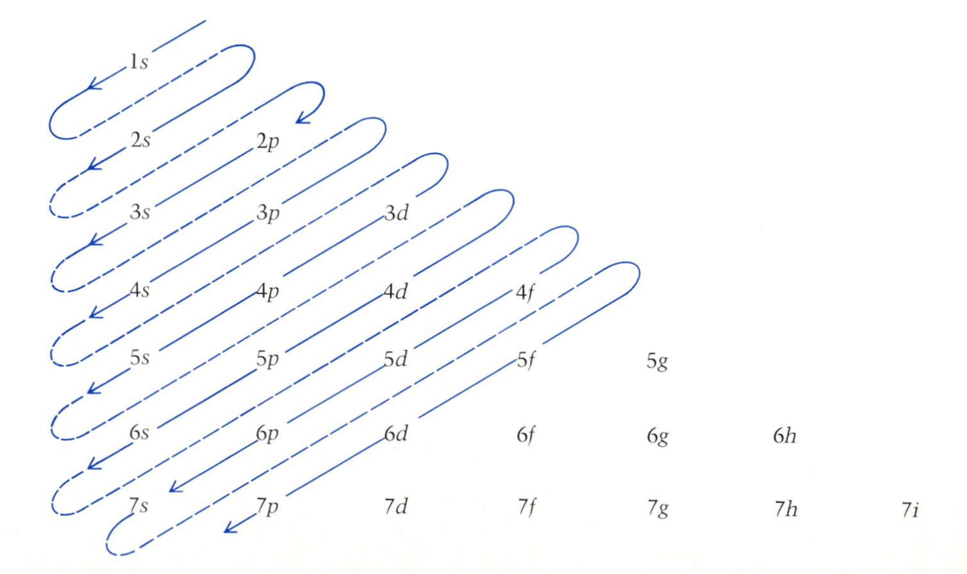

Figure 5.17
The order in which electrons fill energy sublevels

Here are some more examples:

$_3$Li: $1s^2 2s^1$

$_4$Be: $1s^2 2s^2$

$_5$B: $1s^2 2s^2 2p^1$

$_6$C: $1s^2 2s^2 2p^2$

$_{18}$Ar: $1s^2 2s^2 2p^6 3s^2 3p^6$

$_{19}$K: $1s^2 2s^2 2p^6 3s^2 3p^6 4s^1$

$_{20}$Ca: $1s^2 2s^2 2p^6 3s^2 3p^6 4s^2$

$_{21}$Sc: $1s^2 2s^2 2p^6 3s^2 3p^6 4s^2 3d^1$

Because the 4*s* sublevel is lower in energy than the 3*d* sublevel, the last electron in potassium (K) is in the 4*s* sublevel and not the 3*d* sublevel (check Figure 5.17 again). The same is true of calcium (Ca). But beginning with scandium (Sc), the 3*d* sublevel starts to fill. This is because the 4*s* sublevel is now completely filled, and the 3*d* sublevel is now the lowest unfilled energy sublevel. Sublevels continue to fill in this manner. Table 5.3 gives the resulting electron configurations for all elements, and Figure 5.18 shows the pattern of filling in the periodic table.

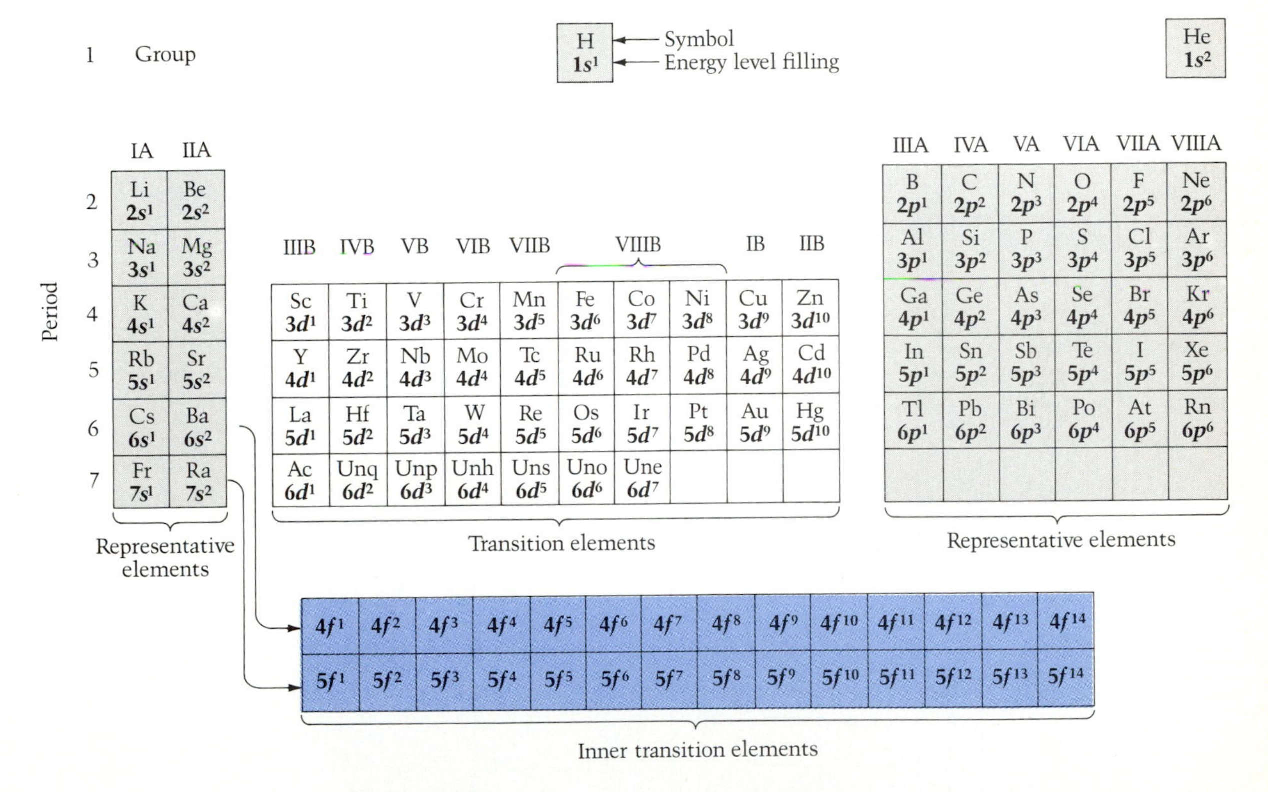

Figure 5.18
The general pattern for filling energy levels of the atoms of each element. (*Note:* There are some exceptions to this general pattern for certain transition and inner-transition elements, as shown in Table 5.3.)

Table 5.3
Electron Configuration of the Elements

Atomic number	Element	1	2		3			4				5				6				7
		s	*s*	*p*	*s*	*p*	*d*	*s*	*p*	*d*	*f*	*s*	*p*	*d*	*f*	*s*	*p*	*d*	*f*	*s*
1	H	1																		
2	He	2																		
3	Li	2	1																	
4	Be	2	2																	
5	B	2	2	1																
6	C	2	2	2																
7	N	2	2	3																
8	O	2	2	4																
9	F	2	2	5																
10	Ne	2	2	6																
11	Na	2	2	6	1															
12	Mg	2	2	6	2															
13	Al	2	2	6	2	1														
14	Si	2	2	6	2	2														
15	P	2	2	6	2	3														
16	S	2	2	6	2	4														
17	Cl	2	2	6	2	5														
18	Ar	2	2	6	2	6														
19	K	2	2	6	2	6		1												
20	Ca	2	2	6	2	6		2												
21	Sc	2	2	6	2	6	1	2												
22	Ti	2	2	6	2	6	2	2												
23	V	2	2	6	2	6	3	2												
24	Cr	2	2	6	2	6	5	1												
25	Mn	2	2	6	2	6	5	2												
26	Fe	2	2	6	2	6	6	2												
27	Co	2	2	6	2	6	7	2												
28	Ni	2	2	6	2	6	8	2												
29	Cu	2	2	6	2	6	10	1												
30	Zn	2	2	6	2	6	10	2												
31	Ga	2	2	6	2	6	10	2	1											
32	Ge	2	2	6	2	6	10	2	2											
33	As	2	2	6	2	6	10	2	3											
34	Se	2	2	6	2	6	10	2	4											
35	Br	2	2	6	2	6	10	2	5											
36	Kr	2	2	6	2	6	10	2	6											

Table 5.3
Electron Configuration of the Elements *(continued)*

Atomic number	Element	1	2		3			4				5				6				7
		s	*s*	*p*	*s*	*p*	*d*	*s*	*p*	*d*	*f*	*s*	*p*	*d*	*f*	*s*	*p*	*d*	*f*	*s*
37	Rb	2	2	6	2	6	10	2	6			1								
38	Sr	2	2	6	2	6	10	2	6			2								
39	Y	2	2	6	2	6	10	2	6	1		2								
40	Zr	2	2	6	2	6	10	2	6	2		2								
41	Nb	2	2	6	2	6	10	2	6	4		1								
42	Mo	2	2	6	2	6	10	2	6	5		1								
43	Tc	2	2	6	2	6	10	2	6	6		1								
44	Ru	2	2	6	2	6	10	2	6	7		1								
45	Rh	2	2	6	2	6	10	2	6	8		1								
46	Pd	2	2	6	2	6	10	2	6	10										
47	Ag	2	2	6	2	6	10	2	6	10		1								
48	Cd	2	2	6	2	6	10	2	6	10		2								
49	In	2	2	6	2	6	10	2	6	10		2	1							
50	Sn	2	2	6	2	6	10	2	6	10		2	2							
51	Sb	2	2	6	2	6	10	2	6	10		2	3							
52	Te	2	2	6	2	6	10	2	6	10		2	4							
53	I	2	2	6	2	6	10	2	6	10		2	5							
54	Xe	2	2	6	2	6	10	2	6	10		2	6							
55	Cs	2	2	6	2	6	10	2	6	10		2	6			1				
56	Ba	2	2	6	2	6	10	2	6	10		2	6			2				
57	La	2	2	6	2	6	10	2	6	10		2	6	1		2				
58	Ce	2	2	6	2	6	10	2	6	10	2	2	6			2				
59	Pr	2	2	6	2	6	10	2	6	10	3	2	6			2				
60	Nd	2	2	6	2	6	10	2	6	10	4	2	6			2				
61	Pm	2	2	6	2	6	10	2	6	10	5	2	6			2				
62	Sm	2	2	6	2	6	10	2	6	10	6	2	6			2				
63	Eu	2	2	6	2	6	10	2	6	10	7	2	6			2				
64	Gd	2	2	6	2	6	10	2	6	10	7	2	6	1		2				
65	Tb	2	2	6	2	6	10	2	6	10	9	2	6			2				
66	Dy	2	2	6	2	6	10	2	6	10	10	2	6			2				
67	Ho	2	2	6	2	6	10	2	6	10	11	2	6			2				
68	Er	2	2	6	2	6	10	2	6	10	12	2	6			2				
69	Tm	2	2	6	2	6	10	2	6	10	13	2	6			2				
70	Yb	2	2	6	2	6	10	2	6	10	14	2	6			2				
71	Lu	2	2	6	2	6	10	2	6	10	14	2	6	1		2				
72	Hf	2	2	6	2	6	10	2	6	10	14	2	6	2		2				

Table 5.3
Electron Configuration of the Elements *(continued)*

Atomic number	Element	1	2		3			4				5				6				7
		s	*s*	*p*	*s*	*p*	*d*	*s*	*p*	*d*	*f*	*s*	*p*	*d*	*f*	*s*	*p*	*d*	*f*	*s*
73	Ta	2	2	6	2	6	10	2	6	10	14	2	6	3		2				
74	W	2	2	6	2	6	10	2	6	10	14	2	6	4		2				
75	Re	2	2	6	2	6	10	2	6	10	14	2	6	5		2				
76	Os	2	2	6	2	6	10	2	6	10	14	2	6	6		2				
77	Ir	2	2	6	2	6	10	2	6	10	14	2	6	7		2				
78	Pt	2	2	6	2	6	10	2	6	10	14	2	6	9		1				
79	Au	2	2	6	2	6	10	2	6	10	14	2	6	10		1				
80	Hg	2	2	6	2	6	10	2	6	10	14	2	6	10		2				
81	Tl	2	2	6	2	6	10	2	6	10	14	2	6	10		2	1			
82	Pb	2	2	6	2	6	10	2	6	10	14	2	6	10		2	2			
83	Bi	2	2	6	2	6	10	2	6	10	14	2	6	10		2	3			
84	Po	2	2	6	2	6	10	2	6	10	14	2	6	10		2	4			
85	At	2	2	6	2	6	10	2	6	10	14	2	6	10		2	5			
86	Rn	2	2	6	2	6	10	2	6	10	14	2	6	10		2	6			
87	Fr	2	2	6	2	6	10	2	6	10	14	2	6	10		2	6			1
88	Ra	2	2	6	2	6	10	2	6	10	14	2	6	10		2	6			2
89	Ac	2	2	6	2	6	10	2	6	10	14	2	6	10		2	6	1		2
90	Th	2	2	6	2	6	10	2	6	10	14	2	6	10		2	6	2		2
91	Pa	2	2	6	2	6	10	2	6	10	14	2	6	10	2	2	6	1		2
92	U	2	2	6	2	6	10	2	6	10	14	2	6	10	3	2	6	1		2
93	Np	2	2	6	2	6	10	2	6	10	14	2	6	10	4	2	6	1		2
94	Pu	2	2	6	2	6	10	2	6	10	14	2	6	10	6	2	6			2
95	Am	2	2	6	2	6	10	2	6	10	14	2	6	10	7	2	6			2
96	Cm	2	2	6	2	6	10	2	6	10	14	2	6	10	7	2	6	1		2
97	Bk	2	2	6	2	6	10	2	6	10	14	2	6	10	8	2	6	1		2
98	Cf	2	2	6	2	6	10	2	6	10	14	2	6	10	10	2	6			2
99	Es	2	2	6	2	6	10	2	6	10	14	2	6	10	11	2	6			2
100	Fm	2	2	6	2	6	10	2	6	10	14	2	6	10	12	2	6			2
101	Md	2	2	6	2	6	10	2	6	10	14	2	6	10	13	2	6			2
102	No	2	2	6	2	6	10	2	6	10	14	2	6	10	14	2	6			2
103	Lr	2	2	6	2	6	10	2	6	19	14	2	6	10	14	2	6	1		2
104	Unq	2	2	6	2	6	10	2	6	10	14	2	6	10	14	2	6	2		2
105	Unp	2	2	6	2	6	10	2	6	10	14	2	6	10	14	2	6	3		2
106	Unh	2	2	6	2	6	10	2	6	10	14	2	6	10	14	2	6	4		2
107	Uns	2	2	6	2	6	10	2	6	10	14	2	6	10	14	2	6	5		2
108	Uno	2	2	6	2	6	10	2	6	10	14	2	6	10	14	2	6	6		2
109	Une	2	2	6	2	6	10	2	6	10	14	2	6	10	14	2	6	7		2

Example 5.3

Using Figure 5.17 write the electron configuration for each of the following elements: *(a)* $_{8}O$ *(b)* $_{30}Zn$

Solution Follow the arrows in Figure 5.17. Remember not to place more than two electrons in an *s* sublevel, six in a *p* sublevel, ten in a *d*, and so on.

(a) $_{8}O$: $1s^22s^22p^4$
(b) $_{30}Zn$: $1s^22s^22p^63s^23p^64s^23d^{10}$

Practice Exercise 5.3 Using Figure 5.17 write the electron configuration for each of the following elements: *(a)* $_{12}Mg$ *(b)* $_{34}Se$

5.9 The Importance of Electron Configuration

Learning Goal 9
Importance of electron configuration

When we know the electron configurations of elements, we can predict how elements will react with each other. The number of electrons in the **outermost shell** (the shell farthest away from the nucleus, by position) controls the chemical properties of an element. Elements with similar outer electron configurations behave in very similar ways. For instance, the elements Ne, Ar, Kr, Xe, and Rn are unreactive; in other words, they are almost chemically inert. They all have *eight* electrons in their outermost energy levels. On the other hand, the elements Li, Na, K, Rb, Cs, and Fr all react violently with water; they all have *one* electron in their outermost energy levels.

The electron configurations of the elements also help us predict how some elements will bond to each other. We will discuss that in Chapter 7.

Summary

When a beam of sunlight passes through a glass prism, the sunlight separates into a rainbow of colors called the visible spectrum. Each color of light is associated with a specific amount of energy. The visible spectrum is part of the electromagnetic spectrum, and both are continuous spectra—their various parts merge into each other. In the 1850s, Kirchhoff and Bunsen showed that when a sample of an element is heated until it glows, it produces a line spectrum of visible light. Each element produces its own characteristic line spectrum.

In 1913 Niels Bohr proposed a model of the atom to explain these line spectra. The model consisted of shells, or imaginary spherical surfaces, roughly concentric with the nucleus. Each shell was associated with a particular energy level. Bohr postulated that electrons travel around the nucleus in their orbits. However, an electron can jump from one shell to another at a higher energy level when it absorbs enough energy. It can fall back to the original shell by emitting this energy in bursts. These separate bursts show up as line spectra.

Revolutionary discoveries occurred in physics in the decades that followed. These discoveries gave rise to the quantum mechanical model of the atom, which is based on a series of mathematical equations. The electrons in an atom were now thought of as occupying energy levels, with each energy level being composed of sublevels. The probability of finding an electron in an orbital, or region of space about the nucleus was postulated.

Using this information, the electron configuration, or arrangement of electrons about the nucleus could be determined. The various energy levels of the atom have been given the letter designations *K* through *Q* and the numbers 1 through 7. (Level *K* or 1 is the lowest energy level.) Associated with each level is a theoretical maximum electron population. Lower energy levels are usually filled before electrons are positioned in higher energy levels. Using the Aufbau principle, electron configurations can be determined and represented using a shorthand notation. The electron configurations endow the elements with many of their properties; elements with similar outer electron configurations exhibit similar chemical behavior.

Key Terms

Aufbau principle **(5.8)**
continuous spectrum **(5.1)**
electromagnetic spectrum **(5.1)**
electron configuration **(5.8)**
energy level **(5.3)**
energy sublevel **(5.6)**
energy subshell **(5.6)**
excited state **(5.3)**
ground state **(5.3)**
line spectrum **(5.1)**
orbital **(5.4)**
outermost shell **(5.9)**
quanta **(5.2)**
quantum mechanics **(5.4)**
shells **(5.3)**

Career Sketch

Life Scientist

As a life scientist, you will look at all aspects of living organisms. You may study a particular animal, a plant, or a specific life process. You may also investigate evolution and the mechanics of heredity. You may decide to concentrate in the area of botany and study plants, or specialize in zoology and study animals, or focus on microbiology and study microorganisms.

Genetic engineering has recently emerged as a new and exciting field in the life sciences. If you choose to pursue a career in genetic engineering, you will join other researchers in searching for ways to transfer hereditary material found within cells to make a whole range of important substances. Researchers in California, for example, have used genetic engineering to produce a chemical compound that is reported to be safe and effective in dissolving blood clots in the coronary arteries of heart attack victims.

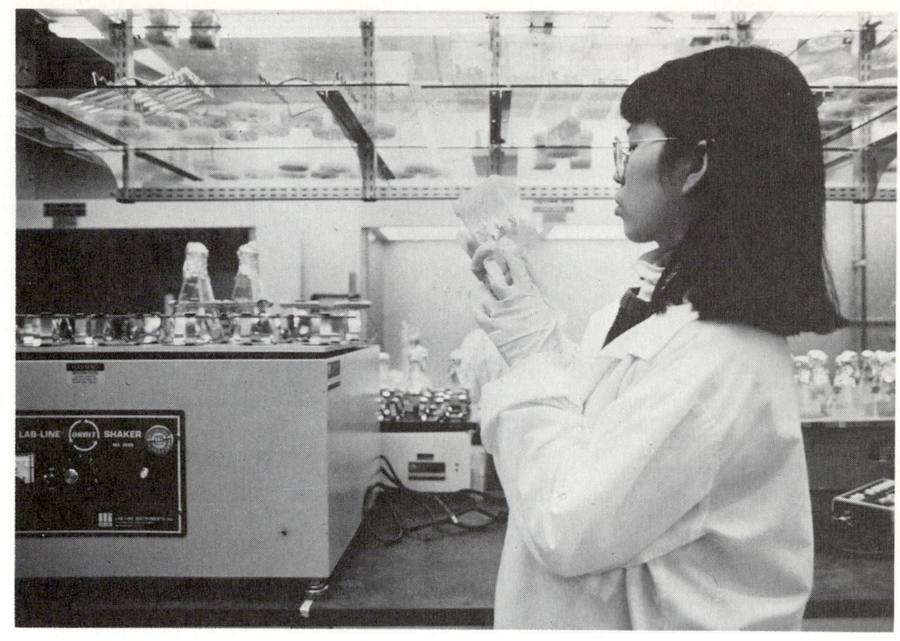

Life scientists engage in many different research projects such as the development of this bioengineered plant reared from cell culture.
(Matt Meadows/Peter Arnold, Inc.)

You will also become familiar with various types of research techniques and learn to use a variety of sophisticated laboratory equipment. Instruments used by life scientists include the transmission electron microscope (TEM) and the scanning tunneling electron microscope (STEM).

If you become a life scientist, you may choose to teach or do research, consult for a government agency, or work for a biotechnology company. To qualify for these positions, you will need to earn at least a baccalaureate degree. Many life scientists continue their education and go on to earn a Ph.D.

CHEMICAL FRONTIERS

CHEMISTRY LEADS TO FRESHER CANNED VEGETABLES*

The process of canning fruits and vegetables to preserve them was originally devised by the French chef and inventor Nicholas Appert nearly 200 years ago. Unfortunately, canned fruits and vegetables are quite mushy, and sales of these canned products have undergone a steady decline. Malcolm Bourne, a Cornell University food science professor, has developed a process that will

*Adapted from "Vegetables 'Fresh' from the Can," *What's Happening in Chemistry* (American Chemical Society, 1990).

allow canned fruits and vegetables to be almost as firm and crisp as fresh vegetables.

Bourne's process "fixes" the thermal stability of canned foods so that they remain firm even if cooked for hours. Bourne and his colleagues discovered that cooking causes vegetables to decompose into simpler molecules. This process is called depolymerization. Large and complex pectin molecules break down when heated, and the food becomes soft.

In traditional canning, two stages are necessary. First, the food is heated or blanched at 200°F. This process drives off gases and allows the food to be packed more tightly. In the second stage, the hermetically sealed can is heated to kill bacteria that might cause food poisoning. The resulting food can be stored in the can safely for a long time.

Bourne knew that pectin methylstearase (PME), an enzyme found naturally in virtually all plant tissue, could reverse the softening process. Unused calcium atoms in vegetables are used by PME to construct salt bridges in the pectin molecules. This reverses food softening. Bourne determined the conditions under which enzymatic activity could be favored to produce more calcium linkages, thus producing more pectin molecules, which leads to firmer foods. The conditions included a cooler blanching temperature and the addition of citric acid to lower pH and favor more salt-bridge construction. A short period of "holding" time was also needed between the blanching and sterilization so that PME could work.

The results were dramatic. Fruits and vegetables that seemed to be "vine-ripened" could be preserved in cans. However, more research is needed. Low blanching temperatures could encourage bacterial growth in the blanching vats, although this can be solved. Commercial food-processing companies are reviewing this new canning method for future use.

Self-Test Exercises

Learning Goal 1: Line Spectra and Continuous Spectra

1. What is a continuous spectrum? a line spectrum?

▲ **2.** What is a spectrum? What is the difference between the electromagnetic spectrum and the visible spectrum?

▲ **3.** How could a spectroscope help scientists analyze the sun and other extraterrestrial bodies?

4. Why must an element be heated until it is glowing before it will exhibit its characteristic spectrum?

5. Can you discover the origin of the word *helium*? (*Hint:* See the *Handbook of Chemistry and Physics*.)

6. Do you think line spectra could be used to identify compounds? Explain.

Learning Goal 2: Bohr Model of the Atom

7. Compare the Rutherford, Thomson, and Bohr models of the atom.

8. Define the terms *ground state* and *excited state*.

9. What is Bohr's most important contribution to our ideas about the structure of the atom?

10. According to the Bohr model of the atom, do electrons "spiral" or "jump" from one energy level to another? What type of spectrum would be produced by each of these types of movements by electrons?

Learning Goal 3: Shells and Energy Levels

11. What subatomic particles produce the line spectrum of an element?

12. Complete the following sentence with the word *higher* or *lower:* The farther away an energy level is from the nucleus, the ______ the energy of an electron in that level.

▲ **13.** What is meant by shells and energy levels? How many shells are known to exist in atoms? State their letter and number designations.

14. Using the $2(n)^2$ rule, calculate the maximum number of electrons that can populate each of the first three energy levels.

Learning Goal 4: Difference Between Ground State and Excited State of an Atom

15. What do we mean by the ground state of an atom? the excited state of an atom?

16. Describe the ground-state configuration of a lithium atom. What changes would take place if the lithium atom were in an excited state?

17. If a neutral hydrogen atom has an electron in the second energy level, is the atom in its ground state or an excited state?

▲ **18.** A particular neutral magnesium atom has two electrons in the first shell, eight electrons in the second shell, and two electrons in the third shell. Would you characterize this atom as in the ground state or an excited state?

Learning Goal 5: Quantum Mechanical Model of the Atom

19. Why was the Bohr model of the atom inadequate to explain the behavior of the atom?

20. What new knowledge regarding the laws of physics was discovered in the early 1900s?

21. Discuss Louis de Broglie's contributions to the field of atomic theory.

22. Discuss Erwin Schrödinger's contributions to the knowledge of the behavior of electrons.

Learning Goal 6: Energy Sublevels

▲ **23.** How many energy sublevels does the first energy level contain?

24. How many energy sublevels does the second energy level contain? Name them.

25. How is the total number of electrons an energy level contains related to the number of electrons a sublevel can hold?

26. The third energy level can hold a maximum of 18 electrons. How many sublevels does this energy level contain? How many electrons does each sublevel contain?

▲ **27.** Determine the following for the fifth energy level of an atom:
(a) number of subshells it contains
(b) maximum number of electrons this energy level can hold

28. Determine the following for the sixth energy level of an atom:
(a) number of subshells it contains
(b) maximum number of electrons this energy level can occupy

Learning Goal 7: Electron Orbitals

▲ **29.** Using an analogy similar to the one in the text, explain what is meant by 95% probability, and describe how it relates to the quantum mechanical model of an atom.

30. What is meant by the "region of highest probability of finding an electron?"

31. For the fifth energy level, determine the number of orbitals in the first two sublevels.

32. For the sixth energy level, determine the number of orbitals in the first three sublevels.

33. Describe the shape of the $2s$, $3s$, and $4s$ electron orbitals.

34. Describe the shape of the $4p$ orbital.

35. Which of the following does not exist? *(a)* $2s$ *(b)* $2p$ *(c)* $2d$ *(d)* $2f$

36. Which of the following does not exist? *(a)* $3s$ *(b)* $3p$ *(c)* $3d$ *(d)* $3f$

Learning Goal 8: Writing Electron Configurations

37. Using Figure 5.17 write the electron configurations for the following elements: *(a)* $_{6}C$ *(b)* $_{22}Ti$ *(c)* $_{36}Kr$ *(d)* $_{53}I$

38. Using Figure 5.17 write the electron configurations for the following elements: *(a)* $_{5}B$ *(b)* $_{16}O$

39. Name the neutral elements with the following electron configurations. (Use Table 5.3.)
(a) $1s^22s^22p^63s^23p^63d^{10}4s^24p^64d^{10}5s^25p^66s^2$
(b) $1s^22s^22p^63s^23p^64s^2$
(c) $1s^22s^22p^63s^23p^63d^{10}4s^24p^65s^2$
(d) $1s^22s^2$

40. Name the neutral elements with the following election configurations. (Use Table 5.3.)
(a) $1s^2$
(b) $1s^22s^22p^3$
(c) $1s^22s^22p^63s^23p^63d^{10}4s^1$
(d) $1s^22s^22p^63s^23p^63d^{10}4s^24p^4$

▲ **41.** A certain neutral element has 2 electrons in the first level, 8 electrons in the second level, 18 electrons in the third level, and 3 electrons in the fourth level. List the following information for this element:
(a) atomic number
(b) total number of *s* electrons
(c) total number of *p* electrons
(d) total number of *d* electrons
(e) number of protons

42. A certain neutral element has 2 electrons in the first level, 8 electrons in the second level, 18 electrons in the third level, and 2 electrons in the fourth level. List the following information for this element:
(a) atomic number
(b) total number of *s* electrons
(c) total number of *p* electrons
(d) total number of *d* electrons
(e) number of protons

43. Write electron configurations to determine which of the following elements belong to the same groups: elements with atomic numbers 6, 19, 17, 14, 3, and 35.

44. Write electron configurations to determine which of the following elements belong to the same groups: elements with atomic numbers 5, 18, 16, 36, 10, and 13.

45. Use the $2n^2$ rule to calculate the maximum number of electrons that the seventh energy level can hold.

46. Use the $2n^2$ rule to calculate the maximum number of electrons that the fourth energy level can hold.

47. Using Figure 5.17, write the electron configurations for each of the following elements: *(a)* $_{31}Ga$ *(b)* $_{51}Sb$ *(c)* $_{82}Pb$ *(d)* $_{88}Ra$

48. Using Figure 5.17 write the electron configurations for the following elements: *(a)* $_{3}Li$ *(b)* $_{12}Mg$ *(c)* $_{56}Ba$

49. Name the neutral elements with the following electron configurations. (Use Table 5.3.)
(a) $1s^22s^22p^63s^2$
(b) $1s^22s^22p^63s^23p^63d^{10}4s^24p^65s^2$
(c) $1s^22s^22p^63s^23p^63d^{10}4s^24p^64d^{10}4f^{14}5s^25p^66s^2$
(d) $1s^22s^22p^63s^23p^63d^{10}4s^24p^64d^{10}4f^{14}5s^25p^65d^{10}6s^26p^1$

50. Name the neutral elements with the following electron configurations. (Use Table 5.3.)
(a) $1s^22s^22p^63s^1$
(b) $1s^22s^22p^63s^23p^63d^{10}4s^24p^64d^{10}4f^{12}5s^25p^66s^2$
(c) $1s^22s^22p^63s^23p^63d^{10}4s^24p^64d^{10}4f^{14}5s^25p^66s^2$
(d) $1s^22s^22p^63s^23p^63d^{10}4s^24p^64d^{10}4f^{14}5s^25p^65d^{10}6s^26p^6$

51. A given neutral element has 2 electrons in the first level, 8 electrons in the second level, 18 electrons in the third level, 22 electrons in the fourth level, 8 electrons in the fifth level, and 2 electrons in the sixth level. List the following information for this element:
(a) atomic number
(b) total number of *s* electrons
(c) total number of *p* electrons
(d) total number of *d* electrons
(e) total number of *f* electrons
(f) number of protons

52. A given neutral element has 2 electrons in the first level, 8 electrons in the second level, 18 electrons in the third level, 32 electrons in the fourth level, 9 electrons in the fifth level, and 2 electrons in the sixth level. List the following information for this element:
(a) atomic number
(b) total number of *s* electrons
(c) total number of *p* electrons
(d) total number of *d* electrons
(e) total number of *f* electrons
(f) number of protons

Learning Goal 9: Importance of Electron Configuration

53. What is the significance of electron configuration?

54. Complete the following sentence: The number of electrons in the ________ shell determines the chemical properties of an element.

55. The elements in Group VIIIA of the periodic table (this group is the vertical column of elements farthest to the right in the periodic table) have eight electrons in their outermost energy level. (The exception to this is helium, which has only two electrons in its outermost energy level.) What is chemically unique about the Group VIIIA elements?

56. Separate the following into groups (vertical columns) of elements that exhibit similar chemical behavior: *(a)* Li *(b)* Ne *(c)* K *(d)* Rb *(e)* Kr *(f)* Na *(g)* Cs *(h)* Rn *(i)* Xe

Extra Exercises

57. Write electron configurations for all elements with atomic numbers between 1 and 20.

58. Write electron configurations for all elements with atomic numbers between 1 and 20 that are isoelectronic with magnesium. *Hint: Isoelectronic* in this instance means "has the same number of electrons in the outermost energy level as."

▲ **59.** What was the reasoning behind replacement of the Rutherford model of the atom with Bohr's model?

60. Explain why the line spectrum of an element is sometimes referred to as its fingerprint.

61. Explain briefly how electrons produce a line spectrum.

62. In Chapter 4 you learned that not all atoms of an element need be identical and that these different atoms are called isotopes. Do isotopes of the same element produce the same line spectrum? Explain.

63. In this chapter we stated that a marble rolling down a flight of stairs is similar to an electron jumping from one energy level to the next. Can you think of other analogous phenomena?

64. What can happen to an electron in an atom when it absorbs too much energy while being excited?

65. What would be the atomic number of an element whose energy levels 1, 2, 3, 4, 5, 6, and 7 were filled with the maximum number of electrons?

6 The Periodic Table: Keeping Track of the Elements

Learning Goals

After you've studied this chapter, you should be able to:

1. Discuss the historical basis and significance of the periodic table.
2. Distinguish between a period and a group in the periodic table.
3. State the difference between an A-group (representative) element and a B-group element (transition metal).
4. Predict trends for properties such as atomic radius, ionization potential, and electron affinity.

Introduction

In preceding chapters, we referred to the periodic table for information about the elements, such as atomic number and mass. The periodic table can be used to predict many properties that can be useful to us. In this chapter, we trace the development of the periodic table and see what kinds of additional information it holds.

6.1 History

Learning Goal 1
Historical basis and significance of periodic table

From the time of the ancient Greeks up to 1800, only 14 elements were known. Then, in a short span of 10 years, from 1800 to 1810, 14 more elements were discovered. By 1830, 45 elements were known. As new elements were discovered, chemists probably began to feel insecure. All these elements had different properties, and there didn't seem to be any relationship among them. And chemists wondered how many elements actually existed. To answer this question, they had to find some relationship among the known elements.

One of the first chemists to notice some order among the elements was the German scientist Johann Döbereiner. He published an account of his observations in 1829. It occurred to Döbereiner that bromine had chemical and physical properties somewhere between those of chlorine and iodine and that

Dmitri Mendeleev (1834–1907). In 1869 he predicted the chemical and physical properties of the element gallium, which he called eka-aluminum.
(Reprinted with permission from *Torchbearers of Chemistry* by Henry Monmouth Smith, ©Academic Press, Inc., 1949)

bromine's atomic mass was almost midway between those of chlorine and iodine (Figure 6.1). Could this be nothing more than coincidence?

Döbereiner searched and found two more groups of similar elements. The first was made up of calcium (Ca), stronium (Sr), and barium (Ba); the second included sulfur (S), selenium (Se), and tellurium (Te) (Figure 6.2). He called these groups *triads* (sets of three).

In 1864 the English chemist John Newlands arranged the known elements in order of increasing atomic mass. He came upon the idea of arranging them in vertical columns. Because he noticed that the eighth element had chemical and physical properties that were similar to those of the first, he let the eighth element start a new column (Figure 6.3). Newlands called his arrangement the *Law of Octaves*. But there were many places in his arrangement where dissimilar elements were next to each other.

In 1869 the German chemist Julius Lothar Meyer devised an incomplete periodic table consisting of 56 elements. In the same year the Russian chemist Dmitri Mendeleev (pronounced "Menduh-LAY-eff") arranged the elements in order of increasing atomic mass. In cases where discrepancies arose, *the chemical properties were used to place the elements*. Mendeleev established horizontal rows, or *periods*. Hydrogen by itself made up the first

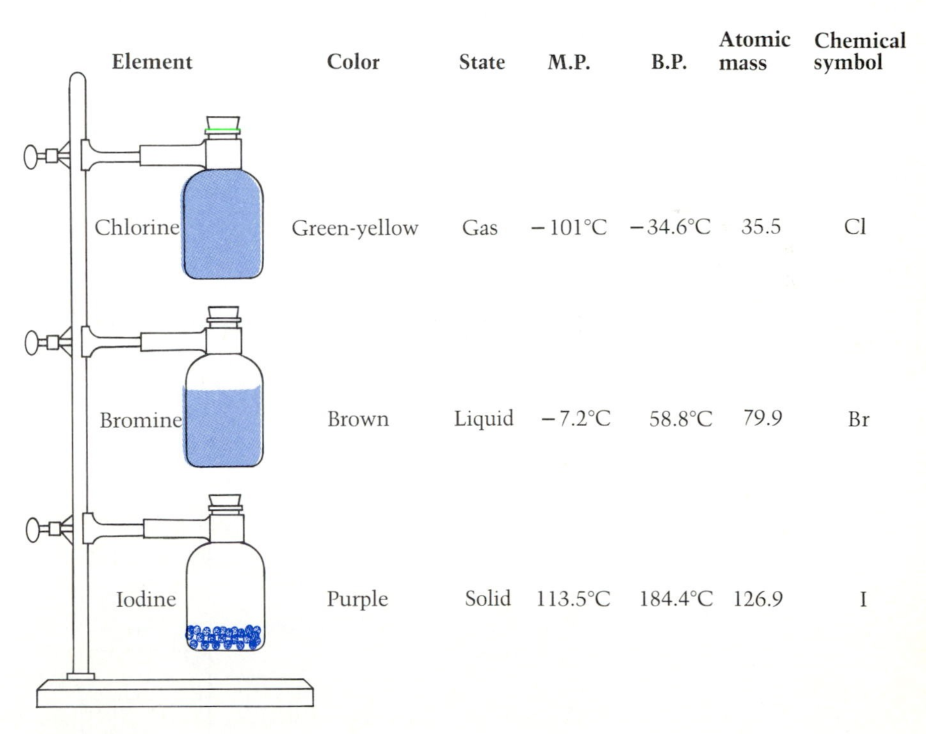

Element	Color	State	M.P.	B.P.	Atomic mass	Chemical symbol
Chlorine	Green-yellow	Gas	−101°C	−34.6°C	35.5	Cl
Bromine	Brown	Liquid	−7.2°C	58.8°C	79.9	Br
Iodine	Purple	Solid	113.5°C	184.4°C	126.9	I

Figure 6.1
Some similar elements

Figure 6.2
More similar elements

Element	Color	State	M.P. (°C)	B.P. (°C)	Atomic mass, amu
Ca	Silver-white	Solid	842.8	1487	40.08
Sr	Silver white to pale yellow	Solid	769	1384	87.62
Ba	Yellow-silver	Solid	725	1140	137.34
S	Yellow	Solid	113	444	32.06
Se	Blue-gray	Solid	217	685	98.96
Te	Silver-white	Solid	452	1390	127.6

period. Each of the next two periods contained seven elements. (The noble gases had not yet been discovered.) The periods after that contained more than seven elements. This is the point at which Mendeleev's table differed from Newlands's table.

In arranging his table, Mendeleev occasionally put heavier elements before lighter ones to keep elements with the same chemical properties in the same vertical column, or *group* (Figure 6.4). Sometimes he left open spaces in the table, where he reasoned that unknown elements should go. This was the case with gallium. In 1869, when he made up his table, the element gallium was unknown; however, Mendeleev *predicted* its existence. He based his prediction on the properties of aluminum (which appeared directly above gallium in the table). Mendeleev even went so far as to predict the melting point, boiling point, and atomic mass of the then-unknown gallium, which he

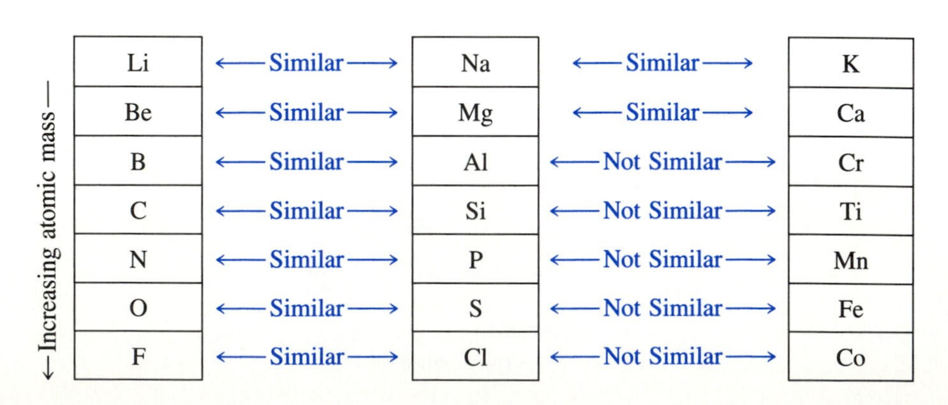

Figure 6.3
Newlands's Law of Octaves

Figure 6.4
Mendeleev's original periodic table arranged in modern form

Period	I	II	III	IV	V	VI	VII	VIII		
1	1 H							2		
2	3 Li	4 Be	5 B	6 C	7 N	8 O	9 F	10		
3	11 Na	12 Mg	13 Al	14 Si	15 P	16 S	17 Cl	18		
4	19 K	20 Ca	21	22 Ti	23 V	24 Cr	25 Mn	26 Fe	27 Co	28 Ni
	29 Cu	30 Zn	31	32	33 As	34 Se	35 Br	36		
5	37 Rb	38 Sr	39 Y	40 Zr	41 Nb	42 Mo	43	44 Ru	45 Rh	46 Pd
	47 Ag	48 Cd	49 In	50 Sn	51 Sb	52 Te	53 I	54		
6	55 Cs	56 Ba	57 La	72	73 Ta	74 W	75	76 Os	77 Ir	78 Pt
	79 Au	80 Hg	81 Tl	82 Pb	83 Bi	84	85	86		
7	87	88	89	104	105					

58 Ce	59	60	61	62	63	64	65 Tb	66	67	68 Er	69	70	71
90 Th	91	92 U	93	94	95	96	97	98	99	100	101	102	103

Paul Emile Lecoq de Boisbaudran (1838–1912). In 1875 he discovered and named the element gallium.
(Reprinted with permission from *Torchbearers of Chemistry* by Henry Monmouth Smith, ©Academic Press, Inc., 1949)

called eka-aluminum. Six years later, while analyzing zinc ore, the French chemist Lecoq de Boisbaudran discovered the element gallium. Its properties were almost identical to those Mendeleev had predicted (Figure 6.5).

The periodic tables proposed by both Mendeleev and Meyer were based on increasing atomic masses. Mendeleev used similarities among chemical properties to account for discrepancies among elements. Certain irregularities in these tables were corrected by the work of Henry G. J. Moseley, a British physicist. Moseley was able to determine the nuclear charge on the atoms of known elements. He concluded that elements should be arranged by increasing atomic number rather than by atomic mass. This research led to the development of the *periodic law,* which states that the chemical properties

Figure 6.5
The predicted element called eka-aluminum and the discovered element, gallium, that it turned out to be

		Eka-aluminum	Gallium
Atomic mass	→	Predicted: 68	Observed: 69.9
Oxide formula	→	Predicted: E_2O_3	Observed: Ga_2O_3
Oxide in acid	→	Predicted: should dissolve in acid	Observed: dissolves in acid
Salt formation	→	Predicted: should form salts as EX_3	Observed: forms salts as GaX_3

of the elements are periodic functions of their atomic numbers. Elements with similar chemical properties recur at regular intervals, and they are placed accordingly on the periodic table.

6.2 The Modern Periodic Table

Learning Goal 2
Periods and groups in the periodic table

The periodic table of today is similar to Mendeleev's but has many more elements—those that have been discovered since 1869. Today's table consists of seven horizontal rows called **periods** and a number of vertical columns called **groups** (or **families**) (Table 6.1).

The groups are numbered with Roman numerals. All the elements in each group have the same number of electrons in their outermost shells, so they all behave similarly. For example, the Group IA elements react violently when they come into contact with water. And all elements in Group IA have one electron in their outermost shell.

Learning Goal 3
A-group and B-group elements

Some of the groups in the periodic table are labeled with a Roman numeral followed by A, others with a Roman numeral followed by B. The A groups are called the **representative elements**. The B groups are called the **transition metals**. You can also see that as we move from the top to the bottom of the periodic table (in other words, from period 1 to period 7), the periods get larger—they have more elements in them. In fact, periods 6 and 7 are so large that to fit the table on one page, we have to write part of each period *below* the rest of the table (the 15 lanthanides and the 15 actinides).

6.3 Periodic Trends

Learning Goal 4
Periodic trends

When we examine the periodic table, we can see certain trends in the properties of the elements. For example, the elements in Group IA (at the left) are all metals. As we move across to the right, we continue to encounter metals through more than half the width of the table. Then we reach elements that have properties intermediate between metals and nonmetals; these are the metalloids. As we move still further to the right, we finally reach the nonmetals. (See Section 3.9 for a review of these terms.)

Table 6.1
Periodic Table of the Elements

Period	Group IA	IIA	IIIB	IVB	VB	VIB	VIIB	VIIIB	VIIIB	VIIIB	IB	IIB	IIIA	IVA	VA	VIA	VIIA	VIIIA
1	1 H 1.00794																	2 He 4.002602
2	3 Li 6.941	4 Be 9.01218											5 B 10.811	6 C 12.011	7 N 14.0067	8 O 15.9994	9 F 18.998403	10 Ne 20.179
3	11 Na 22.98977	12 Mg 24.305											13 Al 26.98154	14 Si 28.0855	15 P 30.97376	16 S 32.066	17 Cl 35.453	18 Ar 39.948
4	19 K 39.0983	20 Ca 40.078	21 Sc 44.9559	22 Ti 47.88	23 V 50.9415	24 Cr 51.9961	25 Mn 54.9380	26 Fe 55.847	27 Co 58.9332	28 Ni 58.69	29 Cu 63.546	30 Zn 65.38	31 Ga 69.723	32 Ge 72.59	33 As 74.9216	34 Se 78.96	35 Br 79.904	36 Kr 83.80
5	37 Rb 85.4678	38 Sr 87.62	39 Y 88.9059	40 Zr 91.22	41 Nb 92.9064	42 Mo 95.94	43 Tc (98)	44 Ru 101.07	45 Rh 102.9055	46 Pd 106.42	47 Ag 107.8682	48 Cd 112.41	49 In 114.82	50 Sn 118.710	51 Sb 121.75	52 Te 127.60	53 I 126.9045	54 Xe 131.29
6	55 Cs 132.9054	56 Ba 137.33	57-71 La–Lu	72 Hf 178.49	73 Ta 180.9479	74 W 183.85	75 Re 186.207	76 Os 190.2	77 Ir 192.22	78 Pt 195.08	79 Au 196.9665	80 Hg 200.59	81 Tl 204.383	82 Pb 207.2	83 Bi 208.9804	84 Po (209)	85 At (210)	86 Rn (222)
7	87 Fr (223)	88 Ra (226)	89-103 Ac–Lr	104 Unq (261)	105 Unp (262)	106 Unh (263)	107 Uns (262)	108 Uno (265)	109 Une (267)									

Transition elements

Lanthanides	57 La 138.9055	58 Ce 140.12	59 Pr 140.9077	60 Nd 144.24	61 Pm (145)	62 Sm 150.36	63 Eu 151.96	64 Gd 157.25	65 Tb 158.9254	66 Dy 162.50	67 Ho 164.9304	68 Er 167.26	69 Tm 168.9342	70 Yb 173.04	71 Lu 174.967
Actinides	89 Ac (227)	90 Th 232.0381	91 Pa (231)	92 U 238.0289	93 Np (237)	94 Pu (244)	95 Am (243)	96 Cm (247)	97 Bk (247)	98 Cf (251)	99 Es (252)	100 Fm (257)	101 Md (258)	102 No (259)	103 Lr (260)

1 ← Atomic number
H ← Symbol
1.00794 ← Atomic mass

Metal	Metalloid	Nonmetal

Other trends in properties that show up in the periodic table have to do with electron configuration, atomic radius, the tendency to lose or to gain electrons, and some physical properties. In the remainder of this chapter, we discuss these trends as well as the properties themselves and their structural basis.

6.4 Periodicity and Electron Configuration

All the elements within each A group have the same number of electrons in their outermost shell. Moreover, for the A-group elements, *the group number tells us how many outer electrons there are*. For example, all the elements in

Group VIA have six electrons in their outermost energy level. All the elements in Group IA have one electron in their outermost energy level.

For the B-group elements, there is no clear connection between group number and the number of electrons in the outermost energy level. For example, Group IVB elements *do not* have four electrons in their outermost energy level.

6.5 Similarities Among Elements in a Group and Period

The configuration of the outermost energy level determines the chemical behavior of an element. Thus all elements in the same group should behave (react) similarly, because they all have the same outer-shell electron configuration. In other words, the properties that these elements exhibit as a result of their interaction with other substances is similar. But what about elements in the same period? In general, the elements in any period have different outer-shell electron configurations, so these elements have different chemical properties.

For example, in period 3, Al (with two $3s$ electrons and one $3p$ electron) is a metal, whereas Cl (with two $3s$ electrons and five $3p$ electrons) is a nonmetal. However, this is not so for the transition metals in periods 4 through 7. Consider the elements Mn, Fe, Co, and Ni in period 4. Their electron configurations differ because they have different numbers of electrons in their $3d$ sublevels. However, level 3 is *not* their outermost energy level. The outermost level for these transition metals is the $4s$ sublevel. We would therefore expect these elements to exhibit somewhat similar behavior, and they do. Moreover, the same is true of the transition elements in periods 5 through 7.

To summarize:

1. Elements in any one group behave in similar ways, because they have (with few exceptions) *identical electron configurations in the outermost shell.*
2. Elements in any one period generally do not behave similarly.
3. However, because of the manner in which electrons fill their shells (filling some inner shells after outer shells), the transition metals—that is, the B groups—show somewhat similar chemical behavior.

6.6 Atomic Radius

The arrangement of the periodic table is helpful in predicting the relative sizes of atoms within any period or group. We shall consider the **atomic radius** (radius of an atom) to be the distance from the center of the nucleus to the outermost electron. Atomic radii are measured with a unit called the *angstrom,* named after a Swedish physicist. One angstrom (1 Å) is equal to 10^{-10} m, or one hundred-millionth of a centimeter.

Figure 6.6
Relative sizes of some Group IIA elements

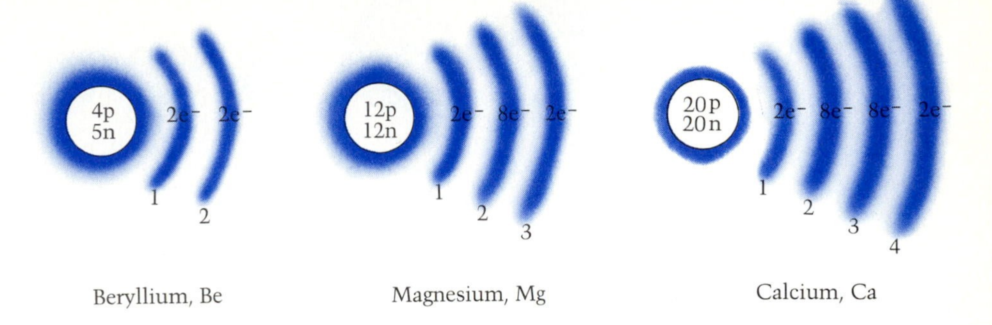

In any group in the periodic table, each element has one more energy level than the element above it (Figure 6.6). Therefore, as we move down through a particular group, the atomic radii of the elements should increase.

But what happens to the atomic radii of the elements as we move from left to right across a period? You might assume that because the number of electrons in the atom is increasing, the size of the atom should also increase. This is not the case. Remember that as we move across a period, the number of protons increases along with the number of electrons. Therefore there is a greater pull on the electrons of the inner energy levels, which usually causes a shrinking of the atomic radius (Figure 6.7).

Using the elements in period 2, we can explain what happens. In lithium (atomic number 3), three protons attract two *K* electrons and one *L* electron. In beryllium (atomic number 4), four protons attract two *K* electrons and two *L* electrons. In boron, five protons attract two *K* electrons and three *L* electrons. This increased positive charge acting on the *K* electrons makes the *K* energy level shrink. The *L* energy level also shrinks, and this results in a decreased atomic radius (Figure 6.8).

Also keep in mind that for the transition metals, atomic radii change very little because electrons are filling the *d* and *f* subshells. Table 6.2 shows the atomic radii of all the elements.

Figure 6.7
The atomic radii of the elements decrease as we move from left to right across a period. The increased nuclear charge results in a greater pull on the electrons of the inner energy levels, so the atom shrinks.

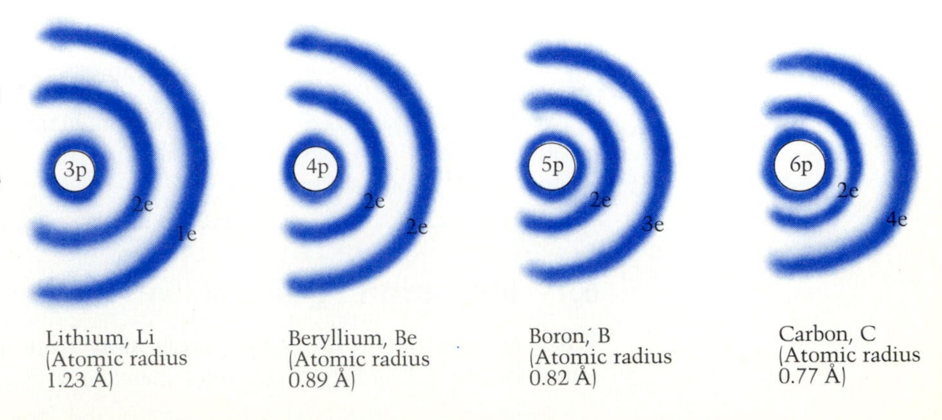

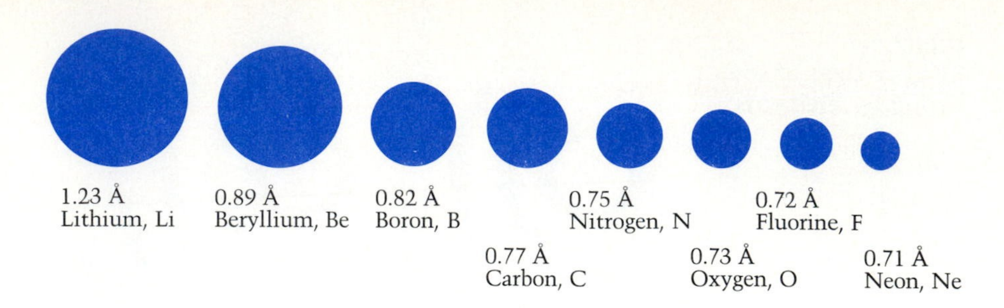

Figure 6.8
The decrease of atomic radii within a period

Example 6.1

List the following elements in order of *increasing* atomic radius: Mg, K, and Ca. Use only the periodic table (Table 6.1); don't look up the atomic radii.

Solution

Understand the Problem

We must order the elements in terms of atomic radius without knowing the specific atomic radius of each element. In other words, we must look at periodic trends.

Devise a Plan

Locate the position of each element in the periodic table, remembering that the atomic radius increases from the top to the bottom of a group of elements and decreases from left to right across a period of elements.

Carry Out the Plan

Remember that magnesium (Mg) is in period 3, whereas potassium (K) and calcium (Ca) are in period 4. Therefore, an atom of Mg has the smallest atomic radius of the three atoms in question. Next we look at K and Ca, the two elements that are in the same period. Since K is in Group IA and Ca is in Group IIA, a K atom has a larger atomic radius than a Ca atom. Therefore, of the three atoms, Mg is the smallest, Ca is larger, and K is the largest.

Look Back

We glance at the periodic table to confirm our reasoning.

Practice Exercise 6.1 List the following elements in order of *decreasing* atomic radius: S, O, N, and B. Use only the periodic table (Table 6.1); don't look up the atomic radii.

6.7 Ionization Potential

When an electron is pulled completely away from a neutral atom, what remains behind is a positively charged particle called an *ion*. A **positive ion**

Table 6.2
Periodic Table of Atomic Radii (in Angstroms)

Period	Group IA	IIA	IIIB	IVB	VB	VIB	VIIB	VIII	VIII	VIII	IB	IIB	IIIA	IVA	VA	VIA	VIIA	VIIIA
1	1 H 0.32																	2 He 0.31
2	3 Li 1.23	4 Be 0.89											5 B 0.82	6 C 0.77	7 N 0.75	8 O 0.73	9 F 0.72	10 Ne 0.71
3	11 Na 1.54	12 Mg 1.36											13 Al 1.18	14 Si 1.11	15 P 1.06	16 S 1.02	17 Cl 0.99	18 Ar 0.98
4	19 K 2.03	20 Ca 1.74	21 Sc 1.44	22 Ti 1.32	23 V 1.22	24 Cr 1.18	25 Mn 1.17	26 Fe 1.17	27 Co 1.16	28 Ni 1.15	29 Cu 1.17	30 Zn 1.25	31 Ga 1.26	32 Ge 1.22	33 As 1.20	34 Se 1.17	35 Br 1.14	36 Kr 1.12
5	37 Rb 2.16	38 Sr 1.91	39 Y 1.62	40 Zr 1.45	41 Nb 1.34	42 Mo 1.30	43 Tc 1.27	44 Ru 1.25	45 Rh 1.25	46 Pd 1.28	47 Ag 1.34	48 Cd 1.48	49 In 1.44	50 Sn 1.40	51 Sb 1.40	52 Te 1.36	53 I 1.33	54 Xe 1.31
6	55 Cs 2.35	56 Ba 1.98	57–71 La–Lu	72 Hf 1.44	73 Ta 1.34	74 W 1.30	75 Re 1.28	76 Os 1.26	77 Ir 1.27	78 Pt 1.30	79 Au 1.34	80 Hg 1.49	81 Tl 1.48	82 Pb 1.47	83 Bi 1.46	84 Po 1.46	85 At 1.45	86 Rn
7	87 Fr	88 Ra 2.20	89–103 Ac–Lr	104 Unq	105 Unp	106 Unh	107 Uns	108 Uno	109 Une									

Transition elements: groups IIIB through IIB.

57 La 1.69	58 Ce 1.65	59 Pr 1.64	60 Nd 1.64	61 Pm 1.63	62 Sm 1.62	63 Eu 1.85	64 Gd 1.62	65 Tb 1.61	66 Dy 1.60	67 Ho 1.58	68 Er 1.58	69 Tm 1.58	70 Yb 1.70	71 Lu 1.56
89 Ac 2.0	90 Th 1.65	91 Pa	92 U 1.42	93 Np	94 Pu	95 Am	96 Cm	97 Bk	98 Cf	99 Es	100 Fm	101 Md	102 No	103 Lr

Key:
6 ⟵ Atomic number
C
0.77 ⟵ Atomic radius

(also called a **cation**), is *an atom (or group of atoms) that has lost one or more electrons*. A **negative ion** (also called an **anion**) is *an atom (or group of atoms) that has acquired one or more electrons*.

The process of removing an electron from a neutral atom is called *ionization*. Here is what happens when a neutral atom of potassium becomes ionized:

$$K \longrightarrow K^{1+} + 1e^{1-}$$

The equation shows the neutral potassium atom (on the left) losing one electron to become a potassium ion (on the right). The superscript 1+ on the potassium shows that it has become a *positively charged particle*. That is, it has lost an electron ($1e^{1-}$), which is negatively charged. Thus the potassium is left with one more positive charge (proton) than negative charge (electron). It is no longer electrically neutral.

The **ionization potential** (also called *ionization energy*) of an element is *the energy needed to pull an electron away from an isolated ground-state atom of that element*. The farther away from the nucleus an electron is, the less it is attracted by the positive charge of the nucleus and the less energy is needed to pull that electron away from the atom. For this reason, the electron is always pulled away from the outermost energy level. In any group of the periodic table, the ionization potential decreases as we move down the group (Figure 6.9).

Ionization energy is usually measured in units called calories. A *calorie* is the amount of heat needed to raise the temperature of one gram of water one degree Celsius. A discussion of the calorie appears in Chapter 12.

Within each period, ionization potential generally increases from left to right (Table 6.3). This results from the decrease in atomic radius (and thus the stronger attraction between nucleus and outer electrons) as we move from left to right in the periodic table. The ionization potentials of the transition metals in each period do not vary much, because their atomic radii do not vary much (see Section 6.6)

Example 6.2

List the following elements in order of *increasing* ionization potential: Ca, Mg, and K. Use only the periodic table (Table 6.1); don't look up the atomic radii.

Solution Locate the position of each element in the periodic table. Remember that the ionization potential decreases from the top to the bottom of a group of elements and increases from left to right across a period of elements. Magnesium (Mg) is in period 3, whereas potassium (K) and calcium (Ca) are in period 4. Therefore, an atom of Mg has the highest ionization potential of the three atoms in question.

Now consider K and Ca, the two elements that are in the same period. Because K is in Group IA and Ca is in Group IIA, a Ca atom has a higher ionization potential than a K atom.

Therefore, of the three atoms, K has the lowest ionization potential, Ca has a higher ionization potential, and Mg has the highest ionization potential. (*Note:* Look back at the solution of Example 6.2. Notice how the ionization potential varies inversely with the atomic radius. In other words, as the atomic radius of an atom gets larger, the ionization potential gets smaller, and vice versa.)

Practice Exercise 6.2 List the following elements in order of *decreasing* ionization potential: S, O, N, and B. Use only the periodic table (Table 6.1); don't look up the atomic radii.

Figure 6.9
The decrease of ionization potential within a group

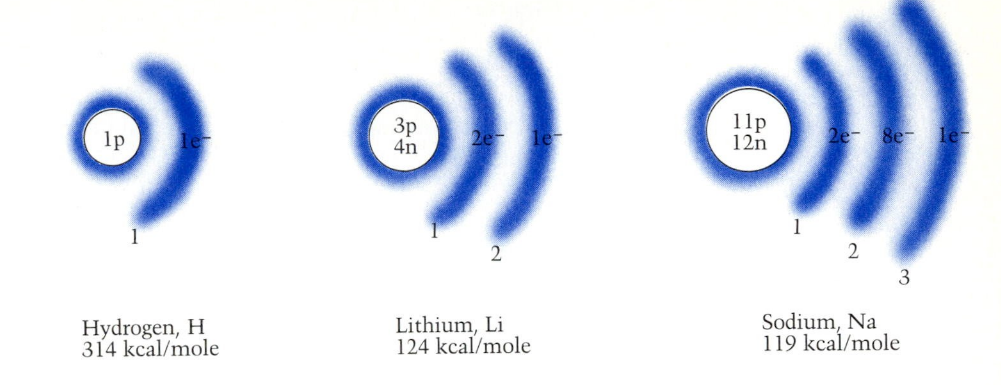

Table 6.3
Periodic Table of Ionization Potentials (in Kilocalories per Mole)

Period	Group IA	IIA	IIIB	IVB	VB	VIB	VIIB	VIII	VIII	VIII	IB	IIB	IIIA	IVA	VA	VIA	VIIA	VIIIA
1	1 H 314																	2 He 566
2	3 Li 124	4 Be 215											5 B 191	6 C $26\bar{0}$	7 N 335	8 O 312	9 F 402	10 Ne 498
3	11 Na 119	12 Mg 176											13 Al 138	14 Si 188	15 P 254	16 S 239	17 Cl $30\bar{0}$	18 Ar 363
4	19 K $10\bar{0}$	20 Ca 141	21 Sc 151	22 Ti 158	23 V 155	24 Cr 156	25 Mn 171	26 Fe 182	27 Co 181	28 Ni 176	29 Cu 178	30 Zn 216	31 Ga 138	32 Ge 187	33 As 242	34 Se 225	35 Br 273	36 Kr 323
5	37 Rb 96	38 Sr 131	39 Y 152	40 Zr $16\bar{0}$	41 Nb 156	42 Mo 166	43 Tc 172	44 Ru 173	45 Rh 178	46 Pd 192	47 Ag 174	48 Cd 207	49 In 133	50 Sn 169	51 Sb 199	52 Te 208	53 I 241	54 Xe $28\bar{0}$
6	55 Cs $9\bar{0}$	56 Ba $12\bar{0}$	57–71 La– Lu	72 Hf 127	73 Ta 140	74 W 184	75 Re 181	76 Os 201	77 Ir 212	78 Pt 208	79 Au 212	80 Hg 241	81 Tl 141	82 Pb 171	83 Bi 184	84 Po 196	85 At	86 Rn 248
7	87 Fr	88 Ra 122	89–103 Ac– Lr	104 Unq	105 Unp	106 Unh	107 Uns	108 Uno	109 Une									

Transition elements: IIIB through IIB.

57 La 129	58 Ce 159	59 Pr 133	60 Nd 145	61 Pm	62 Sm 129	63 Eu 131	64 Gd 142	65 Tb 155	66 Dy 157	67 Ho	68 Er	69 Tm	70 Yb 143	71 Lu 115
89 Ac 162	90 Th	91 Pa	92 U 92	93 Np	94 Pu	95 Am	96 Cm	97 Bk	98 Cf	99 Es	100 Fm	101 Md	102 No	103 Lr

Key	
6	← Atomic number
C	
$26\bar{0}$	← Ionization potential

6.8 Electron Affinity

An atom's ability to acquire an additional electron is called its **electron affinity** and is another important factor that determines its chemical properties. The process in which a neutral atom acquires an electron can be represented in this way:

$$\mathrm{Cl} + \mathrm{e}^{1-} \longrightarrow \mathrm{Cl}^{1-}$$

The equation shows that when a neutral chlorine atom picks up an electron, it becomes a negatively charged chloride ion. When the chlorine atom picks up an electron to form a chloride ion, it gives up something in return: *energy*. The amount of energy the chlorine atom releases is a measure of its electron affinity. And electron affinity depends on the attraction between an electron and the nucleus of an atom.

As we move down a group, the electron affinity of the elements generally decreases. If we look at some Group VIIA elements, we can see a decrease in electron affinity as we move from chlorine to iodine (Table 6.4). This is because in iodine, the additional electron is added to the fifth energy level, whereas in chlorine the additional electron is added to the third energy level. (Remember that the closer the electron is to the nucleus, the more it is attracted to the nucleus, and electron affinity depends on how strongly an electron is attracted to the nucleus.) Fluorine's unexpectedly low electron affinity is caused by the special nature of this element. (Some scientists feel that the electron affinity value for fluorine is reasonable but that the electron affinity value for chlorine is too high.)

If we examine the trend of electron affinity within a period, we find that it *increases* as we go from left to right in a period, and the attraction between the nucleus and the electron becomes greater.

Table 6.5 summarizes what we have discussed about trends in the periodic table. As your study of chemistry progresses, you will find that you rely more and more on information conveyed by the periodic table.

Summary

As more and more elements were discovered during the eighteenth and nineteenth centuries, scientists attempted to find relationships among the properties of the various known elements. Finally, in 1869, Dmitri Mendeleev arranged the elements in a grid or table form, with horizontal rows called periods and vertical columns called groups. Elements with similar chemical properties were positioned in the same group. The table ordered the known elements and their properties so well that Mendeleev was able to use it to predict the properties of yet undiscovered elements. The periodic table of today is similar to Mendeleev's. It consists of seven periods and sixteen groups. For elements in the A groups (representative elements), the group number is the same as the number of electrons in the outermost shell. For elements in the B groups (transition metals), there is not a simple connection between group number and number of electrons in the outermost energy level. However,

Table 6.4
Electron Affinities of Group VIIA Elements

Element	Electron affinity (in electron volts)	Outer electron configuration
F	3.6	$2s^2\,2p^5$
Cl	3.75	$3s^2\,3p^5$
Br	3.53	$4s^2\,4p^5$
I	3.2	$5s^2\,5p^5$

Table 6.5
Summary of Trends in the Periodic Table

Trend of	Top to bottom in a group	Left to right in a period
Atomic radius	Increases	Decreases
Ionization potential	Decreases	Increases
Electron affinity	Decreases	Increases

because the transition metals fill inner shells after outer shells, they show quite similar chemical behavior.

The atomic radius of an atom is the distance from the center of the nucleus to the outermost electron. As we move from the top to the bottom of any group in the periodic table, atomic radius generally increases. And as we move from left to right across a period, atomic radius generally decreases. The ionization potential for an element is the energy needed to pull an electron away from an isolated atom of that element. As we move from the top to the bottom of a group, the ionization potential decreases. And as we move from left to right across a period, the ionization potential generally increases. Electron affinity is a measure of an atom's ability to acquire additional electrons. This measure is related to the energy released when an additional electron is added to a neutral atom. As we move down a group in the periodic table, the electron affinity of the elements generally decreases. And as we move from left to right across a period, the electron affinity increases.

Key Terms

anion **(6.7)**
atomic radius **(6.6)**
cation **(6.7)**
electron affinity **(6.8)**
group (family) **(6.2)**
ionization potential **(6.7)**
negative ion **(6.7)**
period **(6.2)**
positive ion **(6.7)**
representative element **(6.2)**
transition metal **(6.2)**

CAREER SKETCH

FOOD SCIENTIST

As a food scientist, you will investigate the chemical, physical, and biological nature of food and apply this knowledge to the cooking, preserving, packaging, distributing, and storing of food. You may work in the area of research and development. You may also choose to work in quality control or in the processing area of a food-producing plant. Another option is to teach and do research in a college or university setting.

Working in food science means performing basic research in the structure and composition of food and the changes it undergoes in storage and cooking. For example, you may develop new sources of proteins, study the effects of processing on microorganisms, or search for factors that affect flavor, texture, and appearance of foods.

If you choose to work in the area of applied research, you will help create new foods and develop new processing methods. You will also seek to improve existing foods by making them more nutritious and enhancing their flavor, color, and texture.

Food scientists devote themselves to providing us with a safe and abundant food supply. The Food and Drug Administration inspector is sampling coffee beans at the Baltimore docks. (Martha Cooper/Peter Arnold, Inc.)

CHEMICAL FRONTIERS

NEW TREATMENT FOR GROWTH DISORDERS

In the United States there are about 12,000 children with growth disorders, which include dwarfism. In 1982 scientists at the Salk Institute in La Jolla, California, discovered a natural substance called *growth hormone releasing factor* (GHRF). GHRF causes the human body to release proper amounts of human growth hormones (HGHs), which promote normal growth. When children do not produce sufficient HGHs, their growth is stunted.

Until recently, the main sources of additional HGHs were bodies of the deceased, thereby making supplies quite limited. With the advent of genetic engineering, HGHs are produced in the laboratory by genetically engineered bacteria. Although this makes the supply more plentiful, those who need the genetically engineered hormones must be given large amounts by injection up to three times each week. Even so, these injections cannot supply a full spectrum of necessary hormones.

A New Jersey chemist working for Hoffman-La Roche, Inc., reported that GHRF can be produced artificially in the laboratory. Researchers began manufacturing genetically engineered GHRF in late 1988, and it is currently undergoing clinical testing. According to Arthur Felix of Hoffman-La Roche, artificial GHRF could cause the human body to release normal amounts of HGHs.

There are significant advantages to using GHRF instead of HGHs. First, an equal mass of GHRF is between 50 and 150 times more powerful than genetically engineered HGHs, which means smaller doses of GHRF are required. In addition, the smaller, more potent GHRF molecule can be delivered differently. Both skin patches and mist injection are possibilities. Once the body is stimulated to produce its own HGHs, a full spectrum of hormones would be present. With these new chemical discoveries, the future holds much promise for those afflicted with growth disorders.

Self-Test Exercises

Learning Goal 1: Historical Basis and Significance of Periodic Table

▲ **1.** Locate at least two instances in the periodic table (Table 6.1) in which a heavier element appears before a lighter one.

2. What was Mendeleev's reason for placing some heavier elements before lighter ones in his table? Why didn't this upset the "logic" of his table?

3. Using the periodic table (Table 6.1), determine the total number of *(a)* metallic elements, *(b)* metalloids, *(c)* nonmetallic elements.

▲ **4.** Why aren't elements 57 to 71 and 89 to 103 shown with the other elements in the body of Table 6.1?

5. Discuss the contributions to the development of the periodic table made by *(a)* Döbereiner, *(b)* Newlands, *(c)* Mendeleev.

6. How did Mendeleev's table differ from Newlands's table of the elements?

Learning Goal 2: Periods and Groups in the Periodic Table

7. In the periodic table, a horizontal row is called a ________ . A vertical column is called a ________ .

8. Which elements would behave similarly: those within any one group or those within any one period? Why?

▲ **9.** What similarity is there among the outermost energy levels of elements within the same group?

10. "Elements in the B groups (the transition metals) show somewhat similar chemical behavior." Explain why this statement is true.

Learning Goal 3: A-Group and B-Group Elements

11. What does the group number tell us about the A-group elements?

▲ **12.** Where are the A-group elements located on the periodic table? Where are the B-group elements located?

13. The *representative* elements are also known as the ________ (A/B) group elements. The *transition metals* are also known as the ________ (A/B) group elements.

14. "Elements within any one group have the same number of electrons in their outermost energy levels." Is this statement true for the A-group or the B-group elements?

Learning Goal 4: Periodic Trends

▲ **15.** List the following elements in order of increasing ionization potential: Cl, Ca, Sr, Rb, and S. Use only the periodic table (Table 6.1); don't look up the ionization potentials.

16. What is an angstrom, and how big is it?

17. List the following elements in order of increasing atomic radius: In, P, I, Sb, and As. Once again, use only the periodic table.

18. Does atomic radius increase or decrease as we move from the top to the bottom of a group in the periodic table? Why?

19. Define *electron affinity*.

20. As we move from left to right across a period in the periodic table, does atomic radius increase or decrease? Why?

▲ **21.** After you've removed one electron from a neutral atom, would you need more or less energy to remove a second electron? Why?

22. List the following elements in order of increasing atomic radius: In, Tl, H, F, and S. (Use the periodic table.)

23. Where in the periodic table do we find the following?
(a) Elements with the highest electron affinity
(b) Elements with the lowest ionization potential
(c) The most unreactive elements

24. What name is given to the energy needed to pull an electron away from an isolated atom?

25. Complete the following statements:
(a) The ionization potential ________ (increases/decreases) as one moves down a *group* of elements.
(b) The ionization potential ________ (increases/decreases) as one moves from left to right across a *period* of elements.
(c) The atomic radius ________ (increases/decreases) as one moves from left to right across a *period* of elements.

▲ **26.** Explain why it is more difficult to remove an electron from an atom that is located at the top of a group in the periodic table than to remove an electron from an atom that is located at the bottom of a group.

▲ **27.** List the following elements in order of *increasing* atomic radius: Cl, Ca, Sr, Rb, and S. Use only the periodic table (Table 6.1); don't look up the atomic radii.

▲ **28.** The elements sodium and chlorine are both in period 3 of the periodic table. Sodium has an ionization potential of 119 kcal/mole, and chlorine has an ionization potential of $30\overline{0}$ kcal/mole. Why is the ionization potential of chlorine so much greater?

29. List the following elements in order of *increasing* ionization potential: P, I, Sb, and As. Once again, use only the periodic table.

30. List the following elements in order of increasing ionization potential: Br, Ge, Se, Cs, and F. Use only the periodic table.

31. A positive ion is an atom (or group of atoms) that has *(a)* gained a proton, *(b)* lost a proton, *(c)* gained an electron, *(d)* lost an electron.

32. The quantity that depends on the attraction between an electron and the nucleus of an atom is called ________ .

33. A Cl^{1-} ion is a Cl atom that has *(a)* gained a proton, *(b)* lost a proton, *(c)* gained an electron, *(d)* lost an electron.

34. Electron affinity ________ (increases/decreases) as one moves down a *group* of elements.

35. An Mg^{2+} ion is an Mg atom that has *(a)* gained two protons, *(b)* lost two protons, *(c)* gained two electrons, *(d)* lost two electrons.

36. Electron affinity ________ (increases/decreases) as one moves from left to right across a *period* of elements.

Extra Exercises

37. Explain why chlorine is a smaller atom than sodium, even though it contains more subatomic particles.

38. What relationship is there with regard to chemical properties among
(a) all elements in a particular period?
(b) all elements in a particular group?

39. Without using the periodic table, find the electron configurations of the following elements. State which ones are chemically similar. *(a)* $_{11}Na$ *(b)* $_{13}Al$ *(c)* $_{8}O$ *(d)* $_{16}S$ *(e)* $_{5}B$ *(f)* $_{3}Li$

40. Without using the periodic table, find the electron configurations of the following elements. State which ones are chemically similar. *(a)* $_{1}H$ *(b)* $_{3}Li$ *(c)* $_{4}Be$ *(d)* $_{12}Mg$ *(e)* $_{20}Ca$

41. What would you expect the atomic number to be for the noble gas that follows radon?

42. Based on periodic trends, which group of atoms is most likely to form *(a)* 1+ ions? *(b)* 1− ions?

43. Explain the difference between a chlorine atom and a chloride ion.

44. Which member of each of the following pairs should have the larger radius? Why? *(a)* K, Br *(b)* K, Na *(c)* K, K^{1+}

45. Determine how many electrons are in the outermost energy level of a neutral atom of the following elements: *(a)* sodium *(b)* carbon *(c)* nitrogen

46. Which has the higher ionization potential, calcium or magnesium? Why?

47. How many elements are in the *(a)* first period? *(b)* second period? *(c)* third period? *(d)* fourth period? *(e)* fifth period? *(f)* sixth period?

48. Which element of each pair has the smaller ionization potential? *(a)* K, Ca *(b)* O, Se

49. Which element of each pair has the larger atomic radius? *(a)* K, Ca *(b)* O, Se

50. What would be the group number for element 114?

51. Why are the atomic masses of elements 94 through 109 listed in parentheses?

52. If you had the responsibility of naming the next discovered element, what would you name it?

53. We know that the number of electrons in the outermost energy level of the Group IIIB elements is 3 because the group number reflects the number of electrons in the outermost energy level. True or false? Why?

54. We know that the number of electrons in the outermost energy level of the Group IIIA elements is 3 because the group number reflects the number of electrons in the outermost energy level. True or false? Why?

55. By writing electron configurations, determine which of the following elements belong to the same groups: elements with atomic number 15, 38, 50, 56, 14, and 33.

56. By writing electron configurations, determine which of the following elements belong to the same groups: elements with atomic number 16, 36, 39, 8, and 34.

Cumulative Review/Chapters 4–6

1. How are protons, neutrons, and electrons distributed in the atom?

2. How did Eugen Goldstein choose the term *cathode rays?*

3. What experiment did Sir William Crookes perform to determine whether cathode rays were composed of particles or waves?

4. What is the photoelectric effect?

5. What reasoning led physicist J. J. Thomson to develop the "plum-pudding" theory of the atom?

6. When Rutherford tested Thomson's theory of the atom with his own experiments, what results led him to devise a new model of the atom?

7. Why did Rutherford's model of the atom render Thomson's atomic model obsolete?

8. What type of particle adds to the mass of the atom without affecting its charge?

9. John Dalton said that atoms of an element must be identical in size, mass, and shape. How does the existence of isotopes disprove this theory?

▲ **10.** Give the number of protons that are found in each of the following isotopes: *(a)* $^{59}_{27}Co$ *(b)* $^{90}_{40}Zr$ *(c)* $^{74}_{34}Se$

11. Using Table 4.2, calculate the atomic mass of neon.

12. Using Table 4.2, determine how many ^{15}N atoms there are in a sample containing 100,000 nitrogen atoms.

▲ **13.** Write the standard isotopic notation for the following elements. (Use the periodic table to get the symbols of the elements.) *(a)* 17p, 17e, 18n *(b)* 14p, 14e, 14n

14. The hypothetical element Z exists in two isotopic forms, ^{50}Z and ^{52}Z. The atomic mass of Z is 50.5. What is the percentage abundance of each isotope?

15. State the number of protons, electrons, and neutrons in $^{73}_{32}Ge$.

16. How many electrons are there in a neutral $^{60}_{26}Fe$ atom?

17. Of the three isotopes of oxygen, which is the most abundant: oxygen-16, oxygen-17, or oxygen-18?

18. Using Table 4.2, determine how many ^{7}Li atoms there are in a sample containing 2.0×10^3 lithium atoms.

▲ **19.** Using Table 4.2, determine how many atoms of ^{7}Li you would find in a sample of lithium that has 1.000×10^6 atoms of Li.

20. Using Table 4.2, determine how many atoms of Si you would need to get exactly 470,000 atoms of ^{29}Si.

21. Do electrons, protons, or neutrons produce the line spectrum of an element?

22. As an electron travels to an energy level farther from the nucleus, the energy of that electron ________________. *(a)* increases *(b)* decreases *(c)* remains the same

23. How many electrons are in the outermost energy level of an oxygen atom?

24. Without using the periodic table, determine the electron configurations of the following elements: *(a)* $_1H$ *(b)* $_3Li$ *(c)* $_{11}Na$ *(d)* $_{19}K$

25. A neutral atom of magnesium has ________ electrons in its outermost energy level. (You may consult the periodic table.)

26. Name the element that has the following ground-state configuration. You may consult the periodic table.

K, 2 electrons; *L,* 8 electrons; *M,* 6 electrons

▲ **27.** The chemical properties of an element are determined by the number of electrons in the ________.
(a) energy level closest to the nucleus
(b) outermost energy level
(c) second energy level
(d) none of these

28. Why is phosphorus in Group VA?

▲ **29.** Write electron configurations for all elements with atomic numbers between 1 and 20 that are isoelectronic (have the same number of electrons in the outermost energy level) with $_3Li$.

30. What is the atomic number of an element whose energy levels *K, L, M,* and *N* are filled to capacity?

31. When a rainbow is produced naturally, what acts as the prism?

32. How are line spectra used in chemical laboratories today?

33. In a detective story it was discovered that the victim died of arsenic poisoning. This information was made available by analysis of samples of the victim's hair. How could spectral analysis have made this discovery possible?

34. How is radiant energy produced?

35. Waves on a lake travel through water, and sound waves travel through air. Through what medium do electromagnetic waves travel?

36. What types of waves are included in the electromagnetic spectrum? *(a)* visible *(b)* invisible *(c)* both of these *(d)* neither of these

37. How did Heinrich R. Hertz verify the theory of electromagnetic radiation proposed by James Clerk Maxwell?

38. Give an example of light energy being absorbed and re-emitted.

39. Give an example of heat energy being absorbed and re-emitted as light energy.

▲ **40.** Bundles of energy absorbed or emitted by electrons are called ________ .

41. State the periodic law and explain how it is the basis of the modern-day periodic table.

▲ **42.** What was the contribution of Henry G. J. Moseley in terms of the periodic table?

43. The element with atomic number 107 is an artificially produced radioactive element. Both Soviet and German scientists claim to have discovered it, but neither claim has been officially accepted. To produce an isotope of the element with atomic number 107, a sample of bismuth-209 was bombarded with chromium-54. Write the electron configuration of the element with atomic number 107. What is its group number?

44. In 1974 a team of U.S. scientists working under Albert Ghiorso at the Lawrence Berkeley Laboratory in Berkeley, California, bombarded the element californium-249 with a beam of oxygen-18 to produce an isotope of the element with atomic number 106. Write the electron configuration for the element with atomic number 106. What is its group number?

45. Our modern periodic table contains cases where elements are not in proper order according to atomic mass. Where do these cases occur?

46. Ionization potential increases as one moves down a group of elements. True or false?

47. Atomic radius decreases as one moves from left to right across a period of elements. True or false?

▲ **48.** A Ca^{2+} ion is a calcium atom that has gained two protons. True or false?

49. The element with atomic number 108 should behave like what other elements in the periodic table?

50. Name the neutral element with the electron configuration $1s^22s^22p^63s^1$.

51. The elements in the periodic table show properties that are related to their atomic numbers. True or false?

▲ **52.** The element with the electron configuration $1s^22s^22p^6$ belongs in group VIA. True or false?

53. Lithium and fluorine are both in period 2 of the periodic table. Which element has the higher ionization potential? Why?

54. Write the electron configuration for $_{14}Si$.

55. The volume of space occupied by an electron is called an orbital. True or false?

▲ **56.** On the basis of periodic trends, which group of atoms is most likely to form *(a)* 2+ ions? *(b)* 2− ions?

57. Which member of each of the following pairs should have the larger radius? *(a)* Li, Na *(b)* Ca, Ca^{2+}

58. Which element has the lower ionization potential, Mg or Cl?

59. Which element has the larger atomic radius, Li or Cs?

60. Which element has the larger atomic radius, Li or F?

7 Chemical Bonding: How Atoms Combine

Learning Goals

After you've studied this chapter, you should be able to:

1. Write Lewis (electron-dot) structures for the A-group elements.
2. Define and give an example of a covalent bond and a coordinate covalent bond.
3. Name the diatomic elements from memory.
4. Write electron-dot structures for various covalent compounds.
5. Define and give an example of an ionic bond.
6. Use the concept of electronegativity to find the ionic percentage and covalent percentage of a chemical bond.
7. Distinguish between a polar and a nonpolar bond.
8. Distinguish between a polar and a nonpolar molecule.

Introduction

Matter is held together by forces acting between atoms. When chemical reactions occur, atoms can combine to form compounds. In doing so, they join together in such a way as to attain more stable configurations that have lower levels of chemical potential energy.

Atoms have the ability to gain, lose, or share electrons. This enables them to form compounds in two different ways. When electrons are gained and lost, the compounds formed are held together by forces that bind oppositely charged ions. When electrons are shared, the compounds formed are held together by forces that bind atoms together to form molecules.

Whether the binding forces are due to electron loss and gain or to sharing, when atoms combine they are said to be joined together by a **chemical bond.** They have attained a more stable arrangement—usually eight electrons in their outermost shell. In this chapter, we examine the types of bonds that tie atoms together in molecules and unite elements in compounds.

IA	IIA	IIIA	IVA	VA	VIA	VIIA
H·						
Li·	Be·	B·	·C·	·N·	·O:	·F:

Figure 7.1
Electron-dot diagrams of some A-group elements

7.1 Lewis (Electron-Dot) Structures

Learning Goal 1
Lewis structures

Because it is the electrons in the outermost energy levels that form the bonds, we will be concerned only with these electrons in the elements we study. A very useful notation for showing the outermost electrons of an atom (and the bonds they form) was devised in 1916 by G. N. Lewis, an American physical chemist. In this *electron-dot* notation, the outer electrons are shown as dots on the sides of the symbol for the element. (Figure 7.1 provides several examples.) Each dot corresponds to an outer-shell electron. Although dot placement isn't critical, one method is to place the dots one to a side, beginning at the top and moving clockwise. A second dot is then placed on a side only after there is one dot on each of the four sides of the symbol. If you move from left to right in Figure 7.1, you will see how each additional dot (as necessary) is placed on the diagram. Electron-dot structures make it easier to see the role outer shell electrons play in chemical bonding.

Figure 7.2
The electron-dot diagrams of all Group VIIA elements are the same.

Recall now that all the elements in the same A group have the same number of electrons in their outermost energy level. This means that all the elements in any A group have the same electron-dot diagram. Figure 7.2 shows the diagrams for fluorine and bromine, which are both in Group VIIA. Both have seven dots. And Figure 7.3 shows the diagrams for nitrogen and arsenic; both are in Group VA, so both have five dots.

Figure 7.3
The electron-dot diagrams of all Group VA elements are the same.

Now that you know how to write the electron-dot notation for an atom of an element, we'll learn how to use this notation to draw the *electron-dot structure* of a compound. The **electron-dot structure** is *a schematic representation of the bonding in a molecule of a compound*. It does not represent the physical position of the electrons in the molecule. However, the electron-dot notation is useful for understanding how bonding occurs.

7.2 The Covalent Bond: The Octet Rule

Learning Goal 2
Covalent bonds

It has long been known that the Group VIIIA elements tend to be very unreactive and stable. Because of their tendency to keep to themselves, these elements are called the *noble gases*. Researchers believe that this stability is connected with the fact that each of these Group VIIIA elements has eight electrons in its outermost energy level. (The element helium is an exception, because its atom has only two electrons, but these two electrons fill the outermost energy level of the helium atom.) The atoms of other elements can achieve the same type of stability by obtaining eight electrons in their

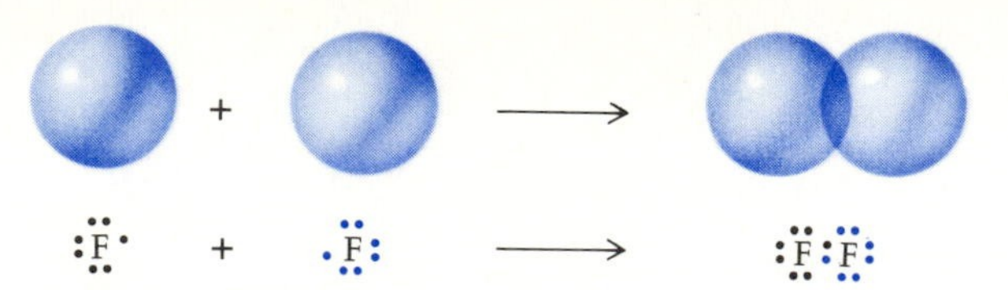

Figure 7.4
Bonding between fluorine atoms

outermost energy levels. We state this relationship formally as the **octet rule,** which tells us that when forming compounds, atoms of elements gain, lose, or share electrons such that a noble gas configuration is achieved for each atom. When atoms share electrons with other atoms, a *covalent bond* is formed. A **covalent bond** is *a chemical bond in which two atoms share a pair of electrons*.

To see how electrons are shared, imagine a sample of fluorine atoms in a closed container. A fluorine atom has seven electrons in its outermost energy level. It needs eight electrons in that energy level to complete its octet. One way to obtain another electron is to share an outer electron belonging to another fluorine atom (Figure 7.4). (Examine Figure 7.4, and think of the arrow as indicating that the items on the left combine to form what is on the right. We will discuss such chemical equations in more detail in Chapter 10.) When a fluorine atom shares one of its electrons with another fluorine atom, a fluorine molecule (F_2) is formed. But now each atom has a stable octet. Evidence of this stability is the fact that fluorine is always found as F_2 molecules.

The type of covalent bond formed between the two fluorine atoms is called a **single covalent bond,** because each atom has shared a *single* electron with the other to form one bond. The single bond is the most common type of covalent bond.

The Diatomic Elements

Learning Goal 3
Diatomic elements

Diatomic elements are elements that are found naturally in molecules with two atoms each. Fluorine is thus a *diatomic element*. In fact, all the Group VIIA elements are diatomic. They all form single covalent bonds (Figure 7.5). Hydrogen, oxygen, and nitrogen also exist in diatomic forms (Figure 7.6), because they share each other's electrons.

An easy way to remember the diatomic elements is with the memory aid HONClBrIF (pronounced "HON kel brif"). Remember that these atoms exist as diatomic molecules because doing so yields a stable octet of electrons or, in the case of hydrogen, a stable duet (two electrons, like helium).

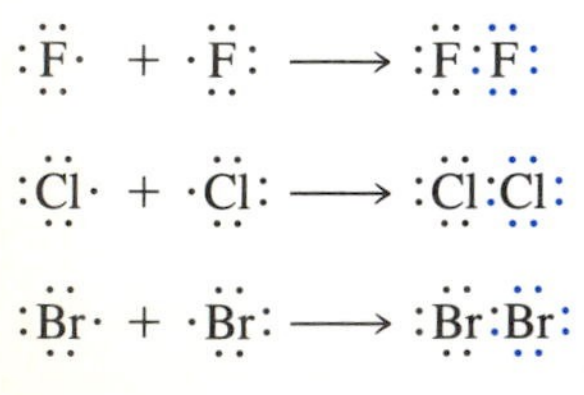

:F· + ·F: ⟶ :F:F:

:Cl· + ·Cl: ⟶ :Cl:Cl:

:Br· + ·Br: ⟶ :Br:Br:

Figure 7.5
Bonding diagrams of Group VIIA diatomic elements

Other Molecules Having Single Covalent Bonds

Now let's see how some elements obtain eight electrons (or, in the case of hydrogen, two electrons) in their outermost energy levels by forming compounds. We will use the electron-dot notation to show the bonding.

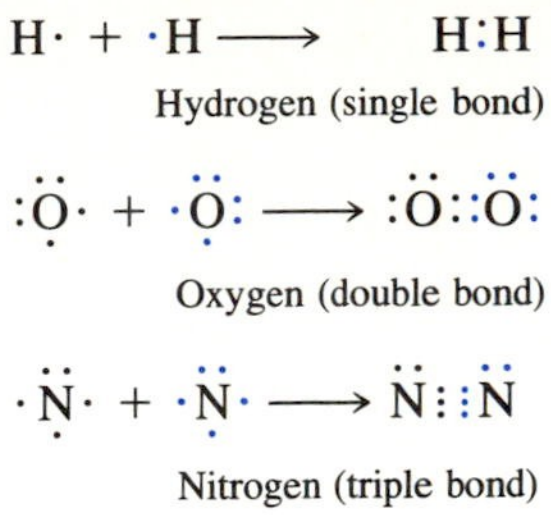

H· + ·H ⟶ H:H

Hydrogen (single bond)

:Ö· + ·Ö: ⟶ :Ö::Ö:

Oxygen (double bond)

·N̈· + ·N̈· ⟶ N⋮⋮N

Nitrogen (triple bond)

Figure 7.6
Bonding diagrams for other diatomic elements. Hydrogen needs only two electrons to complete its outer energy level. In O_2, two electrons from each atom contribute to the bond. Three electrons from each nitrogen atom contribute to the bond.

Water, which has the formula H_2O, bonds in this way:

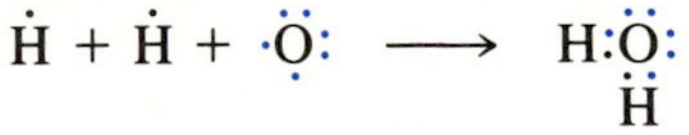

Ḣ + Ḣ + ·Ö: ⟶ H:Ö:
H

The two hydrogens share their single electrons with the oxygen. So the oxygen obtains eight electrons in its outermost energy level, and each hydrogen now has its two electrons.

In ammonia (NH_3), methane (CH_4), and hydrogen chloride (HCl), the bonding looks like this:

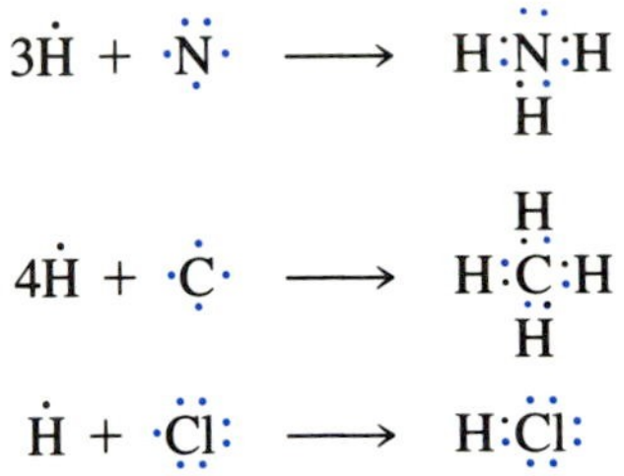

3Ḣ + ·N̈· ⟶ H:N̈:H
H

4Ḣ + ·Ċ· ⟶ H
H:C:H
H

Ḣ + ·C̈l: ⟶ H:C̈l:

The N, C, and Cl atoms now have eight electrons in their outermost energy level. The H atoms complete their outer shells with two electrons. The molecules are more stable than the atoms from which they are formed.

The Use of Dashes

Learning Goal 4
Electron-dot structures of covalent compounds

Instead of using electron-dot notation to show the bonding in compounds, we sometimes use dashes. The dash (—) has the same meaning as a double dot (:), but it simplifies the diagram. The dash, then, represents a single bond, or a shared pair of electrons. The following examples will show you how the dash is used in combination with dot notation.

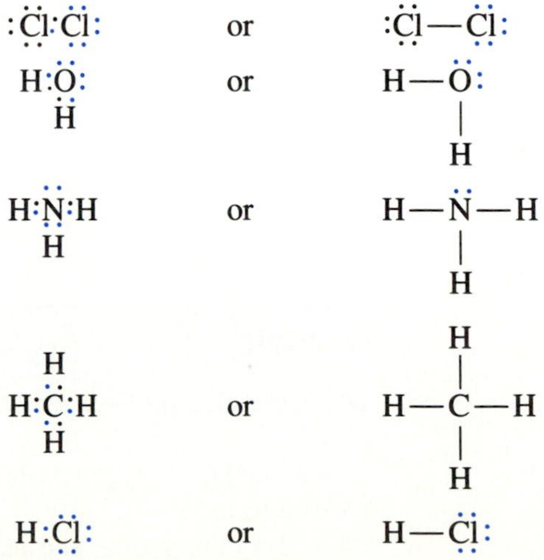

:C̈l:C̈l: or :C̈l—C̈l:

H:Ö: or H—Ö:
H (below O) | H (below O)

H:N̈:H or H—N̈—H
H (below N) | H (below N)

H
H:C:H or H—C—H (with H above and below C, joined by dashes)
H

H:C̈l: or H—C̈l:

This notation is especially useful when you want to represent the structures of compounds that have double and triple bonds.

The Double Bond

A **double covalent bond** is *one in which two pairs of electrons are shared between two atoms*. For example, in oxygen (O_2) the bonding is

$$\cdot\ddot{\underset{\cdot}{O}}: + \cdot\ddot{\underset{\cdot}{O}}: \longrightarrow :\ddot{O}::\ddot{O}: \quad \text{or} \quad :\ddot{O}{=}\ddot{O}:$$

By sharing two electrons, each oxygen atom obtains eight electrons in its outermost energy level. (Actually, this bonding structure for O_2 is simplified and not entirely correct, but the simplified structure communicates the general idea.)

The bonding in the compound ethylene (C_2H_4), which is used to make polyethylene plastic, can be written as

$$\begin{matrix} H & H \\ \overset{\cdot x}{\underset{\cdot x}{C}}: & :\overset{\cdot x}{\underset{\cdot x}{C}} \\ H & H \end{matrix} \quad \text{or} \quad \begin{matrix} H & & H \\ | & & | \\ C & = & C \\ | & & | \\ H & & H \end{matrix}$$

(In this diagram, an x is used to represent the electron in each hydrogen.) Here the hydrogen atoms are satisfied with two electrons, and each of the carbon atoms has the eight electrons it needs.

The Triple Bond

A **triple covalent bond** is *one in which three pairs of electrons are shared between two atoms*. For example, in nitrogen (N_2) the bonding can be diagrammed in this way:

$$\cdot\ddot{\underset{\cdot}{N}}\cdot + \cdot\ddot{\underset{\cdot}{N}}\cdot \longrightarrow \ddot{N}\vdots\vdots\ddot{N} \quad \text{or} \quad \ddot{N}{\equiv}\ddot{N}$$

With the triple bond, each nitrogen atom obtains eight electrons in its outermost energy level. The original N atom has only five electrons, but when two N atoms get together by triple bonding, each has a share in eight electrons.

In the compound acetylene (C_2H_2), the bonding is

$$H\overset{x}{\cdot}C\vdots\vdots C\overset{x}{\cdot}H \quad \text{or} \quad H{-}C{\equiv}C{-}H$$

Example 7.1

Write the electron-dot structures for each of the following covalent compounds.

(a) CO_2 (Each oxygen is bonded to the carbon, but the oxygens aren't bonded to each other.)

(b) CH_3I (The iodine and all the hydrogens are bonded to the carbon.)

(c) $COCl_2$ (This is phosgene. The oxygen and both chlorines are bonded to the carbon.)
(d) $AsCl_3$ (Each chlorine is bonded to the arsenic.)
(e) H_2S (Both hydrogens are bonded to the sulfur.)
(f) CH_2O (This is formaldehyde, which we will talk more about in our discussion of aldehydes in Chapter 20. The oxygen and both hydrogens are bonded to the carbon.)
(g) C_2Cl_2 (One chlorine is bonded to each carbon.)

Solution

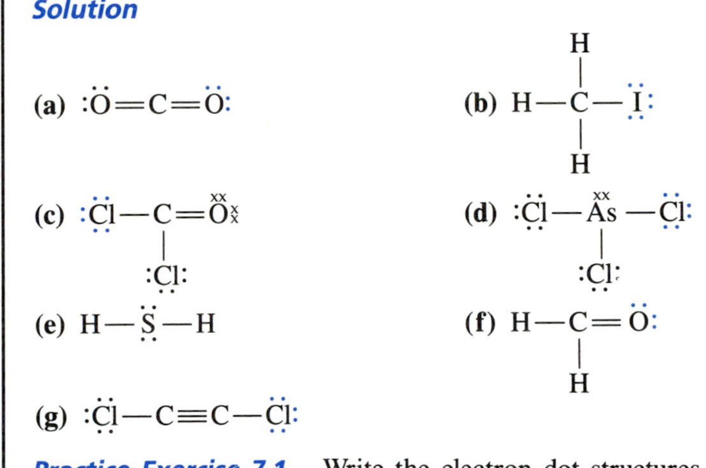

Practice Exercise 7.1 Write the electron-dot structures for the following covalent compounds: *(a)* NH_3 *(b)* HCl *(c)* SO_2

7.3 The Coordinate Covalent Bond

In the examples of covalent bonds that we have discussed so far, both atoms furnished electrons to form the bond. This can be thought of as a "Dutch-treat" bond, in which each atom pays its share of the bill. Another type of covalent bond exists. It is called a **coordinate covalent bond,** *a covalent bond in which only one atom donates the electrons to form the bond.*

A coordinate covalent bond can be thought of as a "you-treat" bond: one atom pays the entire bill. An example of this type of bond is found in the compound phosphoric acid (H_3PO_4). A coordinate covalent bond forms between the phosphorus atom and the single oxygen atom, to which it gives two electrons.

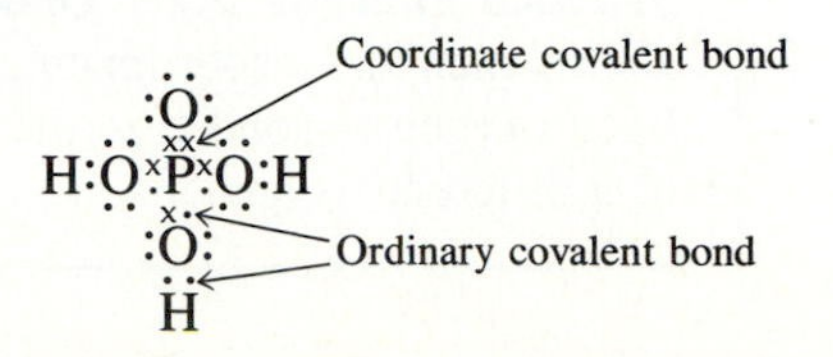

Ordinary (not coordinate) covalent bonds form between the phosphorus atom and the other three oxygen atoms (those with hydrogens attached to them). Once formed, the coordinate covalent bond is indistinguishable from an ordinary covalent bond. However, in drawing bonding diagrams of compounds that have coordinate covalent bonds, we sometimes use an arrow to designate the coordinate covalent bond. For example, H_3PO_4 may be diagrammed as follows:

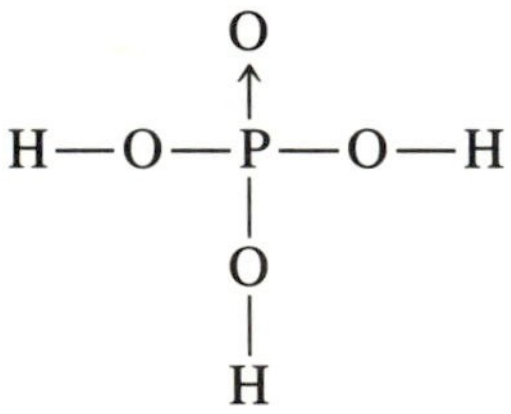

$$\begin{array}{ccccccccc} & & & & O & & & & \\ & & & & \uparrow & & & & \\ H & — & O & — & P & — & O & — & H \\ & & & & | & & & & \\ & & & & O & & & & \\ & & & & | & & & & \\ & & & & H & & & & \end{array}$$

Note that the arrow does not mean that the bond is in any way different.

Example 7.2

Draw bonding diagrams for the following covalent compounds that have coordinate covalent bonds included in their structures.

(a) SO_3 (Each oxygen is bonded to the sulfur.)

(b) H_2SO_4 (Each oxygen is bonded to the sulfur, and each of the hydrogens is bonded to a different oxygen.)

Solution

(a)

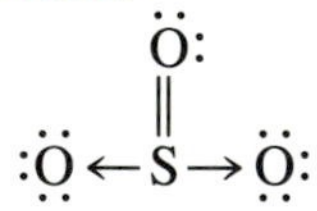

$$\begin{array}{ccccc} & & \ddot{O}: & & \\ & & \| & & \\ :\underset{\cdot\cdot}{\ddot{O}} & \leftarrow & S & \rightarrow & \underset{\cdot\cdot}{\ddot{O}}: \end{array}$$

There are two coordinate covalent bonds in this compound.

(b)

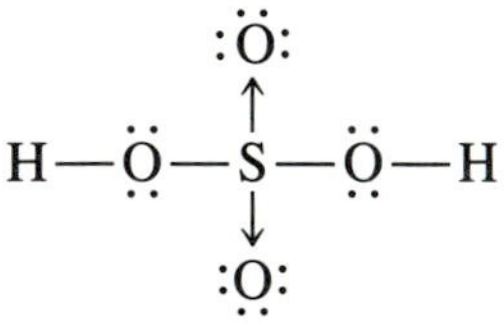

$$\begin{array}{ccccccccc} & & & & :\ddot{O}: & & & & \\ & & & & \uparrow & & & & \\ H & — & \underset{\cdot\cdot}{\ddot{O}} & — & S & — & \underset{\cdot\cdot}{\ddot{O}} & — & H \\ & & & & \downarrow & & & & \\ & & & & :\underset{\cdot\cdot}{O}: & & & & \end{array}$$

There are two coordinate covalent bonds in this compound.

Practice Exercise 7.2 Draw a bonding diagram for H_2SO_3, sulfurous acid, which has a coordinate covalent bond included in its structure. (*Hint:* Each oxygen is bonded to the sulfur, and each of the hydrogens is bonded to a different oxygen.)

7.4 Ionic Bonding

Learning Goal 5
Ionic bonds

One atom can actually *give* electrons to another atom so that they both obtain an octet of electrons in their outermost energy level. This process is called *ionic bonding*. An example of ionic bonding is the formation of sodium fluoride (Figure 7.7). A bond formed by sharing electrons (a covalent bond) would help the fluorine *but not the sodium*. If a sodium atom shared a pair of electrons, it would have only two electrons in its outermost energy level:

Na :F̈:

This does not satisfy the octet rule.

However, note what happens if the sodium *gives* or *transfers* its outermost electron to the fluorine. First, the fluorine obtains eight electrons in its outermost energy level. And, in the sodium atom, the second energy level (which contains eight electrons) now becomes the outermost energy level; so the sodium too has a completed octet.

But what holds the atoms together? Remember that the fluorine now has an extra electron, which used to belong to the sodium. This means that the fluorine now has a total of ten electrons and still has only nine protons. Therefore the fluorine atom *has a net electric charge* of 1− (because it has one extra electron). The sodium atom has given up an electron, so it now has eleven protons and only ten electrons. Therefore the sodium atom *has a net electric charge* of 1+ (because it has one less electron). The two atoms have become *ions*. An **ion** is *an atom or group of atoms that has gained or lost electrons and therefore has a net negative or positive charge*. A **monatomic ion** is *an ion composed of a single atom*. The positively charged sodium atom is called a *sodium ion,* written Na^{1+}. The negatively charged fluorine atom is called a fluor*ide* ion, written F^{1-}. Opposite charges attract, so the Na^{1+} and F^{1-} ions are attracted to each other. This attraction makes them form the neutral salt sodium fluoride (NaF). The positive-negative attraction is what holds these ions together. An **ionic bond** is *a chemical bond in which one atom*

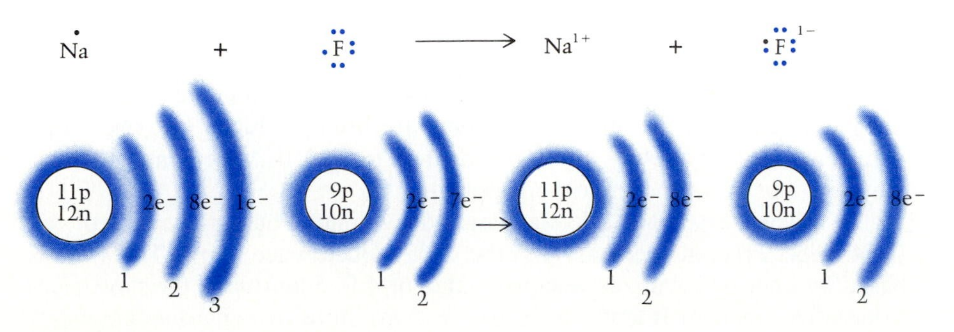

Figure 7.7
Ionic bonding in sodium fluoride

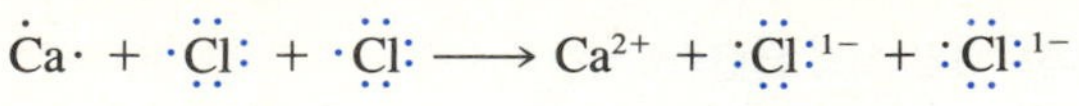

$$\dot{\text{Ca}}\cdot + \cdot\ddot{\underset{\cdot\cdot}{\text{Cl}}}: + \cdot\ddot{\underset{\cdot\cdot}{\text{Cl}}}: \longrightarrow \text{Ca}^{2+} + :\ddot{\underset{\cdot\cdot}{\text{Cl}}}:^{1-} + :\ddot{\underset{\cdot\cdot}{\text{Cl}}}:^{1-}$$

Figure 7.8
Ionic bonding in calcium chloride

transfers one or more electrons to another atom and the resulting ions are held together by the attraction of opposite charges.

Here's another example: Calcium and chlorine combine by forming an ionic bond to make calcium chloride ($CaCl_2$), but it takes two chlorines to do the job (Figure 7.8). In this compound, the calcium atom gives up its two outermost electrons, transferring one electron to each chlorine atom. The calcium atom becomes the positively charged Ca^{2+} ion, because it still has 20 protons and now only 18 electrons. Each chlorine atom now has an extra electron, for a total of 18, but it still has only 17 protons. As a result, each chlorine atom becomes a negatively charged chloride ion. The calcium ion and the chloride ions are attracted to each other, forming $CaCl_2$.

7.5 Exceptions to the Octet Rule

There are exceptions to every rule, and not all elements obey the octet rule. For example, in the compound boron trifluoride (BF_3), the boron atom has only six electrons in its outermost energy level (Figure 7.9).

Arsenic pentachloride is another exception to the rule. (This name means that one arsenic is bonded to five chlorines; *penta* is from the Greek word meaning "five.") In this compound, arsenic has *ten* outer electrons (Figure 7.10). In fact, many of the transition metals fail to follow the octet rule. Still, in the vast majority of cases, the octet rule will give you the correct bonding structure.

7.6 Covalent or Ionic Bonds: The Concept of Electronegativity

Learning Goal 6
Ionic and covalent bonds

At this point you may be asking, "How can I tell whether two elements bond ionically or covalently?" To answer this question, we have to introduce the concept of electronegativity. **Electronegativity** is *the attraction that an atom has for the electrons it is sharing with another atom*. The relative electronegativities for atoms of the different elements are listed in Table 7.1. (Note that no electronegativity values are listed for the Group VIIIA elements, because they tend not to form chemical compounds.)

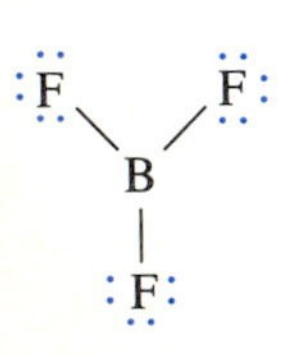

Figure 7.9
Covalent bonding in boron trifluoride

The electronegativity scale was devised by Linus Pauling, winner of the 1954 Nobel Prize in chemistry and the 1962 Nobel Peace Prize. The scale is based on fluorine having an assigned value of 4.0. Fluorine is the most electronegative element. It tends to attract electrons more strongly than any other element does.

The metal elements generally have low electronegativities, whereas the nonmetals have high electronegativities. This means that the atoms of metals

Figure 7.10
Covalent bonding in arsenic pentachloride

tend to lose electrons more readily than the atoms of nonmetals. In fact, the atoms of nonmetals have a strong tendency to pick up electrons. By looking at Table 7.1, you can see the trend for electronegativity, which generally increases as one moves from left to right across a given period of elements and decreases as one moves from the top to the bottom of a group.

To decide whether a bond is ionic or covalent, we find the difference between the electronegativities of the elements forming the bond. Chemists generally agree that if the difference between the electronegativities of the two elements is 2.0 or greater, the bond between the two elements is primarily ionic. If the difference is less than 2.0, the bond is primarily covalent. (Note that 2.0 is simply a convenient number to use as the cutoff point between ionic and covalent bonding. There is nothing special about this

Table 7.1
Periodic Table of Electronegativities

Period	Group IA	IIA	IIIB	IVB	VB	VIB	VIIB	VIII	VIII	VIII	IB	IIB	IIIA	IVA	VA	VIA	VIIA	VIIIA
1	1 H 2.1																	2 He
2	3 Li 1.0	4 Be 1.5											5 B 2.0	6 C 2.5	7 N 3.0	8 O 3.5	9 F 4.0	10 Ne
3	11 Na 0.9	12 Mg 1.2											13 Al 1.5	14 Si 1.8	15 P 2.1	16 S 2.5	17 Cl 3.0	18 Ar
4	19 K 0.8	20 Ca 1.0	21 Sc 1.3	22 Ti 1.5	23 V 1.6	24 Cr 1.6	25 Mn 1.5	26 Fe 1.8	27 Co 1.8	28 Ni 1.8	29 Cu 1.9	30 Zn 1.6	31 Ga 1.6	32 Ge 1.8	33 As 2.0	34 Se 2.4	35 Br 2.8	36 Kr
5	37 Rb 0.8	38 Sr 1.0	39 Y 1.2	40 Zr 1.4	41 Nb 1.6	42 Mo 1.8	43 Tc 1.9	44 Ru 2.2	45 Rh 2.2	46 Pd 2.2	47 Ag 1.9	48 Cd 1.7	49 In 1.7	50 Sn 1.8	51 Sb 1.9	52 Te 2.1	53 I 2.5	54 Xe
6	55 Cs 0.7	56 Ba 0.9	57–71 La Lu–	72 Hf 1.3	73 Ta 1.5	74 W 1.7	75 Re 1.9	76 Os 2.2	77 Ir 2.2	78 Pt 2.2	79 Au 2.4	80 Hg 1.9	81 Tl 1.8	82 Pb 1.8	83 Bi 1.9	84 Po 2.0	85 At 2.2	86 Rn
7	87 Fr 0.7	88 Ra 0.9	89–103 Ac Lr–	104 Unq	105 Unp	106 Unh	107 Uns	108 Uno	109 Une									

Transition elements

57 La 1.1	58 Ce 1.1	59 Pr 1.1	60 Nd 1.1	61 Pm 1.1	62 Sm 1.1	63 Eu 1.1	64 Gd 1.1	65 Tb 1.1	66 Dy 1.1	67 Ho 1.1	68 Er 1.1	69 Tm 1.1	70 Yb 1.1	71 Lu 1.2
89 Ac 1.1	90 Th 1.3	91 Pa 1.5	92 U 1.7	93 Np 1.3	94 Pu 1.3	95 Am 1.3	96 Cm 1.3	97 Bk 1.3	98 Cf 1.3	99 Es 1.3	100 Fm 1.3	101 Md 1.3	102 No 1.3	103 Lr

6 ← Atomic number
C
2.5 ← Electronegativity

number. In fact, there is a more sophisticated way of looking at ionic and covalent bonds, which we will discuss shortly.)

Example 7.3

Determine whether the bonds in the following compounds are ionic or covalent (use Table 7.1): *(a)* H_2S *(b)* KCl *(c)* MgO *(d)* H_2

Solution To solve these problems, we use the specific strategy just described.

(a) There are two H—S bonds in hydrogen sulfide (H_2S). We want to determine whether these are ionic or covalent. From Table 7.1, we find that the electronegativity of H is 2.1, and that of S is 2.5. The difference in electronegativities is

$$2.5 - 2.1 = 0.4$$

Because the difference is only 0.4, the bond between hydrogen and sulfur is covalent. (If the difference is less than 2.0, the bond is considered covalent.)

(b) From Table 7.1, we find that the electronegativity of K is 0.8, and that of Cl is 3.0. The difference in electronegativities is

$$3.0 - 0.8 = 2.2$$

Because the difference is 2.2, the bond between potassium and chlorine is ionic.

(c) From Table 7.1, we find that the electronegativity of Mg is 1.2, and that of O is 3.5. The difference in electronegativities is

$$3.5 - 1.2 = 2.3$$

Because the difference is 2.3, the bond between magnesium and oxygen is ionic.

(d) From Table 7.1, we find that the electronegativity of H is 2.1. The difference between the electronegativities of the two H atoms is

$$2.1 - 2.1 = 0.0$$

The bond between the two H atoms is very definitely a covalent bond. (It's as covalent as it can get!)

Practice Exercise 7.3 Determine whether the bonds in each of the following compounds are ionic or covalent (use Table 7.1): *(a)* H_2O *(b)* KF *(c)* MgS *(d)* Cl_2

7.7 Ionic Percentage and Covalent Percentage of a Bond

So far we have thought of compounds as either ionically bonded or covalently bonded. We decided which *type* of bond by calculating the difference in electronegativities of the elements forming the bond. A difference of less than 2.0 between the electronegativities of the two elements meant that the bond was covalent, whereas a difference of 2.0 or more meant that the bond was ionic. But this is an oversimplified approach.

Chemists find it better to express chemical bonds as a certain percentage ionic and a certain percentage covalent. Table 7.2 relates difference in electronegativities to the ionic and covalent percentages of any bond. With this table, we can state that the hydrogen-oxygen bond in water (with an electronegativity difference of 1.4) is 39% ionic and 61% covalent. (The percentages listed in Table 7.2 come directly from Pauling's work.)

Example 7.4

Determine the ionic and covalent percentages of the bonds in each of the following compounds: *(a)* NH_3 *(b)* SO_2 *(c)* HCl *(d)* NaF

Solution

(a) Look up the electronegativities of nitrogen and hydrogen, determine the difference, and use Table 7.2 to find the ionic and covalent percentages of the bond. The electronegativity of N is 3.0, and that of H is 2.1. The difference between these electronegativities is

$$3.0 - 2.1 = 0.9$$

This corresponds to a bond that is 19% ionic and 81% covalent.
(b) The electronegativity of S is 2.5, and that of O is 3.5. The difference between these electronegativities is

$$3.5 - 2.5 = 1.0$$

This corresponds to a bond that is 22% ionic and 78% covalent.
(c) The electronegativity of H is 2.1, and that of Cl is 3.0. The difference between these electronegativities is

$$3.0 - 2.1 = 0.9$$

This corresponds to a bond that is 19% ionic and 81% covalent.
(d) The electronegativity of Na is 0.9, and that of F is 4.0. The difference between these electronegativities is

$$4.0 - 0.9 = 3.1$$

This corresponds to a bond that is 91% ionic and 9% covalent.

Practice Exercise 7.4 Determine the ionic and covalent percentages of the bonds in each of the following compounds: *(a)* K_2O *(b)* $CaCl_2$ *(c)* Mg_3N_2 *(d)* HF

Example 7.5

List the following compounds in order of decreasing covalence of their bonds: H_2S, H_2O, CO_2, and N_2.

Solution Using Tables 7.1 and 7.2, determine the percentage covalence of each compound. In the compound H_2S, the H—S bond has an electronegativity difference of 0.4, which means it is 96% covalent.

In the compound H_2O, the O—H bond has an electronegativity difference of 1.4, which means it is 61% covalent.

In the compound CO_2, the C=O bond has an electronegativity difference of 1.0, which means it is 78% covalent.

In the compound N_2, the N≡N bond has an electronegativity difference of 0.0, which means it is 100% covalent.

Therefore the order of covalence is

N_2	H_2S	CO_2	H_2O
Most covalent			Least covalent

Practice Exercise 7.5 List the following compounds in order of decreasing covalence of their bonds: CO_2, $AsCl_3$, CS_2, and H_2S.

7.8 Shapes and Polarities of Molecules

Up to now, we have used only two-dimensional diagrams to show how atoms are bonded together in molecules. But our world is *three*-dimensional, and atoms and molecules, like other objects, have three dimensions. In fact, the three-dimensional shapes of some molecules lead to additional properties of covalently bonded compounds.

Three-Dimensional Characteristics of Molecules

If we could enlarge a single molecule of any diatomic element (for instance, H_2, O_2, or N_2), it would appear to be linear (in a line) (Figure 7.11). This would also be true for covalently bonded compounds consisting of two *un*like atoms, such as HCl (Figure 7.12).

But if we could enlarge a single molecule of water, we would see a bent molecule (Figure 7.12). Experiments have shown that the angle between the hydrogen atoms is about 105 degrees.

If we looked at a molecule of methane (CH_4), we would see a pyramid-shaped symmetrical molecule (Figure 7.12). The angle between neighboring hydrogen atoms in methane has been shown to be 109.5 degrees.

Table 7.2
The Relationship Between Electronegativity Difference and the Ionic Percentage and Covalent Percentage of a Chemical Bond

Difference in electronegativity	Ionic percentage	Covalent percentage
0.0	0.0	100
0.1	0.5	99.5
0.2	1.0	99.0
0.3	2.0	98.0
0.4	4.0	96.0
0.5	6.0	94.0
0.6	9.0	91.0
0.7	12.0	88.0
0.8	15.0	85.0
0.9	19.0	81.0
1.0	22.0	78.0
1.1	26.0	74.0
1.2	30.0	70.0
1.3	34.0	66.0
1.4	39.0	61.0
1.5	43.0	57.0
1.6	47.0	53.0
1.7	51.0	49.0
1.8	55.0	45.0
1.9	59.0	41.0
2.0	63.0	37.0
2.1	67.0	33.0
2.2	70.0	30.0
2.3	74.0	26.0
2.4	76.0	24.0
2.5	79.0	21.0
2.6	82.0	18.0
2.7	84.0	16.0
2.8	86.0	14.0
2.9	88.0	12.0
3.0	89.0	11.0
3.1	91.0	9.0
3.2	92.0	8.0

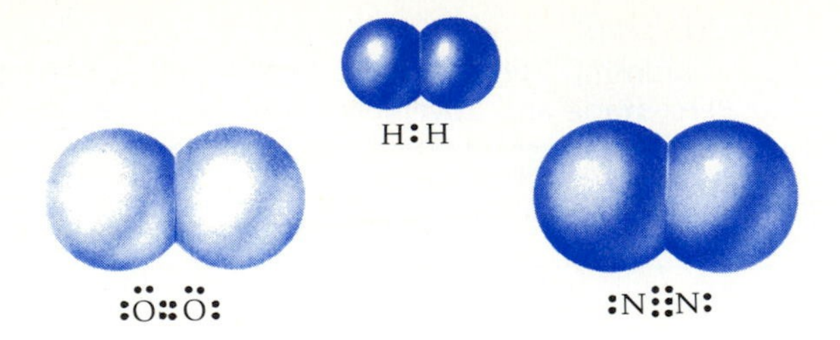

Figure 7.11 Three-dimensional models of some diatomic elements

Molecules of other compounds exhibit many other shapes. The important thing about these shapes is that they can affect bonding and the properties of the molecules.

The Not-So-Covalent Bond

Learning Goal 7
Polar and nonpolar bonds

Recall that when two atoms form a covalent bond, they share electrons. But sometimes they don't share the electrons equally. For example, in the compound hydrogen bromide (HBr), the bromine atom has a greater attraction for the electrons than the hydrogen atom does. In other words, bromine has a greater electronegativity than hydrogen. Because of the bromine's greater attraction for electrons, the two electrons being shared spend more time with the bromine than with the hydrogen. This unequal sharing makes the hydrogen seem to have a partial positive charge, and the bromine a partial negative charge (Figure 7.13). This kind of covalent bond, in which there is unequal sharing of electrons, is called a **polar covalent bond.** This is what we meant when we said a bond was a certain percentage ionic.

Most covalent bonds are polar. However, bonds formed by the diatomic elements (H—H, O=O, and so on) are nonpolar. This is because there is no difference in electronegativity between the two atoms forming the bond. In a **nonpolar covalent bond,** the electrons are shared equally by the atoms.

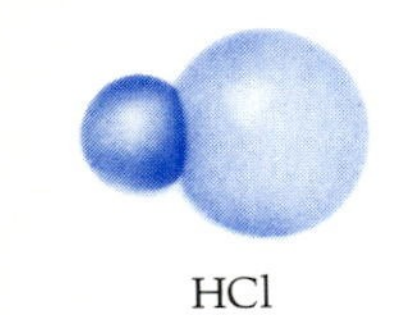

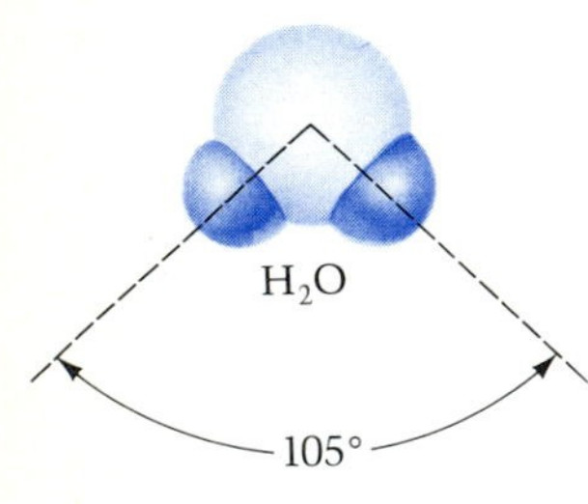

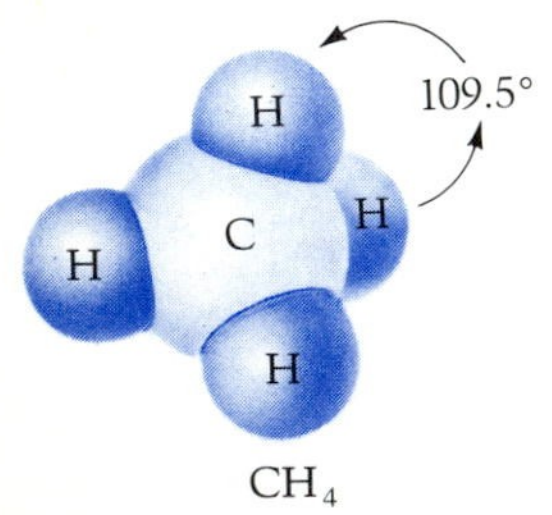

Figure 7.12 Three-dimensional models of hydrogen chloride (HCl), water (H_2O), and methane (CH_4) molecules

Example 7.6

Determine whether each of the following covalent bonds is polar or nonpolar. (*Hint:* Use Table 7.1 to determine the electronegativity of each element.) *(a)* N—O bond *(b)* S—O bond *(c)* Cl—Cl bond *(d)* P—O bond *(e)* C—S bond

Solution We look up the electronegativity values of each element in Table 7.1. If a difference in electronegativity exists between the two atoms forming the bond, then the bond is polar. If no difference in electronegativity exists, then the bond is nonpolar.

(a) N——O
3.0 3.5
0.5 difference

An electronegativity difference exists, so the bond is *polar*.

(b) S——O
2.5 3.5

An electronegativity difference exists, so the bond is *polar*.

1.0 difference

(c) Cl —— Cl
3.0 3.0
0.0 difference

No electronegativity difference exists, so the bond is *nonpolar*.

(d) P —— O
2.1 3.5
1.4 difference

An electronegativity difference exists, so the bond is *polar*.

(e) C —— S
2.5 2.5
0.0 difference

No electronegativity difference exists, so the bond is *nonpolar*.

Practice Exercise 7.6 Determine whether each of the following covalent bonds is polar or nonpolar. (*Hint:* Use Table 7.1.) *(a)* F—F bond *(b)* C—O bond *(c)* B—O bond

Polar and Nonpolar Molecules

Learning Goal 8
Polar and nonpolar molecules

Just as bonds can have polarity, molecules can also have polarity. Even if a molecule contains polar bonds, the molecule itself must not necessarily be a polar molecule. To determine whether or not a molecule is polar we must look at the geometry of the molecule and the polarity of its bonds.

As we mentioned earlier, molecular geometry tells us how the atoms of a molecule are arranged in space in relation to one another. You are familiar with the linear arrangement of the diatomic elements, the bent shape characteristic of a water molecule, and the pyramidal shape of the methane molecule. Such information is necessary when determining molecular polarity. We will look at the polarities of three molecules—HBr, H_2O, and CCl_4—to understand the combined effects of molecular geometry and bond polarity.

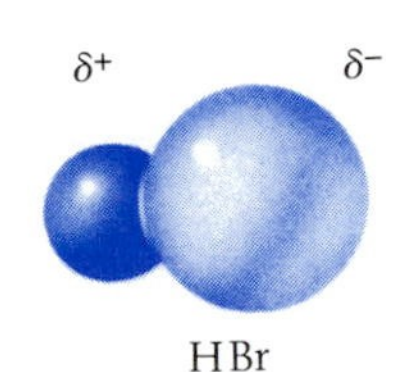

Figure 7.13
A three-dimensional drawing of HBr. The δ^+ indicates that the hydrogen side of the molecule has a partial positive charge. The δ^- shows that the bromine side of the molecule has a partial negative charge. (δ is the lower-case Greek letter delta.)

The linear HBr molecule (Figure 7.13) contains a polar covalent bond. There is a shift in electronic charge such that the hydrogen side of the molecule is positive and the bromine side of the molecule is negative. A **polar molecule** (or **dipole**) is a molecule that is positive on one side and negative on the other side. Thus HBr is a polar molecule.

And then there is water, a compound with two polar bonds. A water molecule is also polar, because there is a center of positive charge between the two hydrogen atoms and a center of negative charge on the oxygen atom (Figure 7.14).

A molecule of carbon tetrachloride (CCl_4) has four carbon-chlorine bonds. Each bond is a *polar* covalent bond. But the molecule itself is *nonpolar* (Figure 7.14). Here's why: The carbon tetrachloride molecule is shaped like a pyramid. The positive charge is on the carbon atom, but so is the center of the negative charges of the four Cl atoms. The two charges coincide and cancel each other, so the molecule is not polar.

Figure 7.14
Three-dimensional models of water and carbon tetrachloride molecules. The water molecule is polar; the carbon tetrachloride is nonpolar.

Example 7.7

Determine whether the bonds in each of the following molecules are polar or nonpolar. Then determine whether the molecule itself is polar or nonpolar.

(a) F_2 (linear in shape)
(b) $CHCl_3$, chloroform (shaped like a pyramid)
(c) *cis*-Dichloroethene (The *cis* means that the two chlorines are on the same side of the double bond.)

```
H         H
 \       /
  C═C
 /       \
Cl        Cl
```

(d) *trans*-Dichloroethene (The *trans* means that the two chlorines are on opposite sides of the double bond.)

```
H         Cl
 \       /
  C═C
 /       \
Cl        H
```

Solution

(a) Fluorine is a diatomic element; therefore the F—F bond is nonpolar and the F_2 molecule is nonpolar.

(b) The bonding in chloroform is similar to the bonding in carbon tetrachloride. Each bond in chloroform is a polar covalent bond. But the molecule of chloroform, unlike that of carbon tetrachloride, is polar. This is because the center of positive charge and the center of negative charge do not coincide. The molecule is not symmetrical:

```
δ+H       Cl
    \    /
      C
    /    \
Cl     δ− Cl
```

(c) In *cis*-dichloroethene, each C—H and C—Cl bond is polar. The C=C bond is nonpolar. The molecule is polar, because it has a center of positive charge between the two hydrogens and a center of negative charge between the two chlorines. Because the centers of positive charge and negative charge do not coincide, the molecule is polar.

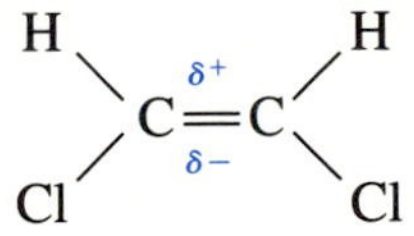

```
H         H
 \  δ+   /
  C===C
 /  δ-   \
Cl        Cl
```

(d) In *trans*-dichloroethene, the C—H and C—Cl bonds are polar and the C=C bond is nonpolar. The molecule is also nonpolar, because it is shaped in such a way that the center of positive charge coincides with the center of negative charge. The charges coincide as shown in the double bond between the two carbon atoms.

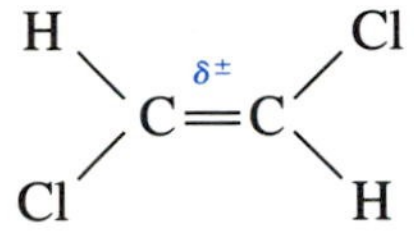

```
H         Cl
 \  δ±   /
  C===C
 /       \
Cl        H
```

Practice Exercise 7.7 Determine whether the bonds in each of the following molecules are polar or nonpolar. Then determine whether the molecule itself is polar or nonpolar.

Molecule	*Shape*
BF_3	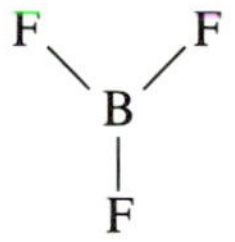 F—B(—F)—F (F\ /F above B, F below B)
Cl_2O	 Cl/O\Cl (O above, Cl at each side below)
CO_2	O=C=O

Summary

In a Lewis (electron-dot) diagram of an element, the symbol for an atom of the element is surrounded by dots. The dots correspond to the electrons in the outermost energy level of an atom of the element. Electron-dot diagrams are used to show how atoms combine to form molecules.

It has long been known that Group VIIIA elements tend to be very stable. Scientists believe this stability is due to the fact that each element in this group (except helium) has eight electrons in its outermost energy level. The atoms of other elements can achieve this same sort of stability by obtaining eight electrons in their outermost energy level. To do so, two atoms may share electrons in a covalent bond. A single covalent bond is formed when each atom donates one electron to the bond, a double covalent bond is formed when each atom donates two electrons, and a triple covalent bond is formed when each atom donates three electrons.

In a coordinate covalent bond, only one atom donates the two electrons required to form the bond. However, both atoms obtain an octet of electrons in their outermost shell.

Elements that are found in molecules with two atoms each (such as H_2) are called diatomic elements; these elements are hydrogen, oxygen, nitrogen, and the Group VIIA elements. The bonds between their atoms are covalent bonds.

An ion is an atom or group of atoms that has gained or lost electrons and therefore has a negative or positive electric charge. In ionic bonding, atoms of one element donate electrons to the atoms of a second element. Both become ions, but they are ions of opposite charges. These opposite charges attract, and this attraction holds the atoms together in a molecule.

Electronegativity is the attraction that an atom has for the electrons it is sharing with another atom. If the difference between the electronegativities of two elements is 2.0 or greater, the bond between the two elements is said to be ionic. If the difference is less than 2.0, the bond is covalent. The concept of electronegativity is also used in determining the ionic and covalent percentages of bonds.

Because of their three-dimensional character, some molecules exhibit unequal sharing of electrons, resulting in polar covalent bonds. Such molecules may have a positive side and a negative side; if so, they are called polar molecules, or dipoles.

Key Terms

chemical bond **(Introduction)**
coordinate covalent bond **(7.3)**
covalent bond **(7.2)**
diatomic element **(7.2)**
double covalent bond **(7.2)**
electron-dot structure **(7.1)**
electronegativity **(7.6)**
ion **(7.4)**
ionic bond **(7.4)**
monatomic ion **(7.4)**
nonpolar covalent bond **(7.8)**
polar covalent bond **(7.8)**
polar molecule (dipole) **(7.8)**
single covalent bond **(7.2)**
triple covalent bond **(7.2)**

Career Sketch

Astronomer

As an astronomer, sometimes called an astrophysicist, you will explore the fundamental nature of the universe. You will search for explanations of how the universe originated and how the solar system evolved. You will use the principles of physics and mathematics to study the behavior of matter and energy in distant galaxies. You will also use this information to support or disprove theories about the nature of matter and energy.

To observe the universe, you will use large earth-based telescopes, space-based telescopes, radio telescopes, and other sophisticated instruments that can detect electromagnetic radiation from distant celestial sources. For example, you may use a spectroscope to analyze the light from a star so that you can determine its chemical composition. You may also use instruments to determine whether a star is emitting radio waves, x rays, or cosmic rays.

As an astronomer, you will also use computers to process astronomical data so that you can calculate the orbits of asteroids and comets or guide spacecraft to specific celestial bodies. Or you may use computers to process and analyze data collected about a star to determine the star's structure or age.

CHEMICAL FRONTIERS

CHEMOPREVENTION LEADS TO DISCOVERY OF A POWERFUL ANTICARCINOGEN FOUND IN ONIONS*

The use of both natural and synthetic chemical compounds to reduce the risk of cancer is called *chemoprevention*. About 550 chemicals found in foods have been identified that are thought to reduce this risk. These include vitamin A and beta-carotene found in green and yellow vegetables, vitamin C found in fruits, and vitamin E found in whole grains and beans as well as vegetable oil. The list includes trace metals such as selenium, which is found in seafoods and organ meats. A new chemical, quercetin, has recently been added to the list. It is found in broccoli, squash, and other vegetables. The best source of this chemical is onions.

Researchers studying quercetin at the National Cancer Institute are working with associates in the People's Republic of China, who have reported that Chinese who consume large amounts of onions and garlic have a rate of stomach cancer that is four times lower than normal.

Terrance Leighton, professor of microbiology and immunology at the University of California at Berkeley, has simulated human digestive biochemistry in the laboratory to establish the process of quercetin liberation. Quercetin is normally bonded to sugar molecules in fruits and vegetables. After a person eats quercetin-rich foods bacterial enzymes in the intestinal tract break the quercetin–sugar bonds and release the protective chemical.

Leighton and his coworkers plan to develop a molecular probe that will allow them to monitor how the activity of bacterial genes changes with different diets. They would like to understand how certain diets high in vegetables provide protection against cancer.

*Adapted from "Powerful Anticarcinogen Found in Onions," *What's Happening in Chemistry* (American Chemical Society, 1990).

Many fruits and vegetables are thought to have anti-cancer properties.
(D. Cavagnaro/Peter Arnold, Inc.)

Self-Test Exercises

Learning Goal 1: Lewis Structures

▲ **1.** Write the electron-dot structure for each of the following elements: *(a)* Sb *(b)* I *(c)* Ba *(d)* Rb *(e)* Kr

▲ **2.** Write the electron-dot structure for each of the following elements: *(a)* F *(b)* Ca *(c)* K *(d)* C *(e)* B

3. Write the electron-dot structure for *(a)* a chlorine atom, *(b)* a chloride ion.

4. Write the electron-dot structure for *(a)* a bromine atom, *(b)* a fluoride ion.

5. Write the electron-dot structure for *(a)* a beryllium atom, *(b)* a beryllium ion.

6. Write the electron-dot structure for *(a)* a barium atom, *(b)* a barium ion.

7. Write the electron-dot structure for each of the following elements: *(a)* S *(b)* Al *(c)* Mg *(d)* Ge

8. Write the electron-dot structure for each of the following elements: *(a)* Si *(b)* Te *(c)* At

9. Write the electron-dot structure for each of the following ions: *(a)* magnesium ion *(b)* sodium ion *(c)* aluminum ion

10. Write the electron-dot structure for each of the following ions: *(a)* calcium ion *(b)* cesium ion *(c)* boron ion

11. Write the electron-dot structure for each of the following ions: *(a)* bromide ion *(b)* selenide ion *(c)* phosphide ion

12. Write the electron-dot structure for each of the following ions: *(a)* sulfide ion *(b)* telluride ion *(c)* nitride ion

Learning Goal 2: Covalent Bonds

▲ **13.** Define the term *covalent bond.*
14. Define the term *coordinate covalent bond.*

15. When each atom donates *two* electrons to form a covalent bond, the bond formed is a ______ (single, double, quadruple) covalent bond.
16. When each atom donates *three* electrons to form a covalent bond, the bond formed is a ______ (single, triple, sextuple) covalent bond.

Learning Goal 3: Diatomic Elements

17. Name the diatomic elements from memory.
▲ **18.** Which of the diatomic elements contain single covalent bonds? Which contain double bonds? Which contain triple bonds?

Learning Goal 4: Electron-Dot Structures of Covalent Compounds

▲ **19.** Write the electron-dot structure for each of the following covalent compounds: *(a)* $CHBr_3$ *(b)* C_2H_2 *(c)* PH_3 *(d)* HCN
20. Write the electron-dot structure for each of the following covalent compounds: *(a)* HCl *(b)* CBr_4 *(c)* SO_3 *(d)* $SbCl_3$

21. Write the electron-dot structure for each of the following covalent compounds: *(a)* CF_4 *(b)* C_2H_4 *(c)* AsH_3 *(d)* H_2S
▲ **22.** Write the electron-dot structure for each of the following covalent compounds: *(a)* CS_2 *(b)* CO_2 *(c)* Cl_2O *(d)* C_3H_8

▲ **23.** Not all compounds follow the octet rule. Draw the bonding diagram for the covalent compound PCl_5. How many electrons surround the P atom in this compound?
24. Not all compounds follow the octet rule. Draw the bonding diagram for the compound BF_3. How many electrons surround the B atom in this compound?

▲ **25.** Draw the bonding diagram for each of the following covalent compounds, which have coordinate covalent bonds: *(a)* HNO_3 *(b)* H_2SO_3
26. Draw the bonding diagram for the compound chloric acid, $HClO_3$, which has a coordinate covalent bond. (*Hint:* The chlorine is bonded to three oxygen atoms, and the hydrogen is bonded to one of the oxygen atoms.)

Learning Goal 5: Ionic Bonds

▲ **27.** Define the term *ionic bond.*
28. How is an ionic bond different from a covalent bond?

Learning Goal 6: Ionic and Covalent Bonds

29. Determine whether each of the following compounds is bonded ionically or covalently: *(a)* KCl *(b)* NH_3 *(c)* CaF_2 *(d)* HBr *(e)* OF_2
30. Determine whether each of the following compounds is bonded ionically or covalently: *(a)* MgO *(b)* $BaCl_2$ *(c)* BaO *(d)* CaF_2 *(e)* SO_2

31. Find the ionic and covalent percentages of the bonds in each of the following compounds: *(a)* CH_4 *(b)* N_2 *(c)* FeO *(d)* Al_2O_3
32. Find the ionic and covalent percentages of the bonds in each of the following compounds: *(a)* HCl *(b)* NH_3 *(c)* KI *(d)* CS_2

▲ **33.** List the following compounds in order of *decreasing* covalence: HI, HCl, HF, and HBr.
34. List the following compounds in order of *decreasing* covalence: Br_2, SO_2, NH_3, HCl.

35. Find the ionic and covalent percentages of the bonds in each of the following compounds: *(a)* CF_4 *(b)* AsH_3 *(c)* CsCl *(d)* CuO
36. Find the ionic and covalent percentages of the bonds in each of the following compounds: *(a)* CO_2 *(b)* F_2 *(c)* $MgCl_2$ *(d)* BaF_2

37. List the following compounds in order of *increasing* covalence: H_2Se, H_2O, H_2Te, and H_2S.
▲ **38.** List the following compounds in order of *increasing* covalence: HF, NH_3, BH_3, CH_4, and H_2O.

39. Which of the following compounds is the *most* ionic? *(a)* LiCl *(b)* NaCl *(c)* KCl *(d)* RbCl *(e)* CsCl
40. Which of the following compounds is the *most* ionic? HF HCl HBr HI

41. Determine whether the bonds in each of the following compounds are ionic or covalent (use Table 7.1): *(a)* SO_3 *(b)* $BiCl_3$ *(c)* LiF *(d)* NO
42. Determine whether the bonds in each of the following compounds are ionic or covalent (use Table 7.1): *(a)* LiH *(b)* OF_2 *(c)* CaI_2 *(d)* MgH_2

Learning Goals 7 and 8: Polar and Nonpolar Bonds and Molecules

▲ **43.** What are polar molecules? What are nonpolar molecules?

44. Determine whether the bond in each of the following molecules is polar or nonpolar: *(a)* H_2 *(b)* NO *(c)* HCl

▲ **45.** Determine whether the *bonds* in each of the following molecules are polar or nonpolar. Then determine whether the *molecule* itself is polar or nonpolar.

Molecule	Shape	Molecule	Shape
(a) PBr_3	```Br \ P / \ Br Br```	*(b)* Cl_2O	```O / \ Cl Cl```
(c) CH_3Cl	```H H \ / C / \ H Cl```	*(d)* O_2	O=O

46. Determine whether the *bonds* in each of the following molecules are polar or nonpolar. Then determine whether the *molecule* is polar or nonpolar.

Molecule	Shape
(a) BeF_2	F—Be—F
(b) BrF_3	```F F \ / Br \ F```
(c) CCl_4	```Cl	Cl—C—Cl	Cl```
(d) Cl_2	Cl—Cl

47. Determine whether the *bonds* in each of the following molecules are polar or nonpolar. Then determine whether the *molecule* itself is polar or nonpolar.

Molecule	Shape	Molecule	Shape
(a) C_2H_2	H—C≡C—H	*(b)* NH_3	```H \ N / \ H H```
(c) CH_4	```H	H—C—H	H```	*(d)* N_2	N≡N

▲ **48.** Determine whether the *bonds* in each of the following molecules are polar or nonpolar. Then determine whether the *molecule* is polar or nonpolar.

Molecule	Shape
(a) H_2	H—H
(b) CO_2	O=C=O
(c) C_6H_6	```H	C / \\ H—C C—H ||	H—C C—H \ // C	H```
(d) CH_2O	```H—C=O	H```

Extra Exercises

49. "In both ionic and covalent bonds, *electron deficiencies are satisfied.*" Explain what is meant by this statement.

▲ **50.** Draw the electron-dot diagram for each of the following: *(a)* $(PH_4)^{1+}$ *(b)* CO_2 *(c)* OF_2

51. Explain why the formula for sodium chloride, NaCl, involves one sodium ion and one chloride ion but the formula for calcium chloride, $CaCl_2$, has two chloride ions with every one calcium ion.

52. Determine the order of increasing bond polarity among the following groups: O=O N—O C—O

53. Which of the following bonds is the most covalent? C—H C—O N—O C—S

▲ **54.** Draw the electron-dot diagram for each of the following: *(a)* $(NH_4)^{1+}$ *(b)* ClBr

55. Predict the compound that is *most likely* to be formed from each of the following pairs of elements: *(a)* Mg and O *(b)* Al and O *(c)* Cs and S

56. Use the octet rule to explain the following:
(a) The formula unit of aluminum chloride is $AlCl_3$.
(b) The formula unit of sodium oxide is Na_2O.
(c) The formula unit of magnesium nitride is Mg_3N_2.

57. Explain how a molecule that has polar covalent bonds can be a nonpolar molecule.

▲ **58.** Use the octet rule to predict the formula unit of each of the following compounds: *(a)* K and S *(b)* Ca and Br *(c)* Li and O *(d)* Al and Cl

59. Explain why a covalently bonded molecule with the formula H_3 should not exist.

60. Using the octet rule, predict the formula and draw the electron-dot diagram for each of the covalently bonded molecules that form between the following atoms: *(a)* C and F *(b)* As and Cl *(c)* I and Cl *(d)* Te and H

8 Chemical Nomenclature: The Names and Formulas of Chemical Compounds

Learning Goals

After you've studied this chapter, you should be able to:

1. Distinguish between the common name of a compound and its systematic name.
2. Write the formula of a binary compound containing two nonmetals when you are given the name of the compound.
3. Write the formula of a binary compound containing a metal with a fixed oxidation number and a nonmetal when you are given the name of the compound.
4. Write the formula of a binary compound containing a metal with a variable oxidation number and a nonmetal when you are given the name of the compound.
5. Write the formula of a polyatomic ion when you are given its name, or write the name of a polyatomic ion when you are given its formula.
6. Write the formula of a ternary or higher compound when you are given the name of the compound.
7. Write the name of a binary compound containing two nonmetals when you are given the formula of the compound.
8. Write the name of a binary compound containing a metal with a fixed or variable oxidation number and a nonmetal when you are given the formula of the compound.
9. Find the oxidation numbers of less familiar elements when you are given chemical formulas of compounds containing these elements.
10. Write the name of a ternary or higher compound when you are given the formula of the compound.
11. Write the formula of an inorganic acid when you are given its name, or write the name of an inorganic acid when you are given its formula.
12. Write the formula of a compound when you are given its common name, or write the common name of a compound when you are given its formula or systematic name.

Introduction

One of the most important topics to master in your study of chemistry is **chemical nomenclature,** which refers to the system for naming chemical compounds and writing their formulas. In any introductory college course, learning the basic vocabulary of the discipline is extremely important. This allows you to discuss the subject matter using the language of the discipline. Learning to write the names and formulas of chemical compounds will build your chemistry vocabulary. In fact, chemical nomenclature is a major part of the language of chemistry. In this chapter, you will learn how to write the formula of a compound when you are given its name. You will also learn how to name a compound when you are given its formula.

Learning Goal 1
Distinguishing between common name and systematic name

As we begin our study of chemical nomenclature, you should be aware that there are two ways of designating chemical compounds: the **systematic chemical name** and the **common name.** The rules governing the systematic names have been developed by the *International Union of Pure and Applied Chemistry* (*IUPAC*). The IUPAC is made up of chemists from all over the world. Members of the IUPAC Commission on the Nomenclature of Inorganic Compounds first met in 1921 to develop a system for naming compounds. This commission meets periodically to refine this system as new types of compounds are synthesized or discovered.

There are no rules that govern the common names of compounds. Most common names, such as table salt (for sodium chloride, NaCl) and water (for H_2O), have been derived from common usage or simply handed down through chemical history. We'll have more to say about common names of compounds later in this chapter. Let's begin our study of nomenclature by learning how to write the formulas of compounds from their systematic chemical names.

8.1 Writing the Formulas of Compounds from Their Systematic Names

In developing the rules for naming inorganic compounds, the IUPAC wanted to make sure that each chemical formula had a unique chemical name and vice versa. The commission also decided that the more *positive portion* of the compound should be written first, and the more *negative portion* last. Therefore, in the formula of a compound composed of two ions, the positive ion precedes the negative ion. That is why the formula for sodium chloride is always written NaCl, not ClNa. In the formula for a covalent compound, the element with the lower electronegativity (the more positive portion) precedes the element with the higher electronegativity (the more negative portion). Thus, the formula for water is written as H_2O, not OH_2.

These simple rules can be used to write the formulas of compounds from their systematic names. We'll begin with **binary compounds** (compounds made up of two different elements) and then move on to **ternary** and higher compounds (those made up of three or more elements).

8.2 Writing the Formulas of Binary Compounds Containing Two Nonmetals

Learning Goal 2
Formulas of binary compounds containing two nonmetals

The formula of a compound containing two nonmetals can easily be deduced from the compound's name, because the name of the compound will contain the name of both elements. For binary compounds, the ending on the second element is always *-ide*. Therefore chlor*ine* becomes chlor*ide*, sul*fur* becomes sulf*ide*, and ox*ygen* becomes ox*ide*. In addition, the number of atoms of each element can be deduced from Greek prefixes that are part of the compound's name. Table 8.1 lists the Greek prefixes used in naming compounds. Note that the Greek prefix *mono-* (meaning "one") is rarely used for the first-named element, though it is used for the second-named element. *The absence of a Greek prefix for the first-named element means that there is only one atom in the formula unit of that compound*. The following example will clarify this point.

Example 8.1

Write the formulas of the following binary compounds composed of two nonmetals:

(a) carbon monoxide
(b) carbon dioxide
(c) phosphorus trichloride
(d) dinitrogen monoxide

Solution

(a) Carbon monoxide is a compound containing carbon and oxygen. Ox*ygen* becomes ox*ide* because it is the second-named element. Notice that there is no prefix with carbon, and the prefix *mono-* appears with oxide. There is one atom of each element in the formula unit: CO.
(b) Carbon dioxide is a compound containing carbon and oxygen. Ox*ygen* becomes ox*ide* because it is the second-named element. Notice that there is no prefix with carbon, and the prefix *di-* appears with oxide. There is one atom of carbon and two atoms of oxygen in the formula unit: CO_2.
(c) Phosphorus trichloride is a compound containing phosphorus and chlorine. Chlor*ine* becomes chlor*ide* because it is the second-named element. Notice that there is no prefix with phosphorus, and the prefix *tri-* appears with chloride. There is one atom of phosphorus and three atoms of chlorine in the formula unit: PCl_3.
(d) Dinitrogen monoxide is a compound containing nitrogen and oxygen. Ox*ygen* becomes ox*ide* because it is the second-named element. Notice the prefix *di-* on carbon and the prefix *mono-* on oxide. There are two atoms of nitrogen and one atom of oxygen in the formula unit: N_2O.

Practice Exercise 8.1 Write the formulas of the following binary compounds composed of two nonmetals:

(a) sulfur dioxide
(b) dinitrogen pentoxide
(c) dinitrogen tetroxide
(d) phosphorus pentabromide

Table 8.1
Greek Prefixes and Their Meanings

Mono-	=	1
Di-	=	2
Tri-	=	3
Tetra-	=	4
Penta-	=	5
Hexa-	=	6
Hepta-	=	7
Octa-	=	8
Nona-	=	9
Deca-	=	10

8.3 Writing the Formulas of Binary Compounds Containing a Metal and a Nonmetal

Before we learn how to write the formulas of binary compounds containing a metal and a nonmetal, we must learn about *oxidation numbers*. Chemists who have studied how elements combine to form compounds have discovered certain trends. It seems that elements tend to form ions with specific charges or they tend to form only a certain number of covalent bonds. To describe this phenomenon, chemists have devised a system to indicate how elements combine to form compounds. This system involves the assignment of oxidation numbers to the substances involved in forming the compound. An **oxidation number** is *the positive or negative number that expresses the combining capacity of an element in a particular compound*. Oxidation numbers are not always real, but are "tools" created by chemists for bookkeeping purposes.

The oxidation number can be positive or negative, depending on whether the element tends to attract electrons strongly or give them up. In carbon tetrachloride (Figure 8.1), the carbon has an oxidation number of 4+ and the chlorine has an oxidation number of 1−. Elements with high electronegativity values usually have negative oxidation numbers, and elements with low electronegativity values usually have positive oxidation numbers.

Tables 8.2 and 8.3 list the oxidation numbers of some important ions. Note that for many monatomic ions, the oxidation number is the same as the charge on the ion. In other words, because the sodium atom tends to give up one electron when forming a chemical compound, it has a charge of 1+ as a sodium ion. Therefore its oxidation number is also 1+. Elements in Group IA, which tend to become ions with 1+ charges, all have oxidation numbers of 1+. Elements in Group IIA, which tend to become ions with 2+ charges, all have oxidation numbers of 2+. Elements in Group IIIA, which tend to become ions with 3+ charges, all have oxidation numbers of 3+.

```
      Cl
      |
Cl—C—Cl
      |
      Cl
```

Figure 8.1
Covalent bonding in carbon tetrachloride

Table 8.2
Charges of Positive Ions Frequently Used in Chemistry

1+		2+	
Hydrogen	H^{1+}	Calcium	Ca^{2+}
Lithium	Li^{1+}	Magnesium	Mg^{2+}
Sodium	Na^{1+}	Barium	Ba^{2+}
Potassium	K^{1+}	Zinc	Zn^{2+}
Mercury(I)*	Hg^{1+} (also called mercurous)	Mercury(II)	Hg^{2+} (also called mercuric)
Copper(I)	Cu^{1+} (also called cuprous)	Tin(II)	Sn^{2+} (also called stannous)
Silver	Ag^{1+}	Iron(II)	Fe^{2+} (also called ferrous)
Ammonium	$(NH_4)^{1+}$	Lead(II)	Pb^{2+} (also called plumbous)
Rubidium	Rb^{1+}	Copper(II)	Cu^{2+} (also called cupric)
Cesium	Cs^{1+}	Strontium	Sr^{2+}
		Nickel(II)	Ni^{2+}
		Chromium(II)	Cr^{2+} (also called chromous)
		Cobalt(II)	Co^{2+} (also called cobaltous)
		Manganese(II)	Mn^{2+} (also called manganous)
3+		**4+**	
Aluminum	Al^{3+}	Tin(IV)	Sn^{4+} (also called stannic)
Iron(III)	Fe^{3+} (also called ferric)	Lead(IV)	Pb^{4+} (also called plumbic)
Bismuth(III)	Bi^{3+}	Manganese(IV)	Mn^{4+}
Chromium(III)	Cr^{3+} (also called chromic)		
Cobalt(III)	Co^{3+} (also called cobaltic)		

* Note that the mercury(I) ion is a diatomic ion. In other words, you never find the Hg^{1+} ion alone, but always as $Hg^{1+}—Hg^{1+}$. The two Hg^{1+} ions are bonded to each other.

Group IVA elements do not usually form ions. However, they still have oxidation numbers, because they do form chemical compounds. The most common oxidation numbers of the Group IVA elements are 4+ and 4−.

The Group VA elements are a bit more difficult to explain. When they form simple ions, they usually do so by gaining three electrons. Therefore they have a charge of 3−. In this case, their oxidation number is also 3−. However, in other compounds in which they bond covalently, their oxidation numbers can be anywhere from 5+ to 5−.

Elements in Group VIA, which tend to become ions with 2− charges, all have oxidation numbers of 2−. (Remember, Group VIA elements tend to gain two electrons to complete their octet.) Elements in Group VIIA, which tend to become ions with 1− charges, all have oxidation numbers of 1−.

Table 8.3
Charges of Negative Ions Frequently Used in Chemistry

1^-		2^-		3^-	
Fluoride	F^{1-}	Oxide	O^{2-}	Nitride	N^{3-}
Chloride	Cl^{1-}	Sulfide	S^{2-}	Phosphide	P^{3-}
Hydroxide	$(OH)^{1-}$	Sulfite	$(SO_3)^{2-}$	Phosphate	$(PO_4)^{3-}$
Nitrite	$(NO_2)^{1-}$	Sulfate	$(SO_4)^{2-}$	Arsenate	$(AsO_4)^{3-}$
Nitrate	$(NO_3)^{1-}$	Carbonate	$(CO_3)^{2-}$	Borate	$(BO_3)^{3-}$
Acetate	$(C_2H_3O_2)^{1-}$	Chromate	$(CrO_4)^{2-}$		
Bromide	Br^{1-}	Dichromate	$(Cr_2O_7)^{2-}$		
Iodide	I^{1-}	Oxalate	$(C_2O_4)^{2-}$		
Hypochlorite	$(ClO)^{1-}$				
Chlorite	$(ClO_2)^{1-}$				
Chlorate	$(ClO_3)^{1-}$				
Perchlorate	$(ClO_4)^{1-}$				
Permanganate	$(MnO_4)^{1-}$				
Cyanide	$(CN)^{1-}$				
Hydrogen sulfite	$(HSO_3)^{1-}$				
Hydrogen sulfate	$(HSO_4)^{1-}$				
Hydrogen carbonate	$(HCO_3)^{1-}$				

Once you know the oxidation numbers of the various ions, you can predict the chemical formulas of any compounds they form. This is because all chemical compounds must be electrically neutral; in other words, they must have no net (overall) charge.

Binary Compounds Containing a Metal with a Fixed Oxidation Number and a Nonmetal

Learning Goal 3
Formulas of binary compounds containing a metal with a fixed oxidation number and a nonmetal

In the examples that follow, we will consider only metals that have fixed oxidation numbers in combination with nonmetals. These metals are the A-group elements of Groups IA, IIA, and IIIA. The nonmetals we will consider are the A-group elements of Groups VA, VIA, and VIIA. Although you can find the oxidation number of many of these elements in Tables 8.2 and 8.3, it is easier to remember that Group IA elements take on a charge of 1+, Group IIA elements take on a charge of 2+, and Group IIIA elements take on a charge of 3+. (*Note:* Although hydrogen is a nonmetal, it is a Group IA element and takes on a charge of 1+ when it enters into most chemical combinations.) For the nonmetals, Group VA elements take on a charge of 3−, Group VIA elements take on a charge of 2−, and Group VIIA elements take on a charge of 1−.

Greek prefixes are not used in the names of binary compounds of this type. Therefore, to write the correct formula, you must choose the correct subscripts for the formula unit by noting the charges of the metal and nonmetal portions. The subscripts you choose must cancel out the charges. The next example will show you how to do this.

Example 8.2

Write the chemical formula for sodium chloride.

Solution Sodium is a Group IA element, and therefore it takes on a charge of 1+ when it enters into chemical combination. Chlorine is a Group VIIA element, and therefore it takes on a charge of 1− when it enters into chemical combination. Thus the formula unit for sodium chloride is

$$Na_1^{1+}\ Cl_1^{1-}$$

The subscripts show that we need one ion of each element to obtain a neutral compound. The subscript 1 is usually not written. Likewise, the charges 1+ and 1− are left out, giving us the chemical formula NaCl.

Practice Exercise 8.2 Write the chemical formula for potassium fluoride.

Let's look more closely at how to choose the proper subscripts for a chemical formula. We know that a formula unit has a positive portion and a negative portion. The oxidation number tells us *how* positive or negative each portion is. Our job is to choose the correct subscripts to balance the charges, so that the formula unit is electrically neutral. Here are some guidelines to help you choose the correct subscripts.

1. Write the symbol of each element of the compound, along with its oxidation number. The element with the positive oxidation number is written first. For example, for the compound calcium chloride, write

$$Ca^{2+}Cl^{1-}$$

2. You want the positive side of the compound to balance the negative side. You can bring this about by crisscrossing the numbers.

$$Ca^{2+}Cl^{1-} \quad \text{or} \quad Ca_1^{2+}Cl_2^{1-} \quad \text{or} \quad CaCl_2$$

3. Note that the subscript numbers are written without charge (that is, without a plus or minus sign). For example, the formula for aluminum oxide is

$$Al^{3+}O^{2-} \quad \text{or} \quad Al_2^{3+}O_3^{2-} \quad \text{or} \quad Al_2O_3$$

4. Also note that subscript numbers should be in least-common-denominator form. For example, the formula for barium oxide is

$$Ba^{2+}O^{2-} \quad \text{or} \quad Ba_2^{2+}O_2^{2-} \quad \text{or} \quad Ba_1^{2+}O_1^{2-} \quad \text{or} \quad BaO$$

Example 8.3

Write the formula for barium chloride.

Solution The periodic table helps us to obtain the charges of barium and chlorine when they enter into chemical combination. We begin by writing $Ba^{2+}Cl^{1-}$. Then we choose the right subscripts to balance the charges:

$$Ba_1{}^{2+}Cl_2{}^{1-} \quad \text{or} \quad BaCl_2$$

Practice Exercise 8.3 Write the formula for magnesium oxide.

Example 8.4

Write the formula for aluminum sulfide.

Solution First we write $Al^{3+}S^{2-}$. Then we choose the right subscripts to balance the charges:

$$Al_2{}^{3+}S_3{}^{2-} \quad \text{or} \quad Al_2S_3$$

Practice Exercise 8.4 Write the formula for strontium sulfide.

Example 8.5

Write the chemical formula for sodium phosphide.

Solution First we write $Na^{1+}P^{3-}$. Then we choose the right subscripts to balance the charges:

$$Na_3{}^{1+}P_1{}^{3-} \quad \text{or} \quad Na_3P$$

Practice Exercise 8.5 Write the formula for potassium oxide.

Example 8.6

Write the chemical formula for aluminum nitride.

Solution First we write $Al^{3+}N^{3-}$. Then we choose the right subscripts to balance the charges:

$$Al_1{}^{3+}N_1{}^{3-} \quad \text{or} \quad AlN$$

Practice Exercise 8.6 Write the formula for aluminum bromide.

Binary Compounds Containing a Metal with a Variable Oxidation Number and a Nonmetal

Learning Goal 4
Formulas of binary compounds containing a metal with a variable oxidation number and a nonmetal

Table 8.2 shows that some atoms have variable oxidation numbers; that is, they can form more than one kind of ion when they enter into chemical combination. This is especially true of the transition metals (the ones in the middle of the periodic table—the B-group elements), because of their unique electron configurations. An element such as copper can combine with other elements in two different ways. It can combine as Cu^{1+} or Cu^{2+}.

There are two methods to name these ions. In the IUPAC *Stock system,* the oxidation number of the metal ion, in the form of a Roman numeral, is used as part of the name. The Roman numeral is placed in parentheses and immediately follows the name of the metal. Therefore Cu^{1+} is called copper(I) (read as "copper-one"), and Cu^{2+} is called copper(II) (read as "copper-two"). Under the Stock system, the oxidation number of a metal that has a variable oxidation number will *always* be part of the compound's name.

The Stock system is now the system of choice, but the older *-ous* and *-ic* system for naming these ions is still used. In the older system, the Latin name of the metal is used along with the suffix *-ous* or *-ic*. The ion with the lower oxidation number is given the suffix *-ous,* and the ion with the higher oxidation number is given the suffix *-ic*. For example, Fe^{2+} is called the ferr*ous* ion, and Fe^{3+} is called the ferr*ic* ion. (Note the suffixes and the use of *ferr-,* not *iron,* as the main stem of the element.) Another example is copper: Cu^{1+} is called the cupr*ous* ion, and Cu^{2+} is called the cupr*ic* ion. The disadvantage of this system is that you have to memorize which ion of a given element has the higher or lower oxidation number to apply the correct suffix. Table 8.2 lists some additional examples of positive ions and their names, using the rules for both systems.

Although you may find the Stock system easier to use, you must also become familiar with the older system, because it is still in use. In the examples that follow, you will practice writing the formulas of compounds when you are given the Stock name or the older *-ous* or *-ic* name.

Example 8.7

Write the formula of each of the following compounds whose metal ions have variable oxidation numbers:
(a) copper(I) chloride
(b) copper(II) chloride
(c) iron(III) sulfide
(d) vanadium(V) oxide
(e) mercuric bromide
(f) ferrous nitride

Solution

(a) The (I) tells us that copper is Cu^{1+}. We already know that chloride is Cl^{1-}, because chlorine is a Group VIIA element. Therefore we have

$$Cu^{1+}Cl^{1-} \quad \text{or} \quad CuCl$$

(b) The (II) tells us that copper is Cu^{2+}. We already know that chloride is Cl^{1-}. Therefore we have

$$Cu_1{}^{2+}Cl_2{}^{1-} \quad \text{or} \quad CuCl_2$$

(c) The (III) tells us that iron is Fe^{3+}. We already know that sulfide is S^{2-}, because sulfur is a Group VIA element. Therefore we have

$$Fe_2{}^{3+}S_3{}^{2-} \quad \text{or} \quad Fe_2S_3$$

(d) The (V) tells us that vanadium is V^{5+}. We already know that oxide is O^{2-}, because oxygen is a Group VIA element. Therefore we have

$$V_2{}^{5+}O_5{}^{2-} \quad \text{or} \quad V_2O_5$$

(e) Table 8.2 tells us that the mercuric ion is Hg^{2+}. We already know that bromide is Br^{1-}, because bromine is a Group VIIA element. Therefore we have

$$Hg_1{}^{2+}Br_2{}^{1-} \quad \text{or} \quad HgBr_2$$

(f) Table 8.2 tells us that the ferrous ion is Fe^{2+}. We already know that nitride is N^{3-}, because nitrogen is a Group VA element. Therefore we have

$$Fe_3{}^{2+}N_2{}^{3-} \quad \text{or} \quad Fe_3N_2$$

Practice Exercise 8.7 Write the formula of each of the following compounds whose metal ions have variable oxidation numbers:
(a) mercury(II) oxide
(b) iron(III) bromide
(c) cobalt(II) iodide
(d) manganese(IV) oxide
(e) cuprous sulfide
(f) ferric sulfide

Before we leave this topic, we should point out that some transition metals do not have variable oxidation numbers. Therefore neither the name of the compound nor the group that it is in will tell you the charge of the ion in that compound. Two important examples are zinc and silver. You should *memorize* that zinc has an oxidation number of 2+ (Zn^{2+}) and silver has an oxidation number of 1+ (Ag^{1+}).

Example 8.8

Write the formulas of the following compounds: *(a)* silver sulfide *(b)* zinc phosphide

Solution

(a) We know that silver is Ag^{1+}. We already know that sulfide is S^{2-}, because sulfur is a Group VIA element. Therefore we have

$$Ag_2{}^{1+}S_1{}^{2-} \quad \text{or} \quad Ag_2S$$

(b) We know that zinc is Zn^{2+}. We already know that phosphide is P^{3-}, because phosphorus is a Group VA element. Therefore we have

$$Zn_3{}^{2+}P_2{}^{3-} \quad \text{or} \quad Zn_3P_2$$

Practice Exercise 8.8 Write the formulas of the following compounds: *(a)* silver chloride *(b)* zinc oxide

8.4 Polyatomic Ions

Learning Goal 5
Formulas and names of polyatomic ions

Over the years, in studying the composition of many compounds, chemists have found that certain groups of covalently bonded atoms appear over and over again. Because these groups are electrically charged, they are not molecules but ions. More specifically, they are *polyatomic ions* (ions that are combinations of many atoms); the prefix *poly-* is from the Greek word for "many." A **polyatomic ion** is a *charged group of covalently bonded atoms.*

Examples of polyatomic ions are nitrate, $(NO_3)^{1-}$; sulfite, $(SO_3)^{2-}$; sulfate, $(SO_4)^{2-}$; carbonate, $(CO_3)^{2-}$; and phosphate, $(PO_4)^{3-}$. Polyatomic ions are common in minerals, plants, animals, and human beings. *They never exist in an uncombined state* but are always part of a chemical compound. For example, one can't isolate nitrate ions and put them in a bottle, but there are hundreds of chemical compounds made up, in part, of nitrate ions. Some examples of nitrate compounds are potassium nitrate, KNO_3 (used in chemical fertilizers); calcium nitrate, $Ca(NO_3)_2$ (used in chemical fertilizers and also in matches); and silver nitrate, $AgNO_3$ (used in photography, mirror manufacturing, hair dyeing, and silver plating and as an external medicine).

Tables 8.2 and 8.3 list the more common polyatomic ions. You should memorize the name, symbol, and charge of each ion. That information will be useful as you learn to write the formulas of compounds containing these ions in the next section.

8.5 Writing the Formulas of Ternary and Higher Compounds

Learning Goal 6
Formulas of ternary and higher compounds

In writing the formulas of compounds containing three or more elements (usually in the form of polyatomic ions), we follow pretty much the same procedure we used to write formulas of binary compounds. The only difference is that we must use the formula and charge of the polyatomic ion in deriving the formulas of ternary and higher compounds. Most of the polyatomic ions that we'll encounter have negative charges and are listed in Table 8.3. The only common polyatomic ion with a positive charge is the ammonium ion, $(NH_4)^{1+}$, which is listed in Table 8.2.

The names of most polyatomic ions end in *-ite* or *-ate*. For example, $(SO_3)^{2-}$ is called the sulf*ite* ion, and $(SO_4)^{2-}$ is called the sulf*ate* ion. Notice that the *-ate* ion has one more oxygen than the corresponding *-ite* ion. There are two polyatomic ions in Table 8.3 that do not end in *-ite* or *-ate*. They are the hydrox*ide* ion, $(OH)^{1-}$, and the cyan*ide* ion, $(CN)^{1-}$.

In the next example, you will practice writing the formulas of compounds containing polyatomic ions. If you've already memorized the polyatomic ions, try writing the formulas using only the periodic table as a reference. Otherwise, look up the formulas and charges for the polyatomic ions in Tables 8.2 and 8.3.

Example 8.9

Write the formula of each of the following ternary compounds:
(a) aluminum sulfate
(b) calcium arsenate
(c) copper(II) phosphate
(d) sodium nitrate

Solution

(a) Aluminum is a Group IIIA element, and therefore it takes on a charge of 3+ when it enters into chemical combination. Sulfate ion has the formula $(SO_4)^{2-}$. Thus the formula for aluminum sulfate is

$$Al_2^{3+}(SO_4)_3^{2-} \quad \text{or} \quad Al_2(SO_4)_3$$

(b) Calcium is a Group IIA element, and therefore it takes on a charge of 2+ when it enters into chemical combination. Arsenate ion has the formula $(AsO_4)^{3-}$. Thus the formula for calcium arsenate is

$$Ca_3^{2+}(AsO_4)_2^{3-} \quad \text{or} \quad Ca_3(AsO_4)_2$$

(c) Copper(II) has a charge of 2+. Phosphate ion has the formula $(PO_4)^{3-}$. Therefore the formula for copper(II) phosphate is

$$Cu_3^{2+}(PO_4)_2^{3-} \quad \text{or} \quad Cu_3(PO_4)_2$$

(d) Sodium is a Group IA element, and therefore it takes on a charge of 1+ when it enters into chemical combination. Nitrate ion has the formula $(NO_3)^{1-}$. Thus the formula for sodium nitrate is

$$Na_1^{1+}(NO_3)_1^{1-} \quad \text{or} \quad NaNO_3$$

Notice that the parentheses are dropped when the polyatomic ion has a subscript of 1.

Practice Exercise 8.9 Write the formula of each of the following ternary compounds:
(a) ferric nitrite
(b) barium phosphate
(c) copper(II) sulfate
(d) potassium chromate

8.6 Writing the Names of Binary Compounds Containing Two Nonmetals

Learning Goal 7
Names of binary compounds containing two nonmetals

Now let's turn our attention to writing the names of compounds when we are given their formulas. We will begin by learning how to name binary compounds composed of two nonmetals.

In Section 8.2, you learned how to write the formulas of these compounds from their names. You need only to reverse the process to write the name of the compound from the formula. Remember that Greek prefixes are used to designate the number of atoms in a formula unit of the compound.

Example 8.10

Write the names of the following binary compounds composed of two nonmetals: *(a)* SO_3 *(b)* NO *(c)* N_2O *(d)* CO_2

Solution

(a) SO_3 is called sulfur trioxide. This compound contains one sulfur atom and three oxygen atoms. There is no prefix used for the sulfur portion of the compound. The prefix *tri-* is used for the oxygen portion of the compound. (Also, remember that in naming a binary compound, you change ox*ygen* to ox*ide* because it is the second-named element.)
(b) NO is called nitrogen monoxide. This compound contains one nitrogen atom and one oxygen atom. There is no prefix used for the nitrogen portion of the compound. The prefix *mono-* is used for the oxygen portion of the compound. (Remember the rule for *mono-*: The prefix *mono-* is used for the second-named element, but rarely appears with the first-named element.)
(c) N_2O is called dinitrogen monoxide. This compound contains two nitrogen atoms and one oxygen atom. The prefix *di-* is used for the nitrogen portion of the compound. The prefix *mono-* is used for the oxygen portion of the compound.
(d) CO_2 is called carbon dioxide. This compound contains one carbon atom and two oxygen atoms. There is no prefix used for the carbon portion of the compound. The prefix *di-* is used for the oxygen portion of the compound.

Practice Exercise 8.10 Write the names of the following binary compounds composed of two nonmetals: *(a)* P_2O_5 *(b)* OF_2 *(c)* SO_2 *(d)* CO

8.7 Writing the Names of Binary Compounds Containing a Metal and a Nonmetal

Learning Goal 8
Names of binary compounds containing a metal and a nonmetal

Learning Goal 9
Determining oxidation numbers

We already know how to write the formulas of binary compounds composed of a metal and a nonmetal when we are given their names (Section 8.3). Once again, we can simply reverse this process to write the name of such a compound from its formula. Remember that for these compounds, *no Greek prefixes are used.* For compounds that contain metals with fixed oxidation numbers, just name the elements. (Don't forget to change the ending of the second-named element to *-ide.*) For compounds that contain metals with variable oxidation numbers, the Roman numeral, which represents the oxidation number of the metal, must be given as part of the compound's name (or the *-ous* and *-ic* suffix system may be used).

Finding Oxidation Numbers

Tables 8.2 and 8.3 list oxidation numbers only for some of the more common ions. However, once you know these, you can use them to find the oxidation numbers of less common ions from the compounds they form. You need only remember that every compound must be neutral. That is, the sum of the oxidation numbers of all the atoms in the compound must be zero.

Example 8.11

What is the oxidation number of cobalt in $CoCl_3$?

Solution We know that each chloride ion has a charge of 1−. The molecule contains three chloride ions, for a total charge of 3−. For the compound to be neutral, the single cobalt ion must have a charge of 3+: $Co_1^{3+}Cl_3^{1-}$.

Practice Exercise 8.11 What is the oxidation number of copper in $CuCl_2$?

Example 8.12

Find the oxidation number of each underlined element: *(a)* $\underline{Mn}O_2$ *(b)* $K\underline{Mn}O_4$ *(c)* $\underline{Cs}_2SO_3$ *(d)* $\underline{Ga}_2O_3$
We follow the same procedure used in the previous example.

Solution

(a) We know that the oxidation number of oxygen is 2−: $Mn_1^{?}O_2^{2-}$

For the compound to be electrically neutral, the manganese must have a charge of 4+: $Mn_1^{4+}O_2^{2-}$

(b) We know that the oxidation number of potassium is 1+, and that of oxygen is 2−: $K_1^{1+}Mn_1^{?}O_4^{2-}$

For the compound to be electrically neutral, the manganese must have a charge of 7+: $K_1^{1+}Mn_1^{7+}O_4^{2-}$

(c) We know that the oxidation number of the sulfite group is 2−: $Cs_2^{?}(SO_3)_1^{2-}$

For the compound to be electrically neutral, the cesium must have a charge of 1+: $Cs_2^{1+}(SO_3)_1^{2-}$

(d) We know that the oxidation number of oxygen is 2−: $Ga_2^{?}O_3^{2-}$

For the compound to be electrically neutral, the gallium must have a charge of 3+: $Ga_2^{3+}O_3^{2-}$

Practice Exercise 8.12 Find the oxidation number of each underlined element: *(a)* $\underline{Sn}F_4$ *(b)* $\underline{W}Cl_5$

Example 8.13

Write the names of the following binary compounds composed of a metal and a nonmetal: *(a)* $AlCl_3$ *(b)* Na_2O *(c)* Mg_3N_2 *(d)* K_3P *(e)* $CoCl_3$ *(f)* Fe_2O_3 *(g)* FeO *(h)* MnO_2

Solution

(a) $AlCl_3$ is called aluminum chloride. All we have to do is name the elements and remember to change the name of chlor*ine* to chlor*ide*.

(b) Na_2O is called sodium oxide. All we have to do is name the elements and remember to change the name of ox*ygen* to ox*ide*.

(c) Mg_3N_2 is called magnesium nitride. All we have to do is name the elements and remember to change the name of nitro*gen* to nitr*ide*.

(d) K_3P is called potassium phosphide. All we have to do is name the elements and remember to change the name of phosph*orus* to phosph*ide*.

(e) $CoCl_3$ is called cobalt(III) chloride (or cobaltic chloride). Cobalt is a metal with a variable oxidation number, and a Roman numeral representing that oxidation number must be part of the compound's name. To determine the oxidation number of cobalt in this compound, we go through the following reasoning: We know that each chloride ion has a charge of 1−. This molecule contains three chloride ions, for a total charge of 3−. For the compound to be neutral, the single cobalt ion must have a charge of 3+:

$$Co_1^{3+}Cl_3^{1-}$$

(f) Fe_2O_3 is called iron(III) oxide (or ferric oxide). Iron is a metal with a variable oxidation number, and a Roman numeral representing that oxidation number must be part of the compound's name. To determine the oxidation number of iron in this compound, we go through the following reasoning: We know that each oxide ion has a charge of 2−. This molecule contains three oxide ions, for a total charge of 6−. For the compound to be neutral, the two iron ions must have a total charge of 6+. Therefore each iron ion must have a charge of 3+:

$$Fe_2^{3+}O_3^{2-}$$

(g) FeO is called iron(II) oxide (or ferrous oxide). Iron is a metal with a variable oxidation number, and a Roman numeral representing that oxidation number must be part of the compound's name. To determine the oxidation number of iron in this compound, we go through the following reasoning: We know that each oxide ion has a charge of 2−. This molecule contains one oxide ion, for a total charge of 2−. For the compound to be neutral, the single iron ion must have a charge of 2+:

$$Fe_1^{2+}O_1^{2-}$$

(h) MnO_2 is called manganese(IV) oxide. Manganese is a metal with a variable oxidation number, and a Roman numeral representing that oxidation number must be part of the compound's name. To determine the oxidation number of manganese in this compound, we go through the following reasoning: We know that each oxide ion has a charge of 2−. This molecule contains two oxide ions, for a total charge of 4−. For the compound to be neutral, the single manganese ion must have a charge of 4+:

$$Mn_1^{4+}O_2^{2-}$$

Practice Exercise 8.13 Write the names of the following binary compounds composed of a metal and a nonmetal: *(a)* RbCl *(b)* Ga_2S_3 *(c)* SrO *(d)* ZnI_2 *(e)* $NiCl_2$ *(f)* FeI_3 *(g)* HgS *(h)* Cu_2O

8.8 Writing the Names of Ternary and Higher Compounds

Learning Goal 10
Names of ternary and higher compounds

You learned how to write the formulas of ternary and higher compounds from their names in Section 8.5. You can just reverse the process to write the name of this type of compound from the formula. Remember that for these compounds, *no Greek prefixes are used,* but you must include the name of the polyatomic ion. For compounds that contain metals with fixed oxidation numbers, simply name the element and the polyatomic ion. For compounds that contain metals with variable oxidation numbers, the Roman numeral, which represents the oxidation number of the metal, must be given as part of the compound's name (or the *-ous* and *-ic* suffix system may be used).

Example 8.14

Write the names of the following ternary and higher compounds: *(a)* $Al_2(CrO_4)_3$ *(b)* Li_2SO_3 *(c)* $Mg(NO_2)_2$ *(d)* $(NH_4)_3PO_4$ *(e)* $Cr(OH)_2$ *(f)* $Fe_2(Cr_2O_7)_3$

Solution

(a) $Al_2(CrO_4)_3$ is called aluminum chromate. All we have to do is name the metal, aluminum, and the polyatomic ion, chromate.
(b) Li_2SO_3 is called lithium sulfite. All we have to do is name the metal, lithium, and the polyatomic ion, sulfite.
(c) $Mg(NO_2)_2$ is called magnesium nitrite. All we have to do is name the metal, magnesium, and the polyatomic ion, nitrite.
(d) $(NH_4)_3PO_4$ is called ammonium phosphate. All we have to do is name the two polyatomic ions, ammonium and phosphate.
(e) $Cr(OH)_2$ is called chromium(II) hydroxide (or chromous hydroxide). Chromium is a metal with a variable oxidation number, and a Roman numeral representing that oxidation number must be part of the compound's name. Therefore we have chromium(II). The polyatomic ion is hydroxide.
(f) $Fe_2(Cr_2O_7)_3$ is called iron(III) dichromate (or ferric dichromate). Iron is a metal with a variable oxidation number, and a Roman numeral representing that oxidation number must be part of the compound's name. Therefore we have iron(III). The polyatomic ion is dichromate.

Practice Exercise 8.14 Write the names of the following ternary and higher compounds: *(a)* $(NH_4)_2O$ *(b)* $Mg(CN)_2$ *(c)* $Al(NO_2)_3$ *(d)* $Zn_3(PO_4)_2$ *(e)* CaC_2O_4 *(f)* $Ni_3(BO_3)_2$

8.9 Writing the Names and Formulas of Inorganic Acids

Learning Goal 11
Formulas and names of inorganic acids

Svante Arrhenius, a well-known Swedish chemist of the late 1800s and early 1900s, developed a classical definition of an acid. According to Arrhenius, an *acid* is a substance that releases hydrogen ions, H^{1+}, in aqueous solutions. Two major classes of inorganic acids exist, and the naming rules differ for the two classes. First we'll examine the rules for naming the non-oxygen-containing acids. Then we'll examine the rules for naming the oxygen-containing acids.

Non-Oxygen-Containing Acids

These acids usually consist of hydrogen plus a nonmetal ion. For example, HCl, which is hydrogen chloride gas in its pure form, becomes hydrochloric acid when dissolved in water. The covalently bonded HCl molecules ionize in water to form hydrogen ions and chloride ions. We derive the names of acids such as HCl by adding the prefix *hydro-* to the name of the nonmetal, which is given an *-ic* suffix. For example,

hydro—	*chlor*—	*ic*—	*acid*
Prefix	Name of nonmetal	Suffix	

Example 8.15

Name each of the following non-oxygen-containing acids: *(a)* HF *(b)* HBr *(c)* H_2S *(d)* HCN

Solution We'll give the name of each substance first as a covalent compound and then as an acid in aqueous solution.

Name as Covalent Compound	Name as Acid in Aqueous Solution
(a) Hydrogen fluoride	Hydrofluoric acid
(b) Hydrogen bromide	Hydrobromic acid
(c) Hydrogen sulfide	Hydrosulfuric acid
(d) Hydrogen cyanide	Hydrocyanic acid

Practice Exercise 8.15 Name each of the following non-oxygen-containing acids: *(a)* HCl *(b)* H_2Se *(c)* HI

Oxygen-Containing Acids

Oxygen-containing acids formed from hydrogen and polyatomic ions (containing oxygen) are named as follows:

1. If the polyatomic ion has an *-ate* suffix, we name the acid by replacing the *-ate* suffix with an *-ic* suffix and adding the word *acid*. For example, HNO_3, composed of a hydrogen ion and nitr*ate* ion, is called nitr*ic* acid in aqueous solution.

Example 8.16

Name each of the following oxygen-containing acids: *(a)* H_2SO_4 *(b)* $HC_2H_3O_2$ *(c)* $HBrO_3$

Solution These three acids are composed of hydrogen plus a polyatomic ion that has the suffix *-ate*.

(a) H_2SO_4 is called sulfur*ic* acid. (SO_4^{2-} is the sulf*ate* ion.)
(b) $HC_2H_3O_2$ is called acet*ic* acid. ($C_2H_3O_2^{1-}$ is the acet*ate* ion.)
(c) $HBrO_3$ is called brom*ic* acid. (BrO_3^{1-} is the brom*ate* ion.)
Note that H_2SO_4 is an exception to our rule, in that the whole name of the element is used rather than the name of the sulfate group.

Practice Exercise 8.16 Name each of the following oxygen-containing acids: *(a)* H_3PO_4 *(b)* $HClO_3$

2. If the polyatomic ion has an *-ite* suffix, we name the acid by replacing the *-ite* suffix with an *-ous* suffix and adding the word *acid*. For example, HNO_2, composed of a hydrogen ion and a nitr*ite* ion, is called nitr*ous* acid.

Example 8.17

Name each of the following oxygen-containing acids: *(a)* H_2SO_3 *(b)* H_3PO_3 *(c)* $HBrO_2$

Solution These three acids are composed of hydrogen plus a polyatomic ion that has the suffix *-ite*.

(a) H_2SO_3 is called sulfur*ous* acid. (SO_3^{2-} is the sulf*ite* ion.)
(b) H_3PO_3 is called phosphor*ous* acid. (PO_3^{2-} is the phosph*ite* ion.)
(c) $HBrO_2$ is called brom*ous* acid. (BrO_2^{1-} is the brom*ite* ion.)

Note that H_2SO_3 is an exception to our rule, in that the whole name of the element is used rather than the name of the sulfite group.

Practice Exercise 8.17 Name each of the following oxygen-containing acids: *(a)* $HClO_2$ *(b)* HNO_2

3. Some elements form more than two oxygen-containing acids. Most notable are the acids formed by the elements chlorine, bromine, and iodine. This occurs because these elements can form four polyatomic ions. For example, bromine forms the following ions:

BrO_4^{1-}, perbrom*ate* ion BrO_3^{1-}, brom*ate* ion
BrO_2^{1-}, brom*ite* ion BrO^{1-}, hypobrom*ite* ion

Note that the first two ions listed have the suffix *-ate* and that the last two ions listed have the suffix *-ite*. Therefore we use rules 1 and 2. Keep the name of the polyatomic ion, but change the *-ate* suffix to *-ic* or the *-ite* suffix to *-ous*.

Example 8.18

Name each of the following oxygen-containing acids: *(a)* $HBrO_4$ *(b)* $HBrO_3$ *(c)* $HBrO_2$ *(d)* $HBrO$

Solution

(a) $HBrO_4$ is called perbrom*ic* acid. (BrO_4^{1-} is the perbrom*ate* ion.)
(b) $HBrO_3$ is called brom*ic* acid. (BrO_3^{1-} is the brom*ate* ion.)

(c) $HBrO_2$ is called brom*ous* acid. (BrO_2^{1-} is the brom*ite* ion.)
(d) HBrO is called hypobrom*ous* acid. (BrO^{1-} is the hypobrom*ite* ion.)

Practice Exercise 8.18 Name each of the following oxygen-containing acids: *(a)* $HClO_4$ *(b)* $HClO_3$ *(c)* $HClO_2$ *(d)* HClO

8.10 Common Names of Compounds

Learning Goal 12
Common names and formulas of compounds

As we stated at the beginning of this chapter, there are no rules governing the common names of compounds. Most common names have been derived from common usage or simply handed down through chemical history. Many common names are still in use, mostly because the systematic names are too complex for everyday use. For example, baking soda is a common substance that has a variety of uses around the house, but hardly anyone would call it sodium hydrogen carbonate. The same could be said of Epsom salts, whose systematic name is magnesium sulfate heptahydrate. Table 8.4 lists the common names, systematic chemical names, and formulas of several familiar substances.

Summary

The system for naming chemical compounds and designating their formulas is called chemical nomenclature. Chemical compounds may be referred to by

Table 8.4
Common Names of Some Chemical Compounds

Common name	Formula	Systematic chemical name
Baking soda	$NaHCO_3$	Sodium hydrogen carbonate
Borax	$Na_2B_4O_7 \cdot 10H_2O$	Sodium tetraborate decahydrate
Dry ice	CO_2	Carbon dioxide
Epsom salts	$MgSO_4 \cdot 7H_2O$	Magnesium sulfate heptahydrate
Gypsum	$CaSO_4 \cdot 2H_2O$	Calcium sulfate dihydrate
Laughing gas	N_2O	Nitrous oxide
Marble	$CaCO_3$	Calcium carbonate
Milk of magnesia	$Mg(OH)_2$	Magnesium hydroxide
Muriatic acid	HCl	Hydrochloric acid
Oil of vitriol	H_2SO_4	Sulfuric acid
Quicklime	CaO	Calcium oxide
Saltpeter	$NaNO_3$	Sodium nitrate

their systematic chemical names or their common names. The rules for the systematic names have been developed by the International Union of Pure and Applied Chemistry (IUPAC). There are no rules that govern the common names. These names have simply been handed down through chemical history.

According to IUPAC rules for writing the formula of a compound, the more positive portion of the compound is written first. In a compound composed of two ions, the positive ion precedes the negative ion. In a covalent compound, the element with the lower electronegativity (the more positive portion) precedes the element with the higher electronegativity (the more negative portion).

Greek prefixes are used to write the names of binary compounds composed of two nonmetals. For binary compounds, the ending of the second element is always *-ide*. Oxidation numbers are used to write the formulas of binary compounds containing a metal and a nonmetal. An oxidation number is the positive or negative number that expresses the combining capacity of an element in a particular compound. Some metals (typically the A-group metals) have fixed oxidation numbers. Other metals (typically the B-group, or transition, metals) have variable oxidation numbers and can combine with a nonmetal element in more than one way. The periodic table can be used to determine the oxidation number of an A-group metal: The group number of the A-group metal is its common oxidation number. The oxidation number of metals with variable oxidation numbers can be determined from the name of the compound, either through the Roman numeral attached to the name of the metal or through the *-ous* or *-ic* suffix attached to the Latin name of the metal.

Ternary and higher compounds contain polyatomic ions. A polyatomic ion is a charged group of covalently bonded atoms. The names of most polyatomic ions end in *-ite* or *-ate*. We use the formula and charge of the polyatomic ion and the rules for writing the formulas of binary compounds in deriving the formulas of ternary and higher compounds.

A number of chemicals are known by their common names, either because of tradition or because the systematic names are too complex for convenient reference. Because there are no rules that govern the common names of compounds, the formulas of these compounds may be associated with their names only through memorization.

Key Terms

binary compound **(8.1)**
chemical nomenclature **(Introduction)**
common name **(Introduction)**
oxidation number **(8.3)**
polyatomic ion **(8.4)**
systematic chemical name **(Introduction)**
ternary compound **(8.1)**

CAREER SKETCH

WASTE-WATER TREATMENT CHEMIST

As a waste-water treatment chemist, you will analyze water samples from streams, examine raw and treated waste water, and look at sludge and other by-products of waste water treatment. You will also work to determine the efficiency of the processes used in waste-water treatment plants to ensure that local, state, and federal standards are met.

Waste-water treatment chemists can work in municipal or privately owned waste water treatment plants. You can also work in a laboratory, where you will work with sophisticated analytical equipment. You will test samples of water and waste water for chemical, bacteriological, and physical components. You will also monitor and regulate the discharge of waste water into sewer and treatment systems.

The proper treatment of waste water is an essential factor in preserving our supply of clean water.
(Steve Delaney/EPA)

Chemical Frontiers

Extending the Life of Organs Used for Transplants*

Recently two researchers, pioneering transplant surgeon Folkert P. Belzer and biochemist James Southard, developed a new solution that extends the life of donor organs used in transplant surgery. Transplant surgeons and health scientists are calling this a major breakthrough in the field.

The new preservative, Belzer UW-Cold Storage Solution®, lengthens the time that tissues can be preserved prior to transplantation. It does this by preventing tissue swelling and cellular damage. Livers and pancreases can be preserved three times longer than in the past. For example, a pancreas or a liver that previously could be kept viable for only 6 hours can now be preserved for 16 to 30 hours. Kidneys that in the past could be kept viable for only 48 hours can now be preserved for up to 72 hours.

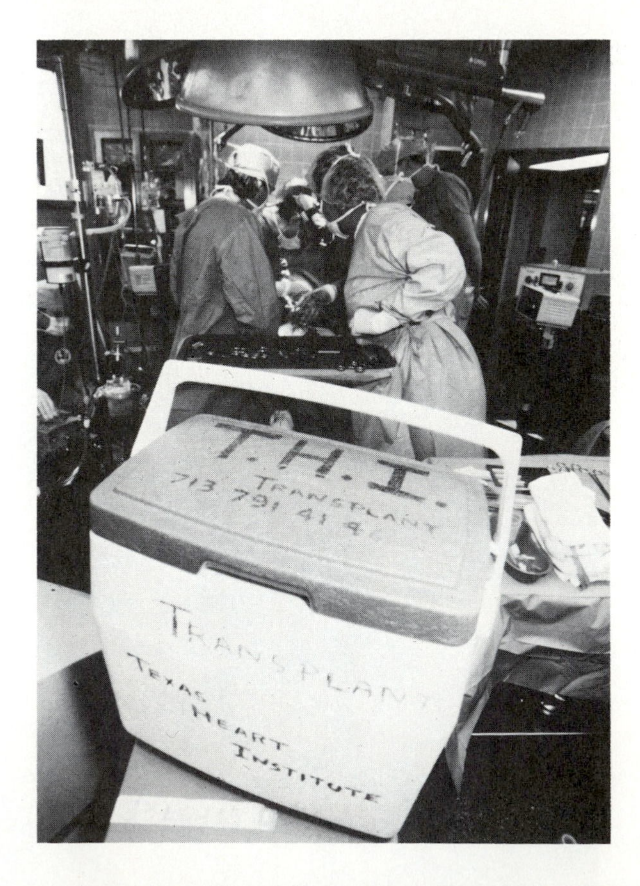

Donor heart is stored in an ice chest for transplant.
(Ken Koge/Texas Heart Institute/Phototake)

*Adapted from "New Life for Organ Transplants," *What's Happening in Chemistry* (American Chemical Society, 1990).

Because of this new solution, the supply of donor organs is increasing. Donor organs that once were wasted because a recipient was too far away can now be placed in this storage solution and flown almost anywhere in the world.

This solution also keeps the organ in better condition, so that a successful transplant is more likely. Organ preservatives of the past were basically simple sugar–salt solutions, containing a sodium or potassium phosphate compound and a high concentration of the sugar glucose. These solutions were used both to flush blood out of the donor organ and to preserve the viability of the cells. A problem with these solutions is that for organs other than kidneys, the glucose solution cannot prevent the organ's cells from gaining water at the low storage temperature of 0° C to 4° C. If the cells of the organ absorb too much water, they die.

To overcome this problem, Belzer and Southard searched for compounds that would prevent cells from gaining water at low temperature. They also searched for compounds capable of suppressing damaging chemical changes that occur in donor organs with no blood supply. Their research succeeded on both counts. Thanks to the collaborative efforts of a biochemist and a physician, this new solution promises to save the lives of numerous transplant patients.

Self-Test Exercises

Learning Goal 1: Distinguishing Between Common Name and Systematic Name

▲ **1.** Explain the difference between the common name and the systematic name of a compound.

▲ **2.** Define or explain the following terms: *(a)* IUPAC *(b)* Stock system *(c)* binary compound *(d)* ternary compound

3. The common name for H_2O is water. What is its systematic name?

4. What is the common name and the systematic name of solid CO_2?

Learning Goal 2: Formulas of Binary Compounds Containing Two Nonmetals

5. Write the number that corresponds to each of the following prefixes: *(a) octa- (b) tri- (c) tetra- (d) nona- (e) penta- (f) hepta-*

6. Write the Greek prefix for each of the following numbers: *(a)* 2 *(b)* 6 *(c)* 1 *(d)* 10

7. Write the formulas of the following binary compounds composed of two nonmetals:
(a) diphosphorus pentasulfide
(b) chlorine dioxide
(c) dinitrogen tetroxide
(d) dichlorine heptoxide

8. Write the formulas of the following binary compounds composed of two nonmetals:.
(a) carbon tetrachloride
(b) tetraphosphorus decoxide
(c) phosphorus pentabromide
(d) selenium dioxide

Learning Goal 3: Formulas of Binary Compounds Containing a Metal with a Fixed Oxidation Number and a Nonmetal

9. Write the formulas of the following binary compounds composed of a metal with a fixed oxidation number and a nonmetal:
(a) aluminum oxide
(b) lithium sulfide

(c) sodium sulfide
(d) calcium nitride
(e) silver iodide
(f) zinc chloride

10. Write the formulas of the following binary compounds composed of a metal with a fixed oxidation number and a nonmetal:
(a) lithium iodide
(b) calcium oxide
(c) strontium bromide
(d) potassium phosphide
(e) rubidium sulfide
(f) barium nitride

Learning Goal 4: Formulas of Binary Compounds Containing a Metal with a Variable Oxidation Number and a Nonmetal

11. Give Roman numerals that correspond to the following Arabic numerals: *(a)* 3 *(b)* 4 *(c)* 5 *(d)* 10

12. Give the Arabic numerals that correspond to the following Roman numerals: *(a)* VI *(b)* IX *(c)* IV *(d)* VI

▲ **13.** Give a definition or an explanation of the following suffixes: *(a) -ic* *(b) -ous* *(c) -ide*

▲ **14.** Explain the difference between *(a)* ferrous and ferric *(b)* cuprous and cupric *(c)* cobaltous and cobaltic *(d)* stannous and stannic

15. Write the formulas of the following binary compounds composed of a metal with a variable oxidation number and a nonmetal:
(a) copper(I) sulfide
(b) mercury(II) chloride
(c) iron(II) oxide
(d) tin(II) iodide
(e) cobaltic bromide
(f) mercuric nitride

16. Write the formulas of the following binary compounds composed of a metal with a variable oxidation number and a nonmetal:
(a) uranium(VI) fluoride
(b) iron(III) nitride
(c) manganous chloride
(d) ferrous sulfide
(e) cuprous oxide
(f) mercurous chloride

Learning Goal 5: Formulas and Names of Polyatomic Ions

▲ **17.** Write the formulas of the following polyatomic ions:
(a) ammonium ion
(b) hydroxide ion
(c) borate ion
(d) hydrogen carbonate ion
(e) oxalate ion
(f) sulfite ion

▲ **18.** Write the formulas of the following polyatomic ions:
(a) cyanide ion
(b) chlorite ion
(c) chlorate ion
(d) chromate ion
(e) dichromate ion
(f) phosphate ion

19. Write the names of the following polyatomic ions: *(a)* $(AsO_4)^{3-}$ *(b)* $(MnO_4)^{1-}$ *(c)* $(SO_3)^{2-}$ *(d)* $(NH_4)^{1+}$ *(e)* $(HSO_4)^{1-}$ *(f)* $(C_2H_3O_2)^{1-}$

20. Write the names of the following polyatomic ions: *(a)* $(NO_2)^{1-}$ *(b)* $(NO_3)^{1-}$ *(c)* $(ClO_4)^{1-}$ *(d)* $(CN)^{1-}$ *(e)* $(BO_3)^{3-}$ *(f)* $(ClO)^{1-}$

Learning Goal 6: Formulas of Ternary and Higher Compounds

21. Write the formulas of the following ternary and higher compounds:
(a) mercury(II) phosphate
(b) tin(II) arsenate
(c) iron(III) acetate
(d) lithium phosphate
(e) aluminum sulfite
(f) zinc nitrite

22. Write the formulas of the following ternary and higher compounds:
(a) cesium hydroxide
(b) copper(I) arsenate
(c) ammonium sulfate
(d) potassium carbonate
(e) ferric cyanide
(f) cuprous sulfate

▲ **23.** Write the formulas of the following ternary and higher compounds:
(a) lead(II) sulfate
(b) cobalt(II) phosphate

(c) ammonium dichromate
(d) calcium oxalate
(e) stannous nitrate
(f) magnesium hydrogen sulfate

▲ **24.** Write the formulas of the following ternary and higher compounds:
(a) potassium chlorate
(b) zinc hydroxide
(c) zinc phosphate
(d) silver nitrite
(e) mercuric nitrate
(f) cuprous sulfite

Learning Goal 7: Names of Binary Compounds Containing Two Nonmetals

25. Write the names of the following binary compounds composed of two nonmetals: (a) P_2S_5 (b) CO (c) SiO_2 (d) ClO_2

26. Write the names of the following binary compounds composed of two nonmetals: (a) N_2O (b) NO_2 (c) SO_3 (d) N_2O_5

Learning Goal 8: Names of Binary Compounds Containing a Metal and a Nonmetal

▲ **27.** Write the names of the following binary compounds containing metals with fixed oxidation numbers: (a) Al_2O_3 (b) NaI (c) $ZnCl_2$ (d) Mg_3N_2 (e) Ag_2S (f) LiI

▲ **28.** Write the names of the following binary compounds containing metals with fixed oxidation numbers: (a) Cs_2O (b) Al_2S_3 (c) BaI_2 (d) $GaCl_3$ (e) K_2O (f) MgS

Learning Goal 9: Determining Oxidation Numbers

29. Find the oxidation number of the underlined element or polyatomic ion: (a) $Na\underline{ClO_4}$ (b) $\underline{Pb_3}O_4$ (c) $\underline{Ge}S_2$ (d) $\underline{V}OCl_3$ (e) $Ca(\underline{HCO_3})_2$

▲ **30.** Find the oxidation number of the underlined element: (a) $H\underline{Cl}O$ (b) $H\underline{Cl}O_2$ (c) $H\underline{Cl}O_3$ (d) $H\underline{Cl}O_4$

▲ **31.** Find the oxidation number of the underlined element or polyatomic ion: (a) $\underline{V_2}O_5$ (b) $\underline{In_2}O$ (c) $\underline{N_2}O$ (d) $\underline{Pb}P_5$ (e) $Mg_3(\underline{BO_3})_2$ (f) $Al(\underline{ClO_3})_3$

32. Find the oxidation number of the underlined element or polyatomic ion: (a) $H_3\underline{B}O_3$ (b) $H\underline{Br}O_3$ (c) $H\underline{I}O_3$ (d) $\underline{Cr}PO_3$ (e) $Al(\underline{HCO_3})_3$ (f) $Li\underline{ClO}$

▲ **33.** Write the names of the following binary compounds containing metals with variable oxidation numbers: (a) OsO_4 (b) Hg_3P (c) FeS (d) $CoCl_2$ (e) Cu_3N (f) Cu_2O

▲ **34.** Write the names of the following binary compounds containing metals with variable oxidation numbers: (a) CuS (b) $AuBr_3$ (c) FeO (d) Cu_3P_2 (e) V_2O_5 (f) MnO_2

Learning Goal 10: Names of Ternary and Higher Compounds

35. Write the names of the following ternary and higher compounds: (a) Ag_2CO_3 (b) $Hg_3(PO_4)_2$ (c) $Fe_2(SO_4)_3$ (d) $NaNO_3$ (e) $CuCrO_4$ (f) $Zn(OH)_2$

36. Write the names of the following ternary and higher compounds: (a) $CaCO_3$ (b) $NaNO_2$ (c) $NaOH$ (d) $Mg(OH)_2$ (e) $K_2Cr_2O_7$ (f) NH_4I

▲ **37.** Write the names of the following ternary and higher compounds: (a) $CaSO_4$ (b) KCN (c) $AlPO_4$ (d) $Cu_2C_2O_4$ (e) $Fe_2(CrO_4)_3$ (f) $Cu(NO_2)_2$

▲ **38.** Write the names of the following ternary and higher compounds: (a) Rb_2SO_4 (b) $Fe(C_2H_3O_2)_2$ (c) $Mg_3(BO_3)_2$ (d) $KMnO_4$ (e) $Bi_2(SO_4)_3$ (f) $(NH_4)_2C_2O_4$

Learning Goal 11: Formulas and Names of Inorganic Acids

39. Write the formulas of the following inorganic acids:
(a) hypochlorous acid
(b) hydrobromic acid
(c) nitric acid
(d) chlorous acid
(e) perbromic acid
(f) sulfuric acid

40. Write the formulas of the following inorganic acids:
(a) acetic acid
(b) sulfurous acid
(c) chloric acid
(d) hydrofluoric acid
(e) phosphoric acid
(f) nitrous acid

▲ **41.** Name the following compounds as inorganic acids: *(a)* HCN *(b)* H_2S *(c)* $HBrO_3$ *(d)* H_2SO_4 *(e)* HClO *(f)* $HBrO_2$

▲ **42.** Name the following compounds as inorganic acids: *(a)* HIO_4 *(b)* $HClO_3$ *(c)* HBr *(d)* HF *(e)* $HClO_2$ *(f)* HIO

Learning Goal 12: Common Names and Formulas of Compounds

43. Write the formula of each of the following compounds designated by their common names:
(a) milk of magnesia
(b) oil of vitriol
(c) saltpeter
(d) laughing gas

44. Write the formulas of each of the following compounds designated by their common names:
(a) Epsom salts
(b) muriatic acid
(c) marble
(d) table salt

45. Write the common name of the compound designated by each formula: *(a)* CaO *(b)* $CaCO_3$ *(c)* $CaSO_4 \cdot 2H_2O$ *(d)* H_2SO_4

46. Write the systematic chemical name of each of the following substances: *(a)* dry ice *(b)* saltpeter *(c)* baking soda *(d)* borax

Extra Exercises

▲ **47.** Predict the compound that is *most likely* to be formed from each of the following pairs of elements: *(a)* Sr and O *(b)* Al and N *(c)* Rb and S

▲ **48.** Write the formula of the compound formed from each of the following pairs of ions:
(a) Al^{3+} and N^{3-}
(b) V^{5+} and O^{2-}
(c) Fe^{3+} and $(OH)^{1-}$
(d) $(NH_4)^{1+}$ and S^{2-}

▲ **49.** Name each of the compounds in Exercise 48.

50. Write the formulas of the following compounds:
(a) tin(II) ion plus sulfide ion
(b) copper(I) ion plus phosphate ion
(c) iron(III) ion plus nitrite ion
(d) magnesium ion plus acetate ion

51. Complete the table below by writing the correct formula of the compound formed by each metal ion in combination with each of the given nonmetal ions.

52. Write the correct name of each of the following compounds: *(a)* $KClO_2$ *(b)* $Fe(CN)_3$ *(c)* P_2S_5 *(d)* CCl_4 *(e)* $HC_2H_3O_2$ (in water) *(f)* SnO_2

53. Write the correct formula of each of the following compounds:
(a) gallium fluoride
(b) palladium(II) nitrate
(c) gold(III) phosphide
(d) lanthanum(III) acetate
(e) plutonium(IV) oxide
(f) ruthenium(VIII) oxide

	Nonmetal ion		
Metal ion	*Bromide*	*Sulfate*	*Phosphate*
Sodium	______	______	______
Calcium	______	______	______
Aluminum	______	______	______

Cumulative Review/Chapters 7–8

1. Write the electron-dot structure for *(a)* a calcium atom, *(b)* a calcium ion.

2. Write the electron-dot structure for *(a)* a chlorine atom, *(b)* a chloride ion.

3. Distinguish among an ionic bond, a covalent bond, and a coordinate covalent bond.

4. Draw the electron-dot diagram for C_3H_8.

▲ **5.** List the following compounds in order of decreasing covalence: HCl, $BaCl_2$, CaF_2, and $CaCl_2$.

6. Which of the following compounds is most ionic? LiF, NaF, RbF, or CsF

7. Write the chemical formula of each of the following compounds:
(a) copper(I) sulfite *(b)* osmium tetroxide
(c) ammonium carbonate *(d)* dinitrogen monoxide

▲ **8.** Find the oxidation number of each underlined element: *(a)* $\underline{W}Cl_5$ *(b)* $\underline{Sn}F_4$ *(c)* $\underline{V}_2O_5$ *(d)* $\underline{In}I_3$

9. Write the name of each of the following compounds: *(a)* NH_4I *(b)* CaC_2 *(c)* Cu_2CrO_4

▲ **10.** Determine the formula of a hypothetical compound that would be formed between elements X and Y if X has three electrons in the outer shell and Y has six electrons in the outer shell.

11. The electron-dot structure for carbon dioxide is

$$:\ddot{\underset{\cdot\cdot}{O}}{}^{\times}_{\times}C{}^{\times}_{\times}\ddot{\underset{\cdot\cdot}{O}}:$$

True or false?

12. An atom having the electron configuration $1s^22s^22p^63s^2$ would form an ion with a 2+ charge. True or false?

13. A Br^{1-} ion, an atom of Kr, and a Rb^{1+} ion have the same electron structure. True or false?

14. The bonds in a molecule of

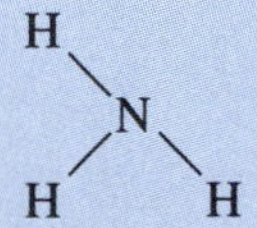

are nonpolar. True or false?

15. A molecule of $CHCl_3$ is polar. True or false?

16. Write the chemical formula for the combination of atoms that have the following electron configuration:

$$1s^22s^22p^4 \quad \text{and} \quad 1s^22s^22p^63s^2$$

▲ **17.** Which of the following bonds is most covalent? C—H, C—Cl, or C—O

18. Write the formula of the compound formed from Na^{1+} and SO_4^{2-}

▲ **19.** Predict the oxidation number that the element with atomic number 16 is most likely to have.

20. Draw the electron-dot diagram for each of the following: *(a)* NO_3^{1-} *(b)* NH_4^{1+}

▲ **21.** Calculate the oxidation number for the element indicated in the following compounds or ions: *(a)* S in $(SO_4)^{2-}$ *(b)* Mn in MnO_2 *(c)* Mn in $(MnO_4)^{-1}$ *(d)* N in HNO_2

22. Use the symbol δ^+ to show a partial positive charge and the symbol δ^- to show a partial negative charge for each of the atoms in the following compounds: *(a)* H_2O *(b)* HI *(c)* Cl_2O *(d)* PCl_5

23. Write the electron-dot structure for each of the following compounds: *(a)* F_2 *(b)* C_2Cl_2 *(c)* PCl_5 *(d)* C_2Cl_4

▲ **24.** State the common oxidation number for each of the following elements when they enter into chemical combination. Base your answer on the element's position in the periodic table. *(a)* cesium *(b)* astatine *(c)* barium *(d)* gallium

25. Write the formulas of the compounds formed by the following ions:
(a) ferrous, Fe^{2+}, and phosphate, $(PO_4)^{3-}$
(b) mercuric, Hg^{2+}, and cyanide, $(CN)^{1-}$
(c) zinc, Zn^{2+}, and bicarbonate, $(HCO_3)^{1-}$
(d) ferric, Fe^{3+}, and dichromate, $(Cr_2O_7)^{2-}$

26. Write the formulas of the following compounds:
(a) indium sulfide
(b) magnesium arsenate
(c) gallium oxide
(d) rubidium selenide
(e) iron(II) oxide
(f) cobalt(II) chloride
(g) copper(I) nitride
(h) copper(I) oxide

27. Write the formulas of the following compounds:
(a) periodic acid *(b)* chloric acid
(c) hydrobromic acid *(d)* chlorous acid

▲ **28.** What is a coordinate covalent bond? Explain.

▲ **29.** Is it possible for nonpolar molecules to contain polar bonds? Explain.

▲ **30.** Element number 116 has not yet been discovered. But using what you know about periodic trends, answer the following questions:

(a) In what family (group) would this element be placed?
(b) How many electrons would be in its outermost energy level?
(c) What would be its most common oxidation number?

9 Calculations Involving Chemical Formulas

Learning Goals

After you've studied this chapter, you should be able to:

1. Calculate the number of moles in a sample of an element when you are given the mass of the sample.
2. Calculate the mass, in grams, of a sample of an element when you are given the number of moles.
3. Calculate the number of atoms of an element in a sample when you are given the mass of the sample.
4. Calculate the empirical formula of a compound when you are given its percentage composition.
5. Calculate the number of moles in a sample of a compound when you are given the mass of the sample.
6. Calculate the mass, in grams, of a sample of a compound when you are given the number of moles.
7. Calculate the number of formula units of a compound in a sample when you are given the mass of the sample.
8. Determine the molecular formula of a compound from its molecular mass and its empirical formula or percentage composition data.
9. Calculate the percentage composition by mass of a compound when you are given its chemical formula.

Introduction

We began our study of chemistry by reviewing the basic structure of matter. We learned about elements, compounds, atoms, and molecules. We discussed the basic structure of the atom and the concept of chemical bonding. We also learned how to name and write the formulas of chemical compounds. We did all of this in a qualitative fashion and didn't get very involved with quantitative calculations. In this chapter, we will revisit the topic of elements and compounds. However, this time we will discuss some quantitative aspects involving these substances.

Amedeo Avogadro
(Reprinted with permission from *Torchbearers of Chemistry*, by Henry Monmouth Smith, © Academic Press, Inc., 1949)

We will begin our study of calculations involving elements and compounds by learning about the *mole,* a useful concept for expressing the amount of a chemical substance. Before proceeding with this chapter, you may find it helpful to review the discussion of atomic mass and formula mass in Chapter 3. Also, because we will use the factor-unit method in our calculations, you may find it helpful to review the discussion of the factor-unit method in Chapter 2.

9.1 Gram-Atomic Mass and the Mole

The **gram-atomic mass** of an element is *its atomic mass expressed in grams.* For example, the atomic mass of gold is 197.0 amu, so gold's gram-atomic mass is 197.0 g. And the gram-atomic mass of carbon is 12.0 g.

One gram-atomic mass of any element contains the same number of atoms as one gram-atomic mass of any other element. For example, 197.0 g of gold contain the same number of atoms as 12.0 g of carbon and the same number of atoms as 1.0 g of hydrogen. In about 1870, scientists discovered that this number is 6.02×10^{23}. It is called **Avogadro's number,** in honor of the scientist whose thinking led to its discovery.

Later scientists took this idea one step further and defined 6.02×10^{23} atoms as **1 mole** of atoms. Thus 1 mole is equal to 6.02×10^{23}; it is simply a particular number of items. We could just as easily talk about 1 mole of people, 1 mole of apples, or 1 mole of dollars. (By the way, if we had 1 mole of dollar bills, we would have $602,000,000,000,000,000,000,000. If we distributed our money equally among the 5 billion people on the earth, each person would get more than 120 trillion dollars:

$$\frac{\$602{,}000{,}000{,}000{,}000{,}000{,}000{,}000}{5{,}000{,}000{,}000} = 1.20 \times 10^{14}$$

A person could spend a million dollars every day, each day of the year, and never run out of money for 330,000 years.)

Learning Goal 1
Moles from mass of an element

A mole of hydrogen atoms weighs 1.0 g; a mole of oxygen atoms weighs 16.0 g; a mole of carbon atoms weighs 12.0 g; and a mole of uranium atoms weighs 238.0 g. (To simplify the mathematics, we are rounding off the atomic masses found in the periodic table to one decimal place. We will do this throughout the text.) Figure 9.1 shows the mass of a mole of some other kinds of atoms.

The mole is very important in chemistry. It gives the chemist a convenient way to describe a large number of atoms or molecules and to relate this number of atoms or molecules to a mass (usually in grams). In fact, the mole has replaced the gram-atomic mass, which is seldom used today. But you do have to be able to convert from moles of atoms of an element to mass in grams, and vice versa. This is like converting inches to feet and feet to inches.

Figure 9.1
One mole of atoms of some common substances

1 mole of silver (Ag) atoms has a mass of 107.9 g.

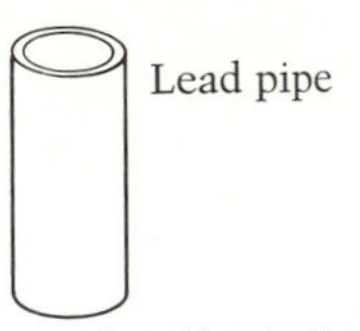

1 mole of lead (Pb) atoms has a mass of 207.2 g.

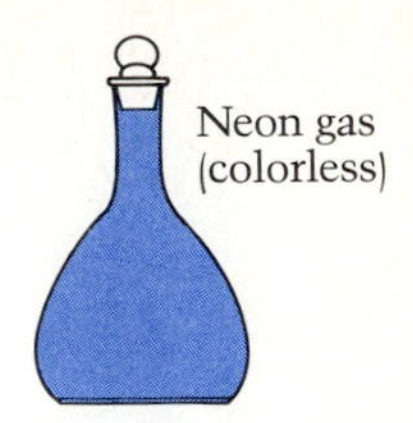

1 mole of neon (Ne) atoms has a mass of 20.2 g.

Example 9.1

Do the following conversions: *(a)* $6\bar{0}$ in. = ? ft *(b)* 8.0 ft = ? in.

Solution

(a) There are 12 in. in 1 ft, so

$$? \text{ ft} = 6\bar{0} \text{ in.} \times \frac{1 \text{ ft}}{12 \text{ in.}} = 5.0 \text{ ft}$$

(b) In the same way,

$$? \text{ in.} = 8.0 \text{ ft} \times \frac{12 \text{ in.}}{1 \text{ ft}} = 96 \text{ in.}$$

Practice Exercise 9.1 Do the following conversions: *(a)* 72.0 in. = ? ft *(b)* 10.0 ft = ? in.

Example 9.2

(a) How many moles of oxygen atoms are there in $8\bar{0}$ g of oxygen?
(b) What is the mass in grams of 0.50 mole of oxygen atoms?

Solution

Understand the Problem

(a) In this example, a specific mass of oxygen atoms is given. This mass must be converted into moles.

Devise a Plan

We use the factor-unit method as described in Chapter 2. (For convenience, we will round all atomic masses to one decimal place.)

Carry Out the Plan

$$\text{Moles of O atoms} = \text{g} \times \frac{\text{moles of O atoms}}{\text{g}}$$

Then, because the atomic mass of oxygen is 16.0 amu,

$$? \text{ moles of O atoms} = 8\bar{0}\ \cancel{\text{g}} \times \frac{1 \text{ mole of O atoms}}{16.0\ \cancel{\text{g}}}$$

$$= 5.0 \text{ moles of O atoms}$$

We find that $8\bar{0}$ g of oxygen atoms equals 5.0 moles.

Look Back

We reason that if 1 mole of oxygen atoms contains 16.0 g, then 5 moles should have five times as many grams. Since 16.0 × 5.0 equals $8\bar{0}$, the answer does make sense.

Understand the Problem

(b) In this case, the number of moles of oxygen is given and it must be converted into a mass (in grams).

Devise a Plan

We use the factor-unit method.

Carry Out the Plan

Using the atomic mass of oxygen, we write

$$\text{g} = \cancel{\text{moles of O atoms}} \times \frac{\text{g}}{\cancel{\text{moles of O atoms}}}$$

$$? \text{ g} = 0.50\ \cancel{\text{mole of O atoms}} \times \frac{16.0 \text{ g}}{1\ \cancel{\text{mole of O atoms}}}$$

$$= 8.0 \text{ g}$$

Look Back

We know that 1 mole of oxygen has a mass of 16.0 g; therefore 0.50 mole of oxygen should have half that mass, or 8.0 g. We conclude that our answer is sensible.

Practice Exercise 9.2 *(a)* How many moles of oxygen atoms are there in $16\bar{0}$ g of oxygen? *(b)* What is the mass in grams of 2.00 moles of oxygen atoms?

Example 9.3

Find the number of moles of atoms in each of the following samples: *(a)* 46 g of Na *(b)* 5.4 g of Al *(c)* 0.12 g of C *(d)* 23.8 g of U

Solution We need to look up the atomic mass of each element in the periodic table inside the front cover.

(a) The atomic mass of Na is 23.0 amu. Therefore

$$? \text{ moles of Na atoms} = 46 \not{g} \times \frac{1 \text{ mole of Na atoms}}{23.0 \not{g}}$$

$$= 2.0 \text{ moles of Na atoms}$$

(b) The atomic mass of Al is 27.0 amu. Therefore

$$? \text{ moles of Al atoms} = 5.4 \not{g} \times \frac{1 \text{ mole of Al atoms}}{27.0 \not{g}}$$

$$= 0.20 \text{ mole of Al atoms}$$

(c) The atomic mass of C is 12.0 amu. Therefore

$$? \text{ moles of C atoms} = 0.12 \not{g} \times \frac{1 \text{ mole of C atoms}}{12.0 \not{g}}$$

$$= 0.010 \text{ mole of C atoms}$$

(d) The atomic mass of U is 238.0 amu. Therefore

$$? \text{ moles of U atoms} = 23.8 \not{g} \times \frac{1 \text{ mole of U atoms}}{238.0 \not{g}}$$

$$= 0.100 \text{ mole of U atoms}$$

Practice Exercise 9.3 Find the number of moles of atoms in each of the following samples: *(a)* 11.50 g of Na *(b)* 2.70 g of Al *(c)* $28\bar{0}$ g of N *(d)* 0.0238 g of U

Example 9.4

Find the mass in grams of each of the following: *(a)* 0.20 mole of Zn atoms *(b)* 4.0 moles of Br atoms *(c)* 1.50 moles of Ca atoms

Solution We need to look up the atomic mass of each element in the periodic table.

(a) The atomic mass of Zn is 65.4 amu. Therefore

$$? \text{ g} = 0.20 \cancel{\text{mole of Zn atoms}} \times \frac{65.4 \text{ g}}{1 \cancel{\text{mole of Zn atoms}}} = 13.1 \text{ g}$$

Learning Goal 2
Mass, in grams, from moles of an element

(b) The atomic mass of Br is 79.9 amu. Therefore

$$? \text{ g} = 4.0 \cancel{\text{moles of Br atoms}} \times \frac{79.9 \text{ g}}{1 \cancel{\text{mole of Br atoms}}} = 320 \text{ g}$$

(c) The atomic mass of Ca is 40.1 amu. Therefore

$$? \text{ g} = 1.50 \cancel{\text{moles of Ca atoms}} \times \frac{40.1 \text{ g}}{1 \cancel{\text{mole of Ca atoms}}} = 60.2 \text{ g}$$

Practice Exercise 9.4 Find the mass in grams of each of the following: *(a)* 15.0 moles of Zn atoms *(b)* 0.200 mole of Br atoms *(c)* 3.00 moles of Ca atoms

Example 9.5

How many *atoms* of each element are there in the samples given in Example 9.4?

Learning Goal 3
Number of atoms of an element from mass of a sample

Solution We know that 1 mole of atoms of any element is 6.02×10^{23} atoms of that element; in other words, there are

$$\frac{6.02 \times 10^{23} \text{ atoms}}{1 \text{ mole of atoms}}$$

for any element. But for this example, we will round off Avogadro's number to 6.0×10^{23}.

(a) For the zinc,

$$? \text{ Zn atoms} = 0.20 \cancel{\text{ mole of Zn atoms}} \times \frac{6.0 \times 10^{23} \text{ Zn atoms}}{1 \cancel{\text{ mole of Zn atoms}}}$$

$$= 1.2 \times 10^{23} \text{ Zn atoms}$$

(b) For the bromine,

$$? \text{ Br atoms} = 4.0 \cancel{\text{ moles of Br atoms}} \times \frac{6.0 \times 10^{23} \text{ Br atoms}}{1 \cancel{\text{ mole of Br atoms}}}$$

$$= 24 \times 10^{23} \text{ Br atoms (or } 2.4 \times 10^{24})$$

(c) For the calcium,

$$? \text{ Ca atoms} = 1.5 \cancel{\text{ moles of Ca atoms}} \times \frac{6.0 \times 10^{23} \text{ Ca atoms}}{1 \cancel{\text{ mole of Ca atoms}}}$$

$$= 9.0 \times 10^{23} \text{ Ca atoms}$$

Practice Exercise 9.5 How many *atoms* of each element are there in the samples given in Practice Exercise 9.4?

9.2 Empirical Formulas

Learning Goal 4
Empirical formula from percentage composition

Chemists prepare hundreds of new compounds in their search for substances that may be beneficial in medicine, agriculture, industry, and the home. A first step in determining the nature of such a new or unknown compound is to obtain its empirical formula. The **empirical formula** of a compound is *the simplest whole-number ratio of the atoms that make up a formula unit of the compound*. For example, the empirical formula of water is H_2O. The subscripts indicate that the ratio of hydrogen atoms to oxygen atoms in this molecule is two to one, often written 2:1. (The lack of a subscript on the O is taken to mean a subscript of 1.)

To find the empirical formula of a compound, the chemist measures the percentage by mass of each element in the compound. From the percentage of each element, the chemist then determines (1) the number of moles of atoms of each element in 100 g of the compound and (2) the ratio of the moles of atoms. (We take 100 g of compound merely as a convenience.)

Example 9.6

Determine the empirical formula of a compound whose composition is 50.05% S and 49.95% O by mass.

Solution If we had 100 g of the compound, 50.05 g would be sulfur and 49.95 g would be oxygen. All we have to do now is convert these masses to moles of atoms and then find their whole-number ratio.

$$? \text{ moles of S atoms} = 50.05 \cancel{\text{g}} \times \frac{1 \text{ mole of S atoms}}{32.1 \cancel{\text{g}}}$$

$$= 1.56 \text{ moles of S atoms}$$

$$? \text{ moles of O atoms} = 49.95 \cancel{\text{g}} \times \frac{1 \text{ mole of O atoms}}{16.0 \cancel{\text{g}}}$$

$$= 3.12 \text{ moles of O atoms}$$

Therefore the formula may be written as $S_{1.56}O_{3.12}$, but this formula does not have whole-number subscripts. One way to obtain a formula with whole-number subscripts is to *divide* all the subscripts by the *smallest* subscript. This gives us

$$S_{1.56/1.56}O_{3.12/1.56} \quad \text{or} \quad SO_2$$

This is an empirical formula, because it has the lowest possible ratio of whole-number subscripts. (The compound is called sulfur dioxide.)

Practice Exercise 9.6 Determine the empirical formula of a compound whose composition is 88.9% oxygen and 11.1% hydrogen by mass.

Example 9.7

Find the empirical formula of a compound whose composition is 3.1% H, 31.5% P, and 65.4% O by mass.

Solution In 100 g of this compound, there are 3.1 g of H, 31.5 g of P, and 65.4 g of O. Therefore there are

$$? \text{ moles of H atoms} = 3.1 \not{g} \times \frac{1 \text{ mole of H atoms}}{1.0 \not{g}}$$
$$= 3.1 \text{ moles of H atoms}$$

$$? \text{ moles of P atoms} = 31.5 \not{g} \times \frac{1 \text{ mole of P atoms}}{31.0 \not{g}}$$
$$= 1.02 \text{ moles of P atoms}$$

$$? \text{ moles of O atoms} = 65.4 \not{g} \times \frac{1 \text{ mole of O atoms}}{16.0 \not{g}}$$
$$= 4.09 \text{ moles of O atoms}$$

This formula may be written $H_{3.1}P_{1.02}O_{4.09}$, but this formula does not have whole-number subscripts. Therefore we divide each subscript by the smallest subscript to obtain

$$H_{3.1/1.02}P_{1.02/1.02}O_{4.09/1.02} \quad \text{or} \quad H_3PO_4$$

Practice Exercise 9.7 Find the empirical formula of a compound whose composition is 51.9% Cr and 48.1% S by mass.

Example 9.8

Determine the empirical formula of a compound whose composition is 23.8% C, 5.9% H, and 70.3% Cl by mass.

Solution In 100 g of this compound, there are 23.8 g of C, 5.9 g of H, and 70.3 g of Cl. Therefore there are

$$? \text{ moles of C atoms} = 23.8 \not{g} \times \frac{1 \text{ mole of C atoms}}{12.0 \not{g}}$$
$$= 1.98 \text{ moles of C atoms}$$

$$? \text{ moles of H atoms} = 5.9 \not{g} \times \frac{1 \text{ mole of H atoms}}{1.0 \not{g}}$$
$$= 5.9 \text{ moles of H atoms}$$

$$? \text{ moles of Cl atoms} = 70.3 \not{g} \times \frac{1 \text{ mole of Cl atoms}}{35.5 \not{g}}$$
$$= 1.98 \text{ moles of Cl atoms}$$

One possible formula is $C_{1.98}H_{5.9}Cl_{1.98}$, but this does not have whole-number subscripts. Again we divide by the smallest subscript to obtain

$$C_{1.98/1.98}H_{5.9/1.98}Cl_{1.98/1.98} \quad \text{or} \quad CH_3Cl$$

Practice Exercise 9.8 Determine the empirical formula of a compound whose composition is 41.5% Zn, 17.8% N, and 40.7% O by mass.

After you divide the subscripts in determining an empirical formula, you may round subscripts to the nearest whole number if they are within 0.1 of the whole number. Otherwise you must search for a factor that will give you whole numbers. For example, suppose a compound contains 68.4% Cr and 31.6% O by mass. This means that 100 g of the compound contains 68.4 g (or 1.32 moles) of Cr and 31.6 g (or 1.98 moles) of O. Therefore the formula is

$$Cr_{1.32}O_{1.98}$$

Following our usual procedure, we divide by the smallest number of moles (the smallest subscript) and obtain

$$Cr_{1.32/1.32}O_{1.98/1.32} \quad \text{or} \quad CrO_{1.5}$$

The formula $CrO_{1.5}$ does not have *whole*-number subscripts. What do we do now? Should we simply round off the 1.5 to the number 2 and make the empirical formula CrO_2? The answer is NO! We may round off only when the number is within 0.1 of the whole number. Therefore we must search for a factor that will give us a whole-number ratio. When we look carefully at $Cr_1O_{1.5}$, we see that if we multiply each subscript by 2, we can get whole-number subscripts. $Cr_1O_{1.5}$ becomes Cr_2O_3, and the empirical formula is Cr_2O_3.

9.3 Gram-Formula Mass and the Mole

Learning Goal 5
Moles from mass of a compound

We're now going to learn how to apply the concept of the *mole* to compounds. Because some compounds are composed of molecules and other compounds are composed of ions, we'll use the term *formula mass* to help us with this application. Recall from Chapter 3 that the formula mass of a compound is *the sum of the atomic masses of all the atoms or ions that make up a formula unit of the compound*. The formula mass of a compound in grams is called the **gram-formula mass.*** This is the mass of a collection of 6.02×10^{23} formula units of the compound. Therefore *1 gram-formula mass of a compound is the mass in grams of 1 mole of molecules for a molecular compound or 1 mole of formula units for an ionic compound.*

Consider some examples. Water (H_2O) is a compound composed of molecules, each containing two hydrogen atoms and one oxygen atom. The

*The term **molar mass** is used in some texts as a general term to describe gram-formula mass and gram-atomic mass of a substance. In this text, we will continue to use the more traditional terms, *gram-formula mass* and *gram-atomic mass*.

Figure 9.2
One mole of molecules of some common substances

1 mole of water (H_2O) molecules has a mass of 18.0 g.

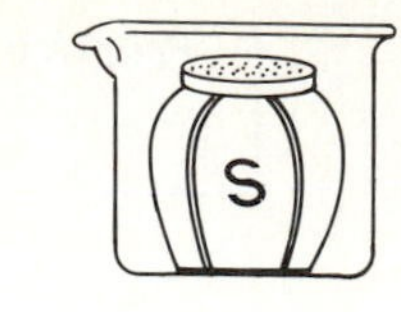

1 mole of glucose sugar ($C_6H_{12}O_6$) molecules has a mass of 180.0 g.

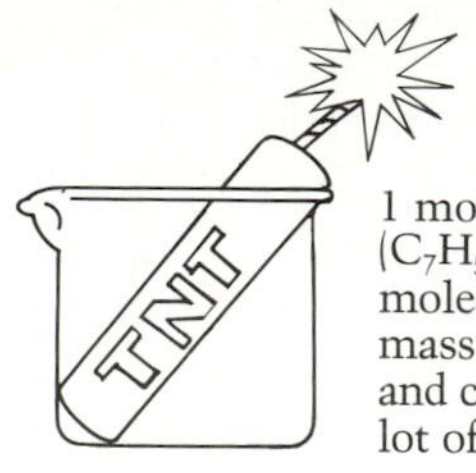

1 mole of TNT ($C_7H_5N_3O_6$) molecules has a mass of 227.0 g and can make a lot of noise.

1 *Scalopus aquaticus* alias 1 mole—has varying mass and can make very little noise.

formula mass of water is 18.0 amu—two times the atomic mass of hydrogen (2 × 1.0) plus the atomic mass of oxygen (16.0). So the mass of 1 mole of water molecules is 18.0 g (Figure 9.2). Calcium chloride ($CaCl_2$) is a compound composed of ions; the formula unit of the compound consists of one calcium ion and two chloride ions. The formula mass of calcium chloride is 111.1 amu—the atomic mass of calcium (40.1) plus two times the atomic mass of chlorine (2 × 35.5). So the mass of 1 mole of calcium chloride ions is 111.1 g.

It is important to be able to convert the number of grams of a compound into the number of moles, and vice versa. You will need to perform these calculations when we study the topic of chemical stoichiometry in Chapter 11. The following examples will give you some practice.

Example 9.9

Find the number of moles in each of the following: *(a)* 32 g of CH_4 *(b)* 0.32 g of CH_4 *(c)* 81 g of H_2O *(d)* 37 g of $Ca(OH)_2$

Solution We first find the formula mass of each compound and then use the factor-units method to determine the number of moles.

(a) The formula mass of CH_4 is 16.0 amu. Therefore we have

$$? \text{ moles of } CH_4 \text{ molecules} = 32 \not{g} \times \frac{1 \text{ mole of } CH_4 \text{ molecules}}{16.0 \not{g}}$$

$$= 2.0 \text{ moles of } CH_4 \text{ molecules}$$

(b) The formula mass of CH_4 is 16.0 amu. Therefore we have

$$? \text{ moles of } CH_4 \text{ molecules} = 0.32 \not{g} \times \frac{1 \text{ mole of } CH_4 \text{ molecules}}{16.0 \not{g}}$$

$$= 0.020 \text{ mole of } CH_4 \text{ molecules}$$

(c) The formula mass of H_2O is 18.0 amu. Therefore

$$? \text{ moles of } H_2O \text{ molecules} = 81 \cancel{g} \times \frac{1 \text{ mole of } H_2O \text{ molecules}}{18.0 \cancel{g}}$$

$$= 4.5 \text{ moles of } H_2O \text{ molecules}$$

(d) The formula mass of $Ca(OH)_2$ is 74.1 amu. Therefore

$$? \text{ moles of } Ca(OH)_2 \text{ formula units} = 37 \cancel{g} \times \frac{1 \text{ mole of } Ca(OH)_2 \text{ formula units}}{74.1 \cancel{g}}$$

$$= 0.50 \text{ mole of } Ca(OH)_2 \text{ formula units}$$

Practice Exercise 9.9 Find the number of moles in each of the following: *(a)* $9\bar{0}$ grams of H_2O *(b)* 0.016 gram CH_4

Learning Goal 6
Mass, in grams, from moles of a compound

Example 9.10

Determine the number of grams in each of the following: *(a)* 6.00 moles of butane (C_4H_{10}) *(b)* 0.025 mole of CO_2 *(c)* 7.00 moles of Al_2O_3 *(d)* 0.400 mole of Cu_3N

Solution We first find the formula mass of each compound and then use the factor-unit method to determine the number of grams.

(a) The formula mass of C_4H_{10} is 58.0 amu. Therefore we have

$$? \text{ g } C_4H_{10} = 6.00 \cancel{\text{moles}} \times \frac{58.0 \text{ g}}{1 \cancel{\text{mole}}} = 348 \text{ g}$$

(b) The formula mass of CO_2 is 44.0 amu. Therefore we have

$$? \text{ g } CO_2 = 0.025 \cancel{\text{mole}} \times \frac{44.0 \text{ g}}{1 \cancel{\text{mole}}} = 1.1 \text{ g}$$

(c) The formula mass of Al_2O_3 is 102.0 amu. Therefore we have

$$? \text{ g } Al_2O_3 = 7.00 \cancel{\text{moles}} \times \frac{102.0 \text{ g}}{1 \cancel{\text{mole}}} = 714 \text{ g}$$

(d) The formula mass of Cu_3N is 204.5 amu. Therefore we have

$$? \text{ g } Cu_3N = 0.400 \cancel{\text{mole}} \times \frac{204.5 \text{ g}}{1 \text{ mole}} = 81.8 \text{ g}$$

Practice Exercise 9.10 Determine the number of grams in each of the following: *(a)* 4.00 moles of NO_2 *(b)* 0.050 mole of $CaCO_3$ *(c)* 2.50 moles of $Ca(C_2H_3O_2)_2$ *(d)* 0.060 mole of $(NH_4)_2SO_4$

Example 9.11

In Example 9.9, how many formula units are present in each sample?

Solution To obtain our factor unit, we recall that there are 6.0×10^{23} formula units in 1 mole of a compound.

(a) For the first sample of CH_4, composed of molecules,

Learning Goal 7
Number of formula units of a compound from mass of a sample

$$? \text{ molecules of } CH_4 = 2.0 \cancel{\text{ moles}} \times \frac{6.0 \times 10^{23} \text{ molecules of } CH_4}{1 \cancel{\text{ mole}}}$$

$$= 12 \times 10^{23} \text{ molecules of } CH_4$$

(b) For the second sample of CH_4, composed of molecules,

$$? \text{ molecules of } CH_4 = 0.020 \cancel{\text{ mole}} \times \frac{6.0 \times 10^{23} \text{ molecules of } CH_4}{1 \cancel{\text{ mole}}}$$

$$= 1.2 \times 10^{22} \text{ molecules of } CH_4$$

(c) For the H_2O, composed of molecules,

$$? \text{ molecules of } H_2O = 4.5 \cancel{\text{ moles}} \times \frac{6.0 \times 10^{23} \text{ molecules of } H_2O}{1 \cancel{\text{ mole}}}$$

$$= 27 \times 10^{23} \text{ molecules of } H_2O$$

(d) For the $Ca(OH)_2$, composed of ions,

? formula units of $Ca(OH)_2$

$$= 0.50 \cancel{\text{ mole}} \times \frac{6.0 \times 10^{23} \text{ formula units of } Ca(OH)_2}{1 \cancel{\text{ mole}}}$$

$$= 3.0 \times 10^{23} \text{ formula units of } Ca(OH)_2$$

Practice Exercise 9.11 In Practice Exercise 9.10, how many *formula units* are present in each sample?

9.4 Molecular Formulas

Learning Goal 8
Molecular formula of a compound

The **molecular formula** of a compound is *a formula that shows the actual number of atoms of each element that are in one molecule of that compound.* The molecular formula is found from the percentages by mass of the elements *and* the molecular mass of the compound.

Example 9.12

A compound is composed of $4\bar{0}\%$ C, 6.6% H, and 53.4% O by mass. The molecular mass of the compound is 180.0. Determine its empirical and molecular formulas.

Solution Find the empirical formula first, by the usual method.

$$? \text{ moles of C atoms} = 4\bar{0}\ \text{g} \times \frac{1 \text{ mole of C atoms}}{12.0 \text{ g}}$$

$$= 3.3 \text{ moles of C atoms}$$

$$? \text{ moles of H atoms} = 6.6\ \text{g} \times \frac{1 \text{ mole of H atoms}}{1.0 \text{ g}}$$

$$= 6.6 \text{ moles of H atoms}$$

$$? \text{ moles of O atoms} = 53.4\ \text{g} \times \frac{1 \text{ mole of O atoms}}{16.0 \text{ g}}$$

$$= 3.3 \text{ moles of O atoms}$$

Hence one possible formula is $C_{3.3}H_{6.6}O_{3.3}$, but this does not have whole-number subscripts. Therefore we divide by the smallest number of moles to get

$$C_{3.3/3.3}H_{6.6/3.3}O_{3.3/3.3} \quad \text{or} \quad CH_2O$$

This is the empirical formula. But it certainly is not the molecular formula of the compound, because CH_2O does not have a molecular mass of 180.0. What is the molecular formula of the compound? It could be any formula wherein the ratio of C:H:O is 1:2:1. Is it $C_2H_4O_2$, $C_3H_6O_3$, $C_4H_8O_4$, $C_5H_{10}O_5$, $C_6H_{12}O_6$, or what? Remember, the correct formula is the one that has a molecular mass of 180.0. Let's determine the molecular mass of each proposed formula.

Molecular Formula	Molecular Mass (amu)
CH_2O	30.0
$C_2H_4O_2$	60.0
$C_3H_6O_3$	90.0
$C_4H_8O_4$	120.0
$C_5H_{10}O_5$	150.0
$C_6H_{12}O_6$	180.0

It is clear from our trial-and-error method that the molecular formula is $C_6H_{12}O_6$. It's the only one that has a molecular mass of 180.0. But is there an easier way to do this? Yes. We simply ask ourselves, "How many times must we take the molecular mass of the empirical formula (30.0 in this problem) to get the true molecular mass (180.0 in this problem)?" The answer is obviously *six* times. It takes six times the mass of the empirical formula, CH_2O, to give the true molecular mass, so the true molecular formula must be $C_6H_{12}O_6$.

Practice Exercise 9.12 A compound has a molecular mass of 180.0 and is composed of 60.0% C, 4.48% H, and 35.5% O by mass. Calculate the empirical and molecular formulas of this common compound, which is called aspirin.

9.5 Percentage Composition by Mass

Learning Goal 9
Percentage composition from chemical formula

You have seen how to determine the empirical formula of a compound from its percentage composition. There will also be times when you will have to do the opposite—find the percentage by mass of each element in a compound. To do so, you need to know the empirical or molecular formula of the compound and the atomic mass of its constituent elements.

Example 9.13

Find the percentage by mass of H and O in water.

Solution In each mole of water (H_2O), there are 2 moles of H, having a mass of 1.0 g/mole, and 1 mole of O, having a mass of 16.0 g/mole. Therefore

$$\text{Percent H} = \frac{\text{mass of hydrogen in 1 mole of water}}{\text{mass of 1 mole of water}} \times 100$$

$$= \frac{2.0 \cancel{\text{g}} \text{ of H}}{18.0 \cancel{\text{g}} \text{ of } H_2O} \times 100 = 11$$

$$\text{Percent O} = \frac{\text{mass of oxygen in 1 mole of water}}{\text{mass of 1 mole of water}} \times 100$$

$$= \frac{16.0 \cancel{\text{g}} \text{ of O}}{18.0 \cancel{\text{g}} \text{ of } H_2O} \times 100 = 89$$

Note that the sum of the percentages must be 100.

Practice Exercise 9.13 Find the percentage by mass of K and Br in potassium bromide (KBr).

Example 9.14

Find the percentage by mass of each element in the following compounds: *(a)* C_8H_{18} *(b)* NH_3 *(c)* FeO *(d)* Fe_2O_3

Solution We must find the formula mass of each compound and the percent contribution of each element to that mass.

(a) C_8H_{18} contains 96.0 g of C + 18.0 g of H = 114.0 g in each mole. Therefore

$$\text{Percent C} = \frac{96.0 \cancel{\text{g}} \text{ of C}}{114.0 \cancel{\text{g}} \text{ of } C_8H_{18}} \times 100 = 84.2$$

$$\text{Percent H} = \frac{18.0 \cancel{\text{g}} \text{ of H}}{114.0 \cancel{\text{g}} \text{ of } C_8H_{18}} \times 100 = 15.8$$

(b) NH_3 contains 14.0 g of N + 3.0 g of H = 17.0 g in each mole. Therefore

$$\text{Percent N} = \frac{14.0 \text{ g of N}}{17.0 \text{ g of } NH_3} \times 100 = 82.4$$

$$\text{Percent H} = \frac{3.0 \text{ g of H}}{17.0 \text{ g of } NH_3} \times 100 = 17.6 \text{ (round to 18)}$$

(c) FeO contains 55.8 g of Fe + 16.0 g of O = 71.8 g in each mole. Therefore

$$\text{Percent Fe} = \frac{55.8 \text{ g of Fe}}{71.8 \text{ g of FeO}} \times 100 = 77.7$$

$$\text{Percent O} = \frac{16.0 \text{ g of O}}{71.8 \text{ g of FeO}} \times 100 = 22.3$$

(d) Fe_2O_3 contains 111.6 g of Fe + 48.0 g of O = 159.6 g in each mole. Therefore

$$\text{Percent Fe} = \frac{111.6 \text{ g of Fe}}{159.6 \text{ g of } Fe_2O_3} \times 100 = 69.9$$

$$\text{Percent O} = \frac{48.0 \text{ g of O}}{159.6 \text{ g of } Fe_2O_3} \times 100 = 30.1$$

Practice Exercise 9.14 Find the percentage by mass of each element in the following compounds: *(a)* $C_{22}H_{44}$ *(b)* NaCl *(c)* C_6H_6

Example 9.15

How many grams of sulfur are there in 256 g of SO_3?

Solution Determine the percentage of sulfur in SO_3 and multiply this by the amount of the sample (256 g). SO_3 contains 32.1 g of S + 48.0 g of O = 80.1 g/mole. Therefore

$$\text{Percent S} = \frac{32.1 \text{ g of S}}{80.1 \text{ g of } SO_3} \times 100 = 40.1$$

So 40.1% of the 256 g of SO_3 is sulfur (S); there are 0.401 × 256 g = 102.7 g of sulfur (or 103 g of sulfur to three significant figures).

Practice Exercise 9.15 How many grams of carbon are there in $\overline{100}$ g of CH_4?

Summary

Each element has its own symbol, and every compound has its own chemical formula. Each element has a unique atomic mass. One mole of anything is equal to 6.02×10^{23} things (Avogadro's number of things): 1 mole of an element contains 6.02×10^{23} atoms of that element, and 1 mole of a compound contains 6.02×10^{23} formula units of that compound.

One mole of atoms of an element has a mass equal to the atomic mass of that element in grams. One mole of formula units of a compound has a mass equal to the formula mass of that compound in grams.

The empirical formula of a compound is the simplest whole-number ratio of the atoms that make up a formula unit of the compound. The molecular formula of a compound is the actual number of atoms of each element that are in one molecule (or formula unit) of the compound. The empirical formula of a compound can be found from percentage composition data. The molecular formula of a compound can be found from its empirical formula and its molecular mass. The percentage composition of a compound may be determined from its empirical or molecular formula.

Key Terms

Avogadro's number **(9.1)**
empirical formula **(9.2)**
formula mass **(9.3)**
gram-atomic mass **(9.1)**
gram-formula mass **(9.3)**
molar mass **(9.3)**
mole **(9.1)**
molecular formula **(9.4)**

Chemical Frontiers

The Use of Epoetin Alfa in the Management of Chronic Renal Failure

Many patients in the end stage of renal (kidney) disease experience severe anemia, a shortage of oxygen-carrying red blood cells. This condition is caused by the failure of the kidneys to produce adequate amounts of a hormone called *erythropoietin* (EPO), which regulates erythropoiesis—red blood cell production. In fact, many of the symptoms that were previously ascribed to uncleared toxins associated with kidney failure are now known to be a result of severe anemia.

Erythropoiesis is dependent on the presence of proper amounts of EPO. The hormone, composed of about 40% carbohydrate and 60% protein, maintains the body's red blood cell mass at the optimum level. EPO is synthesized primarily in the liver of the fetus and in the kidneys of the adult. The rate of red blood cell production is regulated by the concentration of EPO circulating in the bloodstream.

Under normal conditions, the existing number of red blood cells supplies proper amounts of oxygen to the tissues. In this case, the concentration of EPO circulating in the bloodstream is low. When the bloodstream lacks enough oxygen, EPO levels rise up to several hundred times above normal, thus stimulating the production of red blood cells. In contrast, when blood oxygen levels are too high, EPO synthesis decreases. When renal function is impaired, the kidneys are not able to supply adequate levels of EPO, which results in severe anemia.

Recent advances in biotechnology, molecular biology, genetics, immunology, and nucleic acid chemistry have resulted in the production of artificially synthesized human erythropoietin. Recently, the use of epoetin alfa was approved by the Food and Drug Administration for the treatment of anemia associated with chronic renal failure. Epoetin alfa stimulates bone marrow to produce mature red blood cells. This increases the red blood cell level and ends the anemia associated with chronic renal failure in 97% of patients. It also eliminates the dependence on transfusions to replace the hormone, which was the previous method of treatment. By ending the dependence on transfusions, the risk of transmitting infectious disease is also reduced.

Self-Test Exercises

Learning Goal 1: Moles from Mass of an Element

1. How many atoms are there in 1.00 mole of atoms of any element?

2. Twelve items equal one dozen items, but a mole represents a much larger quantity. This quantity is ________ items.

▲ **3.** A sample of the element osmium (Os) has a mass of 400.0 g. How many moles of osmium atoms is this?

4. How many moles are there in 5.62 g of silicon?

▲***5.** The atomic department store is selling uranium atoms at the reduced price of 100,000,000 atoms for 1 cent. What would be the cost in dollars of 1 mg of uranium?

▲ **6.** A sample of sulfur has a mass of 0.963 g. How many moles of sulfur atoms are there in this sample?

7. Determine the number of *moles* of atoms in each of the following samples:
(a) 16.2 g of Al
(b) 239.5 g of Ti
(c) 0.06075 g of Mg
(d) 3,570.0 g of U

▲ **8.** Determine the number of *moles* of atoms in each of the following samples:
(a) 2397.0 g of Cu
(b) 13.79 g of W
(c) 0.00399 g of Ar
(d) 16.19 g of Ag

Learning Goal 2: Mass, in Grams, from Moles of an Element

▲ **9.** Dysprosium (Dy), the sixty-sixth element in the periodic table, has an atomic mass of 162.5. What mass of Dy do you need to obtain 0.5000 mole of Dy?

10. What is the mass in grams of 1.50 moles of rubidium (Rb) atoms?

11. What is the mass in grams of 0.750 mole of cobalt (Co) atoms?

▲ **12.** How many grams of Cr are contained in 0.0250 mole of Cr?

13. Determine the number of grams in each of the following:
(*a*) 4.600 moles of Ni (*c*) 200.0 moles of Ca
(*b*) 0.00300 mole of Br_2 (*d*) 0.04000 mole of S

14. Determine the number of grams in each of the following:
(*a*) 0.400 mole of Rn (*c*) 0.0350 mole of Hg
(*b*) 19.0 moles of F_2 (*d*) 7.200 moles of Yb

Learning Goal 3: Number of Atoms of an Element from Mass of a Sample

▲ **15.** After you've finished an experiment, you find that 69.0 g of sodium (Na) are left in the reaction vessel. How many atoms of sodium are left?

▲ **16.** At the end of an experiment, you find that there are 13.8 g of sodium left in the reaction vessel. How many atoms of sodium are left?

***17.** Diamond is made up of carbon atoms. A 2-carat diamond costs $1,800. What is the cost of 1 carbon atom in the diamond? (Assume that 1 carat weighs 0.2 g.)

***18.** How many atoms are present in a sample of water molecules that has a mass of $1{,}8\overline{0}0$ g?

19. The radius of a silver atom is 1.34 angstroms, or 1.34 Å (1 Å = 10^{-8} cm). If the atoms of 0.1 mole of silver atoms were placed in a straight line, touching each other, what distance (in miles) would they cover?

***20.** How many *atoms* of Cl_2 are present in a sample of Cl_2 that has a mass of $35\overline{0}$ g?

Learning Goal 4: Empirical Formula from Percentage Composition

21. Sodium, the eleventh element in the periodic table (whose name means "headache" in medieval Latin), is a very dangerous chemical when uncombined. But when 39.6% sodium is combined with 60.4% chlorine by mass, a common compound is formed. What is the empirical formula of this compound, and what is its common name?

22. When 15.8% Al is combined with 28.2% S and 56.1% O by mass, a compound is formed. Determine the empirical formula of this compound.

▲ **23.** Cholesterol, a compound suspected of causing hardening of the arteries, has a molecular mass of 386 and the following percentage composition by mass: 84.0% C, 11.9% H, and 4.1% O. What is the molecular formula of cholesterol?

24. When 24.2% Ca is combined with 17.1% N and 58.5% O by mass, a common compound is formed. What are its empirical formula and common name?

25. Determine the empirical formula of a compound that is 80.0% C and 20.0% H by mass. The molecular mass of this compound is 30.0. What is its molecular formula?

▲ **26.** When 60.56% C is combined with 11.18% H and 28.26% N by mass, the compound spherophysine, which is used for treating high blood pressure, is formed. What is the empirical formula of this compound?

27. Determine the empirical formula of a compound that is 54.55% C, 9.09% H, and 36.36% O by mass. The molecular mass of this compound is 88. What is its molecular formula?

28. When 64.81% C is combined with 13.60% H and 21.59% O by mass, a compound that is used as a surgical anesthetic is formed. The name of this compound is diethyl ether. What is its empirical formula?

Learning Goal 5: Moles from Mass of a Compound

29. How many moles of aluminum sulfate are there in 684.0 g of $Al_2(SO_4)_3$?

30. A sample of $CaSO_4$ has a mass of 0.1705 g. How many moles of $CaSO_4$ are there in this sample?

31. How many moles of HCl are there in 100.0 g of this compound?

32. How many moles of O_2 molecules are there in 6.40 g of oxygen gas?

▲ **33.** A sample of laughing gas (N_2O) has a mass of 2.20 g. How many moles of N_2O is this?

▲ **34.** Milk of magnesia, $Mg(OH)_2$, is used as an antacid in small doses and as a laxative in large doses. The typical laxative dose for an adult is about 2.00 g. How many moles of $Mg(OH)_2$ is this?

Learning Goal 6: Mass, in Grams, from Moles of a Compound

35. Determine the number of grams in each of the following:
(*a*) 0.600 mole of H_2O
(*b*) 20.0 moles of SO_2
(*c*) 0.00050 mole of CO_2
(*d*) 0.00600 mole of $Al_2(SO_4)$

36. What is the mass in grams of 10.0 moles of H_2O?

37. How many grams of SO_2 are contained in 0.20 mole of SO_2?
38. How many grams of H_2SO_4 are there in 4.00 moles of this compound?

▲ 39. Portable gas grills use propane (C_3H_8) as a fuel. A cylinder of propane contains 51.6 moles of C_3H_8. How many grams of propane are in the cylinder? How many pounds of propane are in the cylinder?
▲ 40. Baking soda ($NaHCO_3$) has many uses around the house. A box of baking soda contains 5.40 moles of $NaHCO_3$. How many grams is this? How many pounds is this?

Learning Goal 7: Number of Formula Units of a Compound from Mass of a Sample

41. How many *molecules* of CO_2 are there in 198.0 g of CO_2?
42. How many *molecules* of SO_2 are there in 6.40 g of SO_2?

43. How many formula units of NaCl are there in 14.6 g of NaCl?
▲ 44. How many molecules of C_3H_8 are contained in the tank of propane described in Exercise 39?

▲ 45. How many formula units of $NaHCO_3$ are contained in the box of baking soda described in Exercise 40?
46. How many formula units of magnesium nitride are contained in 20.2 g of Mg_3N_2?

Learning Goal 8: Molecular Formula of a Compound

47. According to Nobel Prize winner Linus Pauling, vitamin C, also known as ascorbic acid, is a cure and preventative for colds. The empirical formula of vitamin C is $C_3H_4O_3$, and its molecular mass is 176.0. What is its molecular formula?
48. A compound is found to have the empirical formula HF and a molecular mass of 40.0. What is its molecular formula?

49. The empirical formula of a very important compound, called benzene, is CH. Its molecular mass is 78.0. What is its molecular formula?
50. The compound fumaric acid has the empirical formula CHO. Its molecular mass is 116.0. What is its molecular formula?

51. The plastic melamine, used to make Melmac® dishes, has the empirical formula CH_2N_2. Its molecular mass is 126.0. What is its molecular formula?
52. The compound benzidine, used in the manufacture of dyes, has the empirical formula C_6H_6N. Its molecular mass is 184.0. What is its molecular formula?

▲ 53. Dextrose, a type of sugar, has the empirical formula CH_2O. Acetic acid, which is found in vinegar, shares the same empirical formula as dextrose. The molecular mass of dextrose is 180.0, whereas the molecular mass of acetic acid is 60.0. Write the molecular formula of each compound.
54. A compound of iron and sulfur with a formula mass of 208.0 has an empirical formula of Fe_2S_3. What is its molecular formula?

Learning Goal 9: Percentage Composition from Chemical Formula

▲ 55. Find the percentages of S, H, and O by mass in sulfuric acid (H_2SO_4).
56. Find the percentages by mass of the elements in each of the following compounds: *(a)* C_2H_6 *(b)* NO_2 *(c)* CO_2

57. Determine the percentages by mass of the elements in the following compounds: *(a)* BF_3 *(b)* UF_6 *(c)* $C_3H_8O_3$
58. Determine the percentage by mass of oxygen in each of the following compounds: *(a)* H_2O *(b)* CO *(c)* H_2SO_4

*59. Some sodium chloride (NaCl) and some sugar ($C_{12}H_{22}O_{11}$) are mixed together. The mixture is analyzed and found to contain $2\bar{0}$% chlorine by mass. The mixture has a mass of $5\bar{0}$ g. How much sugar is present?
60. How many moles of nitrogen atoms can be obtained from $1{,}7\bar{0}0$ grams of NH_3?

61. Determine the percentages by mass of the elements in the following compounds: *(a)* SO_2 *(b)* CH_4 *(c)* $C_{12}H_{22}O_{11}$ *(d)* $CaCO_3$
▲ 62. How many grams of hydrogen can be obtained from $18\bar{0}$ g of water?

63. How many grams of copper can be obtained from $1{,}6\bar{0}0$ g of CuO?
64. How many grams of oxygen can be obtained from 320.0 g of SO_2?

65. How many grams of iron can be obtained from 350.0 g of Fe_2O_3?
66. Determine the percentage by mass of sulfur in H_2S.

67. How many grams of oxygen combine with $\overline{200}$ g of sulfur in the compound SO_2? (*Hint:* Do Exercise 61 *(a)* to obtain the percentages by mass of S and O in SO_2.)

68. Determine the percentage by mass of hydrogen in CH_3OH.

***69.** An unknown mixture contains sodium sulfide (Na_2S) and iron(III) oxide (Fe_2O_3). The mixture is analyzed and found to contain 25.0% sulfur by mass. The mixture has a mass of $\overline{500}$ g. How many grams of each compound are present?

70. Determine the percentage by mass of N in HNO_3.

Extra Exercises

71. If you had Avogadro's number of dollars and spent 1 dollar every second of every day, how long would it take you to run out of money?

72. What is the simplest formula for the compound that has the following percentage composition by mass: 24.3% C, 4.1% H, and 71.6% Cl?

73. If the molecular mass of the compound in Exercise 72 is determined to be 99.0, what is its molecular formula?

74. What is the mass of an Fe_3O_4 sample that contains $\overline{100}$ g of oxygen?

▲ **75.** How many carbon atoms are there in a 1.75-carat diamond? (*Hint:* 1.00 carat = $\overline{200}$ mg.)

76. Two ores of iron are hematite (Fe_2O_3) and magnetite (Fe_3O_4). Calculate the percentage by mass of iron in each ore to determine the richer source of iron.

▲ **77.** Two ores of copper have the following mass percentage analyses. What is the empirical formula of each ore?

Ore A	Ore B
Cu, 63.3%	Cu, 34.6%
Fe, 11.1%	Fe, 30.4%
S, 25.6%	S, 34.9%

78. The percentages by mass of oxygen and sulfur are almost the same in the compound SO_2. Calculate the percentage by mass of each.

79. What is the mass in grams of 10 atoms of silver?

80. How many moles of phosphorus are there in 392 g of H_3PO_4?

81. In Exercise 80, how many grams of phosphorus are there?

82. How much is 1 atomic mass unit in grams?

83. Ethyl alcohol, which is found is some popular beverages, has the molecular formula C_2H_6O. You have 9.2 g of ethyl alcohol.
(a) How many moles of ethyl alcohol do you have?
(b) How many molecules of ethyl alcohol?
(c) How many moles of carbon atoms?
(d) How many moles of hydrogen atoms?
(e) How many moles of oxygen atoms?
(f) How many atoms of carbon?
(g) How many atoms of hydrogen?
(h) How many atoms of oxygen?

84. The formula for methanol, also known as wood alcohol, is CH_3OH. You have 3.2 grams of methanol.
(a) How many moles of methanol do you have?
(b) How many molecules of methanol?
(c) How many moles of carbon atoms?
(d) How many moles of hydrogen atoms?
(e) How many moles of oxygen atoms?
(f) How many atoms of carbon?
(g) How many atoms of hydrogen?
(h) How many atoms of oxygen?

85. Determine the number of *moles* of molecules or formula units in each of the following samples:
(a) 7.20 g of H_2O
(b) 260.7 g of MnO_2
(c) 0.070 g of N_2
(d) 1,980.0 g of $(NH_4)_2SO_4$

86. Determine the number of moles of molecules in each of the following: *(a)* 0.34 g of NH_3
(b) 1,335.0 g of MnO_2

▲ **87.** Glycine is the simplest amino acid. It has the molecular formula $C_2H_5O_2N$. You have 300.0 g of glycine.
(a) How many moles of glycine do you have?
(b) How many molecules of glycine?
(c) How many moles of carbon atoms?
(d) How many moles of hydrogen atoms?
(e) How many moles of oxygen atoms?
(f) How many moles of nitrogen atoms?
(g) How many grams of carbon?
(h) How many grams of hydrogen?
(i) How many grams of oxygen?
(j) How many grams of nitrogen?
(k) What should be true about the sum of the numbers of grams of each element in *(g)*, *(h)*, *(i)*, and *(j)* and the number of grams of glycine we started with?

▲ **88.** Determine the number of moles of molecules in each of the following: *(a)* $\overline{720}$ g of H_2O
(b) 13.35 g of MnO_2

10 The Chemical Equation: Recipe for a Reaction

Learning Goals

After you've studied this chapter, you should be able to:

1. Write formula equations from word equations, and vice versa.
2. Balance formula equations.
3. Recognize and give examples of some general types of chemical reactions (combination, decomposition, single-replacement, and double-replacement.)
4. Predict the products for the various types of reactions.
5. Define the terms *acid, base,* and *salt,* and recognize each from its formula.
6. Use the activity series and solubility table to predict reactions.
7. Define the terms *oxidation, reduction, oxidizing agent,* and *reducing agent*.
8. Determine which substance has been oxidized and which substance has been reduced in a redox reaction.
9. Determine which substance is the reducing agent and which substance is the oxidizing agent in a redox reaction.

Introduction

Combine 4 cups of sifted flour and 1 cake of yeast soaked in $1\frac{1}{3}$ cups of 35°C water. Add 2 tablespoons of olive oil and 1 teaspoon of salt. Roll out the dough, add tomato sauce and mozzarella cheese, and bake. When these ingredients are baked, has a chemical reaction occurred? Yes. What's more, this recipe is a set of directions that tells you how to mix certain *reactants* to yield a delectable *product,* a pizza.

In a sense, we can think of a chemical reaction as a set of directions. When we follow these directions and mix the proper chemicals, we get the products that we want. In this chapter, you will study some important classes of chemical reactions. You will also learn how to write chemical equations and how to interpret the information they provide.

10.1 Word Equations

Learning Goal 1
Formula and word equations

One way to represent a chemical reaction is with words. For example, we can say

$$\text{Sulfur trioxide} + \text{water} \longrightarrow \text{sulfuric acid}$$

In a chemical equation such as this one, the reacting substances, or **reactants,** are shown on the left-hand side. The **products** are shown on the right-hand side. The plus sign means *and,* and the arrow means *yields* or *gives*. Our word equation, then, means that "sulfur trioxide and water react to yield sulfuric acid." (By the way, this equation represents part of the process by which "acid rain" develops from one by-product of fuel combustion.)

Our word equation tells us what happens when sulfur trioxide combines with water. But it tells us nothing about the chemical formulas of the reactants and products. Nor does it tell us anything about *how* the reactants com-

Sulfur oxides are formed in the atmosphere when pollutants containing the element sulfur react with oxygen in the air. When these sulfur oxides react with moisture in the air, acids are formed. The final result is acid rain. (Steve Delaney, EPA)

bine to form the products. That's why chemists prefer to use formula notation to write chemical reactions. And, like other people, chemists prefer to use the simpler shorthand notation whenever possible.

10.2 The Formula Equation

Let us substitute the formulas for the reactants and product in our word equation:

$$\text{Sulfur trioxide} + \text{water} \longrightarrow \text{sulfuric acid}$$
$$SO_3 + H_2O \longrightarrow H_2SO_4$$

Learning Goal 2
Balancing equations

Now let's do some "atomic bookkeeping" to see whether we have complied with the Law of Conservation of Mass. On the left-hand side of this equation are one sulfur atom, four oxygen atoms, and two hydrogen atoms. And there are exactly the same numbers of these atoms on the right-hand side. So we have not "created" or "destroyed" atoms in writing the formula equation (see Figure 10.1). The equation is *balanced.* A chemical equation is **balanced** when *the total number of atoms of each element on the left-hand side is exactly equal to the total number of atoms of each element on the right-hand side.*

Let's now look at the chemical reaction in which hydrogen and oxygen form water. Because hydrogen and oxygen are both diatomic elements, we must write the equation for this chemical reaction as follows:

$$H_2 + O_2 \xrightarrow[\text{spark}]{\text{Electric}} H_2O$$

(Many reactions occur only under special conditions. We specify these conditions by writing them above or around the arrow.) Now for the bookkeeping: There are two hydrogens on the left side and two hydrogens on the right side of the equation. However, there are two oxygens on the left side but only *one* oxygen on the right side. This equation is *not balanced,* because the number of atoms of each element on the left *does not* equal the number of atoms of each element on the right.

We can't let the equation stand this way. Every chemical equation must be balanced, because chemical reactions satisfy the Law of Conservation of Mass. As it stands, our reaction falsely implies that an atom of oxygen was lost or destroyed.

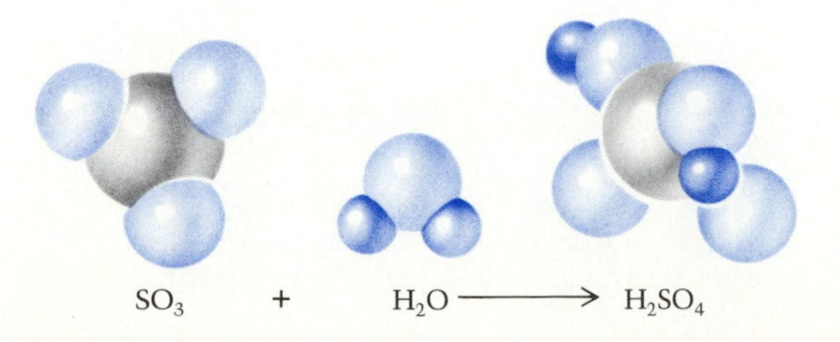

Figure 10.1
Models show how sulfur trioxide and water produce sulfuric acid.

10.3 Balancing a Chemical Equation

To balance a chemical equation, we determine how many molecules of each substance are needed to satisfy the Law of Conservation of Mass. We indicate this by placing *coefficients,* or small whole numbers, before the formulas in the equation. These coefficients are usually obtained by trial and error. For example, in our equation for the formation of water, we have

$$H_2 + O_2 \xrightarrow[\text{spark}]{\text{Electric}} H_2O \qquad \text{(not balanced)}$$

Because this unbalanced equation lacks one oxygen atom on the right, we place the coefficient 2 before the formula for water on the right:

$$H_2 + O_2 \xrightarrow[\text{spark}]{\text{Electric}} 2H_2O \qquad \text{(not balanced)}$$

Now we have two oxygen atoms on the left and two on the right. But the coefficient 2 has unbalanced the hydrogen atoms. There are now four on the right but only two on the left. To remedy this, we use two molecules of H_2:

$$2H_2 + O_2 \xrightarrow[\text{spark}]{\text{Electric}} 2H_2O \qquad \text{(balanced)}$$

Now the equation is balanced: there are four hydrogen atoms and two oxygen atoms on the left and the same numbers on the right. The balanced equation states that two hydrogen molecules and one oxygen molecule react to form two water molecules (Figure 10.2).

There are two rules to keep in mind when you balance a chemical equation:

Rule 1 The equation must be accurate. Correct formulas for all reactants and products must be shown.

Rule 2 The Law of Conservation of Mass must be observed. The same number of atoms of each element must appear on both sides of the balanced equation.

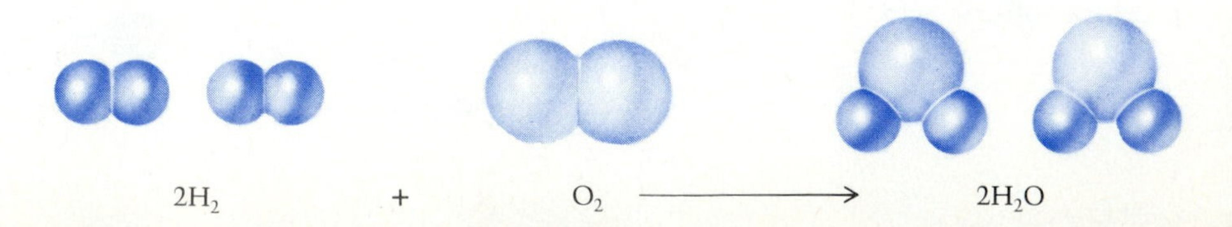

Figure 10.2 The formation of water from hydrogen and oxygen

Example 10.1

Write a balanced chemical equation for the following reaction:

$$\text{Mercury(II) oxide} \xrightarrow{\text{Heat}} \text{mercury metal} + \text{oxygen gas}$$

Solution We begin by writing the formulas for the reactants and products. Remember that oxygen gas is a diatomic element, so we must write O_2.

$$HgO \xrightarrow{\text{Heat}} Hg + O_2 \qquad \text{(unbalanced)}$$

The mercury is balanced, but there is only one oxygen atom on the left and two oxygens on the right. We balance the oxygens by placing the coefficient 2 in front of HgO:

$$2HgO \xrightarrow{\text{Heat}} Hg + O_2 \qquad \text{(unbalanced)}$$

Now the oxygens are balanced, but there are two mercury atoms on the left and only one on the right. We balance the mercury atoms by placing a 2 in front of the Hg:

$$2HgO \xrightarrow{\text{Heat}} 2Hg + O_2 \qquad \text{(balanced)}$$

The balanced chemical equation shows two mercury atoms and two oxygen atoms on each side of the equation.

Practice Exercise 10.1 Write a balanced chemical equation for the following reaction:

$$\text{Copper(II) oxide} \longrightarrow \text{copper metal} + \text{oxygen gas}$$

Here are two important points to remember.

1. You *cannot* balance an equation using *subscripts,* because this would change the formulas of the compounds. For example,

$$HgO \xrightarrow{\text{Heat}} Hg + O_2$$

cannot be balanced by doing this:

$$HgO_2 \xrightarrow{\text{Heat}} Hg + O_2$$

because HgO_2 is not the formula for mercury(II) oxide.

2. You *cannot* insert a coefficient between elements in a chemical formula. For example,

$$HgO \xrightarrow{\text{Heat}} Hg + O_2$$

cannot be balanced by doing this:

$$Hg2O \xrightarrow{\text{Heat}} Hg + O_2$$

because such an expression has no meaning.

After you have balanced a number of equations, you will find it easy to choose the proper coefficients. Here are some suggestions that may help:

1. Balance metal and nonmetal atoms first.
2. Balance hydrogen and oxygen atoms last.
3. Balance complex ions that contain more than one atom *as a group* if they appear unchanged on both sides of the equation.

Example 10.2

Write a balanced chemical equation for the following reaction:

Aluminum + sulfuric acid $\longrightarrow$ aluminum sulfate + hydrogen gas

Solution We first write the correct formula for each compound:

$$Al + H_2SO_4 \longrightarrow Al_2(SO_4)_3 + H_2 \quad \text{(unbalanced)}$$

There is one aluminum atom on the left-hand side but two aluminum atoms on the right-hand side. We balance the aluminums by placing the coefficient 2 in front of the Al on the left-hand side:

$$2Al + H_2SO_4 \longrightarrow Al_2(SO_4)_3 + H_2 \quad \text{(unbalanced)}$$

There is one SO_4 group on the left-hand side and three SO_4 groups on the right-hand side. We balance the SO_4 groups by placing a 3 in front of the H_2SO_4:

$$2Al + 3H_2SO_4 \longrightarrow Al_2(SO_4)_3 + H_2 \quad \text{(unbalanced)}$$

Now there are six hydrogens on the left-hand side and two hydrogens on the right-hand side. We balance the hydrogens by placing a 3 in front of the H_2:

$$2Al + 3H_2SO_4 \longrightarrow Al_2(SO_4)_3 + 3H_2 \quad \text{(balanced)}$$

There are now two aluminum atoms, three sulfate groups, and six hydrogen atoms on each side of the equation. And so we have a balanced chemical equation.

Practice Exercise 10.2 Write a balanced chemical equation for the following reaction:

Zinc + sulfuric acid $\longrightarrow$ zinc sulfate + hydrogen gas

Example 10.3

Write a balanced equation for each of the following reactions:

(a) Hydrogen gas + nitrogen gas $\xrightarrow[\text{Pressure}]{\text{Heat}}$ ammonia(NH_3)

(b) Sodium oxide + water $\longrightarrow$ sodium hydroxide

(c) Calcium bromide + chlorine $\longrightarrow$ calcium chloride + bromine

(*d*) Sodium hydroxide + phosphoric acid (H_3PO_4) $\longrightarrow$ sodium phosphate + water
(*e*) Iron(III) oxide + carbon monoxide $\xrightarrow{\text{Heat}}$ iron + carbon dioxide

Solution We use the same procedure as in the previous examples: Write the formula for each substance, and then balance the equation.

(a) $H_2 + N_2 \xrightarrow[\text{Pressure}]{\text{Heat}} NH_3$ (unbalanced)

$3H_2 + N_2 \xrightarrow[\text{Pressure}]{\text{Heat}} 2NH_3$ (balanced)

(b) $Na_2O + H_2O \longrightarrow NaOH$ (unbalanced)
$Na_2O + H_2O \longrightarrow 2NaOH$ (balanced)
(c) $CaBr_2 + Cl_2 \longrightarrow CaCl_2 + Br_2$ (balanced)
(d) $NaOH + H_3PO_4 \longrightarrow Na_3PO_4 + H_2O$ (unbalanced)
$3NaOH + H_3PO_4 \longrightarrow Na_3PO_4 + 3H_2O$ (balanced)

(e) $Fe_2O_3 + CO \xrightarrow{\text{Heat}} Fe + CO_2$ (unbalanced)

$Fe_2O_3 + 3CO \xrightarrow{\text{Heat}} 2Fe + 3CO_2$ (balanced)

Practice Exercise 10.3 Write a balanced equation for each of the following word equations:
(*a*) Aluminum + chlorine gas $\longrightarrow$ aluminum chloride
(*b*) Sodium + water $\longrightarrow$ sodium hydroxide + hydrogen gas
(*c*) Potassium nitrate $\longrightarrow$ potassium nitrite + oxygen gas
(*d*) Nitric acid + barium hydroxide $\longrightarrow$ barium nitrate + water

10.4 Types of Chemical Reactions

The millions of known chemical reactions may be classified in many ways. We will use a classification scheme that is helpful in predicting the products of many chemical processes. In this scheme, there are four basic classes:

Learning Goal 3
Recognizing types of chemical reactions

1. Combination reactions
2. Decomposition reactions
3. Single-replacement reactions
4. Double-replacement (or double-exchange) reactions

All of these chemical reactions, with the exception of double-replacement reactions, may also be classified as redox reactions. (*Redox* is short for "oxidation–reduction.") We have more to say about these reactions later in the chapter.

Combination Reactions

Combination reactions are *reactions in which two or more substances combine to form a more complex substance* (Figure 10.3). The general formula is

$$A + B \longrightarrow AB$$

Figure 10.3
A combination reaction: iron + sulfur $\longrightarrow$ iron(II) sulfide

In some combination reactions, two *elements* join to form a compound. For example,

$$\underset{\text{Iron}}{4Fe} + \underset{\text{Oxygen gas}}{3O_2} \longrightarrow \underset{\text{Iron(III) oxide}}{2Fe_2O_3}$$

Iron(III) oxide is iron rust. When a product made of iron contacts moist air, the iron rusts, forming a crust of iron(III) oxide. The crust flakes off, a new surface of iron becomes visible, and it too begins to rust. The iron can keep rusting until the entire piece of metal is eaten away.

Another example of a combination reaction that joins two elements is

$$\underset{\text{Carbon}}{C} + \underset{\text{Sulfur}}{2S} \longrightarrow \underset{\text{Carbon disulfide}}{CS_2}$$

Carbon disulfide is an important chemical used in manufacturing rayon and in electronic vacuum tubes. However, it is a highly poisonous substance and should be used only in well-ventilated areas.

In some combination reactions, two *compounds* join to form a more complex compound. For example,

$$\underset{\text{Sulfur trioxide}}{SO_3} + \underset{\text{Water}}{H_2O} \longrightarrow \underset{\text{Sulfuric acid}}{H_2SO_4}$$

$$\underset{\text{Sodium oxide}}{Na_2O} + \underset{\text{Water}}{H_2O} \longrightarrow \underset{\text{Sodium hydroxide}}{2NaOH}$$

Learning Goal 4
Predicting products of chemical reactions

Example 10.4

Complete and balance the following equations for combination reactions:

(a) $H_2 + Cl_2 \longrightarrow ?$
(b) $K + Br_2 \longrightarrow ?$
(c) $Ba + O_2 \longrightarrow ?$

Solution We are told that these are combination reactions, so we know that the products are formed by combination of the elements involved. We first write the correct formula for the product. Then we balance the equation.

(a) $H_2 + Cl_2 \longrightarrow HCl$ (unbalanced)
$H_2 + Cl_2 \longrightarrow 2HCl$ (balanced)

(b) $K + Br_2 \longrightarrow KBr$ (unbalanced)
$2K + Br_2 \longrightarrow 2KBr$ (balanced)

(c) $Ba + O_2 \longrightarrow BaO$ (unbalanced)
$2Ba + O_2 \longrightarrow 2BaO$ (balanced)

Practice Exercise 10.4 Complete and balance the following equations for combination reactions:

(a) $K + O_2 \longrightarrow ?$
(b) $Ca + O_2 \longrightarrow ?$
(c) $H_2 + Br_2 \longrightarrow ?$

Decomposition Reactions

Decomposition reactions are *reactions in which a complex substance is broken down into simpler substances*. They are thus the reverse of combination reactions. The general formula for a decomposition reaction is

$$AB \longrightarrow A + B$$

Some examples of common decomposition reactions follow.

1. The decomposition of a metal oxide (Figure 10.4):

$$\underset{\text{Mercury(II) oxide}}{2HgO} \xrightarrow{\text{Heat}} \underset{\text{Mercury}}{2Hg} + \underset{\text{Oxygen gas}}{O_2}$$

2. The decomposition of a nonmetal oxide:

$$2H_2O \xrightarrow{\text{Electricity}} 2H_2 + O_2$$

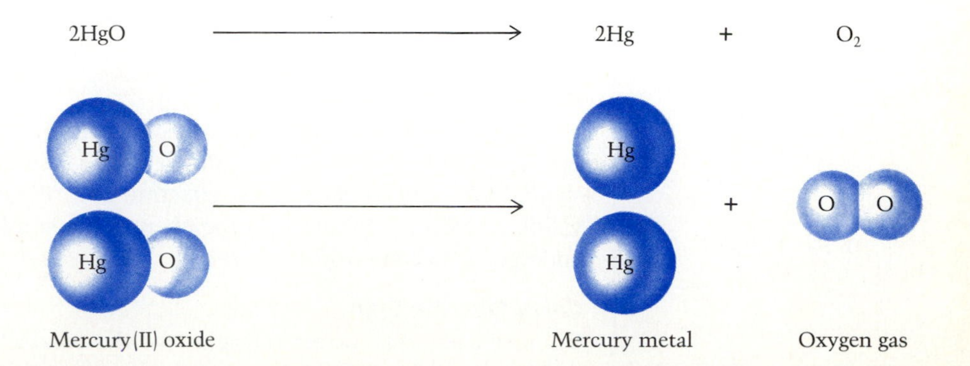

Figure 10.4 The decomposition of mercury(II) oxide molecules

3. The decomposition of a metal chlorate:

$$\underset{\text{Potassium chlorate}}{2KClO_3} \xrightarrow{\text{Heat}} \underset{\text{Potassium chloride}}{2KCl} + 3O_2$$

4. The decomposition of a metal carbonate:

$$\underset{\text{Magnesium carbonate}}{MgCO_3} \xrightarrow{\text{Heat}} \underset{\text{Magnesium oxide}}{MgO} + CO_2$$

5. The decomposition of a metal hydroxide:

$$\underset{\text{Calcium hydroxide}}{Ca(OH)_2} \xrightarrow{\text{Heat}} \underset{\text{Calcium oxide}}{CaO} + H_2O$$

Decomposition reactions are a very important class of reactions. They are the ones that produce pure samples of various elements. For example, very pure hydrogen and oxygen can be obtained from the decomposition of water. Pure copper can be produced by the decomposition of copper(II) oxide.

Example 10.5

Complete and balance the following equations for decomposition reactions:

(a) $Mg(OH)_2 \xrightarrow{\text{Heat}} ? + ?$
(b) $CaCO_3 \xrightarrow{\text{Heat}} ? + ?$
(c) $NaClO_3 \xrightarrow{\text{Heat}} ? + ?$
(d) $HCl \longrightarrow ? + ?$

Solution

Understand the Problem

We must predict the products and balance the equations for these decomposition reactions.

Devise a Plan

We are told that these are decomposition reactions, so we know that the products are formed by a breakdown of the elements in the original substance. We want to balance the equation according to the Law of Conservation of Mass. Look back in the chapter for similar kinds of compounds undergoing decomposition.

Carry Out the Plan

We first write the correct formulas for the products and then balance the equation.

Look Back

We look back and check to be sure we have written an accurate equation.

(a) $Mg(OH)_2 \xrightarrow{\text{Heat}} MgO + H_2O$ (balanced)

(b) $CaCO_3 \xrightarrow{\text{Heat}} CaO + CO_2$ (balanced)

(c) $NaClO_3 \xrightarrow{\text{Heat}} NaCl + O_2$ (unbalanced)

$2NaClO_3 \xrightarrow{\text{Heat}} 2NaCl + 3O_2$ (balanced)

(d) $HCl \longrightarrow H_2 + Cl_2$ (unbalanced)

$2HCl \longrightarrow H_2 + Cl_2$ (balanced)

Practice Exercise 10.5 Complete and balance the following equations for decomposition reactions:

(a) $Sr(OH)_2 \longrightarrow ? + ?$

(b) $SrCO_3 \longrightarrow ? + ?$

(c) $KClO_3 \longrightarrow ? + ?$

(d) $KCl \longrightarrow ? + ?$

Single-Replacement Reactions

A **single replacement reaction** is *one in which an uncombined element replaces another element that is in a compound* (Figure 10.5). The general formula for this type of reaction is

$$A + BC \longrightarrow AC + B$$

Examples of single-replacement reactions follow.

1. The replacement in aqueous solution of one metal by another metal that is more active (the activity of metals is discussed in Section 10.5):

$$Mg + Cu(NO_3)_2 \longrightarrow Mg(NO_3)_2 + Cu$$

2. The replacement of hydrogen in water by a Group IA metal:

$$2Na + 2H_2O \longrightarrow 2NaOH + H_2$$

3. The replacement of hydrogen in an acid by an active metal:

$$Zn + 2HCl \longrightarrow ZnCl_2 + H_2$$

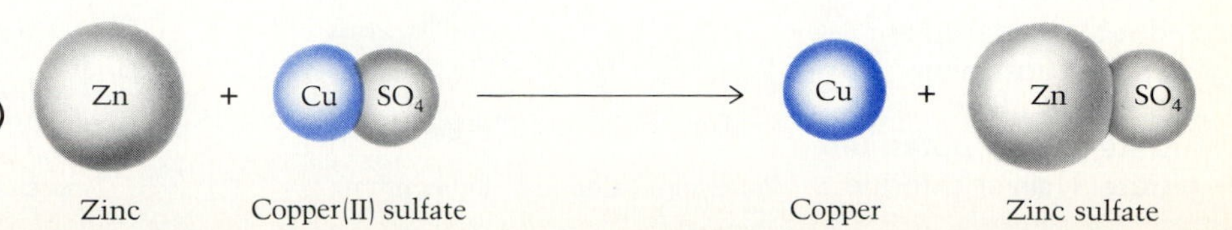

Figure 10.5
A single-replacement reaction: zinc + copper(II) sulfate $\longrightarrow$ copper + zinc sulfate

Example 10.6

Complete and balance the following equations for single-replacement reactions:

(a) $K + H_2O \longrightarrow ? + ?$
(b) $Mg + H_2SO_4 \longrightarrow ? + ?$
(c) $Zn + NiCl_2 \longrightarrow ? + ?$

Solution We first write the correct formulas for the products and then balance the equation.

(a) $K + H_2O \longrightarrow KOH + H_2$ (unbalanced)
$2K + 2H_2O \longrightarrow 2KOH + H_2$ (balanced)
(b) $Mg + H_2SO_4 \longrightarrow MgSO_4 + H_2$ (balanced)
(c) $Zn + NiCl_2 \longrightarrow ZnCl_2 + Ni$ (balanced)

Practice Exercise 10.6 Complete and balance the following equations for single-replacement reactions:

(a) $Li + H_2O \longrightarrow ? + ?$
(b) $Al + H_2SO_4 \longrightarrow ? + ?$
(c) $Mg + Al(NO_3)_3 \longrightarrow ? + ?$

Single-replacement reactions are also a very important class of reactions. Chemists often use single-replacement reactions to obtain various metals from compounds containing these metals. For example, you can obtain pure silver metal from the compound silver nitrate by causing a solution of silver nitrate to react with copper metal.

$$Cu + 2AgNO_3 \longrightarrow Cu(NO_3)_2 + 2Ag$$

Double-Replacement Reactions

A **double-replacement** (or *double-exchange*) **reaction** is *one in which two compounds exchange ions with each other* (Figure 10.6). The general formula is

$$A^+B^- + C^+D^- \longrightarrow A^+D^- + C^+B^-$$

Double-replacement reactions usually take place in water. To show that they do, we use the notation (*aq*), which stands for *aqueous,* or in water.

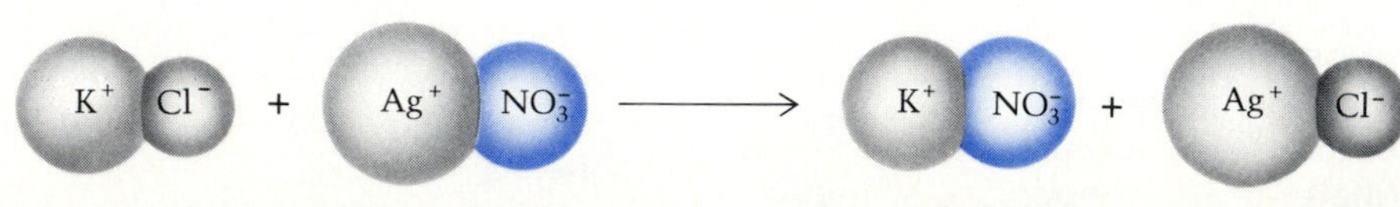

Figure 10.6
A double-replacement reaction: potassium chloride + silver nitrate $\longrightarrow$ potassium nitrate + silver chloride

Learning Goal 5
Acids, bases, and salts

There are two important kinds of double-replacement reactions. One kind takes place between two salts. A **salt** is a compound composed of a positive and a negative ion. The positive ion may be a monatomic ion of a metal (for example, Na^{1+} or Ca^{2+}) or a polyatomic ion such as $(NH_4)^{1+}$. The negative ion may be a monatomic ion of a nonmetal (for example, Cl^{1-}) or a polyatomic ion such as $(SO_4)^{2-}$ or $(NO_3)^{1-}$.

A double-replacement reaction between two salts can occur if one of the *products* separates from the solution—in other words, if it is insoluble in the solution (or if a gas is formed). When a product is insoluble, it separates out of the solution in solid form (called a **precipitate**). To show this, we use the notation (*s*), meaning *solid*. Table B.4 in Supplement B tells you whether a particular product is soluble in water.

When a salt dissolves in water it **dissociates,** or breaks up into its ions. For example, when we dissolve silver nitrate ($AgNO_3$) in water, there are Ag^+ and $(NO_3)^-$ ions in the solution. There are no $AgNO_3$ units as such in the water.

Now let's see what happens when we mix an aqueous solution of silver nitrate and an aqueous solution of potassium chloride. Both salts are soluble in water, so both solutions are clear. But when we mix the two solutions and allow them to react, we get a solid precipitate. This precipitate is silver chloride, which is *not* soluble in water. The reaction is

$$KCl(aq) + AgNO_3(aq) \longrightarrow KNO_3(aq) + AgCl(s)$$

The other product, potassium nitrate, remains in solution [as K^+ and $(NO_3)^-$ ions] because it *is* soluble in water.

The second important kind of double-replacement reaction takes place between an acid and a base. The simplest **acids** are *compounds that dissociate in water into hydrogen ions and negatively charged ions of a nonmetal* (*or negatively charged polyatomic ions*). Acids release hydrogen ions in water. The simplest **bases** are *composed of metallic ions* (*or positively charged polyatomic ions*) *combined with hydroxide ions*. Bases release hydroxide ions in water.

In an acid–base reaction, the acid and base react to form a salt and water. This type of reaction is also called a **neutralization reaction.** Some examples of neutralization reactions are

$$\underset{\text{Hydrochloric acid}}{HCl} + \underset{\text{Sodium hydroxide}}{NaOH} \longrightarrow NaCl + H_2O$$

$$\underset{\text{Sulfuric acid}}{H_2SO_4} + 2NaOH \longrightarrow Na_2SO_4 + 2H_2O$$

$$\underset{\text{Phosphoric acid}}{H_3PO_4} + \underset{\text{Lithium hydroxide}}{3LiOH} \longrightarrow Li_3PO_4 + 3H_2O$$

We make much use of neutralization reactions in our daily lives. For example, if you have an acid stomach (caused by excess hydrochloric acid), you take a product such as milk of magnesia to neutralize the excess acidity. Milk of magnesia contains a base—magnesium hydroxide, $Mg(OH)_2$.

Example 10.7

Complete and balance the following equations for double-replacement reactions:

(a) $BaCl_2(aq) + Na_2SO_4(aq) \longrightarrow ? + ?$
(b) $H_3PO_4 + Ca(OH)_2 \longrightarrow ? + ?$

Solution As usual, we first write the correct formula for each product and then balance the equation.

(a) $BaCl_2(aq) + Na_2SO_4(aq) \longrightarrow NaCl(aq) + BaSO_4(s)$ (unbalanced)

$BaCl_2(aq) + Na_2SO_4(aq) \longrightarrow 2NaCl(aq) + BaSO_4(s)$ (balanced)

(b) $H_3PO_4 + Ca(OH)_2 \longrightarrow Ca_3(PO_4)_2 + H_2O$ (unbalanced)

$2H_3PO_4 + 3Ca(OH)_2 \longrightarrow Ca_3(PO_4)_2 + 6H_2O$ (balanced)

Practice Exercise 10.7 Complete and balance the following equations for double-replacement reactions:

(a) $BiCl_3 + H_2S \longrightarrow ? + ?$
(b) $Fe(OH)_3 + H_2SO_4 \longrightarrow ? + ?$

10.5 The Activity Series

Learning Goal 6
Using the activity series and solubility table

The **activity series** (Table 10.1) is *a list of elements in decreasing order of their reactivity, or ability to react chemically in aqueous solution*. The most active elements (those that react most readily) appear at the top of the table, and reactivity decreases as we move down the list. Under ordinary conditions, each element in the table can replace any element that appears below it by taking part in a single-replacement reaction. For example, aluminum is higher in Table 10.1 than zinc, so it replaces zinc in the compound zinc nitrate:

$$2Al + 3Zn(NO_3)_2 \longrightarrow 2Al(NO_3)_3 + 3Zn$$

The activity series is very helpful in determining whether a particular reaction—especially a single-replacement reaction—can take place under ordinary conditions.

Example 10.8

Predict whether each of the following reactions can occur:

(a) $Cu + 2HCl \xrightarrow{\text{Heat}} CuCl_2 + H_2$

(b) $H_2 + CuO \xrightarrow{\text{Heat}} H_2O + Cu$

Solution

(a) Table 10.1 shows that copper is *below* hydrogen in the activity series. Therefore copper cannot replace hydrogen in the compound HCl, and

$$Cu + HCl \longrightarrow \text{no reaction}$$

(b) Hydrogen is above copper in the activity series, so it can replace copper in the compound CuO. This reaction can occur:

$$H_2 + CuO \longrightarrow H_2O + Cu$$

Practice Exercise 10.8 Predict whether each of the following reactions can occur:

(a) $Cu + H_2SO_4 \longrightarrow CuSO_4 + H_2$
(b) $H_2 + CuCl_2 \longrightarrow 2HCl + Cu$

Table 10.1
The Activity Series

Metals	Nonmetals
Lithium	Fluorine
Potassium	Chlorine
Calcium	Bromine
Sodium	Iodine
Magnesium	
Aluminum	
Zinc	
Chromium	
Iron	
Nickel	
Tin	
Lead	
HYDROGEN*	
Copper	
Mercury	
Silver	
Platinum	
Gold	

*Hydrogen is in capital letters because the activities of the other metals were calculated relative to the activity of hydrogen.

10.6 Oxidation–Reduction (Redox) Reactions

We've just finished classifying reactions according to one scheme—as combination, decomposition, single-replacement, or double-replacement reactions. Now we will introduce another scheme for classifying reactions—as either oxidation–reduction (redox, for short) reactions or nonredox reactions. Let's begin with some definitions.

Oxidation and Reduction

Learning Goal 7
Oxidation and reduction; oxidizing and reducing agents

In the past, chemists defined oxidation as the combining of a substance with oxygen. (That's how the process got its name.) Now, however, the concept of oxidation is expanded to include reactions that do not involve oxygen at all. **Oxidation** is *the loss of electrons by a substance undergoing a chemical reaction. (Alternatively, it is an increase in the oxidation number of a substance.)* (If you don't recall the meaning of *oxidation number,* refer to Section 8.3, and note that the oxidation number of an element in the uncombined state is zero.) As an example, look at the following equation, in which the oxidation number of the iron is shown over its symbol:

$$\overset{0}{4Fe} + 3O_2 \longrightarrow 2\overset{3+}{Fe_2}O_3$$

The oxidation number of the iron changes from 0 to +3. By our definition, this means that the iron is oxidized. But what happens to the oxygen? We say that the oxygen has been *reduced,* which leads us to our definition of reduction. **Reduction** is *the gain of electrons by a substance undergoing a chemical reaction. (Alternatively, it is a decrease in the oxidation number of a substance.)*

In the reaction between iron and oxygen, the oxidation number of the oxygen changes from 0 to −2, because each oxygen atom gains two electrons from the iron:

$$4Fe + 3\overset{0}{O_2} \longrightarrow 2Fe_2\overset{2-}{O_3}$$

Learning Goal 8
Determining oxidized and reduced substances

Note that in this reaction, one substance (iron) is oxidized, while the other substance (oxygen) is reduced. This is not a coincidence: the processes of oxidation and reduction always occur together. In other words, oxidation cannot take place in a reaction unless there is also reduction. And reduction can't take place in a reaction unless there is also oxidation. A reaction in which oxidation and reduction take place is called an **oxidation–reduction** (or **redox**) **reaction.** All other reactions are **nonredox reactions.**

Example 10.9

In each of the following reactions, determine which substance is oxidized and which substance is reduced:

(*a*) $2K + Br_2 \longrightarrow 2KBr$
(*b*) $S + O_2 \longrightarrow SO_2$
(*c*) $Mg + H_2SO_4 \longrightarrow MgSO_4 + H_2$
(*d*) $2H_2O \longrightarrow 2H_2 + O_2$

Solution In each case, we use the ideas of Section 8.3 (and the fact that the oxidation number of an element in its uncombined state is zero) to write in the oxidation numbers for the elements.

(a)
$$\overset{0}{2K} + \overset{0}{Br_2} \longrightarrow 2\,\overset{1+}{K}\,\overset{1-}{Br}$$

The oxidation number of the potassium changes from 0 to +1. This increase means that the potassium is oxidized. The oxidation number of the bromine changes from 0 to −1. This decrease means that the bromine is reduced.

(b)
$$\overset{0}{S} + \overset{0}{O_2} \longrightarrow \overset{4+}{S}\,\overset{2-}{O_2}$$

The oxidation number of the sulfur changes from 0 to +4, so the sulfur is oxidized. The oxidation number of the oxygen changes from 0 to −2, so the oxygen is reduced.

(c)
$$\overset{0}{Mg} + \overset{1+}{H_2}\overset{2-}{(SO_4)} \longrightarrow \overset{2+}{Mg}\overset{2-}{(SO_4)} + \overset{0}{H_2}$$

The oxidation number of the magnesium changes from 0 to +2, so the magnesium is oxidized. The oxidation number of the hydrogen changes from +1 to 0. This decrease means that the hydrogen is reduced. The oxidation number of the sulfate group does not change, so it is neither oxidized nor reduced.

(d)
$$\overset{1+}{2H_2}\,\overset{2-}{O} \longrightarrow \overset{0}{2H_2} + \overset{0}{O_2}$$

The oxidation number of the oxygen changes from −2 to 0, so the oxygen is oxidized. The oxidation number of the hydrogen changes from +1 to 0, so the hydrogen is reduced.

Practice Exercise 10.9 In each of the following reactions, determine which substance is oxidized and which substance is reduced:
(*a*) $Cu + 2AgNO_3 \longrightarrow 2Ag + Cu(NO_3)_2$
(*b*) $2Na + Cl_2 \longrightarrow 2NaCl$
(*c*) $Cl_2 + 2NaBr \longrightarrow 2NaCl + Br_2$
(*d*) $Zn + S \longrightarrow ZnS$

Oxidizing and Reducing Agents

Learning Goal 9
Determining oxidizing and reducing agents

An **oxidizing agent** is *a substance that causes something else to be oxidized.* A **reducing agent** is *a substance that causes something else to be reduced.* As things work out, the oxidizing agent is the substance being reduced, and the reducing agent is the substance being oxidized. Let's take another look at the reaction in which iron and oxygen form iron(III) oxide:

$$4Fe + 3O_2 \longrightarrow 2Fe_2O_3$$

The iron is the substance being oxidized, and the oxygen is the substance being reduced. Therefore the iron is the *reducing agent* and the oxygen is the *oxidizing agent*.

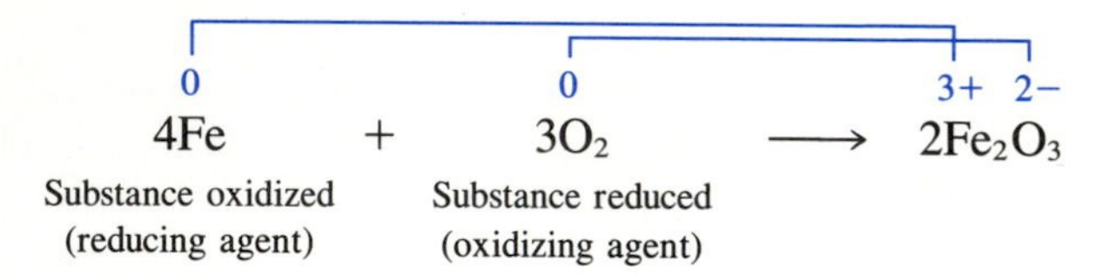

Example 10.10

For each reaction in Example 10.9, state which substance is the oxidizing agent and which substance is the reducing agent.

Solution Just remember that the substance being oxidized is the reducing agent and the substance being reduced is the oxidizing agent.

(a)

$$\overset{0}{2K} + \overset{0}{Br_2} \longrightarrow 2\overset{1+}{K}\overset{1-}{Br}$$

2K: Substance oxidized (reducing agent); Br_2: Substance reduced (oxidizing agent)

(b)

$$\overset{0}{S} + \overset{0}{O_2} \longrightarrow \overset{4+}{S}\overset{2-}{O_2}$$

S: Substance oxidized (reducing agent); O_2: Substance reduced (oxidizing agent)

(c)

$$\overset{0}{Mg} + \overset{1+}{H_2}\overset{2-}{(SO_4)} \longrightarrow \overset{2+}{Mg}\overset{2-}{(SO_4)} + \overset{0}{H_2}$$

Mg: Substance oxidized (reducing agent); $H_2(SO_4)$: Substance reduced (oxidizing agent)

Actually, in this case, the whole compound, H_2SO_4, is considered the oxidizing agent.

(d)

$$2\overset{1+}{H_2}\overset{2-}{O} \longrightarrow \overset{0}{2H_2} + \overset{0}{O_2}$$

The hydrogen is reduced and the oxygen is oxidized. The compound water acts as both oxidizing agent and reducing agent.

Practice Exercise 10.10 For each reaction in Practice Exercise 10.9, state which substance is the oxidizing agent and which substance is the reducing agent.

Summary

A chemical equation is a representation of a chemical reaction. A word equation indicates (in words) which reactants combine (or decompose) and which products are formed as a result of the reaction. In a formula equation, the words are replaced with chemical formulas. In addition, the formula equation must be balanced, so that the number of atoms of each element on the left side of the equation equals the number of atoms of each element on the right side. We balance an equation by placing the proper coefficients before elements or compounds in the equation. Coefficients are generally chosen by trial and error.

Chemical reactions may be classified as combination reactions, decomposition reactions, single-replacement reactions, or double-replacement reactions. Combination reactions are those in which two or more substances combine to form a more complex substance. Decomposition reactions (the reverse of combination reactions) involve the breakdown of a complex substance into simpler substances. A single-replacement reaction is one in which an uncombined element replaces another element in a compound. A double-replacement reaction is one in which two compounds exchange ions with each other.

To predict whether or not a reaction will occur under ordinary conditions, it is important to know the relative activities of the elements involved in the reaction. The activity series is a list of elements in decreasing order of activity or reactivity. It can be used to predict whether or not a single-replacement reaction will occur. That is, each element in the list can replace any element below it in a single-replacement reaction.

Another way of classifying chemical reactions is by dividing them into oxidation–reduction (redox) reactions and non–oxidation–reduction (nonredox) reactions. Oxidation is the loss of electrons by a substance undergoing a chemical reaction (in other words, an increase in the oxidation number of the substance). Reduction is the gain of electrons by a substance undergoing a chemical reaction (in other words, a decrease in the oxidation number of the substance). Oxidation and reduction always occur together. An oxidizing agent is a substance that causes something else to be oxidized, and a reducing agent is a substance that causes something else to be reduced. A redox reaction is a reaction in which oxidation and reduction take place. In such a reaction, the oxidizing agent is reduced, and the reducing agent is oxidized.

Key Terms

acid **(10.4)**
activity series **(10.5)**
balanced equation **(10.2)**
base **(10.4)**
combination reaction **(10.4)**
decomposition reaction **(10.4)**
dissociate **(10.4)**
double-replacement (double-exchange) reaction **(10.4)**
neutralization reaction **(10.4)**
nonredox reaction **(10.4)**
oxidation **(10.6)**
oxidation–reduction (redox) reaction **(10.6)**
oxidizing agent **(10.6)**
precipitate **(10.4)**
product **(10.1)**
reactant **(10.1)**
reducing agent **(10.6)**
reduction **(10.6)**
salt **(10.4)**
single-replacement reaction **(10.4)**

Career Sketch

Occupational Safety and Health Technician

As an occupational safety and health technician, you will work with safety engineers, industrial hygienists, and managers to analyze conditions in the workplace. You will study engineering principles and safety-related legislation, and you will become familiar with techniques used to evaluate the environment. Your aim is to reduce on-the-job accidents and the incidence of work-related illnesses. To that end you will assess physical hazards, air quality, water quality, and hazardous chemicals in the workplace. In addition, you will recommend ways to prevent unsafe working conditions in factories, offices, and other places of business. You may also sketch and lay out safety devices.

Occupational safety and health workers ensure that conditions in the workplace are free of hazards.
(OSHA)

To prepare for this career, you will take courses in the fundamentals of fire protection, safety and health regulations, and first aid. You will study such diverse topics as human relations, power sources and hazards, traffic safety, noise control, hazards of industrial chemicals, environmental health, radiation safety, disaster preparedness, and chemical instrumentation and analysis.

Self-Test Exercises

Learning Goals 1 and 2: Formula and Word Equations; Balancing Equations

▲ **1.** Write a balanced chemical equation for each of the following reactions:

(a) Lithium + water ⟶ lithium hydroxide + hydrogen gas

(b) Nitric acid (HNO_3) + calcium hydroxide ⟶ calcium nitrate + water

(c) Magnesium + zinc nitrate ⟶ magnesium nitrate + zinc

(d) Potassium oxide + water ⟶ potassium hydroxide

(e) Zinc sulfide + oxygen gas $\xrightarrow{\text{Heat}}$ zinc oxide + sulfur dioxide

(f) Aluminum + iron(III) oxide ⟶ aluminum oxide + iron

▲ **2.** Write a balanced chemical equation for each of the following reactions:

(a) Aluminum hydroxide + sulfuric acid ⟶ aluminum sulfate + water

(b) Calcium + oxygen ⟶ calcium oxide

(c) Barium chlorate $\xrightarrow{\text{Heat}}$ barium chloride + oxygen gas

(d) Iron(III) hydroxide + sulfuric acid ⟶ iron(III) sulfate + water

(e) Lithium + oxygen gas ⟶ lithium oxide

(f) Phosphorus + iodine ⟶ phosphorus triiodide

3. Write a balanced chemical equation for each of the following reactions:

(a) Iron + oxygen gas ⟶ iron(III) oxide

(b) Sulfuric acid + aluminum hydroxide ⟶ aluminum sulfate + water

(c) Silver nitrate + barium chloride ⟶ silver chloride + barium nitrate

(d) Copper(I) sulfide + oxygen gas ⟶ copper + sulfur dioxide

4. Write a balanced chemical equation for each of the following reactions:

(a) Hydrogen gas + oxygen gas ⟶ water

(b) Potassium hydroxide + acetic acid ⟶ potassium acetate + water

(c) Nitrogen gas + hydrogen gas ⟶ ammonia

(d) Potassium chlorate ⟶ potassium chloride + oxygen gas

Learning Goal 3: Recognizing Types of Chemical Reactions

▲ **5.** Identify each of the following reactions as a combination, decomposition, single-replacement, or double-replacement reaction:

(a) $2HgO \longrightarrow 2Hg + O_2$

(b) $H_2SO_4 + Ca(OH)_2 \longrightarrow CaSO_4 + 2H_2O$

(c) $2Al + 3ZnCl_2 \longrightarrow 2AlCl_3 + 3Zn$

(d) $2Na + Cl_2 \longrightarrow 2NaCl$

(e) $H_2 + Cl_2 \longrightarrow 2HCl$

(f) $AgNO_3 + HCl \longrightarrow HNO_3 + AgCl$

6. Identify each of the following reactions as a combination, decomposition, single-replacement, or double-replacement reaction:

(a) $H_2 + Br_2 \longrightarrow 2HBr$

(b) $2PbO_2 \xrightarrow{\text{Heat}} 2PbO + O_2$

(c) $2KI + Br_2 \longrightarrow 2KBr + I_2$

(d) $2Fe(OH)_3 + 3H_2SO_4 \longrightarrow Fe_2(SO_4)_3 + 6H_2O$

(e) $2Al + 3Cl_2 \longrightarrow 2AlCl_3$

(f) $2K_3PO_4 + 3BaCl_2 \longrightarrow 6KCl + Ba_3(PO_4)_2$

7. Identify the type (combination, decomposition, single-replacement, or double-replacement) of each reaction in Exercise 13.

8. Identify the type (combination, decomposition, single-replacement, or double-replacement) of each reaction in Exercise 14.

9. Identify each of the reactions in Exercise 3 as a combination, decomposition, single-replacement, or double-replacement reaction.

▲ **10.** Identify each of the reactions in Exercise 2 as a combination, decomposition, single-replacement, or double-replacement reaction.

Learning Goal 4: Predicting Products of Chemical Reactions

▲ **11.** Complete and balance each of the following equations. Use the activity series and solubility table when necessary.

Combination Reactions

(a) $H_2 + Br_2 \longrightarrow ?$
(b) $BaO + H_2O \longrightarrow ?$
(c) $Na + Cl_2 \longrightarrow ?$
(d) Nitrogen gas + oxygen gas ⟶ nitrogen dioxide

Decomposition Reactions

(e) $CaCO_3 \xrightarrow{\text{Heat}} ? + ?$
(f) $KOH \longrightarrow ? + ?$
(g) $Hg(ClO_3)_2 \xrightarrow{\text{Heat}} ? + ?$
(h) $PbCl_2 \longrightarrow ? + ?$

Single-Replacement Reactions

(i) $Li + H_2O \longrightarrow ? + ?$
(j) Zinc + sulfuric acid ⟶ ? + ?
(k) $Ni + Al(NO_3)_3 \longrightarrow ? + ?$
(l) $Al + Hg(C_2H_3O_2)_2 \longrightarrow ? + ?$

Double-Replacement Reactions

(m) Sulfuric acid + ammonium hydroxide ⟶ ? + ?
(n) Silver nitrate + barium chloride ⟶ ? + ?
(o) H_2SO_3 (sulfurous acid) + $Al(OH)_3 \longrightarrow ? + ?$
(p) $NaNO_3(aq) + KCl(aq) \longrightarrow ? + ?$

12. Complete and balance each of the following equations. Use the activity series and solubility table when necessary.

Combination Reactions

(a) $Cu + S \longrightarrow ?$
(b) $CaO + H_2O \longrightarrow ?$
(c) $Na + F_2 \longrightarrow ?$
(d) Phosphorus + iodine ⟶ ?

Decomposition Reactions

(e) $MgCO_3 \xrightarrow{\text{Heat}} ? + ?$
(f) $LiOH \longrightarrow ? + ?$
(g) $Mg(ClO_3)_2 \longrightarrow ? + ?$
(h) $Mg_3N_2 \longrightarrow ? + ?$

Single-Replacement Reactions

(i) $Al + H_3PO_4 \longrightarrow ? + ?$
(j) $Zn + Pb(NO_3)_2 \longrightarrow ? + ?$
(k) $Cl_2 + KBr \longrightarrow ? + ?$
(l) $Br_2 + NaCl \longrightarrow ? + ?$

Double-Replacement Reactions

(m) $AgNO_3 + H_2S \longrightarrow ? + ?$
(n) Barium chloride + ammonium carbonate ⟶ ? + ?
(o) $CaCO_3 + H_3PO_4 \longrightarrow ? + ?$
(p) Silver nitrate + magnesium chloride ⟶ ? + ?

13. Complete and balance each of the following equations. Use the activity series and solubility table when necessary.

(a) $N_2O \xrightarrow{\text{Heat}} ? + ?$
(b) $H_2CO_3 \xrightarrow{\text{Heat}} ? + ?$
(c) Sodium nitrate $\xrightarrow{\text{Heat}}$ sodium nitrite + oxygen gas
(d) $H_2 + F_2 \longrightarrow ?$
(e) $N_2 + H_2 \xrightarrow[\text{High pressure}]{\text{Heat}} ?$
(f) $Ca + HCl \longrightarrow ? + ?$
(g) $Cu + NiCl_2 \longrightarrow ? + ?$
(h) $AgNO_3(aq) + K_3AsO_4(aq) \longrightarrow ? + ?$

▲ **14.** Complete and balance each of the following equations. Use the activity series and solubility table when necessary.

(a) $Na_2O \longrightarrow ? + ?$
(b) $H_2 + I_2 \longrightarrow ?$

(c) Phosphorus + oxygen gas $\longrightarrow$ diphosphorus trioxide
(d) $K_3PO_4 + BaCl_2 \longrightarrow ? + ?$
(e) $KClO_3 \longrightarrow ? + ?$
(f) $Ca + O_2 \longrightarrow ?$
(g) $Cl_2 + NH_4I \longrightarrow ? + ?$
(h) $Ag_2O \xrightarrow{\text{Heat}} ? + ?$

15. Complete and balance each of the following equations:

Combination Reactions

(a) $K + Cl_2 \longrightarrow ?$
(b) $K_2O + H_2O \longrightarrow ?$
(c) $MgO + H_2O \longrightarrow ?$
(d) $CaO + CO_2 \longrightarrow ?$

Decomposition Reactions

(e) $NH_3 \longrightarrow ? + ?$
(f) $SrCO_3 \longrightarrow ? + ?$
(g) $NaClO_3 \longrightarrow ? + ?$
(h) $Mg(OH)_2 \longrightarrow ? + ?$

16. Complete and balance each of the following equations:

Combination Reactions

(a) $K + Br_2 \longrightarrow ?$
(b) $Li_2O + H_2O \longrightarrow ?$
(c) $BaO + H_2O \longrightarrow ?$
(d) $MgO + CO_2 \longrightarrow ?$

Decomposition Reactions

(e) $MgCO_3 \longrightarrow ? + ?$
(f) Copper(I) sulfide $\longrightarrow ? + ?$
(g) Chromium(III) carbonate $\longrightarrow ? + ?$
(h) $Al(ClO_3)_3 \xrightarrow{\text{Heat}} ? + ?$

Learning Goal 5: Acids, Bases, and Salts

▲ **17.** The reaction of an acid and a base produces a salt and water. Find the acid and base needed to produce each of the following salts: *(a)* NaCl *(b)* K_2SO_4 *(c)* $Al(NO_3)_3$

18. The reaction of an acid and a base produces a salt and water. Find the acid and base needed to produce each of the following salts: *(a)* KCl *(b)* $MgSO_4$ *(c)* $Fe_2(SO_4)_3$

Learning Goal 6: Using the Activity Series and Solubility Table

19. Complete and balance each of the following equations. Check the activity series and solubility table to be sure that the reaction can take place.

Single-Replacement Reactions

(a) $Zn + HNO_2 \longrightarrow ? + ?$
(b) $Ag + NiCl_2 \longrightarrow ? + ?$
(c) $Zn + AgNO_3 \longrightarrow ? + ?$
(d) $Cs + H_2O \longrightarrow ? + ?$

Double-Replacement Reactions

(e) $HCl + Al(OH)_3 \longrightarrow ? + ?$
(f) $KNO_3 + ZnCl_2 \longrightarrow ? + ?$
(g) $Al(NO_3)_3 + NaOH \longrightarrow ? + ?$
(h) $K_2CrO_4 + Pb(NO_3)_2 \longrightarrow ? + ?$

▲ **20.** Complete and balance each of the following equations. Check the activity series and solubility table to be sure that the reaction can take place.

Single-Replacement Reactions

(a) $Cu + AgCl \longrightarrow ? + ?$
(b) $Cl_2 + KF \longrightarrow ? + ?$
(c) $Zn + HNO_3 \longrightarrow ? + ?$
(d) $Cu + HCl \longrightarrow ? + ?$

Double-Replacement Reactions

(e) $NH_4NO_3 + H_3PO_4 \longrightarrow ? + ?$
(f) $ZnCl_2 + KOH \longrightarrow ? + ?$
(g) $Ni_3(PO_4)_2 + HCl \longrightarrow ? + ?$
(h) $AgNO_3 + KCl \longrightarrow ? + ?$

Learning Goal 7: Oxidation and Reduction; Oxidizing and Reducing Agents

▲ **21.** Define the following terms and give an example of each: *(a)* oxidation *(b)* reduction *(c)* oxidizing agent *(d)* reducing agent

22. Complete each of the following statements:
(a) A loss of electrons is called ________ .
(b) A gain of electrons is called ________ .
(c) A substance that causes something else to be oxidized is called a(n) ________ .
(d) A substance that causes something else to be reduced is called a(n) ________ .

Learning Goal 8: Determining Oxidized and Reduced Substances

▲ **23.** Balance each of the following reactions and determine which substance is oxidized and which is reduced:
(a) $K + Cl_2 \longrightarrow KCl$
(b) $NH_3 \longrightarrow N_2 + H_2$
(c) $CuO + H_2 \longrightarrow Cu + H_2O$
(d) $Sn + Cl_2 \longrightarrow SnCl_4$

▲ **24.** Balance each of the following reactions and determine which substance is oxidized and which is reduced:
(a) $Na + F_2 \longrightarrow NaF$
(b) $Al + C \longrightarrow Al_4C_3$
(c) $Cu + S \longrightarrow CuS$
(d) $Ag_2O \longrightarrow Ag + O_2$

Learning Goal 9: Determining Oxidizing and Reducing Agents

▲ **25.** For each of the reactions in Exercise 23, determine the oxidizing agent and the reducing agent.

▲ **26.** For each of the reactions in Exercise 24, determine the oxidizing agent and the reducing agent.

27. For each single-replacement reaction in Exercise 19, determine the oxidizing agent and the reducing agent.

28. For each single-replacement reaction in Exercise 20, determine the oxidizing agent and the reducing agent.

Extra Exercises

29. For the reaction

$$5Zn + V_2O_5 \longrightarrow 5ZnO + 2V$$

list *(a)* the oxidizing agent, *(b)* the reducing agent, *(c)* the element oxidized, and *(d)* the element reduced

30. Magnesium carbonate and calcium carbonate, when heated, both decompose to form carbon dioxide and the metal oxides. Write these two reactions.

31. Determine whether the following equations are balanced. If they are not, balance them.
(a) $N_2 + 3H_2 \longrightarrow NH_3$
(b) $Fe + H_2O \longrightarrow Fe_3O_4 + H_2$
(c) $PCl_3 + Cl_2 \longrightarrow PCl_5$

▲ **32.** A student is asked to balance the following:

Calcium hydroxide + hydrochloric acid ⟶ calcium chloride + water

The student works the problem and gets the following solution:

$$CaOH + HCl \longrightarrow CaCl + H_2O$$

This attempt is marked incorrect. Can you explain why the answer is wrong?

33. Write a balanced equation for each of the following. If no reaction occurs, write *no reaction*.
(a) Carbon burning in oxygen gas to form carbon monoxide
(b) Hydrogen gas and chlorine gas reacting to form hydrogen chloride gas
(c) Metallic sodium dropped into water to create sodium hydroxide and hydrogen gas
(d) Silver metal and hydrochloric acid reacting to produce ________

▲ **34.** Consider an experiment in which metal A is immersed in a water solution of ions of metal B. For each of the following pairs A and B, determine whether or not a reaction occurs. If a reaction does occur, write the balanced equation, assuming that the B ions are part of nitrate salts.

	A	B
(a)	Cu	Zn
(b)	Mg	Zn
(c)	Cu	Ni
(d)	Zn	Cu

35. Explain why an oxidizing agent is reduced.

36. Write balanced formula equations for the following word equations and state what type of reaction each is:
(a) Calcium hydroxide + hydrochloric acid ⟶
(b) Zinc metal + sulfuric acid ⟶
(c) Barium metal + oxygen gas ⟶
(d) Cesium oxide + water ⟶

37. Balance the equation

$$NH_3(g) + O_2(g) \longrightarrow N_2O_4(g) + H_2O(g)$$

38. Consider the redox reaction

$$(Cr_2O_7)^{2-} + 14H^{1+} + 6Fe^{2+} \longrightarrow 2Cr^{3+} + 7H_2O + 6Fe^{3+}$$

(a) Determine which substance is oxidized.
(b) Determine which substance is reduced.

(*c*) Determine the oxidizing agent.
(*d*) Determine the reducing agent.

39. Balance the following equation and write the chemical name of each product and reactant in the balanced equation.

$$Na + P_4 \longrightarrow Na_3P$$

▲ **40.** Write a balanced equation for a reaction that produces magnesium chloride. (*Hint:* In other words, magnesium chloride has to be at least one of the products.)

41. Write a balanced equation for an acid–base reaction that produces lithium chloride as one of the products.

42. Are single-replacement reactions also redox reactions? Explain with examples.

43. Write a balanced equation for any reaction that produces carbon dioxide as one of the products.

44. A double replacement reaction produces zinc phosphate. What are the reactants? Write a balanced equation for the reaction.

11 Stoichiometry: The Quantities in Reactions

Learning Goals

After you've studied this chapter, you should be able to:

1. Calculate the quantities of reactants needed or products yielded in a chemical reaction.
2. Determine which starting material is the limiting reactant in a chemical reaction.

Introduction

In Chapter 10, we wrote balanced equations showing how many atoms or molecules of each substance take part in a reaction. For example, we considered the reaction

$$N_2 + 3H_2 \xrightarrow[\text{Pressure}]{\text{Heat}} 2NH_3$$

In this reaction, one molecule of nitrogen combines with three molecules of hydrogen to form two molecules of ammonia (Figure 11.1). Industry uses this reaction to make ammonia for fertilizers, explosives, and plastics, among many other products. The industrial process based on this reaction is called the **Haber process** for the synthesis of ammonia. (To **synthesize** means to combine the parts into the whole.)

Suppose we also wanted to produce ammonia in large quantities for commercial use. To prevent waste and thereby minimize costs, we would like to combine just enough of each reactant. But how much is "just enough of each

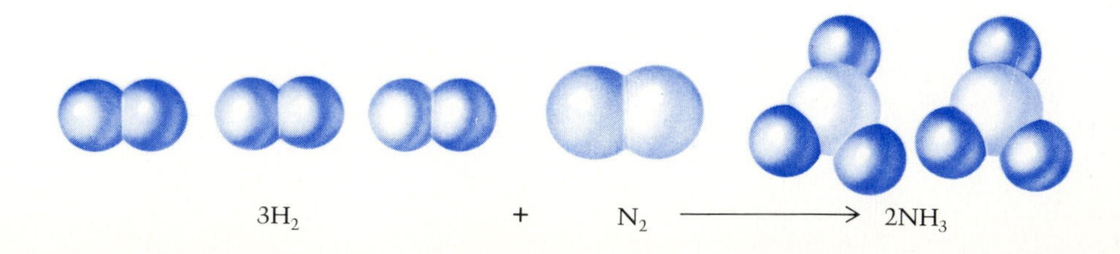

Figure 11.1
The Haber process

reactant"? And how much ammonia would we obtain? Or suppose we wanted to produce just 100 kilograms of ammonia. How much nitrogen and hydrogen would we need?

To answer such questions, we need to "translate" balanced chemical equations into quantities of reactants and products. In this chapter, you will learn how to do so by using *stoichiometry,* which is pronounced "stoy-key-OM-eh-tree."

11.1 The Mole Method

Stoichiometry is defined as *the calculation of the quantities of substances involved in chemical reactions.* An important idea in stoichiometry is that a balanced equation tells us how many *moles* of the various substances take part in the reaction. For example, the balanced equation for the Haber process,

$$N_2 + 3H_2 \xrightarrow[\text{Pressure}]{\text{Heat}} 2NH_3$$

tells us that 1 *mole* of nitrogen and 3 *moles* of hydrogen react to form 2 *moles* of ammonia. (If you've forgotten what a mole is, refer back to Chapter 9.)

The reasoning behind this idea is as follows: The balanced chemical equation tells us the number of molecules or atoms of each substance that is involved in a single instance of the reaction. It also tells us in what *ratio* these molecules or atoms are involved. In the Haber process, the molecules are involved in the ratio one N_2 to three H_2 to two NH_3, or $1N_2 : 3H_2 : 2NH_3$. Using this ratio, we can also say that 1 *Avogadro's number* of nitrogen molecules and 3 *Avogadro's numbers* of hydrogen molecules produce 2 *Avogadro's numbers* of ammonia molecules. Because 1 Avogadro's number of molecules is 1 mole, we can say that 1 mole of nitrogen and 3 moles of hydrogen react to produce 2 moles of ammonia.

This same sort of reasoning holds true for every balanced equation. *The coefficients in a balanced chemical equation indicate the number of moles of each substance that takes part in the reaction.*

Once we know the number of moles of each substance in a reaction, we can find out how many grams (or kilograms or pounds) of each substance are involved. For example, we know that the atomic mass of nitrogen is 14.0 and that the atomic mass of hydrogen is 1.0. Then we can say that since 1 mole of N_2 has a mass of 28.0 grams, 1 mole of H_2 has a mass of 2.0 grams, and 1 mole of NH_3 has a mass of 17.0 grams, 28.0 grams of nitrogen react with 6.0 grams of hydrogen to produce 34.0 grams of ammonia:

N_2	+	$3H_2$	$\xrightarrow[\text{Pressure}]{\text{Heat}}$	$2NH_3$
1 mole		3 moles		2 moles
28.0 grams		6.0 grams		34.0 grams

Note that we have satisfied the Law of Conservation of Mass. The mass of the product is equal to the sum of the masses of the reactants. In the remainder of this chapter, we shall apply this basic idea to several types of stoichiometric problems.

11.2 Quantities of Reactants and Products

Learning Goal 1
Simple stoichiometry

The best way to learn how to solve stoichiometric problems is to work through a number of them. As we do this, remember that the balanced equation for a reaction contains the basic information about the relative quantities of substances involved. Also note that if we know the quantity of even a single reactant or product, we can obtain the quantities of all other reactants or products.

We can begin by asking how many moles of N_2 are needed to produce 5 moles of NH_3? The balanced equation is

$$\begin{array}{ccc} N_2 + 3H_2 & \longrightarrow & 2NH_3 \\ 1\ \text{mole} & & 2\ \text{moles} \\ ? & & 5\ \text{moles} \end{array}$$

Set up the conversion:

$$5\ \cancel{\text{moles}}\ NH_3 \times \frac{1\ \text{mole}\ N_2}{2\ \cancel{\text{moles}}\ NH_3} = 2.5\ \text{moles}\ N_2$$

Next we can ask how many moles of H_2 are needed to produce 5 moles of NH_3? The balanced equation is

$$\begin{array}{ccc} N_2 + 3H_2 & \longrightarrow & 2NH_3 \\ 3\ \text{moles} & & 2\ \text{moles} \\ ? & & 5\ \text{moles} \end{array}$$

Set up the conversion:

$$5\ \cancel{\text{moles}}\ NH_3 \times \frac{3\ \text{mole}\ H_2}{2\ \cancel{\text{moles}}\ NH_3} = 7.5\ \text{moles}\ H_2$$

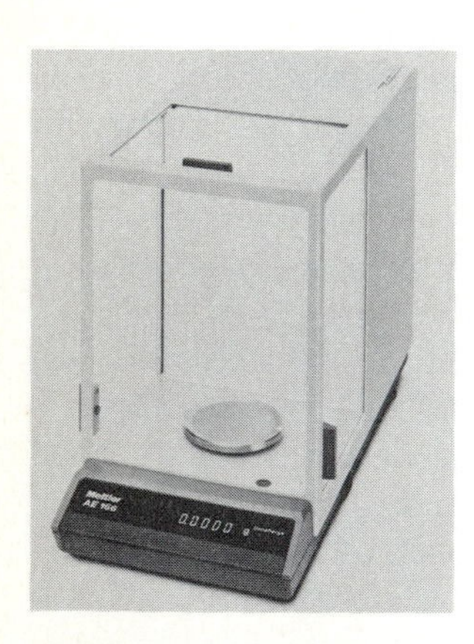

A balance is used to weigh precise quantities of materials in the laboratory.
(Courtesy of Mettler Instrument Corporation)

Let us assume that we want to synthesize 136 grams of ammonia. How many grams of nitrogen and hydrogen do we need? The first thing we do is set up the balanced chemical equation.

$$N_2 + 3H_2 \longrightarrow 2NH_3$$

Next we record what is given and what must be found.

$$\begin{array}{cccc} N_2 & + 3H_2 & \longrightarrow & 2NH_3 \\ ?\ g & ?\ g & & 136\ g \end{array}$$

The next step is to convert grams of ammonia into moles (because the chemical equation relates substances by moles, not grams).

$$? \text{ moles of } HN_3 = (136 \cancel{\text{g}})\left(\frac{1 \text{ mole}}{17.0 \cancel{\text{g}}}\right) = 8.00 \text{ moles of } NH_3$$

$$\begin{array}{lll} N_2 + 3H_2 & \longrightarrow & 2NH_3 \\ ?\text{ g} \quad ?\text{ g} & & 136 \text{ g (which is 8.00 moles)} \end{array}$$

Now we use the coefficients of the balanced equation to calculate the number of moles of hydrogen and nitrogen necessary to form 8 moles of ammonia. We do it in three steps. First, we determine the number of moles of nitrogen:

$$? \text{ moles of } N_2 = (8.00 \cancel{\text{ moles } NH_3})\left(\frac{1 \text{ mole } N_2}{2 \cancel{\text{ moles } NH_3}}\right) = 4.00 \text{ moles of } N_2$$

We obtained the factor

$$\left(\frac{1 \text{ mole } N_2}{2 \text{ moles } NH_3}\right)$$

from the coefficients of the balanced equation. This factor tells us that 1 mole of nitrogen is necessary to produce 2 moles of ammonia. Now we are ready for the second step.

$$? \text{ moles of } H_2 = (8.00 \cancel{\text{ moles } NH_3})\left(\frac{3 \text{ moles } H_2}{2 \cancel{\text{ moles } NH_3}}\right) = 12.0 \text{ moles of } H_2$$

We also obtained the factor

$$\left(\frac{3 \text{ moles } H_2}{2 \text{ moles } NH_3}\right)$$

from the balanced equation, which tells us that 3 moles of hydrogen are necessary to produce 2 moles of ammonia. We now have

$$\begin{array}{ccccc} N_2 & + & 3H_2 & \longrightarrow & 2NH_3 \\ ?\text{ g} & & ?\text{ g} & & 136 \text{ g} \\ 4.00 \text{ moles} & & 12.0 \text{ moles} & & 8.00 \text{ moles} \end{array}$$

Another way to obtain the same information is to remember that the moles of reactants and products must be in the ratio of

$$\begin{array}{ccccc} N_2 & : & H_2 & : & NH_3 \\ 1 & : & 3 & : & 2 \end{array}$$

We could ask ourselves the question,

$$1 : 3 : 2 \qquad \text{is the same as} \qquad ? : ? : 8$$

To answer this, we would note that

$$\begin{array}{c} 1 : 3 : 2 \\ \downarrow \times 4 \\ ? : ? : 8 \end{array}$$

or

$$1 : 3 : 2 \quad \text{is the same as} \quad 4 : 12 : 8$$

Our final step is to change *moles* of nitrogen and hydrogen into *grams*.

$$? \text{ grams of } N_2 = (4.00 \cancel{\text{moles}}) \left(\frac{28.0 \text{ g}}{1 \cancel{\text{mole}}}\right) = 112 \text{ g of } N_2$$

$$? \text{ grams of } H_2 = (12.0 \cancel{\text{moles}}) \left(\frac{2.0 \text{ g}}{1 \cancel{\text{mole}}}\right) = 24 \text{ g of } H_2$$

The problem is solved.

N_2	+	$3H_2$	$\longrightarrow$	$2NH_3$
112 g		24 g		136 g
4.00 moles		12.0 moles		8.00 moles

We can check our mathematics by seeing whether this reaction obeys the Law of Conservation of Mass. The grams of nitrogen and hydrogen we start out with should equal the grams of ammonia produced.

Mass of Reactants	**Mass of Products**
Grams of N_2 = 112 g	Grams of NH_3 = 136 g
Grams of H_2 = 24 g	
Total mass = 136 g	Total mass = 136 g

To practice our stoichiometry, let's solve another problem. Many scientists have suggested using propane gas (C_3H_8) in motor vehicles. Because propane burns cleaner than today's gasolines, these scientists feel that the use of propane would drastically reduce the amount of automobile pollution. When propane reacts with sufficient oxygen, the resulting products are carbon dioxide and water, in the form of water vapor. Our problem is this: How many grams of oxygen are needed for combustion of 22 grams of propane? How many grams of carbon dioxide would be produced? How many grams of water would be produced? (Remember that C_3H_8 is the formula for propane.)

Step 1
Write a balanced equation, listing what is given and what we want to find.

C_3H_8 +	$5O_2$	$\longrightarrow$	$3CO_2$ +	$4H_2O$
22 g	? g		? g	? g

Step 2
Change the grams of C_3H_8 into moles.

$$? \text{ moles of } C_3H_8 = (22 \not{g})\left(\frac{1 \text{ mole}}{44.0 \not{g}}\right) = 0.50 \text{ mole of } C_3H_8$$

C_3H_8	+	$5O_2$	$\longrightarrow$	$3CO_2$	+	$4H_2O$
22 g		? g		? g		? g
0.50 mole						

Step 3
Use the coefficients of the equation to determine the moles of oxygen needed, as well as the moles of carbon dioxide and water produced.

$$? \text{ moles of } O_2 \text{ needed} = (0.50 \cancel{\text{mole } C_3H_8})\left(\frac{5 \text{ moles } O_2}{1 \cancel{\text{mole } C_3H_8}}\right)$$
$$= 2.5 \text{ moles of } O_2 \text{ needed}$$

$$? \text{ moles of } CO_2 \text{ produced} = (0.50 \cancel{\text{mole } C_3H_8})\left(\frac{3 \text{ moles } CO_2}{1 \cancel{\text{mole } C_3H_8}}\right)$$
$$= 1.5 \text{ moles of } CO_2 \text{ produced}$$

$$? \text{ moles of } H_2O \text{ produced} = (0.50 \cancel{\text{mole } C_3H_8})\left(\frac{4 \text{ moles } H_2O}{1 \cancel{\text{mole } C_3H_8}}\right)$$
$$= 2.0 \text{ moles of } H_2O \text{ produced}$$

C_3H_8	+	$5O_2$	$\longrightarrow$	$3CO_2$	+	$4H_2O$
22 g		? g		? g		? g
0.50 mole		2.5 moles		1.5 moles		2.0 moles

Note that you can obtain the same result by the other method.

$$\begin{array}{c} 1 \ : 5 : 3 : 4 \\ 0.50 : ? : ? : ? \end{array}$$

or

$$\begin{array}{c} 1 \ : \ 5 \ : \ 3 \ : \ 4 \\ 0.50 : 2.5 : 1.5 : 2.0 \end{array}$$

Step 4
Change moles of O_2, CO_2, and H_2O into grams.

$$? \text{ grams of } O_2 = (2.5 \cancel{\text{moles}})\left(\frac{32.0 \text{ g}}{1 \cancel{\text{mole}}}\right) = 8\bar{0} \text{ g of } O_2$$

$$? \text{ grams of } CO_2 = (1.5 \cancel{\text{moles}})\left(\frac{44.0 \text{ g}}{1 \cancel{\text{mole}}}\right) = 66 \text{ g of } CO_2$$

$$? \text{ grams of } H_2O = (2.0 \cancel{\text{moles}})\left(\frac{18.0 \text{ g}}{1 \cancel{\text{mole}}}\right) = 36 \text{ g of } H_2O$$

C_3H_8	+	$5O_2$	$\longrightarrow$	$3CO_2$	+	$4H_2O$
22 g		$8\bar{0}$ g		66 g		36 g
0.50 mole		2.5 moles	$\longrightarrow$	1.5 moles		2.0 moles

Step 5
Check your results.

Mass of Reactants		Mass of Products	
Grams of C_3H_8	= 22 g	Grams of CO_2	= 66 g
Grams of O_2	= $8\bar{0}$ g	Grams of H_2O	= 36 g
Total mass	= 102 g	Total mass	= 102 g

Let's summarize the procedure for solving stoichiometric problems:

1. Write a balanced chemical equation for the reaction. Record the given number of grams under the chemical formula of that substance. Place question marks for the unknown numbers of grams under the corresponding chemical formulas.

2. Change the given number of grams into moles.

3. Use the coefficients of the balanced equation and the moles of the given amount of substance to determine the moles of the other substances involved in the reaction.

4. Change the moles of the other substances into grams.

5. Check the results to see whether the mass of the reactants equals the mass of the products.

Another way of summarizing this procedure is to think of it in the following sequence:

Grams known $\longrightarrow$ moles known $\longrightarrow$ moles unknown $\longrightarrow$ grams unknown

Before we continue we would like to show one more way to approach this problem. Some students find it easier to solve a problem such as this one using the mass-mass ratio from the balanced equation. We will find the grams of carbon dioxide and the grams of water produced using this method.

C_3H_8	+ $5O_2$	$\longrightarrow$	$3CO_2$	+	$4H_2O$
1 mole			3 moles		4 moles

44.0 g	3 (44.0 g)	4(18.0 g)
44.0 g	132.0 g	72.0 g

$$22 \cancel{g}\ C_3H_8 \times \frac{132.0\ g\ CO_2}{44.0 \cancel{g}\ C_3H_8} = 66\ g\ CO_2$$

$$22 \cancel{g}\ C_3H_8 \times \frac{72.0\ g\ H_2O}{44.0 \cancel{g}\ C_3H_8} = 36\ g\ H_2O$$

The procedure for solving mass-mass stoichiometric problems can be summarized as follows:

1. First write a balanced chemical equation for the reaction.
2. Underneath the chemical formulas, write the number of moles of the substance given and the number of moles of the substance asked for as indicated by the coefficients.
3. Convert these mole quantities into grams. This may be done underneath the formulas in the equation.
4. Use the gram-gram relationship as a conversion factor with the mass of the given substance to find the mass of the substance.

Practice solving stoichiometric problems by doing the following exercises. Work each problem in steps according to the procedure shown. You may also use an alternate strategy.

Example 11.1

The *thermite* reaction is a chemical reaction that is important to railroads, because railroad workers use it to weld railroad tracks. In this reaction, a mixture of aluminum metal and iron(III) oxide (known as thermite) is heated to yield aluminum oxide and iron metal. Once begun, the reaction releases a tremendous amount of heat—so much that the molten iron produced fuses the railroad tracks (Figure 11.2).

How many grams of aluminum and iron(III) oxide are needed to produce 279 grams of iron metal? How many grams of aluminum oxide will be produced?

Solution

Understand the Problem

We are told how many grams of product (iron metal) are formed, and we can use this information to calculate the grams of reactants [aluminum and iron(III) oxide] necessary to produce this amount of product.

Devise a Plan

Use the same strategy as was described in the previous example.

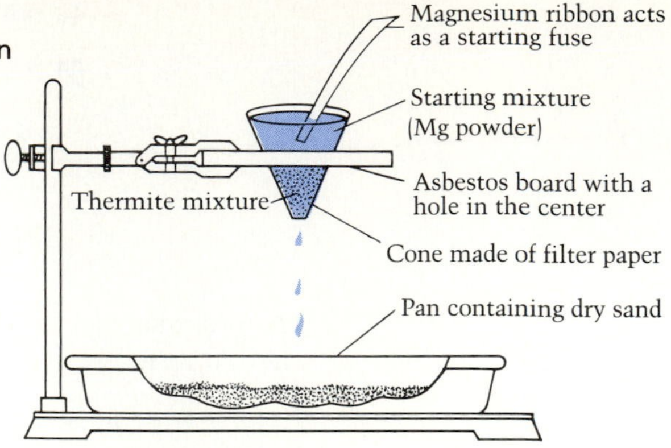

Figure 11.2
The thermite reaction in the lab

Carry Out the Plan

The procedure may be followed in steps. [*Note:* Fe_2O_3 stands for iron(III) oxide, and Al_2O_3 stands for aluminum oxide.]

Step 1

$$\begin{array}{cccc} 2Al\ + & Fe_2O_3 & \longrightarrow & Al_2O_3\ + & 2Fe \\ ?\ g & ?\ g & & ?\ g & 279\ g \end{array}$$

Step 2

$$?\text{ moles of Fe} = (279\ \cancel{g})\left(\frac{1\text{ mole}}{55.8\ \cancel{g}}\right) = 5.00\text{ moles of Fe}$$

$$\begin{array}{cccc} 2Al\ + & Fe_2O_3 & \longrightarrow & Al_2O_3\ + & 2Fe \\ ?\ g & ?\ g & & ?\ g & 279\ g \\ & & & & 5.00\text{ moles} \end{array}$$

Step 3

$$?\text{ moles of Al} = (5.00\ \cancel{\text{moles Fe}})\left(\frac{2\text{ moles Al}}{2\ \cancel{\text{moles Fe}}}\right)$$

$$= 5.00\text{ moles of Al}$$

$$?\text{ moles of }Fe_2O_3 = (5.00\ \cancel{\text{moles Fe}})\left(\frac{1\text{ mole }Fe_2O_3}{2\ \cancel{\text{moles Fe}}}\right)$$

$$= 2.50\text{ moles of }Fe_2O_3$$

$$?\text{ moles of }Al_2O_3 = (5.00\ \cancel{\text{moles Fe}})\left(\frac{1\text{ mole }Al_2O_3}{2\ \cancel{\text{moles Fe}}}\right)$$

$$= 2.50\text{ moles of }Al_2O_3$$

$2Al$	+	Fe_2O_3	$\longrightarrow$	Al_2O_3	+	$2Fe$
? g		? g		? g		279 g
5.00 moles		2.50 moles		2.50 moles		5.00 moles

Step 4

$$? \text{ grams of Al} = (5.00 \cancel{\text{moles}})\left(\frac{27.0 \text{ g}}{1 \cancel{\text{mole}}}\right) = 135 \text{ g of Al}$$

$$? \text{ grams of } Fe_2O_3 = (2.50 \cancel{\text{moles}})\left(\frac{159.6}{1 \cancel{\text{mole}}}\right) = 399 \text{ g of } Fe_2O_3$$

$$? \text{ grams of } Al_2O_3 = (2.50 \cancel{\text{moles}})\left(\frac{102.0 \text{ g}}{1 \cancel{\text{mole}}}\right) = 255 \text{ g of } Al_2O_3$$

$2Al$	+	Fe_2O_3	$\longrightarrow$	Al_2O_3	+	$2Fe$
135 g		399 g		255 g		279 g
5.00 moles		2.50 moles		2.50 moles		5.00 moles

Look Back

We check our results to be sure they satisfy the Law of Conservation of Mass.

Mass of Reactants		Mass of Products	
Grams of Al	= 135 g	Grams of Al_2O_3	= 255 g
Grams of Fe_2O_3	= 399 g	Grams of Fe	= 279 g
Total mass	= 534 g	Total mass	= 534 g

Practice Exercise 11.1 Magnesium oxide is a compound that occurs in nature as the mineral periclase. It is composed of the elements magnesium and oxygen. In industry, magnesium oxide is used in the manufacture of refractory crucibles, firebricks, and casein glue. It is also used as a reflector in optical instruments. It combines with water to form magnesium hydroxide, which acts as an antacid in low concentrations. How many grams of magnesium and oxygen gas are needed to produce 4.032 grams of magnesium oxide?

Example 11.2

Polyvinyl chloride (PVC) is one of the best-known and most widely used plastics in the world today. PVC is used to make food wrappings, luggage, plastic garbage bags, phonograph records, and automobile upholstery. It was originally thought that PVC was an extremely safe compound. But studies now indicate that there may be some hidden hazards. The problem is not with the PVC itself but with the compound vinyl chloride (C_2H_3Cl), from which the PVC is made. Vinyl chloride is a colorless gas, and studies show that it may cause liver damage and some forms of cancer in people who are in constant contact with it. Another problem with PVC and vinyl chloride is that when the compounds are burned, an extremely poisonous gas (hydrogen chloride, HCl) can be produced.

Our problem is this: Assume that the following reaction occurs when we burn vinyl chloride (C_2H_3Cl).

$$2C_2H_3Cl + 5O_2 \longrightarrow 4CO_2 + 2H_2O + 2HCl$$

How many grams of hydrogen chloride (HCl) will be produced if 6.25 g of vinyl chloride are burned? How many grams of O_2 are needed, and how many grams of CO_2 and water are produced?

Solution Follow the procedure we used before. Do it in steps.

Step 1

$$\begin{array}{ccccccccc} 2C_2H_3Cl & + & 5O_2 & \longrightarrow & 4CO_2 & + & 2H_2O & + & 2HCl \\ 6.25\text{ g} & & ?\text{ g} & & ?\text{ g} & & ?\text{ g} & & ?\text{ g} \end{array}$$

Step 2

$$?\text{ moles of } C_2H_3Cl = (6.25\ \cancel{\text{g}})\left(\frac{1\text{ mole}}{62.5\ \cancel{\text{g}}}\right) = 0.100\text{ mole of } C_2H_3Cl$$

$$\begin{array}{ccccccccc} 2C_2H_3Cl & + & 5O_2 & \longrightarrow & 4CO_2 & + & 2H_2O & + & 2HCl \\ 6.25\text{ g} & & ?\text{ g} & & ?\text{ g} & & ?\text{ g} & & ?\text{ g} \\ 0.100\text{ mole} & & & & & & & & \end{array}$$

Step 3

$$?\text{ moles of } O_2 = (0.100\ \cancel{\text{mole } C_2H_3Cl})\left(\frac{5\text{ moles } O_2}{2\ \cancel{\text{moles } C_2H_3Cl}}\right)$$

$$= 0.250\text{ mole of } O_2$$

$$?\text{ moles of } CO_2 = (0.100\ \cancel{\text{mole } C_2H_3Cl})\left(\frac{4\text{ moles } CO_2}{2\ \cancel{\text{moles } C_2H_3Cl}}\right)$$

$$= 0.200\text{ mole of } CO_2$$

$$?\text{ moles of } H_2O = (0.100\ \cancel{\text{mole } C_2H_3Cl})\left(\frac{2\text{ moles } H_2O}{2\ \cancel{\text{moles } C_2H_3Cl}}\right)$$

$$= 0.100\text{ mole of } H_2O$$

$$?\text{ moles of } HCl = (0.100\ \cancel{\text{mole } C_2H_3Cl})\left(\frac{2\text{ moles } HCl}{2\ \cancel{\text{moles } C_2H_3Cl}}\right)$$

$$= 0.100\text{ mole of } HCl$$

$2C_2H_3Cl$	+	$5O_2$	$\longrightarrow$	$4CO_2$	+	$2H_2O$	+	$2HCl$
6.25 g		? g		? g		? g		? g
0.100 mole		0.250 mole		0.200 mole		0.100 mole		0.100 mole

Step 4

$$? \text{ grams of } O_2 = (0.250 \cancel{\text{mole}}) \left(\frac{32.0 \text{ g}}{1 \cancel{\text{mole}}}\right) = 8.00 \text{ g of } O_2$$

$$? \text{ grams of } CO_2 = (0.200 \cancel{\text{mole}}) \left(\frac{44.0 \text{ g}}{1 \cancel{\text{mole}}}\right) = 8.80 \text{ g of } CO_2$$

$$? \text{ grams of } H_2O = (0.100 \cancel{\text{mole}}) \left(\frac{18.0 \text{ g}}{1 \cancel{\text{mole}}}\right) = 1.80 \text{ g of } H_2O$$

$$? \text{ grams of } HCl = (0.100 \cancel{\text{mole}}) \left(\frac{36.5 \text{ g}}{1 \cancel{\text{mole}}}\right) = 3.65 \text{ g of } HCl$$

Step 5

Mass of Reactants		**Mass of Products**	
Grams of C_2H_3Cl =	6.25 g	Grams of CO_2 =	8.80 g
Grams of O_2 =	8.00 g	Grams of H_2O =	1.80 g
		Grams of HCl =	3.65 g
Total mass =	14.25 g	Total mass =	14.25 g

Practice Exercise 11.2 Aluminum nitrate can be produced in a single-replacement reaction. Aluminum replaces zinc in the compound zinc nitrate, because aluminum is higher than zinc in the activity series. The reaction is as follows:

$$2Al + 3Zn(NO_3)_2 \longrightarrow 2Al(NO_3)_3 + 3Zn$$

Aluminum nitrate is used in tanning leather, in antiperspirant preparations, and as a corrosion inhibitor. How many grams of aluminum nitrate are produced when 9.469 g of zinc nitrate react with aluminum? How many grams of aluminum are needed, and how many grams of zinc are produced?

11.3 The Limiting-Reactant Problem

Learning Goal 2
Limiting-reactant problems

Sometimes a chemist who wants to synthesize a compound finds that the quantity of product that can be formed is limited by the amount of one of the starting materials. To find out how much product can be formed, the chemist must first determine which one of the starting materials will be completely used up when the reaction is completed. The other will be in *excess,* so not all of it will be used up to form the product. In other words, the chemist must determine which one of the starting materials is the *limiting reactant*.

To clarify this, let's imagine an automobile manufacturer, the Ajax Automobile Company, which makes a model of a car that is called the Ajax. In order to make just *one* of these automobiles, the assembly line needs (among other things) the following parts: *one* body and *four* wheels. We can represent this statement as follows:

$$1 \text{ body} + 4 \text{ wheels} \longrightarrow 1 \text{ automobile}$$

On a particular day, the assembly line has 50 bodies and 160 wheels. How many automobiles can be produced? If you use the number of bodies as the basis for your answer, you might think Ajax could produce 50 automobiles. However, if you use the number of wheels as the basis for your answer, you'll find that Ajax can produce only 40 automobiles. (Remember, it takes 4 wheels to produce an automobile—not counting spare tires. Therefore $160 \div 4 = 40$.) So the company has 10 extra bodies but not enough wheels to use them. As a result, the most automobiles it can produce is 40. The "limiting reactant" in this case is the wheels, and the bodies are the "reactant" that is in excess.

Let's go through this entire procedure again, this time treating it as if it were a chemical problem.

Step 1
We follow our usual procedure.

1 body	+	4 wheels	$\longrightarrow$	1 automobile
50 bodies		160 wheels		? automobiles

Step 2
In a chemical problem of this type, we would have to change grams into moles at this point. However, in our car problem, we already know the relationship of bodies to wheels. Therefore all we have to do is determine which substance gets used up first (bodies or wheels).

a. Determine how many bodies would be needed to use up 160 wheels.

$$? \text{ bodies needed} = (160 \cancel{\text{wheels}})\left(\frac{1 \text{ body}}{4 \cancel{\text{wheels}}}\right) = 40 \text{ bodies}$$

b. Determine how many wheels would be needed to use up 50 bodies.

$$? \text{ wheels needed} + (50 \cancel{\text{bodies}})\left(\frac{4 \text{ wheels}}{1 \cancel{\text{body}}}\right) = 200 \text{ wheels}$$

But the Ajax Company doesn't have 200 wheels—only 160. Therefore the wheels limit the number of automobiles it can produce.

Step 3

We may use the number of wheels to calculate the number of automobiles Ajax can produce and the number of bodies it will use. (Note that we've already calculated the number of bodies that the company will use in Step 2a.)

$$? \text{ automobiles} = (160 \cancel{\text{wheels}})\left(\frac{1 \text{ automobile}}{4 \cancel{\text{wheels}}}\right) = 40 \text{ automobiles}$$

Now let's do a chemical problem. Assume that a chemist has 203 grams of $Mg(OH)_2$ (magnesium hydroxide) and 164 grams of HCl (hydrogen chloride). The chemist wants to make these substances react to form $MgCl_2$ (magnesium chloride), which can be used for fireproofing wood, or in a disinfectant. How much $MgCl_2$ can the chemist produce, and how much water is formed?

Step 1

We follow our usual procedure.

$$\begin{array}{ccccccc} Mg(OH)_2 & + & 2HCl & \longrightarrow & MgCl_2 & + & 2H_2O \\ 203\text{ g} & & 164\text{ g} & & ?\text{ g} & & ?\text{ g} \end{array}$$

Step 2

Determine the number of moles of each starting substance.

$$? \text{ moles of } Mg(OH)_2 = (203 \cancel{\text{g}})\left(\frac{1 \text{ mole}}{58.3 \cancel{\text{g}}}\right) = 3.48 \text{ moles of } Mg(OH)_2$$

$$? \text{ moles of HCl} = (164 \cancel{\text{g}})\left(\frac{1 \text{ mole}}{36.5 \cancel{\text{g}}}\right) = 4.49 \text{ moles of HCl}$$

Step 3

To find which starting substance is used up, use the following procedure.

a. Write the balanced equation, and place the number of moles of each substance below its formula.

$$\begin{array}{ccccccc} Mg(OH)_2 & + & 2HCl & \longrightarrow & MgCl_2 & + & 2H_2O \\ 203\text{ g} & & 164\text{ g} & & ?\text{ g} & & ?\text{ g} \\ 3.48\text{ moles} & & 4.49\text{ moles} & & & & \end{array}$$

b. Determine how much HCl would be needed to use up all the $Mg(OH)_2$.

$$? \text{ moles of HCl needed} = (3.48 \cancel{\text{moles } Mg(OH)_2})\left(\frac{2 \text{ moles HCl}}{1 \cancel{\text{mole } Mg(OH)_2}}\right)$$

$$= 6.96 \text{ moles of HCl needed}$$

But we have only 4.49 moles of HCl. Therefore:

c. Our limiting reactant must be HCl, and the $Mg(OH)_2$ must be the reactant in excess. We can check this by calculating the amount of $Mg(OH)_2$ needed to react with all the HCl.

$$? \text{ moles of } Mg(OH)_2 \text{ needed} = (4.49 \cancel{\text{ moles HCl}})\left(\frac{1 \text{ mole } Mg(OH)_2}{2 \cancel{\text{ moles HCl}}}\right)$$

$$= 2.25 \text{ moles of } Mg(OH)_2 \text{ needed (or 131 g)}$$

We have 3.48 moles of $Mg(OH)_2$, so we can expect to have $3.48 - 2.25 = 1.23$ moles of $Mg(OH)_2$ left over, or in excess.

Step 4

Using the number of moles of HCl that we have, calculate the moles of $MgCl_2$ and H_2O that will be formed.

$$? \text{ moles of } MgCl_2 = (4.49 \cancel{\text{ moles HCl}})\left(\frac{1 \text{ mole } MgCl_2}{2 \cancel{\text{ moles HCl}}}\right)$$

$$= 2.25 \text{ moles of } MgCl_2$$

$$? \text{ moles of } H_2O = (4.49 \cancel{\text{ moles HCl}})\left(\frac{2 \text{ moles } H_2O}{2 \cancel{\text{ moles HCl}}}\right)$$

$$= 4.49 \text{ moles of } H_2O$$

Step 5

Calculate the number of grams of $MgCl_2$ and H_2O formed.

$$? \text{ grams of } MgCl_2 = (2.25 \cancel{\text{ moles}})\left(\frac{95.3 \text{ g}}{1 \cancel{\text{ mole}}}\right) = 214 \text{ g of } MgCl_2$$

$$? \text{ grams of } H_2O = (4.49 \cancel{\text{ moles}})\left(\frac{18.0 \text{ g}}{1 \cancel{\text{ mole}}}\right) = 80.8 \text{ g of } H_2O$$

$Mg(OH)_2$ +	2HCl	$\longrightarrow$	$MgCl_2$ +	$2H_2O$
131 g	164 g		214 g	80.8 g
2.25 moles	4.50 moles		2.25 moles	4.50 moles

Step 6

Check your mathematics by seeing whether the reaction obeys the Law of Conservation of Mass.

Mass of Reactants		**Mass of Products**	
Grams of $Mg(OH)_2$	= 131 g	Grams of $MgCl_2$	= 214 g
Grams of HCl used	= 164 g	Grams of H_2O	= 80.8 g
Total mass used	= 295 g	Total mass produced	= 294.8 g (or 295)

As we expected, we had 72 g of extra $Mg(OH)_2$. This material remained unchanged during the course of the reaction.

Example 11.3

Chloroform ($CHCl_3$), a quick-acting powerful anesthetic (often used in spy stories to incapacitate the victim), can decompose when it reacts with oxygen. The products formed are HCl and deadly phosgene gas ($COCl_2$), a substance so harmful when inhaled that it was used as a poison gas against enemy troops in World War I. How many grams of $COCl_2$ and HCl can be formed from 35.9 grams of $CHCl_3$ and 6.40 grams of O_2?

Solution Follow the procedure we used before.

Step 1

$$\begin{array}{ccccc} 2CHCl_3 & + \; O_2 & \longrightarrow & 2COCl_2 & + \; 2HCl \\ 35.9\text{ g} & 6.40\text{ g} & & ?\text{ g} & ?\text{ g} \end{array}$$

Step 2

$$? \text{ moles of } CHCl_3 = (35.9 \cancel{\text{g}})\left(\frac{1 \text{ mole}}{119.5 \cancel{\text{g}}}\right) = 0.300 \text{ mole of } CHCl_3$$

$$? \text{ moles of } O_2 = (6.40 \cancel{\text{g}})\left(\frac{1 \text{ mole}}{32.0 \cancel{\text{g}}}\right) = 0.200 \text{ mole of } O_2$$

Step 3

Determine which of the substances is the limiting reactant.

$$\begin{array}{ccccc} 2CHCl_3 & + \; O_2 & \longrightarrow & 2COCl_2 & + \; 2HCl \\ 35.9\text{ g} & 6.40\text{ g} & & ?\text{ g} & ?\text{ g} \\ 0.300\text{ mole} & 0.200\text{ mole} & & & \end{array}$$

$$? \text{ moles of } CHCl_3 \text{ needed} = (0.200 \cancel{\text{mole } O_2})\left(\frac{2 \text{ moles } CHCl_3}{1 \cancel{\text{mole } O_2}}\right)$$

$$= 0.400 \text{ mole of } CHCl_3 \text{ needed}$$

We would need 0.400 mole of $CHCl_3$ to use all the oxygen. However, we do not have 0.400 mole of $CHCl_3$. Therefore the $CHCl_3$ must be the limiting reactant, and the O_2 must be the reactant in excess. Let's check this.

$$? \text{ moles of } O_2 \text{ needed} = (0.300 \cancel{\text{mole } CHCl_3})\left(\frac{1 \text{ mole } O_2}{2 \cancel{\text{moles } CHCl_3}}\right)$$

$$= 0.150 \text{ mole of } O_2 \text{ needed (or 4.80 g)}$$

We have more than enough oxygen present to use all the chloroform, because we have 0.200 mole of O_2 and we need only 0.150 mole of it.

Step 4

Using the moles of $CHCl_3$ that are present, calculate the moles of $COCl_2$ and HCl that will be formed.

$$? \text{ moles of } COCl_2 = (0.300 \cancel{\text{ mole } CHCl_3})\left(\frac{2 \text{ moles } COCl_2}{2 \cancel{\text{ moles } CHCl_3}}\right)$$

$$= 0.300 \text{ mole of } COCl_2$$

$$? \text{ moles of } HCl = (0.300 \cancel{\text{ mole } CHCl_3})\left(\frac{2 \text{ moles } HCl}{2 \cancel{\text{ moles } CHCl_3}}\right)$$

$$= 0.300 \text{ mole of } HCl$$

Step 5

$$? \text{ grams of } COCl_2 = (0.300 \cancel{\text{ mole}})\left(\frac{99.0 \text{ g}}{1 \cancel{\text{ mole}}}\right) = 29.7 \text{ g of } COCl_2$$

$$? \text{ grams of } HCl = (0.300 \cancel{\text{ mole}})\left(\frac{36.5 \text{ g}}{1 \cancel{\text{ mole}}}\right) = 11.0 \text{ g of } HCl$$

$2CHCl_3$	+	O_2	$\longrightarrow$	$2COCl_2$	+	$2HCl$
35.9 g		4.80 g		29.7 g		11.0 g
0.300 mole		0.150 mole		0.300 mole		0.300 mole

Step 6
Check your mathematics by seeing whether the reaction obeys the Law of Conservation of Mass.

Mass of Reactants		Mass of Products	
Grams of $CHCl_3$ used	= 35.9 g	Grams of $COCl_2$	= 29.7 g
Grams of O_2 used	= 4.8 g	Grams of HCl	= 11.0 g
Total mass used	= 40.7 g	Total mass produced	= 40.7 g

Note that in this reaction we had an excess of 0.050 mole (1.6 g) of O_2. This excess oxygen remained unchanged during the course of the reaction.

Practice Exercise 11.3 Potassium acetate, $KC_2H_3O_2$, is a salt with a molecular mass of 98.1. It can be produced by the reaction of potassium hydroxide with acetic acid. Potassium acetate has been used in veterinary medicine, in combating cardiac arrhythmia (irregular heartbeat), and as an expectorant. It acts as a diuretic in animals. How many grams of potassium acetate can be produced from 28.0 g of potassium hydroxide and 120.0 g of acetic acid?

Summary

Stoichiometry is the calculation of the quantities of elements or compounds involved in chemical reactions. Knowledge of stoichiometry enables chemists to determine the quantities of reactants needed and the quantities of products that are formed. It also permits chemists to produce industrial and commercial chemicals with little or no waste.

The coefficients in a balanced equation indicate the ratio in which molecules (and thus moles) of substances take part in a chemical reaction. The mole method for computing the quantities of reactants and products makes use of this fact. First we find the number of moles of each substance involved in the reaction from this ratio. Then, from this, we find the number of grams of each substance. The same method may be used to determine which of several reactants is the limiting reactant. It may also be used to determine the quantity of product that can be formed with the available amount of the limiting reactant.

Key Terms

Haber process **(Introduction)**
stoichiometry **(11.1)**
synthesize **(Introduction)**

CAREER SKETCH

OCEANOGRAPHER

As an oceanographer, you will use the principles of physics; chemistry; mathematics; and engineering to study the oceans, including their movements; physical properties; and plant and animal life. You will explore and study the oceans by means of surface ships, aircraft, and various underwater craft, using specialized instruments to measure and record your findings. You will also learn to use special cameras equipped with underwater lights to photograph marine life and the ocean floor, and you will use sound devices to locate, measure, and map things that lie beneath the vast surfaces of the world's oceans. Your research will not only extend basic knowledge but will also help develop methods for forecasting weather, expand fisheries, mine ocean resources, and improve national defense capabilities.

Many oceanographers perform most of their work at sea, making observations, conducting experiments, and collecting data on tides, currents, and other phenomena. They may study undersea mountain ranges, interactions of the ocean with the atmosphere, or the layers of sediment on and beneath the ocean floor.

Other oceanographers work almost entirely in a laboratory on land where they measure, dissect, and photograph fish; study sea specimens and plankton; and identify, catalog, and analyze different kinds of sea life and minerals. They may also be involved in plotting maps or using computers to test theories about the oceans.

Oceanographers use a sediment trap (the cone-shaped device) to collect samples of sediment at various depths for laboratory analysis. (Peter Wiebe/Woods Hole Oceanographic Institution)

To become an oceanographer you will need to major in either the physical or biological sciences. You will need to take courses in chemistry, biology, and physics leading to a baccalaureate degree. Many oceanographers continue their studies and obtain either a master's degree or Ph.D. in oceanography.

Self-Test Exercises

Learning Goal 1: Simple Stoichiometry

1. In some parts of the world, when a person is convicted of a capital crime and is sentenced to death, the execution is carried out with poison gas. The following chemical reaction is used to produce the poison gas:

$$H_2SO_4 + KCN \longrightarrow K_2SO_4 + \underset{\text{Hydrogen cyanide}}{HCN}$$

(a) Balance the equation.

(b) How many grams of sulfuric acid are needed to react with 13.0 g of potassium cyanide?

(c) How many grams of potassium sulfate are produced? How many grams of hydrogen cyanide are produced?

2. Carbon disulfide can be produced by the reaction

$$C + S \longrightarrow CS_2$$

CS_2 is a very flammable liquid that is also poisonous. It is used in the manufacture of rayon, carbon tetrachloride, and soil disinfectants. Poisoning usually occurs from inhalation, but can also occur due to ingestion or absorption through the skin.
(a) Balance the equation.
(b) How many grams of carbon are needed to react with 0.160 g of sulfur?
(c) How many grams of carbon disulfide are produced?

▲ **3.** Nitrous oxide (N_2O), commonly called laughing gas, can be prepared from the decomposition of ammonium nitrate:

$$NH_4NO_3 \xrightarrow{\text{Heat}} N_2O + H_2O$$

(a) Balance the equation.
(b) How many grams of ammonium nitrate are needed to prepare 2.2 g of nitrous oxide?
(c) How many grams of water are produced?

4. Lithium phosphate is a white crystalline powder that is soluble in dilute acids. It can be prepared by the reaction

$$H_3PO_4 + LiOH \longrightarrow Li_3PO_4 + H_2O$$

(a) Balance the equation.
(b) How many grams of lithium phosphate can be produced from 168.0 g of lithium hydroxide?
(c) How many grams of phosphoric acid are needed, and how many grams of water are produced?

5. We can prepare the chemical explosive trinitrotoluene (TNT) by the reaction of toluene (C_7H_8) with nitric acid:

$$C_7H_8 + HNO_3 \longrightarrow \underset{\text{TNT}}{C_7H_5N_3O_6} + H_2O$$

(a) Balance the equation.
(b) How many grams of toluene and nitric acid are necessary to manufacture $1,\overline{000}$ g of TNT?
(c) How many grams of water will be produced in the reaction?

▲ **6.** Calcium hydroxide, also known as slaked lime, is used in mortar, plaster, cement, and other building and paving materials. When it reacts with phosphoric acid, it produces calcium phosphate and water.
(a) Write a balanced chemical equation for this reaction.
(b) How many grams of calcium phosphate can be produced from 148 g of calcium hydroxide?
(c) How many grams of phosphoric acid are needed, and how many grams of water are produced?

▲ **7.** The compound ammonium sulfate can be used as a local analgesic (pain reliever). This compound can be prepared by a reaction of ammonium chloride and sulfuric acid.
(a) Write a balanced equation for this reaction.
(b) How many grams of sulfuric acid are needed to react with 15.9 g of ammonium chloride?
(c) How many grams of ammonium sulfate and hydrogen chloride are produced?

▲ **8.** Barium chloride is used in manufacturing pigments and paints, in weighting and dyeing textile fabrics, and in tanning and finishing leather. When it reacts with potassium phosphate, the products formed are potassium chloride and barium phosphate.
(a) Write a balanced equation for this reaction.
(b) How many grams of potassium chloride can be produced from 0.208 g of barium chloride?
(c) How many grams of potassium phosphate are needed, and how many grams of barium phosphate are produced?

***9.** Suppose you are a chemist who has been asked to analyze a sample containing potassium chlorate ($KClO_3$) and sodium chloride. You are asked to determine the percentage of potassium chlorate in the sample. You can do this by heating a weighed amount of the sample. The potassium chlorate will decompose to potassium chloride and oxygen gas. You can reweigh the sample after heating, to tell how much oxygen was lost. Then you can backtrack to find the number of grams of potassium chlorate that decomposed to yield this mass of oxygen. From this information, you can calculate the percentage of potassium chlorate in the sample.

Assume that the masses of your sample are as follows, and find the percentage of potassium chlorate in the mixture of potassium chlorate and sodium chloride:

Grams of sample before heating = 5.00 g

Grams of sample after heating = 4.00 g

***10.** A sample containing $NaClO_3$ and KCl must be analyzed to determine the percentage of $NaClO_3$. The sample is weighed and then heated, and the $NaClO_3$ decomposes to NaCl and O_2. The sample is weighed after heating to determine the amount of O_2 lost during decomposition. The number of grams of $NaClO_3$ that decomposed to yield the mass of oxygen is calculated. The percentage of $NaClO_3$ in the sample is then calculated. Perform this analysis using the following information:

Grams of sample before heating = 10.00 g

Grams of sample after heating = 8.00 g

11. Hydrogen peroxide (H_2O_2) is a compound with many uses. In dilute solutions (3% H_2O_2 in water), it can be used as an antiseptic for cuts. In a solution of 90% concentration, it can be used as a fuel for rockets. H_2O_2 is prepared in the following way:

$$\underset{\text{Barium peroxide}}{BaO_2} + H_3PO_4 \longrightarrow H_2O_2 + Ba_3(PO_4)_2$$

(a) Balance the equation.
(b) How many grams of hydrogen peroxide can be made from 338 g of barium peroxide?
(c) How much phosphoric acid is needed to react with 338 g of barium peroxide?

12. Lead(II) iodide is a chemical used in bronzing, printing, and photography, as well as in making gold pencils. It is composed of lead and iodine.
(a) Write a balanced equation for the formation of lead(II) iodide from its elements.
(b) How many grams of lead and iodine are needed to produce 1.38 g of lead(II) iodide?

***13.** Suppose you have a solution containing 588 g of sulfuric acid, and you put $20\bar{0}$ g of magnesium into it. A reaction occurs. After the reaction is complete, 56 g of the original magnesium remain. How many grams of hydrogen were produced?

***14.** Suppose you have a solution containing 216 g of HCl, and you put 100.0 g of magnesium into it. A reaction occurs. After the reaction is complete, 28.00 g of the original magnesium remain. How many grams of hydrogen were produced?

***15.** How many grams of *air* are necessary for a complete combustion of $9\bar{0}.0$ g of ethane (C_2H_6), producing carbon dioxide and water vapor? Assume that the air contains 23.0% oxygen by weight.

***16.** How many grams of *air* are necessary for the complete combustion of 64.0 g of methane (CH_4), producing carbon dioxide and water vapor? Assume that the air contains 23.0% oxygen by weight.

▲ **17.** Butane, C_4H_{10}, is an organic compound used as a fuel in some types of cigarette lighters. When butane is burned with oxygen, the following reaction takes place.

$$C_4H_{10} + O_2 \longrightarrow CO_2 + H_2O \text{ (unbalanced)}$$

(a) Balance the equation.
(b) How many grams of oxygen are needed to burn 23.2 g of butane?
(c) How many grams of carbon dioxide and water are produced?

▲ **18.** Ethane, C_2H_6, is an organic compound that together with 90% propane and 5% butane composes bottled gas. When ethane is burned with oxygen, the following reaction takes place:

$$C_2H_6 + O_2 \longrightarrow CO_2 + H_2O \text{ (unbalanced)}$$

(a) Balance the equation.
(b) How many grams of oxygen are needed to burn $18\bar{0}$ g of ethane?
(c) How many grams of carbon dioxide and water are produced?

19. Ozone, O_3, is a pale-blue gas that can be formed in the atmosphere from O_2. The reaction is

$$3O_2 \longrightarrow 2O_3$$

How many grams of ozone can be formed from 64.0 g of O_2?

▲ **20.** Ozone (O_3) is formed from oxygen gas in the reaction

$$3O_2 \longrightarrow 2O_3$$

How many grams of oxygen gas are needed to form 0.048 g of ozone?

21. Aluminum sulfate has been used to tan leather, to treat sewage, and as an antiperspirant. The compound can be formed by the acid–base reaction

$$Al(OH)_3 + H_2SO_4 \longrightarrow Al_2(SO_4)_3 + H_2O \text{ (unbalanced)}$$

(a) Balance the equation.
(b) How many grams of aluminum hydroxide and sulfuric acid are needed to produce $5\bar{0}0$ g of aluminum sulfate?
(c) How many grams of water are produced?

22. Zinc sulfide is used as a pigment for paints, oilcloths, linoleum, and leather. It is composed of the elements zinc and sulfur.
(a) Write a balanced equation for the formation of zinc sulfide from its elements.
(b) How many grams of zinc and sulfur are needed to produce 2.44 g of zinc sulfide?

23. In a blast furnace, iron(III) oxide is converted to iron metal. The overall reaction for this process is

$$Fe_2O_3 + CO \longrightarrow Fe + CO_2 \text{ (unbalanced)}$$

(a) Balance the equation.
(b) How many grams of iron(III) oxide are needed to produce 454.0 g of iron metal (about one pound)?

(c) How many grams of CO are needed, and how many grams of CO_2 are produced?

24. Silver nitrate, a topical anti-infective agent, reacts with copper to form copper(II) nitrate and silver metal.
(a) Write a balanced equation for this reaction.
(b) How many grams of copper are needed to react with 1.70 g of silver nitrate?
(c) How many grams of silver metal and copper(II) nitrate are formed from the reaction?

25. Nitric acid is prepared commercially by the *Ostwald process*. The first part of this process involves the reaction of ammonia with oxygen to produce nitric oxide, using platinum as a catalyst:

$$NH_3 + O_2 \xrightarrow{Pt} NO + H_2O \text{ (unbalanced)}$$

(a) Balance the equation.
(b) How many grams of oxygen are needed to react with 425 g of ammonia?
(c) How many grams of nitric oxide and water are produced?

26. Glucose, $C_6H_{12}O_6$, reacts with oxygen to produce carbon dioxide and water.
(a) Write a balanced equation for this reaction.
(b) How many grams of glucose are needed to react with 3.20 g of oxygen?
(c) How many grams of carbon dioxide and water are produced?

***27.** Suppose that you have a solution of 500 g of dissolved $Cu(NO_3)_2$ and you add 300 g of Zn metal to it. A reaction occurs. After the reaction is complete, 127 g of Zn remain. How many grams of $Zn(NO_3)_2$ are produced?

28. Hydrogen gas reacts with iodine to produce hydrogen iodide. Hydrogen iodide is a strong irritant that is used in the manufacture of hydroiodic acid.
(a) Write a balanced equation for the reaction between hydrogen and iodine.
(b) How many grams of hydrogen gas and iodine are needed to produce 320 g of hydrogen iodide?

Learning Goal 2: Limiting-Reactant Problems

▲ **29.** The compound acetylene (C_2H_2) is used as a fuel for welding (for example, in the oxyacetylene torch). In years gone by, acetylene was used as a surgical anesthetic. The compound results from the reaction

$$\underset{\text{Calcium carbide}}{CaC_2} + H_2O \longrightarrow Ca(OH)_2 + \underset{\text{Acetylene}}{C_2H_2} \text{ (unbalanced)}$$

(a) Balance this equation.
(b) If you have 128 g of calcium carbide and 144 g of water, how many grams of calcium hydroxide and acetylene can you produce?

▲ **30.** Chloroform, $CHCl_3$, reacts with oxygen to form HCl and deadly phosgene gas, $COCl_2$, according to the equation

$$2CHCl_3 + O_2 \longrightarrow 2COCl_2 + 2HCl$$

How many grams of $COCl_2$ and HCl can be formed from 236 g of $CHCl_3$ and 64.0 g of O_2?

31. Methyl alcohol (CH_3OH), sometimes called wood alcohol, is an extremely poisonous substance when taken internally. Death can result from drinking just 30 mL of it; smaller amounts can produce nausea, convulsions, blindness, and respiratory failure. Methyl alcohol is usually obtained by distilling wood. But it can be synthesized in the laboratory by a reaction of carbon monoxide and hydrogen gas (under high temperature and pressure).

$$CO + H_2 \xrightarrow[\text{3,000 lb/in}^2\text{ pressure}]{\text{300 to 400°C}} CH_3OH \text{ (unbalanced)}$$

(a) Balance the equation.
(b) How many grams of methyl alcohol can be produced from 200 g of carbon monoxide and 40 g of hydrogen?

32. Nitrogen gas reacts with hydrogen gas to form ammonia gas (NH_3). How many grams of ammonia gas can be formed from 0.014 g of N_2 and 0.020 g of H_2?

▲ **33.** Water can be produced from its elements by the reaction

$$H_2 + O_2 \xrightarrow{\text{Electric spark}} H_2O \text{ (unbalanced)}$$

(a) Balance the equation.
(b) How many grams of water can be produced from 10 g of H_2 and 64 g of O_2?

34. Propane, C_3H_8, reacts with oxygen gas to form carbon dioxide and water according to the equation

$$C_3H_8 + 5O_2 \longrightarrow 3CO_2 + 4H_2O$$

How many grams of carbon dioxide and water can be formed from 11.0 g of propane and 32 g of oxygen gas?

35. When calcium phosphide is placed in water, it forms phosphine gas (PH_3). In air, PH_3 usually bursts into flames. In fact, this substance (PH_3) may be responsible for people observing faint flickers of light in marshes. The PH_3 could be produced by the reduction of naturally occuring phosphorus compounds. The reaction for the formation of PH_3 from calcium phosphide is

$$Ca_3P_2 + H_2O \longrightarrow PH_3 + Ca(OH)_2 \text{ (unbalanced)}$$

(a) Balance the equation.
(b) How many grams of PH_3 can be produced from 515 g of calcium phosphide and 216 g of water?
(c) How many grams of calcium hydroxide are produced?

▲ **36.** Aluminum reacts with iron(III) oxide to produce iron and aluminum oxide.
(a) Write a balanced equation for this reaction.
(b) How many grams of aluminum oxide and iron can be produced from 135 g of Al and $8\overline{00}$ g of iron(III) oxide?

37. Elementary phosphorus can be produced by the reaction

$$Ca_3(PO_4)_2 + 3SiO_2 + 5C \longrightarrow 3CaSiO_3 + 5CO + P_2$$

Determine the number of moles of each product formed from 8.0 moles of $Ca_3(PO_4)_2$, $2\overline{0}$ moles of SiO_2, and 45 moles of C.

38. Barium metal reacts with oxygen gas to form barium oxide, a chemical used for drying gases and solvents.
(a) Write a balanced equation for this reaction.
(b) How many grams of barium oxide can be produced from $1\overline{00}$ g of barium and $1\overline{00}$ g of oxygen gas?

39. An inexpensive way of preparing pure hydrogen gas involves the following reaction between iron and steam:

$$Fe + H_2O \longrightarrow Fe_3O_4 + H_2 \text{ (unbalanced)}$$

(a) Balance the equation.
(b) How many grams of hydrogen gas can be prepared from 225 g of Fe and 225 g of H_2O?
(c) How many grams of Fe_3O_4 are produced?

40. Sulfur dioxide is a chemical used in preserving fruits and vegetables and as a disinfectant in breweries and food factories. It is produced from the reaction of sulfur and oxygen gas.
(a) Write a balanced equation for this reaction.
(b) How many grams of sulfur dioxide can be prepared from 100.0 g of sulfur and 200.0 g of oxygen gas?

Extra Exercises

▲ **41.** *(a)* Write the balanced equation for the reaction in which chlorine, Cl_2, reacts with potassium metal, K, to produce potassium chloride. How many moles of chlorine are needed to react with 10.0 g of potassium?
(b) How many grams of potassium chloride are produced?

▲ **42.** Consider the reaction

$$Cu(NO_3)_2 \longrightarrow CuO + NO_2 + O_2 \text{ (unbalanced)}$$

(a) Balance the equation.
(b) What mass of O_2 is produced if $1\overline{00}$ g of copper(II) nitrate react?
(c) How much copper(II) nitrate must react if 54.0 g of NO_2 are formed?

43. Assume that $1\overline{00}$ g of carbon and $15\overline{0}$ g of oxygen are available for the reaction

$$C + O_2 \longrightarrow CO_2$$

(a) What is the limiting reactant?
(b) How much CO_2 is formed?

44. A reaction is carried out with 8.70 g of hydrogen and 64.0 g of oxygen to produce water.
(a) Write a balanced equation for this reaction.
(b) What amount of each substance will be on hand when the reaction is complete?

45. Calculate the number of grams of oxygen gas produced by the decomposition of 1.00 kg of potassium chlorate. (*Hint:* $KClO_3 \longrightarrow KCl + O_2$ (unbalanced))

46. Write the equation for the reaction between zinc metal and sulfuric acid. What mass of zinc is required to produce $6\overline{00}$ g of hydrogen gas?

47. Two experiments are performed. In the first, a 50.0-g plate of silver is immersed in a solution of copper(II) chloride. In the second experiment, a 50.0-g plate of lead is immersed in the same solution.
(a) Determine in which experiment a reaction occurs, and write a balanced equation for this reaction.
(b) Assuming that the metal is the limiting reactant, calculate the masses of all products formed in the reaction that occurs.

48. Aluminum reacts with sulfuric acid to produce aluminum sulfate and hydrogen gas. How many grams of hydrogen are produced from the reaction of 60.0 g of aluminum with excess sulfuric acid?

49. Hydrochloric acid reacts with sodium hydroxide to produce sodium chloride and water. Suppose that 36.5 g of HCl react with 80.0 g of NaOH. How many grams of sodium chloride are produced? How many grams of the excess reactant are left at the end of the reaction?

50. Carbon reacts with oxygen under certain conditions to produce carbon monoxide. Suppose that 48.0 g of carbon react with 64.0 g of oxygen gas to produce *only* carbon monoxide. How many grams of carbon monoxide are produced?

Cumulative Review/Chapters 9–11

1. How many moles are there in 3.43 g of $Al_2(SO_4)_3$?

2. How many moles are there in 4.8 g of C atoms?

3. Determine the number of grams in each of the following:
(a) 0.750 mole of H_2SO_4
(b) 5.00 moles of H_2O
(c) 0.00700 mole of H_2CO_3
(d) 50.0 moles of $HC_2H_3O_2$

4. After completing an experiment, you find 4.6 g of unreacted Na in the vessel. How many atoms does this represent?

5. After a reaction is complete, it is determined that 5.328 g of Sn combined with 1.436 g of O. Calculate the empirical formula of this compound.

6. The empirical formula for the antibiotic nonactin is $C_{10}H_{16}O_3$. If the molecular mass of nonactin is 736, what is its molecular formula?

7. The empirical formula for fructose, the natural sugar found in fruit juice, fruits, and honey, is CH_2O. If the molecular mass of fructose is $18\bar{0}$, what is its molecular formula?

8. The molecular formula for aspirin is $C_9H_8O_4$. You have 36.0 g of aspirin.
(a) How many moles of aspirin are there?
(b) How many molecules of aspirin are there?
(c) How many moles of carbon atoms are there?
(d) How many moles of hydrogen atoms are there?
(e) How many moles of oxygen atoms are there?

9. A 6.08-g sample of nitrogen combines with 13.90 g of oxygen to produce a compound whose molecular mass is 92.0. What is the molecular formula of this compound?

10. Calculate the molecular mass of each of the following compounds: *(a)* $Ca_3(PO_4)_2$ *(b)* CH_4 *(c)* NO_2 *(d)* $Ba(C_2H_3O_2)_2$

11. Calculate the percentage composition of each of the following compounds: *(a)* H_2S *(b)* NaCl *(c)* $Fe(C_2H_3O_2)_3$ *(d)* K_2SO_4

12. Determine the molecular formula of a compound that has a molecular mass of 86.0 and is composed of 83.7% carbon and 16.3% hydrogen by mass.

13. Find the percentage by mass of carbon, hydrogen, and nitrogen in the compound nicotine, given that its empirical formula is C_5H_7N and that its molecular mass is 162.0.

14. How many water molecules are there in a 0.720-g sample of H_2O?

15. A patient has been told to take a $1,\overline{000}$-mg supplement of elemental calcium each day. How many grams of $CaCO_3$ must she take in order to receive $1,\overline{000}$ mg of elemental calcium?

16. Write a balanced chemical equation for the reaction in which calcium bromide + sulfuric acid forms hydrogen bromide + calcium sulfate.

17. When magnesium metal reacts with oxygen, magnesium oxide is formed. Write a balanced equation for this reaction.

18. A flask found in a chemistry laboratory was coated with a thin film that had a cloudy appearance. Chemical analysis of the film showed it to be ammonium chloride formed by reaction of ammonia gas and hydrogen chloride gas. Write the balanced equation for this reaction.

19. Identify each of the following reactions as a combination, decomposition, single-replacement, or double-replacement reaction:
(a) $C_{12}H_{22}O_{11} \longrightarrow 12C + 11H_2O$
(b) $H_2 + Br_2 \longrightarrow 2HBr$
(c) $2KBr + Cl_2 \longrightarrow 2KCl + Br_2$
(d) $2HCl + Ca(OH)_2 \longrightarrow CaCl_2 + 2H_2O$

20. Complete and balance each of the following equations. Use the activity series and solubility table when necessary.
(a) $H_2 + Cl_2 \longrightarrow ?$
(b) $NaCl(aq) + Pb(NO_3)_2(aq) \longrightarrow ? + ?$
(c) $Al + H_2SO_4 \longrightarrow ? + ?$
(d) $SrCO_3 \xrightarrow{\text{Heat}} ? + ?$

21. Complete and balance the following equation:

$$CaCO_3 \xrightarrow{\text{Heat}} ? + ?$$

22. Balance the following equations:
(a) $TiCl_4 + H_2O \longrightarrow TiO_2 + HCl$
(b) $P_4O_{10} + H_2O \longrightarrow H_3PO_4$

23. Complete and balance the following equations. If necessary, check the solubility table to be sure that the reaction can take place.

(a) $Fe(OH)_3 + H_3PO_4 \longrightarrow ? + ?$
(b) $Pb(OH)_2 + HNO_3 \longrightarrow ? + ?$

24. Acid rain is formed in the atmosphere by the reaction of sulfur trioxide (an air pollutant) and water, forming sulfuric acid. Complete and balance the equation describing this process.

25. In an automobile engine, fuel that contains a chemical called octane (C_8H_{18}) reacts with oxygen to form carbon dioxide and water. Write a balanced equation for this reaction.

26. Isopropyl alcohol (C_3H_7OH), or rubbing alcohol, reacts with oxygen to produce carbon dioxide and water. Write a balanced equation for this reaction.

27. Hard water containing $CaSO_4$ can be softened by adding washing soda (Na_2CO_3). Write a balanced chemical equation for this reaction. Which chemical precipitates out, causing the water to soften?

28. When the sugar sucrose ($C_{12}H_{22}O_{11}$) is heated, it decomposes to form carbon and water. Write a balanced chemical equation for this reaction.

29. In the following reaction, the sum of the coefficients is 9. True or false? (*Hint:* Balance the equation before answering.)

$$Al + H_2SO_4 \longrightarrow H_2 + Al_2(SO_4)_3$$

30. In the following reaction, the sum of the coefficients is 4. True or false? (*Hint:* Balance the equation before answering.)

$$NaNO_3 \longrightarrow NaNO_2 + O_2$$

31. In the following reaction, metallic zinc is oxidized and hydrogen ions are reduced. True or false?

$$Zn + H_2SO_4 \longrightarrow ZnSO_4 + H_2$$

32. In the following reaction, sodium is oxidized and chlorine is reduced. True or false?

$$Na + Cl_2 \longrightarrow NaCl$$

33. In the following reaction, tin is the reducing agent and HNO_3 is the oxidizing agent. True or false?

$$Sn + HNO_3 \longrightarrow SnO_2 + NO_2 + H_2O$$

34. The sum of the coefficients for the following oxidation–reduction reaction is 38. True or false?

$$K_2Cr_2O_7 + FeCl_2 + HCl \longrightarrow CrCl_3 + KCl + FeCl_3 + H_2O \text{ (unbalanced)}$$

35. Balance the following redox reaction:

$$C + H_2SO_4 \longrightarrow CO_2 + SO_2 + H_2O$$

36. When copper(II) oxide reacts with hydrogen, copper and water are produced. How many moles of copper oxide are needed to completely react with 5 moles of hydrogen?

37. Consider the following chemical equation:

$$2Na + 2H_2O \longrightarrow 2NaOH + H_2$$

How many moles of sodium hydroxide are produced when 0.46 mole of sodium reacts with 0.20 mole of water?

38. Calculate the number of grams of propane (C_3H_8) needed to produce 6.0 moles of carbon dioxide by the following reaction:

$$C_3H_8 + 5O_2 \longrightarrow 3CO_2 + 4H_2O$$

39. How many grams of hydrochloric acid are needed to produce 340 g of hydrogen sulfide by the following reaction?

$$FeS + HCl \longrightarrow FeCl_2 + H_2S \text{ (unbalanced)}$$

40. A 30.0-g sample of ethane (C_2H_6) is allowed to react with 16.0 g of oxygen to produce carbon dioxide and water. How many grams of carbon dioxide will this reaction produce?

41. How many grams of phosphoric acid are needed to react with 1,360 g of lithium hydroxide in the following reaction?

$$H_3PO_4 + 3LiOH \longrightarrow Li_3PO_4 + 3H_2O$$

42. Calcium hydroxide reacts with phosphoric acid to produce calcium phosphate and water. How many grams of calcium phosphate can be produced from 49.0 grams of phosphoric acid? How many grams of calcium hydroxide are needed?

43. When the compound Epsom salts (magnesium sulfate heptahydrate, $MgSO_4 \cdot 7H_2O$) is heated, 1 mole of magnesium sulfate and 7 moles of water are produced. How many moles of magnesium sulfate are produced when 15.0 moles of Epsom salts are heated? How many grams is this?

44. When hydrogen peroxide (H_2O_2) decomposes, water and oxygen are produced. How many grams of hydrogen peroxide are needed to produce 63.0 grams of water?

45. When sodium chloride and lead(II) nitrate react, lead(II) chloride solid and sodium nitrate result. When 66.4 g of lead(II) nitrate react with 5.85 g of sodium chloride, how many grams of lead(II) chloride form?

46. Carbon dioxide gas reacts with lithium hydroxide to form lithium carbonate and water. How many grams of lithium hydroxide are needed to produce 148 g of lithium carbonate?

47. Magnesium and oxygen react to form magnesium oxide. How many grams of magnesium are required to react with 48.0 g of oxygen?

48. Barium and water form barium hydroxide and hydrogen gas upon reaction. When 137.3 g of barium are mixed with 72.0 g of water, how much barium hydroxide and how much hydrogen are produced? Which reactant is in excess and by how much?

12 Heats of Reaction: Chemistry and Energy

Learning Goals

After you have studied this chapter, you should be able to:

1. Distinguish between exothermic reactions and endothermic reactions.
2. Define the terms *heat of reaction, heat of formation,* and *enthalpy*.
3. Define the terms *calorie* and *specific heat*.
4. Calculate the amount of heat gained or lost by a substance when its temperature changes.
5. Write a thermochemical equation.
6. Explain the significance of positive and negative values of ΔH.
7. Calculate the heat of formation or heat of reaction when you are given a balanced equation and the appropriate information.

Introduction

Turn on a *light,* take a ride in your *car,* cook a roast in your *oven,* or just turn on your portable *radio*. All these devices use energy, and this energy usually comes from chemical reactions. The entire world operates on energy. In fact, the use of energy from chemical reactions is the basis of life itself. In this chapter, we discuss the changes in energy that occur in chemical reactions, and we examine some of the practical uses of energy obtained from various reactions.

12.1 The Release or Absorption of Energy

Learning Goal 1
Exothermic and endothermic reactions

Associated with every substance is a quantity of energy (a sort of chemical potential energy). When a substance reacts to form a new substance, it can either lose energy to the surroundings or gain energy from the surroundings. This is because in chemical reactions, old bonds between atoms are broken and new bonds are formed. Breaking a bond usually requires energy, while forming a bond usually releases energy. Depending on the number of bonds broken and the number formed, and depending on the strengths of the bonds,

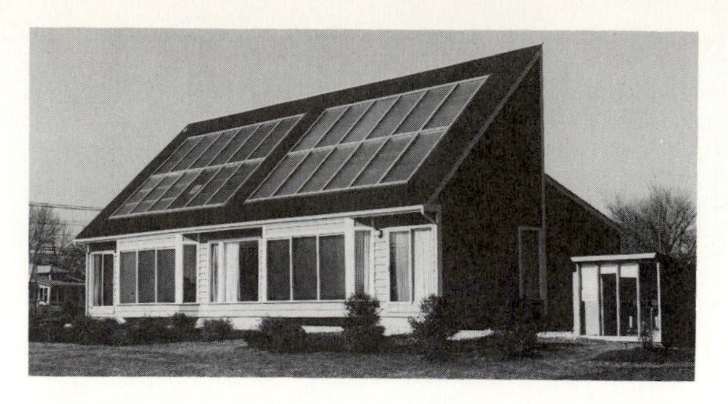

Solar energy is our most powerful energy source. Solar panels on the roof of this experimental house provide the occupants with most of the heat and hot water they need.
(Photograph courtesy of Dr. K. W. Böer)

energy is either released to the surroundings or absorbed from them. This energy is usually in the form of heat.

Reactions that release energy to the surroundings are called **exothermic reactions.** For example, the explosion of a stick of dynamite is a chemical reaction that releases a tremendous quantity of heat energy to the environment. *Reactions that absorb energy from the surroundings* are called **endothermic reactions.** For example, plants live and grow by a process called photosynthesis. In this process, plants use carbon dioxide and water and they absorb *energy from the sun* (sunlight) to make glucose and oxygen.

12.2 Heat and the Chemical Reaction

Learning Goal 2
Definitions of energy terms

Any collection of molecules contains an energy that, if released by chemical reaction, can produce heat effect: a tangible energy effect you can feel with your hands. *The heat content of a specific amount of a particular substance* is called its **enthalpy.** (The symbol for enthalpy is H.) One mole of any substance at a given temperature and pressure has a specific enthalpy, which may be thought of as its heat content. We cannot measure the heat content (enthalpy) of a substance itself. However, we can measure the *change* in heat content that takes place during a chemical reaction. That is, the heat that is absorbed or released during a chemical reaction can be measured. *The change in heat content that results from a chemical reaction* is defined as the **heat of reaction.**

The heat of reaction is usually measured at 25°C and 1 atmosphere (atm) pressure. (The temperature 25°C was probably chosen because it is close to room temperature. The pressure 1 atmosphere—normal atmospheric pressure at sea level—is the pressure in most laboratories. This temperature and pressure are called *standard-state conditions*.) The reactants are kept at a fixed temperature before the reaction, and the products are brought back to the

Learning Goal 3
Definitions of calorie and specific heat

same temperature after the reaction. This allows us to measure only the heat given up or absorbed during the reaction, and not heat involved in temperature or pressure changes. We measure this heat in units of calories. One **calorie** is *the quantity of heat needed to raise the temperature of 1 gram of water by 1 Celsius degree*. If we were to heat 1 gram of water from 20°C to 21°C we might ask how much energy is absorbed. With our knowledge of the definition of a calorie, the answer is 1 calorie. Next we might ask how much energy 1 gram of water absorbs if its temperature changes from 20°C to 25°C. The answer is 5 calories, or 1 calorie for each degree Celsius rise in temperature. Large heat changes are measured in kilocalories (1,000 cal = 1 kcal). (It is the kilocalorie, also known among nutritionists as the large calorie or Calorie, that dieters count when they are trying to lose weight.)

With the adoption of the SI metric system (the revised metric system) in 1960, scientists recognized the *joule* as the basic unit of energy, replacing the calorie. However, the calorie is still used by many scientists to express the quantity of energy something contains, and it may be some time before it disappears from use. In case you find it necessary to convert from calories to joules or vice versa, remember that 1 calorie equals 4.18 joules.

$$1 \text{ calorie} = 4.18 \text{ joules}$$

Example 12.1

One liter of water (1,$\overline{000}$ g) is heated from 2$\overline{0}$°C to 8$\overline{0}$°C. How much energy does the water absorb?

Solution

Understand the Problem

We want to find out how many calories are absorbed by a liter of water when its temperature is raised by 60 Celsius degrees.

Devise a Plan

Given our definition of the calorie, we can say that every gram of water that we heat by 1 Celsius degree absorbs 1 calorie of heat energy. This can be stated as

Heat, in calories = (mass of water, in grams) × (temperature change, in °C)

We can abbreviate this relationship as

$$Q_{\text{cal}} = (m)(\Delta t) \qquad \text{(for water only)}$$

The symbol Δ (the Greek capital letter delta) means "change in." The mass of water heated and the number of degrees that the water was heated can be substituted into this relationship to obtain the number of calories.

Carry Out the Plan

Substitute the given values into the abbreviated relationship.

$$
\begin{aligned}
Q_{\text{cal}} &= (m)(\Delta t) \qquad \text{(for water only)} \\
&= (1,\overline{000}\text{ g})(8\overline{0}^{\circ}\text{C} - 2\overline{0}^{\circ}\text{C}) = (1\overline{000}\text{ g})(6\overline{0}^{\circ}\text{C}) \\
&= 6\overline{0},000 \text{ cal} \qquad \text{or} \qquad 6\overline{0} \text{ kcal}
\end{aligned}
$$

Look Back

We check to see whether the solution makes sense. If 1 calorie is required to raise the temperature of 1 gram of water by 1 Celsius degree, it makes sense that 6,000 times more heat would be required to raise the temperature of 1,000 g of water by 60 Celsius degrees.

Practice Exercise 12.1 You heat $2,\overline{000}$ mL of water ($2,\overline{000}$ g) from $5\overline{0}^{\circ}$C to $7\overline{0}^{\circ}$C. How much energy does the water absorb?

12.3 Specific Heat

The formula used in Example 12.1,

$$Q_{\text{cal}} = (m)(\Delta \text{t})$$

works only for water. However, we can use this formula to perform calculations on other substances once we understand the concept of *specific heat*. The **specific heat** of a substance is *the number of calories needed to raise the temperature of 1 gram of the substance by 1 Celsius degree*. The specific heat of water is taken to be 1.00 calorie per gram per degree Celsius, or 1.00 cal/(g)(°C). Table 12.1 gives the specific heats of some other substances.

Table 12.1
Specific Heat of Some Substances at 25°C

Substance	Specific heat (c) in cal/(*g*)(°C)
Aluminum	0.215
Copper	0.0920
Ethyl alcohol	0.581
Gold	0.0308
Iron	0.107
Nickel	0.106
Zinc	0.0922
Water (liquid)	1.000

Learning Goal 4
Calculating calories

The general formula, then, for calculating the heat (in calories) gained or lost by any substance is

$$Q_{\text{cal}} = (c)(m)(\Delta t)$$

where c is the symbol for specific heat. Let's work some examples using this formula.

Example 12.2

How much heat is absorbed when $50\overline{0}$ g of water are heated from 25.0°C to 75.0°C?

Solution We write the general formula and substitute known values. Remember that the specific heat c for water is 1.00 cal/(g)(°C).

$$\begin{aligned} Q_{\text{cal}} &= (c)(m)(\Delta t) \\ &= \left(1.00 \frac{\text{cal}}{(\cancel{\text{g}})(\cancel{°\text{C}})}\right)(50\overline{0}\ \cancel{\text{g}})(50.0\cancel{°\text{C}}) \\ &= 25{,}\overline{0}00 \text{ cal} \quad \text{or} \quad 25.0 \text{ kcal} \end{aligned}$$

Practice Exercise 12.2 How much heat is absorbed when 250.0 g of water are heated from 10.0°C to 60.0°C?

Example 12.3

How much heat is needed to raise the temperature of a $2\overline{0}0$-g sample of aluminum from 20.0°C to 50.0°C?

Solution Again we write the general formula and then substitute known values. Table 12.1 tells us that c for aluminum is 0.215 cal/(g)(°C). And the change in temperature is 50.0°C − 20.0°C, or 30.0°C. Then,

$$\begin{aligned} Q_{\text{cal}} &= (c)(m)(\Delta t) \\ &= \left(0.215 \frac{\text{cal}}{(\cancel{\text{g}})(\cancel{°\text{C}})}\right)(2\overline{00}\ \cancel{\text{g}})(30.0\cancel{°\text{C}}) \\ &= 1{,}290 \text{ cal} \quad \text{or} \quad 1.29 \text{ kcal} \end{aligned}$$

Practice Exercise 12.3 Find the amount of heat needed to raise the temperature of a 500.0-g sample of aluminum from 25.0°C to 75.0°C.

Example 12.4

If we had heated iron in place of aluminum in Example 12.3, how much heat would we have needed?

Solution We solve this problem exactly as we did Example 12.3. However, the specific heat of iron is 0.107 cal/(g)(°C).

$$Q_{cal} = (c)(m)(\Delta t)$$
$$= \left(0.107 \frac{cal}{(\cancel{g})(^\circ\cancel{C})}\right)(2\overline{0}0\ \cancel{g})(30.0^\circ\cancel{C})$$
$$= 642\ cal \qquad or \qquad 0.642\ kcal$$

(Iron requires about half as much heat as aluminum requires.)

Practice Exercise 12.4 Repeat Practice Exercise 12.3, substituting iron for aluminum.

12.4 The Calorimeter

An instrument called a *calorimeter* can be used to measure changes in the heat content of substances when they undergo chemical reactions. For example, the *bomb calorimeter* shown in Figure 12.1 is frequently used to measure heat-content changes in substances reacting with oxygen.

Suppose we want to measure the heat of reaction of glucose as it reacts with oxygen. We obtain a sample of glucose, say 1.80 g. We place it inside a special heat-transferring container that fits into the calorimeter. Then we position the container in the calorimeter along with an ignition device that will cause a spark, which will allow the sugar and oxygen to react. The calorimeter is filled with a known amount of water, say 1,$\overline{0}$00 g. We measure the temperature of the water and find it to be 25.0°C. We press the ignition switch, and a large quantity of heat is liberated by the reaction between glucose and oxygen.

$$C_6H_{12}O_6(s) + 6O_2(g) \longrightarrow 6CO_2(g) + 6H_2O(l) + \text{energy}$$

The heat passes through the heat-transferring container to the water, the temperature of which rises to 31.7°C. Now we can calculate how much heat is required to raise the temperature of 1,$\overline{0}$00 g of water from 25.0°C to 31.7°C. This is exactly the amount of heat that was liberated by the glucose–oxygen reaction. We proceed as follows:

$$Q_{cal}\ \text{(absorbed by water)} = (c_{water})(m_{water})(\Delta t_{water})$$
$$= \left(1.0 \frac{cal}{(g)(^\circ C)}\right)(1{,}\overline{0}00\ g)(31.7^\circ C - 25.0^\circ C)$$

$$= \left(1.0 \frac{\text{cal}}{(\text{g})(^\circ\text{C})}\right)(1{,}\overline{0}00\ \text{g})(6.7^\circ\text{C})$$

$$= 6{,}700 \text{ cal} \qquad \text{or} \qquad 6.7 \text{ kcal}$$

$$\text{Heat gained by water} = \text{Heat produced by reaction}$$

Thus 6.7 kcal of heat energy must have been available to raise the temperature of the water. But this energy could have come only from the glucose–oxygen reaction. We may therefore say that the reaction of 1.80 g of glucose with oxygen released 6.7 kcal of energy.

It is customary to express heats of reaction in units of kilocalories per mole. To do so for the glucose–oxygen reaction, we first note that the molecular mass of glucose is 180 g/mole. Then,

$$? \frac{\text{kcal}}{\text{mole}} = \left(\frac{6.7 \text{ kcal}}{1.80 \text{ g}}\right)\left(\frac{18\overline{0} \text{ g}}{1 \text{ mole}}\right) = 670 \frac{\text{kcal}}{\text{mole}}$$

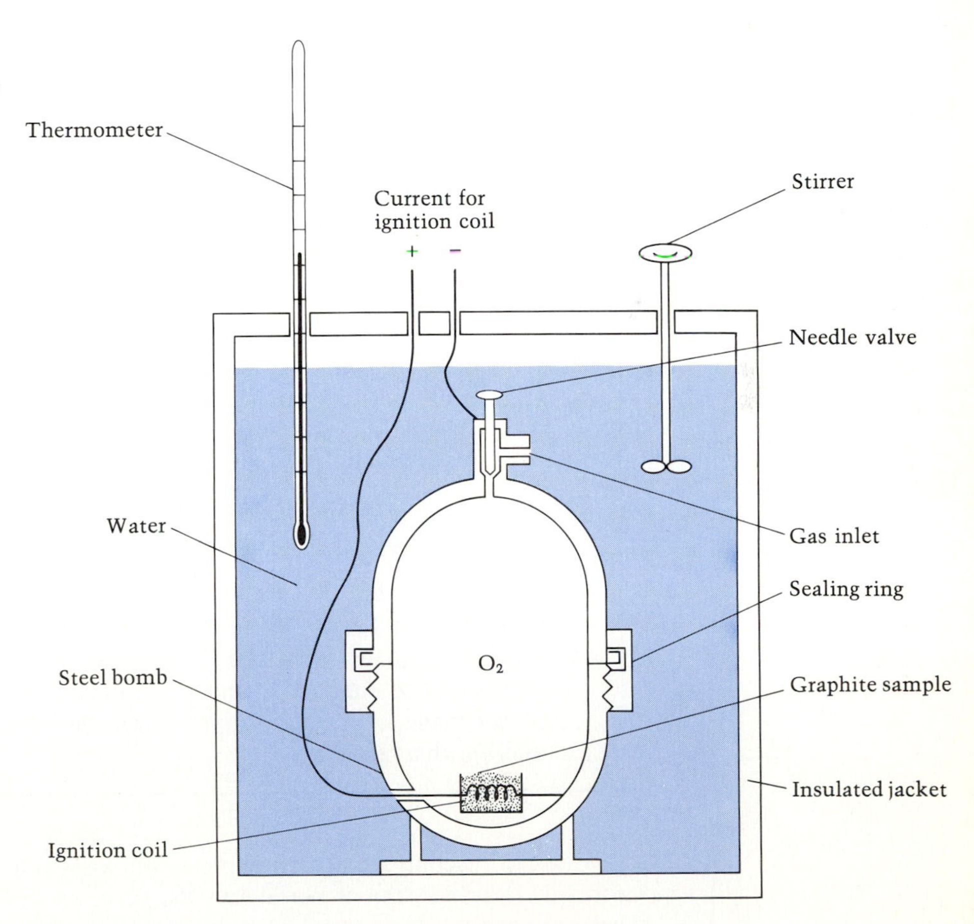

Figure 12.1
A bomb calorimeter to determine the heat of combustion of C(*s*)

Now, in the equation for the glucose–oxygen reaction, we can indicate the exact heat of reaction:

$$C_6H_{12}O_6(s) + 6O_2(g) \longrightarrow 6CO_2(g) + 6H_2O(l) + 670 \text{ kcal}$$

An equation in this form is called a *thermochemical equation*. A **thermochemical equation** is *a chemical equation that is balanced, indicates the physical state of each substance (solid* s, *liquid* l, *or gas* g), *and shows the amount of heat liberated or absorbed during the reaction*. The foregoing equation states that 1 mole of glucose in solid form reacts with 6 moles of gaseous oxygen to form 6 moles of carbon dioxide gas and 6 moles of liquid water, releasing 670 kcal for each mole of glucose that reacts. Note that this also means that 670 kcal are released for every 6 moles of oxygen that react with the glucose, or for every 6 moles of CO_2 or every 6 moles of water that are formed.

12.5 Writing Thermochemical Equations

Learning Goal 5
Writing thermochemical equations

Writing a thermochemical equation is simply a matter of obtaining the necessary information and putting it in the proper form. Here are some examples that illustrate the procedure and a few of the "fine points."

Example 12.5

When 1 mole of carbon (graphite) reacts with oxygen at 25°C and 1 atm pressure, carbon dioxide is formed. In addition, 94.1 kcal of heat are released per mole of carbon dioxide formed. Write the thermochemical equation for this reaction.

Solution

$$\text{C(graphite)} + O_2(g) \longrightarrow CO_2(g) + 94.1 \text{ kcal}$$

The reactants (C and O_2) have a total heat content that is 94.1 kcal greater than the product (CO_2). The heat is given up to the surroundings when the product is formed.

Practice Exercise 12.5 When ethane, $C_2H_6(g)$, reacts with oxygen gas, the products formed are carbon dioxide gas and liquid water. When this reaction is carried out at 25°C and 1 atm pressure, 186 kcal of heat are released per mole of carbon dioxide formed. Write the thermochemical equation for this reaction.

Heats of reaction are usually expressed in kilocalories per mole of product formed. However, this is not always the case, so each equation must be examined carefully. Example 12.6 will help explain this point.

Example 12.6

When hydrogen gas reacts with oxygen gas, water is produced and heat is given off. Experiments have shown that 136.6 kcal of heat are produced when *two* moles of water are formed. Write the thermochemical equation for this reaction.

Solution

$$2H_2(g) + O_2(g) \longrightarrow 2H_2O(l) + 136.6 \text{ kcal}$$

This equation tells us that 136.6 kcal are released to the environment when 2 moles of water are formed. However, you may sometimes see the equation written in the following way:

$$H_2(g) + \tfrac{1}{2}O_2(g) \longrightarrow H_2O(l) + 68.3 \text{ kcal}$$

This equation tells us that 68.3 kcal are released to the environment when 1 mole of water is formed. Because the quantity of heat produced is proportional to the amount of water formed, both equations are correct.

Practice Exercise 12.6 When 2.00 moles of hexane, $C_6H_{14}(l)$, burn in sufficient oxygen gas, the products are carbon dioxide gas and water. In addition, $1{,}98\bar{0}$ kcal of heat are given up to the environment. Write a balanced thermochemical equation for this reaction.

Once you have found the heat of reaction when 1 mole of a substance is formed, you can use this information to find the heat produced or absorbed when *any* amount of that substance is formed.

Example 12.7

How much heat is produced when 144 g of $H_2O(l)$ are formed from the reaction between hydrogen and oxygen?

Solution Because the formation of 1 mole of water produces 68.3 kcal, we must find out how many moles of water there are in 144 g.

$$? \text{ moles of } H_2O = (144 \text{ g})\left(\frac{1 \text{ mole}}{18.0 \text{ g}}\right) = 8.00 \text{ moles of } H_2O$$

$$? \text{ kcal of heat produced} = (8.00 \text{ moles})\left(\frac{68.3 \text{ kcal}}{1 \text{ mole}}\right)$$

$$= 546 \text{ kcal of heat produced}$$

Practice Exercise 12.7 How much heat is produced when 54.0 g of $H_2O(l)$ are formed from the reaction between hydrogen gas and oxygen gas?

So far we have neglected endothermic reactions (reactions that absorb heat from the surroundings). The next example shows you how to write a thermochemical equation for this type of reaction.

Example 12.8

When hydrogen iodide is formed from hydrogen and iodine, heat is *absorbed* from the surroundings. Experiments have shown that 6.2 kcal of heat are absorbed per mole of hydrogen iodide produced. Write the thermochemical equation for the reaction.

Solution There are two ways to write this type of equation.
(a) Write it with the heat of reaction on the left-hand side of the equation. This shows that the reactants (H_2 and I_2) require heat to yield the product (HI).

$$6.2 \text{ kcal} + \tfrac{1}{2}H_2(g) + \tfrac{1}{2}I_2(s) \longrightarrow HI(g)$$

(b) Write it with the heat of reaction on the right-hand side of the equation, as a negative number. Think of this as indicating that the product must *lose* heat for the reaction to be completed.

$$\tfrac{1}{2}H_2(g) + \tfrac{1}{2}I_2(s) \longrightarrow HI(g) - 6.2 \text{ kcal}$$

Chemists usually prefer writing heats of reaction on the right-hand side of the equation.

Practice Exercise 12.8 When $PCl_5(g)$ breaks down to form $PCl_3(g)$ and $Cl_2(g)$, heat is absorbed from the surroundings. Experiments have shown that 22.2 kcal of heat are absorbed per mole of $PCl_5(g)$ broken down. Write the thermochemical equation for this reaction.

12.6 Another Way of Writing Heats of Reaction

Learning Goal 6
Significance of ΔH

The symbol for heat content (enthalpy) is H, so we can represent a *change* in heat content with the symbol ΔH. (Again we are using the Greek letter Δ to mean "change in.") Because a heat of reaction is a change in heat content during a reaction, we represent it with ΔH.

Chemists define ΔH as the sum of the heat contents of the products less the sum of the heat contents of the reactants:

$$\Delta H = H(\text{products}) - (\text{reactants})$$

With this notation, ΔH is negative for an exothermic reaction and positive for an endothermic reaction (Figure 12.2).

In the formation of 1 mole of $CO_2(g)$ from its elements in their standard states, 94.1 kcal of heat are released to the environment. This means that the

Figure 12.2
Reaction diagrams

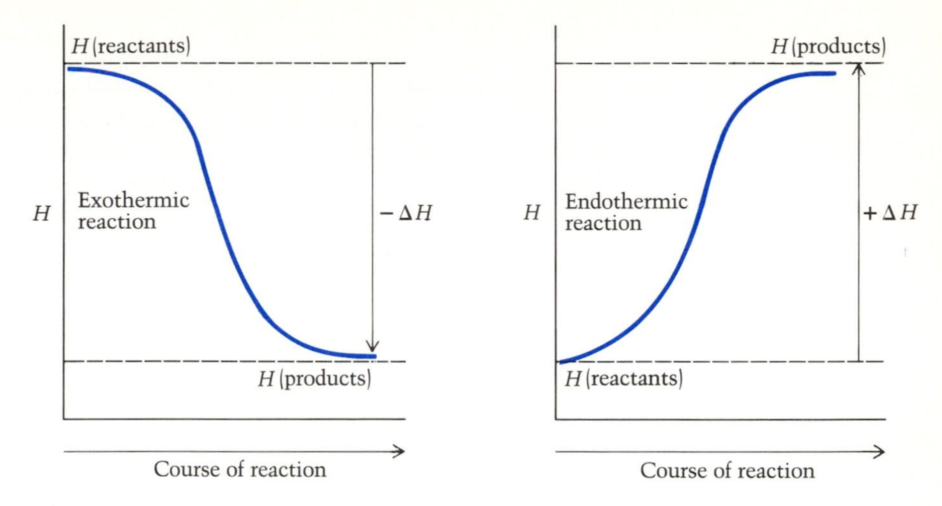

products have a lower heat content than the reactants, and $\Delta H = -94.1$ kcal. This may be indicated by writing

$$C(\text{graphite}) + O_2(g) \longrightarrow CO_2(g) \qquad \Delta H = -94.1 \text{ kcal}$$

or by placing the value of ΔH in the equation and writing

$$C(\text{graphite}) + O_2(g) \longrightarrow CO_2(g) + 94.1 \text{ kcal}$$

In the formation of 1 mole of HI(g) under standard conditions, heat is absorbed, so $\Delta H = +6.2$ kcal. This may be shown as

$$\tfrac{1}{2}H_2(g) + \tfrac{1}{2}I_2(s) \longrightarrow HI(g) \qquad \Delta H = +6.2 \text{ kcal}$$

or by placing the value of ΔH in the equation and writing

$$\tfrac{1}{2}H_2(g) + \tfrac{1}{2}I_2(s) \longrightarrow HI(g) - 6.2 \text{ kcal}$$

12.7 Heats of Formation

Learning Goal 7
Heats of formation and heats of reaction

Up to this point, we have used the term *heat of reaction* for the change in heat content (or enthalpy) during the course of a reaction. However, the examples we used all involved the formation of a compound from its elements:

$$\tfrac{1}{2}H_2(g) + \tfrac{1}{2}I_2(g) \longrightarrow HI(g) \qquad \Delta H = +6.2 \text{ kcal}$$

$$H_2(g) + \tfrac{1}{2}O_2(g) \longrightarrow H_2O(l) \qquad \Delta H = -68.3 \text{ kcal}$$

$$C(\text{graphite}) + O_2(g) \longrightarrow CO_2(g) \qquad \Delta H = -94.1 \text{ kcal}$$

When 1 mole of a compound is formed from its elements in their usual states at 1 atm pressure, the heat of reaction is called the **heat of formation** of the compound and is given the symbol ΔH_f^0. Heats of formation have been determined experimentally for many compounds, and Table 12.2 lists some of them. They can be used to determine ΔH for complex reactions.

Table 12.2
Heats of Formation at 25°C and 1 atm Pressure

Substance	ΔH_f^0 (*kcal/mole*)	Substance	ΔH_f^0 (*kcal/mole*)
$BaCl_2(s)$	−205.56	$H_2O(g)$	−57.8
$BaO_2(s)$	−150.5	$H_2O(l)$	−68.3
$CH_4(g)$	−17.9	$H_2O_2(l)$	−44.8
$C_2H_2(g)$	+54.2	$H_2S(g)$	−4.8
$C_2H_6(g)$	−20.2	$H_2SO_4(aq)$	−217
$C_8H_{18}(l)$	−59.97	$H_3PO_4(aq)$	−308.2
$C_6H_{12}O_6(s)$	−304.4	$SO_2(g)$	−70.96
$CCl_4(l)$	−33.4	$SO_3(g)$	−94.45
$CO(g)$	−26.4	$NH_3(g)$	−11.04
$CO_2(g)$	−94.1	$NO(g)$	+21.5
$CaO(s)$	−151.9	$NO_2(g)$	+8.1
$CaCO_3(s)$	−288.5	$NaCl(aq)$	−97.3
$HF(g)$	−64.2	$NaOH(aq)$	−112
$HCl(g)$	−22.1	$Fe_2O_3(s)$	−197
$HBr(g)$	−8.7	$KOH(aq)$	−115
$HI(g)$	+6.2	$K_2O(s)$	−86.4
$HCl(aq)$	−39.6		

12.8 Determining Heats of Reaction

You will find that Table 12.2 makes it easy to determine heats of reaction. However, you must keep the following rules in mind.

Rule 1 The heat of reaction is equal to the heat of formation of the products less the heat of formation of the reactants:

$$\Delta H = \Delta H_f^0(\text{products}) - \Delta H_f^0(\text{reactants})$$

Rule 2 The heat of formation of an element at standard-state conditions is zero:

$$\Delta H_f^0(\text{element}) = 0$$

Also keep in mind that the values in Table 12.2 are *per mole of compound*. This is important when more or less than 1 mole is involved in the balanced equation.

Example 12.9

Determine ΔH for the reaction

$$CaO(s) + CO_2(g) \longrightarrow CaCO_3(s)$$

Solution We obtain the ΔH_f^0 value for each compound from Table 12.2 and then use Rule 1 above.

Substance	ΔH_f^0(kcal/mole)
$CaO(s)$	−151.9
$CO_2(g)$	−94.1
$CaCO_3(s)$	−288.5

$$\begin{aligned} \Delta H &= \Delta H_f^0(\text{products}) - \Delta H_f^0(\text{reactants}) \\ &= (\Delta H^0_{CaCO_3}) - (\Delta H^0_{CaO} + \Delta H^0_{CO_2}) \\ &= (-288.5 \text{ kcal}) - (-151.9 \text{ kcal} - 94.1 \text{ kcal}) \\ &= (-288.5 \text{ kcal}) - (-246.0 \text{ kcal}) \\ &= -288.5 \text{ kcal} + 246.0 \text{ kcal} \\ &= -42.5 \text{ kcal} \end{aligned}$$

This reaction is exothermic. It produces 42.5 kcal of heat per mole of $CaCO_3$ formed.

Practice Exercise 12.9 Use Table 12.2 to determine ΔH for the reaction

$$K_2O(s) + H_2O(l) \longrightarrow 2KOH(aq)$$

Example 12.10

Determine ΔH for the reaction

$$\underset{\text{Ethane}}{2C_2H_6(g)} + 7O_2(g) \longrightarrow 4CO_2(g) + 6H_2O(l)$$

Solution Again we obtain ΔH_f^0 values from Table 12.2:

Substance	ΔH_f^0(kcal/mole)
$C_2H_6(g)$	−20.2
$O_2(g)$	0.0
$CO_2(g)$	−94.1
$H_2O(l)$	−68.3

Now, because 6 moles of water and 4 moles of CO_2 are formed, we have to multiply ΔH_f^0 for water by 6 and ΔH_f^0 for CO_2 by 4. And because 2 moles of ethane and 7 moles of oxygen are used, we have to multiply ΔH_f^0 for ethane by 2 and ΔH_f^0 for oxygen by 7.

$$\Delta H = \Delta H_f^0(\text{products}) - \Delta H_f^0(\text{reactants})$$
$$= [6(\Delta H^0_{H_2O}) + 4(\Delta H^0_{CO_2})] - [2(\Delta H^0_{C_2H_6}) + 7(\Delta H^0_{O_2})]$$
$$= [6(-68.3 \text{ kcal}) + 4(-94.1 \text{ kcal})]$$
$$- [2(-20.2 \text{ kcal}) + 7(0.0 \text{ kcal})]$$
$$= -786.2 \text{ kcal} + 40.4 \text{ kcal} = -745.8 \text{ kcal}$$

The reaction is exothermic. The combustion of 2 moles of ethane releases 745.8 kcal of heat to the environment.

Practice Exercise 12.10 Use Table 12.2 to determine ΔH for the reaction

$$2CO(g) + O_2(g) \longrightarrow 2CO_2(g)$$

12.9 Practical Applications of Heat Energy from Chemical Reactions

At the beginning of this chapter, we mentioned some uses of energy in our everyday lives. In most cases, this energy is obtained from fuels, substances that are rich in chemical potential energy—for example, oil, natural gas, and coal (often referred to as *fossil fuels*). We usually obtain heat energy from fuels by making them react in the presence of oxygen; in everyday language, we burn them. Then we use the heat energy they produce to do work.

Example 12.11

Modern gas ranges use methane (CH_4) for fuel. Calculate ΔH for the combustion of 1 mole of methane:

$$CH_4(g) + 2O_2(g) \longrightarrow CO_2(g) + 2H_2O(l)$$

Solution Obtain ΔH_f^0 for each compound from Table 12.2, and then use Rule 1:

Substance	ΔH_f^0(kcal/mole)
$CH_4(g)$	−17.9
$O_2(g)$	0.0
$CO_2(g)$	−94.1
$H_2O(l)$	−68.3

$$\begin{aligned}\Delta H &= \Delta H_f^0(\text{products}) - \Delta H_f^0(\text{reactants})\\ &= [2(\Delta H^0_{H_2O}) + 1(\Delta H^0_{CO_2})] - [1(\Delta H^0_{CH_4}) + 2(\Delta H^0_{O_2})]\\ &= [2(-68.3\text{ kcal}) + 1(-94.1\text{ kcal})]\\ &\quad - [1(-17.9\text{ kcal}) + 2(0.0\text{ kcal})]\\ &= -230.7\text{ kcal} + 17.9\text{ kcal} = -212.8\text{ kcal}\end{aligned}$$

This reaction is exothermic. When 1 mole of methane burns, 212.8 kcal of heat are released to the environment.

Practice Exercise 12.11 The first step in the synthesis of nitric acid is the reaction of ammonia gas and oxygen gas to form nitric oxide gas and steam. Calculate ΔH for this reaction:

$$4NH_3(g) + 5O_2(g) \longrightarrow 4NO(g) + 6H_2O(g)$$

Example 12.12

The combustion of methane releases considerable energy to do work. Let's compare this energy with the energy released by the combustion of 1 mole of octane, C_8H_{18} (used in automobile fuels):

$$C_8H_{18}(l) + 12\tfrac{1}{2}O_2(g) \longrightarrow 8CO_2(g) + 9H_2O(l)$$

Solution As usual, obtain ΔH_f^0 for each compound from Table 12.2, and then use Rule 1.

Substance	ΔH_f^0(kcal/mole)
$C_8H_{18}(l)$	−59.97
$O_2(g)$	0.0
$CO_2(g)$	−94.1
$H_2O(l)$	−68.3

$$\begin{aligned}\Delta H &= \Delta H_f^0(\text{products}) - \Delta H_f^0(\text{reactants})\\ &= [8(\Delta H^0_{CO_2}) + 9(\Delta H^0_{H_2O})] - [1(\Delta H^0_{C_8H_{18}}) + 12\tfrac{1}{2}(\Delta H^0_{O_2})]\\ &= [8(-94.1\text{ kcal}) + 9(-68.3\text{ kcal})]\\ &\quad - [1(-59.97\text{ kcal}) + 12\tfrac{1}{2}(0.0\text{ kcal})]\\ &= -1{,}367.5\text{ kcal} + 59.97\text{ kcal} = -1{,}307.5\text{ kcal}\end{aligned}$$

The reaction is exothermic. When 1 mole of octane burns, 1,307.5 kcal of heat are released to the environment. This is more than six times the energy produced by the combustion of 1 mole of methane. Note, however, that because 1 mole of octane weighs seven times as much as 1 mole of methane, we actually get a little more heat from 1 g of *methane* than we do from 1 g of octane.

Practice Exercise 12.12 Iron ore (Fe_2O_3) is reduced in the reaction

$$Fe_2O_3(s) + 3CO(g) \longrightarrow 3CO_2(g) + 2Fe(s)$$

Using Table 12.2, determine ΔH for this reaction.

Example 12.13

We all know that sugar is a source of quick energy. In our bodies, sucrose (sugar) is broken down into carbon dioxide and water by the very complicated process of cellular respiration. Sucrose itself is made up of two simpler sugars, glucose and fructose. Let's determine how much energy our bodies obtain from the oxidation of 1 mole of the simpler sugar, glucose ($C_6H_{12}O_6$), in the reaction

$$C_6H_{12}O_6(s) + 6O_2(g) \longrightarrow 6CO_2(g) + 6H_2O(l)$$

Solution Again we use Table 12.2 and Rule 1.

Substance	ΔH_f^0(kcal/mole)
$C_6H_{12}O_6(s)$	−304.4
$O_2(g)$	0.0
$CO_2(g)$	− 94.1
$H_2O(l)$	− 68.3

$$\begin{aligned}\Delta H &= \Delta H_f^0(\text{products}) - \Delta H_f^0(\text{reactants})\\ &= [6(\Delta H^0_{CO_2}) + 6(\Delta H^0_{H_2O})] - [1(\Delta H^0_{C_6H_{12}O_6}) + 6(\Delta H^0_{O_2})]\\ &= [6(-94.1\text{ kcal}) + 6(-68.3\text{ kcal})]\\ &\quad - [1(-304.4\text{ kcal}) + 6(0.0\text{ kcal})]\\ &= -974.4\text{ kcal} + 304.4\text{ kcal} = -67\overline{0}\text{ kcal}\end{aligned}$$

The reaction is exothermic. Our bodies obtain $67\overline{0}$ kcal of heat from the oxidation of 1 mole of glucose.

Practice Exercise 12.13 Acetylene gas, which is used for welding, reacts with hydrogen to produce ethane gas. Calculate ΔH for this reaction:

$$C_2H_2(g) + 2H_2(g) \longrightarrow C_2H_6(g)$$

Summary

In a chemical reaction, heat is either absorbed from the environment or released to the environment as old bonds are broken and new bonds are formed. A reaction in which energy is absorbed from the surroundings is called endothermic, and a reaction in which energy is released is called exothermic.

Associated with every substance is a heat content called its enthalpy. Although we cannot measure enthalpy directly, we can measure the amount of heat released or absorbed during a reaction. This change in heat content that occurs during a chemical reaction is called the heat of reaction. Heats of reaction are usually measured at standard-state conditions: 25°C and 1 atm pressure. Heat (including heat of reaction) is measured in calories. One calorie is the quantity of heat needed to raise the temperature of 1 gram of water by 1 Celsius degree. The specific heat of a substance is defined as the number of calories needed to raise the temperature of 1 gram of the substance by 1 Celsius degree.

A chemical equation that indicates the heat of reaction and the physical state of each reactant and product is called a thermochemical equation. Heat of reaction ΔH may be defined as the sum of the heat contents of the products less the sum of the heat contents of the reactants. For an exothermic reaction, this difference is negative; for an endothermic reaction, it is positive. When a compound is formed from its elements, the heat of reaction is called the heat of formation ΔH_f^0 for the compound. The heat of formation for an element (at standard-state conditions) is zero.

Key Terms

calorie **(12.2)**
endothermic reaction **(12.1)**
enthalpy **(12.2)**
exothermic reaction **(12.1)**
heat of formation **(12.7)**
heat of reaction **(12.2)**
specific heat **(12.3)**
thermochemical equation **(12.4)**

Career Sketch

Dental Hygienist

If you want to help people achieve and maintain good oral health, you may be interested in becoming a dental hygienist. As a member of the dental health team, you will perform preventive and therapeutic dental services under the supervision of a dentist. Your responsibilities will vary, depending on the laws of the state in which you work. However, your duties will typically include removing deposits and stains from your patients' teeth, instructing patients in how to care for their teeth, giving them dietetic and nutritional counseling, and applying various dental agents to prevent tooth decay. You will also take your patients' medical and dental histories, make impressions of teeth for the dentist to study, and prepare other diagnostic aids (such as x-ray films) for the dentist. In some states, you may also assist the dentist in pain-controlling and restorative procedures.

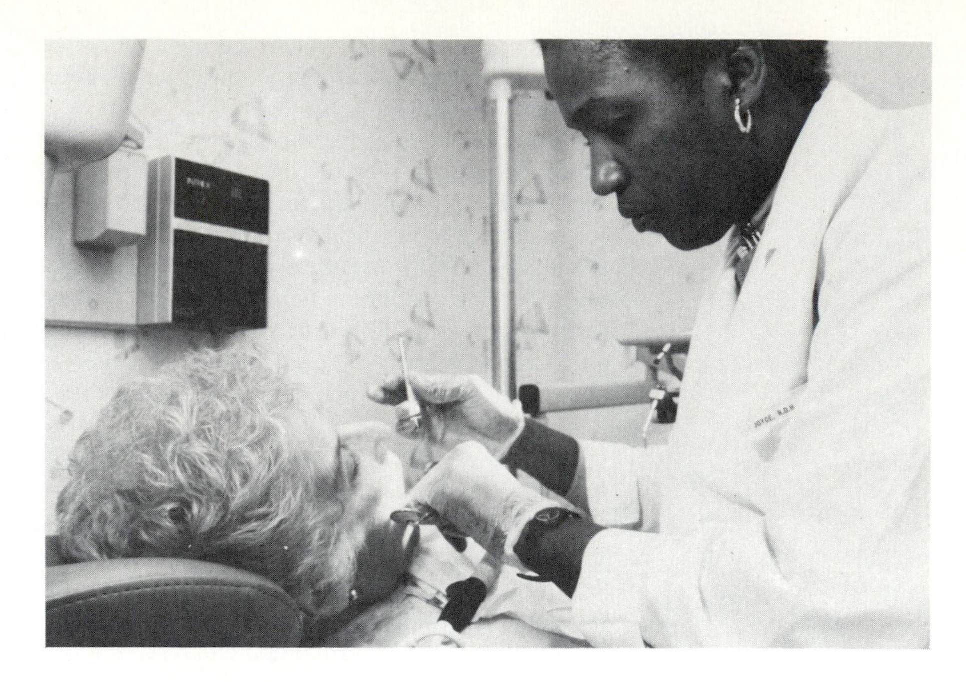

As members of the dental health team, dental hygienists perform preventive and dental therapeutic services under the supervision of a dentist. (Yoav Levy/Phototake)

As a dental hygienist, you may want to work in a school system. There you will examine children's teeth, help the dentist determine the treatment needed, and report findings to parents. You may also scale and polish teeth and instruct children in proper mouth care. You may also present programs on oral health to students.

Another option is to work with a team of dentists performing advanced dental research. With an advanced degree, you may teach in a college that offers a dental hygiene program.

Self-Test Exercises

Learning Goals 1 and 2: Exothermic and Endothermic Reactions; Definitions of Energy Terms

▲ **1.** Define the following terms: *(a)* exothermic reaction *(b)* endothermic reaction *(c)* heat of formation *(d)* heat of reaction

▲ **2.** The symbol H stands for the term ________.

▲ **3.** How does the sum of the heat contents of the products compare with that of the reactants in *(a)* an endothermic reaction, *(b)* an exothermic reaction?

▲ **4.** The change in heat content during a chemical reaction is called the ________.

Learning Goal 3: Definitions of Calorie and Specific Heat

5. Define the terms *calorie* and *specific heat*.

6. What are standard-state conditions, and why is it necessary to adhere to them when calculating changes in heat?

Learning Goal 4: Calculating Calories

▲ **7.** How much heat is needed to raise the temperature of $25\bar{0}$ g of water from 10.0°C to 90.0°C?

▲ **8.** How much heat is needed to raise the temperature of 500.0 g of water from 0°C to 75.0°C?

9. Calculate the amount of heat needed to raise the temperature of 50.0 g of copper from 25.0°C to 100.0°C.
10. Calculate the amount of heat needed to raise the temperature of 100.0 g of copper from 45.0°C to 80.0°C.

*11. How much heat is required to raise the temperature of 100.0 mL of ethyl alcohol from 15.0°C to 60.0°C? The density of ethyl alcohol is 0.800 g/mL.
*12. How much heat is required to raise the temperature of 50.0 cm^3 of ethyl alcohol from 25.0°C to 55.0°C? The density of ethyl alcohol is 0.800 g/cm^3.

13. Suppose that 308 cal of heat are added to a sample of gold that has a mass of $1,\overline{0}00$ g. What is the final temperature of the gold, given that its initial temperature is 20.0°C?
14. Suppose that 5.00 kcal of heat are added to a sample of gold that has a mass of 2.00 kg. What is the final temperature of the gold if its initial temperature is 10.0°C?

*15. A 0.146-g sample of H_2 gas is placed in a calorimeter, where it is allowed to react with oxygen gas. The heat generated by the reaction is transferred to the calorimeter water. The water has a mass of $5\overline{0}0$ g. The temperature of the water increases from 20.0°C to 30.0°C. From this information, calculate the heat of reaction in kilocalories per mole.
16. How much heat is needed to raise the temperature of 1.000 kg of water from 32.0°C to 99.9°C?

Learning Goal 5: Writing Thermochemical Equations

17. Using Table 12.2, write thermochemical equations for the formation of the following compounds from their elements: *(a)* $CO(g)$ *(b)* $HCl(g)$ *(c)* $H_2O_2(l)$
18. Using Table 12.2, write thermochemical equations for the formation of the following compounds from their elements: *(a)* $HF(g)$ *(b)* $HCl(g)$ *(c)* $HBr(g)$

Learning Goal 6: Significance of ΔH

▲ 19. Explain the significance of positive and negative values of ΔH. Use a chemical equation to illustrate your answer.
20. *(a)* What is meant by the term *enthalpy*?
(b) What is the difference between the symbols H and ΔH?

Learning Goal 7: Heats of Formation and Heats of Reaction

21. Calculate the heat released when 102 g of hydrogen sulfide (H_2S) are produced from its elements. *Note:* ΔH_f^0 for H_2S is −4.8 kcal/mole. (*Hint:* Calculate the number of moles of H_2S.)
22. Calculate the heat released when 15.3 g of barium oxide (BaO) are produced from its elements. (ΔH_f^0 for BaO is −133.4 kcal/mole.)

23. Rocket engines often use liquid hydrogen and liquid oxygen for fuel. These reactants are heated, mixed, and united to produce steam.

$$H_2(g) + \tfrac{1}{2}O_2(g) \longrightarrow H_2O(g) + 57.8 \text{ kcal}$$

Calculate the amount of heat released when 1 kg of steam is produced.
24. Hexane is a chemical that is sometimes used in thermometers, instead of mercury. Hexane burns in oxygen to produce carbon dioxide and water according to the equation

$$2C_6H_{14}(l) + 19O_2(g) \longrightarrow 12CO_2(g) + 14H_2O(l) + 1{,}980 \text{ kcal}$$

How many grams of hexane are needed to produce $19,8\overline{0}0$ kcal of heat?

▲ 25. *(a)* Using Table 12.2, calculate ΔH for the reaction

$$2C_2H_2(g) + 5O_2(g) \longrightarrow 4CO_2(g) + 2H_2O(l)$$

(b) Calculate how much heat is produced when $5,2\overline{0}0$ g of C_2H_2 are burned.
26. Carbon dioxide is formed from its elements as follows:

$$C(\text{graphite}) + O_2(g) \longrightarrow CO_2(g) + 94.1 \text{ kcal}$$

How much energy is produced when $88\overline{0}$ g of $CO_2(g)$ are formed?

27. Barium peroxide can be used as a solid propellant for rockets. When it is added to hydrochloric acid, the following reaction takes place:

$$\underset{\text{Barium peroxide}}{BaO_2(s)} + 2HCl(g) \longrightarrow BaCl_2(s) + H_2O_2(l)$$

(a) Using Table 12.2, calculate ΔH for the reaction.
(b) Calculate the number of grams of BaO_2 that must react to produce $5\overline{0}0$ kcal of heat.

28. Liquid water is broken down into its elements by electrolysis. This is an endothermic reaction requiring 68.3 kcal per mole of liquid water decomposed. How much energy is needed to produce $64\bar{0}$ g of oxygen gas?

29. Given the reaction

$$P_2O_5(s) + 3H_2O(l) \longrightarrow 2H_3PO_4(aq) + 228.5 \text{ kcal}$$

calculate ΔH_f^0 for P_2O_5 (use Table 12.2).

▲ **30.** When you burn 1.20 g of carbon (graphite) in the presence of oxygen, carbon dioxide gas is produced. The heat generated by this reaction can raise the temperature of 470.5 g of water from 20.0°C to 40.0°C. Calculate ΔH_f^0 for carbon dioxide from this information, and compare it with the value given in Table 12.2.

▲ **31.** When you burn 3.20 g of sulfur in the presence of oxygen, sulfur trioxide (SO_3) is produced. The heat generated by this reaction can raise the temperature of 1,890 g of water from 25.00°C to 30.00°C. Calculate ΔH_f^0 for sulfur trioxide from this information, and compare it with the value given in Table 12.2.

32. Calculate the heat released when 188.4 g of $K_2O(s)$ are produced from its elements.

33. Eric Cottell, a British-born inventor, has found a method of increasing the burning efficiency of oil and gasoline. He does this simply by mixing the oil or gasoline with water. Cottell has built a device called the Cottell Ultrasonic Reactor, which can actually mix oil and water by breaking down the oil and water droplets into extremely fine particles. By using a mixture containing three parts oil to one part water, Cottell has decreased the fuel consumption in his home furnace by 25%. He claims that cars can be made to run almost pollution-free on a mixture of 18% water and 82% gasoline.

One of the ingredients of home heating oil is $C_{16}H_{34}$. Given the following equation, determine ΔH_f^0 for this ingredient.

$$C_{16}H_{34}(s) + 24\tfrac{1}{2}O_2(g) \longrightarrow 16CO_2(g) + 17H_2O(l) + 2{,}560 \text{ kcal}$$

34. Calculate the heat released when 7.30 g of $HCl(g)$ are produced from its elements.

▲ **35.** Calculate the heat released when 12.8 g of $SO_2(g)$ are produced from its elements.

36. Calculate the heat released when 1.618 kg of $HBr(g)$ are produced from its elements.

37. Calculate the heat absorbed when 51.2 g of $HI(g)$ are produced from its elements.

▲ **38.** Acetylene gas, C_2H_2, burns in oxygen to produce carbon dioxide gas and liquid water. For every mole of acetylene burned, 312.0 kcal of heat are released. From this information, calculate ΔH_f^0 for acetylene, and compare your answer with the value given in Table 12.2.

39. Using Table 12.2, calculate ΔH for the reaction

$$2SO_2(g) + O_2(g) \longrightarrow 2SO_3(g)$$

40. Use the following equation and Table 12.2 to determine ΔH_f^0 for $Ca(OH)_2(s)$:

$$CaO(s) + H_2O(l) \longrightarrow Ca(OH)_2(s) + 15.6 \text{ kcal}$$

41. Hydrogen peroxide (H_2O_2) is sometimes used in diluted form as an antiseptic. Upon standing, hydrogen peroxide decomposes according to the reaction

$$2H_2O_2(l) \longrightarrow 2H_2O(l) + O_2(g)$$

(a) Calculate ΔH for this reaction.
(b) Calculate how much heat is produced when 204 g of H_2O_2 decompose.

42. Calculate the amount of heat released when 2.000 kg of $BaCl_2(s)$ are produced from $Ba(s)$ and $Cl_2(g)$. Use Table 12.2.

43. Calculate ΔH for the decomposition of potassium chlorate:

$$2KClO_3(s) \longrightarrow 2KCl(s) + 3O_2(g)$$

[ΔH_f^0 for $KClO_3(s)$ is -93.50 kcal, and ΔH_f^0 for $KCl(s)$ is -104.18 kcal.]

44. When 0.230 g of $Na(s)$ react with $Cl_2(g)$, $NaCl(s)$ is produced. The heat generated by this reaction can raise the temperature of 24.55 g of water from 10.0°C to 50.0C. Using this information, calculate ΔH_f^0 for $NaCl(s)$.

45. The following equation represents the formation of phosphoric acid from phosphorus pentachloride:

$$PCl_5(s) + H_2O(l) \longrightarrow H_3PO_4(aq) + HCl(aq)$$

(a) Balance the equation.
(b) Calculate ΔH for this reaction. [ΔH_f^0 for $PCl_5(s)$ is -95.35 kcal, and ΔH_f^0 for $HCl(aq)$ is -40.02 kcal.]
(c) How much heat is produced from this reaction if 41.70 g of $PCl_5(s)$ are used?

46. Calculate the amount of heat absorbed when 5.00 moles of $CO_2(g)$ are formed from the reaction

$$C(graphite) + 2H_2O(g) \longrightarrow CO_2(g) + 2H_2(g) - 31.39 \text{ kcal}$$

*47. Calculate ΔH_f^0 for CuO(s) from the following information:

$$3CuO(s) + 2NH_3(g) \longrightarrow 3Cu(s) + 3H_2O(l) + N_2(g) \qquad \Delta H = -71.1 \text{ kcal}$$

[ΔH_f^0 for $NH_3(g)$ is -11.04 kcal.]

48. Calculate the amount of heat released when 5.00 moles of $CO_2(g)$ are formed from the reaction

$$CO(g) + \tfrac{1}{2}O_2(g) \longrightarrow CO_2(g) + 67.64 \text{ kcal}$$

Extra Exercises

49. In a simple calorimeter, a reaction is carried out in a vessel submerged in water. After the reaction occurs, $42\bar{0}$ g of water increase in temperature from 25.0°C to 40.0°C. What is the amount of energy released by the reaction?

50. The heat of reaction for $CaCO_3(s) \longrightarrow CaO(s) + CO_2(g)$ is $\Delta H = 42.5$ kcal. Given that $\Delta H^0_{CaO} = -151.9$ kcal/mole and $\Delta H^0_{CO_2} = -94.1$ kcal/mole, calculate $\Delta H^0_{CaCO_3}$. Compare your calculated value with the value given in Table 12.2.

51. Using Table 12.2, determine whether heat is absorbed or released when hydrogen gas reacts at standard-state conditions with the appropriate elements to form each of the following: *(a)* $C_2H_2(g)$ *(b)* $HCl(g)$ *(c)* $H_2O(l)$ *(d)* $HI(g)$

52. When burned with excess oxygen, does 1.00 mole of methane (CH_4) or 1.00 mole of acetylene (C_2H_2) produce the greater amount of energy? The reactions are

$$CH_4(g) + 2O_2(g) \longrightarrow CO_2(g) + 2H_2O(g)$$

$$C_2H_2(g) + 2\tfrac{1}{2}O_2(g) \longrightarrow 2CO_2(g) + H_2O(g)$$

53. Energy is required to break chemical bonds. Considering this, determine whether the following reaction would be endothermic or exothermic:

$$N_2O_4(g) \longrightarrow 2NO_2(g)$$

54. The heat of reaction for the burning of carbon is 94 kcal/mole. The coal used in a coal-fired power plant is 70.0% carbon. How much energy does 1.00 kg of coal yield?

55. An energy source that is gaining more attention these days is hydrogen gas. When it combines with oxygen, it forms water and releases 58 kcal/mole. Compare the fuel value of hydrogen gas with that of natural gas, which can be approximated as methane (CH_4), having a heat of reaction of 213 kcal/mole, and that of coal, which would yield 94 kcal/mole if it were pure carbon. (*Hint:* The best way to make this comparison is to calculate the heat of formation per gram for each substance.)

▲ **56.** Explain why an exothermic reaction has a negative ΔH and an endothermic reaction has a positive ΔH.

57. Calculate the heat released by the combustion of 50.0 moles of methane.

58. Using Table 12.2, calculate the heat needed to produce the following change in 1.00 mole of water:

$$H_2O(l) \longrightarrow H_2O(g)$$

13 The Gaseous State: Ideal Behavior

Learning Goals

After you have studied this chapter, you should be able to:

1. State and illustrate the kinetic theory of gases.
2. Convert pressures in atmospheres to pressures in torr, and vice versa.
3. Use Boyle's Law to calculate changes in pressure and volume at constant temperature.
4. Use Charles's Law to calculate changes in temperature and volume at constant pressure.
5. Convert from Celsius degrees to kelvins, and vice versa.
6. Use the Combined Gas Law to calculate changes in temperature, pressure, or volume.
7. Use Dalton's Law to calculate the partial pressure of a gas in a mixture.
8. Use Avogadro's Principle to find the number of moles of a gas when you are given the volume (at fixed temperature and pressure), and vice versa.
9. Use the Ideal Gas Law to calculate the pressure, volume, temperature, or number of moles of a gas when you are given the other three.
10. Use the Ideal Gas Law to calculate the molecular mass of a gas.

Introduction

To many people the word *gas* seems to imply mainly the gas we use for heating and cooking. However, every time we breathe we inhale gases; in fact, we inhale 35 pounds of gases each day. Every time we pass a neon sign, use an aerosol spray, or open a refrigerator, we see examples of gas under pressure. A flying airplane is supported by gases.

Gases have deadly as well as useful aspects. A person in a large city inhales in one day as much toxic, or poisonous, material as he or she would by smoking two packs of cigarettes. More than half of this unhealthy mixture is contributed by the combustion of fuel.

Maybe because many gases are colorless and therefore seem invisible, people took a long time to get around to studying them. Until the late 1500s, air was the only vapor-like substance that scientists had studied. Even though alchemists had come into contact with many vapor-like substances, they had found them so mysterious that they called them spirits and ignored them. They used the word *spirits* for all liquids that vaporized, or turned to gas easily.

Jan Baptista van Helmont (1577–1644), a Flemish physician, was the first person to study vapors other than air. He observed that various vapors behaved like air but had different physical and chemical properties. Like air, these other vapors had no definite shape or volume.

In this chapter, we discuss the basic physical properties of gases: pressure, volume, and temperature. Several formulas are presented that allow us to predict the behavior of gases under various conditions.

13.1 Kinetic Theory of Gases

Learning Goal 1
Kinetic theory of gases

We can explain the properties and characteristics of gases with a series of statements called the **kinetic theory of gases.** The kinetic theory is an attempt to account for properties of gases in terms of the forces that exist between the gas molecules and the energy of these molecules. The kinetic theory states:

1. Nearly all gases are composed of molecules. (The so-called noble gases are composed of atoms, not molecules. Remember, atoms of noble gases don't combine readily with other atoms.)

2. The forces of attraction between gas molecules increase as the molecules move closer together. At normal atmospheric pressure and room temperature, the distances between gas molecules are large compared to the size of the molecules and so the forces of attraction are very small.

3. Gas molecules are always in motion. They often collide with other gas molecules or with their container. The molecules do not stick together after collisions occur, and they do not lose any energy on account of the collisions.

4. When the temperature rises, gas molecules move faster. They move more slowly when the temperature drops.

5. All gas molecules (heavy as well as light) have the same average kinetic energy—that is, energy of motion—at the same temperature.

A gas that behaves exactly as predicted by the kinetic theory under all conditions of temperature and pressure is called an *ideal gas*. An **ideal gas** is a gas whose molecules have no attraction for each other. In nature there are no ideal gases, only real gases such as hydrogen (H_2), helium (He), oxygen (O_2), and others. A **real gas** is a gas that does not completely obey the assumptions of the kinetic theory. At very low temperatures and very high pressures, real gases do not exhibit ideal behavior, because the molecules of the gas are very close together, which increases the attractive forces between them. Under moderate conditions of temperature and pressure, the forces of attraction are minimal and real gases behave as ideal gases. This characteristic allows us to use the kinetic theory to predict the behavior of real gases.

13.2 Air Pressure and the Barometer

In the early 1600s, the Italian mathematician and physicist Evangelista Torricelli discovered that air exerts a measurable pressure on the objects it touches. Torricelli filled a glass tube, which was sealed at one end, with liquid mercury. He then turned the tube upside down and placed it in a bowl, also filled with mercury (Figure 13.1). Torricelli noticed that most of the mercury remained in the upright tube. He concluded that the surrounding air exerted pressure on the surface of the mercury in the bowl, which in turn supported the column of mercury in the tube.

At sea level, the height of the mercury supported in the tube averages 760 mm. In honor of Torricelli, the pressure that supports a column of mercury exactly 1 mm high is now called 1 **torr.** Therefore, at sea level, the air pressure averages 760 torr, depending on the weather. (Note that the plural of *torr* is *torr*.)

Figure 13.1
A simple barometer

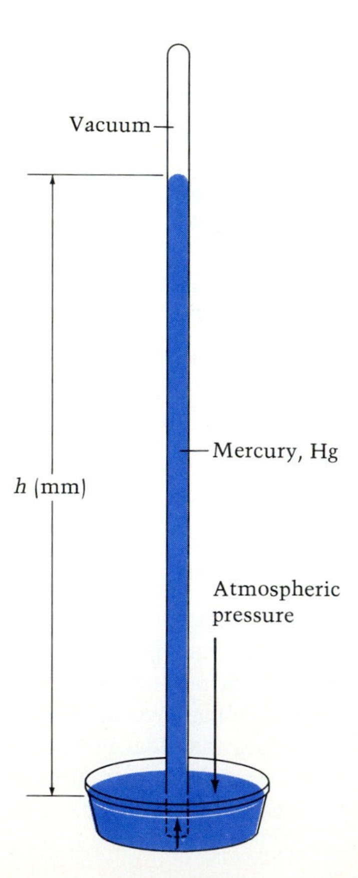

If we take Torricelli's pressure-measuring device, called a **barometer,** to the top of Pikes Peak (which is 14,000 feet above sea level), the height of the mercury column decreases, because the air pressure at the top of the peak is less than it is at sea level. If we take the same barometer to Death Valley (which is 280 feet below sea level), the height of the mercury column increases, because the air pressure below sea level is greater than it is at sea level. We can say that the height of the mercury column is proportional to the atmospheric pressure. In fact, the height of a mercury column supported by *any* gas is proportional to the pressure of that gas. We can thus express the pressure of any gas as a number of torr.

Because we know that the air pressure at sea level supports a column of mercury 760 mm high, we can also define another unit of pressure, called the **atmosphere** (abbreviated *atm*). We have used this unit before, but now we'll define it in terms of torr:

Learning Goal 2
Converting atmospheres to torr, and vice versa

$$1 \text{ atmosphere} = 760 \text{ mm of mercury} = 760 \text{ torr}$$

Example 13.1

Express each of the following pressures in torr: *(a)* 1.50 atm *(b)* 0.150 atm

Solution Use the factor-unit method and the fact that 1 atm = 760 torr.

(a) $? \text{ torr} = (1.50 \cancel{\text{atm}})\left(\frac{76\bar{0} \text{ torr}}{1 \cancel{\text{atm}}}\right) = 1{,}140 \text{ torr}$

(b) $? \text{ torr} = (0.150 \cancel{\text{atm}})\left(\frac{76\bar{0} \text{ torr}}{1 \cancel{\text{atm}}}\right) = 114 \text{ torr}$

Practice Exercise 13.1 Express each of the following pressures in torr: *(a)* 0.075 atm *(b)* 9.700 atm

Example 13.2

Express each of the following pressures in atmospheres: *(a)* 6,460 torr *(b)* 152 torr

Solution Use the factor-unit method and the fact that 1 atm = 760 torr.

(a) $? \text{ atm} = (6{,}460 \cancel{\text{torr}})\left(\frac{1 \text{ atm}}{76\bar{0} \cancel{\text{torr}}}\right) = 8.50 \text{ atm}$

(b) $? \text{ atm} = (152 \cancel{\text{torr}})\left(\frac{1 \text{ atm}}{760 \cancel{\text{torr}}}\right) = 0.200 \text{ atm}$

Practice Exercise 13.2 Express each of the following pressures in atmospheres: *(a)* 7.6 torr *(b)* 9,120 torr

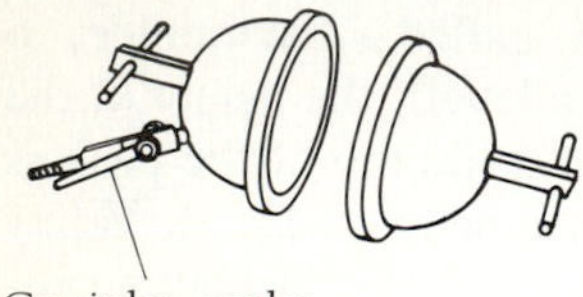

Figure 13.2
von Guericke's hemispheres

Air Pressure: A Strong Force

The German physicist Otto von Guericke developed a pump that could *evacuate,* or empty, the air from a container. In 1657, intrigued by Torricelli's experiments, he performed an experiment in which he put together two tightly fitted metal hemispheres (Figure 13.2). He then used his pump to evacuate the air from inside them. The hemispheres held together. He attached a team of horses to each hemisphere by means of a rope. The horses pulled in opposite directions, but they could not separate the hemispheres, because air was pressing from the outside but *not* from the inside. The air pressure was stronger than the two teams of horses. When the hemispheres were again filled with air, they easily fell apart.

13.3 Boyle's Law

Learning Goal 3
Boyle's Law

After hearing of von Guericke's experiment, an English chemist named Robert Boyle decided to test the effect of pressure on the volume of gases. For his experiments in the mid-1600s, Boyle devised a J-shaped tube. He trapped air in the closed end of the tube by pouring mercury into the open end (Figure 13.3). He changed the pressure on the trapped air by changing the amount of mercury in the tube. He found that if he tripled the pressure on the air, the volume occupied by the air decreased to one-third the original volume. And if he decreased the pressure on the trapped air to one-third the original pressure, the volume occupied by the air tripled.

Boyle failed to realize that the temperature of the air affects this relationship. The French physicist Edme Mariotte, working on his own, also discovered the relationship, and he was the first to state what is now known as **Boyle's Law:** *The volume of a gas increases as the pressure on it decreases, if its temperature remains constant.* Another way to state this principle is to say that at constant temperature, the volume of a given mass of gas varies inversely with its pressure. This is what we mean by an *inverse relationship*.

Boyle's Law can also be stated in terms of the ratios of pressures and volumes:

$$\frac{P_i}{P_f} = \frac{V_f}{V_i} \quad \text{(at constant temperature)}$$

Figure 13.3
Boyle's experiment

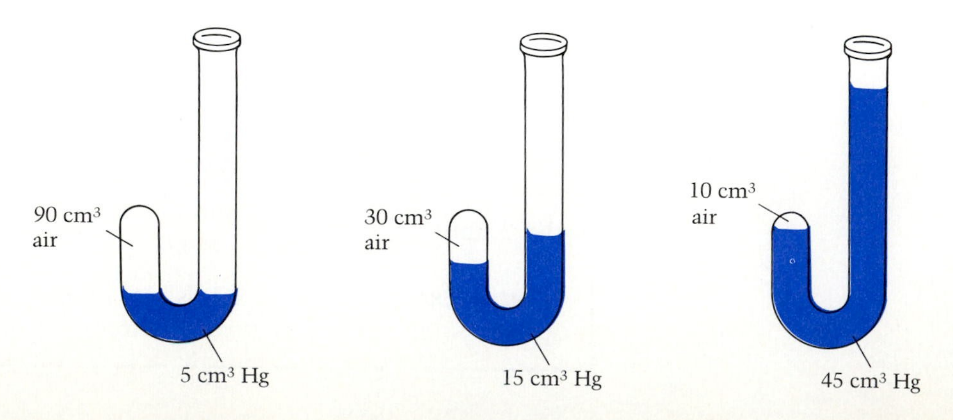

or $$P_i \times V_i = P_f \times V_f \qquad \text{Equation 13.1}$$

where P stands for pressure, V stands for volume, i indicates initial (or original) conditions, and f indicates final conditions. Equation 13.1 can be used to find the final pressure or volume of any gas, given a change in the other quantity.

Example 13.3

Suppose a gas has an initial volume V_i, of $5\bar{0}$ mL at a pressure of 1.0 atm. What will be its final volume, V_f, if the pressure is increased to 2.0 atm (P_f)? (Assume that the temperature remains constant.)

Solution

Understand the Problem

A gas is described as having a particular volume at a certain pressure. We are asked to determine the volume of the gas if the pressure is increased with no change in temperature. We can use Boyle's Law, which states that at constant temperature, the volume of a gas decreases when its pressure increases. Since the pressure is increased, the final volume will decrease, that is, it will be less than the initial volume of $5\bar{0}$ mL.

Carry Out the Plan

First organize the data—that is, the facts that have been given.

$$V_i = 5\bar{0}\text{ mL} \qquad V_f = ?$$

$$P_i = 1.0\text{ atm} \qquad P_f = 2.0\text{ atm}$$

Then write Boyle's Law, solve for V_f, and enter the data in the result.

$$P_i \times V_i = P_f \times V_f \quad \text{so} \quad V_f = \frac{P_i V_i}{P_f}$$

$$V_f = \frac{(1.0\ \cancel{\text{atm}})(5\bar{0}\text{ mL})}{2.0\ \cancel{\text{atm}}} = 25\text{ mL}$$

Look Back

As expected, increasing the pressure on the gas decreases the volume of the gas. Once again, this is an *inverse relationship*.

Practice Exercise 13.3 A gas has an initial volume of 500.0 mL (V_i) at a pressure of 2.50 atm (P_i). What will be its final volume (V_f) if the pressure is increased to 5.00 atm (P_f)? (Assume that the temperature remains constant.)

Example 13.4

Suppose a gas occupies a volume of $50\overline{0}$ mL under a pressure of 3.0 atm at $10\overline{0}°$ C. What will be the volume of the gas when the pressure is decreased to 1.0 atm? (Assume that the temperature remains constant.)

Solution First organize the data:

$$V_i = 50\overline{0} \text{ mL} \qquad V_f = ?$$

$$P_i = 3.0 \text{ atm} \qquad P_f = 1.0 \text{ atm}$$

$$t = 10\overline{0}° \text{ C and is constant}$$

Then write Boyle's Law, solve for V_f, and enter the data in the result.

$$P_i \times V_i = P_f \times V_f \qquad \text{so} \qquad V_f = \frac{P_i V_i}{P_f}$$

$$V_f = \frac{(3.0 \cancel{\text{atm}})(50\overline{0} \text{ mL})}{1.0 \cancel{\text{atm}}} = 1{,}500 \text{ mL}$$

When the pressure was decreased to one-third of its initial value, the volume increased to three times *its* initial value.

Practice Exercise 13.4 A gas occupies a volume of 250.0 mL under a pressure of 4.00 atm at 100.0° C. What will be the volume of the gas when the pressure is decreased to 2.00 atm? (Assume that the temperature remains constant.)

Boyle's Law can also be used to determine the pressure needed to cause a change in volume of a gas by solving the formula $P_i V_i = P_f V_f$ for P_f. See the next example.

Example 13.5

A gas occupies a volume of 0.100 liter (L) under a pressure of $76\overline{0}$ torr at $2\overline{0}$°C. What pressure will cause the gas to occupy a volume of 50.0 mL at the same temperature?

Solution Organize the data:

$$V_i = 0.100 \text{ L} = 10\overline{0} \text{ mL} \qquad V_f = 50.0 \text{ mL}$$

$$P_i = 76\overline{0} \text{ torr} \qquad P_f = ?$$

$$t = 2\overline{0}°\text{C and is constant}$$

Write Boyle's Law, solve for P_f, and enter the data in the result.

$$P_i \times V_i = P_f \times V_f \qquad \text{so} \qquad P_f = \frac{P_i V_i}{V_f}$$

$$P_f = \frac{(76\bar{0}\ \text{torr})(1\bar{0}0\ \text{mL})}{50.0\ \text{mL}} = 1{,}520\ \text{torr}$$

Practice Exercise 13.5 A gas occupies a volume of 0.200 L under a pressure of $76\bar{0}$ torr at 50.0°C. What pressure will cause the gas to occupy a volume of 25.0 mL at the same temperature?

13.4 Charles's Law

Learning Goal 4
Charles's Law

More than a hundred years after Boyle's work, the French physicist Jacques Charles discovered a relationship between the volume and the temperature of a gas. He found that a gas expands when its temperature is increased and that it shrinks when its temperature is decreased. A gas originally at 0°C expands 1/273 of its volume for each 1°C increase in temperature, and it shrinks 1/273 of its volume for each 1°C decrease (if the pressure of the gas is held constant). This is the basis of *Charles's Law*. It stands to reason: Most things do shrink as they get cooler and expand as they get hotter. The old tube-type tires used on automobiles are a good example. These tires expanded as the air inside them got warm, as often happened on a hot day when the car was traveling at high speed. If the tires expanded enough, a blowout would occur.

Learning Goal 5
Converting °C to K, and vice versa

Theoretically, if we took 1 L (or any volume) of a gas at 0°C and kept cooling it (at constant pressure), the volume of the gas would shrink to 0 mL at −273°C (Figure 13.4). (This, of course, has never been accomplished, because all known gases become liquid before reaching that temperature.) The temperature −273°C is called **absolute zero.** It is the basis of a temperature scale called the Kelvin scale, which was named after Lord Kelvin, the British physicist who proposed it. The units of this scale are kelvins, K, not degrees.

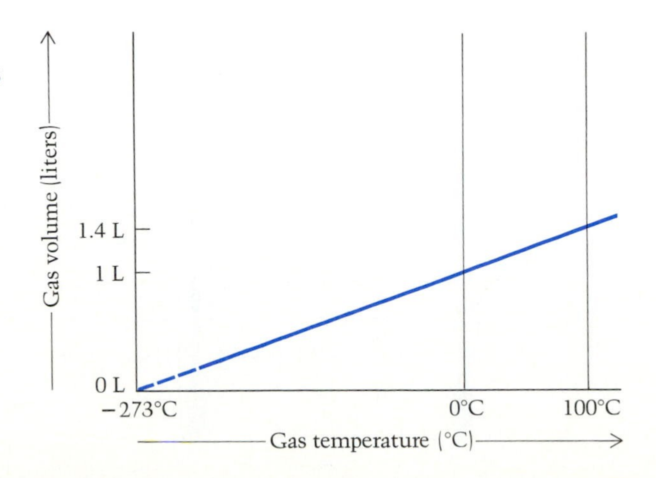

Figure 13.4
A theoretical plot of gas volume versus temperature

One kelvin covers the same temperature interval as one Celsius degree (Figure 13.5). The relationship between the two temperature scales is

$$K = {}^\circ C + 273 \qquad \text{Equation 13.2}$$

Using the Kelvin scale, we can state **Charles's Law** in the following way: *The volume of a gas varies directly with its temperature in kelvins if the pressure remains constant*. Charles's Law can also be stated mathematically:

$$\frac{V_i}{V_f} = \frac{T_i}{T_f} \quad \text{(at constant pressure)} \qquad \text{Equation 13.3}$$

where T is the temperature in kelvins (a lower-case t stands for degrees Celsius).

You must remember to change degrees Celsius to kelvins when you work with the formula for Charles's Law. The direct relationship between the temperature and the volume of a gas, as expressed in Equation 13.3, applies *only* to temperatures measured in kelvins. To convert from one scale to the other, use the formula

$$T = t + 273$$

Figure 13.5
Comparison of kelvins, degrees Celsius, and degrees Fahrenheit

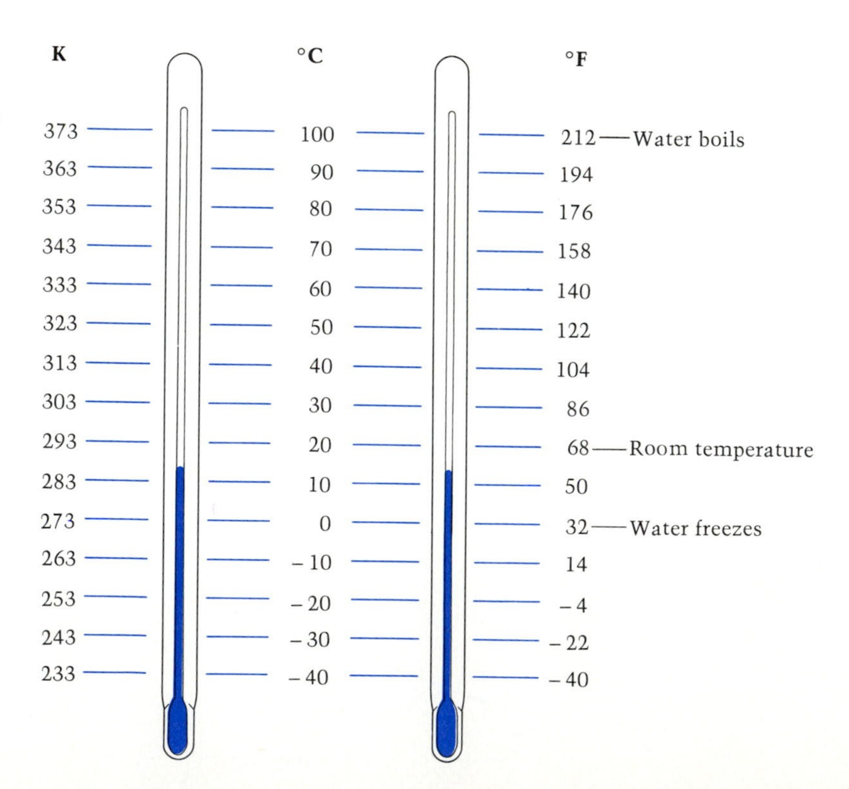

Example 13.6

A gas at 27°C occupies a volume of $1,\overline{000}$ mL. What will be its volume at 127°C? (Assume that the pressure of the gas remains constant.)

Solution First organize the data and change degrees Celsius to kelvins.

$$V_i = 1,\overline{000}\text{ mL} \qquad V_f = ?$$

$$T_i = t_i + 273 \qquad T_f = t_f + 273$$

$$= 27 + 273 = 3\overline{00}\text{ K} \qquad = 127 + 273 = 4\overline{00}\text{ K}$$

Then write Charles's Law, solve for V_f, and enter the data in the result.

$$\frac{V_i}{V_f} = \frac{T_i}{T_f} \quad \text{so} \quad V_f = \frac{V_i T_f}{T_i}$$

$$V_f = \frac{(1,\overline{000}\text{ mL})(4\overline{00}\ \cancel{\text{K}})}{3\overline{00}\ \cancel{\text{K}}}$$

$$= 1,333\text{ mL or } 1,330\text{ mL (three significant figures)}$$

As expected, increasing the temperature increases the volume of the gas. This is what we mean by a *direct relationship*.

You can also solve the problem using this alternate approach. Since Charles's Law predicts an increase in volume with an increase in temperature, the final volume will be greater than the initial volume of $1,\overline{000}$ mL. We may determine the final volume by using the following relationship:

$$\text{Final volume} = \text{initial volume} \times \text{temperature ratio}$$

The correct temperature ratio is the one that will predict the expected volume increase, that is,

$$\text{Final volume} = 1,\overline{000}\text{ mL} \times \frac{4\overline{00}\ \cancel{\text{K}}}{3\overline{00}\ \cancel{\text{K}}}$$

$$= 1,330\text{ mL (three significant figures)}$$

Practice Exercise 13.6 A gas at 127°C occupies a volume of $2,5\overline{00}$ mL. What will be its volume at 27.0°C? (Assume that the pressure of the gas remains constant.)

Charles's Law can also be used to determine the temperature of the gas when the change in volume is known. Solve the formula

$$\frac{V_i}{T_f} = \frac{T_i}{T_f}$$

for either T_i or T_f.

Example 13.7

When a gas is cooled at constant pressure, it contracts from 50.0 mL to 45.0 mL. Its final temperature is $8\overline{0}$°C. What was its initial temperature in degrees Celsius?

Solution First organize the data and change degrees Celsius to kelvins.

$$V_i = 50.0 \text{ mL} \qquad V_f = 45.0 \text{ mL}$$
$$t_i = ? \qquad T_f = t_f + 273$$
$$T_i = ? \qquad = 8\overline{0} + 273 = 353 \text{ K}$$

Then write Charles's Law, solve for T_i, and enter the data in the result.

$$\frac{V_i}{V_f} = \frac{T_i}{T_f} \quad \text{so} \quad T_i = \frac{V_i T_f}{V_f}$$

$$T_i = \frac{(50.0 \text{ mL})(353 \text{ K})}{45.0 \text{ mL}} = 392 \text{ K}$$

Finally, to find the answer in degrees Celsius, proceed as follows:

$$T_i = t_i + 273$$
$$t_i = T_i - 273$$
$$t_i = 392 - 273 = 119°\text{C}$$

Practice Exercise 13.7 When a certain gas is heated at constant pressure, it expands from 50.0 mL to 60.0 mL. Its final temperature is 75.0°C. What was its initial temperature in degrees Celsius?

13.5 The Combined Gas Law

Learning Goal 6
Combined Gas Law

Boyle's Law and Charles's Law can be combined into a single law that includes all three variables—pressure, volume, and temperature. The single law can be used to determine how changes in any two of these variables affect the third variable. This **Combined Gas Law** can be stated as follows:

$$\frac{P_i V_i}{T_i} = \frac{P_f V_f}{T_f} \qquad \text{Equation 13.4}$$

Equation 13.4 incorporates both Charles's Law and Boyle's Law, so it is the only equation you need to memorize. The following examples show why.

Example 13.8

A certain gas at 2°C and 1.00 atm pressure fills a 4.00-L container. What volume will the gas occupy at $10\bar{0}$°C and $78\bar{0}$ torr pressure?

Solution To begin, organize the data, change degrees Celsius to kelvins, and convert all the pressures to the same units.

$$P_i = 1.00 \text{ atm } (76\bar{0} \text{ torr}) \qquad P_f = 78\bar{0} \text{ torr}$$

$$V_i = 4.00 \text{ L} \qquad V_f = ?$$

$$T_i = t_i + 273 \qquad T_f = t_f + 273$$

$$= 2 + 273 = 275 \text{ K} \qquad = 100 + 273 = 373 \text{ K}$$

Then write the Combined Gas Law, solve for V_f, and enter the data in the result.

$$\frac{P_i V_i}{T_i} = \frac{P_f V_f}{T_f} \quad \text{so} \quad V_f = \frac{P_i V_i T_f}{P_f T_i}$$

$$V_f = \frac{(4.00 \text{ L})(76\bar{0} \cancel{\text{torr}})(373 \cancel{\text{K}})}{(78\bar{0} \cancel{\text{torr}})(275 \cancel{\text{K}})} = 5.29 \text{ L}$$

(If you need help solving algebraic equations, see Supplement A, section A.6.)

In this example the final volume depends on both the pressure change (Boyle's Law) and the temperature change (Charles's Law). The Combined Gas Law can be used to solve for the final volume, or the appropriate pressure ratio and temperature ratio can be used according to the following relationship:

Final volume = initial volume × pressure ratio × temperature ratio

From the data given, the pressure change would predict a decrease in the volume of the gas and the temperature change would predict an increase in the volume. Therefore:

$$\text{Final volume} = 4.00 \text{ L} \times \frac{760 \cancel{\text{torr}}}{780 \cancel{\text{torr}}} \times \frac{373 \cancel{\text{K}}}{275 \cancel{\text{K}}}$$

$$= 5.29 \text{ L}$$

Practice Exercise 13.8 A certain gas at 25.0°C and 1.00 atm pressure fills a 5.00-L container. What volume will the gas occupy at 50.0°C and 800.0 torr pressure?

The Combined Gas Law can be used to find the final temperature, as in Example 13.9.

Example 13.9

Suppose a gas is kept in a container (that is, at constant volume), but the pressure of the gas is increased from $76\overline{0}$ torr to $86\overline{0}$ torr. This is done by changing the temperature of the gas from 0°C to some higher temperature. What is this higher temperature? (Report the answer in degrees Celsius.)

Solution To begin, organize the data and change degrees Celsius to kelvins. Note that the volume remains constant.

$$P_i = 76\overline{0} \text{ torr} \qquad P_f = 86\overline{0} \text{ torr}$$

$$T_i = t_i + 273 \qquad t_f = ?$$

$$= 0 + 273 = 273 \text{ K} \qquad T_f = ?$$

$$V \text{ is constant; this means} \qquad V_i = V_f$$

Then write the Combined Gas Law:

$$\frac{P_i V_i}{T_i} = \frac{P_f V_f}{T_f}$$

But because V_i and V_f are equal, they cancel each other, and the equation becomes

$$\frac{P_i}{T_i} = \frac{P_f}{T_f}$$

Solve the equation for T_f and enter the data in the result.

$$T_f = \frac{P_f T_i}{P_i} = \frac{(86\overline{0} \text{ torr})(273 \text{ K})}{76\overline{0} \text{ torr}} = 309 \text{ K}$$

Finally, change kelvins to degrees Celsius.

$$T_f = t_f + 273 \qquad \text{so} \qquad t_f = T_f - 273$$

$$t_f = 309 - 273 = 36°\text{C}$$

Practice Exercise 13.9 A gas is kept in a container at constant volume. The pressure of the gas is decreased from 850.0 torr to 750.0 torr. The temperature of the gas at the start of the experiment is 50.0°C. What is its final temperature in degrees Celsius?

Example 13.10

A certain gas occupies a volume of 3 L at 25°C and 4 atm pressure. What will be its volume at 25°C and 2 atm pressure?

Solution First organize the data and change degrees Celsius to kelvins. Note that the temperature remains constant.

$$P_i = 4 \text{ atm} \qquad P_f = 2 \text{ atm}$$

$$V_i = 3 \text{ L} \qquad V_f = ?$$

$$t_i = t_f = 25°\text{C } (298 \text{ K})$$

Then write the Combined Gas Law.

$$\frac{P_i V_i}{T_i} = \frac{P_f V_f}{T_f}$$

T_i and T_f cancel each other because they are equal (remember that T is constant), and the equation becomes

$$P_i V_i = P_f V_f$$

Finally, solve this equation for V_f and enter the data in the result.

$$V_f = \frac{P_i V_i}{P_f} = \frac{(4\ \cancel{\text{atm}})(3\ \text{L})}{2\ \cancel{\text{atm}}} = 6\ \text{L}$$

Practice Exercise 13.10 A gas occupies a volume of 5.00 L at 298 K and 1.50 atm pressure. What will be its volume at 25°C and 3.00 atm pressure?

13.6 Standard Temperature and Pressure (STP)

For work with gases, scientists have defined a set of standard reference conditions. **Standard temperature and pressure (STP)** are defined as 273 K (0°C) and 1 atm pressure (760 torr). Whenever volumes of gases are reported, they may be corrected to STP so that comparisons may be made. That is, the Combined Gas Law is used to determine what the volume of the gas would be at STP.

Example 13.11

A gas collected at 127°C has a volume of $1\overline{0}$ L and a pressure of $8\overline{0}0$ torr. What would be its volume at STP?

Solution To begin, organize the data and change degrees Celsius to kelvins.

$$P_i = 8\overline{0}0 \text{ torr} \qquad P_f = 7\overline{6}0 \text{ torr}$$

$$V_i = 1\overline{0} \text{ L} \qquad V_f = ?$$

$$T_i = t_i + 273 \qquad T_f = t_f + 273$$

$$= 127 + 273 = 4\overline{0}0 \text{ K} \qquad = 0 + 273 = 273 \text{ K}$$

Then write the Combined Gas Law, solve for V_f, and enter the data in the result.

$$\frac{P_iV_i}{T_i} = \frac{P_fV_f}{T_f} \quad \text{so} \quad V_f = \frac{P_iV_iT_f}{P_fT_i}$$

$$V_f = \frac{(80\bar{0}\ \cancel{\text{torr}})(1\bar{0}\ \text{L})(273\ \cancel{\text{K}})}{(76\bar{0}\ \cancel{\text{torr}})(4\bar{0}\bar{0}\ \cancel{\text{K}})} = 7.2\ \text{L}$$

Practice Exercise 13.11 A gas collected at 25.0°C has a volume of 250.0 mL and a pressure of 780.0 torr. What would be its volume at STP?

13.7 Dalton's Law of Partial Pressures

Learning Goal 7
Dalton's Law

In the laboratory, gases are usually collected by the water-displacement method (Figure 13.6), because many gases are relatively insoluble in water. This is a convenient method for obtaining a sample of a gas, except for one thing: Gas collected by this method always has some water vapor in it. Therefore the pressure in the collection bottle (P_{total}) is due to both the pressure of the gas (P_{gas}) and the pressure of the water vapor (P_{water}). To find out how much gas has been collected, we need to know the pressure of the gas alone—that is, its partial pressure P_{gas}. The **partial pressure** of any gas in a mixture of gases is the pressure that the gas *would* exert at the temperature of the mixture if it were the *only* gas in the container.

About 1800, John Dalton discovered a principle that is useful in this situation. **Dalton's Law of Partial Pressures** states that *the total pressure of a mixture of gases is the sum of the partial pressures of all the gases in the mixture*. In other words, each gas in a mixture exerts a pressure that is independent of the other gases present. This means that we can simply subtract the

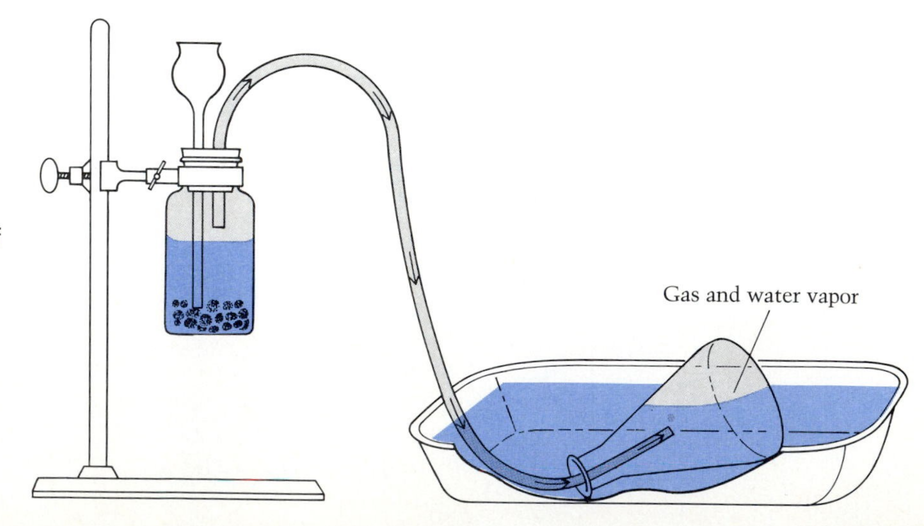

Figure 13.6
Collection of a gas over water. In this example, hydrochloric acid is added, through the thistle tube, to a piece of zinc in the bottle. Hydrogen gas travels out of the bottle through the rubber tube and is collected in the flask. As the hydrogen gas fills the flask, the water that was originally in the flask is pushed out.

pressure of the water vapor from the total pressure to find the pressure exerted by the gas. This is stated mathematically as

$$P_{\text{total}} = P_{\text{gas}} + P_{\text{water}} \qquad \text{(Dalton's Law)}$$

so

$$P_{\text{gas}} = P_{\text{total}} - P_{\text{water}}$$

The partial pressure of water vapor is known at any given temperature. Table 13.1 gives this value for various temperatures.

Example 13.12

Suppose we collect a sample of oxygen gas over water at 27°C and at a pressure of $75\overline{0}$ torr. The gas fills a $5\overline{00}$-mL container but has water vapor mixed in it. What would be the volume of the *dry* gas at STP?

Solution First find the pressure of the dry gas (P_{O_2}). To do this, subtract the pressure of the water vapor (P_{H_2O}) from the total pressure in the collection bottle (P_{total}). Look up the vapor pressure of water at 27°C and find that it is 26.74 torr. Then

$$P_{O_2} = P_{\text{total}} - P_{H_2O}$$

$$P_{O_2} = 75\overline{0}\text{ torr} - 26.74\text{ torr} = 723\text{ torr}$$

Therefore the initial pressure of the dry oxygen gas (P_i) is 723 torr. The next step is to find its volume at STP.

Organize the data and change degrees Celsius to kelvins.

$$P_i = 723\text{ torr} \qquad P_f = 76\overline{0}\text{ torr}$$

$$V_i = 5\overline{00}\text{ mL} \qquad V_f = ?$$

$$T_i = t_i + 723 \qquad T_f = t_f + 273$$

$$= 27 + 273 = 3\overline{00}\text{ K} \qquad = 0 + 273 = 273\text{ K}$$

Then write the Combined Gas Law, solve for V_f, and enter the data in the result.

$$\frac{P_i V_i}{T_i} = \frac{P_f V_f}{T_f} \quad \text{so} \quad V_f = \frac{P_i V_i T_f}{P_f T_i}$$

$$V_f = \frac{(723\ \cancel{\text{torr}})(5\overline{00}\text{ mL})(273\ \cancel{\text{K}})}{(76\overline{0}\ \cancel{\text{torr}})(3\overline{00}\ \cancel{\text{K}})} = 433\text{ mL}$$

The dry oxygen would have a volume of 433 mL at STP.

Practice Exercise 13.12 A sample of oxygen gas is collected over water at 55.0°C and at a pressure of 800.0 torr. The gas occupies a volume of 250.0 mL. What would be the volume of the *dry* gas at STP?

Table 13.1
Pressure of Water Vapor, P_{H_2O}, at Various Temperatures

Temperature (°C)	Pressure (torr)	Temperature (°C)	Pressure (torr)
0	4.58	32	35.66
5	6.54	33	37.73
10	9.21	34	39.90
15	12.79	35	42.18
16	13.63	36	44.56
17	14.53	37	47.07
18	15.48	38	49.69
19	16.48	39	52.44
20	17.54	40	55.32
21	18.65	45	71.88
22	19.83	50	92.51
23	21.07	55	118.04
24	22.38	60	149.38
25	23.76	65	187.54
26	25.21	70	233.7
27	26.74	75	289.1
28	28.35	80	355.1
29	30.04	85	433.6
30	31.82	90	525.8
31	33.70	95	633.9
		100	760.0

Example 13.13

Suppose 2.0 L of hydrogen gas is collected over water at 45°C and at a pressure of 700 torr. What would be the volume of the dry hydrogen gas at STP?

Solution Again, first find the pressure of the dry hydrogen gas (P_{H_2}). To do this, subtract the pressure of the water vapor (P_{H_2O}) from the total pressure in the collection bottle (P_{total}). The vapor pressure of water at 45°C is 71.88 torr, so

$$P_{H_2} = P_{total} - P_{H_2O}$$

$$P_{H_2} = 7\overline{0}0 \text{ torr} - 71.88 \text{ torr} = 628 \text{ torr}$$

Therefore the initial pressure of the dry hydrogen gas (P_i) is 628 torr. Next organize the data and change degrees Celsius to kelvins.

$$P_i = 628 \text{ torr} \qquad P_f = 76\overline{0} \text{ torr}$$

$$V_i = 2.0 \text{ L} \qquad V_f = ?$$

$$T_i = t_i + 273 \qquad T_f = t_f + 273$$

$$= 45 + 273 = 318 \text{ K} \qquad = 0 + 273 = 273 \text{ K}$$

Write the Combined Gas Law, solve for V_f, and enter the data in the result.

$$\frac{P_i V_i}{T_i} = \frac{P_f V_f}{T_f} \quad \text{so} \quad V_f = \frac{P_i V_i T_f}{P_f T_i}$$

$$V_f = \frac{(628 \text{ torr})(2.0 \text{ L})(273 \text{ K})}{(76\overline{0} \text{ torr})(318 \text{ K})} = 1.4 \text{ L}$$

Practice Exercise 13.13 A sample of hydrogen gas is collected over water at 20.0°C and 715 torr. It occupies a volume of 5.50 L. What would be the volume of the *dry* gas at STP?

13.8 Gay-Lussac's Law of Combining Volumes

In the early 1800s, the French chemist Joseph Louise Gay-Lussac (pronounced "Gay-lew-SACK") investigated the reaction in which hydrogen gas and oxygen gas form water vapor. He found that if the gases were at the same temperature and pressure, 2 L of hydrogen gas plus 1 L of oxygen gas produced 2 L of water vapor. He also studied the reaction between hydrogen and chlorine. He found that 1 L of hydrogen gas plus 1 L of chlorine gas reacted to form 2 L of hydrogen chloride gas.

Joseph Louis Gay-Lussac (Reprinted with permission from *Torchbearers of Chemistry* by Henry Monmouth Smith, © Academic Press, Inc., 1949)

$$\underset{2\text{ L}}{\text{Hydrogen gas}} + \underset{1\text{ L}}{\text{oxygen gas}} \longrightarrow \underset{2\text{ L}}{\text{water vapor}}$$

$$\underset{1\text{ L}}{\text{Hydrogen gas}} + \underset{1\text{ L}}{\text{chlorine gas}} \longrightarrow \underset{2\text{ L}}{\text{hydrogen chloride gas}}$$

What Gay-Lussac had discovered is now called **Gay-Lussac's Law of Combining Volumes,** which states that *when gases at the same temperature and pressure combine, the volumes of the reactants and products are in the same ratio as the coefficients of the balanced equation for the reaction.*

The Law of Combining Volumes states, for example, that two volumes of hydrogen combine with one volume of oxygen to form two volumes of water vapor

$$\underset{2\text{ volumes}}{2H_2(g)} + \underset{1\text{ volume}}{O_2(g)} \longrightarrow \underset{2\text{ volumes}}{2H_2O(g)}$$

whether "volume" means liter, cubic inch, gallon, or even coffee-can-full, as long as all three gases are at the same temperature and pressure. This law provides an easy way to work stoichiometric problems involving gases.

Example 13.14

How many liters of carbon dioxide can be produced with 10 L of carbon monoxide? How much oxygen is required (provided that all three gases are at the same temperature and pressure)?

Solution First write the balanced equation and organize the data.

$$\underset{10\ \text{L}}{2\text{CO}(g)} + \underset{?\ \text{L}}{\text{O}_2(g)} \longrightarrow \underset{?\ \text{L}}{2\text{CO}_2(g)}$$

Then use the coefficients of the balanced equation, just as in Chapter 11 (except that here liters can be substituted for moles), to calculate the amount of O_2 needed and the amount of CO_2 formed.

$$?\ \text{L of CO}_2 = (10\ \cancel{\text{L CO}})\left(\frac{2\ \text{L CO}_2}{2\ \cancel{\text{L CO}}}\right)$$

$$= 10\ \text{L of CO}_2$$

$$?\ \text{L of O}_2 = (10\ \cancel{\text{L CO}})\left(\frac{1\ \text{L O}_2}{2\ \cancel{\text{L CO}}}\right)$$

$$= 5\ \text{L of O}_2$$

Practice Exercise 13.14 How many liters of hydrogen fluoride gas can be produced from 20.0 L of hydrogen gas and 20.0 L of fluorine gas? (Assume that all gases are at the same temperature and pressure.)

13.9 Molar Gas Volume and Avogadro's Principle

In 1811, the Italian physicist Amedeo Avogadro, whose name we associate with Avogadro's number, advanced one of the most important hypotheses in chemistry. In an attempt to explain Gay-Lussac's Law of Combining Volumes, Avogadro said that *at the same temperature and pressure, equal volumes of all gases contain the same number of molecules*. This is called **Avogadro's Principle.**

Learning Goal 8
Avogadro's Principle

To understand this principle, consider the following experiment: Suppose exactly 1 mole of hydrogen gas (H_2) is placed in a container whose volume can be changed and whose temperature can be kept at 0°C (273 K). The volume of the container is adjusted so that the pressure of the hydrogen gas is exactly 1 atm. We find that the volume of the container (and therefore of the gas) is 22.4 L.

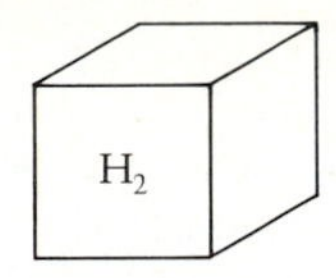

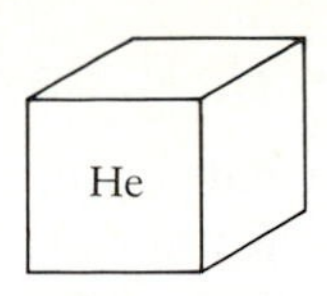

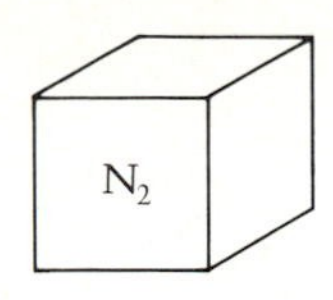

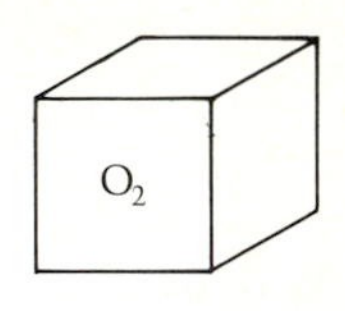

1 mole H_2 at STP, V = 22.4 liters; 1 mole He at STP, V = 22.4 liters; 1 mole N_2 at STP, V = 22.4 liters; 1 mole O_2 at STP, V = 22.4 liters

Figure 13.7
Molar gas volume

We now remove the hydrogen gas and repeat the experiment with exactly 1 mole of oxygen (O_2). Again we find that the volume of the container (and the oxygen) is 22.4 L. And every time the experiment is repeated with exactly 1 mole of a different gas, we find that its volume is 22.4 L at standard temperature and pressure (Figure 13.7).

Our experiment enables us to state the following conclusion: *Exactly 1 mole of any gas at STP has a volume of 22.4 L.* This volume is called the **molar volume of a gas,** and this conclusion can be used to solve problems like those in the next two examples.

Example 13.15

A certain gas occupies a volume of 179.2 L at STP. How many moles of gas are in this sample?

Solution One mole of any gas has a volume of 22.4 L at STP. Therefore

$$? \text{ moles of gas} = (179.2 \cancel{\text{L}})\left(\frac{1 \text{ mole}}{22.4 \cancel{\text{L}}}\right) = 8.00 \text{ moles of gas}$$

Practice Exercise 13.15 How many moles of gas occupy a volume of 448 mL at STP?

Example 13.16

Determine the volume of 5.00 moles of a gas at STP.

Solution Again, use the idea that 1 mole of a gas has a volume of 22.4 L at STP. Therefore

$$? \text{ L} = (5.00 \cancel{\text{moles}})\left(\frac{22.4 \text{ L}}{1 \cancel{\text{mole}}}\right) = 112 \text{ L}$$

Practice Exercise 13.16 What is the volume of 0.200 mole of a gas at STP?

13.10 The Ideal Gas Law

Learning Goal 9
Ideal Gas Law

We can now combine Boyle's Law, Charles's Law, and Avogadro's Principle into a relationship among the gas variables (P, V, T) and the number n of moles of gas. This relationship, called the **Ideal Gas Law,** is represented by the equation

$$\underset{\text{in atm}}{\text{Pressure}} \longrightarrow PV = nRT \longleftarrow \underset{\text{in K}}{\text{Temperature}} \qquad \text{Equation 13.5}$$

(Labels: n — Number of moles; V — Volume in liters; R — Ideal gas constant)

If this equation is used with the pressure expressed in atmospheres, the volume in liters, the temperature in kelvins, and the amount of gas in moles, then the value for R, the ideal gas constant, is

$$0.0821 \frac{(\text{L})(\text{atm})}{(\text{K})(\text{mole})}$$

Notice the units of R. These units will remind you that when you use the Ideal Gas Law, pressure must be in atmospheres, volume in liters, temperature in kelvins, and gas in moles. We can show that R has this value by solving the ideal gas equation for R and substituting known values for the other variables in the equation.

$$PV = nRT \qquad \text{so} \qquad R = \frac{PV}{nT}$$

We know that at STP (273 K, 1 atm), 1 mole of a gas has a volume of 22.4 L. These values may be substituted into the equation.

$$R = \frac{PV}{nT} = \frac{(1 \text{ atm})(22.4 \text{ L})}{(1 \text{ mole})(273 \text{ K})} = 0.0821 \frac{(\text{L})(\text{atm})}{(\text{K})(\text{mole})}$$

The following examples show how the Ideal Gas Law is used.

Example 13.17

Two moles of a gas exert a pressure of 1,520 torr at a temperature of 127°C. What volume does the gas occupy under these conditions?

Solution First organize the data. Change the 1,520 torr to atmospheres and the degrees Celsius to kelvins. To convert torr to atmospheres, remember that 1 atm = $76\overline{0}$ torr.

$$? \text{ atm} = (1{,}520 \cancel{\text{torr}})\left(\frac{1 \text{ atm}}{76\bar{0} \cancel{\text{torr}}}\right) = 2.00 \text{ atm}$$

$$P = 1{,}520 \text{ torr } (2.00 \text{ atm})$$

$$V = ?$$

$$n = 2.00 \text{ moles}$$

$$R = 0.0821 \frac{(\text{L})(\text{atm})}{(\text{K})(\text{mole})}$$

$$T = t + 273 = 127 + 273 = 4\overline{00} \text{ K}$$

Next write the ideal gas equation, solve for V, and enter the data in the result.

$$PV = nRT \qquad \text{so} \qquad V = \frac{nRT}{P}$$

$$V = \frac{(2.00 \cancel{\text{moles}})\left(0.0821 \dfrac{(\text{L})(\cancel{\text{atm}})}{(\cancel{\text{K}})(\cancel{\text{mole}})}\right)(4\overline{00} \cancel{\text{K}})}{2.00 \cancel{\text{atm}}} = 32.8 \text{ L}$$

Practice Exercise 13.17 Three moles of a gas exert a pressure of 1,555 torr at a temperature of 27.0°C. What volume does the gas occupy under these conditions?

Example 13.18

A gas has a pressure of $38\bar{0}$ torr at a temperature of $5\bar{0}$°C. The volume of the gas is $8\overline{00}$ mL. How many moles of gas are present?

Solution Organize the data. Change torr to atmospheres and degrees Celsius to kelvins.

$$? \text{ atm} = (38\bar{0} \cancel{\text{torr}})\left(\frac{1 \text{ atm}}{76\bar{0} \cancel{\text{torr}}}\right) = 0.500 \text{ atm}$$

$$P = 0.500 \text{ atm}$$

$$V = 8\overline{00} \text{ mL } (0.800 \text{ L})$$

$$n = ?$$

$$R = 0.0821 \frac{(\text{L})(\text{atm})}{(\text{K})(\text{mole})}$$

$$T = t + 273 = 5\bar{0} + 273 = 323 \text{ K}$$

Write the ideal gas equation, solve for n, and enter the data in the result.

$$PV = nRT \qquad \text{so} \qquad n = \frac{PV}{RT}$$

$$n = \frac{(0.500 \text{ atm})(0.800 \text{ L})}{\left(0.0821 \frac{(\text{L})(\text{atm})}{(\text{K})(\text{mole})}\right)(323 \text{ K})} = 0.0151 \text{ mole}$$

Practice Exercise 13.18 A gas exerts a pressure of 444 torr at a temperature of 55.0°C. The volume of the gas is 1.00 L. How many moles of gas are present?

13.11 Determining Molecular Mass with the Ideal Gas Law

Learning Goal 10
Ideal Gas Law and molecular mass

The number of moles n of a sample of a gas is equal to the mass of the sample (g) in grams divided by the molecular mass of the gas (MM).

$$n = \frac{\text{g}}{\text{MM}}$$

If g/MM is used in place of n in the Ideal Gas Law, we have a valuable tool for determining the molecular masses of gases:

$$PV = \frac{\text{g}RT}{\text{MM}} \qquad \text{Equation 13.6}$$

For an unknown gas, we can weigh a sample of it into a container, measure its pressure, volume, and temperature, and use this form of the Ideal Gas Law to calculate its molecular mass.

Example 13.19

What is the molecular mass of the gas for which the following experimental data apply?

$$\text{Mass} = 1.00 \text{ g} \qquad V = 82.0 \text{ mL } (0.0820 \text{ L})$$
$$P = 1.50 \text{ atm} \qquad T = 3\overline{0}0 \text{ K}$$

Solution Write Equation 13.6, solve for MM, and enter the data in the result.

$$PV = \frac{\text{g}RT}{\text{MM}} \qquad \text{so} \qquad \text{MM} = \frac{\text{g}RT}{PV}$$

$$\text{MM} = \frac{(1.00 \text{ g})\left(0.0821 \frac{(\text{L})(\text{atm})}{(\text{K})(\text{mole})}\right)(3\overline{0}0 \text{ K})}{(1.50 \text{ atm})(0.0820 \text{ L})} = 2\overline{0}0 \text{ g/mole}$$

> ***Practice Exercise 13.19*** Determine the molecular mass of the gas to which the following hypothetical data apply: mass = 2.50 g, V = 100.0 mL, P = 2.00 atm, T = 400.0 K.

13.12 The Kinetic Theory Revisited

If the gas laws are considered in light of the kinetic theory presented at the beginning of this chapter, it is clear why gases behave as they do. For example, the kinetic theory explains Boyle's Law. Remember, Boyle's Law says that the volume of a gas increases as its pressure decreases and that the pressure of a gas increases as its volume decreases. When we assume that gases are composed of molecules that are constantly in motion, colliding with each other and with the walls of their container, it seems reasonable that:

1. Keeping these molecules in a smaller space causes them to strike the walls of their container more often; this means that more force is exerted against the walls of the container, and therefore the pressure is greater (Figure 13.8).
2. Allowing these molecules more space to move around in causes them to strike the walls less often; therefore the pressure is decreased (Figure 13.9).

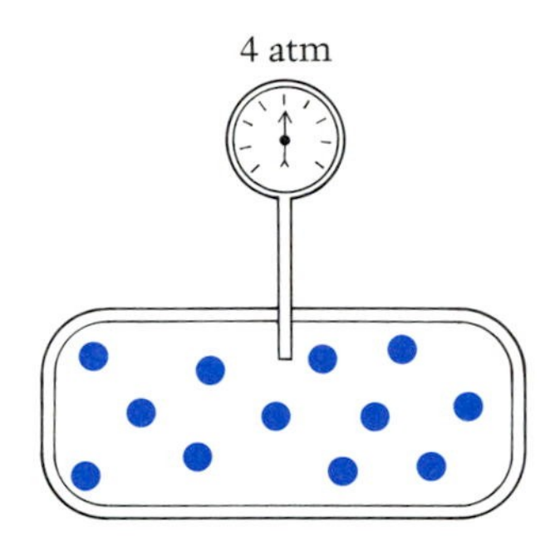

Figure 13.8
A sample of gas in a small container. The pressure gauge shows the pressure that is exerted by the gas.

The kinetic theory also explains Charles's Law, which says that the volume of a gas increases as its temperature increases and that the volume of a gas decreases as its temperature decreases. Consider a balloon filled with air (Figure 13.10). The molecules of air constantly move around, hitting the walls of the balloon. Suppose a candle is held near the balloon, which warms the air inside. The gas gets warmer because the molecules begin to move more rapidly, hitting the walls of the balloon with more force. Because the balloon is made of flexible material, it expands. As the molecules of air move more slowly, hitting the walls of the balloon more gently, the balloon contracts, and the gas cools to room temperature.

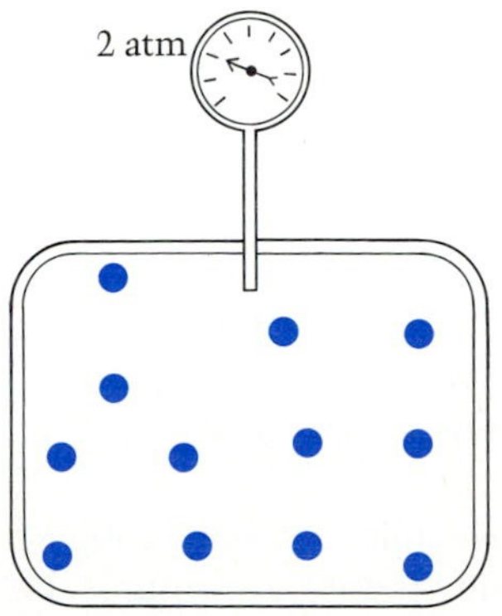

Figure 13.9
The same sample of gas in a container twice the size of that shown in Figure 13.8

Summary

The kinetic theory of gases is concerned with the motion of gas molecules and the forces of attraction between molecules. It can be used to explain the various physical properties of gases.

In the early 1600s, Torricelli devised a pressure-measuring device called a barometer and discovered that air exerts pressure on objects it touches. The units of pressure we use are atmospheres and torr, and air exerts a pressure of about 760 torr (1 atm) at sea level.

The experiments of Robert Boyle in the mid-1600s led to the formulation of Boyle's Law, which states that at constant temperature, the volume of a given mass of gas varies inversely with the pressure. In the 1780s, Jacques Charles discovered a relationship between the volume and the temperature of

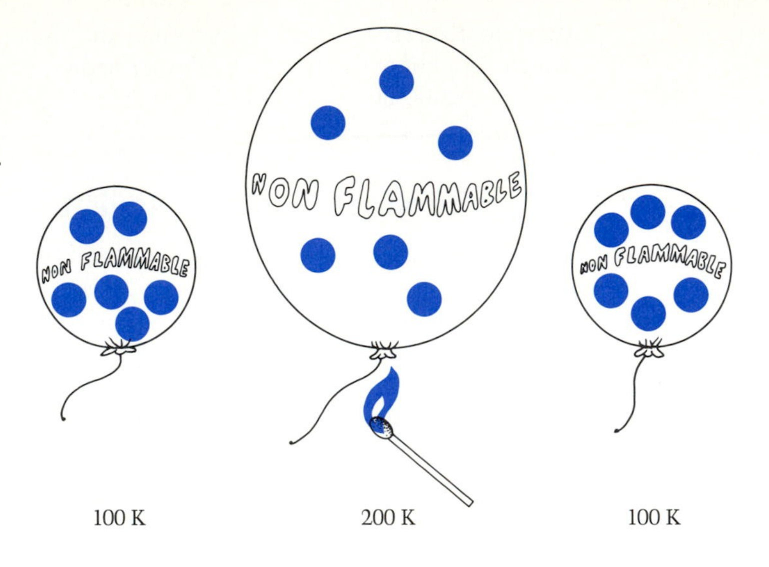

Figure 13.10
The balloon experiment shows the temperature–volume relationship expressed by Charles's Law.

a gas. Charles's Law states that at constant pressure, the volume of a given amount of gas varies directly with its temperature in kelvins. On the Kelvin scale, the temperature in kelvins is equal to the Celsius temperature plus 273. Boyle's Law and Charles's Law can be combined into a single law, the Combined Gas Law, which shows the interrelationship of changes in the temperature, pressure, and volume of a gas.

Standard temperature and pressure (STP) for gases are defined as 273 K (0°C) and 1 atm pressure (760 torr). The Combined Gas Law may be used to determine what the volume of a gas would be at STP. Dalton's Law of Partial Pressures states that the total pressure exerted by a mixture of gases is the sum of the partial pressures that each of the gases in the mixture would exert on its own at the same temperature as the mixture.

In the early 1800s, Gay-Lussac discovered the Law of Combining Volumes. This law states that when gases at the same temperature and pressure combine, the volumes of the reactants and products are in the same ratio as the coefficients of the balanced equation for the reaction. In an attempt to explain the Law of Combining Volumes, Avogadro hypothesized that equal volumes of all gases at the same temperature and pressure contain the same number of molecules. From the hypothesis, called Avogadro's Principle, scientists determined that 1 mole of any gas at STP has a volume of 22.4 L. This is called the molar volume of a gas.

Boyle's Law, Charles's Law, and Avogadro's Principle can be combined into a single relationship among the pressure P, temperature T, volume V, and number of moles n of a gas. This relationship, called the Ideal Gas Law, is represented by the equation $PV = nRT$, where R is the ideal gas constant and is equal to 0.0821 (L) (atm)/(K) (mole). The Ideal Gas Law can be used to determine the molecular mass of a gas when n is replaced by the fraction that represents mass/molecular mass (g/MM) and the equation is rearranged.

Key Terms

absolute zero **(13.4)**
atmosphere **(13.2)**
Avogadro's Principle **(13.9)**
barometer **(13.2)**
Boyle's Law **(13.3)**
Charles's Law **(13.4)**
Combined Gas Law **(13.5)**
Dalton's Law of Partial Pressures **(13.7)**
Gay-Lussac's Law of Combining Volumes **(13.8)**
ideal gas **(13.1)**
Ideal Gas Law **(13.10)**
Kelvin scale **(13.4)**
kinetic theory of gases **(13.1)**
molar volume of a gas **(13.9)**
partial pressure **(13.7)**
real gas **(13.1)**
standard temperature and pressure (STP) **(13.6)**
torr **(13.2)**

CHEMICAL FRONTIERS

BIOCHEMICAL RESEARCH BENEFITS ALLERGY SUFFERERS*

After more than a decade and a half of research, biochemists have determined the structure of the immunoglobulin E (IgE) receptor, a key protein involved in triggering allergic reactions. All allergic reactions—sneezing, sniffling, developing an inflammation, coughing, wheezing, itching—involve IgE receptors.

There are five groups of antibodies, called *immunoglobulins,* produced by the human immune system as a reaction to foreign substances, or *antigens*. These immunoglobulins are labeled IgE, IgD, IgM, IgG, and IgA, and each has a different function. IgE is involved in the protection against parasitic infections and in the allergic response. Allergic people produce large amounts of IgE when exposed to *allergens,* or substances that trigger an allergic reaction. For example, people who are allergic to wool produce IgE in response to that allergen.

As soon as the allergen enters the body, IgE antibodies are produced and attach themselves to IgE receptors, which are on specialized cells called *basophils* and *mast cells*. Basophils circulate in the blood, while mast cells line the membranes of the nasal passages, bronchial tubes, and other sites where symptoms of allergic reactions are present. As the IgE–mast cell complex comes in contact with the allergen, the mast cell spews forth a mass of chemicals that cause allergic symptoms. Among these chemicals is the compound histamine.

*Adapted from "New Hope for Allergy Sufferers," *What's Happening in Chemistry* (American Chemical Society, 1990).

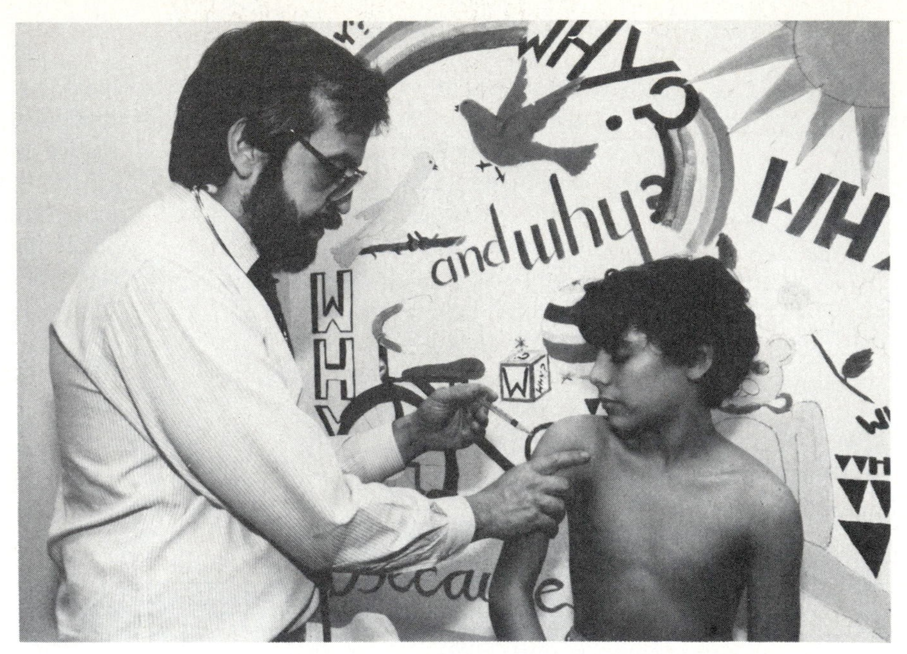

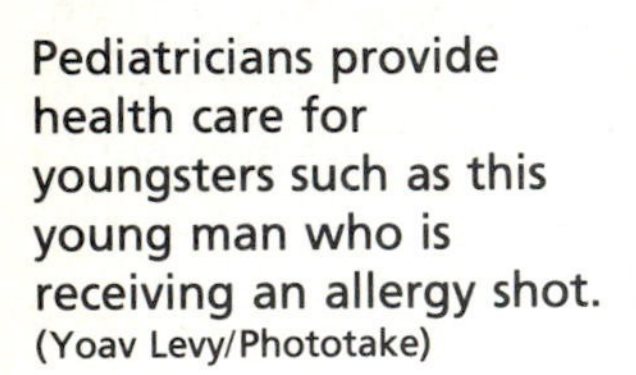
Pediatricians provide health care for youngsters such as this young man who is receiving an allergy shot. (Yoav Levy/Phototake)

The most common allergy drugs contain antihistamines, which act *after* the allergic reaction occurs. Researchers hope that their new knowledge of the structure of IgE will enable them to develop drugs that prevent the allergic reactions.

Self-Test Exercises

Learning Goal 1: Kinetic Theory of Gases

▲ **1.** State the five basic principles of the kinetic theory of gases. Explain how each of these principles is related to the gas laws discussed in this chapter.

▲ **2.** Explain how the kinetic theory of gases accounts for the properties of gases in terms of the forces between gas molecules and the energy these molecules have.

Learning Goal 2: Converting Atmospheres to Torr, and Vice Versa

▲ **3.** Express the following pressures in atmospheres: *(a)* $25\overline{0}$ torr *(b)* 1,550 torr *(c)* 7,600 torr

▲ **4.** Express the following pressures in atmospheres: *(a)* $35\overline{0}$ torr *(b)* $2{,}5\overline{0}0$ torr *(c)* $1{,}96\overline{0}$ torr

5. Express the following pressures in torr: *(a)* 3.00 atm *(b)* 0.300 atm *(c)* 0.150 atm

6. Express the following pressures in torr: *(a)* 2.00 atm *(b)* 0.250 atm *(c)* 0.123 atm

7. Express the following pressures in atmospheres: *(a)* 76 torr *(b)* 2,600 torr *(c)* 3.80 torr

8. Express the following pressures in atmospheres: *(a)* 152 torr *(b)* 304 torr *(c)* 0.152 torr

9. Express the following pressures in torr: *(a)* 0.0100 atm *(b)* 12.0 atm *(c)* 5.50 atm

10. Express the following pressures in torr: *(a)* 0.200 atm *(b)* 15.2 atm *(c)* 7.6 atm

Learning Goal 3: Boyle's Law

11. Given the following data for a gas, use Boyle's Law to determine V_f: $P_i = 0.60$ atm, $V_i = 30.0$ L, $P_f = 0.10$ atm.

12. Given the following data for a gas, use Boyle's Law

to determine V_f: $P_i = 0.500$ atm, $V_i = 40.0$ L, $P_f = 0.200$ atm.

▲ **13.** Given the following data for a gas, use Boyle's Law to determine V_f: $P_i = 3.00$ atm, $V_i = 5\overline{0}0$ mL, $P_f = 3\overline{8}0$ torr.

▲ **14.** Given the following data for a gas, use Boyle's Law to determine V_f: $P_i = 2.00$ atm, $V_i = 25\overline{0}$ mL, $P_f = 4\overline{0}0$ torr.

15. Given the following data for a gas, use Boyle's Law to determine P_f: $P_i = 1{,}9\overline{0}0$ torr, $V_i = 25\overline{0}$ mL, $V_f = 50.0$ mL.

16. Given the following data for a gas, use Boyle's Law to determine P_f: $P_i = 1{,}52\overline{0}$ torr, $V_i = 455$ cm^3, $V_f = 75.0$ cm^3.

Learning Goal 4: Charles's Law

▲ **17.** Given the following data for a gas, use Charles's Law to determine t_f (that is, express your answer in °C): $t_i = 77.0$°C, $V_i = 2.00$ L, $V_f = 4.00$ L.

▲ **18.** Given the following data for a gas, use Charles's Law to determine t_f (that is, express your answer in °C): $t_i = 90.0$°C, $V_i = 4.00$ L, $V_f = 5.50$ L.

19. Given the following data for a gas, use Charles's Law to determine t_f (that is, express your answer in °C): $t_i = -73$°C, $V_i = 8.00$ L, $V_f = 32.0$ L.

20. Given the following data for a gas, use Charles's Law to determine t_f (that is, express your answer in °C): $t_i = 85.0$°C, $V_i = 7.50$ L, $V_f = 4.50$ L.

21. Given the following data for a gas, use Charles's Law to determine V_f: $t_i = 327$°C, $V_i = 2\overline{0}0$ L, $t_f = -173$°C.

22. Given the following data for a gas, use Charles's Law to determine V_f: $t_i = 400.0$°C, $V_i = 45\overline{0}$ L, $t_f = -73$°C.

Learning Goal 5: Converting °C to K, and Vice Versa

▲ **23.** Change the following temperatures to °C:
(a) $5\overline{0}0$ K *(b)* 4 K

▲ **24.** Change the following temperatures to °C:
(a) 353 K *(b)* 3 K

25. Change the following temperatures to kelvins:
(a) 10°C *(b)* −10°C *(c)* −73°C

26. Change the following temperatures to kelvins:
(a) 15°C *(b)* −15°C *(c)* −50°C

27. Change the following temperatures to °C:
(a) 0 K *(b)* $1\overline{0}0$ K *(c)* $2\overline{0}0$ K *(d)* $3\overline{0}0$ K

28. Change the following temperatures to °C:
(a) 5 K *(b)* 150 K *(c)* 250 K *(d)* 350 K

29. Change the following temperatures to kelvins:
(a) 20°C *(b)* 0°C *(c)* $-1\overline{0}0$°C *(d)* 212°F

30. Change the following temperatures to kelvins:
(a) 25°C *(b)* −5°C *(c)* −75°C *(d)* 424°F

Learning Goal 6: Combined Gas Law

▲ **31.** Given the following data for a gas, use the Combined Gas Law to determine P_f:

$P_i = 0.5$ atm $V_f = 5$ L
$V_i = 10$ L $t_f = 327$°C
$t_i = 27$°C

32. Given the following data for a gas, use the Combined Gas Law to determine P_f:

$P_i = 0.20$ atm $V_f = 10.0$ L
$V_i = 8.0$ L $t_f = 35\overline{0}$°C
$t_i = 43$°C

33. A tire is filled with air at a pressure of 1,520 torr at 20.0°C. What will be the pressure in the tire after the car is driven and the temperature of the air in the tire rises to 34.0°C? (Assume that the volume of the tire remains constant, which is the case in a tubeless tire.)

▲ **34.** A tubeless tire (whose volume remains constant) is filled with air at a pressure of $76\overline{0}$ torr at 30.0°C. What will be the pressure in the tire after the car is driven and the temperature of the air in the tire rises to 35.0°C?

35. A balloon is filled with a gas to a volume of $4\overline{0}0$ mL, at a pressure of 1.00 atm and a temperature of 25.0°C. The temperature is increased to 50.0°C, and the pressure is increased to $78\overline{0}$ torr. What will be the volume of the balloon?

36. A balloon is filled with air to a volume of 350.0 mL, at a pressure of 1.20 atm and a temperature of 28.0°C. The temperature is increased to 55.0°C, and the pressure is increased to 2.00 atm. What is the new volume of the balloon?

37. Suppose that you collect a gas over water at 30.0°C. The volume of the gas is $8\overline{0}0$ mL, and the total pressure is $78\overline{0}$ torr. What is the volume of *dry* gas at STP?

38. A gas is collected over water at 28.0°C. The volume of the gas is 750.0 mL, and the total pressure is 790.0 torr. What is the volume of the *dry* gas at STP?

39. Assume that your lungs can hold about $5\overline{0}0$ mL of air at STP. You take a deep breath and hold it, then dive into the ocean to a depth at which the temperature is 10.0°C and the pressure is $1,00\overline{0}$ torr. What is the volume of the air in your lungs now?

40. A gas is collected over water at 35.0°C. The volume of the gas is 1.00 L, and the total pressure is 800.0 torr. What is the volume of the *dry* gas at STP?

41. The Gulf of Mexico puffer fish can live in deep water, where the water pressures are enormous. These fish have the ability to compress and expand their bodies to compensate for changes in pressure. They do this by holding in or letting out air. If a fish of this type decided to move from an area in which the water pressure was 4.0 atm to an area in which it was 1.5 atm, by what ratio would the volume of its body increase?

42. Assume that the Gulf of Mexico puffer fish described in Exercise 41 moves from an area in which the water pressure is 2.00 atm to an area in which it is 4.50 atm. What happens to the volume of its body?

43. An airtight chamber is filled with a gas at a pressure of 25.0 atm and a temperature of 50.0°C. The temperature inside the chamber is increased to $5\overline{0}0$°C. What happens to the pressure of the gas? (The volume of the chamber remains constant.)

44. An airtight chamber is filled with a gas to a pressure of 15.0 atm and a temperature of 350.0°C. If the temperature inside the chamber is decreased to 50.0°C, what will happen to the pressure of the gas? (The volume of the chamber remains constant.)

45. A gas has a volume of $4\overline{0}0$ mL at a pressure of 3.00 atm and a temperature of 127°C. What would its volume be at STP?

46. A gas has a volume of 40.0 mL at a pressure of 2.50 atm and a temperature of 100.0°C. What would its volume be at STP?

47. A sample of oxygen gas is collected over water. The volume of gas collected is $3\overline{0}0$ mL, at a pressure of $75\overline{0}$ torr and a temperature of 15.0°C. What is the volume of the *dry* gas at STP?

48. A sample of hydrogen gas is collected over water. The volume of gas collected is 250.0 mL, at a pressure of 815 torr and a temperature of 27.0°C. What is the volume of the *dry* gas at STP?

49. Complete the following table for each data set:

Data Set	P	V	n	T
1	constant	increased	constant	?
2	?	decreased	constant	constant
3	decreased	?	increased	constant

50. Complete the following table for each data set:

Data Set	P	V	n	T
1	decrease	?	constant	constant
2	?	increase	constant	constant
3	increase	constant	constant	?

Learning Goal 7: Dalton's Law

▲ **51.** Find the total pressure in an enclosed vessel that contains the following gases: nitrogen at $2\overline{0}0$ torr, hydrogen at $3\overline{0}0$ torr, oxygen at $5\overline{0}$ torr, and helium at $15\overline{0}$ torr.

▲ **52.** Find the total pressure in an enclosed vessel that contains the following gases: helium at $2\overline{0}0$ torr, hydrogen at $15\overline{0}$ torr, nitrogen at $9\overline{0}$ torr, and oxygen at $15\overline{0}$ torr.

53. Nitrogen gas makes up about $8\overline{0}$ volume percent of the atmosphere. What is the pressure exerted by this nitrogen gas when the total atmospheric pressure is $76\overline{0}$ torr?

54. Suppose we collect a $75\overline{0}$-mL sample of oxygen gas over water at 22.0°C and $77\overline{0}$ torr. What would be the volume of the *dry* gas at STP?

Learning Goal 8: Avogadro's Principle

55. Methane (CH_4) burns in air to produce carbon dioxide (CO_2) and water vapor. How many liters of water vapor are produced from 12 L of CH_4?

56. Ethane (C_2H_6) burns in air to produce carbon dioxide and water vapor. How many liters of water vapor are produced from 15.0 L of ethane?

▲ **57.** Determine the volume of 4.00 moles of oxygen gas at STP.

58. What is the volume of 3.00 moles of hydrogen gas at STP?

59. You are given the reaction

$$2Na + 2H_2O \longrightarrow 2NaOH + H_2$$

How many grams of sodium would be needed to release 10.0 L of hydrogen gas at STP?

***60.** Consider the reaction

$$2Li + 2H_2O \longrightarrow 2LiOH + H_2$$

How many grams of lithium are needed to release 10.0 L of hydrogen gas at STP?

***61.** Suppose that $8\bar{0}$ mL of hydrogen gas react with $8\bar{0}$ mL of oxygen gas. How many milliliters of oxygen gas remain uncombined? How many milliliters of water vapor are formed?

***62.** Suppose that 1.00 L of hydrogen gas reacts with 1.00 L of oxygen gas at STP. How many grams of oxygen gas remain uncombined? How many grams of water vapor are formed?

63. Sulfur dioxide can combine with oxygen in the air to produce sulfur trioxide:

$$SO_2(g) + O_2(g) \longrightarrow SO_3(g)$$

(a) Balance the equation.
(b) How many liters of oxygen gas are needed to combine with $15\bar{0}$ L of SO_2?
(c) How many liters of SO_3 are produced?

64. Nitric oxide can combine with oxygen gas in the air to produce nitrogen dioxide:

$$NO(g) + O_2(g) \longrightarrow NO_2(g)$$

(a) Balance the equation.
(b) How many liters of oxygen gas are needed to combine with 10.0 L of NO?
(c) How many liters of NO_2 are formed?

65. Butane burns in air to produce carbon dioxide and water vapor:

$$C_4H_{10}(g) + O_2(g) \longrightarrow CO_2(g) + H_2O(g)$$

(a) Balance the equation.
(b) How many liters of oxygen gas are needed to burn 15.0 L of butane?
(c) How many liters of carbon dioxide and water are produced?

66. Propane burns in air to produce carbon dioxide and water vapor:

$$C_3H_8(g) + O_2(g) \longrightarrow CO_2(g) + H_2O(g)$$

(a) Balance the equation.
(b) How many liters of oxygen gas are needed to burn 25.0 L of propane?
(c) How many liters of carbon dioxide gas are formed?

***67.** Nitrogen gas can combine with oxygen gas to produce nitrogen dioxide:

$$N_2(g) + O_2(g) \longrightarrow NO_2(g)$$

(a) Balance the equation.
(b) Suppose that 75.0 mL of nitrogen gas react with 75.0 mL of oxygen gas. How many milliliters of nitrogen gas remain uncombined?
(c) How many milliliters of $NO_2(g)$ are formed?

***68.** Nitrogen gas can combine with hydrogen gas to produce ammonia:

$$N_2(g) + H_2(g) \longrightarrow NH_3(g)$$

(a) Balance the equation.
(b) Suppose that 120.0 mL of N_2 can react with 120.0 mL of H_2 gas. How many milliliters of nitrogen gas remain uncombined?
(c) How many milliliters of ammonia gas are formed?

Learning Goal 9: Ideal Gas Law

69. Determine the number of grams of nitrogen gas present in a 3.00-L container at a temperature of 27°C and a pressure of 1,520 torr.

70. Determine the number of grams of hydrogen gas present in a 2.00-L container at a temperature of 127°C and a pressure of $2{,}28\bar{0}$ torr.

▲ **71.** Part of Avogadro's Principle is that 1.00 mole of any gas at STP has a volume of 22.4 L. Using this information, solve the ideal gas equation for R.

72. What volume will 0.75 mole of hydrogen gas occupy at STP?

73. Determine the volume of 0.500 mole of $N_2(g)$ at STP.

▲ **74.** How many grams of chlorine gas are present in a 1,500.0-mL flask at a temperature of 150.0°C and a pressure of 1.50 atm?

75. Determine the number of grams of NO_2 gas present in a $5\overline{00}$-mL flask at a temperature of 227°C and a pressure of $3{,}04\bar{0}$ torr.

76. How many moles of fluorine gas are present in a 2.00-L flask at a temperature of 127°C and a pressure of 1.00 atm?

77. Determine the pressure of 12.20 moles of H_2 gas occupying a volume of $2{,}\overline{000}$ mL at a temperature of −73°C.

78. What is the pressure of 4.00 moles of oxygen gas that occupy a volume of 5.00 L at a temperature of 0°C?

79. A 128-g sample of SO_2 has a pressure of $38\bar{0}$ torr and a volume of $50\bar{0}$ mL. What is the temperature of this gas?

80. A 92.0-g sample of NO_2 gas has a pressure of $77\bar{0}$ torr and a volume of $95\bar{0}$ mL. What is the temperature of this gas?

Learning Goal 10: Ideal Gas Law and Molecular Mass

***81.** A gaseous compound is made up of 83% carbon and 17% hydrogen by mass. When 1.2 g of this gas are placed in a $50\bar{0}$-mL container at 25°C, the pressure of the gas is found to be 1.0 atm.
(a) What is the empirical formula of the gas?
(b) What is the molecular mass of the gas?
(c) What is the molecular formula of the gas?

82. Given that 4.00 g of a certain gas have a volume of 500.0 mL at a pressure of 2,005 torr and a temperature of 27.0°C, calculate the molecular mass of the gas.

83. Calculate the molecular mass of a gas, given that 2.0 g of the gas have a volume of $40\bar{0}$ mL at a pressure of 3,040 torr and a temperature of 127°C.

***84.** Using the Ideal Gas Law equation and remembering that density equals mass divided by volume, calculate the density of hydrogen gas at STP. (Report your answer in grams per liter.)

***85.** Using the ideal gas equation and remembering that density equals mass divided by volume, calculate the density of oxygen gas at STP. (Report the answer in grams per liter.)

86. A 20.0-g sample of a gas has a pressure of 4.00 atm and a volume of 1,500.0 mL at a temperature of 30.0°C. What is the molecular mass of the gas?

▲ **87.** A 14.63-g sample of a gas has a pressure of 6.00 atm and a volume of $2{,}50\bar{0}$ mL at a temperature of 27°C. What is the molecular mass of this gas?

▲ **88.** Given that 150.0 g of a certain gas have a volume of 5.00 L at a pressure of $2{,}00\bar{0}$ torr and a temperature of 150.0°C, calculate the molecular mass of the gas.

***89.** A gaseous compound is 69.57% oxygen and 30.43% nitrogen by mass. When 1.44 g of this gas are placed in a $25\bar{0}$-mL flask at 20.0°C, the pressure of the gas is found to be 1.50 atm.
(a) What is the empirical formula of the gas?
(b) What is the molecular mass of the gas?
(c) What is the molecular formula of the gas?
(Use 16.00 as the molecular mass of oxygen and 14.01 as the molecular mass of nitrogen.)

90. Given that 75.0 g of a certain gas have a volume of 4.50 L at a pressure of $1{,}56\bar{0}$ torr and a temperature of 127°C, calculate the molecular mass of the gas.

Extra Exercises

▲ **91.** Explain Boyle's Law and Charles's Law using the kinetic theory of gases.

92. One-half of a 10.0-L tank of hydrogen gas at 25°C and 10.0 atm is released into a balloon at constant temperature. The atmospheric pressure is $76\bar{0}$ torr. Determine the volume of the balloon.

93. The balloon in Problem 92 is released. It rises 10.0 km into the atmosphere, where it bursts. What was its final volume, given that the temperature was −50.0°C and the pressure was $23\bar{0}$ torr?

94. An industrial process for the production of ammonia involves reacting hydrogen gas with nitrogen gas. How many liters of ammonia gas could be produced from 1.00 kg of nitrogen gas, given that the process operates at a temperature of 30.0°C and a pressure of $75\bar{0}$ torr?

95. Change the following temperatures to kelvins:
(a) $4\bar{0}$°F *(b)* $18\bar{0}$°F *(c)* 212°F

96. Calculate the volume occupied by 1.00 mole of oxygen gas at STP (density = 1.429 g/L).

97. Calculate the density of CH_4 gas at STP.

▲ **98.** Here is a very interesting experiment (see if your instructor will do it): Boil a small amount of water in a tin can. While the water is boiling, seal the can and remove the heat. As the can cools, it collapses. Explain this in terms of the gas laws you are now familiar with.

99. A volume of 272 mL of oxygen is collected at 27.0°C. What volume would the oxygen occupy at $2\bar{0}\bar{0}$°C, assuming that the pressure is held constant?

100. The following data are collected on a gas sample: P = 1.00 atm, V = 2.00 L, t = 27.0°C, mass = 2.56 g. What is the molecular mass of the gas? Suggest an element or compound that the gas might be if it is known to be diatomic.

14 The Liquid and Solid States

Learning Goals

After you've studied this chapter, you should be able to:

1. Discuss the properties that distinguish gases, liquids, and solids from one another.
2. Apply the kinetic theory to describe the behavior of liquids and solids.
3. Define and illustrate the terms *evaporation, equilibrium vapor pressure,* and *boiling point*.
4. Explain what happens as a solid changes to liquid and the liquid changes to gas.
5. Explain the processes of distillation and fractional distillation, and define the relevant terms.
6. Describe what is meant by crystalline solids and amorphous solids.
7. Discuss the three kinds of crystalline solids and the forces of attraction in each.
8. Define and give an example of a crystal lattice and a unit cell.
9. List some chemical and physical properties of water.
10. Write equations for the reaction of water with oxides of metals and nonmetals.
11. Explain what a hydrate is, give examples of hydrates, and calculate the percentage of water in a hydrated salt.
12. Define the terms *anhydrous, hygroscopic, deliquescence,* and *efflorescence*.

Introduction

Matter can exist in any of three physical states: gas, liquid, and solid. The early chemists—the alchemists—had trouble dealing with gases, and it was only in the seventeenth century that research into gases really began. The alchemists were, however, able to study liquids and solids extensively, and most of what we know of their work is related to these two states of matter.

In Chapter 13, we discussed the special properties of gases; in this chapter, we do the same for liquids and solids. We begin by using the kinetic theory (which was applied to gases in Section 13.1) to explain some of the differences among the three states.

14.1 The Kinetic Theory Extended to Liquids and Solids

Learning Goal 1
Properties of gases, liquids, and solids

Learning Goal 2
Kinetic theory of liquids and solids

Table 14.1 lists three physical properties of gases, liquids, and solids. In Chapter 13, we explained these properties of gases according to the kinetic theory of gases. We noted that gas molecules are relatively far apart, are always in motion, and thus show little attraction for each other. It is for these reasons that a gas has no shape of its own, will move through and fill any volume, and is easily compressed.

In a liquid, however, the molecules are very close to each other. Therefore, according to the kinetic theory, the forces of attraction between liquid molecules are much greater than those between gas molecules. The liquid molecules tend to "stick together" and have a definite volume, but they do not maintain a specific shape. And, because the molecules in a liquid are so close together, liquids tend to be incompressible. When an ocean liner goes through the locks of a canal, it does not crush the water down. Water molecules can't be squeezed any closer together than they already are.

The particles composing a solid are not only close together but are also held together in a fixed position in relation to one another. Later in this chapter, we will discuss the forces that hold the particles together. For now, we'll just say that the fact that the particles of a solid *are held in a fixed position* explains why a solid has a definite shape, has a definite volume, and is practically incompressible.

Table 14.1
Properties of Gases, Liquids, and Solids

Property	Gases	Liquids	Solids
Volume	Have no definite volume (gases expand to fill their containers)	Have a definite volume	Have a definite volume
Shape	Have no definite shape	Have no definite shape (liquids assume the shape of their containers)	Have a definite shape (most solids have a well-defined crystalline arrangement)
Compressibility	Are easily compressed	Tend to be incompressible	Cannot be compressed (they are practically incompressible)

14.2 Evaporation of Liquids

Learning Goal 3
Evaporation, equilibrium vapor pressure, and boiling point

If a liquid is placed in an open container, it slowly disappears. That is, the liquid **evaporates** as its molecules change from the liquid state to the gaseous (or vapor) state. For any liquid, the speed of evaporation depends on the temperature of the liquid and its exposed surface area.

Before we discuss the process of evaporation, let us perform an experiment. We place a cube of sugar in a container of water and leave the container undisturbed for a few hours. Later, when we examine the container, we see that the cube of sugar has disappeared. It has dissolved in the water. If we taste the solution, we discover that the molecules of sugar have been distributed evenly throughout the water (Figure 14.1). All parts of the solution have the same sweetness. This could happen only if the molecules had moved around. From experiments like this, we conclude that *the molecules of a liquid are constantly in motion*. We can use this knowledge to explain evaporation.

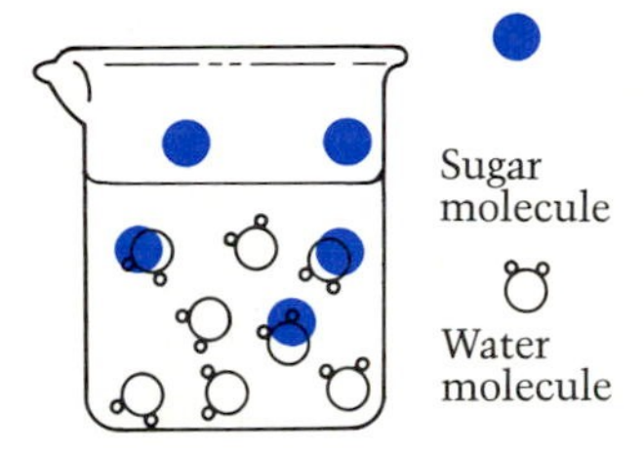

Figure 14.1
A solution of sugar and water

Because liquid molecules are constantly in motion, each must have a specific amount of energy. And when the molecules collide, energy can be transferred from one molecule to another. If a molecule on the surface of a liquid collides with a molecule that is on its way up toward the surface, the surface molecule may obtain enough energy from the collision to escape from the liquid entirely (Figure 14.2). That molecule has then changed state, from liquid to vapor. This is similar to what happens in a game of pool. When the cue ball strikes another ball, it passes energy to that other ball. The struck ball then starts to move. And if it is struck hard enough, it can leave the table entirely.

A Dynamic Equilibrium

If we place a cover on a container that is partly filled with water, the evaporation process begins, but it seems to stop after a time. Here is what happens.

As Figure 14.3 shows, water molecules begin to escape from the surface of the liquid. These molecules occupy the space above the water in the partly filled container. As evaporation continues, more and more escaping molecules occupy this same space. The molecules are now closer together, they

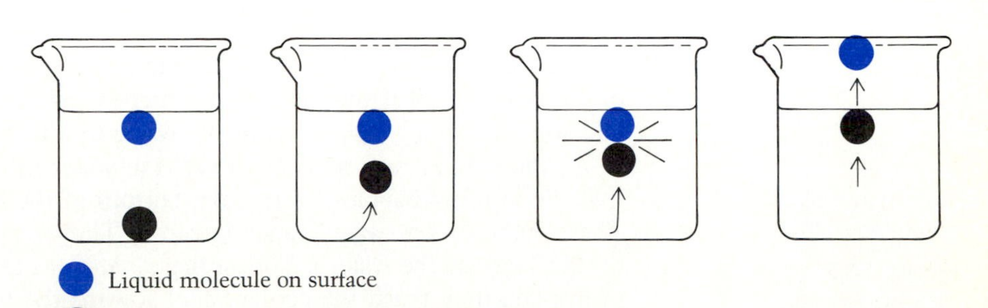

Figure 14.2
How liquids evaporate

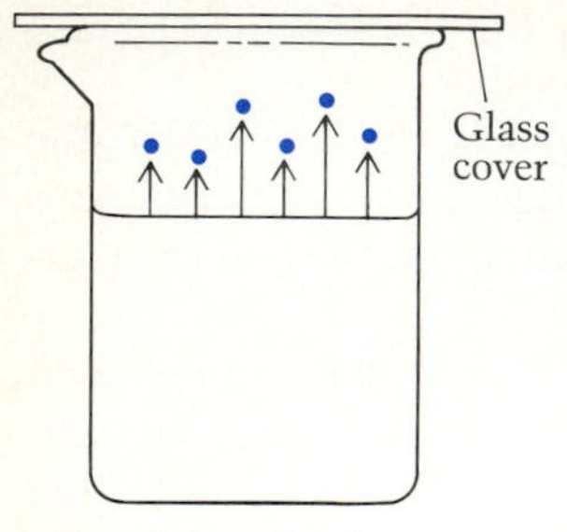

Figure 14.3
Liquid molecules escaping from the surface of the liquid

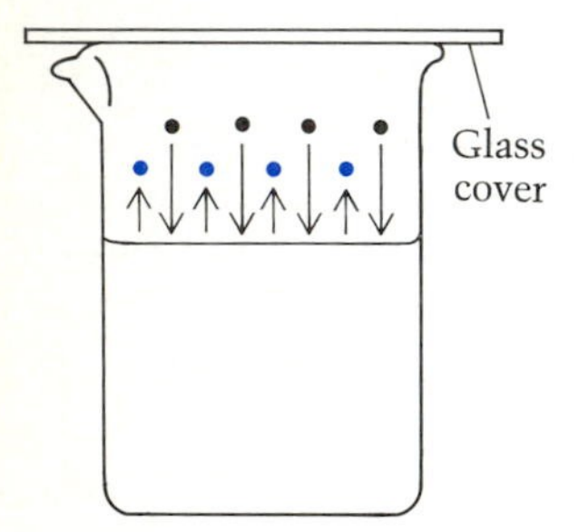

Figure 14.4
A dynamic equilibrium

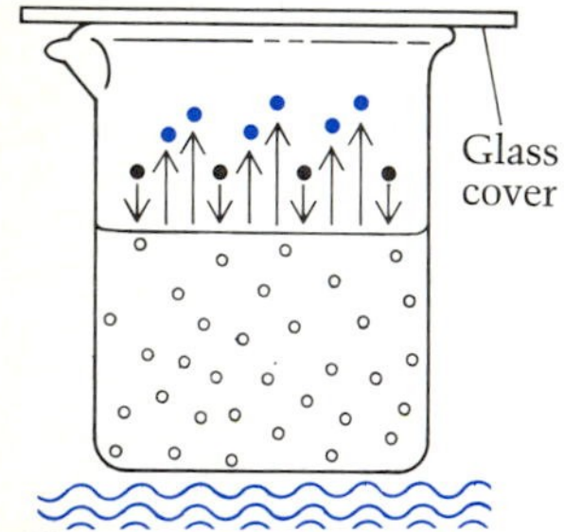

Figure 14.5
The higher the temperature, the more molecules escape.

collide with one another more often, and some of them are bumped back into the water. At first, many molecules escape and few return to the water. But after a while, as the concentration of molecules in the vapor state increases, the number of molecules returning to the liquid also increases. Eventually, the number of molecules escaping from the water equals the number of molecules returning to it (Figure 14.4). The net result is that no additional liquid seems to be evaporating from the container. But in reality, there is a *dynamic equilibrium* between evaporation and return to the liquid state. *These two opposing processes are occurring at the same rate.*

Equilibrium Vapor Pressure

The molecules that have escaped the liquid in Figure 14.4 exert a pressure on the container, the cover, and the surface of the liquid. This pressure is called the *equilibrium vapor pressure*. We can define **equilibrium vapor pressure** as *the pressure exerted by a vapor when it is in equilibrium with its liquid at a given temperature*.

Every liquid has its own characteristic equilibrium vapor pressure, and this equilibrium pressure depends on the temperature. To see this, consider what happens when we raise the temperature of the water in Figure 14.4. Because we are adding heat to raise the temperature, we are also supplying additional energy to the liquid molecules. As a result, a greater number of molecules obtain sufficient energy to escape from the surface of the liquid and become vapor (Figure 14.5). The equilibrium that had existed is upset, but a new equilibrium is soon established at the new temperature. And, because there are now more molecules in the space above the liquid, the new equilibrium vapor pressure is higher than the original one. That is, the higher the temperature, the greater the equilibrium vapor pressure for any substance.

The more easily a liquid evaporates, the more **volatile** it is said to be. Liquids that have high equilibrium vapor pressures tend to evaporate easily and so are considered volatile. Some examples are alcohol, gasoline, and ether. Liquids that have low equilibrium vapor pressures tend to evaporate less easily. Examples include motor oil, water, and glycerin.

14.3 Boiling of Liquids

If a beaker of water is heated over a flame, bubbles soon form on the bottom of the beaker. These first bubbles are merely dissolved air (oxygen, nitrogen, and so forth) being driven out of the water by the heat.

As the temperature rises, however, the water molecules become more active. Soon other bubbles form at the bottom of the beaker and begin to rise. These bubbles are water vapor (steam). They form on the *bottom* of the beaker because the water is hottest there. At first, these steam bubbles disappear when they reach the cooler water toward the top of the beaker (Figure 14.6). After a while, however, they begin to rise all the way to the surface. At this point, we say that the water is *boiling*. What we mean is that the vapor

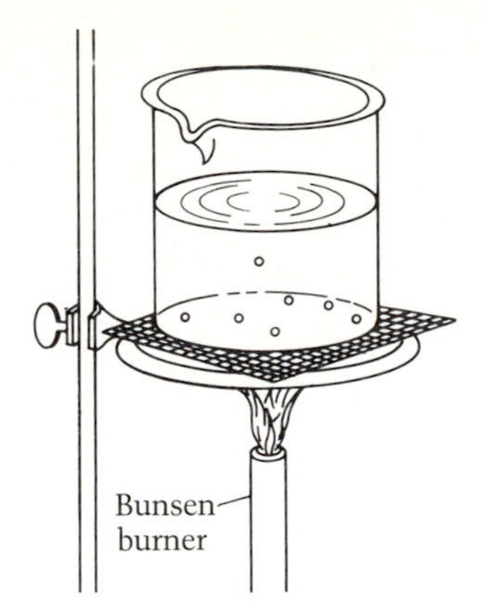

Figure 14.6
The first bubbles of water vapor

pressure exerted by the bubbles of steam has become equal to the atmospheric pressure (Figure 14.7). So the bubbles of steam are able to rise up and out of the water.

The **boiling point** of a liquid is *the temperature at which the equilibrium vapor pressure of the liquid equals the atmospheric pressure*. This means that the boiling point depends on the atmospheric pressure: increasing the atmospheric pressure increases the boiling point, and decreasing the atmospheric pressure decreases the boiling point. So, when we discuss the boiling points of liquids, we mean the *normal* boiling points, at the *normal* pressure of 1 atm, unless we state otherwise.

The temperature of a liquid remains constant as it boils at constant pressure. For example, if you heat a container of water, the temperature of the water increases to 100°C (at 1 atm pressure), and the water then starts to boil. But the temperature of the liquid does not increase beyond 100°C *no matter how much heat you add*. What happens is this: The heat you add goes first to raising the temperature of the water to 100°C. After that, all the heat that is added is used to vaporize the water molecules. This heat it called the **heat of vaporization.** That is, energy is required for the water molecules to move fast enough to escape the liquid, and this energy is supplied by the heat. The more heat you add, the faster the water is vaporized. But the temperature of the water remains at the boiling point until the last molecule escapes.

14.4 Freezing of Liquids

Learning Goal 4
Change of state

When heat is removed from a liquid, its temperature decreases. At a certain temperature, the liquid begins to **freeze,** or change to a solid. This temperature is called the **freezing point** of the liquid. As more heat (or energy) is removed, more and more of the liquid freezes *at this same temperature*. During the change of state from liquid water to solid ice, all of the heat removed from the system is used to turn the liquid into solid. This is called the **heat of crystallization (solidification).** Once all the liquid has been transformed to the solid state, its temperature will decrease if more heat is removed.

For example, the freezing point of pure water is 0°C. As liquid water is cooled, its temperature decreases. At 0°C, the liquid begins to turn to ice—or solid water. As more heat is removed by cooling, more of the liquid turns to ice, but the temperature stays at 0°C. Finally, when all the water has been frozen, the temperature of the ice will drop below 0°C if it is cooled further.

When a solid is heated, its temperature increases until its freezing point is reached. At that temperature, the solid begins to **melt,** or change to a liquid. (For this reason, the freezing point of a substance is also called its **melting point**.) As more heat is added, more of the solid turns to liquid *at this same temperature*. Once all the solid has become liquid, its temperature will increase if more heat is added. Figure 14.8 shows the complete process—solid to liquid to gas—for ice at 1 atm pressure.

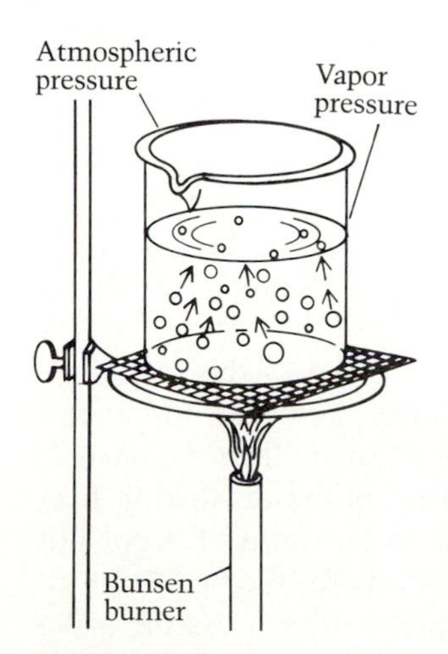

Figure 14.7
Boiling water

Some substances can change directly from the solid state to the gaseous state without first becoming a liquid. This process is called **sublimation.** An

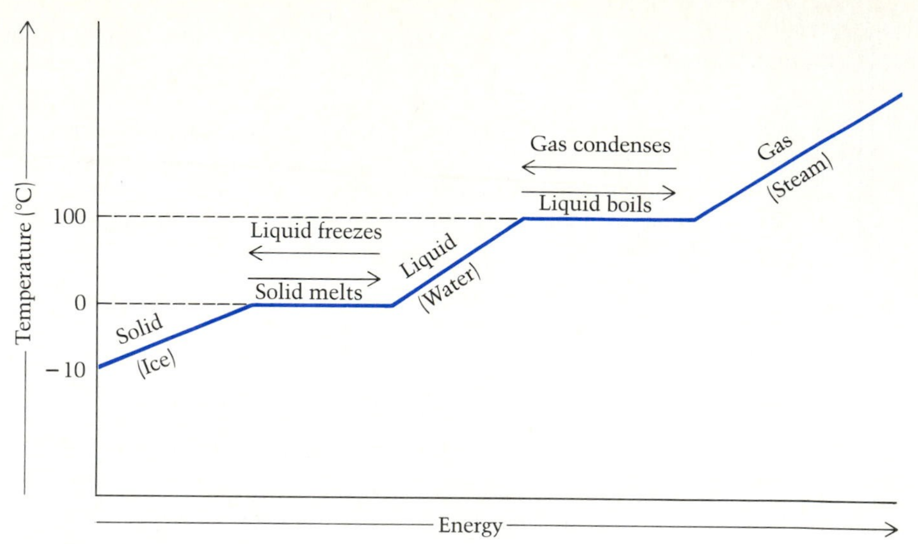

Figure 14.8
How water at 1 atm pressure changes from a solid to a liquid to a gas. Below 0°C, water is in the solid state—ice. In this diagram, we begin with ice at −10°C. Energy (heat) is supplied so that the temperature of the ice rises. When it reaches 0°C, the ice starts to melt into water. As more energy is supplied, more ice melts into water. This water remains at 0°C until all the ice melts. After all the ice has melted, the energy supplied starts to warm the water. This continues until the water temperature reaches 100°C. At this point, the water starts to vaporize into steam. The temperature of the water remains constant at 100°C as the liquid is turned into gas.

example is dry ice, which is solid carbon dioxide. It goes directly from the solid state to the gaseous state at 1 atm pressure.

14.5 The Process of Distillation

Learning Goal 5
Distillation and fractional distillation

The fact that liquids boil gives rise to an interesting method for separating one liquid from another, or even separating a solid from a liquid in which it is dissolved. This process is called *distillation*.

Figure 14.9 shows a simple distillation apparatus. It works this way: Suppose you have a solution of sodium chloride and water. You want to separate the sodium chloride from the water and at the same time collect the purified water. You place the sodium chloride–water solution in the distillation flask and heat it to boiling. As the solution boils, water in the form of vapor fills the flask and travels toward the distilling head. From there, the hot water vapor travels down the double-jacketed, water-cooled condenser. As the vapor travels down within the inner walls of the condenser, it is cooled by tap water

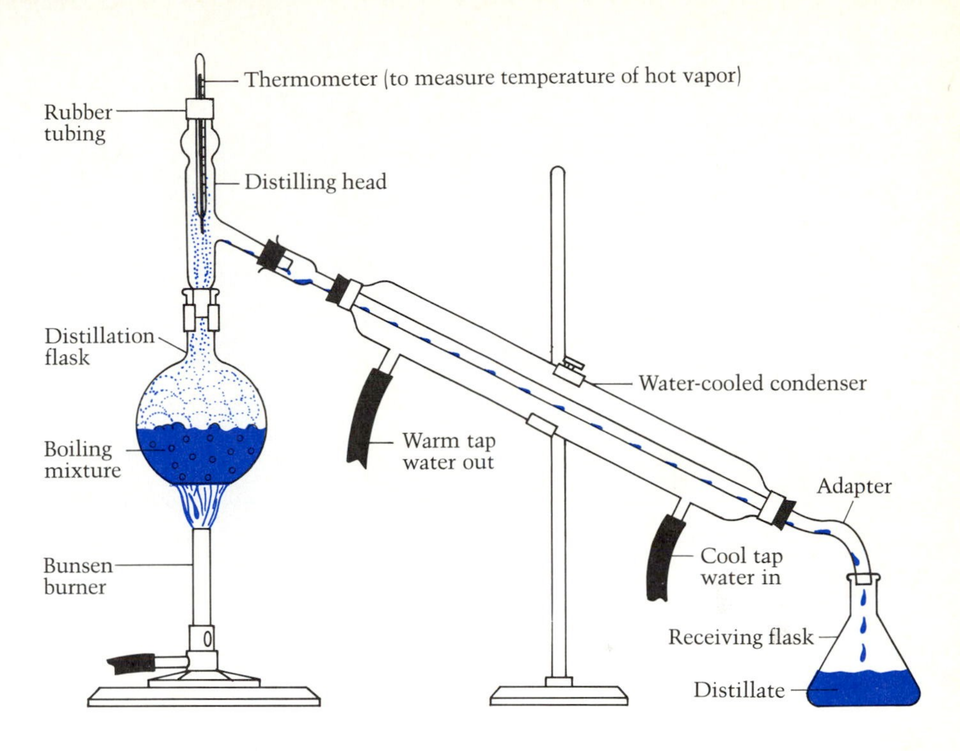

Figure 14.9
A distillation apparatus

running between the outer walls of the condenser. The water vapor **condenses,** or changes back to its liquid form. The condensed liquid, known as the *distillate,* is collected in a receiving flask. Because only the water, and not the sodium chloride, vaporizes and passes through the condenser, the distillate contains only pure water. The sodium chloride remains behind in the distillation flask.

Miscible liquids are those that form solutions when mixed. Two or more miscible liquids that have different boiling points may also be separated by the process of distillation. For example, when a mixture of ethyl alcohol (boiling point, 78.5°C) and water (boiling point, 100°C) is heated, the component with the lower boiling point (ethyl alcohol) vaporizes more readily than the component with the higher boiling point (water). Therefore the distillate is richer in ethyl alcohol than in water. And it can be distilled again to yield even purer ethyl alcohol. If this process is repeated enough times, a distillate of essentially pure ethyl alcohol is obtained.

This technique, called *fractional distillation,* is extremely useful in petroleum refining. The different components of crude petroleum all have different boiling points, so they can be separated by distillation. Petroleum companies use huge distilling columns, known as *fractionating columns,* to separate the various petroleum products (Figure 14.10). When the crude oil is heated, the components (or fractions) with lower boiling points rise to the top of the column, those with higher boiling points rise to the middle of the column, and solids remain at the bottom of the column. As the components rise up

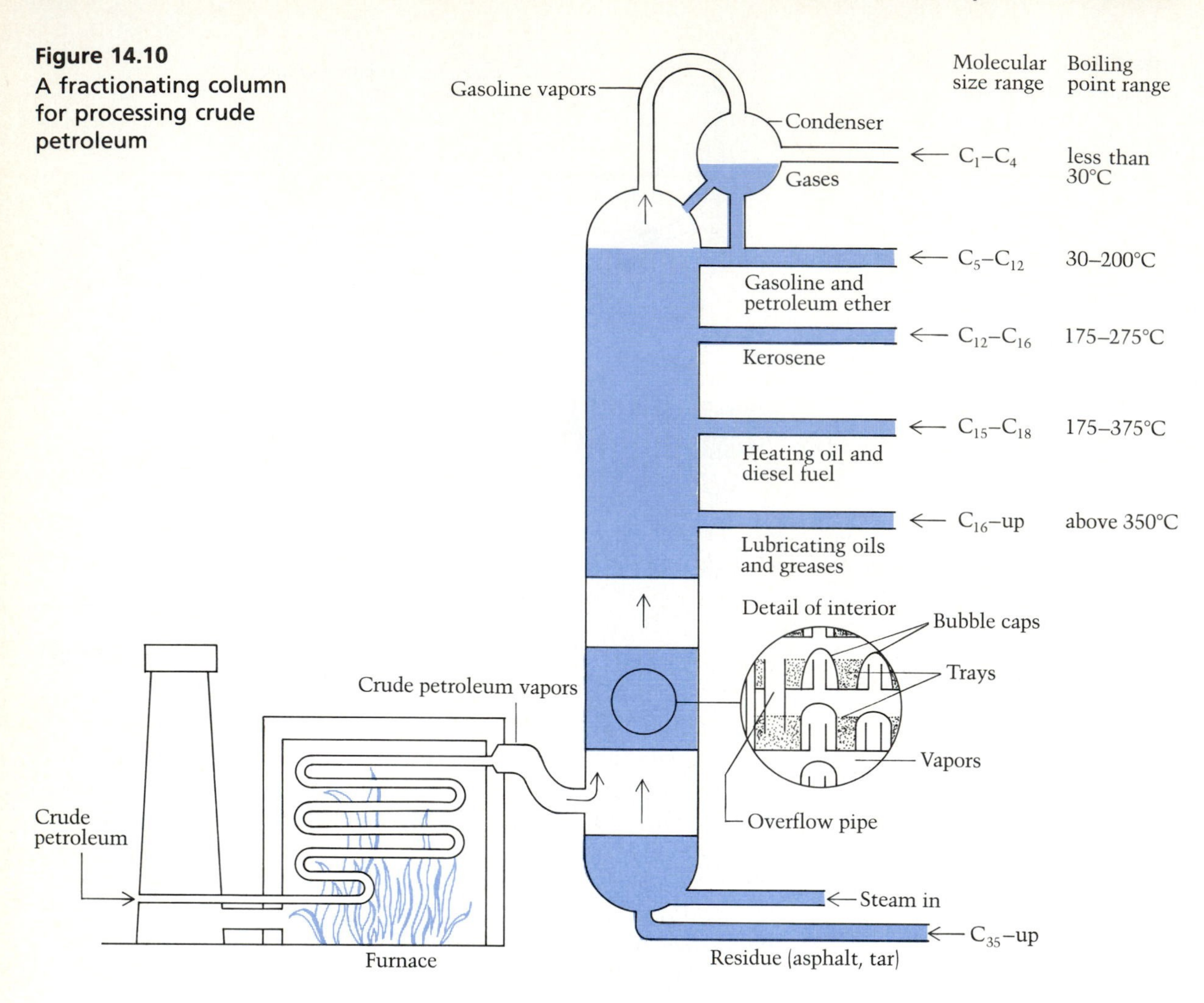

Figure 14.10 A fractionating column for processing crude petroleum

the column and condense, they are drawn off, to be purified further if necessary.

14.6 Solids

Learning Goal 6 Crystalline and amorphous solids

Solids have two main characteristics that distinguish them from gases and liquids: *fixed shape* and *fixed volume*. Most solids owe these unique properties to their well-defined crystal structure (Figure 14.11). (We'll explain the term *crystal* shortly.) However, a few solids, such as glass and paraffin, do not have a well-defined crystal structure, although they do have definite shape and volume. *Such solids are called* amorphous, *which means that they have no definite internal structure or form.* This makes them interesting to research scientists. However, we shall concentrate on crystalline solids.

Figure 14.11
Examples of different types of crystalline solids

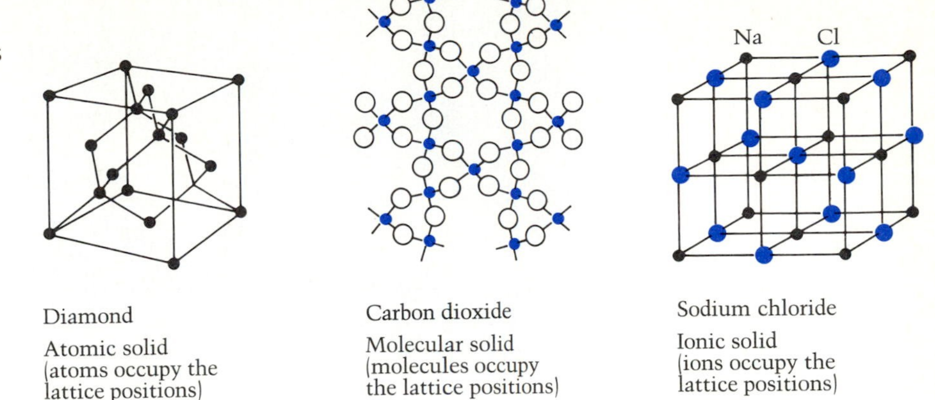

Diamond
Atomic solid
(atoms occupy the lattice positions)

Carbon dioxide
Molecular solid
(molecules occupy the lattice positions)

Sodium chloride
Ionic solid
(ions occupy the lattice positions)

Particles That Make Up Solids

Solids can be composed of three types of particles: atoms, molecules, and ions. Solids such as carbon, silicon, copper, and gold are composed of atoms. Solids such as dry ice (CO_2), iodine, and water (ice) are composed of molecules. Solids such as sodium chloride, calcium chloride, and sodium bromide are composed of ions. The three types of solids have very different properties. Table 14.2 lists some of the differences.

Learning Goal 7
The kinds of crystalline solids

What Is a Crystal?

A **crystal** is *a solid that has a fixed, regularly repeating, symmetrical internal structure. This symmetrical structure, formed by the particles in a crystal, is called the* **crystal lattice.** A crystal lattice is not an actual object, but a *struc-*

Learning Goal 8
Crystal lattice and unit cell

Table 14.2
Some Properties of Crystalline Solids

Type	Melting point	Hardness	Conductivity	Example
Ionic	High	Hard and brittle	Nonconductors of electricity	Sodium chloride, NaCl
Molecular	Low	Soft	Nonconductors of electricity	Dry ice, CO_2
Atomic				
Metallic	High	Varying hardness	Good conductors of electricity	Copper, Cu
Nonmetallic	High	Hard and brittle	Nonconductors of electricity	Diamond, C

Figure 14.12
A quartz crystal
(Courtesy of the American Museum of Natural History)

ture or pattern. The particles that make up the pattern are different in different types of crystals, but in each case they occupy specific *lattice positions*. In an atomic solid, *atoms* occupy the lattice positions; in a molecular solid, *molecules* occupy them; and in an ionic solid, *ions* occupy them (Figure 14.11). It is this internal structure that results in the plane (flat) surfaces exhibited by crystals (see Figure 14.12).

Each repeated unit in a crystal lattice is called a **unit cell** (Figure 14.13), and the unit cell is repeated over and over again in the crystal. Because the whole crystal is made up of repetitions of the lattice structure, we know the

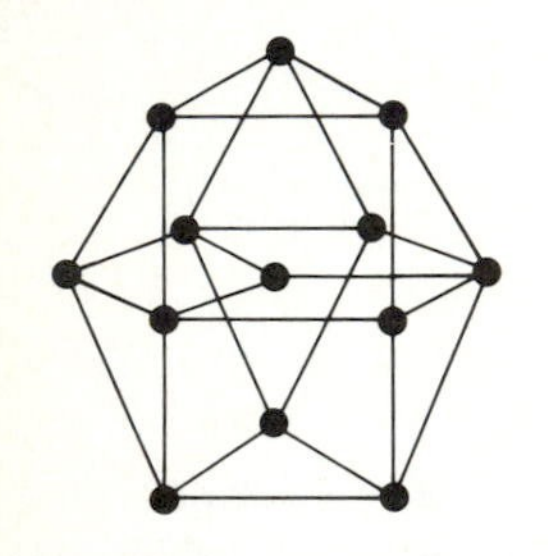

Unit cell of magnesium (Mg)

Figure 14.13
The unit cell of magnesium

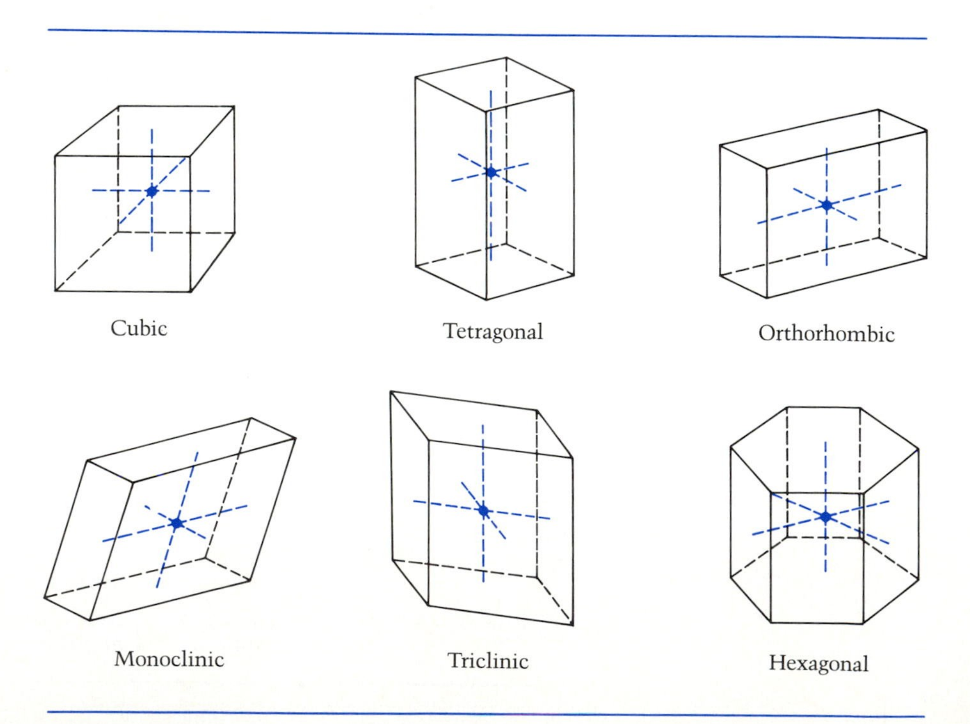

Figure 14.14
Six types of crystals

Figure 14.15
Sodium chloride crystals
(Courtesy Morton Salt)

three-dimensional pattern of the crystal when we know the lattice. Figure 14.14 shows six different types of crystals—each precise, orderly, and capable of existing side by side with exact copies of itself. Figure 14.15 is a photograph taken with a microscope, showing the cubic structure of sodium chloride crystals.

What Holds a Crystal Together?

The properties of a crystalline solid are partly the result of the forces that bind together the particles that make up the crystal. Let's look briefly into the bonding forces within the three types of crystals (ionic, molecular, and atomic).

In an *ionic* crystal, the attractive forces between oppositely charged ions hold the crystal together. This kind of bonding force is very strong, so ionic solids have very high melting points. For example, table salt ($NaCl$) has a melting point of 800°C. A great deal of heat energy is required to force those ions apart.

Molecular crystals, on the other hand, are usually held together by weak electrical forces. Only molecular crystals whose molecules are polar have binding forces that have some degree of strength. The forces binding most molecular crystals together are so weak that these crystals frequently have much lower melting points than ionic crystals. For example, ice (a molecular crystal) melts at 0°C.

Atomic crystals have strong binding forces—Table 14.3 shows just *how* strong. When the atoms are nonmetallic, they are held together by covalent bonds. An example is diamond, made up of carbon atoms that bond covalently. It takes a temperature of 3,500°C to break the bonds that unite carbon atoms in diamond.

In atomic crystals that are composed of metallic atoms, it is the attraction of opposite charges that makes the bonds so strong. Scientists think of the metallic lattice as composed of positive ions surrounded by a cloud of electrons. The electrons are given off by the metallic atoms, but they are considered part of the entire crystal—electrons at large, you might say. Because the electrons are free to wander throughout the crystal, they are able to conduct electricity. This is why metals are good electrical conductors.

Table 14.3
Melting Points of Some Atomic Crystals

Substance	Melting point
Iron	1,535°C
Copper	1,083°C
Diamond	3,500°C
Silicon	1,410°C

CHEMICAL FRONTIERS

NEW SYNTHETIC DIAMONDS HAVE UNUSUAL PROPERTIES

Synthetic diamonds consisting of almost pure carbon-12 have been produced by General Electric scientists. These diamonds have unusual properties that suggest a variety of applications. For example, the new diamonds have 50% higher heat conductivities than natural diamonds. In addition, they remain electrical insulators and absorb less light at most wavelengths than do natural diamonds. They also have a threshold to laser damage that is ten times higher than that of their natural counterparts.

Production of the new diamonds is made possible by an interesting application of chemical knowledge. The process begins with a mixture of hydrogen gas and ^{12}C-enriched methane gas. The enriched methane is 99.97% ^{12}C, while ordinary methane is 98.89% ^{12}C. Microwave energy is used to ionize the mixture of gases to form a plasma. The plasma deposits a layer of polycrystalline diamond, which is then crushed to a powder. The powder is added to molten iron and aluminum in the presence of a diamond seed crystal on which the structure builds. Finally, at a temperature of 1,500°C and a pressure of 55,000 atm, a single-crystal diamond forms. The largest diamond produced in this way weighed 1.3 carats; this diamond has been sliced for use in a variety of devices.

General Electric scientists expect that these synthetic diamonds will be useful in dissipating heat from electronic integrated circuits and fiber-optic systems. One promising application is in the circuitry of orbiting satellites and undersea electronic instruments. It is also hoped that these diamonds may be used in laser cutting and optical devices.

14.7 Water

Water is the world's most important liquid. You might be able to live a month without food, but you could survive only two or three days without water. All vegetation disappears without water: In West Africa, an area the size of the United States is becoming a desert because there has been almost no rain for some years. The human body is composed of about two-thirds water, and many fruits and vegetables are as much as 80% or 90% water. Milk, one of our basic foods, is more than 90% water.

Yet the supply of water is *not* unlimited. Today the entire population of the United States has only about the same quantity of drinkable water that was available when Columbus landed in 1492. Americans use an estimated 2,000 gallons per person day, and the trend is to use still more. (This figure of 2,000 gallons includes water used by industry, by power plants, and so forth.)

Physical Properties of Water

Learning Goal 9
Chemical and physical properties of water

Pure water is a tasteless, odorless, colorless substance that has a melting point of 0°C and a boiling point of 100°C at 1 atm pressure. These two temperatures serve as reference points on the Celsius temperature scale.

The density of water is another reference standard. It relates metric units of volume (milliliters) to metric units of mass (grams). At 4°C, water has a density of 1 gram per milliliter (1 g/mL). This means that 1 gram of water at 4°C occupies a volume of 1 milliliter. If we measured out exactly 1 mL of 4°C water, we would find that it has a mass of 1 g (Figure 14.16).

Water has intrigued humanity since the days of the ancient Greeks. One of its most interesting properties is the way in which its volume changes when its temperature is changed. Suppose we start with a certain amount of water at room temperature. If we warm it, the volume of the water increases. If we cool the water below room temperature, its volume at first decreases. This is exactly the behavior we would expect from a liquid. However, as we cool the water below 4°C, the volume of the water starts to *increase*. And its volume continues to increase until it freezes at 0°C. This behavior is very unusual, because the volumes of most liquids decrease continuously as the liquids are cooled.

One result of this strange behavior of water is that the density of ice is less than the density of water. Remember that density D = mass/volume. At 4°C, the volume of, say, 10 g of water is 10 mL. But at 0°C, the volume of the ice (solidified water) has increased to 10.9 mL. Therefore

$$D \text{ (at 4°C)} = 10 \text{ g}/10 \text{ mL} = 1 \text{ g/mL}$$

$$D \text{ (at 0°C)} = 10 \text{ g}/10.9 \text{ mL} = 0.92 \text{ g/mL}$$

Because the density of ice is less than the density of water, ice floats on water. Most other substances reach their minimum volume (and maximum density) at the freezing point, not above it. This causes the solid form to sink in the liquid form.

It is fortunate for us that ice floats on water, If it didn't, when lakes, rivers, and seas freeze and water turns to ice in the winter, the ice would sink to

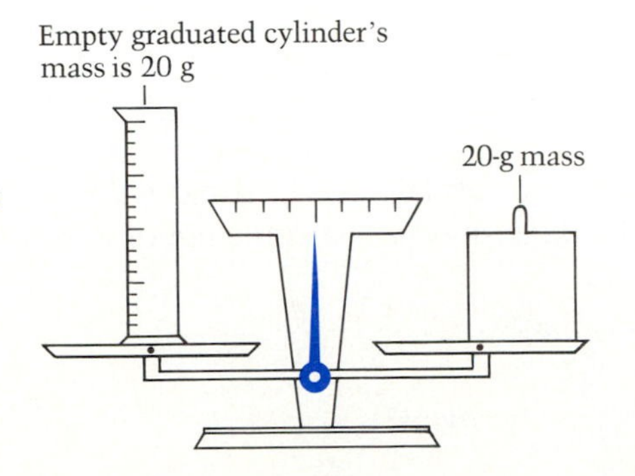

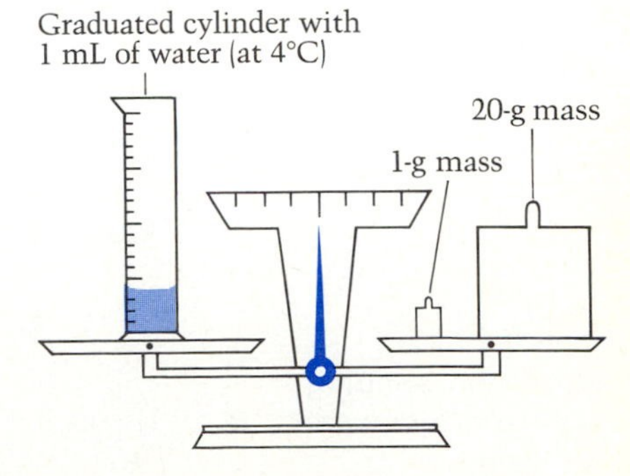

Figure 14.16 Establishing the density of water (at 4°C). The experiment shows that 1 mL of water has a mass of 1 g (in other words, the density of water is 1 g/mL).

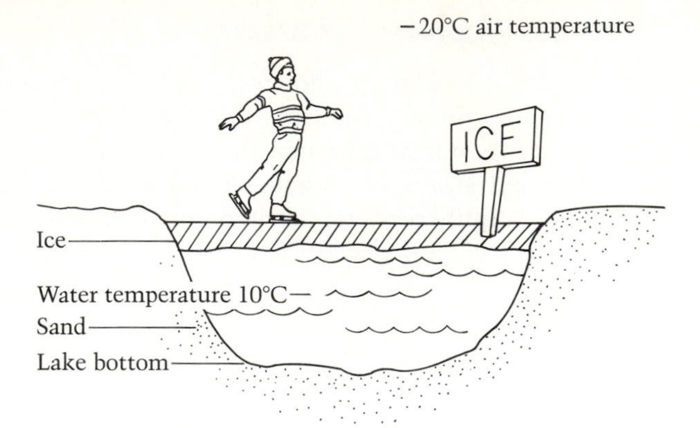

Figure 14.17
Floating ice insulates the lake water below.

the bottom. The water above the ice would protect it from the sun in the summer, so the ice would never melt. Eventually, the entire body of water would become frozen, and all life forms in it would be killed. And, if it is true that life on earth originated in the sea, then human beings would never have evolved. However, ice *does* float on water, and the layer of floating ice insulates the rest of the water and prevents it from freezing. This explains why an entire body of water does not freeze, even though the air temperature above it stays below freezing for many months (Figure 14.17).

Water also has a much higher freezing point and boiling point than would normally be expected. The unusual behavior of water results from its bonding and the three-dimensional shape of the water molecule. If we examine a single water molecule, we find that:

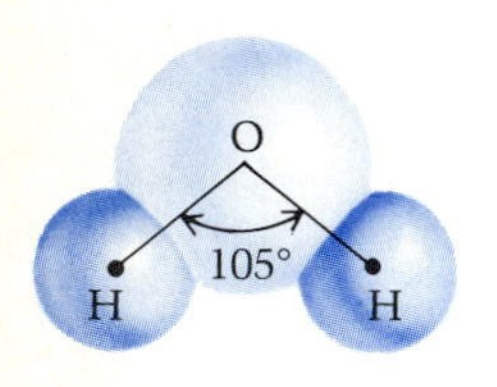

Figure 14.18
The structure of a water molecule

1. The bonds between the hydrogen and oxygen atoms are polar. Each hydrogen atom seems to have a slight positive charge, and the oxygen atom seems to have a slight negative charge.
2. The bond angle between the hydrogen atoms is about 105° (Figure 14.18).

The slightly positively charged hydrogen atom of one water molecule can attract the slightly negatively charged oxygen atom of another water molecule. The two molecules then are held together by a force called a **hydrogen bond** (Figure 14.19). Thousands of water molecules may bond together in this way to form long chains (Figure 14.20). It is the hydrogen bonds, which hold the water molecules together, that give water its unusually high boiling point.

But how do we explain the fact that water expands as it freezes? If we looked at the three-dimensional structure of a crystal of ice, it would seem to

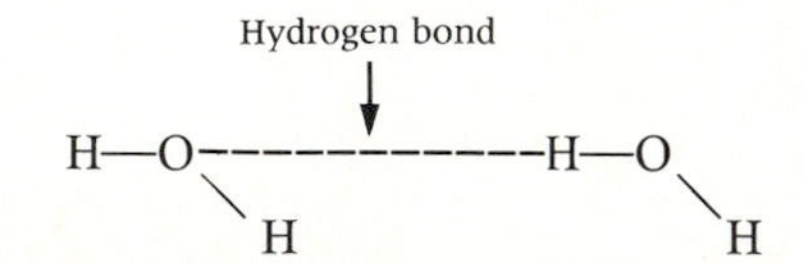

Figure 14.19
Hydrogen bonding between two water molecules

Figure 14.20
Hydrogen bonding among many water molecules. (*Note:* True bond angles are not shown.)

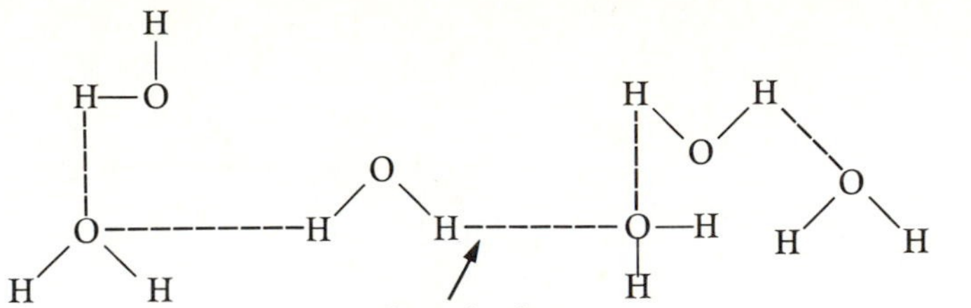

Hydrogen bond

be hexagonal, or six-sided (Figure 14.21). To attain this type of geometry while maintaining their 105° bond angle, the molecules have to move farther apart in ice than they are in liquid water. Therefore water that is in the form of ice has a greater volume than water that is in the form of a liquid.

Chemical Properties of Water

Water is one of the most stable compounds—so stable that for many years it was considered one of the basic *elements* of matter. Even when heated as high as 2,000°C, water shows little decomposition. However, if an electric current is passed through it, water breaks down.

$$2H_2O \xrightarrow{\text{Electricity}} 2H_2 + O_2$$

Learning Goal 10
Reactions of water

Water reacts with oxides of *non*metals to produce *acids*. For example, in the case of sulfur trioxide,

$$SO_3 + H_2O \longrightarrow \underset{\text{Sulfuric acid}}{H_2SO_4}$$

and in the case of sulfur dioxide,

$$SO_2 + H_2O \longrightarrow \underset{\text{Sulfurous acid}}{H_2SO_3}$$

Water reacts with oxides of metals to produce *bases*. For example,

$$CaO + H_2O \longrightarrow Ca(OH)_2$$

$$Na_2O + H_2O \longrightarrow 2NaOH$$

Learning Goal 11
Hydrates and water of hydration

Many salts combine with water to form compounds called *hydrates*. A **hydrate** is *a crystalline compound that contains a definite proportion of water molecules loosely held*. Some common hydrates are

$Na_2B_4O_7 \cdot 10H_2O$	Borax	$MgSO_4 \cdot 7H_2O$	Epsom salts
$CaSO_4 \cdot 2H_2O$	Gypsum	$Na_2CO_3 \cdot 10H_2O$	Washing soda
$KAl(SO_4)_2 \cdot 12H_2O$	Alum	$CuSO_4 \cdot 5H_2O$	Blue vitriol

The centered dot between the formula for the salt and the formula for the water is not a decimal point or a multiplication sign. It is a way of showing that

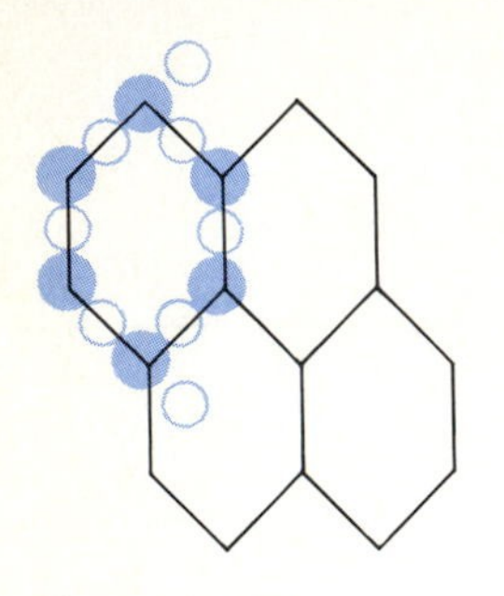

Figure 14.21
The crystal structure of ice. (*Note:* Not all the hydrogen atoms are shown, because some of them would be behind the plane of the paper.)

the water molecules maintain their integrity in the crystal structure. When a hydrated salt is heated, the water is driven off. For example,

$$MgSO_4 \cdot 7H_2O \xrightarrow{\text{Heat}} MgSO_4 + 7H_2O$$

$$\underset{\substack{\text{A blue}\\\text{compound}\\\text{(hydrated salt)}}}{CuSO_4 \cdot 5H_2O} \xrightarrow{\text{Heat}} \underset{\substack{\text{A white}\\\text{compound}\\\text{(anhydrous salt)}}}{CuSO_4} + 5H_2O$$

The salt is left in its **anhydrous** form—the form in which the water of hydration is no longer attached to the salt.

Example 14.1

Anhydrous nickel chloride, a golden-yellow salt, is used as an absorbent for ammonia in gas masks. When the anhydrous salt picks up water, it changes color to form nickel chloride hexahydrate, which is green. The reaction is as follows:

$$\underset{\text{Golden-yellow}}{NiCl_2} + 6H_2O \longrightarrow \underset{\text{Green}}{NiCl_2 \cdot 6H_2O}$$

Calculate the percentage of water (by mass) in nickel chloride hexahydrate.

Solution

Understand the Problem

We are asked to determine the percentage of water (by mass) in a hydrated salt.

Devise a Plan

Our strategy will be to determine the molecular mass of the hydrated salt, then divide the mass of the water molecules in the hydrated salt by the molecular mass of the hydrated salt itself. Multiplying the result by 100 will give the percentage of water (by mass).

Carry Out the Plan

Calculate the molecular mass of the hydrated salt as follows:

$$NiCl_2 \cdot 6H_2O$$

$$58.7 + 71.0 + 6(18.00)$$

$$129.7 + 108.00 = 237.7$$

Now determine the percentage of water (by mass):

$$\text{Percent } H_2O \text{ (by mass)} = \frac{\text{mass of water's contribution}}{\text{mass of hydrated salt}} \times 100$$

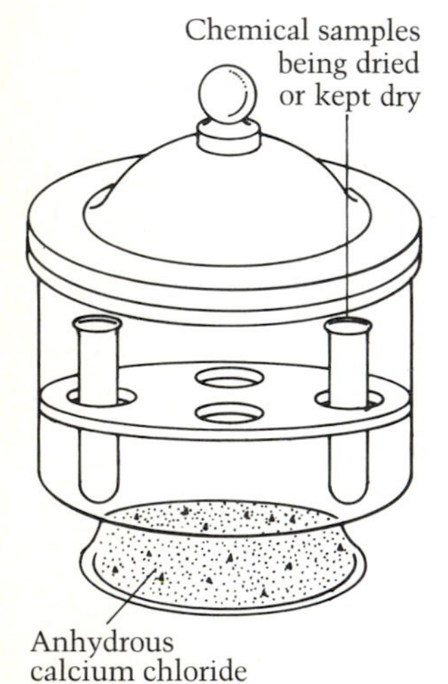

Figure 14.22
A desiccator is used to dry chemical samples and to keep them in a moisture-free atmosphere.

$$= \frac{108.0}{237.7} \times 100$$

$$= 45.44\%$$

Look Back

We check to see whether the solution makes sense. Since 108 is slightly less than half of 237, our answer of 45.44% is reasonable.

Practice Exercise 14.1 Pale-green anhydrous chromium(III) nitrate also exists in a deep-violet nonahydrate form, as shown in the following reaction:

$$\underset{\text{Pale green}}{Cr(NO_3)_3 + 9H_2O} \longrightarrow \underset{\text{Deep violet}}{Cr(NO_3)_3 \cdot 9H_2O}$$

Calculate the percentage of water (by mass) in chromium(III) nitrate nonahydrate.

Learning Goal 12
Definitions of anhydrous, hygroscopic, deliquescence, and efflorescence

On exposure to air, some anhydrous salts (as well as other substances) can absorb water from the atmosphere. Such a substance is said to be **hygroscopic.** Some substances are so hygroscopic that they absorb enough water from the atmosphere to form a solution. These substances are said to be **deliquescent.** One such substance is sodium hydroxide (NaOH). A small piece of solid sodium hydroxide, if left in an open beaker, soon absorbs enough water from the atmosphere to form a solution.

Some hygroscopic substances, such as calcium chloride ($CaCl_2$), are used in the laboratory as drying agents. Such substances are called **desiccants.** If you want to keep some chemicals in a dry atmosphere, you can place them in a container called a **desiccator** (Figure 14.22). Anhydrous calcium chloride in the desiccator ensures that the atmosphere in the desiccator remains moisture-free. (Of course, in time, the calcium chloride absorbs as much water as it can, and it must be changed.)

Desiccant pellets are sometimes placed in vitamin containers. The pellets absorb any moisture that might accumulate in the containers possibly causing the vitamins to decompose.

We have described hygroscopic substances as those that can spontaneously absorb water from the atmosphere. There are also substances that can do the opposite. A few hydrated salts can spontaneously lose their water of hydration when exposed to the atmosphere. This process is known as **efflorescence.** For example,

$$Na_2SO_4 \cdot 10H_2O \xrightarrow[\text{temperature}]{\text{Room}} Na_2SO_4 + 10H_2O$$

Efflorescence can occur when the vapor pressure of water in the atmosphere is less than the vapor pressure of the hydrate. In the case of the sodium sulfate decahydrate, shown above, it is possible to watch the crystals lose their water and form a noncrystalline-looking powder.

Summary

Gases have no definite shape or volume and are easily compressed. Liquids have a definite volume but no definite shape, and they tend to be incompressible. Solids have a definite volume and shape and are practically incompressible.

The kinetic theory helps explain the properties of liquids and solids. The forces of attraction between liquid molecules and between solid molecules are much greater than those between gas molecules. As a result, the molecules of a liquid are very close to each other. This gives the liquid a definite volume but no specific shape. The particles composing a solid are not only close together but are also held in a fixed position. This gives the solid both a definite volume and a definite shape.

The molecules of a liquid are constantly in motion. When molecules collide, energy is transferred from one molecule to another. A molecule near the surface of the liquid can obtain enough energy in a collision to evaporate, or escape from the liquid entirely. Molecules that evaporate (or vaporize) have changed from the liquid state to the gaseous state.

When the liquid is in a covered container, the molecules that have escaped the liquid remain above its surface. Some of these molecules return to the liquid and others escape it. Eventually, a dynamic equilibrium is attained in which the rate of escape equals the rate of return to the liquid. The pressure exerted by the escaped molecules at this point is called the equilibrium vapor pressure of the liquid.

The boiling point of a liquid is the temperature at which its equilibrium vapor pressure equals the atmospheric pressure. A heated liquid will continue to boil at its boiling point until it has evaporated completely. The boiling phenomenon is used in the process of distillation, in which liquids are separated by being evaporated in turn.

The melting point (or freezing point) of a substance is the temperature at which it begins to change from the solid state to the liquid state (or from the liquid state to the solid state). The temperature of a substance remains at its melting point (or freezing point) as long as it is melting (or freezing).

Most solids owe their fixed shape and volume to their crystal structure, although a few solids, such as glass and paraffin, are amorphous (do not have a definite internal structure) rather than crystalline. A crystal is a solid that has a definite, repeating, symmetrical internal structure. The form of this internal structure is called the crystal lattice.

In ionic crystals, ions occupy the fixed lattice positions, and the attractive forces between oppositely charged ions hold the crystal together. Ionic solids have high melting points because these attractive forces are so strong. Molecular crystals, in which molecules occupy the lattice positions, are usually held together by weak electrical forces. Most molecular crystals have much lower melting points than those of ionic crystals. Only molecular crystals whose molecules are polar have binding forces that are fairly strong. Atomic crystals, in which atoms occupy the lattice positions, have strong binding forces. These crystals can be composed of either nonmetallic or metallic ions.

Water is the world's most important liquid. Pure water is a colorless, odorless, tasteless substance that has a melting point of 0°C and a boiling point of 100°C at 1 atm pressure. At 4°C, water has a density of 1 g/mL. Because of the way in which hydrogen and oxygen atoms bond to form water, and because of the three-dimensional shape of the water molecule, the volume of water *increases* as water is cooled below 4°C. For this reason, ice is less dense than liquid water and floats on water. Water is a polar molecule: the positively charged hydrogen atom of one water molecule can attract the negatively charged oxygen atom of another water molecule to form a *hydrogen bond*. Such bonds can hold together long strings of water molecules, which accounts for the high boiling point of water. Water reacts with oxides of nonmetals to produce acids and with oxides of metals to produce bases. Water combines chemically with many salts to form hydrates.

Key Terms

amorphous **(14.6)**
anhydrous **(14.7)**
boiling point **(14.3)**
condense **(14.5)**
crystal **(14.6)**
crystal lattice **(14.6)**
deliquescent **(14.7)**
desiccant **(14.7)**
desiccator **(14.7)**
efflorescence **(14.7)**
equilibrium vapor pressure **(14.2)**
evaporation **(14.2)**
freeze **(14.4)**
freezing point **(14.4)**
heat of crystallization **(14.4)**
heat of vaporization **(14.3)**
hydrate **(14.7)**
hydrogen bond **(14.7)**
hygroscopic **(14.7)**
melt **(14.4)**
melting point **(14.4)**
miscible **(14.5)**
sublimation **(14.4)**
unit cell **(14.6)**
volatile **(14.2)**

Self-Test Exercises

Learning Goal 1: Properties of Gases, Liquids, and Solids

▲ **1.** *(a)* How is the boiling point of a liquid affected by an increase in atmospheric pressure?
(b) How is the melting point of a solid affected by an increase in atmospheric pressure?

2. A solid has a definite shape but no definite volume. True or false?

3. How is the boiling point of a liquid affected by a decrease in atmospheric pressure?

4. Gases are easily compressed, but solids and liquids are not. True or false?

5. State whether each of the following statements refers to a solid, liquid, or gas:
(a) It has definite volume but no definite shape.
(b) It has no definite volume or shape.
(c) It has both definite volume and definite shape.
6. Solids, liquids, and gases all have a definite volume. True or false?

Learning Goal 2: Kinetic Theory of Liquids and Solids

▲ 7. Explain each of the following properties of liquids in terms of the kinetic theory: *(a)* no definite shape *(b)* definite volume *(c)* incompressibility
8. Explain each of the following properties of solids in terms of the kinetic theory: *(a)* definite shape *(b)* definite volume *(c)* incompressibility

9. A man heats a pot of potatoes. When the water starts to boil, he turns up the heat "to cook the potatoes faster." Is he wasting gas, or will the potatoes actually cook faster? Explain.
*10. The directions on a box of fudge brownie mix say, "High-Altitude Directions (3,500 to 6,500 feet): Stir 2 tablespoons of all-purpose flour into brownie mix (dry). Increase baking time to 35 to 40 minutes." Why are these high-altitude directions necessary? If you wanted to bake cake-like brownies, no adjustment would be necessary. Why?

11. It takes 15 minutes to cook a hard-boiled egg at sea level. Explain why it takes more time to cook a hard-boiled egg on a mountain top.
▲ 12. A kettle full of water is boiling on a stove. If you turn up the heat, will the water get hotter? Explain.

*13. Adding salt to water raises the boiling point of water. Give a possible explanation for this phenomenon.
14. You have a bottle of rubbing alcohol on the shelf and you forget to put the cap on the bottle. What will happen to the alcohol in the bottle?

15. As liquid evaporates from an open container, the temperature of the remaining liquid drops. Give a possible explanation for this occurrence.
16. A house heated by a forced-hot-air system has air that is very dry. How could the situation be remedied?

17. Will an equilibrium vapor pressure be reached in the space above a liquid in an open container (for example, a half-filled glass of water)? Explain.

18. Name a liquid that evaporates slowly.

▲ 19. Describe an experiment that you could perform to show that the molecules of a liquid are constantly in motion.
▲ 20. A thermometer containing mercury, a toxic liquid metal, breaks. Why is it dangerous to collect the mercury and let it sit in an open container?

21. If you place equal amounts of gasoline and water in separate glasses, the gasoline evaporates first. Explain why.
22. Which system would have a higher equilibrium vapor pressure: a covered jar half-filled with water at 25°C or a covered jar half-filled with water at 75°C?

23. Adding salt to water lowers its freezing point. Give an explanation for this occurrence.
24. Have you ever been inside a vehicle whose fuel tank is being filled with gasoline and smelled the fumes from the gasoline? If a quart of oil were being added to this same vehicle, you would not smell any fumes from the oil. Why is this so?

Learning Goal 3: Evaporation, Equilibrium Vapor Pressure, and Boiling Point

▲ 25. Define each of the following: *(a)* normal boiling point *(b)* equilibrium vapor pressure
(c) evaporation
26. Which liquid do you think would boil first, ethyl alcohol or salad oil? Explain your answer.

▲ 27. Define the term *dynamic equilibrium* as it applies to vapor pressure.
28. What happens to the boiling point of a liquid if the weather changes from sunny and dry to rainy? (*Hint:* Keep in mind that sunny and dry weather usually means a high atmospheric pressure system is in the area, while rainy weather usually means a low atmospheric pressure system is in the area.)

29. Explain the difference between the terms *boiling point* and *normal boiling point*.
30. A beaker of water is boiling at 1 atm pressure. The temperature of the water is 100°C. What is the temperature of the water vapor at the boiling point?

Learning Goal 4: Change of State

▲ 31. Explain what happens to molecules of water as they go from solid (ice) to liquid (water) to gas (steam).

*32. When dry ice (solid carbon dioxide) is placed in a beaker of water, vigorous bubbling is seen and a white vapor leaves the beaker. Explain what is happening.

Learning Goal 5: Distillation and Fractional Distillation

33. Explain the following processes: *(a)* distillation *(b)* fractional distillation

▲ **34.** Water and isopropyl alcohol are mixed and distilled. Which liquid is collected first?

35. Define each of the following: *(a)* condensation *(b)* distillate *(c)* distilling head *(d)* condenser

36. What is the difference between distillation and fractional distillation?

Learning Goal 6: Crystalline and Amorphous Solids

37. Solids that have no well-defined crystalline structure are said to be ________.

38. Name two solids with a well-defined crystalline structure.

39. Crystalline solids are said to exist in a ________ structure.

40. Name two amorphous solids.

Learning Goal 7: The Kinds of Crystalline Solids

41. List the properties associated with each of the following:
(a) solids composed of oppositely charged ions
(b) solids composed of atoms
(c) solids composed of molecules

42. Compare the melting points of ionic solids, molecular solids, and atomic solids.

43. The compound naphthalene is a soft solid substance with a low melting point. What type of solid (ionic, molecular, or atomic) is naphthalene? Explain.

44. Ionic solids are hard and brittle with low melting points. True or false?

45. The compound cesium chloride has a melting point of 646°C. It is a brittle substance. What type of crystalline solid is it?

46. Molecular solids are nonconductors of electricity. True or false?

▲ **47.** For each of the following statements, classify the solid as ionic, molecular, or atomic (metallic or nonmetallic):
(a) The solid has a high melting point and is a good conductor of electricity.
(b) The solid has a high melting point, does not conduct electricity, and can be decomposed into simpler substances.

48. Which types of solids can be good conductors of electricity? Which types can be good nonconductors? What determines whether a solid is a conductor or a nonconductor?

49. Fill in the blanks.
(a) In atomic solids, ________ occupy the lattice positions.
(b) In molecular solids, ________ occupy the lattice positions.
(c) In ionic solids, ________ occupy the lattice positions.

50. Which types of solids are hard and brittle?

51. Describe the forces that hold the crystals together in *(a)* atomic solids, *(b)* molecular solids, *(c)* ionic solids.

52. A certain compound is a soft solid with a low melting point and does not conduct electricity. Describe the particles that occupy its lattice positions.

Learning Goal 8: Crystal Lattice and Unit Cell

53. Trace the unit cell of sodium chloride from the drawing in Figure 14.11, and show how several unit cells can build up into a crystalline structure.

▲ **54.** Name six types of crystals.

55. Define each of the following: *(a)* crystal lattice *(b)* unit cell

56. Which type of crystal structure is exhibited by sodium chloride?

Learning Goal 9: Chemical and Physical Properties of Water

57. Why is it incorrect to say that rainwater is "pure" water?

58. The density of ice at 0°C is 0.92 g/cm^3. What is the volume of a cube of ice that has a mass of 18.4 g?

▲ **59.** A glass is filled to capacity with ice cubes and water. When the ice cubes melt, will the water overflow? (Assume that there is no evaporation.) Explain.

60. The density of water at 4°C is 1.00 g/cm³. What is the mass of 535 cm³ of water?

61. The density of ice at 0°C is 0.92 g/cm³. What is the mass of an ice cube that measures 4.0 cm on each side?

62. What would happen to our rivers, lakes, and seas if ice did not float on water?

63. Draw a picture of a hydrogen bond between two water molecules.

64. How large is the bond angle between the hydrogen atoms in a water molecule?

65. The density of water at 4°C is 1.00 g/cm³. What is the volume of 1.00 kg of water?

66. What is a hydrogen bond?

Learning Goal 10: Reactions of Water

67. Predict the products formed when water combines with *(a)* MgO, *(b)* SO_3, *(c)* Na.

▲ **68.** Predict the products formed when water combines with *(a)* CaO, *(b)* SO_2, *(c)* K.

69. Complete the following equations, and then balance them:
(a) $SrO + H_2O \longrightarrow$
(b) $SO_2 + H_2O \longrightarrow$
(c) $Li + H_2O \longrightarrow$

70. Complete the following equations, and then balance them:
(a) $Na_2O + H_2O \longrightarrow ?$
(b) $CO_2 + H_2O \longrightarrow ?$
(c) $Rb + H_2O \longrightarrow ?$

Learning Goal 11: Hydrates and Water of Hydration

▲ **71.** Determine the percentage of water (by mass) in $CuSO_4 \cdot 5H_2O$.

72. Determine the percentage of water (by mass) in $CaSO_4 \cdot 2H_2O$.

73. A commercial for a well-known cleaning powder says it "shakes out white, then turns blue." What type of reaction is occurring? What compound could be used to produce this effect?

74. What is the difference between a hydrated salt and an anhydrous salt?

75. Determine the percent composition by mass of water in each of the following hydrated salts:
(a) $MgSO_4 \cdot 7H_2O$
(b) $Na_2CO_3 \cdot 10H_2O$
(c) $KAl(SO_4)_2 \cdot 12H_2O$

▲ **76.** Determine the percentage by mass of water in each of the following hydrated salts:
(a) $Ca(ClO_3)_2 \cdot 2H_2O$
(b) $CaCO_3 \cdot 6H_2O$
(c) $CaCl_2 \cdot 6H_2O$

77. Complete the following equations, and then balance them:

(a) $CaSO_4 \cdot 2H_2O \xrightarrow{\text{Heat}}$

(b) $KAl(SO_4)_2 \cdot 12H_2O \xrightarrow{\text{Heat}}$

78. Complete the following equations, and then balance them:

(a) $Mg(C_2H_3O_2)_2 \cdot 4H_2O \xrightarrow{\text{Heat}} ? + ?$

(b) $MgBr_2 \cdot 6H_2O \xrightarrow{\text{Heat}} ? + ?$

***79.** There are three hydrated forms of the salt $CoSO_4$. They are $CoSO_4 \cdot 1H_2O$, $CoSO_4 \cdot 6H_2O$, and $CoSO_4 \cdot 7H_2O$. From the following data, determine which form of $CoSO_4 \cdot nH_2O$ is being analyzed: When a chemist heats 52.58 g of hydrated salt, 21.60 g of water are produced.

***80.** The two hydrated forms of the salt $Mg(NO_3)_2$ are $Mg(NO_3)_2 \cdot 2H_2O$ and $Mg(NO_3)_2 \cdot 6H_2O$. From the following data, determine which of the hydrated salts is being analyzed: When 25.64 g of hydrated salt are heated, 10.80 g of water are produced.

Learning Goal 12: Definitions of Anhydrous, Hygroscopic, Deliquescence, and Efflorescence

81. Define and give an example of each of the following: *(a)* anhydrous salt *(b)* hygroscopic substance *(c)* deliquescent substance *(d)* substance that undergoes efflorescence

▲ **82.** Complete the following sentences:
(a) A salt that has lost its water of hydration is called a(n) ________ salt.
(b) A substance that can absorb water from the atmosphere is ________.

(c) A substance that absorbs enough water from the atmosphere to form a solution is ________.
(d) A salt that can spontaneously lose its water of hydration to the atmosphere is a(n) ________ salt.

Extra Exercises

83. Explain, in terms of the kinetic theory, why the temperature of a pure substance remains constant during a change of state.

84. In what way is a unit cell like a molecule?

85. Is it possible for a substance *not* to have a normal boiling point? Explain.

86. Compare the freedom of motion of a molecule of a gas with that of a molecule of a liquid and with that of a molecule of a solid.

87. Using Figure 14.8, explain why steam at 100°C produces more severe burns than liquid water at 100°C.

88. Explain what is happening when a glass rod is heated and begins to sag.

89. Calcium chloride, $CaCl_2$, is often used as a desiccant (drying agent) for wet chemicals. Explain.

90. Certain weather indicators operate by color changes. For example, one indicator turns red if the air is moist and turns blue in dry weather. Explain how this happens.

91. Compare and contrast the characteristics of the three types of solids.

92. State whether each of the following characteristics refers to a solid, a liquid, or a gas. (*Note:* Each characteristic may refer to more than one state of matter.)
(a) has definite volume
(b) has definite shape
(c) has no definite volume

93. Describe an experiment that you could perform to show that the molecules of a gas are constantly in motion.

94. Do liquids that evaporate easily at a particular temperature have *higher* or *lower* equilibrium vapor pressures than liquids that do not evaporate easily at the same temperature? Explain.

95. Even though water is a relatively small molecule, it has a very high boiling point compared to molecules of similar size. Can you explain this in terms of the structure of water?

Cumulative Review/Chapters 12–14

1. Distinguish between endothermic and exothermic reactions.

▲ **2.** How much heat is needed to raise the temperature of 500.0 g of water from 15.0°C to 35.0°C?

3. How much heat is required to raise the temperature of 25.0 g of copper from 30.0°C to 90.0°C?

4. How much heat is required to raise the temperature of 25.0 mL of ethyl alcohol from 25.0°C to 50.0°C? (The density of ethyl alcohol is 0.800 g/mL.)

5. Why must we use standard-state conditions when calculating changes in heat?

6. How much heat is needed to raise the temperature of 1.000 kg of water from 50.0°C to 75.0°C?

▲ **7.** Write the thermochemical equation for the formation of 1 mole of water from hydrogen gas and oxygen gas.

8. Use Table 12.2 to determine ΔH for the reaction

$$H_2O(l) + SO_3(g) \longrightarrow H_2SO_4(aq)$$

9. Calculate the heat released when 254 g of $SO_2(g)$ are produced from its elements. Use Table 12.2.

▲ **10.** Calculate the amount of heat released when 10.2 g of $H_2O_2(l)$ decompose to form $H_2O(l)$ and $O_2(g)$. Use Table 12.2.

11. A gas has an initial volume of 55 mL at a pressure of 2.0 atm. What will be its final volume if the pressure of the gas is increased to 4.5 atm and the temperature remains constant?

▲ **12.** Suppose a gas occupies a space of 15.05 mL under a pressure of 2.50 atm at 110.0°C. What will be the volume of the gas when the pressure is decreased to 2.00 atm? (Assume the temperature remains constant.)

13. A gas occupies a volume of 0.200 L under a pressure of $77\bar{0}$ torr at 50.0°C. What pressure will cause the gas to occupy a volume of 250.0 mL at the same temperature?

▲ **14.** A gas occupies a volume of 1,500.0 mL at 27°C. What will be its volume at 227°C, assuming its pressure remains constant?

15. When a gas is cooled at constant pressure, it contracts from 100.0 mL to 55.0 mL. Its final temperature is 180.0°C. What was its initial temperature?

16. A gas at 50.5°C and 1.50 atm pressure fills a 2.50-L container. What volume will the gas occupy at 250.0°C and 775 torr?

17. A gas collected at 127°C has a volume of 10.5 L and a pressure of 775 torr. What will be its volume at STP?

18. Suppose a sample of oxygen is collected over water at 20.0°C and a pressure of 775 torr. The gas fills a $4\bar{0}0$-mL container but has water vapor mixed in with it. What would be the volume of the *dry* gas at STP?

19. Suppose 5.00 L of hydrogen gas are collected over water at 28.0°C and a pressure of 725 torr. What would be the volume of the *dry* hydrogen gas at STP?

▲ **20.** A gas occupies a volume of 358.4 L at STP. How many moles of gas are in this sample?

21. Determine the volume of 15.0 moles of a gas at STP.

22. Twenty (20.0) moles of a gas exert a pressure of 2,280 torr at a temperature of 227°C. What volume does the gas occupy under these conditions?

23. A gas exerts a pressure of 550.0 torr at a temperature of 155°C. If the volume of the gas is 2.00 L, how many moles of gas are present?

24. Determine the molecular mass of a gas, given the following experimental data: mass = 3.55 g, volume = 255 mL, pressure = $1{,}52\bar{0}$ torr, and T = 227°C.

▲ **25.** When ethane (C_2H_6) reacts with oxygen, carbon dioxide and water vapor are formed. How many liters of carbon dioxide are produced from 5.00 L of ethane?

26. Using the Ideal Gas Law and the formula for density ($D = m/V$), calculate the density of chlorine gas at STP. Report your answer in grams per liter.

27. Once released, a weather balloon travels to the upper atmosphere. Why does its volume increase there?

▲ **28.** An old tube tire has a volume of 2.55 L at 25°C and 1.39 atm pressure. During a trip, the temperature of the tire increases by 12°C and the pressure increases to 1.41 atm. What is the new volume of the tire?

29. A 25.0-g sample of a gas has a pressure of 4.50 atm and a volume of 1.45 L at a temperature of 25.0°C. What is the molecular mass of the gas?

30. Examining the physical states of the halogens at room temperature, we find that fluorine is a gas, bromine a liquid, and iodine a solid. Use the kinetic theory of gases to explain these properties.

▲ **31.** A balloon is filled with 4.0 L of air at standard pressure. What will happen to the volume of air in the balloon if it is placed in a hyperbaric chamber at 3.5 atm pressure? (Assume the temperature doesn't change.)

32. A sample of intestinal gas is analyzed and found to have the following approximate percentage composition by volume: 42.0% CO_2, 40.0% H_2, 16.9% N_2, and 0.00300% CH_4. If the total pressure in the intestine is 815 torr, what is the partial pressure of each gas?

33. Match each of the following reactions with its description:

(a) $SO_2 + H_2O \longrightarrow H_2SO_3$	1. Electrolysis of water
(b) $CaO + H_2O \longrightarrow Ca(OH)_2$	2. Metal oxide + water
(c) $2H_2O \longrightarrow 2H_2 + O_2$	3. Nonmetal oxide + water

34. A 20.8-g sample of calcium carbonate is heated until all the water is driven off. The remaining salt weighs 10.00 g. Determine the formula of the hydrate.

▲ **35.** Calculate the percentage of water in calcium chloride hexahydrate.

36. Calculate the percentage of water in calcium sulfate dihydrate.

▲ **37.** A hydrate of barium chloride is analyzed and found to contain 14.7% water by mass. Determine the formula of this hydrate.

38. Calculate the formula for the hydrate of sodium carbonate that is 63% water by mass.

39. How many grams of water are needed to produce 25.00 g of oxygen by electrolysis?

40. Match each of the following formulas with the name of the proper hydrate:

(a) $MgSO_4 \cdot 7H_2O$	1. Alum
(b) $KAl(SO_4)_2 \cdot 12H_2O$	2. Gypsum
(c) $CaSO_4 \cdot 2H_2O$	3. Epsom salts

41. Match each of the following properties with the correct term:

(a) Has no definite shape but has a definite volume	1. Solid
(b) Is easily compressed	2. Liquid
(c) Has definite shape and volume	3. Gas

42. Eggs are being cooked in boiling water. The heat is increased. Which of the following statements is true in this situation?

(a) The eggs cook faster.
(b) The water evaporates more quickly.

▲ **43.** On a sunny day following a winter snowstorm, the volume of snow decreases, yet the temperature has not allowed melting to take place. What is the most probable explanation for this occurrence?

44. A crystalline solid has a low melting point and is soft. Is it more likely to be a molecular or an atomic solid?

45. A crystalline solid has a high melting point and is hard and brittle. Is it more likely to be an ionic or a molecular solid?

46. Is a solid with no definite shape more likely to be a crystalline or an amorphous solid?

▲ **47.** Using the kinetic theory, explain what happens to water as it changes from ice to vapor.

▲ **48.** Benzene evaporates more easily than water. Which liquid has the higher equilibrium vapor pressure?

49. Why are some solids capable of conducting electricity and others are not?

50. Why are atomic crystals that are composed of metallic atoms so strong?

15 The Chemistry of Solutions

Learning Goals

After you have studied this chapter, you should be able to:

1. Define the terms *solution, solute,* and *solvent*.
2. Predict whether substances will form solutions using the idea that "like dissolves like."
3. Define the terms *saturated solution, unsaturated solution,* and *supersaturated solution*.
4. Calculate concentrations of solutions in percent by mass, percent by volume, and percent by mass–volume.
5. Calculate the molarity, number of moles, or volume of a solution when you are given two of the quantities.
6. Explain the concept of equivalents for an acid or base, and calculate the number of equivalents when you are given the number of grams or moles.
7. Calculate the normality, number of equivalents, or volume of a solution when you are given the other two quantities.
8. Calculate concentrations of normal and molar solutions after they have been diluted.
9. Explain what electrolytic and nonelectrolytic solutions are, and give examples of each.
10. Define the terms *ion* and *ionization*.
11. Explain what is meant by the molality of a solution.
12. Calculate the molality of a solution, given mass of solute and mass of solvent.
13. Define the term *colligative properties,* and explain how a solute affects the boiling point and freezing point of a solution.
14. Define the terms *freezing-point depression constant* (K_f) and *boiling-point elevation constant* (K_b).

15. Calculate the boiling and freezing points of a solution when you are given the molality of the solution, its K_f, and its K_b.
16. Define the following terms: *diffusion, osmosis, osmotic pressure,* and *osmometry*.

Introduction

Solutions are all around us. We drink them, we breathe them, we swim in them, we are even composed of them. Every time you drink a cup of tea or a soda, you swallow a solution. Each time you breathe, you inhale a solution—air. When you swim in the ocean, you're swimming in a solution of salt in water. Even your blood is a solution.

There is a lot more to solutions than just two things put together. In this chapter, we look at the types of solutions, some of their properties, and some special units that apply to solutions.

15.1 Like Dissolves Like

Learning Goal 1
Definitions of solution, solute, and solvent

In Chapter 3, we defined a *solution* as a homogeneous mixture. A solution can contain two substances, three substances, or more. The most common types of solutions are made by dissolving a solid in a liquid—for example, a salt in water. However, solutions can be made up from combinations of all three states of matter. Table 15.1 shows eight possible types of solutions and gives an example of each.

Because solutions are always composed of at least two substances, we need to be able to identify the role that each substance plays. The **solute** is *the substance that is being dissolved,* whereas the **solvent** is *the substance that is doing the dissolving*. For example, in a solution of salt water, salt is the solute and water is the solvent.

Table 15.1
Different Types of Solutions

	Solvent		
Solute	**Solid**	**Liquid**	**Gas**
Solid	Copper metal dissolved in silver metal (for example, coins)	Salt dissolved in water	———
Liquid	Mercury in silver (dental fillings)	Ethyl alcohol dissolved in water	Water vapor in air
Gas	Hydrogen dissolved in platinum metal	Carbon dioxide dissolved in water (soda water)	Oxygen gas dissolved in nitrogen gas

When a solution is composed of two substances in the same state (as is a liquid–liquid solution), it is difficult to establish which substance is the solute and which is the solvent. One rule of thumb is that the substance present in the larger amount is the solvent. In a solution of 10 mL of ethyl alcohol and 90 mL of water, the ethyl alcohol is the solute and the water is the solvent.

Example 15.1

Determine which is the solute and which is the solvent in each of the following solutions:
(a) sugar and water
(b) hydrogen chloride gas and water
(c) 75.0 mL of ethyl alcohol and 25.5 mL of water
(d) 80 mL of nitrogen and 20 mL of hydrogen
(e) soda water

Solution

(a) Sugar is the solute; water is the solvent.
(b) Hydrogen chloride is the solute; water is the solvent.
(c) Water is the solute; ethyl alcohol is the solvent.
(d) Hydrogen is the solute; nitrogen is the solvent.
(e) Carbon dioxide is the solute; water is the solvent.

Practice Exercise 15.1 Which is the solute and which is the solvent in each of the following solutions?
(a) salt and water
(b) sulfur trioxide gas and water
(c) $3\bar{0}$ mL of isopropyl alcohol and 25 mL of water
(d) 40 mL of hydrogen gas and 50 mL of oxygen gas
(e) carbonated mineral water

Learning Goal 2
Substances that form solutions

Before we go on, we should note that some pairs of liquids do not form solutions when mixed; such liquids are said to be **immiscible.** For example, gasoline and water do not readily form a solution. The same holds true for oil and vinegar, as in salad dressing (Figure 15.1). There are also substances that do mix, but only to a slight degree. (The opposite of immiscible is miscible. Pairs of liquids that form solutions when mixed are said to be *miscible.*)

How can we determine whether two substances will form a solution? The answer is found in the statement "like dissolves like." To understand its significance, recall that a molecule may be either polar or nonpolar. "Like dissolves like" means that polar substances dissolve other polar substances and that nonpolar substances dissolve other nonpolar substances. But a polar and a nonpolar substance generally do not form a solution. Alcohol and water mix because both are composed of polar molecules. Gasoline and water do

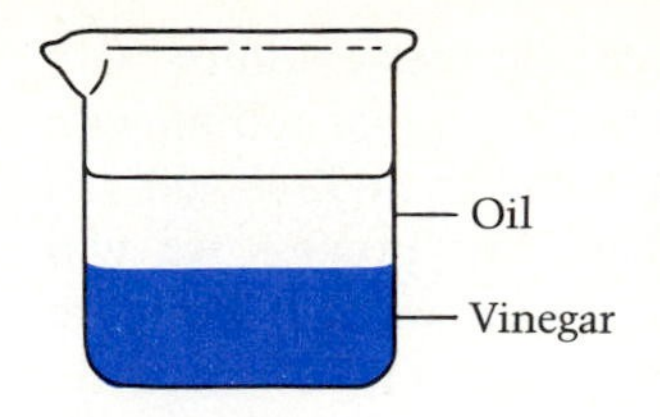

Figure 15.1
Two immiscible liquids

not mix, because gasoline is composed of nonpolar molecules and water is composed of polar molecules.

Example 15.2

State whether the following pairs of substances form solutions. (*Hint:* Refer to Chapter 7.)

(*a*) methane (CH_4) and water
(*b*) HBr and water
(*c*) *trans*-dichloroethene and carbon tetrachloride

Solution

(a) Methane is nonpolar; water is polar. They do not form a solution.
(b) Both HBr and water are polar. They do form a solution.
(c) *trans*-dichloroethene is nonpolar, and so is carbon tetrachloride. They do form a solution.

Practice Exercise 15.2 State whether the following pairs of substances form solutions. (*Hint:* You may want to refer to Chapter 7.)

(*a*) oil and water
(*b*) hydrogen chloride gas and water
(*c*) benzene and carbon tetrachloride

15.2 Saturated, Unsaturated, and Supersaturated Solutions

Learning Goal 3
Saturated, unsaturated, and supersaturated solutions

Suppose we take a glass of water and decide to add salt. We begin with a teaspoonful of salt. As the salt and water mix, the salt begins to dissociate. Here's how we think the process happens. We know that water molecules are polar, and the negative end of the water molecule (the oxygen portion) can align itself toward the positive sodium ions. The positive end of the water molecule (the hydrogen portion) can align itself toward the negative chloride ions. Thus, the polar water molecules surround ions on the surface of the sodium chloride crystal, forcing them to break away from the actual surface of the crystal. The first ions to break away become hydrated sodium ions and hydrated chloride ions, meaning that each ion is surrounded by a group of water molecules. A newly exposed surface is formed and is ready to be acted on by other water molecules. The process continues until all of the sodium chloride crystals become hydrated sodium ions and hydrated chloride ions. The salt is then said to have dissolved in the water. We can stir the salt water to increase the rate at which the salt dissolves. Stirring helps distribute the hydrated ions uniformly throughout the solution.

We can add another teaspoonful of salt, stir, and watch the salt crystals disappear into solution. As the number of hydrated ions in solution increases, the chance of ions colliding with each other increases. Besides colliding with each other, ions can collide with solvent molecules, and they can collide with any remaining undissolved solid. When they collide with the undissolved solute, they leave the solution. If we continue to repeat the process we will eventually reach the point at which, no matter how much we stir, no more salt will dissolve. For any given temperature there is a point at which no more solute can dissolve in a particular quantity of solvent. Once this point is reached the solution is said to be **saturated.** When a solution is saturated, the rate at which solid dissolves into the solution is the same as the rate at which dissolved solid crystallizes out of the solution. Ions leave the solution to return to the undissolved solute at the same rate at which ions enter the solution.

An **unsaturated** solution contains less solute than can be dissolved at a particular temperature. In an unsaturated solution, more solute can be dissolved into the solvent without changing the temperature. This happens when the number of ions in the solution is low, making the chance of collision between hydrated ions and the undissolved solute low. The result is that the undissolved solute won't have much opportunity to recapture ions, so they stay in solution.

Although it appears as if nothing is happening once a solution is saturated, we know that a dynamic process is occurring at the molecular level. The undissolved solute is in equilibrium with the dissolved solute. This dynamic equilibrium can be represented as follows:

$$\text{Undissolved solute} \rightleftharpoons \text{dissolved solute}$$

The double arrow means the reaction is reversible.

Table 15.2 gives the solubility of several substances in water. **Solubility** is the amount of a solute that dissolves in a particular amount of solvent to give a saturated solution. Note that solubility varies according to the temperature, that is, a particular solution that is saturated at one temperature is not neces-

Table 15.2
Solubility of Various Substances in Water at 0°C and 50°C

Solute	Solubility (g solute/100 g solvent)	
	0°C	*50°C*
$CuSO_4$	14.3	33.3
$NaCl$	35.7	37.0
$AgNO_3$	122.0	455.0
$CsCl$	161.4	218.5
$C_{12}H_{22}O_{11}$	203.9 (20°C)	260.4

sarily saturated at another temperature. In general, an increase in temperature causes more solute to be dissolved.

It is possible to prepare solutions that contain a greater concentration of solute than is needed for saturation. Suppose we start with a teaspoonful of sugar and a glass of water at room temperature. We add the sugar to the water and stir. The sugar dissolves in the water. We repeat the process until no more sugar will dissolve. At this point we find sugar resting on the bottom of the glass, and we know that the solution is saturated. We decide to heat the solution. The sugar that was resting on the bottom of the glass now dissolves. We add more sugar, and it also dissolves when stirred. We allow the solution to cool slowly to room temperature. The solid does not precipitate out of the solution, and what we have is a **supersaturated** solution. Such a solution is unstable and will revert to a saturated solution if disturbed. Such action will cause the excess solute to crystallize rapidly from the solution. If we were to tie a string around a tiny sugar crystal and suspend the string in the supersaturated sugar-water solution, we could grow rock candy, which is made of large sugar crystals.

Example 15.3

Determine whether each of the following solutions is unsaturated, saturated, or supersaturated (see Table 15.2).

(a) 200.0 g of $AgNO_3$ dissolved in 200.0 g of water at 0°C.
(b) 50.0 g of $CuSO_4$ dissolved in 150.0 g of water at 50°C.
(c) 220.0 g sucrose ($C_{12}H_{22}O_{11}$) dissolved in 75.00 g of water at 50°C.

Solution

Understand the Problem

In each case, we are given an amount of substance dissolved in a given amount of water at a certain temperature. We are asked whether the resulting solution is saturated (that is, there is just enough solute to achieve dynamic equilibrium), unsaturated (more of the solute could dissolve in the solution), or supersaturated (the solution contains more dissolved substance than a saturated solution).

Devise a Plan

For each example use the factor-unit method, and calculate the grams of solute per 100 g of water. Compare this value to the value in Table 15.2.

Carry Out the Plan

(a) $$? \text{ g } AgNO_3 = (100.0 \text{ g } \cancel{H_2O})\left(\frac{200.0 \text{ g } AgNO_3}{200.0 \text{ g } \cancel{H_2O}}\right)$$

$$= 100.0 \text{ g } AgNO_3 \text{ (at 0°C)}$$

This solution is *unsaturated;* Table 15.2 states that 122.0 g $AgNO_3$ can be dissolved in 100.0 g H_2O at 0°C.

(b) $? \text{ g CuSO}_4 = (100.0 \text{ g } \cancel{H_2O})\left(\frac{50.0 \text{ g CuSO}_4}{150.0 \text{ g H}_2\text{O}}\right)$

$= 33.3 \text{ g CuSO}_4$ (at 50°C)

This solution is *saturated;* Table 15.2 states that 33.3 g of $CuSO_4$ can be dissolved in 100.0 g H_2O at 50°C.

(c) $? \text{ g sucrose} = (100.0 \text{ g } \cancel{H_2O})\left(\frac{220.0 \text{ g sucrose}}{75.00 \text{ g } \cancel{H_2O}}\right)$

$= 293.3 \text{ g sucrose}$ (at 50°C)

This solution is *supersaturated;* Table 15.2 states that 260.4 g of sucrose can be dissolved in 100.0 g H_2O at 50°C.

Look Back

Recheck your calculations, and see if your answers are reasonable.

Practice Exercise 15.3 Determine whether each of the following solutions is unsaturated, saturated, or supersaturated (see Table 15.2).

(a) 60.0 g of CsCl dissolved in 25.0 g of water at 50°C.
(b) 75.0 g of CsCl dissolved in 50.0 g of water at 0°C.
(c) 185.0 g of NaCl dissolved in 500.0 g of water at 50°C.

15.3 Concentrations of Solutions by Percent

Learning Goal 4
Percent concentrations

Because solutions are mixtures, their components can be present in different ratios. For example, we can make many different salt-and-water solutions, each with a different **concentration,** or ratio of solute to solvent. The concentrations or relative quantities of solutions can vary widely, so we must have a way of describing them. (It is important to remember, however, that at any given temperature there is a limit to the amount of solute that can be dissolved in a solution. This concentration, called the **saturation point,** is the point at which no more solute dissolves to form a stable solution.)

One method of defining the concentrations of solutions is based on the percent of solute in the solution. This method can cause confusion, because there can be three types of percent concentrations:

1. Percent by mass
2. Percent by volume
3. Percent by mass–volume

A 30%-by-mass solution of alcohol and water is not the same as a 30%-by-volume solution of alcohol and water. Therefore a label that reads "30% alcohol in water" tells us nothing. The solution could be a 30%-by-mass solution, a 30%-by-volume solution, or a 30%-by-mass–volume solution. Chemists seldom use the percent method of defining concentrations, however, biologists and medical workers do use percent by mass–volume, so we will discuss all three types.

Figure 15.2
Preparing a 30%-by-mass alcohol–water solution

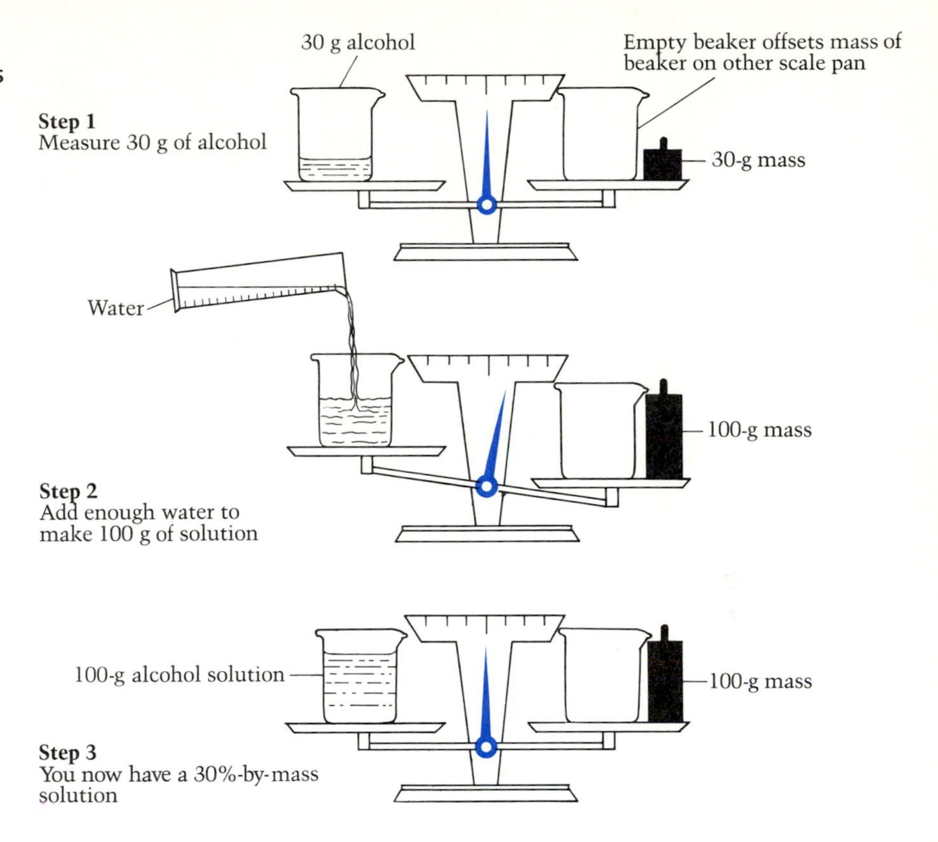
30 g alcohol
Empty beaker offsets mass of beaker on other scale pan
Step 1
Measure 30 g of alcohol
30-g mass
Water
100-g mass
Step 2
Add enough water to make 100 g of solution
100-g alcohol solution
100-g mass
Step 3
You now have a 30%-by-mass solution

Percent-by-Mass Solutions

To find percent by mass, we divide the *mass of the solute* by the *total mass of the solution* and multiply the result by 100. (The mass of the solution equals the mass of the solute plus the mass of the solvent.) For example, suppose we made a solution of alcohol in water by mixing 30 g of alcohol with enough water to make 100 g of solution (Figure 15.2). The concentration of our solution would be 30% alcohol by mass.

$$\text{Percent alcohol by mass} = \frac{\text{mass of alcohol}}{\text{total mass of solution}} \times 100$$

$$= \frac{30 \not{g}}{100 \not{g}} \times 100 = 30\%$$

Example 15.4

Find the concentrations of the following solutions in percent by mass.
(a) $2\bar{0}$ g of salt with enough water to make $6\bar{0}$ g of solution
(b) $5\bar{0}$ g of sugar with enough water to make $35\bar{0}$ g of solution

Solution

(a) Percent salt by mass $= \dfrac{\text{mass of salt}}{\text{total mass of solution}} \times 100$

$$= \frac{2\bar{0}\ \text{g}}{6\bar{0}\ \text{g}} \times 100 = 33\%$$

(b) Percent sugar by mass $= \dfrac{\text{mass of sugar}}{\text{total mass of solution}} \times 100$

$$= \frac{5\bar{0}\ \text{g}}{35\bar{0}\ \text{g}} \times 100 = 14\%$$

Practice Exercise 15.4 Find the concentrations of the following solutions in percent by mass:

(a) 40.0 g of salt with enough water to make 90.0 g of solution
(b) 75.0 g of sugar with enough water to make 250.0 g of solution

Percent-by-Volume Solutions

To find percent by volume, we divide the *volume of the solute* by the *total volume of the solution* and multiply the result by 100. For example, suppose we mixed 3$\bar{0}$ mL of alcohol with enough water to make 1$\overline{00}$ mL of solution (Figure 15.3). The concentration of the solution would be 3$\bar{0}$% alcohol by volume.

$$\text{Percent alcohol by volume} = \frac{\text{volume of alcohol}}{\text{total volume of solution}} \times 100$$

$$= \frac{3\bar{0}\ \text{mL}}{1\overline{00}\ \text{mL}} \times 100 = 3\bar{0}\%$$

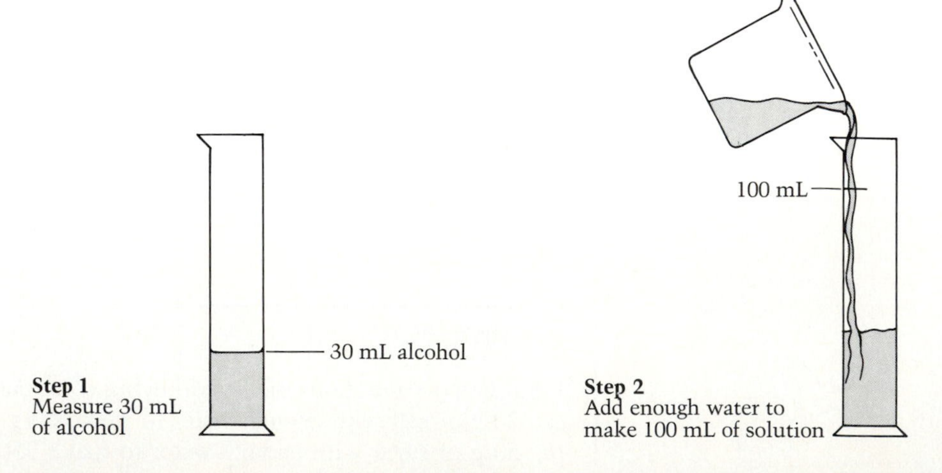
30 mL alcohol
Step 1
Measure 30 mL
of alcohol
100 mL
Step 2
Add enough water to
make 100 mL of solution

Figure 15.3
Preparing a 30%-by-volume alcohol–water solution

Example 15.5

Find the concentrations of the following solutions in percent by volume:
(a) 50.0 mL of alcohol with enough water to make $4\bar{0}0$ mL of solution
(b) $1\bar{0}$ mL of benzene is added to $4\bar{0}$ mL of carbon tetrachloride to make $5\bar{0}$ mL of solution

Solution

(a) $$\text{Percent alcohol by volume} = \frac{\text{volume of alcohol}}{\text{total volume of solution}} \times 100$$

$$= \frac{50.0\ \cancel{\text{mL}}}{4\bar{0}0\ \cancel{\text{mL}}} \times 100 = 12.5\%$$

(b) $$\text{Percent benzene by volume} = \frac{\text{volume of benzene}}{\text{total volume of solution}} \times 100$$

$$= \frac{1\bar{0}\ \text{mL}}{(1\bar{0}\ \text{mL} + 4\bar{0}\ \text{mL})} \times 100$$

$$= \frac{1\bar{0}\ \cancel{\text{mL}}}{5\bar{0}\ \cancel{\text{mL}}} \times 100 = 2\bar{0}\%$$

Practice Exercise 15.5 Find the concentrations of the following solutions in percentage by volume:
(a) 60.0 mL of alcohol with enough water to make 300.0 mL of solution
(b) 40.0 mL of carbon tetrachloride in 50.0 mL of benzene

Percent-by-Mass–Volume Solutions

To find percent by mass–volume, we divide the *mass of the solute* in grams by the *volume of the solution* in milliliters and multiply the result by 100. For example, suppose that we mix $3\bar{0}$ g of alcohol with enough water to make $1\bar{0}\bar{0}$ mL of solution (Figure 15.4). The concentration of the solution would be $3\bar{0}\%$ alcohol by mass–volume.

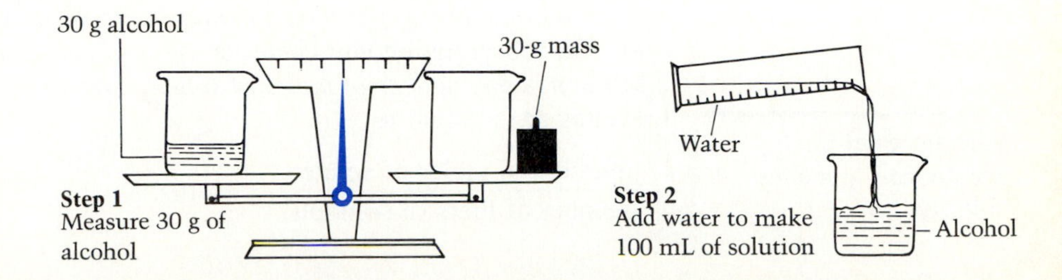
30 g alcohol
30-g mass
Water
Step 1
Measure 30 g of alcohol
Step 2
Add water to make 100 mL of solution
Alcohol

Figure 15.4 Preparing a 30%-by-mass–volume alcohol–water solution

$$\text{Percent alcohol by mass–volume} = \frac{\text{mass of alcohol (g)}}{\text{volume of solution (mL)}} \times 100$$

$$= \frac{3\bar{0}\ \text{g}}{1\bar{0}0\ \text{mL}} \times 100 = 3\bar{0}\%$$

Example 15.6

Find the concentrations of the following solutions in percent by mass–volume:

(a) 2.0 g of iodine in enough carbon tetrachloride to make $8\bar{0}$ mL of solution

(b) $13\bar{0}$ g of sugar in enough water to make $65\bar{0}$ mL of solution

Solution

(a) $\text{Percent iodine by mass–volume} = \frac{\text{mass of iodine}}{\text{volume of solution}} \times 100$

$$= \frac{2.0\ \text{g}}{8\bar{0}\ \text{mL}} \times 100 = 2.5\%$$

(b) $\text{Percent sugar by mass–volume} = \frac{\text{mass of sugar}}{\text{volume of solution}} \times 100$

$$= \frac{13\bar{0}\ \text{g}}{65\bar{0}\ \text{mL}} \times 100 = 20.0\%$$

Practice Exercise 15.6 Find the concentrations of the following solutions in percent by mass–volume:

(a) 4.5 g of *trans*-dichloroethene in enough carbon tetrachloride to make 100.0 mL of solution

(b) 150.0 g of sucrose in enough water to make 750.0 mL of solution

15.4 Molarity

Chemists run many reactions in solutions. Because most of these reactions take place between the solute *particles,* we need a concentration unit that expresses the *number of particles of solute* present in a given amount of solution. The concentration unit chemists use is called *molarity*. The **molarity** of a solution is *the number of moles of solute per liter of solution*. Molarity (abbreviated M) is equal to

Learning Goal 5
Calculations involving molarity

$$\frac{\text{number of moles of solute}}{\text{number of liters of solution}} \quad \text{so that} \quad M = \frac{\text{moles}}{\text{liter}} \quad \text{or } M = \frac{n}{V}$$

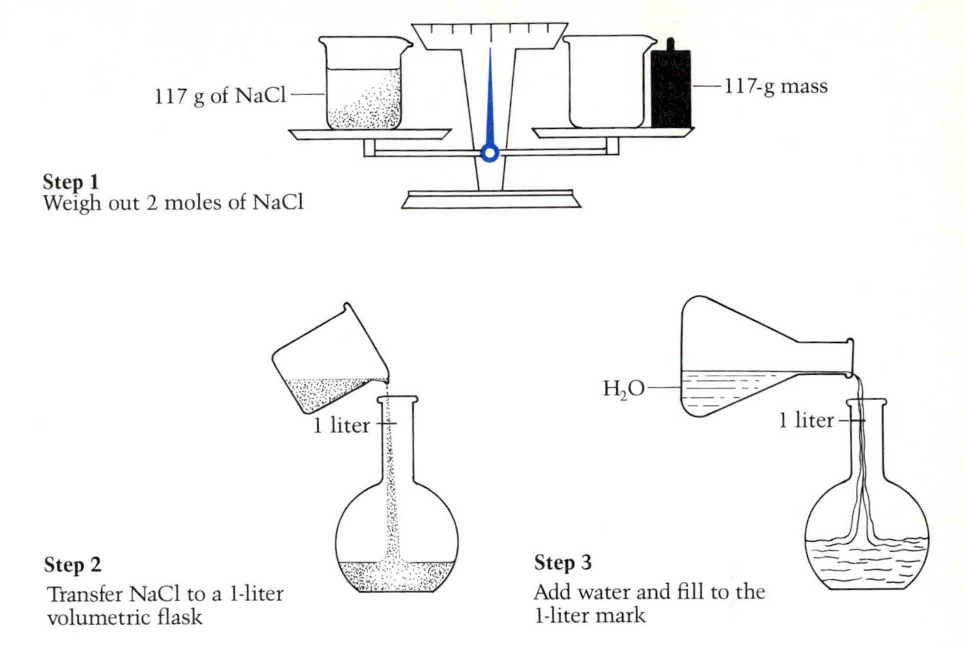
117 g of NaCl
117-g mass
Step 1
Weigh out 2 moles of NaCl
1 liter
Step 2
Transfer NaCl to a 1-liter volumetric flask
H2O
1 liter
Step 3
Add water and fill to the 1-liter mark

Figure 15.5
Preparing 1 L of a 2 *M* NaCl solution

There can be no confusion when the concentrations of solutions are expressed in these terms. A solution made by dissolving 117 g (2.00 moles) of NaCl in enough water to obtain 1 L of solution (Figure 15.5) would be a 2-molar NaCl solution (usually written as 2 *M* NaCl). (For a review of moles, see Chapter 9.) To calculate the molarity of this solution, we would write

$$M = \frac{\text{moles}}{\text{liter}} \qquad ?\,M = \frac{2\text{ moles}}{1\text{ liter}} = 2\,M$$

Chemists often need to prepare solutions that have particular concentrations. *Volumetric flasks* are helpful in preparing such solutions. These flasks are calibrated to contain a specific volume of a solution at a particular temperature (usually 20°C), as shown in Figure 15.5.

Example 15.7

Calculate the molarity of each of the following solutions:
(a) 3.65 g of HCl in enough water to make $5\overline{0}0$ mL of solution
(b) 160 g of NaOH in enough water to make 6.0 L of solution
(c) 36.8 g of ethyl alcohol (C_2H_6O) in enough water to make $1{,}6\overline{0}0$ mL of solution

Solution

(a) First determine the number of moles of HCl.

$$3.65 \text{ g of HCl} \times \frac{1 \text{ mole of HCl}}{36.5 \text{ g of HCl}} = 0.100 \text{ mole of HCl}$$

Then convert the volume of solutions to liters.

$$5\overline{0}0 \text{ mL} \times \frac{1 \text{ L}}{1000 \text{ mL}} = 0.500 \text{ L}$$

Then,

$$M = \frac{n}{V} = \frac{0.100 \text{ mole}}{0.500 \text{ L}} = 0.200\ M$$

(b) We have 6.0 L of solution, and we first calculate that

$$160 \text{ g of NaOH} = 4.0 \text{ moles}$$

Then,

$$M = \frac{\text{moles}}{\text{liter}} = \frac{n}{V} \qquad ?\ M = \frac{4.0 \text{ moles}}{6.0 \text{ L}} = 0.67\ M$$

(c) We first calculate that

$$36.8 \text{ g of } C_2H_6O = 0.800 \text{ mole}$$

$$1{,}6\overline{0}0 \text{ mL of solution} = 1.600 \text{ L}$$

Then,

$$M = \frac{\text{moles}}{\text{liter}} = \frac{n}{V} \qquad ?\ M = \frac{0.800 \text{ mole}}{1.600 \text{ L}} = 0.500\ M$$

Practice Exercise 15.7 Calculate the molarity of each of the following solutions:
(a) 9.80 g of H_2SO_4 in enough water to make 0.500 L of solution
(b) 4.60 g of KOH in enough water to make 5.00 L of solution
(c) 73.6 g of ethyl alcohol (C_2H_6O) in enough water to make 1.500 L of solution

Example 15.8

Suppose that you have a supply of 3.0 *M* NaOH solution in your laboratory. You take $2\overline{0}0$ mL of this solution and evaporate it to dryness. How many grams of *solid* NaOH remain?

Solution First solve the molarity formula,

$$M = \frac{n}{V}$$

for n, the number of moles,

$$n = MV$$

Then substitute the values.

$$n = \frac{3.0 \text{ mole}}{\cancel{\text{liter}}} \times 0.200 \cancel{\text{liter}}$$

$$= 0.60 \text{ mole of NaOH}$$

Convert the moles NaOH to grams.

$$0.60 \cancel{\text{mole}} \text{ NaOH} \times \frac{40.0 \text{ g NaOH}}{1 \text{ mole NaOH}} = 24 \text{ g of NaOH}$$

Practice Exercise 15.8 Suppose that you have a supply of 2.00×10^{-3} M $Ca(OH)_2$ solution in your laboratory. You evaporate 250.0 mL of this solution to dryness. How many grams of *solid* $Ca(OH)_2$ do you obtain?

Example 15.9

Suppose that you have a supply of 0.500 M $Ca(NO_3)_2$ solution in your laboratory. How many milliliters of this solution must be evaporated to obtain 32.8 g of solid $Ca(NO_3)_2$?

Solution Change the 32.8 g of $Ca(NO_3)_2$ to moles, and then solve for the number of liters of 0.50 M $Ca(NO_3)_2$ solution needed to obtain this number of moles.

$$32.8 \cancel{\text{g of Ca(NO}_3)_2} \times \frac{1 \text{ mole of Ca(NO}_3)_2}{164.1 \cancel{\text{g of Ca(NO}_3)_2}} = 0.200 \text{ mole of Ca(NO}_3)_2$$

Solve the molarity formula

$$M = \frac{n}{V}$$

for V, the number of liters.

$$V = \frac{n}{M}$$

Then substitute the values.

$$V = \frac{0.200 \cancel{\text{mole}}}{0.500 \text{ mole/L}}$$

$$= 0.400 \text{ L} \quad \text{or} \quad 4\overline{00} \text{ mL}$$

You would need to evaporate 0.400 L ($4\overline{0}0$ mL) of a 0.500 *M* $Ca(NO_3)_2$ solution to obtain 32.8 g of solid $Ca(NO_3)_2$.

Practice Exercise 15.9 Suppose that you have a supply of 0.200 *M* $Mg(NO_3)_2$ solution in your laboratory. How many milliliters of this solution must be evaporated to obtain 1.48 g of solid $Mg(NO_3)_2$?

15.5 Normality

Chemists have developed another important concentration concept: *normality*. The definition of normality is similar to that of molarity, but normality is easier to use in certain chemical calculations. It is most often applied to solutions of acids and bases. The **normality** of a solution *is the number of equivalents of solute per liter of solution*. Normality (abbreviated *N*) is equal to

$$\frac{\text{number of equivalents of solute}}{\text{number of liters of solution}} \quad \text{so that} \quad N = \frac{\text{equivalents}}{\text{liter}}$$

$$\text{or } N = \frac{\text{Eq}}{\text{L}}$$

Learning Goal 6
Equivalents of acids and bases

Of course, the question "What is an equivalent?" immediately comes to mind. Because we'll use normality for calculations involving acids and bases, we will define an equivalent in terms of an acid or a base. An **equivalent**

Table 15.3
The Mass of 1 Mole and of 1 Equivalent for Some Common Acids and Bases

Substance	Mass of 1 mole	Mass of 1 equivalent	Number of replaceable H^+ or OH^- ions
HCl (hydrochloric acid)	$\frac{36.5\text{ g}}{1\text{ mole}}$	$\frac{36.5\text{ g}}{1\text{ equivalent}}$	1
$HC_2H_3O_2$ (acetic acid)	$\frac{60.0\text{ g}}{1\text{ mole}}$	$\frac{60.0\text{ g}}{1\text{ equivalent}}$	1
H_2SO_4 (sulfuric acid)	$\frac{98.1\text{ g}}{1\text{ mole}}$	$\frac{49.1\text{ g}}{1\text{ equivalent}}$	2
H_3PO_4 (phosphoric acid)	$\frac{98.0\text{ g}}{1\text{ mole}}$	$\frac{32.7\text{ g}}{1\text{ equivalent}}$	3
NaOH (sodium hydroxide)	$\frac{40.0\text{ g}}{1\text{ mole}}$	$\frac{40.0\text{ g}}{1\text{ equivalent}}$	1
$Ca(OH)_2$ (calcium hydroxide)	$\frac{74.1\text{ g}}{1\text{ mole}}$	$\frac{37.1\text{ g}}{1\text{ equivalent}}$	2

of an acid is *the mass of the acid that contains Avogadro's number of hydrogen ions.* An **equivalent of a base** is *the mass of the base that contains Avogadro's number of hydroxide ions.*

To find the number of grams in 1 equivalent of an acid or a base, first obtain the molecular mass of the substance. Then divide the molecular mass in grams by the number of replaceable hydrogen ions (for an acid) or the number of replaceable hydroxide ions (for a base). This is called the equivalent mass. For example, sulfuric acid (H_2SO_4) has a molecular mass of 98.1. Because H_2SO_4 contains two replaceable hydrogen ions, the equivalent mass or the grams per equivalent is as follows:

Learning Goal 7
Calculations involving normality

$$\text{Equivalent mass of } H_2SO_4 = \frac{\text{molecular mass}}{\text{number of replaceable hydrogen ions}}$$

$$= \frac{98.1 \text{ g}}{2 \text{ Eq}} = \frac{49.1 \text{ g}}{\text{Eq}}$$

This means that a 1 N solution of H_2SO_4 contains 49.1 g of H_2SO_4 per liter of solution. Table 15.3 lists the masses of 1 mole and of 1 equivalent for some common acids and bases.

Example 15.10

What is the normality of the solution formed by dissolving 196 g of H_2SO_4 in enough water to make 800 mL of solution? What is the molarity of the solution?

Solution First convert the grams of H_2SO_4 to equivalents, and then use the normality formula. H_2SO_4 has

$$\frac{49.1 \text{ g}}{1 \text{ equivalent}} \quad \text{or} \quad \frac{1 \text{ equivalent}}{49.1 \text{ g}}$$

Therefore

$$? \text{ equivalents of } H_2SO_4 = (196 \text{ g})\left(\frac{1 \text{ equivalent}}{49.1 \text{ g}}\right)$$

$$= 3.99 \text{ equivalents of } H_2SO_4$$

$$N = \frac{\text{Eq}}{\text{L}} = \frac{\text{equivalents}}{\text{liter}} \qquad ?\,N = \frac{3.99 \text{ equivalents}}{0.800 \text{ L}} = 4.99\,N$$

To find the molarity of this solution, we first convert grams of H_2SO_4 to moles, and then find the number of moles per liter. (The molecular mass of H_2SO_4 is 98.1.)

$$? \text{ moles of } H_2SO_4 = (196 \text{ g})\left(\frac{1 \text{ mole}}{98.1 \text{ g}}\right) = 2.00 \text{ moles of } H_2SO_4$$

$$M = \frac{\text{moles}}{\text{liter}} \qquad ?\,M = \frac{2.00\text{ moles}}{0.800\text{ L}} = 2.50\,M \text{ solution of } H_2SO_4$$

Practice Exercise 15.10 What is the normality of the solution formed by dissolving 60.0 g of $HC_2H_3O_2$ in enough water to make 500.0 mL of solution? What is the molarity of the solution?

Example 15.11

What are the normality and the molarity of the solution formed by dissolving 19.6 g of H_3PO_4 in enough water to make $3\overline{00}$ mL of solution?

Solution Change the 19.6 g of H_3PO_4 to equivalents, and then use the normality formula. H_3PO_4 has

$$\text{Equivalent mass of } H_3PO_4 = \frac{\text{molecular mass}}{\text{number of replaceable } H^+ \text{ ions}} = \frac{98.0\text{ g}}{3\text{ Eq}} = \frac{32.7\text{ g}}{\text{Eq}}$$

$$\frac{32.7\text{ g}}{1\text{ equivalent}} \quad \text{or} \quad \frac{1\text{ equivalent}}{32.7\text{ g}}$$

Therefore

$$?\text{ equivalents of } H_3PO_4 = (19.6\text{ g})\left(\frac{1\text{ equivalent}}{32.7\text{ g}}\right) = 0.599\text{ equivalent of } H_3PO_4$$

$$N = \frac{\text{equivalents}}{\text{liter}} \qquad ?\,N = \frac{0.599\text{ equivalent}}{0.300\text{ L}} = 2.00\,N$$

To find the molarity of the solution, change grams of H_3PO_4 to moles, and then find the number of moles per liter.

$$?\text{ moles of } H_3PO_4 = (19.6\text{ g})\left(\frac{1\text{ mole}}{98.0\text{ g}}\right) = 0.200\text{ mole of } H_3PO_4$$

$$M = \frac{\text{moles}}{\text{liter}} = \frac{n}{V} \qquad ?\,M = \frac{0.200\text{ mole}}{0.300\text{ L}} = 0.667\,M \text{ solution of } H_3PO_4$$

Practice Exercise 15.11 What are the normality and the molarity of the solution formed by dissolving 7.23 g of HCl in enough water to make 450.0 mL of solution?

Example 15.12

How many grams of $Ca(OH)_2$ are there in $5\overline{0}$ mL of a 3.0 N solution?

Solution Solve the normality formula,

$$N = \frac{\text{Eq}}{\text{L}}$$

for Eq, the number of equivalents,

$$\text{Eq} = N\,\text{L}$$

Then substitute the values.

$$\text{Eq} = 3.0\ \text{Eq}/\cancel{\text{L}} \times 0.050\cancel{\text{L}}$$
$$= 0.15\ \text{Eq of Ca(OH)}_2$$

Next, find the equivalent mass of $Ca(OH)_2$.

$$\text{Equivalent mass of Ca(OH)}_2 = \frac{\text{molecular mass}}{\text{number of replaceable OH}^{1-}\text{ ions}}$$
$$= \frac{74.1\ \text{g}}{2\ \text{Eq}}$$
$$= 37.1\ \frac{\text{g}}{\text{Eq}}$$

Convert the 0.15 Eq of $Ca(OH)_2$ to grams.

$$0.15\ \text{Eq}\ \cancel{\text{Ca(OH)}_2} \times \frac{37.1\ \text{g Ca(OH)}_2}{1\ \text{Eq}\ \cancel{\text{Ca(OH)}_2}} = 5.6\ \text{g Ca(OH)}_2$$

Practice Exercise 15.12 How many grams of NaOH are there in 75.0 mL of a 4.00 N NaOH solution?

15.6 Dilution of Solutions

Learning Goal 8
Calculations for dilutions of solutions

Solutions of acids and bases are usually purchased from chemical supply houses. For the sake of economy, the suppliers furnish most of these solutions in highly concentrated form. (This is like buying concentrated soups and adding water to them.) Chemists have to know how to dilute concentrated solutions to the strengths they want. If the concentration is expressed as a normality, the formula for doing this is

$$N_c V_c = N_d V_d$$

where N_c is the normality of the concentrated solution, V_c is the volume of the concentrated solution, N_d is the normality of the dilute solution, and V_d is the volume of the dilute solution. If the concentration is expressed in terms of molarity, the formula becomes

$$M_c V_c = M_d V_d$$

The volume V can be expressed in any convenient unit. But the units used, such as liters or milliliters, must be the same on both sides of the equation. This formula can be applied when solving practical laboratory problems like those in the next two examples.

Example 15.13

How would you prepare 2 L of 3 *M* HCl from a 12 *M* HCl solution?

Solution You need to find the amount V_c of concentrated HCl that, when diluted with water, will give 2 L of 3 *M* HCl.

$$M_c V_c = M_d V_d \qquad \text{so} \qquad V_c = \frac{M_d V_d}{M_c}$$

$$= \frac{(3\ \cancel{M})(2\ \text{L})}{12\ \cancel{M}} = 0.5\ \text{L}$$

Therefore, to make the final solution, measure 0.5 L of concentrated HCl into a flask and dilute with water until you have 2 L. (But see the cautionary note in Example 15.14.)

Practice Exercise 15.13 How would you prepare 3.00 L of a 4.00 *M* HCl solution from a stock solution of 6.00 *M* HCl?

Example 15.14

How would you prepare 300 mL of a 0.6 *N* H_2SO_4 solution from a 36 *N* solution?

Solution You need the amount V_c of concentrated H_2SO_4 that, when diluted with water, will give 300 mL of 0.6 *N* H_2SO_4.

$$N_c V_c = N_d V_d \qquad \text{so} \qquad V_c = \frac{N_d V_d}{N_c}$$

$$= \frac{(0.6\ \cancel{N})(300\ \text{mL})}{36\ \cancel{N}} = 5\ \text{mL}$$

Therefore, to make the final solution, measure 5 mL of concentrated H_2SO_4 into a flask and dilute with water to 300 mL. This will be a 0.6 *N*

H_2SO_4 solution. (A cautionary note: *Always add acid to water, never the reverse. In the case of H_2SO_4, adding water to the concentrated acid will make the acid react violently. In the problem we just solved, it would be wise to put most of the water in the flask before adding the H_2SO_4.*)

Practice Exercise 15.14 How would you prepare 500.0 mL of a 0.400 *N* H_3PO_4 solution from a stock solution of 12.0 *N* H_3PO_4?

15.7 Ionization in Solutions

Learning Goal 9
Electrolytic and nonelectrolytic solutions

Learning Goal 10
Ions and ionization

For centuries, scientists have tried to find out what happens when one substance dissolves in another. The chemists of the 1800s were puzzled by the fact that some solutions would conduct electric current and others would not. The English physicist Michael Faraday tried to explain this phenomenon. He classified solutions generally in the following way: **Electrolytic solutions** are *solutions that conduct electric current*. A solute that produces ions in solution is an electrolyte. **Nonelectrolytic solutions** are *solutions that do not conduct electric current*. A solute that does not produce ions in solution is a nonelectrolyte.

Faraday said that electrolytic solutions—such as sodium chloride in water—contain charged particles. He called these particles *ions* (from a Greek word that may be translated as "wanderer"). He said that ions wander through the solution carrying electric current (Figure 15.6). But questions arose about the nature of these ions. For example, why are there ions in a salt-and-water solution but no ions in a sugar-and-water solution? Another scientist, the Swedish chemist Svante Arrhenius, came up with the answer.

In 1884, in his Ph.D. thesis, Arrhenius advanced his ideas about ions. He suggested that Faraday's ions were really simple atoms (or groups of atoms) carrying a positive or a negative charge. He said that when some substances dissolve in solution, they break up into ions:

$$NaCl(s) \xrightarrow{H_2O} Na^{1+}(aq) + Cl^{1-}(aq)$$

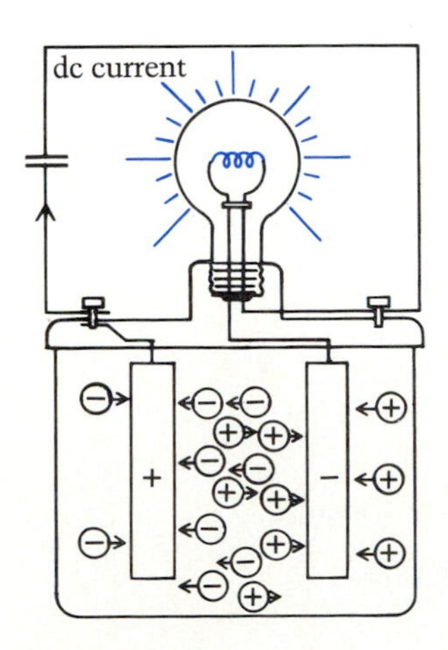

Figure 15.6
An electrolytic solution conducts electricity.

This process is what we call **ionization** (Figure 15.7). Actually, the solid sodium chloride *dissociates* when placed in water, because it is composed of ions. When placed in water, the ions separate from each other and are free to move about in the solution. Solutions that contain ions are electrolytic because they contain charged particles that can carry electric current. Substances in solution that conduct electricity are called *electrolytes*. A solution of sodium chloride actually contains a 1:1 mixture of sodium *ions* and chloride *ions*. A bottle label reading "1 *M* NaCl" refers to how the solution was made and not what is really in the bottle.

On the other hand, some substances dissolve in solution and *do not ionize*. They simply break up into their neutral molecules and become surrounded by the molecules of solvent:

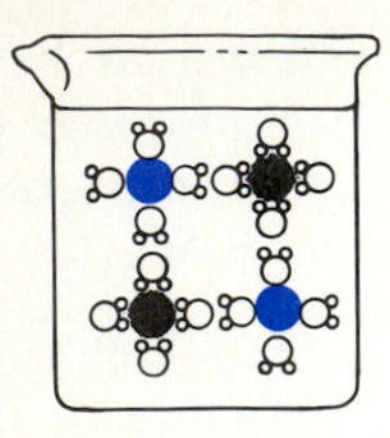

Sodium ion, Na^{1+}

Chloride ion, Cl^{1-}

Water molecule

Figure 15.7
Dissociation of NaCl in water

$$\underset{\text{(Glucose)}}{C_6H_{12}O_6(s)} \xrightarrow{H_2O} C_6H_{12}O_6(aq)$$

These solutions are nonelectrolytic because they contain no charged particles to carry electric current (Figure 15.8). Substances that do not ionize in solution are called *nonelectrolytes*.

Electrolytes and nonelectrolytes acting as solutes have some very interesting effects on the boiling points and freezing points of solutions. However, before we can examine these effects, we must discuss another way of expressing the concentration of solutes in solutions.

15.8 Molality

Learning Goal 11
Definition of molality

We can express the concentration of a solute in a solution in terms of the *number of moles of solute per kilogram of solvent*. This concentration unit is known as **molality** (m).

$$\text{Molality } (m) = \frac{\text{number of moles of solute}}{\text{kilogram of solvent}} = \frac{n}{\text{kg}}$$

We can determine the molality of a solution if we know either the number of moles or the mass of a solute dissolved in a given mass of solvent.

Example 15.15

Calculate the molality of each of the following solutions:
(a) 36 g of glucose ($C_6H_{12}O_6$) dissolved in $5\overline{0}0$ g of water
(b) 1.03 g of sodium bromide (NaBr) dissolved in $25\overline{0}$ g of water
(c) 3.40 g of ammonia (NH_3) dissolved in $30\overline{0}$ g of water

Solution

(a) First change the 36 g of $C_6H_{12}O_6$ (molecular mass = 180.0) to moles.

$$?\ \text{moles} = (36\ \not{g})\left(\frac{1\ \text{mole}}{180.0\ \not{g}}\right) = 0.20\ \text{mole}$$

Next change $5\overline{0}0$ g of water to kilograms.

$$?\ \text{kg} = (5\overline{0}0\ \not{g})\left(\frac{1\ \text{kg}}{1{,}000\ \not{g}}\right) = 0.500\ \text{kg}$$

Finally, substitute these numbers into the molality formula.

$$m = \frac{\text{number of moles of solute}}{\text{kilogram of solvent}}$$

$$?\ m = \frac{0.20\ \text{mole of glucose}}{0.500\ \text{kg of water}} = 0.40\ m$$

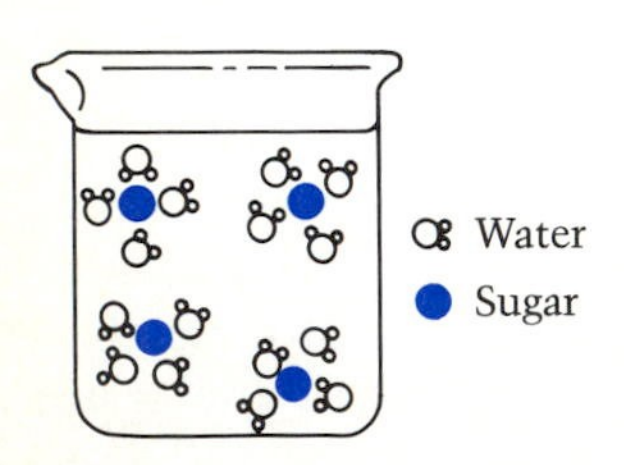

Figure 15.8
Sugar dissolves in water without ionizing.

Learning Goal 12
Calculation of molality

(b) First change 1.03 g of NaBr (molecular mass = 102.9) to moles.

$$? \text{ moles} = (1.03 \text{ g})\left(\frac{1 \text{ mole}}{102.9 \text{ g}}\right) = 0.0100 \text{ mole}$$

Next change $25\bar{0}$ g of water to kilograms.

$$? \text{ kg} = (25\bar{0} \text{ g})\left(\frac{1 \text{ kg}}{1{,}000 \text{ g}}\right) = 0.250 \text{ kg}$$

Finally, substitute these numbers into the molality formula.

$$m = \frac{\text{number of moles of solute}}{\text{kilogram of solvent}}$$

$$? \, m = \frac{0.0100 \text{ mole of NaBr}}{0.250 \text{ kg of } H_2O} = 0.0400 \, m$$

(c) First change the 3.40 g of NH_3 (molecular mass = 17.0) to moles.

$$? \text{ moles} = (3.40 \text{ g})\left(\frac{1 \text{ mole}}{17.0 \text{ g}}\right) = 0.200 \text{ mole}$$

Next change the $3\overline{00}$ g of water to kilograms.

$$? \text{ kg} = (3\overline{00} \text{ g})\left(\frac{1 \text{ kg}}{1{,}000 \text{ g}}\right) = 0.300 \text{ kg}$$

Finally, substitute these numbers into the molality formula.

$$m = \frac{\text{number of moles of solute}}{\text{kilogram of solvent}}$$

$$? \, m = \frac{0.200 \text{ mole of } NH_3}{0.300 \text{ kg of } H_2O} = 0.667 \, m$$

Practice Exercise 15.15 Calculate the molality of each of the following solutions:
(a) 72 g of glucose ($C_6H_{12}O_6$) dissolved in $4\overline{00}$ g of water
(b) 1.17 g of NaCl dissolved in $5\overline{00}$ g of water
(c) 0.680 g of NH_3 dissolved in $25\bar{0}$ g of water

Chemists find that molality is a useful concept when they must deal with the effect of a solute on the boiling point and freezing point of a solution. The next section explains what this means.

15.9 Colligative (Collective) Properties of Solutions

Certain properties of solutions depend more on the number of solute particles present than on the type of solute particles. In other words, these properties

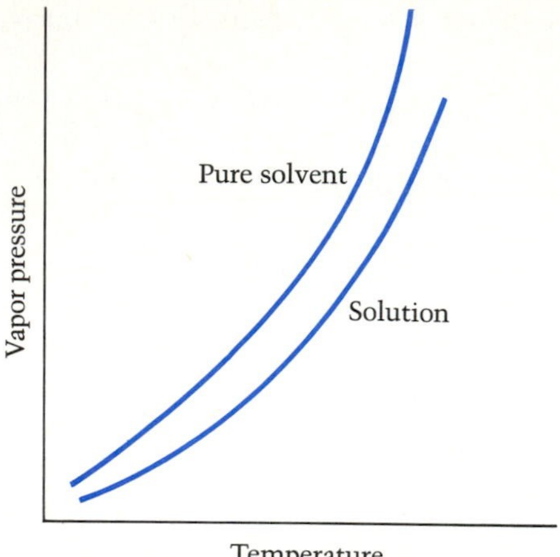

Figure 15.9
At a given temperature, the vapor pressure of the pure solvent is always greater than that of the solution of a nonvolatile solute.

Learning Goal 13
Definition of colligative properties and effect of solute on boiling and freezing points

are the same regardless of what substance is the solute. *These properties, called* **colligative properties,** *depend on the number of solute particles present in a given quantity of solvent*. For example, a mole of sugar molecules will have the same effect on the physical properties of a solution as will a mole of soluble starch or glycerol molecules.

One important colligative property of solutions is related to the vapor pressure of the solution. When a solute is dissolved in a solvent, the vapor pressure of the solution *decreases*. In other words, if we compared the vapor pressure of pure water at 50°C with the vapor pressure of a sugar–water solution at 50°C, we would find that the sugar–water solution had a lower vapor pressure than the pure water (Figure 15.9).

Two other colligative properties of solutions result from this vapor-pressure-reduction property.

1. The boiling point of a solution is *higher* than the boiling point of the pure solvent alone. This is called the **boiling-point-elevation** property.

2. The freezing point of a solution is *lower* than the freezing point of the pure solvent alone. This is called the **freezing-point-depression** property.

For example, when salt is added to water, the boiling point of the water is raised, and food cooks in the water at a higher temperature. And, by adding salt to the ice that is used to freeze ice cream, we lower its freezing point. We both raise the boiling point and lower the freezing point of automobile radiator water when we add antifreeze/coolant.

15.10 Boiling-Point Elevation, Freezing-Point Depression, and Molality

Learning Goal 14
Definition of K_f and K_b

The amount by which the boiling point of a solution is elevated or the freezing point lowered depends on (1) the solvent and (2) the concentration of the solute particles. For water, the **boiling-point-elevation constant** K_b is 0.52°C per mole of solute particles per kilogram of water. That is, a 1-molal (1 *m*) solution of sugar in water raises the boiling point of the water from 100°C to 100.52°C. The **freezing-point-depression constant** K_f for water is 1.86°C per mole of solute particles per kilogram of water. So a 1 *m* solution of sugar in water would have a freezing point of −1.86°C. Other solvents have different values for the constants K_b and K_f.

The values of K_f and K_b for water are the same for all non-ionic solutes. Thus a 1 *m* solution of glycerin in water also raises the boiling point of water to 100.52°C and lowers the freezing point to −1.86°C. (However, if we dissolve 1 mole of NaCl in 1,000 g of water, the effect on the boiling point and freezing point would be about double, because the NaCl ionizes into Na^{1+} ions and Cl^{1-} ions. In other words, 1 mole of NaCl yields 2 moles of solute particles. However, here we'll discuss only non-ionizing solutes.) For a solute that does not ionize in water, we can make this statement: *A 1-molal solution of molecular solute lowers the freezing point of 1 kg of water by 1.86°C and raises the boiling point by 0.52°C.*

Each additional mole of solute per kilogram of solvent lowers the freezing point of the solution by an additional 1.86°C and raises the boiling point of the solution by an additional 0.52°C. We can state this mathematically as

$$\Delta t = mK_f \quad \text{(for freezing-point depression)}$$

$$\Delta t = mK_b \quad \text{(for boiling-point elevation)}$$

where Δt is the boiling-point elevation or freezing-point depression in Celsius degrees.

Learning Goal 15
Calculations involving boiling and freezing points of solutions

Example 15.16

Calculate the boiling point of each of the following situations:
(a) 36 g of glucose dissolved in $5\bar{0}0$ g of water
(b) 9.2 g of glycerol ($C_3H_8O_3$) in 250 g of water

Solution

(a) First we need to determine the molality of a solution of 36 g of glucose in $5\bar{0}0$ g of water. If you look back at Example 15.15(a), you'll see that we calculated this molality to be 0.40 *m*.

Now use the formula for boiling-point elevation to calculate the change in temperature Δt.

$$\Delta t = mK_b = (0.40\ \cancel{m})(0.52°\text{C}/\cancel{m}) = 0.21°\text{C}$$

If pure water boils at 100°C, then this 0.40 *m* solution boils at 100.21°C.
(b) First determine the molality of a solution of 9.2 g of $C_3H_8O_3$ (molecular mass = 92.0) in $25\bar{0}$ g of water.

$$?\ \text{moles } C_3H_8O_3 = (9.2\ \cancel{\text{g}})\left(\frac{1\ \text{mole}}{92.0\ \cancel{\text{g}}}\right) = 0.10\ \text{mole}$$

$$? \text{ kg water} = (25\bar{0} \text{ g})\left(\frac{1 \text{ kg}}{1{,}000 \text{ g}}\right) = 0.250 \text{ kg}$$

$$? \, m = \frac{0.10 \text{ mole}}{0.250 \text{ kg}} = 0.40 \, m$$

Now use the formula for boiling-point elevation to calculate the change in temperature Δt.

$$\Delta t = mK_b = (0.40 \, m)(0.52°\text{C}/m) = 0.21°C$$

If pure water boils at 100°C, then this 0.40 m solution boils at 100.21°C.

Practice Exercise 15.16 Calculate the boiling point of each of the following solutions:
(a) 72.0 g of glucose ($C_6H_{12}O_6$) dissolved in $25\bar{0}$ g of water
(b) 1.84 g of glycerol ($C_3H_8O_3$) dissolved in $5\overline{00}$ g of water

Example 15.17

Calculate the freezing point of each of the solutions in Example 15.16.

Solution

(a) We already know that the molality of this solution is 0.40 m. We also know that K_f for water is 1.86°C. Therefore

$$\Delta t = mK_f = (0.40 \, m)(1.86°\text{C}/m) = 0.74°\text{C}$$

Because pure water freezes at 0°C, this 0.40 m solution freezes at −0.74°C.

(b) We already know that the molality of this solution is 0.40 m. Therefore

$$\Delta t = mK_f = (0.40 \, m)(1.86°\text{C}/m) = 0.74°\text{C}$$

This 0.40 m solution freezes at −0.74°C.

Practice Exercise 15.17 Calculate the freezing point of each of the solutions in Practice Exercise 15.16.

15.11 The Processes of Diffusion and Osmosis

Learning Goal 16
Diffusion and osmosis

Diffusion and osmosis are important colligative properties in living systems and in the chemistry laboratory as well. In this section, we'll take a brief look at each of these processes.

If you place a few drops of dye in a glass of water, what happens? Very quickly the dye disperses through the water. Given enough time, the dye will distribute itself evenly throughout the water (Figure 15.10). This process is

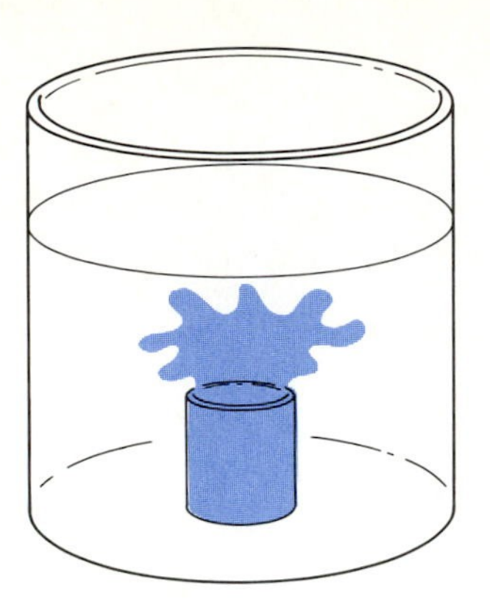
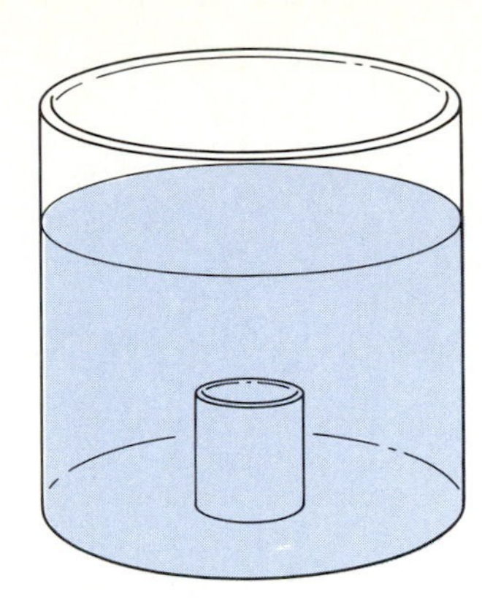

Figure 15.10
A dye will diffuse through a solution from areas of high dye concentration to areas of low dye concentration. Eventually, the dye will distribute itself evenly throughout the solution.

known as **diffusion.** It occurs because the dye moves from an area of high dye concentration to an area of low dye concentration. In other words, there is a *concentration gradient,* or gradually changing difference in concentration, between the place in the water where the dye is dropped and the rest of the water. The diffusion process stops when the dye is evenly distributed throughout the water, so that the concentration gradient no longer exists.

It is this process of diffusion that moves most of the substances in our body. For example, the breathing process is based on the diffusion of gases from areas of high concentration to areas of low concentration. And the process of diffusion is the system by which electrolytes in the body are carried to where they are needed.

A special type of diffusion process called **osmosis** involves the *passage of water through a semipermeable membrane, like cellophane or cell walls.* In this process, *water* moves from an area of low solute concentration to an area of high solute concentration. The semipermeable membrane allows only the water, not the solute, to pass through it.

For example, consider a beaker of water having two compartments separated by a semipermeable membrane (Figure 15.11). Compartment A is filled with a 5% sucrose solution. Compartment B is filled with an equal volume of a 20% sucrose solution. Soon after the solutions are placed in the beaker, we notice a shift in the water levels of the compartments. Water moves from the compartment containing the 5% sucrose solution to the compartment containing the 20% sucrose solution, in an attempt to equalize the sucrose concentrations in both compartments (Figure 15.12). This movement of water is the process of osmosis.

You may be thinking, "Why doesn't the sucrose in the 20% solution move through the semipermeable membrane to equalize the concentrations of the two solutions?" This doesn't happen because an osmotic membrane does not

Figure 15.11
A beaker with two compartments separated by a semipermeable membrane

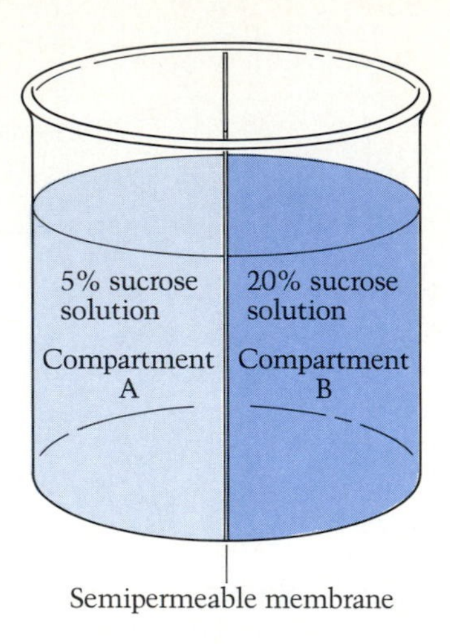

allow solute particles to pass through it but does allow solvent particles to pass freely.

Osmotic Pressure

Consider the two sucrose solutions that we've just discussed. After the process of osmosis has proceeded for a while, it finally stops. Osmosis ends when the rate of water molecules passing from compartment A to compartment B equals the rate of water molecules passing from compartment B to compartment A (Figure 15.13). In other words, a dynamic equilibrium is in progress. But why does this occur?

The answer lies in looking at the height of the water in each compartment. The height of the water in compartment B is greater than that in A (Figure

Figure 15.12
This is what happens to our two solutions after osmosis has proceeded for a while. Water has left compartment A, increasing the concentration of that sucrose solution. The water has gone into compartment B, diluting that sucrose solution.

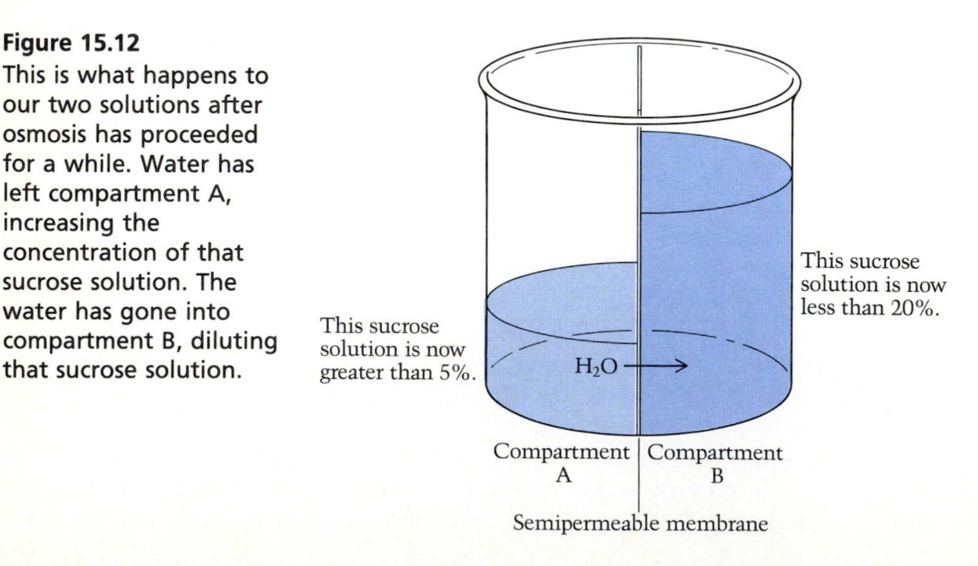

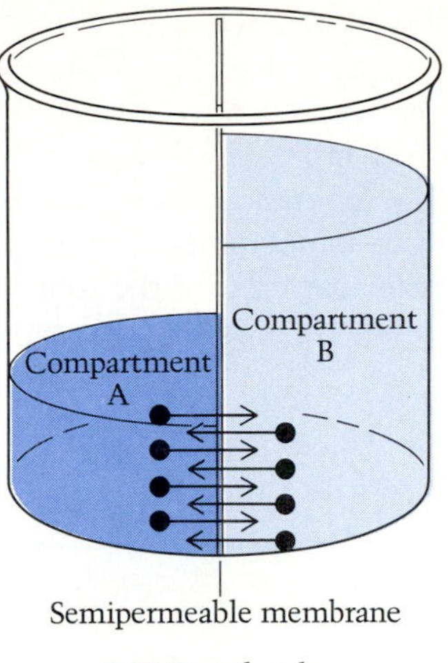

Figure 15.13
The process of osmosis stops when the number of water molecules moving from compartment A to compartment B equals the number of water molecules moving from compartment B to compartment A.

15.14). At first, water molecules moved from A to B in an attempt to equalize the concentrations of the two sucrose solutions. But eventually the height of the water in compartment B caused a pressure to be exerted on the water molecules in that compartment, pushing them back into compartment A. When these two opposing forces became equal, the process of osmosis stopped. The pressure exerted by the water in compartment B at this point is called the *osmotic pressure* and is related to the concentration of the solute in the solution. **Osmotic pressure** can be defined as *the amount of pressure that must be applied to prevent the flow of water through a membrane*. This concept is extremely important in the transport of body fluids.

Like other colligative properties, osmotic pressure can be used to determine the molecular mass of a solute in a solution. A Dutch chemist named

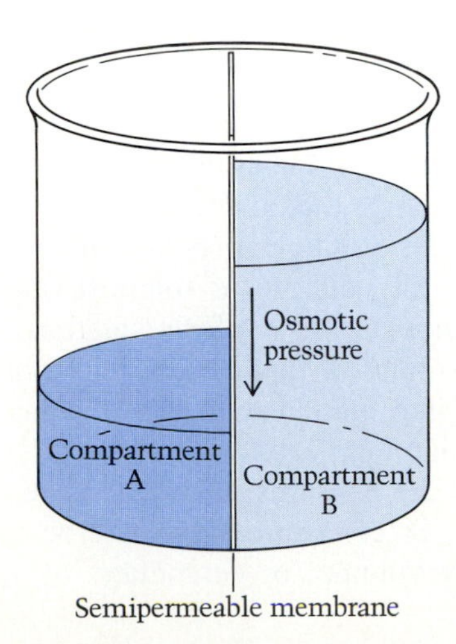

Figure 15.14
Osmotic pressure is due to the height of the water column in compartment B.

van't Hoff developed a mathematical expression that relates the osmotic pressure to the molar concentration of the solution. A technique based on this equation, called **osmometry,** has proved to be far more sensitive for molecular mass determinations than freezing-point depression or boiling-point elevation. In this method, the difference in the height of the water columns in the two compartments is used to obtain the osmotic pressure of the system. From this information the molecular mass of the solute is determined. Although we won't pursue this calculation here, chemists have found this technique quite useful in the laboratory.

Example 15.18

Two solutions of equal volume are placed in a beaker similar to the one in Figure 15.11. Compartment A of the beaker contains a 10% NaCl solution. Compartment B of the beaker contains a 5% NaCl solution. Which way does the water move, from A to B or from B to A?

Solution The water moves from compartment B to compartment A. The solvent moves from the side of low solute concentration to the side of high solute concentration.

Practice Exercise 15.18 Cells in the body have both intracellular and extracellular fluid. The intracellular fluid has an osmotic pressure equivalent to 5% glucose. If an individual is given an intravenous solution that has an osmotic pressure equivalent to 10% glucose, would water move in or out of the cell? (*Hint:* The intravenous solution replaces the extracellular fluid.)

Summary

A solution is a homogeneous mixture composed of at least two substances. The solute is the substance that is dissolved, and the solvent is the substance in which it is dissolved. Solutions may be saturated, unsaturated, or supersaturated depending on the quantity of solute dissolved in a given quantity of solvent. Pairs of liquids that do not form solutions when mixed are said to be immiscible, whereas pairs of liquids that form solutions are said to be miscible. In general, polar substances dissolve polar substances, and nonpolar substances dissolve nonpolar substances.

Because solutions are mixtures, they may have various ratios of solvent to solute. These ratios, or concentrations, may be measured as percentage by mass, percentage by volume, or percentage by mass–volume. In addition, the concentration of a solution may be measured as its molarity: the number of moles of solute per liter of solution. Concentration may also be measured as normality: the number of equivalents of solute per liter of solution.

An equivalent of any acid is the mass of that acid that contains Avogadro's number of hydrogen ions. An equivalent of a base is the mass of that base that contains Avogadro's number of hydroxide ions. The number of grams in one equivalent of an acid or a base is computed by dividing the molecular mass of the substance by the number of replaceable hydrogen ions (for an acid) or replaceable hydroxide ions (for a base) in one molecule of the substance.

Electrolytic solutions are solutions that conduct electricity, and nonelectrolytic solutions are solutions that do not conduct electricity. Electrolytic solutions are capable of conducting electricity because they contain ions, which are charged atoms or groups of atoms. Substances that ionize, or break up into ions, when they dissolve are called electrolytes.

The molality of a solution is the number of moles of solute per kilogram of solvent. Molality is useful in dealing with colligative properties of solutions, which are properties that depend mainly on the number of solute particles present (in a given quantity of solvent) rather than on the type of solute particles. Vapor-pressure reduction, boiling-point elevation, and freezing-point depression are examples of colligative properties. In general, a solution exhibits a lower vapor pressure, a higher boiling point, and a lower freezing point than the pure solvent alone.

Diffusion is the movement of a solute in a solution from an area of high solute concentration to an area of low solute concentration. Osmosis is the passage of water through a semipermeable membrane from the less concentrated to the more concentrated solution.

Key Terms

boiling-point elevation **(15.9)**
boiling-point-elevation constant **(15.10)**
colligative properties **(15.9)**
concentration **(15.3)**
diffusion **(15.11)**
electrolytic solution **(15.7)**
equivalent (of an acid or a base) **(15.5)**
freezing-point depression **(15.9)**
freezing-point-depression constant **(15.10)**
immiscible **(15.1)**
ionization **(15.7)**
molality **(15.8)**
molarity **(15.4)**
nonelectrolytic solution **(15.7)**
normality **(15.5)**
osmometry **(15.11)**
osmosis **(15.11)**
osmotic pressure **(15.11)**
saturated solution **(15.2)**
saturation point **(15.3)**
solubility **(15.2)**
solute **(15.1)**
solvent **(15.1)**
supersaturated solution **(15.2)**
unsaturated solution **(15.2)**

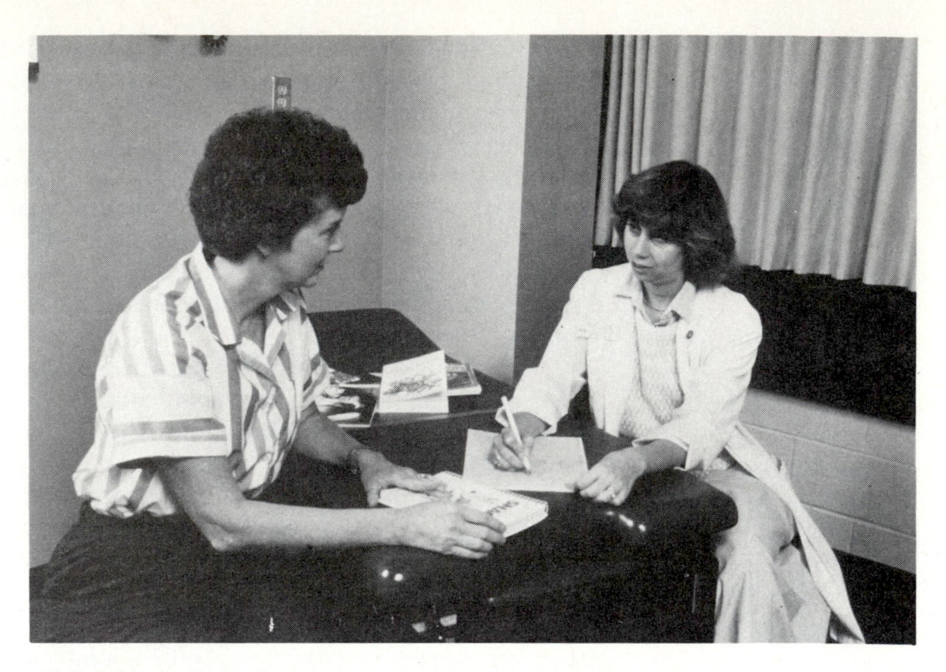

The dietitian recommends a nutritious eating plan for this patient.
(Paul Fry/Peter Arnold)

CAREER SKETCH

DIETITIAN

As a dietitian, you will plan balanced and appetizing meals to ensure that healthy people stay healthy and sick people recover. You may choose to specialize in one of three areas: clinical (the largest area), administrative, or research.

As a *clinical dietitian,* sometimes called a *therapeutic dietitian,* you will plan diets and supervise the service of meals to patients in a hospital, nursing home, or clinic. You will also confer with physicians about patients' nutrition, instruct patients and their families on diet requirements, and suggest ways for patients to follow their diets after they return home.

As an *administrative dietitian,* you will supervise the planning and preparation of meals for large numbers of people in an institution such as a hospital, university, or school. You will also select, train, and direct food-service workers; purchase food, equipment, and supplies; enforce sanitary and safety regulations; and prepare records and reports. If you become the director of a dietetic department, you will also coordinate dietetic services with the activities of other departments. You will be responsible for your department's budget, which in a large organization may amount to millions of dollars annually.

As a *research dietitian,* you will conduct research aimed at improving the nutrition of both healthy and sick people. You may specialize in nutrition science, education, food management, or food-service systems and equipment. You may study how the body uses food, or you may investigate the nutritional needs of very young or very old people, of individuals with chronic diseases, or even of space travelers. Most research dietitians work in a medical center or at a college or university.

Self-Test Exercises

Learning Goal 1: Definitions of Solution, Solute, and Solvent

▲ **1.** Define the following and give an example of each: *(a)* solute *(b)* solvent *(c)* solution

2. A solution is a heterogeneous mixture. True or false?

▲ **3.** Which is the solute and which the solvent in each of the following solutions?
(a) 5%-by-mass sodium hydroxide in water
(b) 3%-by-volume ether in carbon tetrachloride
(c) 1%-by-mass–volume iodine in alcohol

4. A coin composed of two metals that were melted, mixed together, and solidified is an example of a solution. True or false?

5. Give an example of *(a)* a solid–solid solution, *(b)* a solid–liquid solution *(c)* a solid–gas solution, *(d)* a liquid–liquid solution.

▲ **6.** Which is the solute and which the solvent in each of the following solutions?
(a) 10%-by-mass sodium chloride in water
(b) 5%-by-mass benzene in carbon tetrachloride
(c) 10%-by-mass–volume methyl alcohol in ethyl alcohol

Learning Goal 2: Substances That Form Solutions

7. "Like dissolves like." True or false? What does this statement refer to in terms of molecular structure?

▲ **8.** You are served a cup of coffee in a restaurant. Assuming that no sugar has been added to the coffee, you add 2 teaspoons of sugar, stir well, and notice a layer of sugar at the bottom of the cup that won't dissolve. What type of solution have you produced?

9. The point at which no more solute dissolves in a solution is called the ______ point.

10. What are the meanings of the terms *miscible* and *immiscible*?

Learning Goal 3: Saturated, Unsaturated, and Supersaturated Solutions

11. *(a)* A solution that contains less solute than can be dissolved in a given quantity of solvent at a particular temperature is a ______ solution.
(b) A solution that contains more solute than is needed for saturation for a given quantity of solvent at a particular temperature is a ______ solution.

12. Determine whether each of the following solutions is unsaturated, saturated, or supersaturated. (See Table 15.2.)
(a) 15.0 g of NaCl in 50.0 g of water at 50°C
(b) 45.5 g of $AgNO_3$ in 50.0 g of water at 50°C

▲ **13.** You are given a saturated solution that contains undissolved solute. The solute is constantly dissolving, yet the concentration of the solute in the solution remains the same. Explain.

14. You are given a solution by your instructor who asks you to determine whether or not it is supersaturated. Explain the procedure you would use to test for supersaturation.

Learning Goal 4: Percent Concentrations

▲ **15.** What is the concentration in percentage by mass of each of the following solutions?
(a) $4\bar{0}$ g of sodium nitrate with enough water to make $2\bar{0}0$ g of solution

(b) 3.0 g of ethyl alcohol with enough water to make $2\bar{0}0$ g of solution
(c) $3\bar{0}$ g of potassium chloride *plus* $17\bar{0}$ g of water

▲ **16.** What is the concentration in percentage by mass of each of the following solutions?
(a) 80.0 g of sodium chloride with enough water to make $25\bar{0}$ g of solution
(b) 7.50 g of methyl alcohol with enough water to make $4\bar{0}0$ g of solution
(c) 50.0 g of potassium nitrate *plus* $15\bar{0}$ g of water

17. How many grams of KOH are there in $2\bar{0}\bar{0}$ g of $1\bar{0}$%-by-mass KOH solution?
18. How many grams of NaOH are there in 500.0 g of a 15.0%-by-mass NaOH solution?

19. What is the percentage by volume of the solute in each of the following solutions?
(a) $2\bar{0}\bar{0}$ mL of ether in enough carbon tetrachloride to make 1.0 L of solution
(b) 5 mL of alcohol in enough water to make 80 mL of solution
20. What is the percentage by volume of the solute in each of the following solutions?
(a) 300.0 mL of ether in enough carbon tetrachloride to make 2.00 L of solution
(b) 25.0 mL of alcohol in enough water to make 100.0 mL of solution

***21.** Find the volume of alcohol present in $1\bar{0}\bar{0}$ g of 10.0%-by-mass alcohol–water solution. (*Hint:* The density of alcohol is 0.800 g/mL.)
***22.** Find the volume of alcohol present in 250.0 g of a 10.0%-by-mass alcohol–water solution. (*Hint:* The density of alcohol is 0.800 g/mL.)

23. How many milliliters of chloroform are there in $5\bar{0}\bar{0}$ mL of a 4.0%-by-volume chloroform–ether solution?
24. How many milliliters of carbon tetrachloride are there in 250.0 mL of a 5.00%-by-mass–volume carbon tetrachloride–benzene solution?

▲ **25.** What are the percentages by mass–volume of the solutes in the following solutions?
(a) $5\bar{0}$ g of sodium chloride in 1.5 L of solution
(b) $1\bar{0}$ g of ammonia dissolved in $2\bar{0}\bar{0}$ g of solution (Assume that the density of the solution is 1.00 g/cm^3.)

▲ **26.** What is the percentage by mass–volume of the solute in each of the following solutions?
(a) 25.0 g of sodium nitrate in 1.00 L of solution
(b) 15.0 g of sodium hydroxide dissolved in 100.0 g of solution (Assume that the density of the solution is 1.00 g/cm^3.)

27. How many grams of sugar are there in $1\bar{0}\bar{0}$ mL of a 15%-by-mass–volume sugar–water solution?
28. How many grams of lactose (milk sugar) are there in 150.0 mL of a 10%-by-mass–volume lactose–water solution?

***29.** If a person's blood-sugar level falls below 60 mg per 100 mL, insulin shock can occur. What is the percentage by mass of sugar in the blood at this level? (Assume that the density of blood is 1.2 g/mL.)

▲ **30.** A person's blood-sugar level is calculated to be 120.0 mg per $1\bar{0}\bar{0}$ mL blood. What is the percentage by mass of sugar in the blood at this level? (The density of blood is 1.20 g/mL.)

31. What are the concentrations in percentage by mass of the following solutions?
(a) 25.0 g of NaCl in enough water to make $2\bar{0}\bar{0}$ g of solution
(b) 8.00 g of sugar in enough water to make $16\bar{0}$ g of solution
32. What is the concentration in percentage by mass of each of the following solutions?
(a) 55.0 g of Na_2CO_3 in enough water to make 400.0 g of solution
(b) 14.5 g of sugar in enough water to make 290.0 g of solution

33. How many grams of KCl are there in $5\bar{0}\bar{0}$ g of a 6.0%-by-mass KCl solution?
34. How many grams of KNO_3 are there in 750.0 g of a 7.50%-by-mass KNO_3 solution?

35. What are the percentages by volume of the solutes in the following solutions?
(a) $15\bar{0}$ mL of ethyl alcohol in enough water to make 3.0 L of solution
(b) $5\bar{0}\bar{0}$ mL of methyl alcohol (dry gas) in enough gasoline to make 60.0 L of solution (This represents a can of dry gas in a nearly full tank of gasoline in a standard-size automobile.)
36. What is the percentage by volume of the solute in each of the following solutions?
(a) 175 mL of isopropyl alcohol in enough water to make 4.00 L of solution
(b) 600.0 mL of methyl alcohol (dry gas) in enough gasoline to make 60.0 L of solution

▲ **37.** Determine the volume of ether present in $5\bar{0}0$ g of a 4.0%-by-mass ether–carbon tetrachloride solution. (*Hint:* The density of ether is 0.714 g/cm^3.)

***38.** Determine the volume of ether in 750.0 g of a 10.0%-by-mass ether–carbon tetrachloride solution. (*Hint:* The density of ether is 0.714 g/cm^3.)

39. How many milliliters of ethyl alcohol are there in 2.00 L of a 10.0%-by-volume ethyl alcohol–water solution?

40. How many milliliters of isopropyl alcohol are there in 4.50 L of a 20.0%-by-volume isopropyl alcohol–water solution?

41. What are the percentages by mass–volume of the solutes in the following solutions?
(*a*) 25.0 g of sodium nitrate in $8\bar{0}0$ mL of solution
(*b*) 0.800 g of potassium chloride in 16.0 mL of solution

42. What is the percentage by mass–volume of the solute in each of the following solutions?
(*a*) 50.0 g of sodium carbonate in 500.0 mL of solution
(*b*) 0.160 g of potassium nitrate in 80.0 mL of solution

43. How many grams of sodium chloride are there in $25\bar{0}$ mL of a 5.00%-by-mass–volume solution of sodium chloride and water?

44. How many grams of sodium chloride are there in 100.0 mL of a 9.00%-by-mass–volume solution of sodium chloride and water?

Learning Goal 5: Calculations Involving Molarity

▲ **45.** What is the molarity of each of the following solutions?
(*a*) 5.4 g of HCl in enough water to make $5\bar{0}0$ mL of solution
(*b*) 117 g of sodium chloride in enough water to make 4 L of solution

46. What is the molarity of each of the following solutions?
(*a*) 1.18 g of HCl in enough water to make 2.00 L of solution
(*b*) 23.40 g of sodium chloride in enough water to make 5.00 L of solution

***47.** What is the molarity of a 10.0%-by-mass HCl solution? (Assume that the density of the solution is 1.00 g/mL.)

***48.** What is the molarity of a 5.00%-by-mass HCl solution? (Assume that the density of the solution is 1.00 g/cm^3.)

49. How many grams of KOH are obtained by evaporating to dryness $5\bar{0}$ mL of a 3.0 *M* KOH solution?

50. How many grams of NaOH are obtained by evaporating to dryness 25.0 mL of a 2.00 *M* NaOH solution?

***51.** The *Merck Manual* reports that a blood-alcohol level of $15\bar{0}$ to $2\bar{0}0$ mg per $1\bar{0}0$ mL produces intoxication and that a level of $3\bar{0}0$ to $4\bar{0}0$ mg per $1\bar{0}0$ mL produces unconsciousness. At a blood-alcohol level above $5\bar{0}0$ mg per $1\bar{0}0$ mL, a person may die. What is the molarity of the blood with respect to alcohol at the level of $5\bar{0}0$ mg per $1\bar{0}0$ mL? (*Hint:* The molecular mass of alcohol is 46.0.)

▲***52.** A person has been drinking ethyl alcohol. What is the molarity of the alcohol in this person's blood if the alcohol concentration is $4\bar{0}0$ mg per $1\bar{0}0$ mL of blood?

▲ **53.** Salt (NaCl) is a necessary ingredient of our diets, and the body has a system to maintain a delicate balance of salt. In certain illnesses, this salt balance can be lost, and a physician or nurse must administer salt intravenously, using a 0.85% solution. What is the molarity of a 0.85%-by-mass–volume sodium chloride solution?

▲ **54.** How many milliliters of water are needed to make a 2.00 *M* NaCl solution if 5.85 g of NaCl are present?

55. What is the molarity of each of the following solutions?
(*a*) 24.5 g of H_2SO_4 in enough water to make 1.50 L of solution
(*b*) 10.1 g of KNO_3 in enough water to make $5\bar{0}0$ mL of solution

56. What is the molarity of each of the following solutions?
(*a*) 49.0 g of H_2SO_4 in enough water to make 2.00 L of solution
(*b*) 2.02 g of KNO_3 in enough water to make 200.0 mL of solution

***57.** What is the molarity of a 5.00%-by-mass magnesium chloride solution? (Assume that the density of the solution is 1.00 g/cm^3.)

***58.** What is the molarity of a 7.50%-by-mass sodium nitrate solution? (Assume that the density of the solution is 1.00 g/cm^3.)

59. How many grams of Na_3PO_4 are obtained by evaporating to dryness $25\bar{0}$ mL of a 0.400 *M* Na_3PO_4 solution?

60. How many grams of $NaC_2H_3O_2$ are obtained by evaporating to dryness 400.0 mL of a 0.500 *M* $NaC_2H_3O_2$ solution?

Learning Goal 6: Equivalents of Acids and Bases

▲ **61.** Determine the number of equivalents in each of the following samples: *(a)* 4.9 g of H_2SO_4 *(b)* 7.2 g of HCl *(c)* 32.1 g of $Fe(OH)_3$

62. Determine the number of equivalents in each of the following samples: *(a)* 9.8 g of H_2SO_4 *(b)* 3.6 g of HCl *(c)* 6.42 g of $Fe(OH)_3$

63. For each of the following acids and bases, determine the number of grams per mole and the number of grams per equivalent: *(a)* H_2SO_3 *(b)* $Fe(OH)_3$ *(c)* $Fe(OH)_2$ *(d)* H_3BO_3

64. For each of the following acids and bases, determine the number of grams per mole and the number of grams per equivalent: *(a)* $HC_2H_3O_2$ *(b)* NaOH *(c)* $Mg(OH)_2$ *(d)* HNO_3

65. Determine the number of equivalents in each of the following: *(a)* 1.23 g of H_2SO_3 *(b)* 26.7 g of $Fe(OH)_3$

66. Determine the number of equivalents in each of the following: *(a)* 246 g of H_2SO_3 *(b)* 4.00 g of NaOH

▲ **67.** Determine the number of grams in each of the following: *(a)* 25.0 equivalents of $Fe(OH)_2$ *(b)* 0.450 equivalent of H_3BO_3

▲ **68.** Determine the number of grams in each of the following: *(a)* 0.250 equivalent of $Mg(OH)_2$ *(b)* 2.00 equivalents of $HC_2H_3O_2$

Learning Goal 7: Calculations Involving Normality

▲ **69.** Determine the normality of each of the following solutions:
(a) $97\bar{0}$ g of H_3PO_4 in enough water to make 10.0 L of solution
(b) 3.7 g of $Ca(OH)_2$ in enough water to make $8\overline{00}$ mL of solution
(c) Calculate the molarities of the solutions in *(a)* and *(b)*.

▲ **70.** Determine the normality of each of the following solutions:
(a) 465 g of H_3PO_4 in enough water to make 8.00 L of solution
(b) 740.0 g of $Ca(OH)_2$ in enough water to make 740.0 mL of solution
(c) Calculate the molarities of the solutions in *(a)* and *(b)*.

71. How many grams of $Ca(OH)_2$ are there in $8\overline{00}$ mL of a 2.0 *N* $Ca(OH)_2$ solution?

72. How many grams of $Mg(OH)_2$ are there in $75\bar{0}$ mL of a 3.00 *N* $Mg(OH)_2$ solution?

73. How many milliliters of a 0.60 *N* H_3PO_4 solution do you need to get 4.9 g of H_3PO_4?

74. How many milliliters of a 0.450 *N* H_3PO_4 solution do you need to get 98.0 g of H_3PO_4?

75. Determine the normality of each of the following solutions:
(a) 11.23 g of $Fe(OH)_2$ in enough water to make $25\bar{0}$ mL of solution
(b) 4.12 g of H_3BO_3 in enough water to make $4\overline{00}$ mL of solution
(c) Calculate the molarities of the solutions in *(a)* and *(b)*.

76. Determine the normality of each of the following solutions:
(a) 2.246 g of $Fe(OH)_2$ in enough water to make 500.0 mL of solution
(b) 16.48 g of H_3BO_3 in enough water to make 500.0 mL of solution
(c) Calculate the molarities of the solutions in *(a)* and *(b)*.

▲ **77.** How many grams of $Al(OH)_3$ are there in $25\bar{0}$ mL of a 0.500 *N* $Al(OH)_3$ solution?

▲ **78.** How many grams of $Fe(OH)_3$ are there in 500.0 mL of a 0.250 *N* $Fe(OH)_3$ solution?

79. How many milliliters of 0.100 *N* H_2SO_4 do you need to obtain 2.94 g of H_2SO_4?

80. How many milliliters of a 0.200 *N* H_3PO_4 solution do you need to obtain 29.40 g of H_3PO_4?

Learning Goal 8: Calculations for Dilutions of Solutions

▲ **81.** How would you make $5\overline{00}$ mL of 4.0 *N* H_2SO_4 solution from a 16 *N* solution?

▲ **82.** How would you make 250.0 mL of 8.00 *N* H_3PO_4 solution from a 10.0 *N* solution?

83. How would you make 3L of a 0.1 *N* NaOH solution from a 6 *N* solution?
84. How would you make 4.00 L of 0.200 *N* $Ca(OH)_2$ solution from 2.00 *N* solution?

85. How would you prepare $25\bar{0}$ mL of a 5.00 *M* HCl solution from a 12.0 *M* solution?
86. How would you prepare 450.0 mL of a 7.50 *M* HNO_3 solution from an 11.0 *M* solution?

87. How would you prepare 5.00 L of 0.0500 *N* H_2SO_4 solution from a 3.00 *M* solution?
88. How would you prepare 4.00 L of a 0.0200 *N* $HC_2H_3O_2$ solution from a 2.00 *N* solution?

Learning Goals 9 and 10: Electrolytic and Nonelectrolytic Solutions; Ions and Ionization

▲ **89.** Define the terms *ion* and *ionization,* and give examples of each.
90. True or false: *(a)* Ions are groups of charged atoms. *(b)* Charged elements are not ions.

▲ **91.** Explain why some solutions conduct electric current and others do not.
▲ **92.** Solution A conducts electricity, but solution B does not. If solutions A and B are mixed and no chemical reaction occurs, will the resulting solution conduct electricity?

▲ **93.** Define and give an example of an electrolytic solution and of a nonelectrolytic solution.
94. Which will conduct electricity, a glass of salt water or a cup of tea with sugar? Explain.

Learning Goal 11: Definition of Molality

▲ **95.** Define *molality*. What information is necessary to calculate the molality of a solution?
▲ **96.** How is *molality* different from *molarity?*

Learning Goal 12: Calculation of Molality

***97.** Calculate the molality of the intravenous salt solution in Self-Test Exercise 53. (Assume that there is 0.85 g of NaCl per $1\overline{00}$ g of water.)
▲ **98.** A solution contains 0.373 g of KCl per $1\overline{00}$ g of water. What is its molality?

▲ **99.** Determine the molality of each of the following solutions:
(a) 6.4 g of methanol (CH_3OH) in $25\bar{0}$ g of water
(b) 90.0 g of glucose ($C_6H_{12}O_6$) in 1,$\overline{500}$ g of water
100. Determine the molality of each of the following solutions:
(a) 0.128 g of CH_3OH in $\overline{500}$ g of water
(b) $45\bar{0}$ g of $C_6H_{12}O_6$ in 2,$\overline{000}$ g of water

Learning Goal 13: Definition of Colligative Properties and Effect of Solute on Boiling and Freezing Points

▲ **101.** Explain how a solute affects the boiling point and the freezing point of a solution.
▲ **102.** What would happen to the engine of an automobile if antifreeze were not added to the water in the radiator during a cold winter?

Learning Goal 14: Definition of K_f and K_b

▲ **103.** Define the following terms, and give an example of each: *(a)* K_f *(b)* K_b
▲ **104.** What are the values of K_f and K_b for water? What do they mean?

Learning Goal 15: Calculations Involving Boiling and Freezing Points of Solutions

▲ **105.** You can winterize a typical car radiator system by adding 1.00 gal of antifreeze to 1.00 gal of water. At what temperature does this water–antifreeze solution begin to freeze? (Assume that the antifreeze is pure ethylene glycol, $C_2H_6O_2$, which has a density of 1.10 g/cm^3.) (*Hint:* You must change the 1.00 gal of ethylene glycol solute to grams and then to moles. Also, the 1.00 gal of water must be changed to kilograms.)
106. Calculate the boiling point of a solution prepared by dissolving 5.40 g of glucose in $\overline{500}$ g of water. Glucose is $C_6H_{12}O_6$.

107. Calculate the boiling point of the solution in Self-Test Exercise 105.
▲ **108.** Calculate the boiling point of a solution prepared by dissolving 0.184 g of glycerol ($C_3H_8O_3$) in 1,$\overline{000}$ g of water.

109. Calculate the boiling point and the freezing point of a solution prepared by dissolving 27.0 g of glucose, $C_6H_{12}O_6$, in $1\overline{00}$ g of water.

110. Calculate the freezing point of the solution in Self-Test Exercise 108.

Learning Goal 16: Diffusion and Osmosis

111. *(a)* The passage of water through a semipermeable membrane is the process of ________.
(b) The movement of a solute in a solvent from an area of high solute concentration to an area of low solute concentration is the process of ________.

▲ **112.** A 50.0-g sample of a freshly peeled raw potato is placed in a beaker containing a saturated NaCl solution. The potato is left in the beaker for two hours. It is then removed, dried, and weighed. Will the mass of the potato be lower, higher, or the same? Explain.

▲ **113.** Two solutions of equal volume are placed in a beaker similar to the one in Figure 15.11. Compartment A of the beaker contains a 5% KCl solution. Compartment B of the beaker contains a 20% KCl solution. Which way does the water move, from A to B or from B to A?

114. Cells in the body have both intracellular and extracellular fluid. The intracellular fluid has an osmotic pressure equivalent to 0.9% sodium chloride. If an individual is given an intravenous solution that has an osmotic pressure equivalent to 2.0% sodium chloride, would water move in or out of the cell? (*Hint:* The intravenous solution replaces the extracellular fluid.)

Extra Exercises

▲***115.** Use the kinetic theory to explain how and why a solute affects the boiling and freezing points of a solution.

116. Calculate the molarity of a solution of HCl that has a hydrogen-ion concentration of 10^{-2} *M*.

117. Calculate the molarity and the normality of a solution of sulfuric acid that has a $[H^{1+}]$ of 10^{-2} *M*.

118. How many grams of H_2SO_4 are necessary to make 10.0 L of a 5.00%-by-mass sulfuric acid solution? (Assume that the density of the solution is 1.00 g/mL.)

119. Find the molarity and the normality of the solution in Extra Exercise 118.

120. How does a solution differ from a mixture?

▲ **121.** In Chicago, salt is sometimes used on roads to keep the roads free of ice. In northern Minnesota, however, people use sand to avoid slipping on ice. Why don't the Minnesota residents try to use salt to get rid of their ice?

▲ **122.** Which of the following would lower the freezing point of $1,\overline{000}$ g of water by the greatest amount? Explain your choice.
(a) $\overline{100}$ g of glucose ($C_6H_{12}O_6$)
(b) $\overline{100}$ g of sucrose ($C_{12}H_{22}O_{11}$)
(c) $\overline{100}$ g of ethanol (C_2H_6O)

***123.** Concentrated hydrochloric acid is delivered as 37.00% HCl by mass. The density of this solution is 1.19 g/mL. What volume of this acid is needed to produce 1.00 L of a 0.500 *M* HCl solution?

124. You dissolve a 15.0-g sample of an unknown substance (nonelectrolyte) in $\overline{100}$ g of water. The solution freezes at −1.86°C. What is the molecular mass of the substance?

125. A chemist prepares a solution by taking 1.00 L of water and adding 2.00 moles of solute. Is this solution a 2.00 *M* solution or a 2.00 *m* solution? Explain.

***126.** You add a 10.0-g sample of a substance to $1,\overline{000}$ g of water. You find that the freezing point of the solution is −5.00°C. What is the molecular mass of the substance?

127. You add 5.85 g of NaCl to $1,\overline{000}$ g of water. What is the approximate boiling point of the solution?

128. Determine the number of equivalents in each of the following: *(a)* 73.0 g of HCl *(b)* 98.1 g of H_2SO_4

129. Determine the number of equivalents in each of the following: *(a)* 1.00 mole of HCl *(b)* 1.00 mole of H_2SO_4

16 Acids, Bases, and Salts

Learning Goals

After you have studied this chapter, you should be able to:

1. Define the terms *acid* and *base* in the way Arrhenius did.
2. Define the terms *acid* and *base* according to the Brønsted–Lowry concept.
3. List the important characteristics of acids and bases.
4. Write equations for the preparation of acids and bases.
5. Write equations for important chemical reactions of acids and bases.
6. Define the terms *strong acid* and *strong base,* and give examples of each.
7. Define the terms *weak acid* and *weak base,* and give examples of each.
8. Explain the concept of dynamic equilibrium.
9. Write an equation for the ionization of water.
10. Define pH and explain the different values on the pH scale.
11. Determine the pH of a solution, given its hydrogen-ion or hydroxide-ion concentration in moles per liter.
12. State the relationship between the hydrogen-ion concentration and the hydroxide-ion concentration in aqueous solutions.
13. Explain what the term *salt* means, and give some examples.
14. Cite the most important properties of salts.
15. List methods and equations for the preparation of various salts.
16. Define the term *titration*.

Introduction

Three important groups of compounds—acids, bases, and salts—are the subject of this chapter. You are probably familiar with many compounds found in each of these groups. Acetic acid is the substance that gives vinegar its characteristic taste and odor, and citric acid is responsible for the tartness of lemons and grapefruits. Many popular household cleansers contain the base ammonia. Bases are the substances that make soaps and cleansers feel slippery. In this chapter, we also discuss titration, an analytical procedure that enables the chemist to determine how much of a substance is present in solution. And finally, we examine salts, which are widely used in both home and industry.

16.1 Acids

Three different definitions of acids and bases have been developed in the last 100 years, and each has its place in explaining observed behavior. Two of these definitions will serve as the basis for our discussion of the properties, reactions, and uses of acids.

The Arrhenius Definition

Learning Goal 1
Arrhenius acids

In 1884 the Swedish scientist Svante Arrhenius described an **acid** as a substance that releases hydrogen ions (H^{1+}) when it is dissolved in water. This description was part of his ionic theory of solutions, discussed in Chapter 15.

Let's look at how an acid dissolves in water. In Chapter 7, we saw that water is a polar molecule: its oxygen end is negatively charged and its hydrogen end is positively charged. When we bubble hydrogen chloride gas (HCl) into a beaker of water, the gas seems to dissolve in the water, but actually much more is happening. The HCl gas is reacting with the water (Figure 16.1). The covalently bonded hydrogen chloride molecules are being broken apart (ionized) by the water into hydrogen ions (H^{1+}) and chloride ions (Cl^{1-}), and the hydrogen ions are adding on to the negative end of the water molecules, forming an ion. This ion, $(H_3O)^{1+}$, is called a **hydronium ion.** The reaction is

$$HCl + H_2O \longrightarrow (H_3O)^{1+} + Cl^{1-}$$

Similar reactions occur when the other acids listed in Table 16.1 are dissolved in water. For example,

$$H_2SO_4 + 2H_2O \longrightarrow 2(H_3O)^{1+} + (SO_4)^{2-}$$

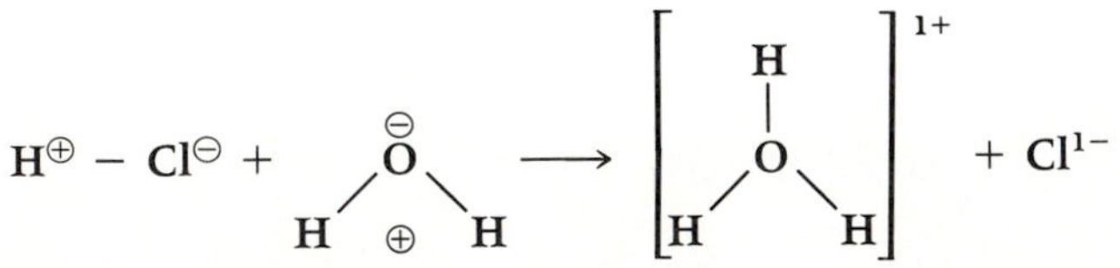

$$H^{\oplus} - Cl^{\ominus} + \underset{H\ \ \oplus\ \ H}{\overset{\ominus}{O}} \longrightarrow \left[\underset{H\qquad H}{\overset{H}{O}} \right]^{1+} + Cl^{1-}$$

Figure 16.1
The positive end of the HCl molecule is attracted by the negative end of the water molecule. The attraction is strong enough to break the bond between the H and the Cl, forming a hydronium ion, $(H_3O)^{1+}$, and a chloride ion, Cl^{1-}.

Table 16.1
Some Common Arrhenius Acids

Name of acid	Formula
Hydrochloric acid	HCl
Nitric acid	HNO_3
Sulfuric acid	H_2SO_4
Phosphoric acid	H_3PO_4
Acetic acid	$HC_2H_3O_2$

According to Arrhenius, HCl is an acid because it undergoes the following reaction when placed in water:

$$HCl + H_2O \longrightarrow (H_3O)^{1+} + Cl^{1-}$$

or, in a simpler form,

$$HCl \xrightarrow{\text{Water}} H^{1+} + Cl^{1-}$$

Table 16.1 lists some additional Arrhenius acids. When we confine ourselves to the Arrhenius definition of an acid, then, our discussion is limited to aqueous (water) solutions and to substances that liberate hydrogen ions in solution. Because of these limitations, other definitions for the term *acid* were proposed. However, the Arrhenius definition is adequate for most of our purposes.

The Brønsted–Lowry Definition

In 1923 two chemists, J. N. Brønsted and T. M. Lowry, working independently, proposed a much broader definition of the term *acid*. They described an acid as any substance that is a proton donor. According to this definition, an **acid** is any substance that can donate a proton (an H^{1+} ion) to another substance. This definition includes all substances that qualify as acids under the Arrhenius definition and many other compounds and ions as well (Table 16.2). For example, the ammonium ion is a Brønsted–Lowry acid because

$$(NH_4)^{1+} + H_2O \rightleftharpoons NH_3 + (H_3O)^{1+}$$

Learning Goal 2
Brønsted–Lowry acids

or, more simply,

$$\underset{\text{Acid}}{(NH_4)^{1+}} \xrightleftharpoons{\text{Water}} NH_3 + H^{1+}$$

Properties of Acids

Learning Goal 3
Characteristics of acids

The acids listed in Table 16.1 have many characteristics in common:

1. *Each has a sour taste.* (The word *acid* comes from the Latin *acidus*, which means "sour.")

Table 16.2
Some Common Brønsted–Lowry Acids

Name of acid	Formula
Hydrochloric acid	HCl
Nitric acid	HNO_3
Sulfuric acid	H_2SO_4
Phosphoric acid	H_3PO_4
Acetic acid	$HC_2H_3O_2$
Hydronium ion	$(H_3O)^{1+}$
Hydrogen sulfate ion	$(HSO_4)^{1-}$
Ammonium ion	$(NH_4)^{1+}$
Water	H_2O

2. *Each can change the color of certain dyes.* For example, blue litmus dye or blue litmus paper turns red in the presence of an acid. These dyes are called *indicators* because they indicate whether a substance is an acid or a base (Figure 16.2).

3. *Each can react with certain active metals, such as zinc and magnesium, to produce hydrogen gas.*

$$Zn + 2HCl \longrightarrow ZnCl_2 + H_2(g)$$

$$Zn + H_2SO_4 \longrightarrow ZnSO_4 + H_2(g)$$

4. *Each can neutralize bases.* (We will discuss this property later.)

One of the important characteristics of acids is the one we used in the Arrhenius definition: *They dissolve in water to produce hydrogen ions* (H^{1+}). These ions are responsible for the other properties of acids.

Preparation of Acids

Learning Goal 4
Preparation of acids

There are many different ways of preparing acids. One of the most important methods involves the reaction of a nonmetal oxide with water.

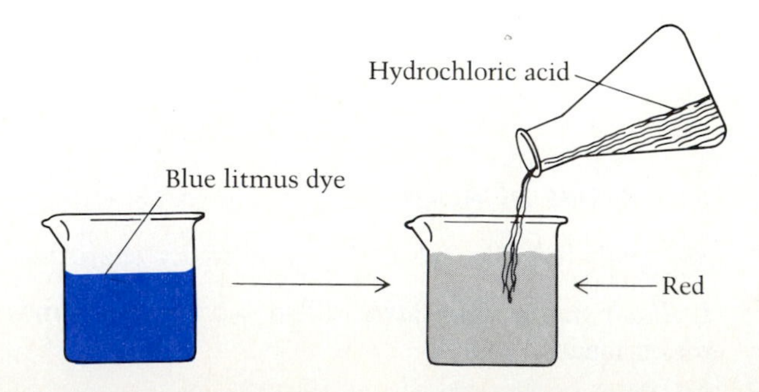

Figure 16.2
The reaction of an acid–base indicator. Hydrochloric acid turns blue litmus red.

$$SO_3 + H_2O \longrightarrow H_2SO_4$$

Sulfur trioxide → Sulfuric acid

$$N_2O_5 + H_2O \longrightarrow 2HNO_3$$

Nitrogen (V) oxide → Nitric acid

The acids that result from some of these reactions are called *mineral* acids (because they are prepared from minerals). Other acids, called *organic* acids, are obtained from living things such as citrus fruits or fermented grapes or are found in animal matter.

Reactions of Acids

Learning Goal 5
Chemical reactions of acids

Acids undergo many types of reactions. Two of these are important for our purposes:

1. The reactions of acids with metals
2. The reactions of acids with bases

We have already discussed the reaction of an acid with a metal. You learned in Chapter 10 that this type of reaction is a *single-replacement reaction* involving the replacement of hydrogen atoms in an acid by atoms of a metal (which must lie above hydrogen in the activity series, Table 10.1). For example,

$$Mg + H_2SO_4 \longrightarrow MgSO_4 + H_2$$

$$Mg + 2HBr \longrightarrow MgBr_2 + H_2$$

Hydrobromic acid

We have also already discussed the reaction of an acid with a base (in Section 10.4). We'll come back to this type of reaction, sometimes called a neutralization reaction, later in this chapter when we cover the reactions of bases.

Uses of Some Common Acids

Acids are important substances with many household and industrial uses. Some acids are very toxic or can burn the skin and must be handled carefully. Let's look at some of the modern uses of acids.

Hydrochloric acid, a strong mineral acid known commercially as muriatic acid, is used industrially to clean metals that are about to be coated or plated. (This cleaning is known as *pickling*.) Hydrochloric acid is also used to clean brick and cement.

Nitric acid, another mineral acid, is highly reactive. When a fair-skinned person happens to spill some on his or her skin, it turns the skin yellow. This is because of a reaction between the nitric acid and the protein in skin. Nitric acid is used mainly in the production of fertilizers but is also used to make dyes, plastics, and explosives.

Sulfuric acid is another very strong mineral acid that is used in fertilizers. In addition, it is used to make smokeless powder (a type of gunpowder) and nitroglycerin. Automobile batteries contain a diluted form of sulfuric acid. Such batteries work because there is a reaction between their metal plates and the acid, creating a movement of electrons and therefore an electric current. The Soviets are manufacturing sulfuric acid from the oxides of sulfur that result when fossil fuels are burned to produce electricity. In other words, sulfuric acid can be made as a by-product of electric power plants.

Phosphoric acid is a weak mineral acid. Its uses range from flavoring soft drinks to producing superphosphates for fertilizers. And the false teeth of millions of people are held in place by dental cements containing phosphoric acid.

Acetic acid is found in vinegar. According to the pure-food laws of the United States, vinegar must contain no less than 4% acetic acid. Concentrated acetic acid is used in making synthetic fibers called *acetates*. It is also used in manufacturing various plastics. (Remember, though, that acetic acid—unlike the other acids we have just discussed—is an organic acid.)

CHEMICAL FRONTIERS

FUELING CARS WITH HYDROGEN*

In an effort to decrease our dependence on fossil fuels, researchers are exploring the use of hydrogen as an alternative fuel to power cars, trucks, and buses. Hydrogen fuel has several advantages. It burns very cleanly in air and can be easily obtained from water by electrolysis.

But several problems must be solved before hydrogen fuel can become a practical alternative. First, it must be available at a reasonable cost. Natural gas is the cheapest source of liquid hydrogen used to fuel NASA rockets, but it costs 4.4 times more than gasoline. Solar energy to produce electricity for the electrolysis process is even more expensive. Researchers hope that special photovoltaic cells, which produce voltage when exposed to light, can be used to convert sunlight to electricity more cheaply, leading to more affordable solar-generated hydrogen.

The design of hydrogen-powered vehicles is another challenge. Tanks storing hydrogen gas are heavy and bulky, and vehicles are restricted in the distances they can travel by the limited amount of fuel they can carry. One possible innovation is to replace the heavy tanks with lightweight metal "hydride" beds that release pressurized hydrogen gas as needed. Liquid hydrogen fuel may one day be practical once the problems associated with storing liquid hydrogen at −423°F are overcome.

Another promising possibility is the use of hydrogen-powered fuel cells,

*Adapted from "Hydrogen Dreams," *Technology Review,* MIT: Cambridge, Mass. August/September, 1990.

which can convert 60% of hydrogen into useful energy. This technology is more efficient than gasoline engines, but the fuel cells required are quite expensive.

Converting from fossil-fuel to hydrogen-powered systems is an ambitious undertaking. But over the next few years, we can expect to see significant developments in this area as the need for alternative fuels intensifies.

16.2 Bases

As in the case of acids, three different definitions of bases have been developed over the last 100 years. Once again, we will look at two useful definitions—those of Arrhenius and of Brønsted and Lowry.

The Arrhenius Definition

Arrhenius's definition of an acid applies only to aqueous solutions, and the same is true of his definition of a base. A **base** is a substance that releases hydroxide ions, $(OH)^{1-}$, when that substance is dissolved in water. According to Arrhenius, NaOH is a base because it undergoes the following reaction when placed in water:

$$NaOH \xrightarrow{\text{Water}} Na^{1+} + (OH)^{1-}$$

Learning Goal 1
Arrhenius bases

Many of the common bases found in the laboratory are composed of metallic ions plus hydroxide ions—as is sodium hydroxide (NaOH). Table 16.3 lists some additional Arrhenius bases.

Incidentally, there is another word that people often use for a base: **alkali.** An alkali is any substance that dissolves to yield a basic (or alkaline) solution.

The Brønsted–Lowry Definition

Just as Brønsted and Lowry defined an acid as a proton donor, they defined a base as a proton acceptor. In other words, according to Brønsted and Lowry,

Table 16.3
Some Common Arrhenius Bases

Name of base	Formula
Lithium hydroxide	$LiOH$
Sodium hydroxide	$NaOH$
Potassium hydroxide	KOH
Calcium hydroxide	$Ca(OH)_2$
Magnesium hydroxide	$Mg(OH)_2$
Iron(III) hydroxide	$Fe(OH)_3$

Table 16.4
Some Common Brønsted–Lowry Bases

Name of base	Formula	Name of base	Formula
Ammonia	NH_3	Chloride ion	Cl^{1-}
Hypochlorite ion	ClO^{1-}	Carbonate ion	$(CO_3)^{2-}$
Hypobromite ion	BrO^{1-}	Sulfate ion	$(SO_4)^{2-}$
Cyanide ion	CN^{1-}	Acetate ion	$(C_2H_3O_2)^{1-}$
Phosphate ion	PO_4^{3-}	Hydroxide ion	$(OH)^{1-}$
Hydride ion	H^{1-}	Water	H_2O

a **base** is any substance that accepts protons (H^{1+} ions). This important definition includes all substances that qualify as bases under the Arrhenius definition, since the OH^{1-} ion from the base accepts an H^{1+} ion from an acid to form water. Many other compounds and ions act as bases as well (Table 16.4). For example, the compound ammonia is a Brønsted–Lowry base because

Learning Goal 2
Brønsted–Lowry bases

$$NH_3 + H^{1+} \longrightarrow (NH_4)^{1+}$$

Properties of Bases

Learning Goal 3
Characteristics of bases

The bases listed in Table 16.3 have many characteristics in common:

1. *Each has a bitter taste.*
2. *Each can change the color of certain dyes.* For example, red litmus turns blue in the presence of a base (Figure 16.3).
3. *Each feels slippery or soapy on the fingers.*
4. *Each can neutralize acids.* (This property is discussed later.)

One of the important characteristics of bases is the one we used in the Arrhenius definition: *They dissolve in water to produce hydroxide ions,* $(OH)^{1-}$. These hydroxide ions are responsible for the other properties of bases.

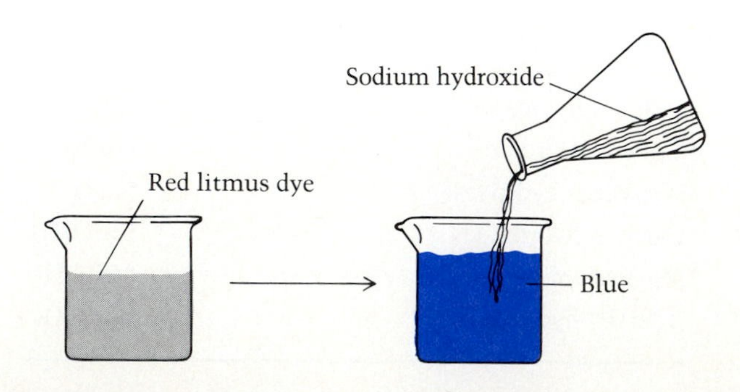

Figure 16.3
The reaction of an acid–base indicator. Sodium hydroxide turns red litmus blue.

How do bases dissolve in water to produce hydroxide ions? Consider, for example, solid sodium hydroxide (NaOH). Sodium hydroxide is an ionic compound made up of sodium ions and hydroxide ions, or Na^{1+} and $(OH)^{1-}$. The ions are held together by the attraction of their opposite charges. When we add sodium hydroxide to water, the negative end of each water molecule (the oxygen end) attracts the sodium ion, and the positive end (the hydrogen end) attracts the hydroxide ion. This attraction breaks the bonds between the sodium ions and the hydroxide ions (Figure 16.4). The result is that there are now free sodium ions and free hydroxide ions in solution. The reaction can be represented by the equation

$$NaOH \xrightarrow{\text{Water}} Na^{1+} + (OH)^{1-}$$

Preparation of Bases

Learning Goal 4
Preparation of bases

Bases can be prepared in many ways. One major method involves the reaction of a metal oxide with water.

$$\underset{\text{Sodium oxide}}{Na_2O} + H_2O \longrightarrow \underset{\text{Sodium hydroxide}}{2NaOH}$$

$$\underset{\text{Calcium oxide}}{CaO} + H_2O \longrightarrow \underset{\text{Calcium hydroxide}}{Ca(OH)_2}$$

Reactions of Bases

Bases undergo many types of reactions. Two that are important for our purposes are:

1. The reactions of bases with nonmetal oxides
2. The reactions of bases with acids

Learning Goal 5
Chemical reactions of bases

When a base reacts with a nonmetal oxide, it produces a salt and water:

$$2NaOH + CO_2 \longrightarrow Na_2CO_3 + H_2O$$

$$Ca(OH)_2 + CO_2 \longrightarrow CaCO_3 + H_2O$$

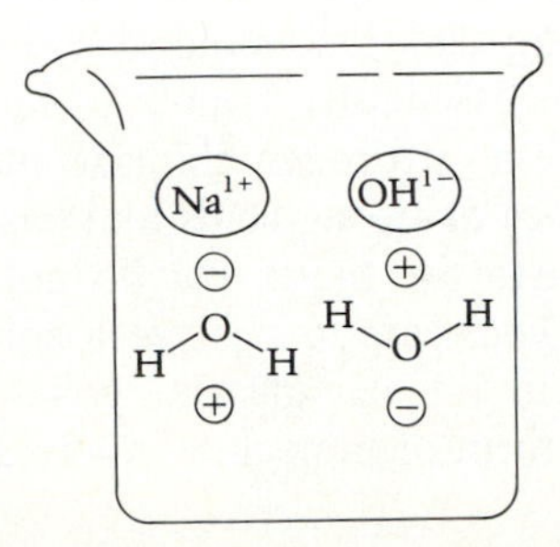

Figure 16.4
The dissociation of NaOH

When a base reacts with an acid, it again produces a salt and water:

$$NaOH + HCl \longrightarrow NaCl + H_2O$$

$$2NaOH + H_2SO_4 \longrightarrow Na_2SO_4 + 2H_2O$$

$$3Ca(OH)_2 + 2H_3PO_4 \longrightarrow Ca_3(PO_4)_2 + 6H_2O$$

These latter reactions are sometimes called *neutralization reactions,* because when we slowly add an acid to a base, the H^{1+} and $(OH)^{1-}$ ions neutralize each other by forming water.

Uses of Some Common Bases

Like acids, bases play an important role in our daily lives, and they are just as dangerous to our skin and clothes. They are used in our homes, in industrial processes, and in manufacturing drugs and medicines.

Sodium hydroxide (NaOH) is a base used in drain cleaners, or what we call lye. It is also used to make soaps and cellophane. The petroleum industry uses sodium hydroxide to neutralize acids in the refining process.

Calcium hydroxide, $Ca(OH)_2$, sometimes called slaked lime, is important in the construction industry. It is used in mortar, plaster, cement, and many other building and paving materials.

Ammonia (NH_3) dissolved in water is the household cleaner we know as ammonia water, and it has an intense, almost suffocating odor. Ammonia is the chief ingredient of smelling salts. It is excellent for removing stains, and it is also used in manufacturing textiles (such as rayon), plastics, and fertilizers.

Magnesium hydroxide, $Mg(OH)_2$, dispersed in water, commonly called milk of magnesia, has many uses as a medicine. In small doses (a few hundred milligrams), magnesium hydroxide acts as an antacid (bases counteract acid just as reliably in the stomach as in the laboratory). In large doses (two to four grams), it acts as a laxative.

16.3 The Strengths of Acids and Bases

Learning Goal 6
Strong acids and bases

Acids and bases can both be subdivided into two categories: (1) *strong* acids and *strong* bases and (2) *weak* acids and *weak* bases. The **strong acids and bases** are those that ionize completely in dilute solution. For example, hydrochloric acid is a strong acid. One mole of hydrogen chloride gas dissolved in 1 L of water contains essentially 1 mole of hydrogen ions and 1 mole of chloride ions. There are no hydrogen chloride molecules to speak of. Table 16.5 gives some examples of strong acids and bases.

Learning Goal 7
Weak acids and bases

Weak acids and bases are those that do not ionize completely. Acetic acid is a weak acid. If you were to analyze a solution containing 1 mole of acetic acid ($HC_2H_3O_2$) in 1 L of solution, you would find that most of the acetic acid stays in the form of molecules. Only about 0.4% ionizes to form

Table 16.5
Strong Acids and Bases

	Name	Formula
Acids	Hydrochloric acid	HCl
	Sulfuric acid	H_2SO_4
	Nitric acid	HNO_3
Bases	Sodium hydroxide	$NaOH$
	Potassium hydroxide	KOH

hydrogen ions and acetate ions. We can represent the dissociation of this weak acid in the following manner:

$$HC_2H_3O_2 \xrightleftharpoons{\text{Water}} H^{1+} + (C_2H_3O_2)^{1-}$$

The *double arrow* indicates that in a given solution of acetic acid, an *equilibrium* (or balance) exists among the acetic acid molecules, the hydrogen ions, and the acetate ions. This equilibrium lies in the direction of (provides more of) the undissociated $HC_2H_3O_2$. That's why the arrow that points to the left is longer than the arrow that points to the right. (If the situation were the other way around, the arrow pointing to the right would be longer.) In weak acids and weak bases, the equilibrium concentration of the undissociated species is favored. Table 16.6 offers some examples of weak acids and bases. (Also see Figure 16.5.)

Since they produce ions in aqueous solution, acids and bases are electrolytes. Strong acids and strong bases are considered to be strong electrolytes, because they are about 100% dissociated into ions when placed in aqueous solution. Weak acids and weak bases are considered to be weak elec-

Table 16.6
Weak Acids and Bases

	Name	Formula
Acids	Acetic acid	$HC_2H_3O_2$
	Carbonic acid	H_2CO_3
	Boric acid	H_3BO_3
Bases	Ammonia (in water)	NH_3
	Aluminum hydroxide	$Al(OH)_3$

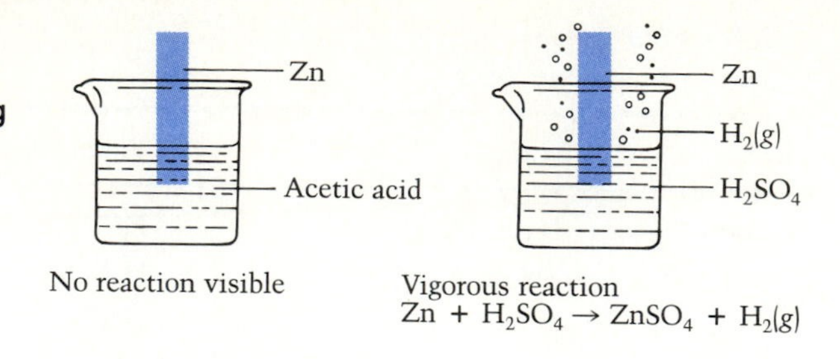

Figure 16.5 One difference between a weak acid and a strong acid

trolytes, because they dissociate slightly into ions when placed in aqueous solution. The strong electrolytes contain mostly ions in aqueous solution, whereas the weak electrolytes contain mostly covalently bonded molecules in equilibrium with a small number of ions in aqueous solution.

16.4 Weak Acids and Bases and Dynamic Equilibrium

Learning Goal 8
Dynamic equilibrium

The equilibrium that exists in a solution of a weak acid or weak base is not a static (unchanging) equilibrium but rather a *dynamic* equilibrium. That is, some molecules are always dissociating to form ions, and some ions are always recombining to form molecules. But these dissociations and recombinations take place at the same rate, so the proportion of molecules and ions remains constant—they remain in equilibrium.

As an analogy, consider a popular restaurant during the busy lunch hour. The restaurant can seat 200 people, and it is full at noon. But some people are leaving the restaurant and others are entering. A quarter of an hour later, the situation is the same: 200 people in the restaurant, some people entering, and others leaving. And again, 15 minutes later, the same situation prevails. During the lunch hour, those who enter the restaurant take the places of those who have eaten and left, maintaining a population equilibrium of 200 diners. But the equilibrium is dynamic (changing), because individual diners are constantly being replaced by other diners.

A chemical equilibrium is dynamic in the same sense. For example, in a solution of acetic acid, there are some hydrogen ions (H^{1+}), some acetate ions $(C_2H_3O_2)^{1-}$, and many acetic acid molecules ($HC_2H_3O_2$). But at any given time, some hydrogen ions are combining with acetate ions to form acetic acid molecules. And some acetic acid molecules are ionizing to form hydrogen ions and acetate ions. As long as these two processes go on at the same rate, the concentrations of the various substances don't change, and a dynamic equilibrium is maintained.

Chemists use various methods to measure the extent of dissociation of acids and bases. There are also quantitative ways of describing the extent of dissociation.

16.5 The Ionization of Water

Learning Goal 9
Ionization of water

Like weak acids and bases, water has a tendency to ionize. One way to represent the ionization of water is as follows:

$$\text{H—OH} + H_2O \rightleftharpoons \underset{\text{Hydronium ion}}{(H_3O)^{1+}} + \underset{\text{Hydroxide ion}}{(OH)^{1-}} \qquad \text{Equation 16.1}$$

Here the formula for the first water molecule is written as H—OH so that we can see how the ionization occurs. Equation 16.1 shows that a hydronium ion and a hydroxide ion are formed.

We can also represent the ionization of water in this simpler way:

$$H_2O \rightleftharpoons H^{1+} + (OH)^{1-} \qquad \text{Equation 16.2}$$

This equation shows that a hydrogen ion and a hydroxide ion are formed. Even though Equation 16.1 is more accurate (technically), Equation 16.2 is easier to use. However, as the double arrow indicates, the ionization of water occurs only to a slight extent. In fact, in exactly 1 L of water, which is 55.6 moles of water, only 0.0000001 mole (which can be written as 10^{-7} mole) actually dissociates into hydrogen and hydroxide ions. In other words, there are 10^{-7} mole of hydrogen ions and 10^{-7} mole of hydroxide ions present in 1 L of water. Because there are equal amounts of hydrogen ions and hydroxide ions in pure water, it is said to be neutral.

16.6 The pH Scale

Learning Goal 10
pH and pH scale

The pH scale is used to measure and express the acidity or basicity (acid or base strength) of a solution. The **pH** of a solution is *equal to the negative of the logarithm of its hydrogen-ion concentration in moles per liter*. This equation is written as

$$pH = -\log [H^{1+}]$$

where the brackets, [ ], mean *concentration in moles of* H^{1+} *per liter of solution*. The pH scale runs from 0 to 14 (Figure 16.6).

Recall from our discussion of the ionization of water that the concentration of hydrogen ions in pure water is 10^{-7} mole per liter. It happens that this con-

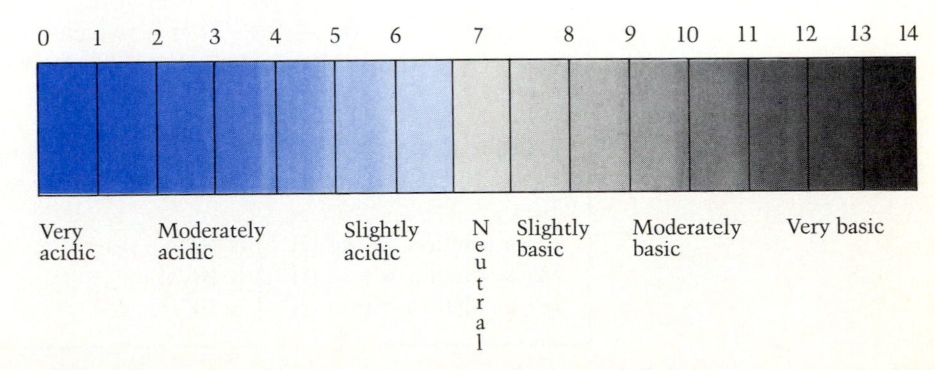

Figure 16.6
The pH scale

centration corresponds to a pH of 7. Moreover, we noted that pure water is neutral (neither acidic nor basic). Therefore any solution with a pH of 7 is considered neutral.

A solution that has a pH between 0 and 2 is considered strongly acidic. A solution that has a pH between 2 and 5 is moderately acidic, and one with a pH between 5 and 7 is slightly acidic. A solution with a pH of 7 is, as we saw, neutral. A solution that has a pH between 7 and 9 is slightly basic, one with a pH between 9 and 11 is moderately basic, and one with a pH between 12 and 14 is strongly basic. Table 16.7 lists the pH values of some common substances.

We can determine the pH of a solution by many methods. One simple method is to place a few drops of a chemical dye or a piece of paper tape that contains the dye into the solution. The dye or paper turns a color that depends on the pH of the solution. There is also a device called a pH meter, which directly reads the pH of a solution.

Learning Goal 11
Determining pH from hydrogen-ion or hydroxide-ion concentration

It is easy to find the pH of a solution when its hydrogen-ion concentration (in moles per liter) is given as a power of 10. For example, for pure water, with a hydrogen-ion concentration of 10^{-7} mole/liter,

$$pH = -\log [H^{1+}] = -\log 10^{-7} = -(-7) = 7$$

Note that the logarithm of a power of 10 is simply that power of 10. For example, $\log 10^{-2} = -2$. (*Note:* pH values don't have to be whole numbers. However, in this text we will deal only with whole-number pH values and with concentrations to the nearest power of ten.)

Example 16.1

Determine the pH of each of the following solutions. Then use Figure 16.6 to determine how acidic or basic the solution is.
(a) a solution whose $[H^{1+}]$ is 10^{-3}
(b) a solution whose $[H^{1+}]$ is 10^{-13}
(c) a solution whose $[H^{1+}]$is 10^{0}

Solution $pH = -\log [H^{1+}]$. Therefore,
(a) $pH = -\log 10^{-3} = -(-3) = 3$, which is moderately acidic.
(b) $pH = -\log 10^{-13} = -(-13) = 13$, which is very basic.
(c) $pH = -\log 10^{0} = 0$, which is very acidic.

Practice Exercise 16.1 Determine the pH of each of the following solutions. Then use Figure 16.6 to determine how acidic or basic the solution is.
(a) a solution whose $[H^{1+}]$ is 10^{-5}
(b) a solution whose $[H^{1+}]$ is 10^{-10}
(c) a solution whose $[H^{1+}]$ is 10^{-1}

Table 16.7
The pH Values of Some Common Substances

Substance	pH	Substance	pH
1 *M* HCl	0.1	Rainwater	6.2
0.5 *M* H_2SO_4	0.3	Milk	6.5
Lemons	2.3	Pure water	7.0
Vinegar	2.8	Sea water	8.5
Soft drinks	3.0	Milk of magnesia	10.0
Oranges	3.5	0.1 *M* NH_4OH	11.1
Tomatoes	4.2	1 *M* NaOH	14.0

16.7 Relationship Between the Hydrogen-Ion and Hydroxide-Ion Concentrations

Learning Goal 12
Relationship between hydrogen-ion and hydroxide-ion concentrations

There is an interesting relationship between the hydrogen-ion concentration and the hydroxide-ion concentration in aqueous solutions: The product of $[H^{1+}]$ and $[OH^{1-}]$ is always equal to 10^{-14}. In other words,

$$[H^{1+}][OH^{1-}] = 10^{-14}$$

Remember that pure (neutral) water has an H^{1+} concentration of $10^{-7}\ M$ and an $(OH)^{1-}$ concentration of $10^{-7}\ M$. The product of these two numbers is $10^{-14}\ M^2$:

$$[H^{1+}][OH^{1-}] = (10^{-7}\ M)(10^{-7}\ M) = 10^{-14}\ M^2$$

The value 10^{-14} is the *dissociation constant* of water. It is usually given the symbol K_w. Therefore,

$$[H^{1+}][OH^{1-}] = K_w = 10^{-14}$$

(We generally don't bother writing the M^2 after 10^{-14}, but remember that it is understood to be there all the same.) A solution that has a hydrogen-ion concentration of $10^{-5}\ M$ has to have a hydroxide-ion concentration of $10^{-9}\ M$, because

$$(10^{-5}\ M)[OH^{1-}] = 10^{-14}\ M^2 \quad \text{so} \quad [OH^{1-}] = \frac{10^{-14}\ \cancel{M}^2}{10^{-5}\ \cancel{M}} = 10^{-9}\ M$$

Table 16.8 summarizes the relationships among hydrogen-ion concentration, hydroxide-ion concentration, and pH.

Table 16.8
The Relationship Between Hydrogen-Ion Concentration and Hydroxide-Ion Concentration

$[H^{1+}]$	pH	$[OH^{1-}]$
$10^{0} = 1$	0	10^{-14}
$10^{-1} = 0.1$	1	10^{-13}
$10^{-2} = 0.01$	2	10^{-12}
$10^{-3} = 0.001$	3	10^{-11}
$10^{-4} = 0.0001$	4	10^{-10}
$10^{-5} = 0.00001$	5	10^{-9}
$10^{-6} = 0.000001$	6	10^{-8}
$10^{-7} = 0.0000001$	7	10^{-7}
$10^{-8} = 0.00000001$	8	10^{-6}
$10^{-9} = 0.000000001$	9	10^{-5}
$10^{-10} = 0.0000000001$	10	10^{-4}
$10^{-11} = 0.00000000001$	11	10^{-3}
$10^{-12} = 0.000000000001$	12	10^{-2}
$10^{-13} = 0.0000000000001$	13	10^{-1}
$10^{-14} = 0.00000000000001$	14	10^{0}

Example 16.2

Determine the pH of each of the following strong electrolytes:
(a) a 0.001 *M* HCl solution
(b) a 0.01 *M* HNO_3 solution
(c) a solution whose $[OH^{1-}]$ is 10^{-3}
(d) a solution whose $[OH^{1-}]$ is 10^{-11}
(e) a 0.0001 *M* NaOH solution

Solution

(a) 0.001 *M* HCl solution releases 0.001 mole of hydrogen ions per liter of solution, so its $[H^{1+}]$ is 10^{-3}. Therefore its pH is 3.

$$pH = -\log [H^{1+}] = -\log 10^{-3} = 3$$

(b) A 0.01 *M* HNO_3 solution releases 0.01 mole of hydrogen ions per liter of solution, so its $[H^{1+}]$ is 10^{-2}. Therefore its pH is 2.

$$pH = -\log [H^{1+}] = -\log 10^{-2} = 2$$

(c) A solution whose $[OH^{1-}]$ is 10^{-3} has $[H^{1+}] = 10^{-11}$ (see Table 16.8). Therefore its pH is 11.

$$pH = -\log [H^{1+}] = -\log 10^{-11} = 11$$

(d) A solution whose $[OH^{1-}]$ is 10^{-11} has $[H^{1+}] = 10^{-3}$ (see Table 16.8). Therefore its pH is 3.

$$pH = -\log [H^{1+}] = -\log 10^{-3} = 3$$

(e) A 0.0001 *M* NaOH solution releases 0.0001 mole of hydroxide ions per liter of solution, so its $[OH^{1-}]$ is 10^{-4}. A solution whose $[OH^{1-}]$ is 10^{-4} has $[H^{1+}] = 10^{-10}$ (see Table 16.8). Therefore its pH is 10:

$$pH = -\log [H^{1+}] = -\log 10^{-10} = 10$$

Practice Exercise 16.2 Determine the pH of each of the following solutions:

(a) a 0.0001 *M* HCl solution
(b) a 0.001 *M* HCl solution
(c) a solution whose $[OH^{1-}]$ is 10^{-5}
(d) a solution whose $[OH^{1-}]$ is 10^{-11}
(e) a 0.10 *M* KOH solution

16.8 Salts

Definition of Salt

We have seen that acids and bases react with each other in what are called neutralization reactions. The product of these reactions is a salt. Water is also produced if hydroxide ion is in the formula of the base.

$$HCl + NaOH \longrightarrow \underset{\text{Sodium chloride}}{NaCl} + H_2O$$

$$HC_2H_3O_2 + NaOH \longrightarrow \underset{\text{Sodium acetate}}{NaC_2H_3O_2} + H_2O$$

$$HCl + NH_3 \longrightarrow \underset{\text{Ammonium chloride}}{NH_4Cl}$$

Learning Goal 13
Definition of salt

In these reactions, the hydrogen ions from the acid combine with the hydroxide ions from the base to produce water. At the same time, the positive ions of the base combine with the negative ions of the acid to produce a compound that we call a salt (Figure 16.7). A *salt* is a compound composed of the positive ion of a base and the negative ion of an acid.

Most salts are composed of a metal ion combined with a nonmetal (or polyatomic) ion that has a negative charge. Examples of such salts are sodium chloride (NaCl), which is table salt, silver bromide (AgBr), potassium sulfate (K_2SO_4), and iron(III) phosphate ($FePO_4$). Some salts are composed of a polyatomic ion that has a positive charge combined with a nonmetal (or polyatomic) ion that has a negative charge. Examples of such salts are ammonium chloride (NH_4Cl) and ammonium nitrate (NH_4NO_3).

Figure 16.7
A neutralization reaction

$$HCl + NaOH \longrightarrow NaCl + H_2O$$

Properties, Preparations, and Some Uses of Salts

Learning Goal 14
Important properties of salts

Different salts have different properties, but *with few exceptions salts are ionic substances* (which means that they are bonded ionically). Their solubilities in water may vary, but their aqueous solutions always contain ions. For example, sodium chloride dissociates into sodium ions and chloride ions when it is dissolved in water. Most salts are strong electrolytes.

$$NaCl \xrightarrow{\text{Water}} Na^{1+} + Cl^{1-}$$

Learning Goal 15
Preparation of salts

There are many ways to prepare salts. However, we'll limit our discussion to three methods: (1) direct combination of two elements, (2) reaction of an active metal with an acid, and (3) neutralization of an acid by a base.

1. When we heat sodium metal in the presence of an excess of chlorine gas, a vigorous reaction takes place. The product is pure sodium chloride. Many salts can be prepared by heating elements together.

$$2Na + Cl_2 \longrightarrow 2NaCl$$

2. In Chapter 10, we noted that hydrogen could be produced by the reaction of an active metal with an acid. The other product formed in such a reaction is a salt. We can use this type of reaction to produce the salt we want by choosing the appropriate metal and acid. If we want to prepare magnesium chloride, for example, we use magnesium metal and hydrochloric acid.

$$Mg + 2HCl \longrightarrow MgCl_2 + H_2$$

An important fact to keep in mind is that the metal we choose must be able to replace the hydrogen in the acid. Table 10.1, the activity series, reveals which metals can replace hydrogen.

3. To form a salt by a neutralization reaction, we choose a base that contains the positive ion of the salt, and an acid that contains the negative ion of the salt. For example, if we want to prepare potassium bromide, we can use the reaction between potassium hydroxide and hydrobromic acid:

$$KOH + HBr \longrightarrow KBr + H_2O$$

Salts have many different uses. Many salts, such as sodium chloride, are necessary for life itself. Calcium phosphate is the main ingredient of our bones. Other salts, such as calcium phosphate dihydrate [$Ca_3(PO_4)_2 \cdot 2H_2O$], are used for building materials. The variety of uses for salts is matched only by the number of salts known.

16.9 Titration

When explaining double-replacement reactions in Section 10.4, we discussed neutralization as one kind of double-replacement reaction. *Neutralization* is the reaction between an acid and a base to form water and a salt. A neutralization reaction is often represented by the general equation

$$HX + MOH \longrightarrow MX + H_2O$$

where HX (an acid) and MOH (a base) form a salt (MX) and water.

Learning Goal 16
Titration

The most common use that chemists make of neutralization reactions is called titration. **Titration** is *a volumetric analytical procedure that enables the chemist to find out how much of a solute is present in solution*. In the laboratory, titration is sometimes used to find the concentrations of chemicals in urine, blood, plasma, and drinking water. When a neutralization titration is performed, the amount of acid or base can be determined.

In an acid–base titration, a measured volume of a solution of acid or base of known concentration (called the titrant) is added to a measured volume of a solution of acid or base whose concentration is to be determined. A device called a buret is used for dispensing the measured quantities of acid or base (Figure 16.8). The buret enables the chemist to stop adding titrant by means of a stopcock at the bottom of the buret. The stopcock can be closed to stop the flow of solution at any time. Calibration marks etched onto the surface of the buret reveal exactly how much titrant has been added. The titrant used is generally a *standard solution* of acid or base, which is a solution whose concentration is accurately known.

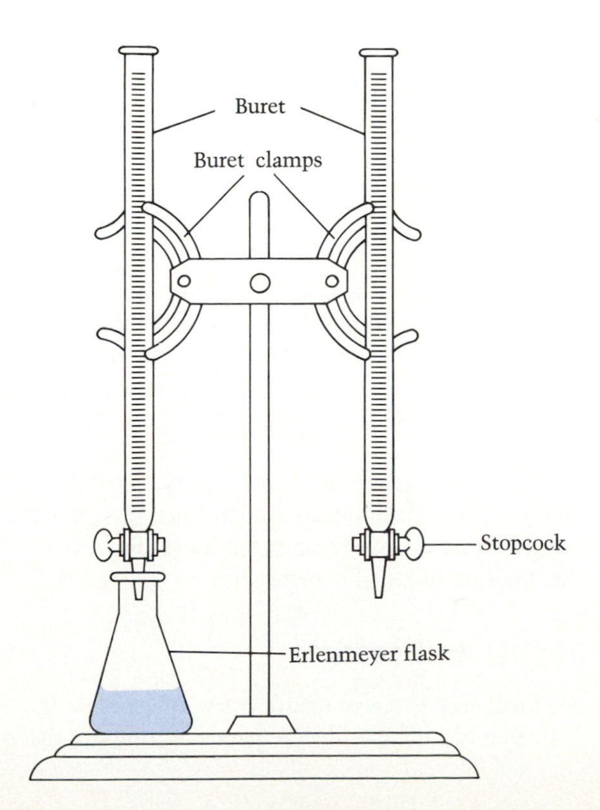

Figure 16.8
A titration setup. A measured volume of one solution is placed in the flask. The buret above it contains another solution. The second solution is added gradually until the end point is reached. The stopcock controls the rate of flow of the solutions into the flask.

How does the chemist know when enough titrant has been added? In other words, when is it evident that the neutralization is complete? *When equal numbers of equivalents of acid and base have been added to the reaction vessel (which is usually an Erlenmeyer flask), the* **equivalence point** *has been reached.* At this point, the volume and concentration of the titrant that has been added can be used to calculate the concentration of the acid or base being titrated.

In order to determine whether the equivalence point has been reached, the chemist must know (1) the pH of a solution of the salt formed by complete neutralization of the particular acid and base involved in the titration and (2) its concentration. In general, the pH at the equivalence point varies with the salt formed by the particular acid and base used in the neutralization reaction, and with the concentration of the salt formed.

For example, if we are neutralizing a solution of a strong base, such as sodium hydroxide, with a standard solution of a strong acid, such as nitric acid, we begin with a solution of sodium hydroxide of unknown concentration and quite high pH. As the nitric acid is added, the pH decreases, because $(OH)^{1-}$ is removed by the reaction $H^{1+} + (OH)^{1-} \rightarrow H_2O$. When the equivalence point is reached, the numbers of equivalents of acid and base are the same and there are equal amounts of H^{1+} and $(OH)^{1-}$. The pH at the equivalence point should be 7, because when $[H^{1+}] = [OH^{1-}]$, each value is 10^{-7} *M*. When a strong acid and a strong base take part in a titration reaction, the equivalence point is reached when the pH reaches 7. At that point, neutralization is complete and no more titrant should be added. The salt sodium nitrate is neutral.

What happens when a strong acid and a weak base take part in a titration? At the equivalence point, when the numbers of equivalents of acid and base are the same, is the pH equal to 7? No, it is not, because when the salt of a strong acid and a weak base forms, it further reacts with water to produce a solution with a pH less than 7. For example, a solution of NH_4Cl is slightly acidic.

Similarly, when a strong base and a weak acid take part in a titration, at the equivalence point the numbers of equivalents of acid and base are the same, but the solution is slightly basic. When the salt of a strong base and a weak acid forms, it further reacts with water to produce a solution with a pH greater than 7. Sodium acetate is an example of a salt that gives a slightly basic solution.

Chemicals called indicators can be carefully selected to show the point at which all the available hydrogen ions have reacted with all the available hydroxide ions in a solution. One must, however, take into account the pH of the solution of salt formed.

16.10 Indicators

An **indicator** is *a dye whose color depends on its pH*. Paper saturated with an indicator is called indicator paper. Litmus is the most common indicator. It

Table 16.9
Common Indicators

Indicator and pH range			
Cresol red 0.2–1.8	Red	Color change	Yellow
Methyl orange 3.1–4.4	Red	Color change	Yellow
Methyl red 4.2–6.2	Red	Color change	Yellow
Litmus 5.5–7.5	Red	Color change	Blue
Bromthymol blue 6.0–7.6	Yellow	Color change	Blue
Cresol purple 7.4–9.0	Yellow	Color change	Purple
Phenolphthalein 8.0–9.8	Colorless	Color change	Red-violet

pH scale: 0 1 2 3 4 5 6 7 8 9 10 11 12 13 14

turns blue when the solution in which it is placed is basic; it turns red when the solution is acidic. *In general, indicators change color within a certain pH range, and the range of pH values within which the color change takes place depends on the particular indicator*. For example, litmus changes from red to blue within the pH range of 5.5 to 7.5. It is blue above 7.5 and red below 5.5. Another common indicator, phenolphthalein, is colorless in acid solution and red-violet in alkaline solution. It changes color within the pH range of 8.0 to 9.8. Table 16.9 lists common indicators and the pH ranges within which their color changes take place.

When an indicator is added to solution that is being titrated and the indicator just changes color, what we call the **end point** has been reached. To do an acid–base titration, chemists estimate the pH at the equivalence point, select an indicator that changes color in the appropriate pH range, and add a few drops of that indicator to the solution being titrated.

16.11 The Mathematics of Acid–Base Titrations

In the laboratory, units of normality are used to simplify calculations. Equal numbers of equivalents of acid and base react with each other. That is,

$$\text{Number of equivalents of acid} = \text{number of equivalents of base}$$

Remember, normality (N) is the number of equivalents of acid or base present in one liter of solution, so

$$\text{Number of equivalents} = \text{normality } (N) \times \text{volume (liters)}$$

At the equivalence point,

$$N_a \times V_a = N_b \times V_b$$

where N_a is the concentration of the acid, V_a is the volume of the acid, N_b is the concentration of the base, and V_b is the volume of the base. Liters do not have to be used as the unit of volume. Any unit may be used as long as it is the same for both V_a and V_b. This equation is often used in calculations involving titrations.

Example 16.3

Calculate the normality (or molarity; see the alternative solution that follows) of an HNO_3 solution if 25.0 mL of the solution are neutralized by 100.0 mL of 0.200 N (0.100 M) $Ca(OH)_2$ solution.

Solution

Understand the Problem

We are asked to calculate the normality of an acidic solution, a given volume (25.0 mL) of which is neutralized by 100.0 mL of 0.200 N basic solution. Organizing the data will help you see what is given and what must be found. Let N_a refer to the normality of the acid and N_b to the normality of the base.

$$N_a = ? \qquad N_b = 0.200\ N$$
$$V_a = 25.0\text{ mL} \qquad V_b = 100.0\text{ mL}$$

Devise a Plan

We can use the equation given in the text to solve for the unknown, N_a.

Carry Out the Plan

Rewrite the formula

$$N_a \times V_a = N_b \times V_b$$

in terms of N_a:

$$N_a = \frac{N_b V_b}{V_a}$$

Then solve the equation.

$$N_a = \frac{(0.200\ N)(100.0\ \cancel{\text{mL}})}{25.0\ \cancel{\text{mL}}} = 0.800\ N$$

The concentration of the acid is 0.800 *N*.

Look Back

We can reason that since 25.0 mL of acidic solution is neutralized by 100.0 mL of 0.200 *N* basic solution, we can expect a higher concentration for the acid. In fact, we see that its concentration is four times greater than that of the base.

Alternative Solution Using Molarity You may also perform titration calculations by using only molarity. Here's how it's done. First calculate how many moles of $Ca(OH)_2$ solution it took to neutralize the HNO_3 solution.

$$M_{Ca(OH)_2} = 0.100\ \frac{\text{mole}}{\text{L}} \qquad V_{Ca(OH)_2} = 100.0\text{ mL, or } 0.1000\text{ L}$$

$$M = \frac{\text{moles}}{\text{liter}} \quad \text{or} \quad M = \frac{n}{V}$$

$$\text{moles} = (M)(\text{liter}) \quad \text{or} \quad n = MV$$

$$?\text{ moles} = \left(0.100\frac{\text{mole}}{\cancel{\text{L}}}\right)(0.1000\ \cancel{\text{L}})$$
$$= 0.0100\text{ mole } Ca(OH)_2$$

Next write the balanced equation for the reaction between the HNO_3 and the $Ca(OH)_2$.

$$2HNO_3 + Ca(OH)_2 \longrightarrow Ca(NO_3)_2 + 2H_2O$$

The balanced equation tells us that 2 moles of HNO_3 are neutralized by 1 mole of $Ca(OH)_2$. Use this information and the factor-unit method to calculate the moles of HNO_3 neutralized by the 0.0100 mole of $Ca(OH)_2$.

$$? \text{ moles } HNO_3 = (0.0100 \text{ mole } \cancel{Ca(OH)_2})\left(\frac{2 \text{ moles } HNO_3}{1 \text{ mole } \cancel{Ca(OH)_2}}\right)$$

$$= 0.0200 \text{ mole } HNO_3$$

Finally, calculate the molarity of the HNO_3 solution from this information and the volume of the HNO_3 (25.0 mL, or 0.0250 L).

$$M = \frac{\text{moles}}{\text{liter}} \quad \text{or} \quad M = \frac{n}{V}$$

$$? M = \frac{0.0200 \text{ mole}}{0.0250 \text{ L}} = 0.800 M$$

Practice Exercise 16.3 Calculate the normality (or molarity) of an HCl solution if 150 mL of the solution are neutralized by $5\bar{0}$ mL of 1.0 N (1.0 M) NaOH solution.

Example 16.4

A quality-control laboratory is analyzing a sample of vinegar to determine its acetic acid content. The technicians find that it requires 14.0 mL of 1.25 N (1.25 M) NaOH solution to neutralize a 25.0-mL sample of vinegar. What is the normality (or molarity) of the vinegar solution?

Solution Organize the data.

$$N_a = ? \qquad N_b = 1.25\ N$$

$$V_a = 25.0 \text{ mL} \qquad V_b = 14.0 \text{ mL}$$

Substitute the values into the equation and solve it for N_a.

$$N_a \times V_a = N_b \times V_b$$

$$N_a = \frac{N_b V_b}{V_a} = \frac{(1.25\ N)(14.0\ \cancel{\text{mL}})}{25.0\ \cancel{\text{mL}}} = 0.700\ N$$

Alternative Solution Using Molarity First calculate how many moles of NaOH it took to neutralize the vinegar.

$$M_{NaOH} = 1.25\ \frac{\text{moles}}{\text{L}} \qquad V_{NaOH} = 14.0\ \text{mL, or } 0.0140\ \text{L}$$

$$M = \frac{\text{moles}}{\text{liter}} \quad \text{or} \quad M = \frac{n}{V}$$

$$\text{moles} = (M)(\text{liter}) \quad \text{or} \quad n = MV$$

$$?\ \text{moles} = \left(1.25\frac{\text{moles}}{\cancel{\text{L}}}\right)(0.0140\ \cancel{\text{L}})$$

$$= 0.0175\ \text{mole NaOH}$$

Next write the balanced equation for the reaction between the $HC_2H_3O_2$ and the NaOH.

$$HC_2H_3O_2 + NaOH \longrightarrow NaC_2H_3O_2 + H_2O$$

The balanced equation reveals that 1 mole of $HC_2H_3O_2$ is neutralized by 1 mole of NaOH. Use this information and the factor-unit method to calculate the moles of $HC_2H_3O_2$ neutralized by the 0.0175 mole of NaOH.

$$?\ \text{moles } HC_2H_3O_2 = (0.0175\ \text{mole } \cancel{\text{NaOH}})\left(\frac{1\ \text{mole } HC_2H_3O_2}{1\ \text{mole } \cancel{\text{NaOH}}}\right)$$

$$= 0.0175\ \text{mole } HC_2H_3O_2$$

Finally, calculate the molarity of the vinegar from this information and the volume of the vinegar solution (25.0 mL, or 0.0250 L).

$$M = \frac{\text{moles}}{\text{liter}}$$

$$?\ M = \frac{0.0175\ \text{mole}}{0.0250\ \text{L}} = 0.700\ M$$

Practice Exercise 16.4 Calculate the normality (or molarity) of a solution of acetic acid if 43.50 mL of vinegar are neutralized by 25.00 mL of 0.15 N (0.15 M) NaOH.

Example 16.5

A 50.00-mL sample of $Mg(OH)_2$ is titrated with 65.00 mL of 0.500 N (0.500 M) HCl. What is the normality (or molarity) of the $Mg(OH)_2$ solution?

Solution Organize the data.

$$N_a = 0.500\ N \qquad N_b = ?$$

$$V_a = 65.00\ \text{mL} \qquad V_b = 50.00\ \text{mL}$$

Substitute the values into the equation and solve it for N_b.

$$N_a \times V_a = N_b \times V_b$$

$$N_b = \frac{N_a V_a}{V_b} = \frac{(0.500\ N)(65.00\ \cancel{\text{mL}})}{50.00\ \cancel{\text{mL}}} = 0.650\ N$$

Alternative Solution Using Molarity First calculate how many moles of HCl it took to neutralize the $Mg(OH)_2$ solution.

$$M_{\text{HCl}} = 0.500\ \frac{\text{mole}}{\text{L}} \qquad V_{\text{HCl}} = 65.00\ \text{mL, or}\ 0.06500\ \text{L}$$

$$M = \frac{\text{moles}}{\text{liter}}$$

$$\text{moles} = (M)(\text{liter})$$

$$?\ \text{moles} = \left(0.500\ \frac{\text{mole}}{\cancel{\text{L}}}\right)(0.06500\ \cancel{\text{L}})$$

$$= 0.0325\ \text{mole HCl}$$

Next write the balanced equation for the reaction between the HCl and the $Mg(OH)_2$.

$$2HCl + Mg(OH)_2 \longrightarrow MgCl_2 + 2H_2O$$

The balanced equation reveals that 2 moles of HCl are neutralized by 1 mole of $Mg(OH)_2$. Use this information and the factor-unit method to calculate the number of moles of $Mg(OH)_2$ neutralized by the 0.0325 mole of HCl.

$$?\ \text{moles Mg(OH)}_2 = (0.0325\ \cancel{\text{mole HCl}})\left(\frac{1\ \text{mole Mg(OH)}_2}{2\ \cancel{\text{moles HCl}}}\right)$$

$$= 0.0163\ \text{mole Mg(OH)}_2$$

Finally, calculate the molarity of the $Mg(OH)_2$ solution from this information and the volume of the $Mg(OH)_2$ solution (50.00 mL, or 0.05000 L).

$$M = \frac{\text{moles}}{\text{liter}}$$

$$?\ M = \frac{0.0163\ \text{mole}}{0.05000\ \text{L}} = 0.326\ M$$

Practice Exercise 16.5 A 25.65-mL sample of NaOH solution is titrated with 100.0 mL of 0.105 N (0.105 M) hydrochloric acid solution. What is the normality (or molarity) of the NaOH solution?

Example 16.6

A chemistry student was determining the normality of an $Fe(OH)_3$ solution. It required 30.00 mL of a 0.200 *N* (0.200 *M*) HCl solution to neutralize 10.00 mL of the $Fe(OH)_3$ solution. Calculate the normality (or molarity) of the $Fe(OH)_3$ solution.

Solution Organize the data.

$$N_a = 0.200\ N \qquad N_b = ?$$

$$V_a = 30.00\ \text{mL} \qquad V_b = 10.00\ \text{mL}$$

Substitute the values into the equation and solve it for N_b.

$$N_a \times V_a = N_b \times V_b$$

$$N_b = \frac{N_a V_a}{V_b} = \frac{(0.200\ N)(30.00\ \cancel{\text{mL}})}{10.00\ \cancel{\text{mL}}} = 0.600\ N$$

Alternative Solution Using Molarity First calculate how many moles of HCl it took to neutralize the $Fe(OH)_3$ solution.

$$M_{HCl} = 0.200 \frac{\text{mole}}{\text{L}} \qquad V_{HCl} = 30.00\ \text{mL, or } 0.03000\ \text{L}$$

$$M = \frac{\text{moles}}{\text{liter}}$$

$$\text{moles} = (M)(\text{liter})$$

$$?\ \text{moles} = \left(0.200\ \frac{\text{mole}}{\cancel{\text{L}}}\right)(0.03000\ \cancel{\text{L}})$$

$$= 0.00600\ \text{mole HCl}$$

Next write the balanced equation for the reaction between HCl and $Fe(OH)_3$.

$$3HCl + Fe(OH)_3 \longrightarrow FeCl_3 + 3H_2O$$

The balanced equation reveals that 3 moles of HCl are neutralized by 1 mole of $Fe(OH)_3$. Use this information and the factor-unit method to calculate the number of moles of $Fe(OH)_3$ neutralized by the 0.0600 mole of HCl.

$$?\ \text{moles } Fe(OH)_3 = (0.00600\ \cancel{\text{mole HCl}})\left(\frac{1\ \text{mole } Fe(OH)_3}{3\ \cancel{\text{moles HCl}}}\right)$$

$$= 0.00200\ \text{mole } Fe(OH)_3$$

Finally, calculate the molarity of the $Fe(OH)_3$ solution from this information and the volume of the $Fe(OH)_3$ solution (10.00 mL, or 0.01000 L).

$$M = \frac{\text{moles}}{\text{liter}}$$

$$?\,M = \frac{0.00200 \text{ mole}}{0.01000 \text{ L}} = 0.200\,M$$

Practice Exercise 16.6 In a titration, it took 20.4 mL of 0.1330 *N* (0.0665 *M*) sulfuric acid solution to neutralize 26.5 mL of potassium hydroxide solution of unknown concentration. Calculate the normality (or molarity) of the potassium hydroxide solution.

Summary

Arrhenius defined an acid as a substance that releases hydrogen ions, H^{1+}, when dissolved in water. Brønsted and Lowry defined an acid as a proton or H^{1+} donor. Some common Arrhenius acids are HCl, HNO_3, H_2SO_4, and $HC_2H_3O_2$. These are also Brønsted–Lowry acids, as are such substances as $(H_3O)^{1+}$, $(HSO_4)^{1-}$, $(NH_4)^{1+}$, and H_2O. Generally, acids taste sour, change the color of certain dyes, react with certain metals to produce hydrogen gas, and neutralize bases. Most important, they dissolve in water to increase the hydrogen-ion concentration. Acids are commonly prepared by the reaction of a nonmetal oxide with water. Acids react with metals as well as with bases. They are used in the manufacturing of a variety of products.

Bases, like acids, can be defined in more than one way. Arrhenius defined a base as a substance that releases hydroxide ions, $(OH)^{1-}$, when dissolved in water. Brønsted and Lowry defined a base as a proton or H^{1+} acceptor. Some common Arrhenius bases are LiOH, KOH, NaOH, $Ca(OH)_2$, and $Mg(OH)_2$. These are also Brønsted–Lowry bases, as are such substances as chloride ions, carbonate ions, sulfate ions, acetate ions, and water. Generally, bases taste bitter, change the color of certain dyes, feel soapy or slippery, and neutralize acids. Most important, they dissolve in water to produce hydroxide ions. Bases can be produced by the reaction of a metal oxide with water. They react with nonmetal oxides and with acids. Like acids, they have a number of uses in industry and in our daily lives.

Strong acids and bases are those that essentially ionize completely in a solution. Weak acids and bases are those that don't ionize completely. In solution, the ions and molecules of a weak acid or a weak base are in dynamic equilibrium.

Water ionizes partially, just like a very weak acid or a very weak base. When a water molecule ionizes, a hydronium ion $(H_3O)^{1+}$ and a hydroxide ion are formed. In 1 L of water, only 10^{-7} mole of water actually dissociates, forming 10^{-7} mole of hydronium ions and 10^{-7} mole of hydroxide ions.

The pH scale provides a means of expressing the very dilute concentrations of acids and bases. The pH of any solution is equal to the negative of the logarithm of its hydrogen-ion concentration in moles per liter. The pH of pure water and that of any other neutral solution is 7. An acidic solution has a pH below 7, and a basic solution has a pH above 7. The product of the hydrogen-ion concentration and the hydroxide-ion concentration in an aqueous solution is always 10^{-14}.

A salt is a compound composed of the positive ion of a base and the negative ion of an acid. In the solid state the positive and negative ions are held together by an ionic bond. In aqueous solution, they are not held together at all. Salts can be prepared by direct combination of two elements, by reaction of an active metal with an acid, or by neutralization of an acid by a base.

Titration is an analytical procedure that enables the chemist to determine the concentration of a substance present in solution. A measured volume of a solution of unknown concentration is found. Acid–base titrations involve a neutralization reaction.

Key Terms

alkali **(16.2)**
Arrhenius acid **(16.1)**
Arrhenius base **(16.2)**
Brønsted–Lowry acid **(16.1)**
Brønsted–Lowry base **(16.2)**
end point **(16.10)**
equivalence point **(16.9)**
hydronium ion **(16.1)**
indicator **(16.10)**
pH **(16.6)**
strong acid **(16.3)**
strong base **(16.3)**
titration **(16.9)**
weak acid **(16.3)**
weak base **(16.3)**

Career Sketch

Radiologic (X-ray) Technologist

As a radiologic (or x-ray) technologist, you will operate x-ray equipment and take x-ray pictures (called radiographs) under the supervision of a radiologist, a physician who specializes in the use of x-rays for diagnosis and treatment of illnesses or injuries. You may work in a hospital, a private physician's office, a clinic, an x-ray manufacturing firm, or an x-ray supply company.

There are three specialties in the field of radiologic technology. *X-ray technology* involves taking x-ray pictures of parts of the human body. *Nuclear medicine technology* uses radioactive isotopes and sophisticated detector equipment to obtain precise images of various internal organs. Radiation therapy is the use of radiation-producing machines to administer therapeutic treatments recommended by radiologists, especially for cancer patients.

To prepare for this career, you will study chemistry, anatomy and physiology, radiologic physics, radiation biology, and radiography in a two- or four-year college program. In most states, once you have completed 2,000 hours of clinical practice, you may take the American Registry Examination and then apply for a license.

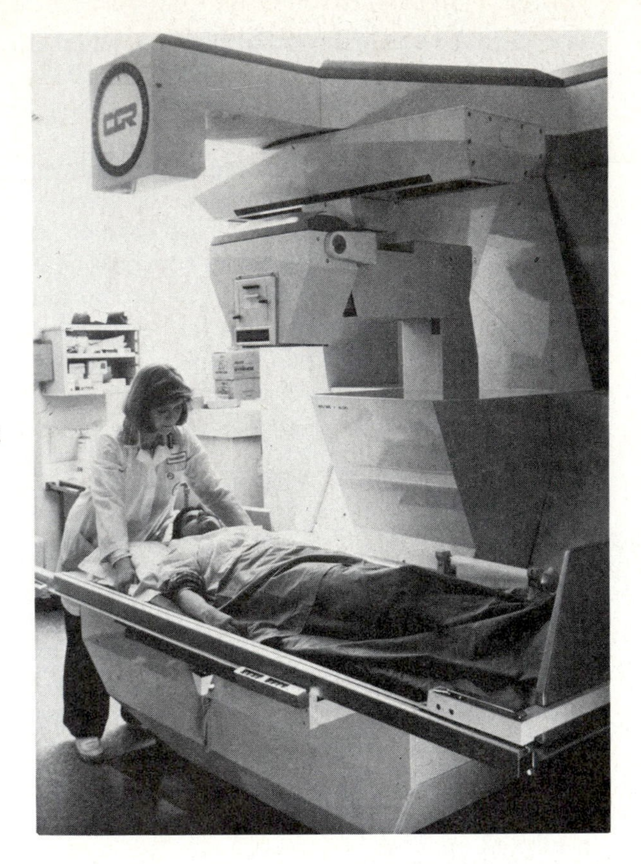

This x-ray technologist uses state-of-the art equipment for diagnostic purposes.
(Yoav Levy/Phototake)

Self-Test Exercises

Learning Goal 1: Arrhenius Acids and Bases

▲ **1.** Using the Arrhenius definition, define *(a)* an acid, *(b)* a base.

2. Name two Arrhenius acids and two Arrhenius bases.

3. What is the acid found in vinegar?

▲ **4.** What are the limitations of the Arrhenius definition of acids and bases?

Learning Goal 2: Brønsted–Lowry Acids and Bases

▲ **5.** Using the Brønsted–Lowry definition, define *(a)* an acid, *(b)* a base.

6. Why is water both an acid and a base according to the Brønsted–Lowry definitions?

***7.** Complete the following equation, which shows why ammonia (NH_3) is a Brønsted–Lowry base:

$$NH_3 + H_2O \rightleftharpoons$$

▲***8.** Complete the following equations, which show why each substance is a Brønsted–Lowry acid or base:

(a) $\underset{\text{Acid}}{(HSO_4)^{1-}} \longrightarrow ? + ?$

(b) $\underset{\text{Base}}{Cl^{1-}} + ? \longrightarrow ?$

Learning Goal 3: Characteristics of Acids and Bases

▲ **9.** What are the characteristic properties of acids, bases, and salts?

10. What do we call a substance that is used to reveal whether a chemical is an acid or a base?

▲ **11.** What is the difference between hydrogen chloride gas and hydrochloric acid?

▲ **12.** What is a hydronium ion?

13. Why do people store acids in containers made of nonmetal materials?

14. A substance that dissolves in water to produce hydroxide ions is a(n) ________.

15. The two types of acids are mineral acids and ________ acids.

16. A substance that dissolves in water to produce hydrogen ions is a(n) ________.

17. Place the words *red* and *blue* in the proper spaces.
(a) ________ litmus paper turns ________ in an acid.
(b) ________ litmus paper turns ________ in a base.

18. An acid feels soapy or slippery. True or false?

19. Write a reaction showing the formation of a hydronium ion from HBr.

20. Write an equation for the formation of hydronium ions from HNO_3.

21. Write a reaction showing the formation of hydroxide ions when calcium hydroxide dissolves in water.

22. Write an equation for the formation of hydroxide ions when magnesium hydroxide dissolves in water.

Learning Goal 4: Preparation of Acids and Bases

▲ **23.** Write a reaction for the preparation of each of the following bases from a metal oxide and water: *(a)* KOH *(b)* $Mg(OH)_2$

24. Write a reaction for the preparation of each of the following bases from a metal oxide and water: *(a)* LiOH *(b)* $Ca(OH)_2$

25. Write a reaction for the preparation of each of the following acids from a nonmetal oxide and water: *(a)* H_2SO_3 (sulfurous acid) *(b)* H_2CO_3 (carbonic acid)

▲ **26.** Write a reaction for the preparation of each of the following acids from a nonmetal oxide and water: *(a)* HNO_3 *(b)* H_3PO_4

Learning Goal 5: Chemical Reactions of Acids and Bases

27. Complete and balance the following equations:
(a) $H_3PO_4 + Mg(OH)_2 \longrightarrow$
(b) $P_2O_5 + H_2O \longrightarrow$
(c) $K_2O + H_2O \longrightarrow$
(d) $Mg(OH)_2 + CO_2 \longrightarrow$
(e) $Zn + H_3PO_4 \longrightarrow$

28. Complete and balance the following equations:
(a) $HC_2H_3O_2 + Ca(OH)_2 \longrightarrow$
(b) $Na_2O + H_2O \longrightarrow$
(c) $Mg + HCl \longrightarrow$
(d) $Sr(OH)_2 + CO_2 \longrightarrow$
(e) $N_2O_5 + H_2O \longrightarrow$

▲ **29.** Suppose you spill some sodium hydroxide on your clothes. What could you do to keep this chemical from ruining them?

▲ ***30.** Suppose that you accidentally drank a liquid containing sodium hydroxide. What should you do to neutralize its effect?

31. Do vinegar and lye react? If a reaction does take place, write a chemical equation for it.

***32.** Power plants that burn coal with a high sulfur content send huge quantities of SO_2 into the atmosphere. When the SO_2 reacts with the moisture in the clouds, acid rain results. Write an equation for the formation of acid rain.

▲ **33.** Paper mills emit huge quantities of sulfur dioxide (SO_2) into the atmosphere at the mill site. What can happen to exposed metal machinery when rain falls near an operating paper mill?

34. Write a balanced chemical equation for each of the following word equations:
(a) magnesium + hydrochloric acid ⟶ ? + ?
(b) sulfuric acid + sodium hydroxide ⟶ ? + ?
(c) ? + ? ⟶ magnesium sulfate + hydrogen gas
(d) ? + ? ⟶ zinc phosphate + water

35. Write balanced chemical equations for the following word equations:

(a) zinc + sulfuric acid $\longrightarrow$ zinc sulfate + hydrogen gas
(b) hydrochloric acid + lithium hydroxide $\longrightarrow$? + ?
(c) ? + ? $\longrightarrow$ calcium sulfate + H_2
(d) ? + ? $\longrightarrow$ calcium phosphate + water

36. Complete and balance the following equations:
(a) $Zn + H_3PO_4 \longrightarrow$
(b) $Zn + HCl \longrightarrow$
(c) $Zn + H_2SO_4 \longrightarrow$

37. Complete and balance the following equations:
(a) $Zn + HCl \longrightarrow$
(b) $HCl + Ca(OH)_2 \longrightarrow$
(c) $Mg(OH)_2 + CO_2 \longrightarrow$

38. All of the reactions in Exercise 36 can be classified as ________ reactions.

Learning Goal 6: Strong Acids and Bases

▲ **39.** Define and give an example of *(a)* a strong acid *(b)* a strong base

▲ **40.** What is the difference between a strong acid and a strong base?

41. Write formulas for the following acids:
(a) nitric acid *(b)* hydrofluoric acid
(c) hydroiodic acid

42. Write formulas for the following acids:
(a) phosphoric acid *(b)* sulfuric acid *(c)* boric acid

▲ **43.** Give names for the following bases: *(a)* NaOH *(b)* $Ca(OH)_2$ *(c)* $Fe(OH)_2$ *(d)* $Al(OH)_3$

▲ **44.** Give names for the following bases:
(a) $Mg(OH)_2$ *(b)* $Zn(OH)_2$ *(c)* KOH *(d)* LiOH

45. Give names for the following acids: *(a)* HNO_3 *(b)* H_3PO_4 *(c)* HI *(d)* H_2SO_4

46. Name the following acids: *(a)* HCl *(b)* H_2SO_3 *(c)* H_2CO_3 *(d)* $HC_2H_3O_2$

47. Write formulas for the following bases:
(a) lithium hydroxide *(b)* barium hydroxide
(c) iron(II) hydroxide *(d)* iron(III) hydroxide

48. Write formulas for the following bases:
(a) potassium hydroxide *(b)* sodium hydroxide
(c) zinc hydroxide *(d)* calcium hydroxide

Learning Goal 7: Weak Acids and Bases

▲ **49.** Define and give an example of *(a)* a weak acid, *(b)* a weak base.

▲ **50.** A weak acid or a weak base ionizes almost completely. True or false?

Learning Goal 8: Dynamic Equilibrium

***51.** Explain what is meant by dynamic equilibrium. Show how it is related to weak acids and bases.

▲***52.** Explain how a dynamic equilibrium comes to exist when carbon dioxide is dissolved in water to produce carbonic acid, a weak acid.

Learning Goal 9: Ionization of Water

53. Write an equation for the ionization of water.

54. How many moles of hydrogen ions and how many moles of hydroxide ions are present in 2 L of water?

Learning Goal 10: pH and pH Scale

55. State how acidic or basic each of the following substances is (very acidic, moderately acidic, very basic, and so on):
(a) vinegar, pH = 2.8
(b) sea water, pH = 8.5
(c) blood, pH = 7.4
(d) saliva, pH = 7.0

56. State how acidic or basic each of the following substances is (very acidic, moderately acidic, very basic, and so on):
(a) oranges, pH = 3.5
(b) milk, pH = 6.5
(c) lemons, pH = 2.3
(d) rainwater, pH = 6.2

▲ **57.** Why is it important to measure the pH of the water in a swimming pool?

▲ **58.** Why must the water in a fish tank be the proper pH?

***59.** The pH scale is a logarithmic scale. This means that a difference of one pH unit indicates a multiple of 10 between the actual acidities or basicities of two substances.
(a) The pH of grapefruit juice is 3; the pH of beer is 5. How many times more acidic is grapefruit juice than beer?
(b) The pH of lemon juice is 2; the pH of pure water is 7. How many times more acidic is lemon juice than water?

*60. The pH of soft drinks is 3.0; the pH of pure water is 7.0. How many times more acidic is the soft drink than the water?

61. In your own words, give a working definition of pH.
62. What is "pH-balanced shampoo"?

Learning Goal 11: Determining pH, Hydrogen-Ion or Hydroxide-Ion Concentration

▲ 63. Determine the pH of each of the following aqueous solutions:
(a) $[H^{1+}] = 10^{-3}$
(b) $[H^{1+}] = 10^{-4}$
(c) $[H^{1+}] = 10^{-2}$
(d) $[H^{1+}] = 10^{-11}$
(e) $[OH^{1-}] = 10^{-4}$
(f) $[OH^{1-}] = 10^{-9}$
64. Determine the pH of each of the following aqueous solutions:
(a) $[H^{1+}] = 10^{-5}$
(b) $[H^{1+}] = 10^{-10}$
(c) $[H^{1+}] = 10^{-8}$
(d) $[OH^{1-}] = 10^{-5}$
(e) $[OH^{1-}] = 10^{-10}$
(f) $[OH^{1-}] = 10^{-8}$

▲ 65. Determine the $[H^{1+}]$ and $[OH^{1-}]$ of each of the aqueous solutions with the following pH values:
(a) pH = 5
(b) pH = 9
(c) pH = 2
(d) pH = 12
66. Determine the $[H^{1+}]$ and $[OH^{1-}]$ of each of the aqueous solutions with the following pH values:
(a) pH = 4
(b) pH = 7
(c) pH = 1
(d) pH = 13

Learning Goal 12: Relationship Between Hydrogen-Ion and Hydroxide-Ion Concentrations

▲ 67. For each of the following aqueous solutions, determine the hydrogen-ion concentration and the pH:
(a) $[OH^{1-}] = 10^{-5}$
(b) $[OH^{1-}] = 10^{-8}$
(c) $[OH^{1-}] = 10^{-10}$
(d) $[OH^{1-}] = 10^{-2}$
*68. For each of the following aqueous solutions, determine the hydrogen-ion concentration and the pH:
(a) $[OH^{1-}] = 10^{-4}$
(b) $[OH^{1-}] = 10^{-11}$
(c) $[OH^{1-}] = 10^{-9}$
(d) $[OH^{1-}] = 10^{-3}$

*69. Determine the $(OH)^{1-}$ concentration for each of the following aqueous solutions:
(a) $[H^{1+}] = 10^{-3}$
(b) $[H^{1+}] = 10^{-4}$
(c) $[H^{1+}] = 10^{-8}$
(d) pH = 9
(e) pH = 6
(f) pH = 2
*70. Determine $[OH^{1-}]$ for each of the following aqueous solutions:
(a) $[H^{1+}] = 10^{-2}$
(b) $[H^{1+}] = 10^{-9}$
(c) $[H^{1+}] = 10^{-5}$
(d) pH = 3
(e) pH = 4
(f) pH = 7

Learning Goal 13: Definition of Salt

▲ 71. Name each of the following salts: *(a)* Na_2S *(b)* KBr *(c)* Na_2SO_4 *(d)* KNO_3 *(e)* $CaCO_3$ *(f)* BaI_2
▲ 72. Name each of the following salts: *(a)* Na_2CO_3 *(b)* KCl *(c)* $BaCl_2$ *(d)* $NaNO_3$ *(e)* $Ca(NO_3)_2$ *(f)* CaI_2

▲ 73. Write the formula for each of the following salts: *(a)* silver nitrate *(b)* cobalt(II) chloride *(c)* copper(I) sulfite *(d)* mercury(II) phosphate
▲ 74. Write the formula for each of the following salts: *(a)* silver chloride *(b)* iron(III) nitrate *(c)* iron(II) nitrate *(d)* copper(I) chloride

Learning Goal 14: Important Properties of Salts

▲ 75. List at least two important properties of salts.
▲ 76. What kind of bonding is found in all salts?

Learning Goal 15: Preparation of Salts

▲ **77.** List at least two ways in which salts can be prepared by chemical reactions.

▲ **78.** How would you prepare the salt silver nitrate?

79. Write an acid–base reaction showing the preparation of the salt lithium sulfate (Li_2SO_4).

80. Write an acid–base reaction showing the preparation of the salt lithium nitrate ($LiNO_3$).

81. Write a single-replacement reaction showing the preparation of the salt zinc bromide ($ZnBr_2$).

82. Write a single-replacement reaction showing the preparation of the salt zinc chloride ($ZnCl_2$).

83. Write a single-replacement reaction showing the preparation of each of the following salts:
(a) magnesium chloride *(b)* zinc sulfate

84. Write a single-replacement reaction showing the preparation of each of the following salts:
(a) magnesium bromide *(b)* copper(II) sulfate

85. Write an acid–base reaction showing the preparation of each of the following salts: *(a)* potassium phosphate *(b)* calcium nitrate

86. Write an acid–base reaction showing the preparation of each of the following salts: *(a)* lithium phosphate *(b)* calcium chloride

87. Write a reaction for the preparation of each of the following salts by the direct combination of their elements: *(a)* KCl *(b)* $CaBr_2$

88. Write a reaction for the preparation of each of the following salts by the direct combination of their elements: *(a)* KI *(b)* $CaCl_2$

Learning Goal 16: Titration

▲ **89.** Calculate the normality of an HNO_3 solution if 30.0 mL of the solution are neutralized by 75.0 mL of 0.25 *N* NaOH.

▲ **90.** Calculate the normality and the molarity of an HCl solution if 125 mL of the solution are neutralized by 25.0 mL of 1.5 *N* NaOH.

91. Calculate the normality and the molarity of a solution of acetic acid if 45.00 mL of the solution are neutralized by 50.00 mL of 0.2500 *N* (0.2500 *M*) NaOH.

92. A 35.0-mL sample of NaOH solution is titrated with 75.0 mL of 0.150 *N* lactic acid. Calculate the normality of the NaOH solution.

93. A 25.0-mL sample of vinegar is titrated with 30.00 mL of 0.150 *N* (0.150 *M*) NaOH solution. Calculate the normality and the molarity of the vinegar.

94. A 35.00-mL sample of KOH solution required 25.00 mL of 1.000 *N* H_2SO_4 to neutralize it. Calculate the concentration of the KOH solution.

95. If 50.0 mL of 1.00 *M* HCl solution neutralize 75.0 mL of $Ca(OH)_2$ solution, what is the molarity of the $Ca(OH)_2$ solution?

96. Calculate the molarity of an NaOH solution if 25.0 mL of 0.150 *M* HCl are needed to titrate 30.0 mL of the NaOH solution.

Extra Exercises

97. Consider the reaction $PH_3 + NaH \longrightarrow NaPH_2 + H_2$.
(a) Identify PH_3 as a Brønsted–Lowry acid or base.
(b) Identify NaH as a Brønsted–Lowry acid or base.

98. Consider the reaction $NH_3 + HCl \longrightarrow (NH_4)^{1+} + Cl^{1-}$.
(a) Identify NH_3 as a Brønsted–Lowry acid or base.
(b) Identify HCl as a Brønsted–Lowry acid or base.

99. Consider the reaction $(HSO_4)^{1-} + (H_3O)^{1+} \longrightarrow H_2SO_4 + H_2O$.
(a) Identify $(HSO_4)^{1-}$ as a Brønsted–Lowry acid or base.
(b) Identify $(H_3O)^{1+}$ as a Brønsted–Lowry acid or base.

100. Calculate the pH and $[OH^{1-}]$ of a 0.001 *M* solution of NaOH.

101. Complete and balance the following equations:
(a) $HCl + NaOH \longrightarrow$
(b) $BaO + HCl \longrightarrow$
(c) $Zn + H_2SO_4 \longrightarrow$

▲ **102.** Show by a reaction with water that each of the following is a base: *(a)* $(OH)^{1-}$ *(b)* NH_3 *(c)* $(HSO_4)^{1-}$

▲ **103.** Show by a reaction with water that each of the following is an acid: *(a)* H_2O *(b)* $(HCO_3)^{1-}$ *(c)* HNO_3

▲ **104.** Many midwestern industries give off large amounts of sulfur trioxide, SO_3, through their smokestacks. Explain why the pH of lakes in New England is decreasing.

105. Explain why phosphorus trichloride, PCl_3, might be a Brønsted–Lowry base.

▲ **106.** Suggest a way to prepare each of the following via a correctly balanced equation: *(a)* Na_2SO_4 *(b)* $ZnCl_2$ *(c)* KCl *(d)* H_2SO_3 (*Hint:* Choose your reactants, and then use them in a combination reaction, single-replacement reaction, or double-replacement reaction.)

17 Chemical Kinetics and Chemical Equilibrium

Learning Goals

After you have studied this chapter, you should be able to:

1. Define the terms *chemical kinetics* and *chemical equilibrium,* and state LeChatelier's principle.
2. Define the terms *reaction mechanism, activated complex,* and *activation energy*.
3. List the factors that affect the rate of a chemical reaction and those that affect chemical equilibrium, and describe the effect of each.
4. Explain what is meant by a reversible reaction.
5. Explain how the rates of the forward and reverse reactions change as a chemical reaction approaches equilibrium.
6. Write the equilibrium expression for a chemical reaction, and calculate the equilibrium constant K_{eq}, given the necessary information.
7. Calculate the ionization constant K_a for a weak acid or K_b for a weak base, given the appropriate data.
8. Calculate the equilibrium concentrations of reactants and products, given the ionization constant K_a or K_b and the initial concentrations of the reactants.
9. Write the expression for the solubility product constant K_{sp} for a slightly soluble salt, and calculate K_{sp}, given the appropriate data.
10. Calculate the solubility of a salt when you are given its solubility product constant K_{sp}.
11. Explain what is meant by the term *buffer*.
12. Explain how the carbonic acid/bicarbonate buffer system maintains the pH of the blood at a fairly constant level.

Introduction

In water, hydrogen ions, H^{1+}, react with hydroxide ions, $(OH)^{1-}$, to form water, H_2O. This reaction occurs in less than a millionth of a second. Water is also formed when hydrogen gas, H_2, reacts with oxygen gas, O_2. But this reaction is so slow that if we mixed the two gases together today and left them, we could come back in the year 2000 and find hardly any water. Obviously, these two reactions occur at vastly different rates.

As another example of very different reaction rates, compare the oxidation of gunpowder (an explosion) with the oxidation of iron (rusting). Gunpowder explodes almost instantaneously, but the rusting of iron is a slow reaction that takes place over a long period of time.

In this chapter, we discuss several ideas that are related to chemical reaction rates. We look at the process by which reactants combine to form products. We discuss the dynamic equilibrium that exists between reactants and products in many chemical reactions. And we talk about the factors that affect both this equilibrium and the rate at which the reaction takes place.

17.1 Chemical Kinetics: Rates of Reaction and the Reaction Mechanism

Learning Goal 1
Chemical kinetics

Chemical kinetics is *the study of how chemical reactions occur and the rates at which they occur*. One theory of chemical kinetics explains how molecules and atoms react. According to this **collision theory,** a chemical reaction can occur only if reactant molecules collide with each other. However, a simple collision is not enough. The molecules must possess a certain amount of energy. We can say that there is an energy barrier—a certain minimum amount of energy—that is necessary for a reaction to take place. As Figure 17.1

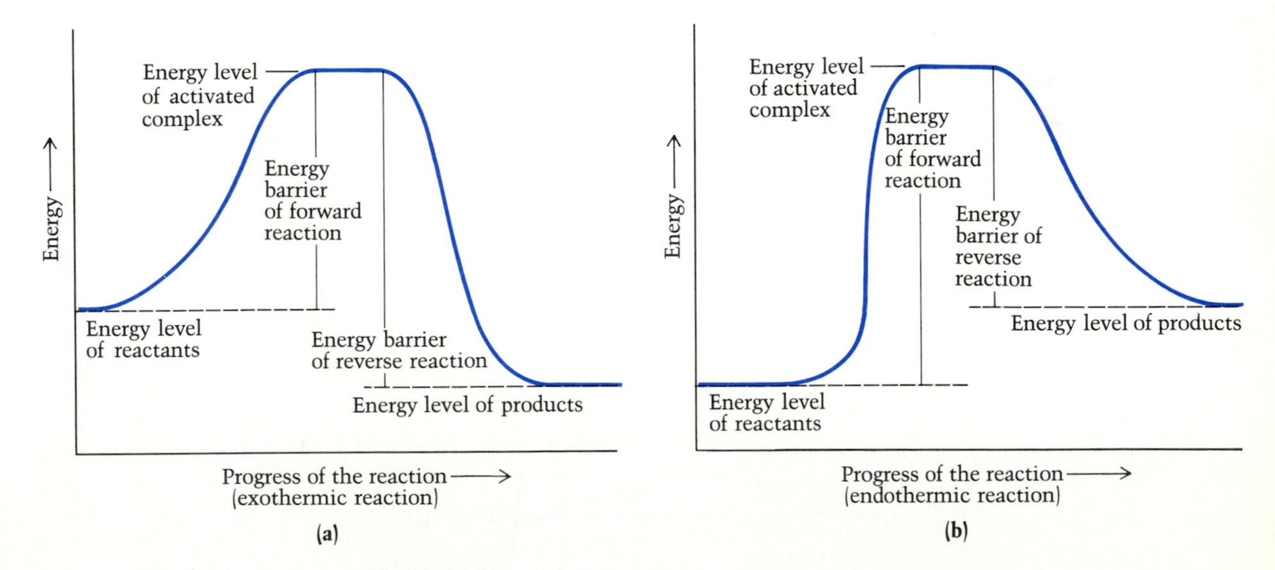

Figure 17.1
Reactant molecules must attain enough energy to overcome this energy barrier. Once the energy barrier is overcome, reaction can take place in either direction.

shows, this energy barrier is larger than the energy of the reactant molecules or the energy of the product molecules.

Thus the reactant molecules—if they are to react—must pick up some extra energy. Molecules that have this extra energy are said to be *activated*. And the total energy needed to overcome the energy barrier is called the **activation energy.** When two activated reactant molecules (for instance, one activated molecule X_2^* and one activated molecule Y_2^*) collide, they join to form an activated complex, which may be represented by the symbol $X_2^*Y_2^*$. This activated complex is not very stable (that is, it does not last long). It is simply part of the path of a chemical reaction: the **reaction mechanism.** However, the molecules that make up the activated complex do possess the required energy of activation. So when the activated complex breaks down, it can form either the reaction product XY or the reverse, the reactants X_2 and Y_2 again (Figure 17.2).

Learning Goal 2
Mechanism of reactions

Sometimes two colliding reactant molecules have enough energy to form an activated complex, but they do not have the proper *orientation*. That is, the wrong parts of the molecules are touching, so the X—Y bond cannot form. In that case, there is no activated complex formed that can lead to the product XY. A collision that does lead to a chemical reaction is called an *effective* collision.

Learning Goal 3
Factors that affect the rate of a chemical reaction

On the basis of this collision theory, we can see that several factors influence the rate at which a chemical reaction occurs. Of course, the *reactants* themselves affect their reaction rate. That is, different sets of reactants react at different rates, depending on their physical and chemical properties. But for a given set of reactants, the *concentration,* the *temperature,* and the *presence of catalysts* influence the rate at which the reaction proceeds.

Concentration

The greater the *concentration* of a reactant, the more molecules are present and the more often collisions take place. Therefore, increasing the concentra-

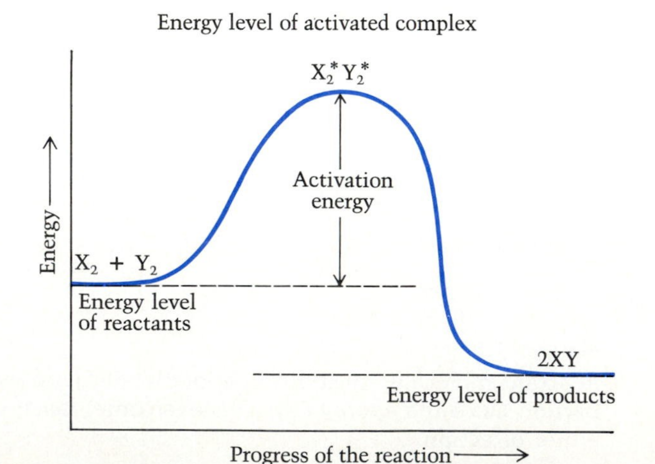

Figure 17.2
The activation energy is the amount of energy needed to form the activated complex through which the forward or reverse reaction occurs.

tion increases the reaction rate. We can often speed up a slow reaction by adding more of one reactant. This makes it easier for collisions to occur. Collisions between molecules and atoms may be likened to the conditions on a crowded roller-skating rink. When many skaters are present on the rink, collisions occur often. When just a few skaters are present, collisions are less likely to occur.

Temperature

By increasing the *temperature* at which a reaction takes place, we increase the energy of the reactant particles. This has three main effects: (1) it causes the reactant particles to move about more rapidly, so collisions occur more frequently; (2) it causes more reactant particles to attain activation energy; (3) it causes more product particles to attain activation energy. For these reasons, more collisions occur in a given time and more of them are effective collisions, so the reaction rate increases. *(It has been found that an increase of about 10°C often doubles the rate at which a reaction takes place.)* Returning to our skating-rink analogy, we note that when the skaters are moving about on the rink at greater speeds, they are more likely to collide with one another.

Presence of Catalysts

A **catalyst** is a substance that increases the rate at which a reaction occurs. A catalyst remains unchanged at the end of a reaction. It does not change the concentrations or the temperature of the reactants. It does, however, present a different path for a reaction, a path with a lower energy barrier. It makes possible a new activated complex whose lower activation energy increases the chance of effective collisions (Figure 17.3). Because more reactant molecules can attain this lower activation energy, the reaction rate increases.

The rate of reaction between hydrogen gas and oxygen gas, for example, has a very high activation energy; it proceeds very slowly. The addition of a

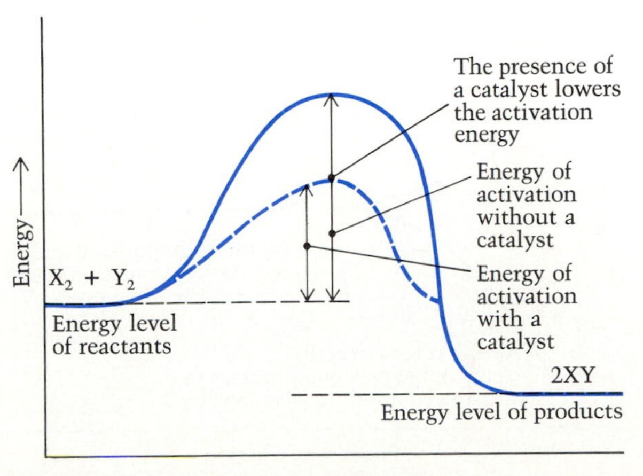

Figure 17.3
The presence of a proper catalyst presents a new reaction path with a lower activation energy.

catalyst consisting of finely divided platinum lowers the activation energy and allows the reaction to take place at a much faster rate. The platinum is unchanged when the reaction ends.

17.2 Reversible Chemical Reactions

Learning Goal 4
Reversible reactions

In previous chapters, we have considered mainly irreversible reactions. An **irreversible reaction** is *a chemical reaction in which one or more reactants form one or more products—but the products cannot then react to form the reactants.* An irreversible reaction stops when one of the reactants is used up. For example, when a raw egg is cooked and has turned into a hard-boiled egg, or when ingredients are mixed and baked to form a cake, an irreversible reaction has occurred. It is not possible to unboil the egg or unbake the cake. When an irreversible reaction stops, we say that it has *gone to completion.*

But many reactions are reversible. A **reversible reaction** is *a chemical reaction in which the products, once they are formed, can react to yield the original reactants.* We can represent the general case of a reversible reaction as

$$A + B \rightleftharpoons C + D$$

where A and B are the original reactants, and C and D are the original products formed. The reaction $A + B \longrightarrow C + D$ is called the *forward* reaction, and $C + D \longrightarrow A + B$ is called the *reverse* reaction. The double arrow tells us that the reaction is reversible. The reaction *does not* go to completion, because products C and D are reacting to form reactants A and B at the same time that A and B are reacting to form C and D.

Learning Goal 1
Chemical equilibrium

However, the system eventually attains an equilibrium condition. In a **chemical equilibrium system,** *the forward reaction and the reverse reaction occur at the same rate.* Figure 17.4 shows how this equilibrium condition comes about. The forward reaction starts off at a fast rate, because the con-

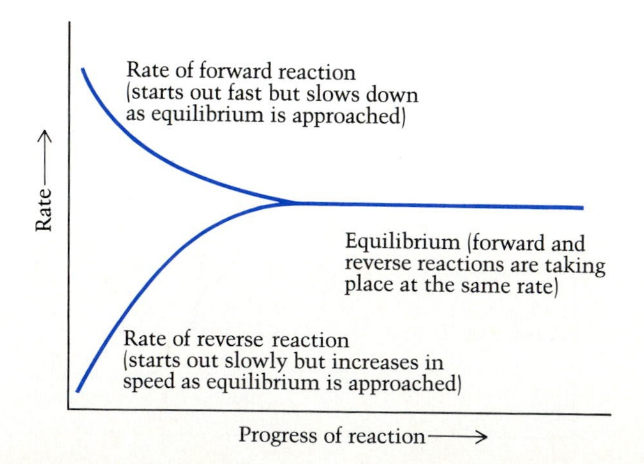
Rate of forward reaction
(starts out fast but slows down
as equilibrium is approached)
Rate ⟶
Equilibrium (forward and
reverse reactions are taking
place at the same rate)
Rate of reverse reaction
(starts out slowly but increases in
speed as equilibrium is approached)
Progress of reaction ⟶

Figure 17.4
When the forward and reverse reactions are proceeding at the same rate, equilibrium has been achieved.

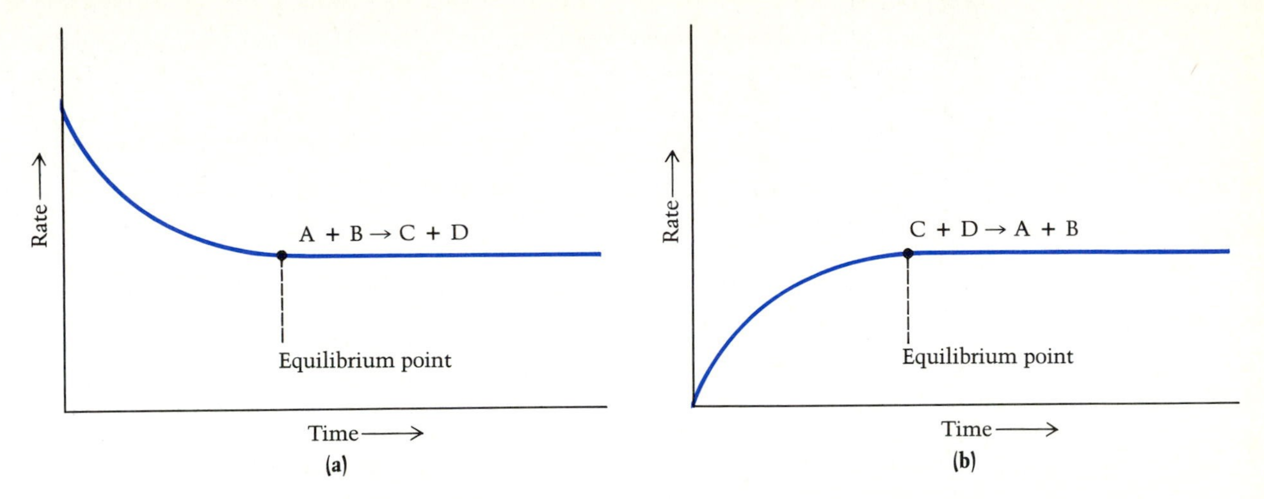

Figure 17.5
Rates of formation of products and reactants before and at equilibrium. (*a*) In the forward reaction, the reactants combine to form products. Note that the forward reaction starts out fast but slows down as equilibrium is approached. (*b*) In the reverse reaction, the products recombine to form reactants. Note that the reverse reaction starts out slowly but increases in speed as equilibrium is approached.

centration of reactant particles is high. However, as the reaction proceeds, products are formed and the concentration of reactants decreases, so the forward reaction rate decreases. At first, the concentration of products is zero, so the reverse reaction rate is zero. As products are formed by the forward reaction, the concentration of products increases, so the reverse reaction rate increases. Eventually (at equilibrium), the two reaction rates become and remain equal (Figure 17.5).

Learning Goal 5
Rates of reaction and equilibrium

In a chemical equilibrium, both products and reactants are being formed, so a chemical equilibrium is a *dynamic* equilibrium. However, equilibrium does not mean that equal *amounts* of products and reactants are present—only that they are being formed at equal *rates*.

In Chapter 16, we explained dynamic equilibrium by using the analogy of the people eating in a restaurant during a busy lunch hour. The restaurant can seat 200 people, and all the seats are taken by noon. All the seats are still taken 15 minutes later, but people are constantly entering and leaving the restaurant. This same situation continues over the next 45 minutes. In other words, during lunchtime at this restaurant, there is a population equilibrium. As a given number of people finish eating and leave the restaurant, the same number of hungry people enter it.

As a specific example of a reversible reaction, consider a solution containing 1 mole of acetic acid ($HC_2H_3O_2$) in a liter of solution at 25°C. Analysis of the solution would show that most of the acetic acid is in the form of molecules and that only about 0.4% is in the form of hydrogen ions and acetate ions. That is, the forward reaction, in which the acetic acid dissociates into ions, is only about 0.4% complete.

$$HC_2H_3O_2 \xrightarrow{H_2O} H^{1+} + (C_2H_3O_2)^{1-} \qquad \text{(about 0.4\% complete)}$$

If you begin with 1 *M* ions of H^{1+} and $(C_2H_3O_2)^{1-}$, however, the reaction is 99.6% complete.

$$H^{1+} + (C_2H_3O_2)^{1-} \xrightarrow{H_2O} HC_2H_3O_2 \qquad \text{(about 99.6\% complete)}$$

Thus an equilibrium system exists. It can be represented by a single equation:

$$HC_2H_3O_2 \rightleftharpoons H^{1+} + (C_2H_3O_2)^{1-}$$

Once again, the double arrow indicates that the system is *reversible* and that an *equilibrium* exists. Here the equilibrium is among the acetic acid molecules, the hydrogen ions, and the acetate ions. (Notice that the arrows in this equation are like those we used in Chapter 16 to show the side of the equation that the equilibrium favors. In this case, the equilibrium lies in favor of the undissociated $HC_2H_3O_2$ molecules.)

We have examined a few reversible chemical systems in previous chapters. For example, we discussed the *dissociation* of sodium chloride. The forward reaction may be represented as

$$NaCl(s) + H_2O \longrightarrow Na^{1+}(aq) + Cl^{1-}(aq) \qquad \text{(dissociation)}$$

The reverse reaction may be represented as

$$Na^{1+}(aq) + Cl^{1-}(aq) \longrightarrow NaCl(s) + H_2O \qquad \text{(crystallization)}$$

In a *saturated* solution of NaCl, dissociation and crystallization occur simultaneously. At equilibrium, these two reactions may be represented by one equation:

$$NaCl(s) \underset{}{\overset{H_2O}{\rightleftharpoons}} Na^{1+}(aq) + Cl^{1-}(aq)$$

(Notice that the arrows in this equation are of equal size. We'll use this notation when we're not interested in showing the side of the equation that the equilibrium favors.)

An example of a reversible reaction that is *visible* is the conversion of nitrogen dioxide (NO_2), a colored gas, to dinitrogen tetraoxide (N_2O_4), a colorless gas. This reaction is very sensitive to temperature. When pure NO_2 gas (which is yellow-brown) is sealed in a container, the intensity of its color changes with temperature, indicating that different amounts of $NO_2(g)$ and $N_2O_4(g)$ exist in equilibrium (Figure 17.6). The equilibrium reaction may be represented as follows:

$$\underset{\text{Colored}}{2NO_2(g)} \underset{\text{Heating}}{\overset{\text{Cooling}}{\rightleftharpoons}} \underset{\text{Colorless}}{N_2O_4(g)}$$

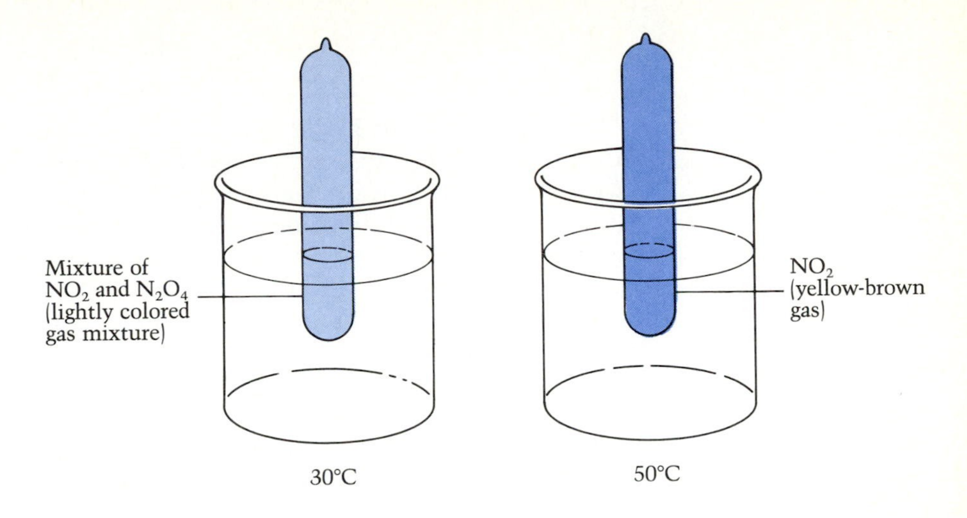

Figure 17.6
The conversion of colored NO_2 to colorless N_2O_4 is temperature-sensitive.

17.3 The Equilibrium Constant

Learning Goal 6
Equilibrium expression and equilibrium constant

In 1862 Marcellin Berthelot and Jean de Saint Gilles were experimenting with the reaction in which ethyl alcohol and acetic acid combine to produce ethyl acetate and water:

$$\text{Ethyl alcohol} + \text{acetic acid} \longrightarrow \text{ethyl acetate} + \text{water}$$

The two scientists mixed various amounts of the reactants and then measured the concentrations of the reactants and products. (We now know that they were measuring *equilibrium concentrations*.) They found that in every case, no matter how much of each reactant they started with, the concentrations of the products and reactants had the same relation to each other. They then mixed ethyl acetate and water and found that the reaction was reversible. And again they found the same kind of relation among the concentrations of the reactants and products.

In 1864 two Norwegian chemists, Cato Guldberg and Peter Waage, discovered that for an equilibrium system, the ratio of the product of the *product concentrations* to the product of the *reactant concentrations* is a constant. That is,

$$\frac{\text{Product of } \textit{product concentrations}}{\text{Product of } \textit{reactant concentrations}} = \text{constant } K_{eq} \qquad \text{(at temperature } T\text{)}$$

regardless of the initial concentrations of the reactants. This is called the **Law of Mass Action.** The concentrations are usually expressed in moles per liter, and the constant K_{eq} is called the **equilibrium constant** for the reaction.

For the general reversible reaction

$$A + B \rightleftharpoons C + D$$

we would write the equilibrium constant as

$$K_{eq} = \frac{[C][D]}{[A][B]}$$

where, as usual, the brackets mean "concentration in moles per liter." This expression for K_{eq} is called the **equilibrium expression.** If we measured the concentrations of the reactants and products at equilibrium at a given temperature, we could substitute them into the equilibrium expression and find the value of K_{eq} for that temperature. (The equilibrium constant changes with the temperature but is not affected by the initial concentrations of the reactants.)

Suppose we wanted to find the equilibrium expression for the reaction

$$H_2(g) + I_2(g) \rightleftharpoons 2HI(g)$$

How would we handle the coefficient 2 on the right side of the equation? One way to do so is to note that we can write the chemical equation without coefficients:

$$H_2(g) + I_2(g) \rightleftharpoons HI(g) + HI(g)$$

Then, as above, we can write the equilibrium expression as

$$K_{eq} = \frac{[HI][HI]}{[H_2][I_2]}$$

Now, because [HI][HI] is $[HI]^2$, we write

$$K_{eq} = \frac{[HI]^2}{[H_2][I_2]}$$

There really is no need to go through these intermediate steps. Instead, we simply write the coefficients of the balanced equation as exponents in the equilibrium expression. Thus, for the general reversible reaction

$$aA + bB \rightleftharpoons cC + dD$$

the equilibrium expression is

$$K_{eq} = \frac{[C]^c[D]^d}{[A]^a[B]^b}$$

Example 17.1

Write the equilibrium expression for the reaction

$$N_2(g) + 3H_2(g) \rightleftharpoons 2NH_3(g)$$

Solution Apply the method we just discussed, being sure to place the product concentration in the numerator of the ratio and the reactants in the denominator.

$$K_{eq} = \frac{[NH_3]^2}{[N_2][H_2]^3}$$

Practice Exercise 17.1 Write the equilibrium expression for the reaction

$$CO(g) + Cl_2(g) \rightleftharpoons COCl_2(g)$$

Example 17.2

Calculate the equilibrium constant for the reaction

$$2NO(g) + O_2(g) \rightleftharpoons 2NO_2(g)$$

given that the equilibrium molar concentrations (at a particular temperature) are as follows: $[NO_2]$ = 8.8 moles/liter, $[NO]$ = 1.2 moles/liter, and $[O_2]$ = 1.6 moles/liter.

Solution

Understand the Problem

We are asked to calculate K_{eq} for a certain reaction with given equilibrium molar concentrations. Organizing the data will help you see what is given and what must be found.

$$[NO_2] = 8.8 \text{ moles/L}$$

$$[NO] = 1.2 \text{ moles/L}$$

$$[O_2] = 1.6 \text{ moles/L}$$

Devise a Plan

We can write the equilibrium expression from the balanced equation for the reaction and enter the given concentrations to calculate K_{eq}.

Carry Out the Plan

The equilibrium expression is

$$K_{eq} = \frac{[NO_2]^2}{[NO]^2[O_2]}$$

Enter the concentrations and solve for K_{eq}.

$$K_{eq} = \frac{(8.8)^2}{(1.2)^2(1.6)} = 34$$

Look Back

We check to be sure our answer is sensible. We have a large equilibrium constant (K_{eq} = 34), as we should expect from the numbers used.

Practice Exercise 17.2 When carbohydrates are broken down in the body, one important step is the conversion of glucose-1-phosphate to glucose-6-phosphate. The reaction may be represented as follows:

$$\text{Glucose-1-phosphate} \rightleftharpoons \text{glucose-6-phosphate}$$

Calculate K_{eq} for this reaction, given that the concentrations at equilibrium at body temperature are as follows: glucose-1-phosphate, 0.0050 mole/liter; and glucose-6-phosphate, 0.10 mole/liter.

Example 17.3

Calculate K_{eq} for the reaction

$$PCl_5(g) \rightleftharpoons PCl_3(g) + Cl_2(g)$$

given that the equilibrium concentrations at 25°C are as follows: PCl_5, 0.0096 mole/liter; Cl_2, 0.0247 mole/liter; and PCl_3, 0.0247 mole/liter.

Solution First write the equilibrium expression for the reaction from the balanced equation.

$$K_{eq} = \frac{[PCl_3][Cl_2]}{[PCl_5]}$$

Next enter the concentrations and find K_{eq}.

$$K_{eq} = \frac{(0.0247)(0.0247)}{0.0096} = 0.064$$

Practice Exercise 17.3 Use your result from Practice Exercise 17.2 to calculate the concentration of glucose-6-phosphate at equilibrium if the concentration of glucose-1-phosphate is 0.10 mole/liter. (Assume that the temperature does not change.)

The magnitude of the equilibrium constant K_{eq} is a good indication of how far the reaction goes toward completion. A large equilibrium constant means that the concentrations of the products are much larger than the concentrations of the reactants at equilibrium. Therefore the reaction tends toward completion. In other words, the forward reaction is favored. A small equilibrium constant means that the concentrations of the products are much smaller than the concentrations of the reactants. Therefore the reverse reaction is favored. In Example 17.2, K_{eq} is relatively large. This means that the forward reaction is favored. In Example 17.3, K_{eq} is relatively small. This means that the reverse reaction is favored. The larger the equilibrium constant at a given temperature, the more highly favored is the formation of products. The smaller the equilibrium constant, the more highly favored is the formation of reactants.

17.4 Equilibrium Concentrations

Chemical equilibrium can occur only in a reaction that does not go to completion. Equilibrium is established when the rate of the forward reaction is equal to the rate of the reverse reaction. This means that products are being formed at the same rate as reactants. It also means that products are being formed as fast as they are used up to yield reactants. And reactants are being formed as fast as they are used up to yield products. So the concentrations of products and the concentrations of reactants do not change as long as the equilibrium condition is maintained. *Again, we stress the fact that this does not mean that these concentrations are equal. It means only that the concentrations remain unchanged.*

If we know how far a reversible reaction proceeds toward completion, we can determine the equilibrium concentrations of the reactants and products. (The reactants and products are called the *species* taking part in the reaction.) For example, we saw earlier that at 25°C, a solution of acetic acid molecules is about 0.4% ionized. Then, at equilibrium and 25°C, a 1 *M* solution of acetic acid would contain 0.996 mole of $HC_2H_3O_2$ molecules, 0.004 mole of H^{1+} ions, and 0.004 mole of $(C_2H_3O_2)^{1-}$ ions in each liter of solution:

$$\underset{0.996\text{ mole}}{HC_2H_3O_2} \xrightarrow{H_2O} \underset{0.004\text{ mole}}{H^{1+}} + \underset{0.004\text{ mole}}{(C_2H_3O_2)^{1-}} \quad \text{(at equilibrium, 25°C)}$$

The concentrations are then 0.996 *M*, and 0.004 *M*, respectively. To find these concentrations, we simply find the amounts of the various species that are present at equilibrium, as in the example that follows.

Example 17.4

Into a 1-liter vessel are introduced 1 mole of H_2 gas and 1 mole of I_2 gas at 440°C. These gases are allowed to react to produce HI gas, according to the reaction

$$H_2(g) + I_2(g) \rightleftharpoons 2HI(g)$$

How many moles of each gas are present at equilibrium? The forward reaction is 78.2% complete at equilibrium.

Solution We started with 1 mole of H_2 and 1 mole of I_2. If the reaction is 78.2% complete at equilibrium, then

$$? \text{ moles of } H_2 \text{ have reacted} = (1.000 \text{ mole})(0.782) = 0.782 \text{ mole } H_2$$

Because H_2 and I_2 react in the ratio of 1 mole to 1 mole, 0.782 mole of H_2 and 0.782 mole of I_2 have reacted. Then the number of moles of H_2 (and I_2) that *have not reacted* (that is, the number that remain) is $1.000 - 0.782 = 0.218$ mole.

But what about the amount of HI that is produced? According to the balanced equation, for each mole of H_2 and I_2 that reacts, 2 moles of HI are produced. Because 0.782 mole of H_2 and 0.782 mole of I_2 have reacted,

$$? \text{ moles of HI} = (0.782 \cancel{\text{ mole } H_2})\left(\frac{2 \text{ moles HI}}{1 \cancel{\text{ mole } H_2}}\right) = 1.564 \text{ moles HI}$$

So 1.564 moles of HI are produced. We can now summarize the equilibrium condition as follows:

$$\underset{0.218\text{ mole}}{H_2(g)} + \underset{0.218\text{ mole}}{I_2(g)} \rightleftharpoons \underset{1.564\text{ moles}}{2HI(g)} \quad \text{(at equilibrium)}$$

Practice Exercise 17.4 Into a 1-liter vessel at 400°C are introduced 1 mole of the hypothetical gas A_2 and 1 mole of the hypothetical gas B_2. The gases react to produce AB, according to the reaction

$$A_2(g) + B_2(g) \rightleftharpoons 2AB(g)$$

How many moles of each gas are present at equilibrium? The forward reaction is 60.0% complete at equilibrium.

17.5 LeChatelier's Principle

More than a century ago, the French chemist Henri LeChatelier experimented with solutions containing reactants and products at equilibrium. He observed what happened when the equilibrium was disturbed in some way—that is, when a *stress* was applied to the system. He found that the system adjusted itself to a new equilibrium in such a way as to relieve the stress. This observation is now known as **LeChatelier's principle:** *When a stress is applied to a system at equilibrium, the system adjusts to a new equilibrium position, if possible, in such a way as to reduce the effect of the stressful condition.*

Learning Goal 1
LeChatelier's principle

A chemical system adjusts its equilibrium only by shifting to new equilibrium concentrations of products and reactants. Depending on the nature of the stress, the formation of products may be favored (the equilibrium shifts to the right through a net forward reaction) or the formation of reactants may be favored (the equilibrium shifts to the left through a net reverse reaction). In each case, a new equilibrium system results. We shall look at the effects of three types of stresses on chemical systems at equilibrium.

Effect of a Change in Concentration

When a system is at equilibrium and we change the concentration of one of its components, a stress is placed on the system. For example, suppose that the system

$$CO_2(g) + H_2(g) \rightleftharpoons CO(g) + H_2O(g)$$

is at equilibrium. According to LeChatelier's principle, increasing the concentration of one of the reactants (for example, adding CO_2) would shift the equilibrium to the right. That is, the system would act to reduce the stress—here, the excess CO_2—by using it to form more product. The result would be a new equilibrium with greater concentrations of the products and a lower concentration of H_2. The concentration of CO_2 would be greater than the original concentration but lower than the increased concentration that caused the shift in equilibrium.

According to LeChatelier's principle, if the concentration of one of the *products* is increased, the equilibrium shifts to the left. That is, the system acts to relieve the excess of product by using it up to form more reactant. At the new equilibrium point, the concentrations of reactants are higher and the concentrations of products are lower than at the start of the reaction. The equilibrium constant for the reaction is unchanged.

Example 17.5

Consider the equilibrium system

$$A(g) + B(g) \rightleftharpoons C(g) + D(g)$$

Use LeChatelier's principle to predict the direction in which the equilibrium shifts when:

(a) The concentration of species A is increased.
(b) The concentration of species B is decreased.
(c) The concentration of species C is decreased.
(d) The concentration of species D is increased.

Solution

(a) When more A is added, the equilibrium shifts to the *right*.
(b) When some B is removed, the equilibrium shifts to the *left*. Decreasing the concentration of a reactant has the same effect as increasing the concentration of a product.
(c) When the concentration of C is decreased, the equilibrium shifts to the *right*. Decreasing the concentration of a product has the same effect as increasing the concentration of a reactant.
(d) When the concentration of D is increased, the equilibrium shifts to the *left*.

Practice Exercise 17.5 Consider the equilibrium system

$$4HCl(g) + O_2(g) \rightleftharpoons 2H_2O(g) + 2Cl_2(g)$$

Use LeChatelier's principle to predict the direction in which the equilibrium shifts when:

(a) The concentration of $HCl(g)$ is decreased.
(b) The concentration of $O_2(g)$ is increased.
(c) The concentration of $H_2O(g)$ is increased.
(d) The concentration of $Cl_2(g)$ is decreased.

Effect of a Change in Pressure or Volume

Recall from your study of the gas laws in Chapter 13 that the pressure and volume of a gas are inversely related. If we increase the pressure on a system of gas molecules, the volume is decreased. And if we decrease the pressure, the volume is increased. Therefore pressure changes and volume changes should have opposite effects on an equilibrium system composed entirely of gases.

The fewer molecules there are in a given volume of gas, the smaller the stress produced by an increase in pressure. And, according to LeChatelier's principle, an equilibrium system shifts in such a way as to reduce a stress imposed on the system. Therefore, if you *increase* the pressure on a system composed of gas molecules, *the equilibrium shifts in the direction that yields the smaller number of gas molecules*. If you *decrease* the pressure on a system composed of gas molecules, the equilibrium shifts in the direction that yields the greater number of gas molecules. Changes in the volume of a gaseous equilibrium system have the exact opposite effects.

Three additional facts are important in predicting equilibrium shifts due to changes in pressure or volume:

1. The numbers of reactant and product molecules that take part in a reaction (or that are present in the reaction vessel) are proportional to the coefficients of the balanced equation.
2. If the balanced equation for a reaction shows the same number of gas molecules on the right and on the left, then the equilibrium does *not* shift in response to a change in pressure. Shifting either way would do nothing to reduce the stress.
3. In a system consisting of gases along with solids or liquids or both, only the gas molecules are significantly affected by changes in pressure.

Example 17.6

Consider the equilibrium system

$$N_2(g) + 3H_2(g) \rightleftharpoons 2NH_3(g)$$

Predict the direction in which the equilibrium will shift when:
(a) The pressure is increased.
(b) The volume is increased.

Solution

(a) When the pressure is increased, the equilibrium shifts in the direction of *fewer* gas molecules. Because the coefficients on the left side of the balanced equation add up to 4, and the coefficients on the right side add up to 2, the equilibrium will shift to the *right*.

(b) When the volume is increased, the pressure is decreased. And a decrease in pressure shifts the equilibrium in the direction of the *greater* number of gas molecules. Therefore the equilibrium will shift to the *left*.

Practice Exercise 17.6 Consider the equilibrium system

$$CO_2(g) + 2NH_3(g) \rightleftharpoons H_2O(g) + CO(NH_2)_2(s)$$

Predict the direction in which the equilibrium will shift when:
(a) The pressure is increased.
(b) The volume is increased.

Example 17.7

For each of the following equilibrium systems, predict the direction in which the equilibrium will shift when the pressure is increased:
(a) $SiF_4(g) + 2H_2O(g) \rightleftharpoons SiO_2(s) + 4HF(g)$
(b) $H_2(g) + Cl_2(g) \rightleftharpoons 2HCl(g)$

Solution

(a) The equilibrium shifts in the direction of fewer gas molecules. The coefficients of the *gaseous* molecules on the left side of the equation add up to 3. The coefficient for the *gaseous* molecules on the right side of the equation is 4. Therefore the equilibrium will shift to the *left*.
(b) Here the sum of the coefficients of the gases on the left is equal to the "sum" of the coefficients on the right. An equilibrium shift would not reduce the stress caused by the increased pressure, and no shift will occur.

Practice Exercise 17.7 For each of the following equilibrium systems, predict the direction in which the equilibrium will shift when the pressure is increased:
(a) $CO(g) + H_2O(g) \rightleftharpoons CO_2(g) + H_2(g)$
(b) $COCl_2(g) \rightleftharpoons CO(g) + Cl_2(g)$

Effect of a Change in Temperature

Chemical reactions that require heat in order to proceed are called *endothermic* reactions. Chemical reactions that produce heat are called *exothermic* reactions (Figure 17.7). When the temperature of an equilibrium system is raised, heat is added to the system. The equilibrium then shifts in the direction that absorbs the added heat—to relieve the stress caused by the heat. For *endothermic* reactions, an increase in temperature (heat) favors the *forward* reaction. For *exothermic* reactions, an increase in temperature (heat) favors the *reverse* reaction.

An easy way to remember this is to think of heat as a reactant in endothermic reactions or as a product in exothermic reactions. Then an increase in temperature increases the "concentration" of heat. And the equilibrium shifts in such a way as to reduce this concentration.

Example 17.8

For each of the following reactions, predict the direction in which the equilibrium will shift when heat is added:

(a) $C(s) + CO_2(g) + \text{heat} \rightleftharpoons 2CO_2(g)$
(b) $H_2(g) + Cl_2(g) \rightleftharpoons 2HCl(g) + \text{heat}$

Solution

(a) This is an *endothermic* reaction, so the addition of heat will shift the equilibrium to the right. (Increasing the concentration of the *reactant* heat shifts the equilibrium to the right.)
(b) This is an *exothermic* reaction, so the addition of heat will shift the equilibrium to the *left*. (Increasing the concentration of the *product* heat shifts the equilibrium to the left.)

Practice Exercise 17.8 For each of the following reactions, predict the direction in which the equilibrium will shift when heat is added:
(a) $4HCl(g) + O_2(g) \rightleftharpoons 2H_2O(g) + 2Cl_2(g) + \text{heat}$
(b) $CO_2(g) + 2NH_3(g) \rightleftharpoons H_2O(g) + CO(NH_2)_2(s) + \text{heat}$
(c) $CO_2(g) + H_2(g) + \text{heat} \rightleftharpoons CO(g) + H_2O(g)$
(d) $N_2(g) + 3H_2(g) \rightleftharpoons 2NH_3(g) + \text{heat}$

Effect of a Catalyst

Recall that catalysts are substances that increase the speed of a chemical reaction but are not themselves chemically changed in the reaction. Experiments have shown that a catalyst has no effect on chemical equilibrium. That is, it does not shift the equilibrium in either direction. Because catalysts change re-

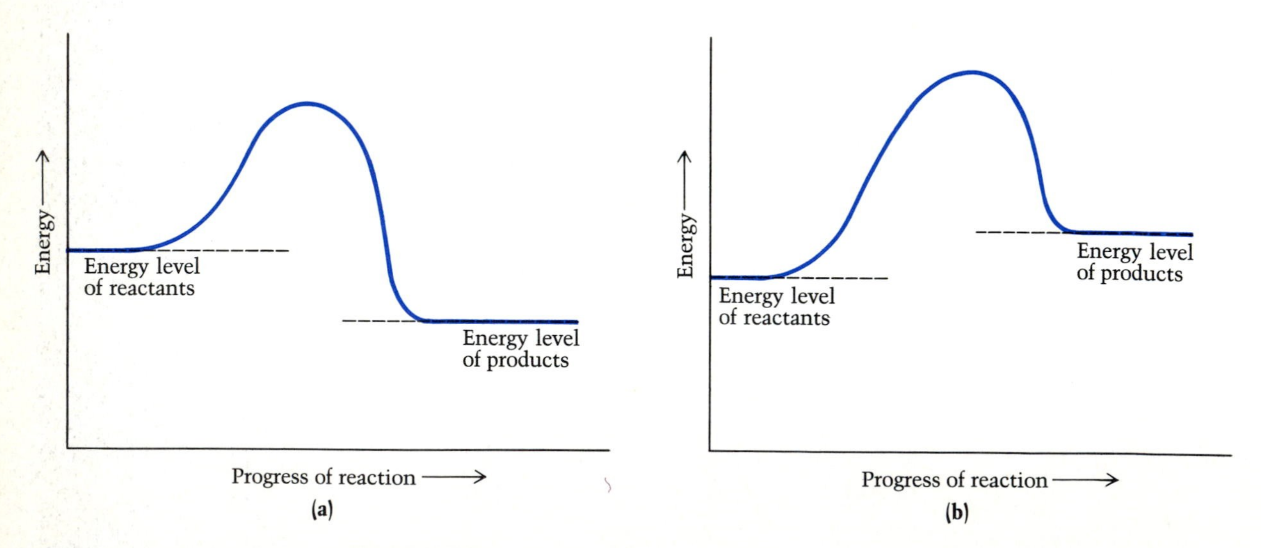

Figure 17.7
Energy diagrams for (*a*) an exothermic reaction (energy is released to the environment) and (*b*) an endothermic reaction (energy is absorbed from the environment).

action paths, catalysts must increase the forward reaction rate and the reverse reaction rate to exactly the same degree.

Catalysts are used to a great extent, in both the laboratory and commercial preparation of chemicals. In addition, more than 1,000 biological catalysts, or *enzymes,* are known. Enzymes work in our bodies to help us digest our food. They conduct nerve impulses throughout our bodies and break down food components into smaller molecules. If even one enzyme is missing in the body, the results can be disastrous.

An example of a disease caused by a missing enzyme is *phenylketonuria,* or PKU for short. This genetically transmitted disease results in a deficiency of the enzyme phenylalanine hydrolase, which is produced in the liver and which converts the amino acid phenylalanine in protein-containing foods. A baby who is born with PKU may be unusually irritable, may have epileptic seizures, and may vomit often. Phenylacetic acid appears in the urine and sweat and can be detected by its odor. About two-thirds of the babies who have elevated levels of phenylalanine show severe mental retardation. This disease is treated by lowering the amount of phenylalanine in the diet, not by eliminating it altogether. Babies are fed Lofenelac (a formula fortified with vitamins, iron, and other minerals) instead of milk, which contains phenylalanine. Low-protein foods such as fruits, vegetables, and cereals are also included in the diet. With such treatment, mental retardation can be avoided and many other problems associated with this condition can be reversed.

How LeChatelier's Principle Can Be Used Practically

Perhaps the best example of how LeChatelier's principle has been put to practical use in controlling chemical equilibria is the synthesis of ammonia (NH_3). The industrial synthesis of ammonia, which is the fourth-most-important major chemical produced commercially in the United States, is as follows:

$$3H_2(g) + N_2(g) \rightleftharpoons 2NH_3(g) + 22.08 \text{ kcal}$$

If the pressure is increased, the equilibrium shifts to the right, favoring the formation of ammonia. Because ammonia is formed in an exothermic reaction, adding heat would shift the equilibrium to the left, which would *not* favor the formation of ammonia. Therefore, if chemists want to increase the maximum yield of ammonia, the reaction should take place at low temperature and high pressure, in accordance with LeChatelier's principle. Manufacturing plants produce ammonia at pressures of about 1,000 atm and moderate temperatures of about 500°C. At temperatures below 500°C, the reaction occurs too slowly. In addition, a catalyst is added to speed up the reaction.

The story of the production of ammonia is an interesting one that illustrates the relationship between science and society. Thomas Malthus theorized that although food supply grows at a steady rate, population grows many, many times faster. He predicted that population would eventually outstrip agricultural production. In 1890 Sir William Crookes, addressing the British Association for the Advancement of Science, discussed the problem of the *fixed nitrogen* supply. Nitrogen, although it composes 79% of the air, is not useful when in the form of the unreactive N_2 molecule. Atmospheric nitrogen must be *fixed* in order to be used by plants. To *fix* nitrogen means to stabilize it and change it from a gaseous to a solid or liquid form, in which it is bound with another element or with more than one element. Some forms of fixed nitrogen are ammonia, nitrite ions, and nitrate ions. Certain bacteria and algae are capable of fixing atmospheric nitrogen. These organisms are found free in nature. Some live in colonies on the roots of plants called legumes. Crookes maintained that if people were to have adequate food supplies, scientists must figure out a way to fix atmospheric nitrogen, because nitrogen-containing fertilizers were the only hope for growing enough food. The source of nitrogen used for fertilizers at the time was limited to manures and to sodium nitrate, which is found in Chile.

The Germans were interested in producing fixed nitrogen. A large chemical company called Badische Anilin-und-Sodafabrik appealed to Fritz Haber, a professor of physical chemistry at Karlsruhe, for help. Haber had read a paper published by LeChatelier in 1901, which stated that increased pressure resulted in definite amounts of ammonia being formed from its elements. Haber pursued this path. By 1911 the German chemical industry had built a plant to produce ammonia from its elements by using high pressure, moderate temperatures, and the proper catalyst.

In August 1914, when World War 1 broke out, the German chemical industry was able to supply ammonium nitrate for use as an explosive. A discovery that was intended for use in fertilizer, to grow food, was used to produce explosives. In 1918 Haber was awarded the Nobel Prize in chemistry for his discovery.

17.6 Ionization Constants

Learning Goal 7
Calculation of K_a and K_b

As we saw in Chapter 16, weak acids and weak bases ionize only partially in solution. An equilibrium is eventually established between the undissociated acid or base (the "reactant") and its ions (the "products"). We can write an equilibrium expression for the dissociation, just as we do for a reaction involving molecules. Then, if we can obtain the equilibrium concentrations of the various species at a particular temperature, we can calculate the equilibrium constant for the acid or base at that temperature. Such an equilibrium constant is called the **ionization constant** or **dissociation constant** of the acid or base. It is usually given the symbol K_a for an acid and K_b for a base.

Example 17.9

Calculate K_a for acetic acid at 25°C if measured equilibrium concentrations are as follows:

$$HC_2H_3O_2 \overset{H_2O}{\rightleftharpoons} H^{1+} + (C_2H_3O_2)^{1-}$$
$$0.09986\ M \qquad 0.00134\ M \qquad 0.00134\ M$$

Solution Write the equilibrium expression for the reaction, enter the concentrations of the species, and calculate K_a.

$$K_a = \frac{[H^{1+}][(C_2H_3O_2)^{1-}]}{[HC_2H_3O_2]}$$

$$= \frac{(0.00134)(0.00134)}{0.09986} = 1.80 \times 10^{-5}$$

This small value of K_a tells us that the equilibrium favors the reverse reaction—the formation of acetic acid.

Practice Exercise 17.9 Calculate K_a for nitrous acid, HNO_2, given the following equilibrium concentrations:

$$HNO_2 \rightleftharpoons H^{1+} + NO_2^{1-}$$
$$0.0935\ M \qquad 6.5 \times 10^{-3}\ M \qquad 6.5 \times 10^{-3}\ M$$

Learning Goal 8
Calculation of equilibrium concentrations

Table 17.1 lists the ionization constants of some weak acids and bases at 25°C. Once you know K_a or K_b for a substance at a particular temperature, you can use it to calculate the concentrations of ions and molecules in an equilibrium solution of that substance.

Table 17.1
Ionization Constants of Weak Acids and Bases at 25°C

Substance	Formula	K_a or K_b
Acetic acid	$HC_2H_3O_2$	1.8×10^{-5}
Benzoic acid	$HC_7H_5O_2$	6.3×10^{-5}
Cyanic acid	$HCNO$	2.0×10^{-4}
Formic acid	$HCHO_2$	1.8×10^{-4}
Hydrocyanic acid	HCN	4.0×10^{-10}
Hypochlorous acid	$HClO$	3.5×10^{-8}
Nitrous acid	HNO_2	4.5×10^{-4}
Ammonium hydroxide	NH_4OH	1.8×10^{-5}
Copper(II) hydroxide	$Cu(OH)_2$	1.0×10^{-8}
Zinc hydroxide	$Zn(OH)_2$	4.0×10^{-5}

Example 17.10

Calculate the equilibrium concentrations of H^{1+}, $(C_2H_3O_2)^{1-}$, and $HC_2H_3O_2$ in a 0.50 *M* acetic acid solution at 25°C ($K_a = 1.8 \times 10^{-5}$).

Solution The reaction is

$$HC_2H_3O_2 \rightleftharpoons H^{1+} + (C_2H_3O_2)^{1-}$$

The initial concentration of $HC_2H_3O_2$ is 0.50 *M*. Let us suppose that x moles of $HC_2H_3O_2$ dissociate. Then, according to the stoichiometry of the reaction, x moles of H^{1+} ions and x moles of $(C_2H_3O_2)^{1-}$ ions are produced. Summarize this as follows:

	$HC_2H_3O_2$	$\rightleftharpoons$	H^{1+}	$+ (C_2H_3O_2)^{1-}$
Init. conc.	0.50 mole/liter		0 mole/liter	0 mole/liter
Conc. at equil.	$(0.50 - x)$ mole/liter		x moles/liter	x moles/liter

Next derive the equilibrium expression from the balanced equation and enter all known values.

$$K_a = \frac{[H^{1+}][(C_2H_3O_2)^{1-}]}{[HC_2H_3O_2]} \quad \text{so} \quad 1.8 \times 10^{-5} = \frac{(x)(x)}{0.50 - x}$$

At this point we could solve the equation for x. However, we would obtain a quadratic equation. Instead, we simplify the mathematics by assuming that x is very small compared to 0.50, so that $(0.50 - x)$ is still about 0.50. (We can safely make this simplifying assumption because the value of K_a, 1.8×10^{-5}, tells us that the acetic acid is only slightly dissociated.) Our expression becomes

$$1.8 \times 10^{-5} = \frac{(x)(x)}{0.50}$$

so that

$$9.0 \times 10^{-6} = x^2 \quad \text{and} \quad 3.0 \times 10^{-3} = x$$

Therefore the equilibrium concentration of H^{1+} is 3×10^{-3} mole/liter. The equilibrium concentration of $(C_2H_3O_2)^{1-}$ is also 3×10^{-3} mole/liter. The concentration of $HC_2H_3O_2$ is $0.50 - (3 \times 10^{-3})$, which is $0.50 - 0.003$, or 0.497, mole/liter. Rounded to the proper number of significant figures, this is 0.50 mole/liter. The initial assumption was sensible.

Practice Exercise 17.10 Citric acid is found in citrus fruits and is responsible for their characteristic taste. Calculate the equilibrium concentrations of hydrogen ion, dihydrogen citrate ion, and citric acid for a 1.00 *M* citric acid solution ($K_a = 8.4 \times 10^{-4}$).

Example 17.11

Morphine, $C_{17}H_{19}NO_3$, a naturally occurring base, is administered medically to relieve pain. The ionization constant K_b for morphine is 1.60×10^{-6} at 25°C. We can represent the ionization of morphine in water by the following equation (let's abbreviate the formula of morphine as Mor):

$$\text{Mor}(aq) + H_2O \rightleftharpoons (\text{HMor})^{1+} + (\text{OH})^{1-}$$

Or, because the concentration of the water changes very little, we can simplify the equation to

$$\text{Mor} \xrightleftharpoons{H_2O} (\text{HMor})^{1+} + (\text{OH})^{1-}$$

If the initial concentration of Mor is 0.1000 mole/liter, what is the concentration of each species at equilibrium?

Solution The initial concentration of Mor is 0.1000 mole/liter. If x moles of Mor dissociate, then according to the stoichiometry of the reaction, x moles of $(\text{HMor})^{1+}$ and x moles of $(\text{OH})^{1-}$ ions are produced. This can be summarized as follows:

	Mor	$\xrightleftharpoons{H_2O}$ $(\text{HMor})^+$	+ $(\text{OH})^{1-}$
Init. conc.	0.1000 mole/liter	0 mole/liter	0 mole/liter
Conc. at equil.	$(0.1000 - x)$ mole/liter	x moles/liter	x moles/liter

Derive the expression K_b from the balanced equation and enter all known values.

$$K_b = \frac{[(\text{HMor})^{1+}][(\text{OH})^{1-}]}{[\text{Mor}]} \quad \text{so} \quad 1.60 \times 10^{-6} = \frac{(x)(x)}{0.1000 - x}$$

If we assume that x is small compared to 0.1000 (on the basis of the value of K_b), then the expression simplifies to

$$1.60 \times 10^{-6} = \frac{(x)(x)}{0.1000}$$

so that

$$1.60 \times 10^{-7} = x^2 \quad \text{and} \quad 4.00 \times 10^{-4} = x$$

Therefore, the equilibrium concentrations of $(\text{HMor})^{1+}$ and $(\text{OH})^{1-}$ are 4.00×10^{-4} mole/liter. The equilibrium concentration of Mor is $0.1000 - (4.00 \times 10^{-4})$ which is $0.1000 - 0.0004$, or 0.0996, mole/liter.

Practice Exercise 17.11 The ionization constant for the reaction

$$NH_3 + H_2O \rightleftharpoons (NH_4)^{1+} + (OH)^{1-}$$

is $K_b = 1.8 \times 10^{-5}$. The initial concentration of NH_3 is 1.0 mole/liter. Calculate the concentration of each species at equilibrium.

17.7 The Solubility Product

Learning Goal 9
Writing K_{sp} expressions and calculating K_{sp}

Many salts are only slightly soluble in water. For example, the salt silver chloride has a solubility of 1.3×10^{-5} mole/liter at 25°C. In other words, if we place some silver chloride in a liter of water at 25°C, only 1.3×10^{-5} mole of it will dissociate. The rest of the silver chloride will remain undissolved. However, there will (as usual) be a dynamic equilibrium between the ions and the undissolved salt. We can represent this equilibrium with the equation

$$AgCl(s) \rightleftharpoons Ag^{1+} + Cl^{1-}$$

We could write an expression for the equilibrium constant K_{eq} for this reaction. However, the experiment shows that the product of the ionic concentrations in solution is found to be constant, regardless of how much solid is present. Therefore, we eliminate the denominator. We define the **solubility product** constant K_{sp} of a slightly soluble salt as the product of the concentrations of its ions in a saturated solution at a particular temperature. For silver chloride, then,

$$K_{sp} = [Ag^{1+}][Cl^{1-}]$$

We have already noted that the solubility of AgCl at 25°C is 1.3×10^{-5} mole/liter. Therefore, on the basis of the stoichiometry of the ionization reaction, $[Ag^{1+}]$ is 1.3×10^{-5} and $[Cl^{1-}]$ is 1.3×10^{-5}. From these concentrations, we obtain

$$K_{sp} = [Ag^{1+}][Cl^{1-}]$$
$$= (1.3 \times 10^{-5})(1.3 \times 10^{-5}) = 1.7 \times 10^{-10}$$

Generally speaking, the higher the solubility product constant of a salt, the more soluble the salt is in water. For example, the solubility product constants given in Table 17.2 show that barium sulfate ($BaSO_4$) is more soluble than silver chloride at 25°C. And silver acetate is more soluble than barium sulfate.

Suppose we have 1 liter of a saturated aqueous solution of silver chloride. We know that the concentration of Ag^{1+} ions in this solution is 1.3×10^{-5} mole/liter and that the concentration of Cl^{1-} ions is the same. What would happen to the concentration of Cl^{1-} ions if we increased the concentration of Ag^{1+} ions to 0.010 mole/liter? Would it change? We can use the solubility product to find out.

Table 17.2
Solubility Product Constants, K_{sp}, at 25°C

Substance	K_{sp}	Substance	K_{sp}
$AgCl$	1.7×10^{-10}	$BaSO_4$	1.5×10^{-9}
$AgBr$	5.0×10^{-13}	CaF_2	3.9×10^{-11}
AgI	8.5×10^{-17}	CuS	9.0×10^{-45}
$AgC_2H_3O_2$	2.0×10^{-3}	$Fe(OH)_3$	6.0×10^{-38}
Ag_2CrO_4	1.9×10^{-12}	$Mn(OH)_2$	2.0×10^{-13}
$BaCrO_4$	8.5×10^{-11}		

Because K_{sp} for silver chloride is 1.7×10^{-10}, we have

$$K_{sp} = [Ag^{1+}][Cl^{1-}]$$

$$1.7 \times 10^{-10} = (0.010)[Cl^{1-}]$$

Then

$$[Cl^{-1}] = \frac{1.7 \times 10^{-10}}{0.010} = 1.7 \times 10^{-8}$$

Therefore the concentration of the chloride ion *decreases* to 1.7×10^{-8} mole/liter. (Remember, before we added the excess Ag^{1+}, the Cl^{1-} concentration was much higher: 1.3×10^{-5} mole/liter.) *This result, in which the concentration of one ion is reduced when the concentration of another ion is increased at equilibrium, is called the* **common ion effect.**

Example 17.12

Write the K_{sp} expressions for the following slightly soluble salts: *(a)* $AgBr$ *(b)* $BaCrO_4$ *(c)* CaF_2

Solution First write the equilibrium equation. Then use the equation to write the K_{sp} expression.

(a) $AgBr \rightleftharpoons Ag^{1+} + Br^{1-}$
$K_{sp} = [Ag^{1+}][Br^{1-}]$

(b) $BaCrO_4 \rightleftharpoons Ba^{2+} + (CrO_4)^{2-}$
$K_{sp} = [Ba^{2+}][(CrO_4)^{2-}]$

(c) $CaF_2 \rightleftharpoons Ca^{2+} + 2F^{1-}$
$K_{sp} = [Ca^{2+}][F^{1-}]^2$

Practice Exercise 17.12 Write the K_{sp} expressions for the following salts: *(a)* $PbSO_4$ *(b)* Ag_2CrO_4 *(c)* AgI

Example 17.13

Calculate the solubility of the salt $BaSO_4$ in grams per liter, given that $K_{sp} = 1.5 \times 10^{-9}$.

Solution Write the equilibrium equation. Then write the expression for K_{sp} and insert its value.

Learning Goal 10
Calculation of the solubility of a salt

$$BaSO_4 \rightleftharpoons Ba^{2+} + (SO_4)^{2-}$$

$$K_{sp} = [Ba^{2+}][(SO_4)^{2-}]$$

$$1.5 \times 10^{-9} = [Ba^{2+}][(SO_4)^{2-}]$$

The equilibrium equation tells us that the concentrations of Ba^{2+} and $(SO_4)^{2-}$ in solution are equal. We let x represent these concentrations in *moles per liter of solution*. Then

$$1.5 \times 10^{-9} = [Ba^{2+}][(SO_4)^{2-}] \qquad \text{or} \qquad 1.5 \times 10^{-9} = (x)(x)$$

Therefore

$$1.5 \times 10^{-9} = x^2 \qquad \text{and so} \qquad 3.9 \times 10^{-5} = x$$

Therefore 3.9×10^{-5} moles of $BaSO_4$ can dissolve per liter of solution. If we want to express the solubility in *grams per liter,* we have to multiply this result by the molecular mass of $BaSO_4$.

$$\frac{?\ \text{g}}{\text{liter}} = \left(\frac{3.9 \times 10^{-5}\ \cancel{\text{mole}}}{1\ \text{liter}}\right)\left(\frac{233.4\ \text{g}}{1\ \cancel{\text{mole}}}\right)$$

$$= 9.1 \times 10^{-3}\ \text{g/liter}$$

Practice Exercise 17.13 Calculate the solubility in grams per liter of $PbSO_4$, given that $K_{sp} = 1.3 \times 10^{-8}$ for this salt at 25°C.

17.8 Buffer Solutions and the Control of pH

Learning Goal 11
Buffers

A **buffer solution** is *a solution that resists changes in pH when small amounts of strong acids or bases are added to it or when it is diluted with water*. A buffer solution may be made up of one of the following:

1. A weak acid and its salt
2. A weak base and its salt

Pure water has a pH of 7. But if we added even a small amount of an acid or a base to pure water, its pH would change drastically. The top part of Table 17.3 gives two examples of such changes. However, if the water contained a buffer, the pH would change much less, as shown in the bottom part of Table 17.3.

Table 17.3
Changes in pH As a Result of Adding HCl and NaOH to Pure Water and to an Aqueous Acetic Acid–Sodium Acetate Buffer Solution

Solution	pH	Change in pH
Pure H_2O (1 liter)	7	—
Add: 0.01 mole HCl	2	−5
Add: 0.01 mole NaOH	12	+5
Buffer solution (1 liter)	4.74	—
0.5 M $HC_2H_3O_2$ + 0.5 M $NaC_2H_3O_2$		
Add: 0.01 mole HCl	4.72	−0.02
Add: 0.01 mole NaOH	4.76	+0.02

Learning Goal 12
Carbonic acid/bicarbonate buffer system

To appreciate how a buffer system works, consider the carbonic acid/bicarbonate buffer system, which maintains the pH of our blood at a fairly constant level. Carbon dioxide and water, which are present in the blood, combine to form carbonic acid.

$$H_2O + CO_2 \rightleftharpoons \underset{\text{Carbonic acid}}{H_2CO_3} \qquad \text{Equation 17.1}$$

As Equation 17.1 indicates, an equilibrium exists among the water, carbon dioxide, and carbonic acid. However, some of the carbonic acid ionizes to form an equilibrium with hydrogen ions and bicarbonate ions.

$$H_2CO_3 \rightleftharpoons H^{1+} + \underset{\text{Bicarbonate ion}}{(HCO_3)^{1-}} \qquad \text{Equation 17.2}$$

The system composed of Equations 17.1 and 17.2 can be represented as follows:

$$H_2O + CO_2 \rightleftharpoons H_2CO_3 \rightleftharpoons H^{1+} + (HCO_3)^{1-} \qquad \text{Equation 17.3}$$

Equation 17.3 indicates that the blood contains water, carbon dioxide, carbonic acid, hydrogen ions, and bicarbonate ions in equilibrium with each other. The fact that they are in equilibrium with each other is very important, because under these conditions LeChatelier's principle applies: *If the concentration of any one of these substances changes, the equilibrium shifts in a direction to minimize this change*. For example, if too many hydrogen ions form in the blood, the equilibrium shifts to the left, reducing the number of hydrogen ions. If too much carbon dioxide is present in the blood, the equilibrium shifts to the right, and the concentration of CO_2 is reduced.

The normal pH of blood is 7.4. However, suppose a person ingests a basic substance or has an internal disorder that causes the concentration of hydroxide ions in the blood to increase (a condition known as *metabolic alkalosis*). If the blood had *no* buffer system, its pH would increase greatly. This could prove fatal to the individual. But because the blood has the benefit of the carbonic acid/bicarbonate buffer system, the following reaction occurs:

$$H_2CO_3 + (OH)^{1-} \longrightarrow H_2O + (HCO_3)^{1-}$$

The carbonic acid reacts with the hydroxide ions, producing water and bicarbonate ion and maintaining the pH at about 7.4. The excess bicarbonate ion is eventually eliminated via the kidneys and urine.

Now suppose that the concentration of hydrogen ions in a person's blood were to increase suddenly, as the result of the person's ingesting an acidic substance, experiencing kidney failure, or having diabetes mellitus. (This condition is known as *metabolic acidosis*.) The excess hydrogen ions react with the bicarbonate ions, producing carbonic acid. The excess carbonic acid then decomposes to form water and carbon dioxide, which is eventually eliminated from the blood. The reactions can be summarized in this way:

$$(HCO_3)^{1-} + H^{1+} \longrightarrow H_2CO_3 \longrightarrow H_2O + CO_2$$

Again, the pH of the blood remains almost constant.

Buffer systems work very well—within their limits. It is, of course, possible to overpower a buffer system by adding huge amounts of a strong acid or base. This uses up all the buffer ions and causes a drastic change in pH.

Example 17.14

The blood buffer system can be represented, in part, as follows:

$$H_2O + CO_2 \rightleftharpoons H_2CO_3 \rightleftharpoons H^{1+} + (HCO_3)^{1-}$$

In *hypoventilation*, the body does not take in enough oxygen. As a result, the level of CO_2 in the blood increases. Describe the effect of hypoventilation on the blood buffer system. What condition would it lead to?

Solution Hypoventilation leads to a buildup of CO_2, which shifts the equilibrium to the *right*. This leads to an acidosis condition. (In medical terms, this is known as *respiratory acidosis*.)

Practice Exercise 17.14 In terms of the blood buffer system described in Example 17.14, explain which way the equilibrium would shift if a person *hyperventilated*. (A person who is hyperventilating is breathing at an above-normal rate.) What condition would this lead to?

Summary

According to the collision theory, for two species to react, their molecules must collide. In addition, the colliding molecules must be activated. That is, they must possess a certain amount of energy called the activation energy. If they do, they will form an activated complex. And if the reactant molecules are oriented properly, the activated complex can form product molecules.

Increased concentrations of reactants increase the reaction rate by producing more frequent collisions. Increased temperature increases the reaction rate by increasing the velocity and energy of reactant molecules. Catalysts increase the reaction rate by offering a new path for reaction, a path that involves a lower activation energy.

A reversible reaction is one in which the products can react to yield the reactants. At first, the formation of products (the forward reaction) proceeds at a rapid rate. This rate decreases as products are formed and the concentration of reactants decreases. The formation of reactants (the reverse reaction) proceeds slowly at first. However, this rate increases as more product is formed in the forward reaction. Eventually, the forward and reverse reaction rates become equal, and the reaction is said to be at equilibrium. This chemical equilibrium is a dynamic one, in which products and reactants are both being formed. The concentrations of products and reactants remain constant as long as this equilibrium condition is maintained.

According to LeChatelier's principle, if a stress is applied to a chemical system at equilibrium, the equilibrium shifts in the direction that tends to relieve the stress. Thus, for a chemical system at equilibrium, an increase in the concentration of one species shifts the equilibrium in such a way as to use up more of that species. An increase in temperature shifts the equilibrium in such a way as to favor an endothermic reaction. An increase in the pressure on (or a decrease in the volume of) gaseous species shifts the equilibrium in such a way as to favor the production of fewer molecules. The introduction of a catalyst has no effect on the equilibrium.

For any reaction, the equilibrium constant K_{eq} is the ratio of the product of the *product concentrations* to the product of the *reactant concentrations* at equilibrium. The larger the equilibrium constant, the more the formation of products is favored at equilibrium. The value of K_{eq} for a given reaction depends on the temperature but not on the initial concentrations of the reactants.

The equilibrium constant for a weak acid or base is called its ionization (or dissociation) constant, represented as K_a or K_b. The ionization constant indicates the degree of dissociation in aqueous solution. The solubility product constant K_{sp} for a slightly soluble salt is equal to the product concentrations of its ions (in moles/liter) at equilibrium. This constant is useful in computing solubilities.

A buffer system or buffer solution is a solution that resists changes in pH when small amounts of strong acids or bases are added. Buffer solutions are solutions of either a weak acid and its salt or a weak base and its salt.

Key Terms

activation energy **(17.1)**
buffer solution **(17.8)**
catalyst **(17.1)**
chemical equilibrium system **(17.2)**
chemical kinetics **(17.1)**
collision theory **(17.1)**
common ion effect **(17.7)**
dissociation constant **(17.6)**
equilibrium constant **(17.3)**
equilibrium expression **(17.3)**
ionization constant **(17.6)**
irreversible reaction **(17.2)**
Law of Mass Action **(17.3)**
LeChatelier's principle **(17.5)**
reaction mechanism **(17.1)**
reversible reaction **(17.2)**
solubility product **(17.7)**

Career Sketch

Medical Laboratory Worker

As a medical laboratory worker, you will analyze body tissues, blood, and other body fluids by using precision techniques and instruments. You will work under the supervision of a physician or a scientist who specializes in clinical chemistry, microbiology, or some other biological science. Your analysis of body fluids and tissues will help determine the presence of bacte-

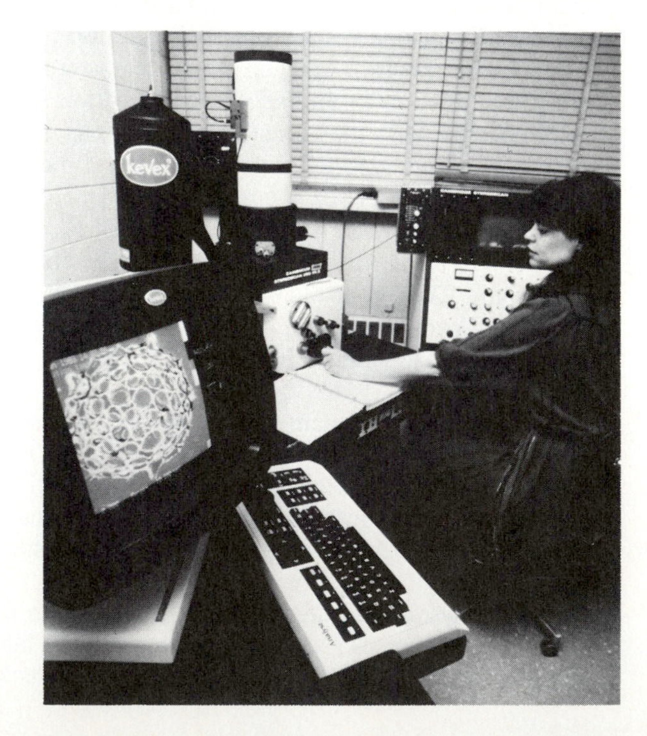

This medical laboratory worker uses a scanning electron microscope to gather data.
(Yoav Levy/Phototake)

ria, parasites, or other microorganisms; assess the chemical content or reaction of substances you are testing; or determine an individual's blood type.

There are three kinds of medical laboratory workers. *Medical technologists* conduct sophisticated chemical, microscopic, and bacteriological tests, such as testing cholesterol levels and examining blood under a microscope to detect diseases. They also make cultures of body fluid and tissue for analysis. This work requires four years of college training.

Medical laboratory technicians perform a wide range of laboratory procedures and tests that require a high level of skill. These technicians generally receive two years of college training.

Medical laboratory assistants help medical technologists and technicians perform routine tests. They may store and label plasma, clean and sterilize laboratory equipment, prepare solutions, keep records of tests, and identify specimens. One year of post–high school training is required.

Self-Test Exercises

Learning Goal 1: Chemical Kinetics, Chemical Equilibrium, and LeChatelier's Principle

1. Match each term on the left with the proper definition on the right.

(a) Chemical kinetics	1. A system under stress will move in a direction to relieve that stress ______
(b) Chemical equilibrium	2. The study of the rate of speed at which a chemical reaction occurs ______
(c) LeChatelier's principle	3. A dynamic state in which two or more opposing processes are taking place simultaneously at the same rate ______

▲ **2.** Define and give an example of a chemical equilibrium.

3. Explain how LeChatelier's principle is related to chemical equilibrium.

4. Define the term *reaction mechanism.*

Learning Goal 2: Mechanism of Reactions

▲ **5.** Use a specific example to explain a reaction mechanism. Identify the activated complex.

▲ **6.** What is meant by the term *energy barrier?*

7. Draw a graph showing the progress of a reaction versus changes in energy for an exothermic reaction. Label the activation energy.

▲ **8.** Define the term *activated complex.*

▲ **9.** How does a catalyst affect the activation energy of a reaction?

▲ **10.** What is an effective collision?

Learning Goal 3: Factors That Affect the Rate of a Chemical Reaction

▲ **11.** Name the four factors that affect the rate of a chemical reaction.

▲ **12.** How can a change in concentration increase the speed of a reaction?

13. For a reversible reaction, explain what happens to the rate of the forward reaction as the reaction proceeds. Explain what happens to the rate of the reverse reaction.

▲ **14.** What is the rate of the reverse reaction at the start of a chemical reaction?

▲ **15.** Explain how temperature affects reaction rate. What is the general rule?

▲ **16.** What effect does a decrease in temperature have on reaction rate?

***17.** For the gas-phase reaction

$$2H_2(g) + 2NO(g) \longrightarrow 2H_2O(g) + N_2(g)$$

the initial rate of reaction at 800°C is 20 torr per minute. What would be the approximate rate at 820°C?

▲ **18.** Consider the reaction

$$A(g) + B(g) \xrightarrow{\text{Catalyst}} AB(g)$$

Explain what would happen to the reaction rate if:
(a) The concentration of A were doubled.
(b) The catalyst were removed.

▲ **19.** Explain how a catalyst affects the rate of a reaction.

20. Explain the following observation: On a windy day a brush fire burns more rapidly than on a calm day.

Learning Goal 4: Reversible Reactions

▲ **21.** Write the forward and reverse reactions for the ionization of acetic acid.

▲ **22.** Write the forward and reverse reactions for the ionization of carbonic acid.

23. Write the forward and reverse reactions for the dissociation of sodium chloride and water.

24. Write the forward and reverse reactions for the dissociation of sodium sulfate and water.

Learning Goal 5: Rates of Reaction and Equilibrium

▲ **25.** Explain how the rates of the forward and reverse reactions change as a chemical reaction approaches equilibrium.

▲ **26.** What is meant by the phrase "reaction goes to completion"?

27. Which of the following statements are correct?
(a) In a reaction at equilibrium, the concentrations of the reactants and those of the products are equal.
(b) A catalyst increases reaction rates by changing reaction mechanisms.
(c) The amount of product available at equilibrium is proportional to the speed at which equilibrium is attained.
(d) A catalyst lowers the activation energy of a reaction by equal amounts for both forward and reverse reactions.

▲ **28.** How do catalysts affect reaction rates?

Learning Goal 6: Equilibrium Expression and Equilibrium Constant

▲ **29.** Write an equilibrium expression for each of the following equations:
(a) $2A(g) + 3B(g) \rightleftharpoons 5C(g) + 4D(g)$
(b) $H_2(g) + I_2(g) \rightleftharpoons 2HI(g)$
(c) $H_2(g) + Cl_2(g) \rightleftharpoons 2HCl(g)$
(d) $2NO_2(g) \rightleftharpoons N_2O_4(g)$

30. Write an equilibrium expression for each of the following equations:
(a) $A(g) + 2B(g) \rightleftharpoons C(g) + 4D(g)$
(b) glucose-1-phosphate $\rightleftharpoons$ glucose-6-phosphate
(c) $N_2(g) + 2O_2(g) \rightleftharpoons 2NO_2(g)$
(d) $H_2O(g) + CO(g) \rightleftharpoons H_2(g) + CO_2(g)$

31. Write an equilibrium expression for each of the following equations:
(a) $N_2(g) + 3H_2(g) \rightleftharpoons 2NH_3(g)$
(b) $2NO(g) + O_2(g) \rightleftharpoons 2NO_2(g)$
(c) $PCl_5(g) \rightleftharpoons PCl_3(g) + Cl_2(g)$
(d) $2CO_2(g) \rightleftharpoons 2CO(g) + O_2(g)$

32. Write an equilibrium expression for each of the following equations:
(a) $2SO_2(g) + O_2(g) \rightleftharpoons 2SO_3(g)$
(b) $C(s) + H_2O(g) \rightleftharpoons CO(g) + H_2(g)$
(c) $A(g) + 3B(g) \rightleftharpoons 2C(g)$
(d) $H_2(g) + Br_2(g) \rightleftharpoons 2HBr(g)$

33. Calculate K_{eq} at a given temperature for the reaction

$$2NO(g) + O_2(g) \rightleftharpoons 2NO_2(g)$$

The concentrations at equilibrium are as follows: $NO(g)$, 0.600 mole/liter; $O_2(g)$, 0.800 mole/liter; and $NO_2(g)$, 4.40 moles/liter.

▲ **34.** Calculate K_{eq} at 440°C for the reaction

$$H_2(g) + I_2(g) \rightleftharpoons 2HI(g)$$

given the following data: equilibrium concentration of $H_2(g)$, 0.436 *M;* equilibrium concentration of $I_2(g)$, 0.436 *M;* and equilibrium concentration of $HI(g)$, 3.128 *M*.

35. Calculate K_{eq} at 440°C for the reaction

$$H_2(g) + I_2(g) \rightleftharpoons 2HI(g)$$

The concentrations at equilibrium are as follows: $H_2(g)$, 0.218 mole/liter; $I_2(g)$, 0.218 mole/liter; and $HI(g)$, 1.564 moles/liter.

36. Consider the reaction

$$\text{Glucose-1-phosphate} \rightleftharpoons \text{glucose-6-phosphate}$$

Calculate K_{eq} if the concentrations at equilibrium at a certain temperature are as follows: glucose-6-phosphate, 1.00 mole/liter; and glucose-1-phosphate, 0.050 mole/liter.

Learning Goal 7: Calculation of K_a and K_b

▲ **37.** Calculate K_a for benzoic acid at 25°C for the reaction

$$HC_7H_5O_2 \rightleftharpoons H^{1+} + (C_7H_5O_2)^{1-}$$

The equilibrium concentrations are as follows: $HC_7H_5O_2$, 0.0975 mole/liter; H^{1+}, 0.0025 mole/liter; and $(C_7H_5O_2)^{1-}$, 0.0025 mole/liter.

▲ **38.** Calculate K_b for NH_3 at 20°C for the reaction

$$NH_3(aq) + H_2O(aq) \rightleftharpoons (NH_4)^{1+}(aq) + OH^{1-}(aq)$$

The equilibrium concentrations are as follows: $(NH_4)^{1+}$, 0.0013 mole/liter; $(OH)^{1-}$, 0.0013 mole/liter; and $NH_3(aq)$, 0.0939 mole/liter.

39. Calculate K_a for cyanic acid at 25°C for the reaction

$$HCNO(aq) \rightleftharpoons H^{1+}(aq) + (CNO)^{1-}(aq)$$

The equilibrium concentrations are as follows: HCNO, 0.0955 mole/liter; H^{1+}, 0.0045 mole/liter; and $(CNO)^{1-}$, 0.0045 mole/liter.

40. Calculate K_a for acetic acid at 25°C for the reaction

$$HC_2H_3O_2(aq) \rightleftharpoons H^{1+}(aq) + (C_2H_3O_2)^{1-}(aq)$$

The equilibrium concentrations are as follows: $HC_2H_3O_2$, 0.3990 mole/liter; H^{1+}, 0.00268 mole/liter; and $(C_2H_3O_2)^{1-}$, 0.00268 mole/liter.

Learning Goal 8: Calculation of Equilibrium Concentrations

▲ **41.** Calculate the equilibrium concentrations of H^{1+} and $(C_2H_3O_2)^{1-}$ at 25°C for a 0.100 *M* acetic acid solution ($K_a = 1.80 \times 10^{-5}$).

42. Calculate the equilibrium concentrations of H^{1+} and F^{1-} for a 0.400 *M* HF solution if K_a is 8.29×10^{-4}.

43. Calculate the equilibrium concentrations of H^{1+} and $(CN)^{1-}$ for a 0.500 *M* hydrocyanic acid solution ($K_a = 4.00 \times 10^{-10}$).

▲ **44.** The vitamin niacin, which we can represent as HNiac, dissociates to form H^{1+} and $(Niac)^{1-}$ ions. K_a for niacin is 1.40×10^{-5}. Calculate the equilibrium concentrations of H^{1+} and $(Niac)^{1-}$ for a 1.00 *M* HNiac solution.

45. Calculate the equilibrium concentrations of $(CuOH)^{1+}$ and $(OH)^{1-}$ for a 0.500 *M* copper(II) hydroxide solution ($K_b = 1.00 \times 10^{-8}$).

46. Calculate the equilibrium concentrations of Ag^{1+} and $(OH)^{1-}$ for a 1.50 *M* silver hydroxide solution ($K_b = 1.1 \times 10^{-4}$).

47. Calculate the equilibrium concentrations of $(NH_4)^{1+}$ and $(OH)^{1-}$ for a 0.100 *M* ammonium hydroxide solution ($K_b = 1.80 \times 10^{-5}$).

48. Calculate the equilibrium concentrations of $(CaOH)^{1+}$ and $(OH)^{1-}$ ions for a 1.00 *M* calcium hydroxide solution at 25°C ($K_b = 3.74 \times 10^{-3}$).

Learning Goal 9: Writing K_{sp} Expressions and Calculating K_{sp}

▲ **49.** Write K_{sp} expressions for these slightly soluble salts: *(a)* AgI *(b)* $Fe(OH)_3$ *(c)* Ag_2CrO_4

50. Write K_{sp} expressions for these slightly soluble salts: *(a)* $BaCO_3$ *(b)* BaF_2 *(c)* PbI_2

51. Write K_{sp} expressions for these slightly soluble salts: *(a)* $BaCrO_4$ *(b)* CaF_2 *(c)* CuS

52. Write K_{sp} expressions for these slightly soluble salts: *(a)* ZnS *(b)* CuCl *(c)* $Al(OH)_3$

53. Calculate K_{sp} for $BaSO_4$ at 25°C if the concentrations of Ba^{2+} and $(SO_4)^{2-}$ ions are both 3.9×10^{-5} mole/liter. (*Hint:* The equilibrium equation for the ionization of $BaSO_4$ is $BaSO_4 \rightleftharpoons Ba^{2+} + (SO_4)^{2-}$.)

▲ **54.** Calculate K_{sp} for silver chloride if the concentrations of the Ag^{1+} and Cl^{1-} ions are both 1.30×10^{-5} mole/liter at 25°C.

55. Calculate K_{sp} for Ag_2CrO_4 at 25°C if the concentration of Ag^{1+} ions is 1.56×10^{-4} mole/liter and the concentration of $(CrO_4)^{2-}$ ions is 7.80×10^{-5} mole/liter. (*Hint:* The equilibrium equation for the ionization of Ag_2CrO_4 is $Ag_2CrO_4 \rightleftharpoons 2Ag^{1+} + (CrO_4)^{2-}$.)

56. Calculate K_{sp} for calcium fluoride if the concentration of Ca^{2+} is 2.14×10^{-4} mole/liter and that of F^{1-} is 4.28×10^{-4} mole/liter at 25°C.

Learning Goal 10: Calculation of the Solubility of a Salt

▲ **57.** Calculate the solubility of CuS at 25°C in grams per liter, given that $K_{sp} = 9.0 \times 10^{-45}$.

▲ **58.** Calculate the solubility of AgI at 25°C in grams per liter, given that K_{sp} is 8.5×10^{-17}.

59. Calculate the solubility of Ag_2CrO_4 at 25°C in grams per liter, given that $K_{sp} = 1.9 \times 10^{-12}$.

60. Calculate the solubility of AgBr at 18°C in grams per liter, given that K_{sp} is 4.1×10^{-13}.

Learning Goals 11 and 12: Buffers; Carbonic Acid/Bicarbonate Buffer System

▲ **61.** How does one prepare a buffer solution?

62. Write the equation for the equilibrium among $(HCO_3)^{1-}$, H^{1+}, H_2CO_3, and CO_2 in water. What effect would an increase in the $(HCO_3)^{1-}$ concentration have on the hydrogen-ion concentration?

63. Write the equation(s) that represent the carbonic acid/bicarbonate buffer system in the blood.

▲ **64.** What are the effects of hypoventilation and of hyperventilation on the carbonic acid/bicarbonate buffer system in the blood?

65. Given the following reactions at equilibrium,

$$H_2O + CO_2 \rightleftharpoons H_2CO_3 \rightleftharpoons H^{1+} + (HCO_3)^{1-}$$

decide which way the equilibrium shifts when:
(*a*) Excess H^{1+} ion is added.
(*b*) Excess CO_2 is added.
(*c*) Bicarbonate ion, $(HCO_3)^{1-}$, is removed.

▲ **66.** State whether each of the following statements is true or false, and explain your answers:
(*a*) Carbonic acid is capable of neutralizing any weak base present in the blood.
(*b*) Bicarbonate ion is capable of neutralizing any weak acid that is present in the blood.

▲ **67.** Show by the use of a chemical equation how the carbonic acid/bicarbonate buffer system in the blood works in the event of a sudden increase in the concentration of H^{1+} ions.

▲ **68.** A person who hyperventilates and goes into alkalosis is told to breathe into a paper bag. Why? Explain your answer in terms of the carbonic acid/bicarbonate buffer system.

Extra Exercises

▲ **69.** For each of the following reactions, in which direction does the equilibrium shift if the pressure is increased?
(*a*) $H_2(g) + I_2(g) \rightleftharpoons 2HI(g)$
(*b*) $2SO_2(g) + O_2(g) \rightleftharpoons 2SO_3(g)$
(*c*) $N_2(g) + 3H_2(g) \rightleftharpoons 2NH_3(g)$

▲ **70.** For each of the following reactions, in which direction does the equilibrium shift if the temperature is increased?
(*a*) $H_2(g) + Cl_2(g) \rightleftharpoons 2HCl(g) + 44.2$ kcal
(*b*) $2CO_2(g) + 135.2$ kcal $\rightleftharpoons 2CO(g) + O_2(g)$
(*c*) $H_2(g) + I_2(g) + 12.4$ kcal $\rightleftharpoons 2HI(g)$

71. Write an equilibrium expression for each of the following reactions:
(*a*) $N_2(g) + O_2(g) \rightleftharpoons 2NO(g)$
(*b*) $2CO_2(g) \rightleftharpoons 2CO(g) + O_2(g)$
(*c*) $2SO_2(g) + O_2(g) \rightleftharpoons 2SO_3(g)$

72. For the reaction

$$A(g) + 2B(g) + 3C(g) \rightleftharpoons 4D(g)$$

calculate K_{eq} if the concentrations at equilibrium are [A] = 4.0 moles/liter, [B] = 3.0 moles/liter, [C] = 2.0 moles/liter, [D] = 1.0 mole/liter.

73. Calculate K_a for formic acid, $HCHO_2$, at 25°C, given the following equilibrium concentrations: $[HCHO_2] = 0.0958$ mole/liter, $[H^{1+}] = 0.0042$ mole/liter, and $[(CHO_2)^{1-}] = 0.0042$ mole/liter.

74. Calculate the equilibrium concentration of H^{1+} and $(NO_2)^{1-}$ at 25°C for a 1.0 *M* HNO_2 solution ($K_a = 4.5 \times 10^{-4}$).

▲ **75.** Which of the following indicates the reaction that goes farthest toward completion?
(*a*) $K_{eq} = 1$ (*b*) $K_{eq} = 10$ (*c*) $K_{eq} = 100$
(*d*) $K_{eq} = 0.1$

76. Write the K_{sp} expression for each of these salts: *(a)* $Mn(OH)_2$ *(b)* $AgCl$ *(c)* A_2X_3

77. Calculate the solubility of CaF_2 in grams per liter at 25°C, given that $K_{sp} = 3.9 \times 10^{-11}$.

▲ **78.** Which compound, Ag_2CrO_4 or $BaCrO_4$, has the greater *molar* solubility? (*Hint:* Check Table 17.2.)

79. Write an equilibrium equation for each of the following systems:
(a) a mixture of ice and water at 0°C
(b) a mixture of water and steam in a closed system at 100°C
(c) carbonic acid (H_2CO_3) in equilibrium with carbon dioxide and water
(d) ammonia and water in equilibrium with ammonium ion and hydroxide ion

Cumulative Review/Chapters 15–17

▲ **1.** Which is the solute and which is the solvent in each of the following solutions?
(a) 75 mL of nitrogen and 35 mL of oxygen
(b) sulfur dioxide gas and water
(c) lemon juice, sugar, and water in lemonade

▲ **2.** State whether each of the following pairs of substances form a solution:
(a) hydrogen fluoride gas and water
(b) benzene and *trans*-dichloroethane

3. Find the concentration of each of the following solutions in percentage by mass:
(a) 50.0 g of salt in enough water to make 75.0 g of solution
(b) 175 g of sugar in enough water to make $1{,}5\overline{0}0$ g of solution

4. Find the concentration of each of the following solutions in percentage by volume:
(a) 25.0 mL of alcohol in enough water to make 455 mL of solution
(b) 35.0 mL of benzene dissolved in enough carbon tetrachloride to make 95.0 mL of solution

▲ **5.** If 5.0 g of benzene are dissolved in enough carbon tetrachloride to make 75 mL of solution, what is the concentration of the solution in percentage by mass–volume?

6. What is the molarity of the solution formed by dissolving 1.84 g of ethyl alcohol (C_2H_5OH) in enough water to make 1.5 L of solution?

▲ **7.** When $25\overline{0}$ mL of 1.33 *M* NaOH solution are evaporated to dryness, how many grams of NaOH remain?

8. How many milliliters of 1.00 *M* $CaCl_2$ solution must one evaporate to obtain 26.0 g of $CaCl_2$?

▲ **9.** Distinguish between normality and molarity.

10. How many grams of H_3PO_4 are present in 1.00 L of a 2.00 *N* solution of H_3PO_4?

11. Calculate the normality and the molarity of a solution of calcium hydroxide formed by dissolving 18.5 g of calcium hydroxide in enough water to make $5\overline{0}0$ mL of solution.

12. How would 5.50 L of 1.33 *N* NaOH solution be prepared from a 36.0 *N* solution of NaOH?

▲ **13.** A photographer is preparing 2.00 L of 0.150 *M* silver nitrate solution. How many grams of silver nitrate are needed?

14. A solution is a heterogeneous mixture. True or false?

15. Calculate the grams of solute in $1\overline{0}0$ mL of 18.0 *M* H_2SO_4 solution.

16. A solution of sugar in water contains 25.0% sugar, by mass, and has a density of 1.09 g/mL. The molecular formula for sugar is $C_{12}H_{22}O_{11}$.
(a) Calculate the number of grams of sugar present in 2.00 L of solution.
(b) Calculate the molarity of the solution.

17. Calculate the grams of solute present in 100.0 g of 0.90% by mass NaCl solution.

18. To what volume must a solution of 75.0 g of H_2SO_4 in $4\overline{0}0$ mL of water be diluted to give a 0.200 *M* solution?

▲ **19.** Calculate the boiling point of a solution prepared by dissolving 100.0 g of ethylene glycol, $C_2H_6O_2$, in 500.0 g of water. K_b for water is 0.520°C/*m*.

20. A solution is prepared by dissolving 12.8 g of naphthalene, $C_{10}H_8$, in 1.00 L of water. Is this solution 0.100 *M* or 0.100 *m?*

21. State the difference between the Arrhenius and the Brønsted–Lowry definitions of an acid.

▲ **22.** A solution of pH 4 is 100 times more acidic than a solution of pH 7. True or false?

23. Complete and balance each of the following equations:

(a) $HC_2H_3O_2 + Ca(OH)_2 \longrightarrow ?$
(b) $N_2O_5 + H_2O \longrightarrow ?$

24. Write an equation for the formation of acid rain in the atmosphere.

25. Determine the pH of a solution that has a hydrogen-ion concentration of 1.0×10^{-2} *M*.

26. Calculate the pH of black coffee if its hydrogen ion-concentration is 1.0×10^{-6} *M*.

27. How many milliliters of 0.500 *M* HCl can one prepare by diluting $75\overline{0}$ mL of 12.0 *M* HCl?

28. Calculate the molarity of each of the ions present in a solution prepared by combining 50.0 mL of 2.00 *M* HCl with 50.0 mL of 2.00 *M* NaOH.

29. If 40.0 mL of 1.50 *M* NaOH solution are needed to titrate 50.0 mL of HCl, what is the molarity of the HCl?

30. Calculate the molarity of NaOH if 15.0 mL of 0.150 *M* HCl are needed to titrate 50.00 mL of NaOH.

31. Differentiate between the terms *ionization* and *dissociation*.

32. Which solution is more acidic, 1 *M* H_2SO_4 or 1 *N* H_2SO_4?

33. Which solution is more acidic, 1 *M* HCl or 1 *M* $HC_2H_3O_2$?

34. Calculate the molarity of the sum of all the ions present in 0.02 *M* NaCl if the solution is 100% ionized.

35. Calculate the concentration of Sr^{2+} ions in a solution of $SrCl_2$ that has a Cl^{1-} ion concentration of 0.500 *M*. (Assume 100% ionization.)

36. If 50.00 mL of 0.500 *M* HCl neutralize 100.0 mL of $Ca(OH)_2$ solution, what is the molarity of the $Ca(OH)_2$ solution? The reaction is

$$Ca(OH)_2(aq) + 2HCl(aq) \longrightarrow CaCl_2(aq) + 2H_2O$$

▲ **37.** Do the terms *strong acid* and *strong base* refer to concentration or ionization?

▲ **38.** When a mole of $SrCl_2$ is dissolved in water, are there more Sr^{2+} ions or more Cl^{1-} ions present?

39. Calculate the pH of tomato juice whose H^{1+} concentration is 1.0×10^{-4}.

40. What is the $(OH)^{1-}$ concentration of a solution whose H^{1+} concentration is 10^{-8}?

▲ **41.** Cite four factors that affect the rate of a chemical reaction.

42. As a chemical reaction approaches equilibrium, what happens to the rates of the forward and reverse reactions?

43. Express the following reaction in equation form: Water and water vapor are both present at 100°C in a pressurized system.

44. Is the following reaction endothermic or exothermic?

$$2NO(g) + O_2(g) \longrightarrow 2NO_2(g) + 27 \text{ kcal}$$

45. A large equilibrium constant for a reaction means that at equilibrium, the formation of reactants is favored over the formation of products. True or false?

▲ **46.** Consider the carbonic acid/bicarbonate buffer system in the blood.

$$H_2O + CO_2 \rightleftharpoons H_2CO_3 \rightleftharpoons H^{1+} + (HCO_3)^{1-}$$

When excess acid is added to the blood, this system accommodates to the "stress" in such a way as to maintain the normal pH of the blood. Explain how.

47. Write the equilibrium expression for the following reaction:

$$CaCO_3(s) \rightleftharpoons CaO(s) + CO_2(g)$$

48. Calculate the solubility product constant for $Pb(IO_3)_2$ if the concentration of the saturated solution of the salt is 4.0×10^{-5} *M*.

49. Calculate the solubility of Ag_2S in moles per liter, given that $K_{sp} = 6.0 \times 10^{-50}$.

50. Calculate the solubility of $PbSO_4$ in grams per liter, given that $K_{sp} = 6.0 \times 10^{-8}$.

51. Calculate the equilibrium concentrations of H^{1+}, $(C_2H_3O_2)^{1-}$, and $HC_2H_3O_2$ in a 0.75 *M* acetic acid solution at 25°C, given that $K_a = 1.8 \times 10^{-5}$.

52. Write the K_{sp} expression for $Ba_3(PO_4)_2$.

53. Write the equilibrium expression for the following reaction:

$$HClO_2(aq) \rightleftharpoons H^{1+}(aq) + (ClO_2)^{1-}(aq).$$

54. Using Table 17.2, determine which compound has the greater molar solubility, AgCl, AgBr, or AgI.

55. Predict the direction in which the equilibrium will shift when heat is added to the following reaction:

$$4NH_3(g) + 3O_2(g) \rightleftharpoons 2N_2(g) + 6H_2O(g) + 366 \text{ kcal}$$

18 Special Topics: Nuclear Chemistry, Electrochemistry, and Environmental Chemistry

Learning Goals

After you have studied this chapter, you should be able to:

1. Explain what is meant by radioactivity.
2. List the three major types of nuclear radiation and the characteristics of each.
3. Write balanced nuclear equations.
4. Define the term *half-life*.
5. Perform calculations involving the half-life of a substance.
6. Define the terms *nuclear fission* and *nuclear fusion*.
7. Discuss the emergence of nuclear power plants.
8. Explain how radiation affects human health.
9. Cite several methods of detecting radiation.
10. Explain how radiation is measured.
11. Give an example of a radioactive isotope that is used in medicine, and discuss how it is used.
12. Describe a voltaic cell and an electrolytic cell.
13. Discuss the operation of a lead storage battery and a fuel cell.
14. List the normal ingredients of air and the common pollutants it may contain.
15. Describe some common reactions that produce air pollutants.
16. Discuss the health problems of pollution presented by the automobile engine and other pollution sources.
17. Describe some common ways of reducing air pollution.
18. Discuss the problem of hazardous wastes.
19. Define the term *potable,* and list sources of potable water.
20. List some common water pollutants.

Introduction

It is the year 1800. In his laboratory, the Italian physicist Alessandro Volta is experimenting with solutions and metals that are able to conduct electricity. He finds that two metals and such a solution can be arranged to *produce* an electric current. Volta has invented the first *electrochemical battery*—a device that produces electrical energy from a chemical reaction.

It is the year 1895 and New Zealand physicist Ernest Rutherford performs experiments on the magnetic deflection of radioactivity at the Cavendish laboratory of Cambridge University. This research leads to the discovery of two kinds of particles. He calls them alpha particles and beta particles. He publishes his results in 1899. Shortly thereafter the Frenchman P. V. Villard discovers the gamma ray.

It is late October 1948. The place is Donora, Pennsylvania, a small industrial town located at the bottom of a river valley. An unusually thick smog has settled over the town. People look up at the leaden sky and sniff uneasily. The smog is so thick that it makes people's eyes sting and their throats become raw. Sulfur dioxide, dust, and waste products from the nearby zinc, iron, and steel mills all contribute to the smog. Local weather conditions have created an *atmospheric inversion,* which keeps the polluted air near the ground (Figure 18.1). By the time the smog lifts some 5 days later, 17 people have died as a direct result of it, and almost half the population of Donora has been affected to some degree. The nightmare of environmental pollution has made its first widely recognized appearance in the United States.

These three events, which took place at different times in history, all involve chemical processes. They represent, however, aspects of three different areas of chemistry: nuclear chemistry, electrochemistry, and environmen-

Figure 18.1
(*Left*) Shenandoah National Park on a pollution-free day. (*Right*) The same vantage point as shown in the photo on the left, during an atmospheric inversion on a day when pollution from other areas has settled in.
(National Park Service photographs)

Antoine Henri Becquerel
(Reprinted with permission from Torchbearers of Chemistry by Henry Monmouth Smith, © Academic Press, Inc., 1949)

tal chemistry. In this chapter, we briefly survey these three important areas of study.

18.1 Nuclear Chemistry

Although the explosion of the atomic bomb over Hiroshima in 1945 is the most dramatic sign that the nuclear age had begun, the study of nuclear chemistry actually began much earlier. In 1895 the German physicist Wilhelm Konrad Röntgen accidentally discovered x rays. He found that x rays have a very great penetrating power—so great that they could easily pass through wood and plaster walls.

In 1896 the French physicist Antoine Becquerel placed a salt containing uranium on a photographic plate that was wrapped in black paper. When he developed the photographic plate, he found the image of the pile of salt. The uranium salt had taken a picture of itself! Becquerel concluded that the uranium was giving off some type of penetrating rays and that these rays must be very strong to be able to pass through the black paper and expose the photographic plate. When Becquerel placed a thick barrier of lead between the salt and the photographic plate, the lead absorbed the rays and no image was formed. Becquerel thought that these penetrating rays were probably Röntgen's x rays.

Learning Goal 1
Radioactivity and atomic energy

At the same time, in Paris, a young Polish chemist named Marie Curie was working with her husband Pierre in the laboratories at the Sorbonne. The Curies became interested in Becquerel's discovery. In fact, it was Marie Curie who defined **radioactivity** as the ability of a substance to emit penetrating rays from its nucleus. The Curies found that a radioactive substance seems to keep on emitting these powerful penetrating rays, year after year. In 1898 they discovered that the element thorium is radioactive. So are polonium and radium.

Learning Goal 2
Types of nuclear radiation

When early experimenters allowed the radiation produced by uranium to pass through a magnetic field, they found that the radiation was deflected in different directions (Figure 18.2). Three types of radiation were detected,

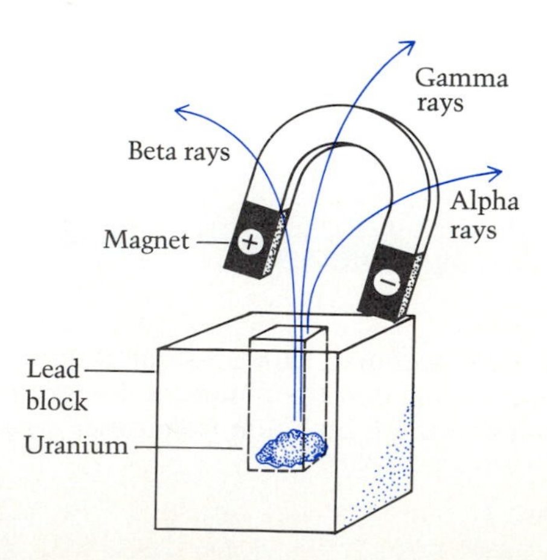

Figure 18.2
Analyzing the radiation from uranium

Marie Curie
(Reprinted with permission from Torchbearers of Chemistry by Henry Monmouth Smith, © Academic Press, Inc., 1949)

and they were called *alpha particles, beta particles,* and *gamma rays,* from the first three letters of the Greek alphabet: α, β, and γ.

Gamma Rays (γ Rays)

Gamma (γ) rays *are made up of energy. They have no mass and carry no electrical charge.* Gamma rays travel straight through a magnetic field—they are not deflected. For this reason, the early experimenters assumed that they carried no charge. Gamma rays seemed to have properties like those of x rays, but they were more energetic. In fact, gamma rays have a very high energy content and strong penetrating ability. Today we know that because of this penetrating ability, a high concentration of gamma rays can harm human genes. The nucleus of an atom gives off gamma rays only under certain conditions, when the nucleus loses energy. And the atom, of course, has to be of a radioactive element.

Beta Particles (β Particles)

Beta (β) particles *are identical to an electron in mass and size. These particles have a negative electrical charge.* When Becquerel passed beta particles through a magnetic field, he found that they behaved like cathode rays—that is, like the rays that emanate from a negative electrode. Therefore Becquerel concluded that these rays were composed of speeding particles, which he assumed to be electrons. However, beta particles, like gamma rays and alpha particles, originate in the *nucleus* of the atom, and electrons do not exist in the nucleus.

How, then, are beta particles formed? Scientists believe they are formed when neutrons decompose during a nuclear reaction:

$$\text{Neutron} \longrightarrow \text{proton} + \beta \text{ particle}$$

Like an electron, a beta particle has a zero mass number and a charge of -1. We can represent it in the following way:

$${}_{-1}^{0}\text{e} = \beta \text{ particle}$$

Beta particles have less penetrating power than gamma rays. They can be stopped by a thin sheet (a few centimeters thick) of almost any metal.

Alpha Particles (α Particles)

Alpha (α) particles *are helium nuclei. These particles carry a double positive charge.* Scientists of the early 1900s found that in a magnetic field, a stream of alpha particles was deflected in the opposite direction from beta particles. They therefore assumed that alpha radiation was made up of positively charged particles. Experiments showed that these particles have a mass number of 4 and a charge of +2. Werner Heisenberg, a German physicist, suggested that we picture an alpha particle as a helium-4 atom with its electrons removed. In other words, we can imagine it as composed of two protons and two neutrons. We can thus represent an alpha particle in this way:

$$^{4}_{2}\text{He} = \alpha \text{ particle}$$

Alpha particles have little penetrating ability. They can be stopped by a thin sheet of paper, but their effects on the human body can be horrendous.

Nuclear Transformations

Learning Goal 3
Writing balanced nuclear equations

Table 18.1 summarizes the basic properties of the three types of radioactivity. It also indicates the changes that take place in atoms that emit alpha or beta particles. Thus we can use the table to write the equations for nuclear reactions, and we will do so shortly.

An element that is emitting radiation is said to be undergoing **radioactive** or **nuclear decay.** Some elements do so spontaneously—naturally—without help from scientists or laboratory equipment. An example is radium-226, $^{226}_{88}\text{Ra}$. When it undergoes nuclear decay, a radium atom emits an alpha particle and actually becomes a different element! We call this a **nuclear transformation** or **transmutation,** and we can represent it with the nuclear equation

$$^{226}_{88}\text{Ra} \longrightarrow \underset{\substack{\alpha \\ \text{particle}}}{^{4}_{2}\text{He}} + \ ?$$

But what element is formed in this transformation? To find out, we note that in a balanced nuclear equation, the sum of the superscripts (mass numbers) and the sum of the subscripts (protons) must be the same on both sides of the equation. Subtraction then tells us that the element formed has the atomic number 86 and the mass number 222. In the periodic table (inside the front cover), we find that element 86 is radon, Rn. So we write the nuclear equation as

$$^{226}_{88}\text{Ra} \longrightarrow {}^{4}_{2}\text{He} + {}^{222}_{86}\text{Rn}$$

Here's another example: The isotope thorium-234 produces radiation by emitting beta particles. We can represent this nuclear reaction as

$$^{234}_{90}\text{Th} \longrightarrow \underset{\substack{\beta \\ \text{particle}}}{^{0}_{-1}\text{e}} + \ ?$$

What is the element that is formed? Table 18.1 tells us that the emission of a beta particle increases an element's atomic number by 1. Therefore the unknown element must be the the element whose atomic number is 91. That element is protactinium, so we can write our equation as

$$^{234}_{90}\text{Th} \longrightarrow {}^{0}_{-1}\text{e} + {}^{234}_{91}\text{Pa}$$

Again the sum of the superscripts and the sum of the subscripts are the same on both sides of the equation.

Table 18.1
Summary of Nuclear Particles

			Changes caused by emission	
Type of emission	**Mass number of particle emitted**	**Charge of particle emitted**	*Mass number*	*Atomic number*
Alpha	4	+2	Decreases by 4	Decreases by 2
Beta	0	−1	No change	Increases by 1
Gamma	0	0	No change*	No change*

* Gamma rays may accompany some radioactive decay reactions.

Example 18.1

Complete the following nuclear equations:

(a) $^{14}_{6}C \longrightarrow {}^{0}_{-1}e + ?$
(b) $? \longrightarrow {}^{0}_{-1}e + {}^{24}_{12}Mg$
(c) $? \longrightarrow {}^{4}_{2}He + {}^{234}_{90}Th$

Solution

Understand the Problem

We are given partial equations for three nuclear reactions. To complete them, we must analyze what happens to particular atoms when they undergo nuclear change.

Devise a Plan

We can use the Law of Conservation of Mass and the periodic table to find the missing atom that will balance the equation.

Carry Out the Plan

According to the Law of Conservation of Mass we know that the total mass on the left side of the equation equals the total mass on the right side of the equation.

(a) $^{14}_{6}C \longrightarrow {}^{0}_{-1}e + {}^{14}_{7}N$
(b) $^{24}_{11}Na \longrightarrow {}^{0}_{-1}e + {}^{24}_{12}Mg$
(c) $^{238}_{92}U \longrightarrow {}^{4}_{2}He + {}^{234}_{90}Th$

Look Back

Check that in each case the sum of the mass numbers (superscripts) on the right side of the equation equals the mass number on the left side of the equation. Also be sure that the sum of the atomic numbers (subscripts) on the right side of the equation equals the atomic number on the left side of the equation.

Practice Exercise 18.1 Complete the following nuclear equations:

(a) $^{87}_{36}Kr \longrightarrow ^{1}_{0}n + ?$

(b) $^{212}_{84}Po \longrightarrow ^{4}_{2}He + ?$

(c) $^{32}_{15}P \longrightarrow ^{0}_{-1}e + ?$

The Half-life of a Radioactive Element

Learning Goal 4
Half-life

No one really knows why some elements are radioactive—that is, why they undergo nuclear decay. Certain combinations of protons and neutrons make the atomic nucleus unstable. These nuclei may become more stable by emitting alpha or beta particles.

And no one can tell when a particular atom of a radioactive element will decay. But, for a large number of atoms of any radioactive element, scientists can predict how many atoms will decay in a given time. This decay rate is a characteristic of the radioactive isotope and is expressed as the time required for half of the nuclei in a sample of the radioactive isotope to decay. It is called the *half-life* of the element. The **half-life** of a radioactive element is *the time required for the decay of half of the atoms in a sample of that element*. Half-lives of the radioactive isotopes vary in length. For example, the half-life of $^{238}_{92}U$ is 4.5 billion years, whereas the half-life of $^{257}_{103}Lr$ is 8 seconds.

Example 18.2

The half-life of strontium-90 is 28 years. Suppose we have 100 g of strontium-90 today. How many grams of strontium-90 will be left in the sample in 84 years?

Learning Goal 5
Calculations using half-life

Solution Because the half-life of strontium-90 is 28 years, half of the strontium-90 will decay every 28 years. After 84 years, the sample will contain 12.5 g of strontium-90.

Time in Years	Grams of Strontium-90 Remaining
Now	100 g
28	50 g
56	25 g
84	12.5 g

Practice Exercise 18.2 The half-life of lawrencium-257 is 8 seconds. If we have 100 g of lawrencium-257 right now, how many grams of this radioactive element will be left in the sample in 64 seconds?

Radioactive isotopes react chemically exactly as nonradioactive ones do, and therefore they cannot be distinguished from each other by chemical means. While discussing strontium-90, we should note that it has a chemical similarity to calcium (Ca and Sr are in the same chemical family). It will replace the calcium in bone and stay there, undergoing its radioactive decay. When nations test atomic bombs in the atmosphere, strontium-90 gets into the atmosphere and begins to float around the world. Gradually it falls to earth—on fields, for example. When cows eat the grass that grows in the fields, this strontium-90 gets into milk and other dairy products; it also gets into vegetables planted in the fields. People consume these farm products, and the radioactive isotope, strontium-90, accumulates in their bones, side by side with calcium. It then bombards nearby tissues and organs with beta particles. The isotope's 28-year half-life means that it stays around for a long time. And if a person is exposed to a large dose, it can cause bone cancer.

Artificially Produced Radiation—the Alchemists' Dream Come True

In 1919 Ernest Rutherford bombarded nitrogen-14 with a stream of alpha particles. The results were astounding. Rutherford found that he had produced oxygen-17, a nonradioactive isotope of oxygen.

$$\underset{\substack{\alpha \\ \text{particle}}}{{}^{4}_{2}\text{He}} + {}^{14}_{7}\text{N} \longrightarrow {}^{17}_{8}\text{O} + \underset{\substack{\text{A} \\ \text{proton}}}{{}^{1}_{1}\text{H}}$$

The dreams of the alchemists had come true! By triggering the first nuclear reaction produced by humans, Rutherford had at last transmuted an element (Figure 18.3).

Fifteen years later, the French physicists Irène Joliot-Curie and Frédéric Joliot (daughter and son-in-law of Marie and Pierre Curie) bombarded boron-10 with alpha particles. They obtained nitrogen-13, an artificially produced radioactive isotope.

$${}^{10}_{5}\text{B} + \underset{\substack{\alpha \\ \text{particle}}}{{}^{4}_{2}\text{He}} \longrightarrow {}^{13}_{7}\text{N} + \underset{\substack{\text{A} \\ \text{neutron}}}{{}^{1}_{0}\text{n}}$$

By the 1930s, then, scientists were unraveling the atom's deepest secrets and were about to discover the power of the atom. They were, in fact, on the

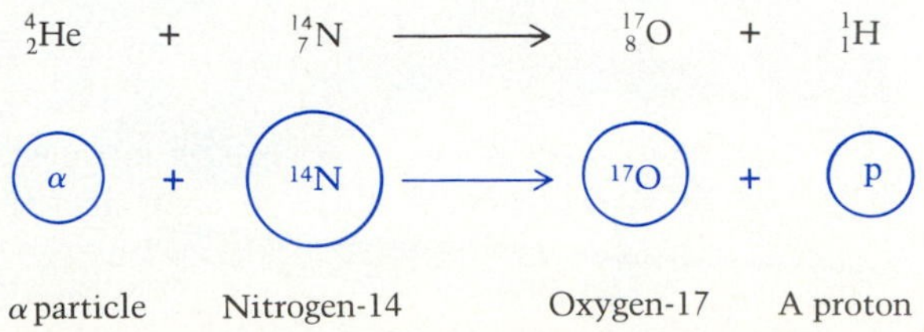

$${}^{4}_{2}\text{He} + {}^{14}_{7}\text{N} \longrightarrow {}^{17}_{8}\text{O} + {}^{1}_{1}\text{H}$$

$$\alpha + {}^{14}\text{N} \longrightarrow {}^{17}\text{O} + \text{p}$$

α particle   Nitrogen-14   Oxygen-17   A proton

Figure 18.3
Nitrogen-14, bombarded with an alpha particle, takes on an entirely new identity and becomes oxygen-17.

verge of producing a nuclear chain reaction, which could turn very small quantities of mass into huge quantities of energy.

The Chain Reaction and Nuclear Fission

Learning Goal 6
Nuclear fission

In 1934 the Italian physicist Enrico Fermi (who won the Nobel Prize in 1938) bombarded uranium (element 92) with neutrons in an attempt to produce element 93 (neptunium). But instead, Fermi found himself with an isotope of barium (element 56), a mystifying outcome.

In 1938 the German physical chemist Otto Hahn, who was also working on atomic power, proposed an explanation. He said that uranium atoms split into different atoms when they are bombarded by neutrons. This splitting process is called **nuclear fission.** Furthermore, when these uranium atoms split, they produce *more* neutrons, which in turn are able to split still more uranium atoms in what is called a **chain reaction.** It can be likened to placing dominoes in a triangular pattern. When you tip the first domino, it causes two others to tip over, and they can tip over several more, and so on down the line.

In September 1939, Europe exploded into World War II and all nations looked for new weapons. Many scientists who were studying nuclear reactions knew that nuclear fission could be turned into a decisive weapon of war. It was only a matter of time before someone perfected the technique. Would the Germans be first? Were they racing toward that goal? Everyone knew that Germany's scientists were second to none.

The United States, though not yet at war, recognized the danger and launched a research program to produce an atomic (fission) bomb. Under the leadership of Enrico Fermi, the so-called Manhattan Project (a code name) was started. It operated out of a windy, makeshift laboratory beneath the bleachers of Stagg Field at the University of Chicago. On December 2, 1942, Enrico Fermi and his team achieved the first sustained nuclear chain reaction.

Figure 18.4
An atomic bomb of the World War II variety
(Courtesy National Air and Space Museum, Smithsonian Institution)

By July 1945, through hard work and perseverance, the scientists had managed to scrape together enough uranium-235 to make a fission bomb (Figure 18.4). To test it, the United States had to fire it; the first explosion of an A-bomb took place in the desert near Alamogordo, New Mexico. Most people were in bed when the bomb exploded in the middle of the night. By the location of their cattle's singed hair, ranchers many miles away could determine the orientation of their cattle during the explosion. With feelings of deep foreboding, the scientists set about making more bombs.

By August 1945, they had two more bombs prepared. Just two! Yet such was the crisis (Americans believed that they would lose a million soldiers if they tried to storm the Japanese mainland) that the United States went ahead and exploded them: the first at Hiroshima and the second, three days later, at Nagasaki, Japan. The Japanese surrendered shortly thereafter, which ended World War II.

Nuclear Power Plants

Although the first energy generated by nuclear fission was used in making war, scientists quickly learned to apply this energy to peaceful purposes. In 1946 Congress established the Atomic Energy Commission (AEC) to control the development of nuclear energy in the United States. Private industry soon became involved in the development of commercial nuclear power, and the 1950s saw the emergence of full-scale nuclear power plants in the United States and around the world. This trend escalated in the decades that followed.

Learning Goal 7
Emergence of nuclear power plants

Nuclear power plants generate electricity in much the same way as conventional fossil-fueled power plants. In a fossil-fueled power plant, the heat obtained from the burning of coal or oil is used to change water into steam, which is then used to turn a turbine and produce electricity. In a nuclear power plant, the heat obtained from the naturally occurring fission (splitting) of uranium-235 atoms is used to change water into steam, which is then used to turn a turbine and produce electricity (Figure 18.5). Control rods are used in the nuclear reactor to absorb neutrons and regulate the rate of the fission reaction. Boron is commonly used in control rods, because it absorbs neutrons well and does not undergo fission. The fuel rods that compose the reactor's core are submerged in water. This not only cools the core but also helps moderate the chain reaction.

The products of nuclear fission include free neutrons and fission fragments that produce alpha particles, beta particles, and gamma rays. Plutonium, one of the by-products of atomic fission, has a half-life of 24,000 years. Some argue that even though plutonium remains radioactive for so long, its most harmful high-level radiation dissipates after a few hundred years. Nevertheless, such radioactive waste products must be handled very carefully for a very long time.

Several problems are associated with the use of nuclear fission to produce energy. Many of the issues involved are universally acknowledged as prob-

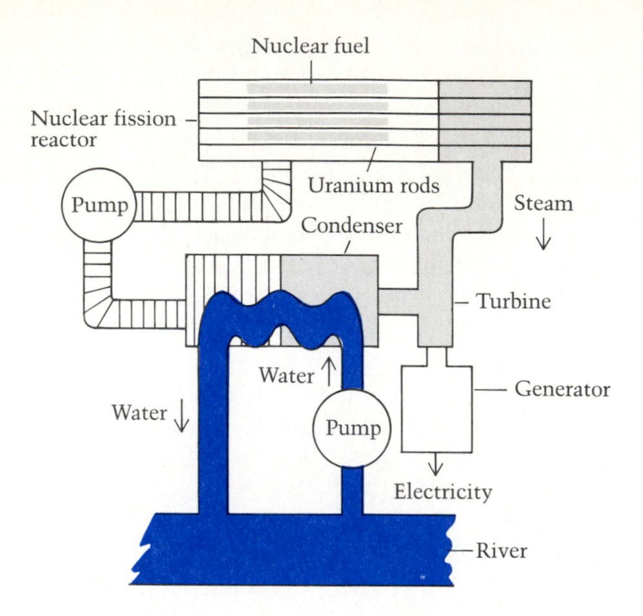

Figure 18.5
A nuclear reactor

lems that must be solved; the significance of others is widely debated. These issues involve the impact of radiation on health, safety, and the environment. The problems that no one denies concern the safe disposal of radioactive waste, the matter of security (uranium and plutonium must be kept out of the hands of unauthorized people), and the high cost of operating nuclear power plants. More controversial are the questions of whether functioning nuclear power plants routinely add significant amounts of radiation to the environment and whether a colossal nuclear accident could take place that would destroy life and property.

On both sides of the nuclear power question we find well-informed citizens. Those who argue for it cite the need for nuclear energy to maintain our standard of living and point to the relatively good safety record of the nuclear power industry. Those who oppose atomic energy mention the unknown effects of low levels of radiation on the incidence of such diseases as cancer. They argue that no safe method for disposing of nuclear waste exists and that no one can guarantee that nuclear materials will never fall into the hands of irresponsible people, terrorists, or criminals.

In addition, several recent accidents at nuclear power plants have tarnished the overall safety record of the industry. On March 28, 1979, a faulty water pump at the Three Mile Island Nuclear Power Plant near Harrisburg, Pennsylvania, failed. This was followed by a chain of mechanical and human errors. As a result, radioactive water was released from the emergency cooling system into the Susquehanna River. Radioactive steam was vented into the environment, and water was converted into hydrogen gas, making an explosion possible and raising the specter of a reactor meltdown. Although experts believe that the low levels of radiation released could never threaten human

The explosion at the Chernobyl nuclear power station in the USSR caused damage to the building, destroyed the reactor, killed 31 people, and raised radiation levels throughout Europe.
(Wide World)

health, others fear that long-term effects of the radiation will emerge in years to come.

On April 26, 1986, the worst accident in the history of nuclear power took place at the Chernobyl plant about 110 km north of Kiev in the Soviet Union. At the Chernobyl plant, the uranium was contained in fuel rods surrounded by graphite bricks that served to moderate the nuclear reaction. The accident was a result of engineers turning off most of the reactor's automatic safety and warning systems to keep them from interfering with an unauthorized safety experiment. The cooling water was one of the systems turned off. The remaining water in the reactor turned to steam, and the steam reacted with the nuclear fuel and the graphite bricks. An explosive mixture of gases was formed and ignited. The reactor was destroyed, the roof was blown off of the building, and the graphite bricks caught fire. Soviet officials claimed the fire was extinguished on April 29. According to Soviet reports, 500 people were hospitalized and the acknowledged death count stood at 31. Incidences of thyroid cancer, leukemia, and other radiation-related illnesses are higher among people living near the power plant. More ominously, the radioactive particles produced by the explosion have been dispersed all over the planet by the natural circulation of air. It will be years before all the effects of the Chernobyl disaster can be assessed.

Nuclear Fusion

Learning Goal 6
Nuclear fusion

Nuclear fusion is *the fusing or combining of small atoms into larger atoms.* The sun and other stars get their energy through a process of nuclear fusion in which hydrogen atoms are combined to form a helium atom. This process creates huge amounts of heat and high temperatures—on the order of tens of millions of degrees Celsius—and no waste products are left over. However, acute problems have so far prevented the use of this process to generate electricity. How do you contain such fantastic heat? What vessel or tools can handle it? Moreover, to duplicate the sun's fusion process on earth, you would need a great amount of energy to start with. So far, the only device that scientists have developed from their knowledge of nuclear fusion is the hydrogen bomb. This bomb is now possessed by the United States, the Soviet Union, Great Britain, France, China, and India.

Because it uses atomic fusion rather than fission, the hydrogen bomb is much more powerful than the "old-fashioned" atomic bomb. (In fact, in the hydrogen bomb, an atomic bomb acts as the initiator or detonator.) It is clear that nuclear fusion, used unwisely, could lead to our doom. However, once we learn how to control the process and tame it for peaceful uses, fusion may provide an unlimited power supply that burdens us with no radioactive waste products to pollute our environment.

Physicists at Princeton University's Plasma Physics Laboratory are conducting a series of major experiments as part of a national effort in that direction. One experiment utilizes a *tokamak* device, in which a very hot ionized gas called *plasma* is confined in a large stainless steel vacuum vessel shaped like a doughnut. The plasma must reach a temperature of about 100,000,000°C for 1 second with a density of about one one-hundred-thousandth that of air at sea level. Under these special conditions, a sizable number of atomic nuclei within the plasma collide and fuse, producing helium nuclei and high-speed neutrons. The energy of motion of these neutrons can be converted to electricity. Experiments such as these will, it is hoped, provide a safe way of producing energy for all human needs even thousands of years after conventional fuels have been used up. The fusion reactions being studied are

$$\underset{\text{Tritium}}{{}^{3}_{1}\text{H}} + \underset{\text{Deuterium}}{{}^{2}_{1}\text{D}} \longrightarrow {}^{4}_{2}\text{He} + {}^{1}_{0}\text{n} + \text{energy}$$

$$\underset{\text{Tritium}}{{}^{3}_{1}\text{H}} + \underset{\text{Protium}}{{}^{1}_{1}\text{H}} \longrightarrow {}^{4}_{2}\text{He} + \text{energy}$$

Radiation's Effects on Health

When an atom or molecule is hit by an alpha particle, beta particle, or gamma ray, there may be some transfer of energy. As a result, the particle or ray may cause the target to ionize when an electron is knocked out of its orbital, producing a cation (positively charged ion).

Learning Goal 8
How radiation affects human health

Radioactive particles and rays are commonly called ionizing radiation (visible light does not commonly cause atoms or molecules to ionize). Sometimes ionizing radiation strikes a molecule in such a way that a bond is broken. If that bond lies inside a gene within a chromosome, the cell may undergo mutation, changing the genetic message it carries. The cell may even die. If either a sperm cell or egg cell undergoes mutation via ionizing radiation, the organism formed by that sperm and egg may contain a genetic defect.

Substances that emit radioactive particles are also dangerous if inhaled or ingested. The radioactive particles often settle in the body and damage a certain body part. This is how workers who used paint containing the element radium were harmed in the early 1900s. The paint was used to produce watch dials that glowed in the dark. To create a fine point at the end of the paintbrush, workers used their lips and mouths, thereby ingesting the radioactive paint. A large proportion of those workers fell victim to bone cancer, leukemia, and other bone marrow diseases. Because radium is chemically similar to calcium, the radium ions were metabolized as though they were calcium ions, and the ionizing radiation they gave off caused illness and death.

Detection of Radiation

Learning Goal 9
Methods of detecting radiation

Several devices are commonly used to detect ionizing radiation. The best-known device is the **Geiger–Müller counter,** which consists of a detecting tube and a counter. The detecting tube is made of a pair of electrodes surrounded by a gas that can be ionized. As radiation enters the tube, the gas is ionized and the ions move toward an area of high voltage that surrounds the electrodes. Pulses of electric current caused by the presence of ions at the electrodes are detected and recorded by the counter. The Geiger–Müller tube is most sensitive to beta radiation; it is not useful for detecting gamma rays (which pass right through the tube undetected) or alpha radiation (which generally cannot pass through the window of the tube).

The Geiger–Müller tube records pulses of current entering the window per minute, but it gives no indication of the energy of the radiation. A device that records the amount *and* the energy of radiation produced is a *scintillation counter*. This device operates when the radiation reacts with a crystal containing sodium iodide and thallium iodide, producing a series of flashes of varying intensity. The intensity of the flashes is proportional to the energy of the radiation.

People working in areas where radiation exists commonly wear badges that contain photographic film. Each month the film is developed to indicate the degree of that individual's exposure to radiation that month.

Measurement of Radiation

Several different kinds of measuring units are available to express the quantity of radiation emitted by a source or received by a living being. Some of the units express the activity of the radiation to its impact on living tissue.

Learning Goal 10
Measurement of radiation

The simplest unit is the **curie** (symbol Ci), which is equal to 3.7×10^{10} disintegrations per second. This unit reveals little about the impact of radiation on human tissue because, in order to determine that, one must know the quantity of radiation received by the tissues.

Other units for the measurement of radiation include the *röentgen,* abbreviated *r,* and the *rad,* which stands for "radiation absorbed dose." The **röentgen** measures the ionizing ability of x rays and gamma rays. The amount of radioactivity that produces 2×10^9 ion pairs in 1 cubic centimeter of air is equivalent to 1 röentgen. A measurement taken in **rads** reveals the amount of radiation absorbed by living tissue, independent of the type of radiation.

Lethal dose is a term used to express the toxicity of radiation. A **lethal dose** is *the amount of radiation that will kill 50% of all exposed organisms within 30 days*. (This is written LD^{30}_{50}.) The most useful unit, however, is the **rem,** which stands for "röentgen equivalent man." The rem indicates the biological effect of radiation regardless of the source. Hence it includes the effect of medical x rays. The rem is calculated by taking the number of rads measured and multiplying it by a weighting factor. Each type of radiation has its own weighting factor.

In general, the recommended dose limit for the general public is 0.5 rem/year. Background radiation from cosmic rays, natural uranium and radon, and other isotopes ranges from 0.08 to 0.20 rem/year. A chest x ray is equivalent to 0.20 rem, whereas a series of x rays of the gastrointestinal tract, using fluoroscopy, produces about 22 rems. Nuclear power plants are estimated to add 0.001 rem/year to the general population.

18.2 Radioisotopes and Medicine

Learning Goal 11
Radioisotopes and medicine

Medical practitioners take advantage of the behavior of radioisotopes to diagnose and treat millions of patients each year. Specific radioactive isotopes, once introduced into a system, find their way to specific parts of the body. This behavior allows practitioners to follow an element with precision in order to diagnose a disease or target cancer cells for destruction. In medical diagnosis, the technique of radioactive tracing is routinely applied. Commonly used isotopes are chromium-51, iron-59, iodine-131, phosphorus-32, and technetium-99m.

A patient with an overactive thyroid can be administered iodine-131, a radioactive isotope of iodine. The isotope localizes in the thyroid gland, and its radiation destroys some of the thyroid cells, thereby decreasing the size of the thyroid gland and reducing its activity.

Because the body treats radioisotopes, also called tagged isotopes, the same way it treats nonradioactive elements, the destinations of element and isotope are the same. Iodine, for example, radioactive or not, travels to the thyroid gland where it is incorporated into the amino acid thyroxine. This is why iodine-131 is perfect for monitoring the function of the thyroid gland, as well as for reducing its activity. A scanning device can determine the rate at which the isotope is taken into the thyroid gland after it is administered.

The isotope technetium-99m emits only gamma rays, which can be detected by a scintillation counter. This isotope is used to create "images" of the brain, heart, kidneys, liver, spleen, lungs, and bones. Brain scans and bone scans both utilize technetium-99m, along with other elements.

Chromium, in the form of sodium chromate, attaches strongly to the hemoglobin of red blood cells. This makes radioactive chromium-51 an excellent isotope for determining the flow of blood through the heart. This isotope is also useful for determining the lifetime of red blood cells, which can be of great importance in the diagnosis of anemias.

Radioactive cobalt (cobalt-59 or cobalt-60) is used to study defects in vitamin B_{12} absorption. Cobalt is the metallic atom at the center of the B_{12} molecule. By injecting a patient with vitamin B_{12} that is labeled with (or contains) radioactive cobalt, a physician can study the path of the vitamin through the body and discover any irregularities.

Radiation can also be used to destroy cancer cells. The radiation can be delivered to the malignant area in three ways. In **teletherapy,** a high-energy beam of radiation is aimed at the cancerous tissue. In **brachytherapy,** a radioactive isotope is placed in the area to be treated. The radioactive source is usually enclosed in a seed (it could be a glass bead containing the isotope) that is implanted under the skin. In **radiopharmaceutical therapy,** the isotope is administered either orally or intravenously. Then the isotope travels along normal body pathways to seek its target. This method is used to get iodine-131 to the thyroid gland.

CAT Scans

Medical practitioners have long used x rays to diagnose problems with bones and some organs in the body. These photographs compress three-dimensional information into two dimensions and thereby lose a great deal of detail.

In 1979 the Nobel Prize in physiology and medicine was given to Godfrey N. Hounsfield and Allan M. Corwin for developing a three-dimensional x-ray scanning technique that uses a computer along with a source of x rays. The **CAT scan** (CAT stands for "computerized axial tomography") rotates a source of x rays around a patient and uses a computer to come up with a three-dimensional picture of the patient's internal organs.

PET Scan

In contrast to CAT scans, which show the different densities of human tissues, the **PET scan** (PET stands for "positron emission tomography") uses computers to analyze radiation coming from radioisotopes. This process informs medical practitioners of exactly when a radioisotope enters an organ and enables them to trace its mode of action.

PET scans have taught researchers a great deal about brain chemistry. They have revealed how the brain of a normal individual functions differently from the brain of an individual with schizophrenia. Degenerative brain diseases like Alzheimer's disease are also being studied with this technique.

18.3 Electrochemistry

The Voltaic Cell

Learning Goal 12
Voltaic and electrolytic cells

As we noted in the chapter introduction, Volta devised the first electrochemical battery, or **voltaic cell,** in 1800. In his battery, Volta used the two metals silver and zinc, separated from each other by a piece of paper soaked in a salt-and-water solution. Wires were attached to the silver and zinc strips. When these wires were connected in a circuit, an electric current ran through them (Figure 18.6). Many scientists immediately tried to see how other combinations of metals would react when they were set up similarly.

The Electrolytic Cell

A few weeks after Volta's discovery was made known, two English chemists, William Nicholson and Anthony Carlisle, performed the reverse of Volta's experiment. Volta had used a chemical reaction to produce electricity; Nicholson and Carlisle used electricity to produce a chemical reaction. They ran electric current through water, using an experimental setup similar to the one shown in Figure 18.7. The current caused the water to decompose slowly into hydrogen gas and oxygen gas. What Nicholson and Carlisle had developed is called an **electrolytic cell.** It uses electricity to produce a chemical reaction.

The Lead Storage Battery

Learning Goal 13
The lead storage battery and the fuel cell

Today many batteries incorporate the principles of both the voltaic cell (when they are discharging) and the electrolytic cell (when they are being charged). An example is the *lead storage battery,* which supplies energy for automo-

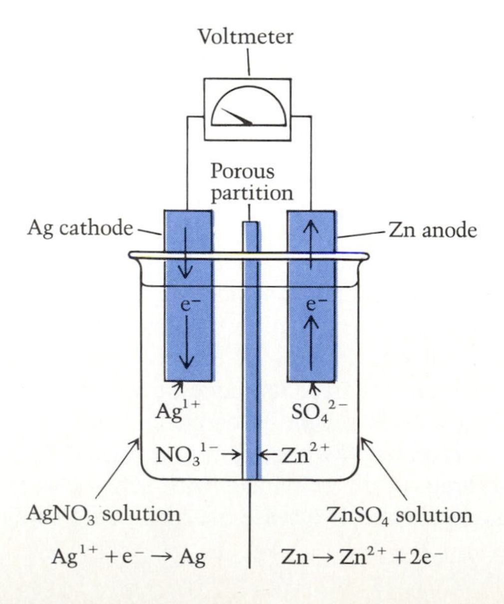

Figure 18.6
An example of a voltaic cell. Although this cell has a more advanced design, it is very much like Volta's original cell.

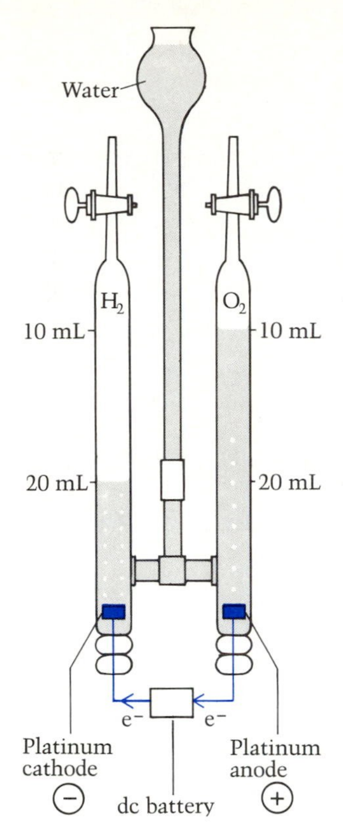

Figure 18.7
An electrolytic cell

biles, trucks, and other vehicles. The lead storage battery is actually a device that changes chemical energy into electrical energy.

All batteries operate through the transfer of electrons from one substance to another substance. (An electric current is simply a flow of electrons.) For example, when a piece of zinc metal is dropped into a solution of copper(II) sulfate, a single-replacement reaction occurs. The products are copper metal and zinc sulfate.

$$Zn + CuSO_4 \longrightarrow Cu + ZnSO_4$$

In this reaction, electrons move from the zinc to copper ions. We can show this by writing the reaction in the following way:

$$Zn + Cu^{2+} + SO_4^{2-} \longrightarrow Cu + Zn^{2+} + SO_4^{2-}$$

The Zn gives up two electrons and the Cu^{2+} picks them up. If we arrange these substances so that they can't make direct contact with each other but are indirectly connected, we have a voltaic cell (Figure 18.8), which can produce electricity.

The lead storage battery in an automobile uses lead(IV) oxide and lead, separated by a sulfuric acid solution (Figure 18.9). Electricity flows from the battery when the following reaction takes place:

$$Pb + PbO_2 + 2H_2SO_4 \longrightarrow 2PbSO_4 + 2H_2O \qquad \text{(battery discharging)}$$

In this reaction, lead metal reacts with sulfate ions to produce lead(II) sulfate, with the release of two electrons.

$$Pb + SO_4^{2-} \longrightarrow PbSO_4 + 2e^{-1}$$

While this is happening, lead(IV) oxide is reacting with sulfuric acid to produce lead(II) sulfate, with the gain of two electrons.

$$PbO_2 + 4H^{1+} + SO_4^{2-} + 2e^{-1} \longrightarrow PbSO_4 + 2H_2O$$

Figure 18.8
A zinc–copper voltaic cell

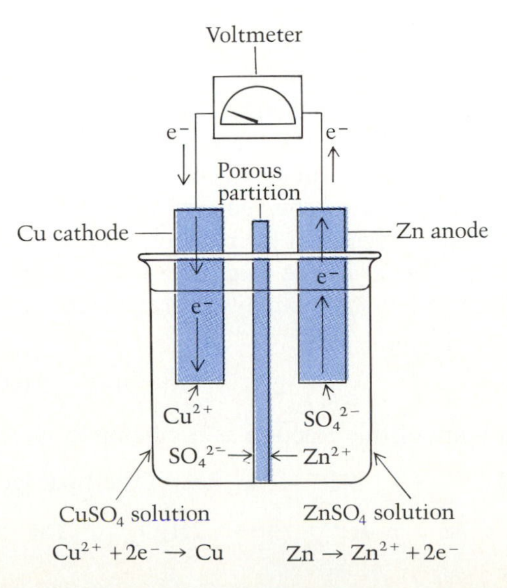

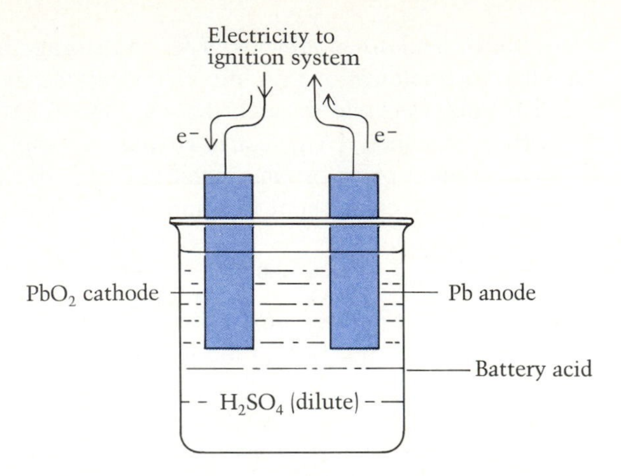

Figure 18.9
One cell of a lead storage battery

Electrons are thus transferred from the lead to the lead(IV) oxide. This produces electricity.

Each time a car is started, current is drained from its battery. The battery discharges as lead(II) sulfate is formed. Why, then, does the battery last so long? Because the lead storage battery has another characteristic. As the engine runs, the battery is recharged by the automobile's alternator (or generator). A current flows through the battery, moving in the opposite direction from the discharging current. The battery now becomes an electrolytic cell. The reaction that takes place during the charging of the battery is just the reverse of the discharging reaction:

$$2PbSO_4 + 2H_2O \longrightarrow Pb + PbO_2 + 2H_2SO_4 \quad \text{(battery charging)}$$

Lead and lead(IV) oxide are formed, and the battery is again ready to produce electricity.

Fuel Cells

When the United States put its space program into high gear, it needed a battery that would offer high efficiency, have few moving parts, and be light in weight. The **fuel cell** was the answer. This type of battery works like the voltaic cell except that in a fuel cell the reactants are introduced continuously and the products are removed continuously. One kind of fuel cell used in space vehicles depends on the reaction between hydrogen and oxygen (Figure 18.10). The nickel in one electrode and a combination of nickel and nickel oxide in the other act as *catalysts* for the reaction. The overall chemical reaction is

$$2H_2 + O_2 \longrightarrow 2H_2O$$

We can think of this reaction as occurring in two steps:

$$2H_2 \longrightarrow 4H^{1+} + 4e^{-1} \quad \text{(The hydrogen releases electrons.)}$$

$$O_2 + 4e^{-1} + 4H^{1+} \longrightarrow 2H_2O \quad \text{(The oxygen gains electrons.)}$$

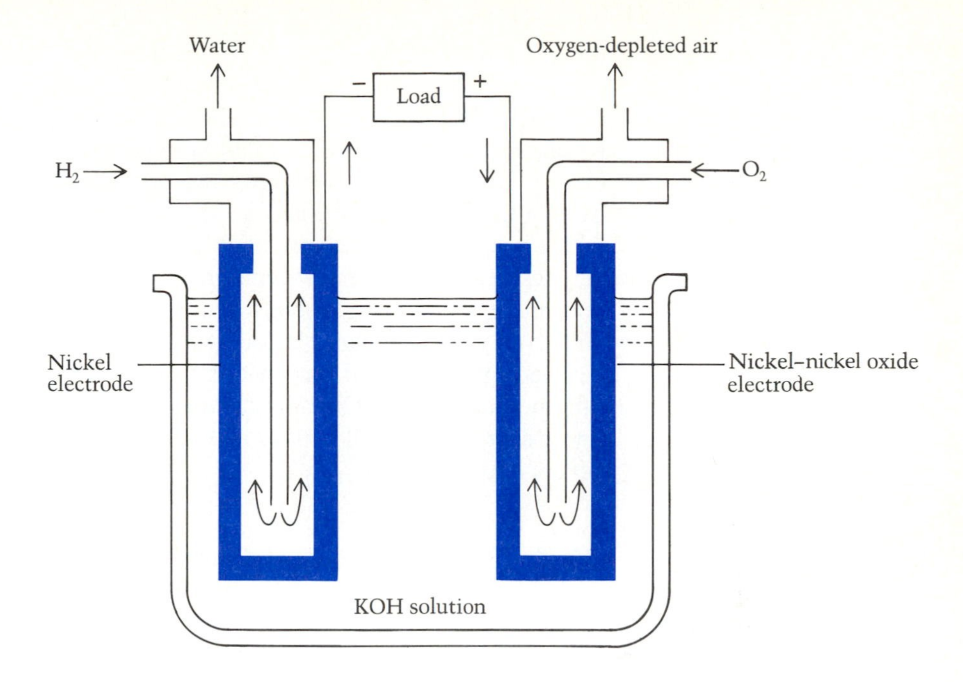

Figure 18.10 The hydrogen–oxygen fuel cell

An advantage of this fuel cell is that it produces a consumable end product—water! More important, it does not pollute the air. For this reason, research scientists are looking at fuel cells as a possible means of powering automobiles.

CHEMICAL FRONTIERS

ENERGY-SAVING CHEMICAL INNOVATIONS

In the United States, electricity accounts for almost 40% of the energy consumed. By the year 2010, the Department of Energy estimates that this level will rise to about 60%—at a cost of about $250 billion in today's dollars. But there are other costs, especially the damage to the environment caused by the burning of fossil fuels to produce electricity. Air pollution, acid rain, and global warming, or the *greenhouse effect,* that results from the production of carbon dioxide gas all result when coal, oil, and natural gas are burned. There are also the problems associated with the production of nuclear waste from nuclear power plants.

A variety of technological innovations, many of them based on chemical research, point the way to tremendous energy savings. Among these innovations are solid-state voltage regulators that have been developed to make fluorescent lights more efficient. Nonconductive chemical coatings that absorb less light as heat are also available for fluorescent lights. These coatings reduce the amount of waste heat generated by lights, thus lowering the need for air conditioning. In a "smart" building, a central energy-management computer can be connected to multiple thermostats located throughout the building. The computer decides whether the air conditioner or furnace should be turned on, and it uses ventilation fans to distribute cool air or heat more evenly. Low-emissivity windows are specially insulated to keep heat from passing through them. Filling the space between the panes of these windows with argon gas improves insulation another 25%.

These and other energy-saving innovations will become even more important in the future. One high-efficiency light bulb eliminates the need for 524 pounds of coal or 1 barrel of oil, which translates to a reduction of between 220 and 382 pounds of carbon-containing environmental pollutants over the lifetime of the bulb. Ongoing chemical research and development will help lower energy costs still further and help preserve our environment for future generations.

18.4 Environmental Chemistry and Air Pollution

Never before has our world been faced with so many environmental problems, because never before has the population been so large. The air we breathe, the water we drink, the oceans and lakes, and the land we live on are rapidly being polluted. We ask ourselves, "How can we stop this pollution?" To stop it we must first understand how it is caused. In this section, we'll look briefly at the mechanisms of air pollution from the chemist's point of view. We can define **air pollution** as *the presence of a contaminant in the outdoor atmosphere in a concentration large enough to injure human, plant, or animal life or to interfere with the enjoyment of life or property.*

The Air We Breathe

Learning Goal 14
Composition of the air and air pollutants

Air consists, by volume, of 78% nitrogen, 21% oxygen, and 1% other gases (argon, helium, hydrogen, carbon dioxide, water vapor, and other inert gases). These percentages are fairly accurate, but unfortunately they don't tell the whole story. A number of substances that we often find in air are missing from this list. And although their concentrations may be extremely small, they have a devastating effect on human health. These unnamed ingredients include **gaseous air pollutants,** such as carbon monoxide, hydrocarbons, sulfur dioxide, sulfuric acid, nitrogen oxide, and nitrogen dioxide, and **particulate matter,** or tiny particles of smoke, dust, fumes, and aerosols in the atmosphere.

Reactions in the Air

Air pollution has been a problem since ancient times, but until quite recently the causes were natural. Active volcanoes spewed out huge amounts of ash, and swamps emitted foul-smelling gases. This natural pollution, since it was isolated and infrequent, wasn't much of a problem for human beings or the atmosphere. However, as the world population increased and people advanced technologically, a problem arose.

In England, for example, coal was the main source of fuel for centuries. Coal is basically carbon, along with some sulfur and other impurities. When the English burned coal to protect themselves from the cold and damp air, the smoke from their millions of fires threw out huge amounts of particulates and gases, and some of these gases were poisonous. It was only in 1956 (after a disastrous 4-day smog had killed 4,000 Londoners in 1952) that Britain passed the Clean Air Act and made it illegal to burn coal in the cities. Britain even switched its railroads from coal to electricity and diesel fuel.

Learning Goal 15
Reactions that produce air pollution

Here are some of the reactions that occur when coal is burned:

$$C + O_2 \longrightarrow \underset{\text{Carbon dioxide}}{CO_2} \qquad S + O_2 \longrightarrow \underset{\text{Sulfur dioxide}}{SO_2}$$

$$2SO_2 + O_2 \longrightarrow \underset{\text{Sulfur trioxide}}{2SO_3} \qquad SO_3 + H_2O \longrightarrow \underset{\text{Sulfuric acid}}{H_2SO_4}$$

The sulfur in the coal is oxidized to sulfur dioxide, which is eventually oxidized to sulfuric acid. The sulfuric acid then remains as a mist in the atmosphere. When it is inhaled, this acid mist can irritate the nasal passages. It can penetrate deep into the lungs, destroying parts of the lung tissue. Sulfuric acid can also corrode many metals and building materials. (This is happening, for example, to the beautiful old buildings and statues in Venice, Italy.) **Acid rain,** which is rain containing sulfurous and sulfuric acids, can ruin trees and farm land and destroy the aquatic life of ponds and lakes.

Coal isn't the only offender. Oil, the main source of fuel in the United States today, contains sulfur impurities also. Industry uses immense amounts of oil, and some power companies use it to produce electricity.

The Internal Combustion Engine

Learning Goal 16
Health problems caused by pollution

The automobile is another major source of air pollution. In this case, the problem is caused by flaws in the internal combustion engine:

1. Not all the gasoline that enters the engine is burned completely. This leads to the emission of carbon monoxide and hydrocarbons, which are chemical compounds that contain the elements carbon and hydrogen. (Gasoline is a mixture of hydrocarbons.)

2. The nitrogen and oxygen in air react at the high operating temperatures of auto engines. This leads to the emission of oxides of nitrogen.

Incompletely burned gasoline travels out the exhaust pipe and into the air. Once in the air, the hydrocarbons can react with nitrogen oxides to produce *photochemical smog*. *Smog* is a mixture of smoke and fog. **Photochemical smog** is smog that results from atmospheric reactions in which sunlight is a catalyst. Both can be deadly to people with respiratory problems. And CO is poisonous to everything that breathes oxygen.

The other major by-products of the combustion of gasoline—oxides of nitrogen—are produced in a series of steps. First, nitrogen and oxygen produce nitric oxide within the engine.

$$N_2 + O_2 \longrightarrow \underset{\text{Nitric oxide}}{2NO}$$

The nitric oxide gets out into the atmosphere, where some of it reacts with oxygen in the air to produce nitrogen dioxide.

$$2NO + O_2 \longrightarrow \underset{\text{Nitrogen dioxide}}{2NO_2}$$

Nitrogen dioxide, which is unstable in the presence of sunlight, can decompose into nitric oxide and atomic oxygen.

$$NO_2 + \text{sunlight} \longrightarrow NO + \underset{\text{Atomic oxygen}}{O}$$

In the presence of sunlight, the atomic oxygen can react with unburned hydrocarbons to produce other harmful chemicals, but more commonly the atomic oxygen (O) reacts with molecular oxygen (O_2) to produce ozone (O_3). High concentrations of ozone near ground level pose a severe health problem, because ozone irritates the eyes and the respiratory system. It can also react with many different substances. For example, ozone can destroy rubber by causing it to harden and crack.

Ozone has another role—one of protecting human health. Despite the fact that ozone formed near the surface of the earth is a pollutant, a protective layer of ozone is found about 15 miles above the earth's surface. The *ozone layer,* as it is called, shields the earth from harmful ultraviolet radiation produced by the sun's rays. Unfortunately chemicals called chlorofluorocarbons are capable of reacting with ozone to produce atomic chlorine and oxygen. This reaction destroys the ozone layer and leads to a higher incidence of skin cancer in humans.

How Can Air Pollution Be Controlled?

Learning Goal 17
Ways of reducing air pollution

The major drawbacks of both coal and oil as fuels are emissions of particulate matter and sulfur oxides. Emission of sulfur oxides can be reduced by using coal or oil that has a low sulfur content. We can also reduce emissions by passing exhaust gases through a **scrubber,** a device that absorbs gaseous pollutants. A scrubber for removing sulfur dioxide in a smokestack usually consists of a fine spray of water. Gas rising through the stack passes through the scrubber, where the water absorbs the sulfur dioxide (Figure 18.11). This process can even produce a useful by-product: sulfurous or sulfuric acid.

Figure 18.11
A scrubber

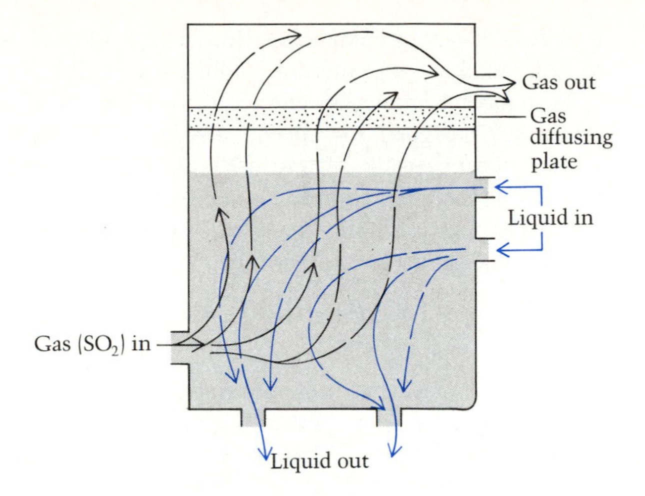

Particulate emissions can be reduced by using mechanical devices. One such device is the **electrostatic precipitator** (Figure 18.12), in which dust particles that have been electrically charged are collected between highly charged electrical plates. This eliminates the dust particles from the gases leaving the smokestack.

Because cars and trucks contribute half of all air pollution, eliminating their emissions is a top-priority problem for automobile manufacturers. One solution would be to switch from the internal combustion engine to electrically powered cars. However, for many technical reasons, this is not practical at the present time. Moreover, in a sense, it would merely shift the blame for

Figure 18.12
An electrostatic precipitator

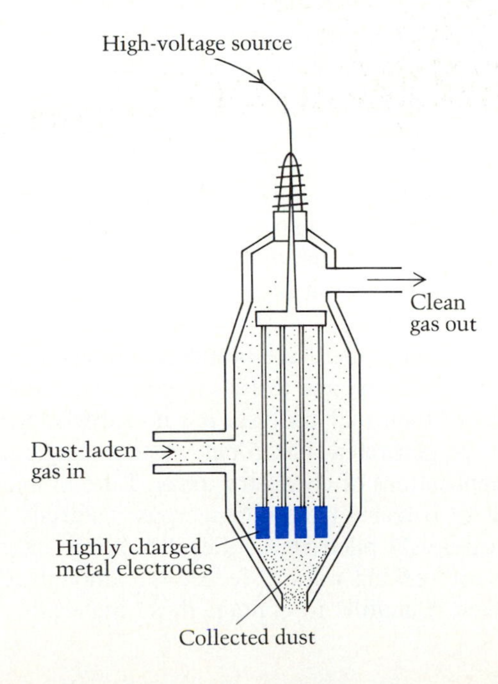

pollution to power companies. So, at least for the near future, improvements must be made to the internal combustion engine.

A number of pollution-control devices are used in today's motor vehicles. Some are designed to reduce emissions of hydrocarbons by reducing the evaporation of gasoline from fuel tanks. Others aim at more nearly complete burning of gasoline; complete combustion would produce only carbon dioxide and water. For example, here is the equation for the complete combustion of octane, one ingredient of gasoline:

$$\underset{\text{Octane}}{2C_2H_{18}} + 25O_2 \longrightarrow 16CO_2 + 18H_2O$$

Catalytic converters were developed to reduce the emission of nitrogen oxides. Exhaust gases are passed through the converter, where the proper catalyst enhances the conversion of these oxides into less troublesome substances. Some converters also enhance the oxidation of partially burned hydrocarbons, to produce carbon dioxide and water.

18.5 The Problem of Hazardous Wastes

As technology and industry have developed new processes, the problem of handling chemical waste has become crucial. Countless barrels of hazardous-waste materials have been produced, threatening the integrity of our land, air, and waterways.

Learning Goal 18
Hazardous wastes

Hazardous waste is any waste that may cause death or irreversible or incapacitating illness and that threatens human health and the environment because of its composition. Many hazardous wastes do not decompose into harmless substances that cycle back into our ecological system. They will remain a threat to future generations unless we develop ways to dispose of them safely.

Most plastics, for example, don't decompose. A toy composed of polyethylene or a bottle made of PVC (polyvinyl chloride) may last for thousands of years. And when PVC is burned, it produces hydrogen chloride gas, which is toxic. Some plastics contain additives called plasticizers to make them more pliable. Among these are the polychlorinated biphenyls (PCBs) and phthalates. PCBs can leach out of plastics and, once free in the environment, can enter the food chain.

Besides plastics, there are pesticides, medicines, paints, petroleum products, metals, leathers, and textiles that all visit hazardous wastes upon us as a by-product of their manufacture. It's no surprise that with all the toxic waste that has been generated, hazardous-waste management is one of the most important applications of chemistry today. Like time bombs waiting to explode, thousands of barrels of these dangerous materials lie buried in illegal dump sites or carelessly piled up in illegal holding areas (Figure 18.13a,b). Researchers are seeking ways to reduce the amount of hazardous waste that has accumulated. Landfills to contain these materials are being developed, and

(a) (b) (c)

Figure 18.13
Chemical Control Corporation in Elizabeth, New Jersey, the former site of 40,000 drums of hazardous chemical wastes, (*a*) before the fire and explosion in April 1980, and (*b*) after the fire. The U.S. Environmental Protection Agency's Mobile Incineration System (*c*), developed for the purpose of destroying hazardous wastes on-site, before such accidents take place.
(Courtesy Environmental Protection Agency)

injection wells are being considered in geologically stable regions. Other researchers are devising ways to treat the waste products through incineration, thermal destruction, high-temperature decomposition, and chemical stabilization (Figure 18.13c). The objective of these scientists is to rid the earth of these deadly materials. The quality of life available to our descendants may depend on their success.

18.6 Water Pollution

Learning Goal 19
Potable water

Water—a vital substance for human survival—must be free of contamination from deadly chemicals and bacteria in order to be *potable* (drinkable). In the United States, each American uses over 340 liters (90 gallons) of water per day for domestic purposes, and statistics show that the demand is increasing.

The world's supply of water is enormous, and water is found as vapor, liquid, and solid. Over 97% of the earth's water is located in the oceans, however, and is unfit to drink. Just 3% of all water is found as fresh water fit for human consumption.

Our sources of potable water include lakes, rivers, springs, and wells. As our wells and springs are rapidly drained, many large cities are turning to rivers and lakes as sources of potable water. To be considered drinkable, water must meet standards set by the U.S. Public Health Service (Table 18.2). Because of the pollutants that have been dumped into them over the years, very few of our rivers and lakes can now meet these standards.

In the past, a nearby river was commonly used as the discharge point for waste water. In the United States today, despite legislation to control this form of pollution, millions of gallons of raw, untreated waste are still being dumped into our waterways.

Learning Goal 20
Water pollutants

There are many ways in which water pollution occurs today. Nearly half of the water withdrawn from rivers, lakes, and estuaries each year in the United States is used to cool electric power plants. When the water is returned it contains no pollutants, but its temperature is increased significantly, which can adversely affect the ecological relationships in the body of water.

Another widespread problem called **eutrophication** occurs when erosion and run-off introduce plant nutrients from the land into lakes. The algae nor-

Table 18.2
U.S. Public Health Service Standards for Potable Water

Contaminating ion(s)	Maximum concentration (milligrams/liter)	Contaminating ion(s)	Maximum concentration (milligrams/liter)
Arsenic	0.05	Lead	0.05
Barium	1.00	Manganese	0.05
Cadmium	0.01	Nitrate	45
Chloride	250	Organics	0.20
Chromium	0.05	Selenium	0.01
Copper	1.00	Silver	0.05
Cyanide	0.20	Sulfate	250
Fluoride	2.00	Zinc	5.00
Iron	0.30	Total dissolved solids	500

mally present in a healthy lake feed on phosphates and nitrates, nutrients that originated in agricultural or urban centers. However, populations of algae grow out of control when the lake becomes overloaded with phosphates and nitrates derived from fertilizers, animal waste, or sewage treatment plants. The vital oxygen content of the lakes then decreases, and the water tastes bad and has a foul odor. It is the decay of the resulting algae that causes eutrophication.

The process of eutrophication has destroyed the natural balance of life in the Great Lakes. The effect on Lake Erie has been particularly serious. Located near major population centers, the Great Lakes are sinks for large amounts of phosphates and toxic chemicals. However, as a result of pollution-control measures, conditions have continued to improve at the Great Lakes for more than two decades.

Ground-water contamination is yet another problem of growing proportions. Accidental chemical spills, leaking sewer pipelines, leaks from the hundreds of thousands of underground storage tanks and abandoned hazardous-waste sites, and agricultural and industrial run-off—all contribute to ground-water pollution. This is the most difficult type of pollution to control, and there are just a few thousand people trained in the ground-water field who can work effectively on the problem.

Oil spills constitute a continuing water-pollution hazard. Besides natural seepage of crude petroleum from deposits below the ocean bottom, accidents, urban run-off, and ruptures of offshore wells contribute to the problem.

When water pollutants get into the bodies of birds, fish, and shellfish, their life cycles can be altered or death can occur. Once these substances enter the food chain, their effects can be far-reaching. When humans consume fish that have ingested harmful pollutants, for example, these unwanted substances build up in human tissues, creating potential health problems (Figure 18.14).

Legislation has been enacted in an effort to put an end to our waste-water contamination. America's two major water-pollution control laws are the Federal Water Pollution Act of 1972 and the Clean Water Act of 1977 (which was amended in 1981). These laws required the Environmental Protection Agency to set a system of standards for waste water leaving a sewage treatment or industrial plant. They also required all municipalities to treat sewage biologically by 1988. In addition, goals were set so that by 1983, all U.S. waters would be safe for fishing and swimming. Elimination of the discharge of 129 highly toxic pollutants into U.S. waters was to end by 1985, and the discharge of any pollutants into waterways would be allowed by EPA permit only.

Because of this legislation, the amount of pollution dumped into our waterways by industry was reduced by one-half between 1972 and 1980. Problems still exist today; for example, 37 states could not meet the goal of making all bodies of water safe for fishing and swimming by 1983. Much pollution from agricultural and urban run-off could not be effectively controlled, and the slow pace at which bureaucratic machinery turns, construction delays, and lack of funds made compliance impossible in many areas.

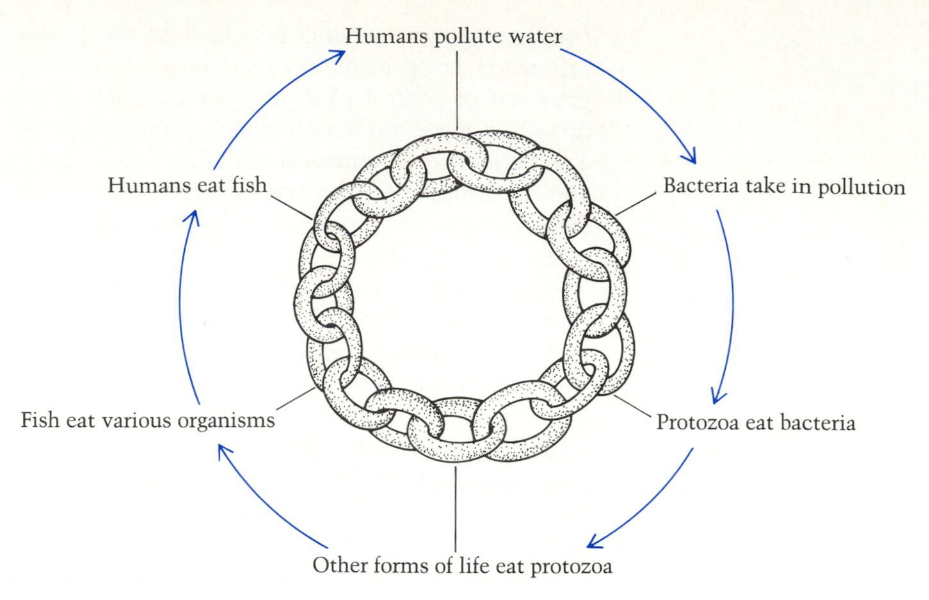

Figure 18.14 The path of water pollution through the food chain

Even so, and despite the fact that our waterways continue to be polluted, conditions are much better than they were a decade or two ago.

Summary

Radioactivity is the ability of an atom to emit penetrating rays from its nucleus spontaneously. There are three types of radiation: (1) alpha particles, which have a mass number of 4, an electrical charge of +2, and little penetrating ability; (2) beta particles, which have a mass number of zero, a charge of −1, and only slight penetrating power; and (3) gamma rays, which have neither mass nor charge but have very high energy content and strong penetrating ability.

Some elements produce radiation naturally. When an atom of such an element emits an alpha or beta particle, it undergoes a nuclear transformation and actually becomes a different element. The half-life of a radioactive element is the time required for half the atoms in a sample of the element to decay in this manner.

Radioactive isotopes can be produced artificially by bombarding certain elements with particles such as neutrons. It is also possible to "split" atoms by bombarding them in a process called nuclear fission. Such a process, which results in a chain reaction leading to the splitting of more and more atoms, is the basis of the atomic (fission) bomb. Controlled nuclear fission is used in power plants to produce electricity.

The sun and stars get their energy through nuclear fusion, a process in which nuclei are combined, or fused together. This process is the basis of the hydrogen (fusion) bomb. Nuclear fusion may one day supply us with almost limitless amounts of energy for conversion to electricity.

Radioactive isotopes are used in medicine to diagnose and treat a variety of diseases. Radiation is also commonly used to destroy cancer cells.

Electrochemistry involves the production of electrical energy from chemical reactions. In the voltaic cell, electrons move from one metallic substance to another, producing electricity. In the electrolytic cell, electricity is used to cause a chemical reaction. The common lead storage battery acts as a voltaic cell when it is discharging and as an electrolytic cell when it is being charged. The fuel cell is a battery that works like a voltaic cell, except that the reactants are introduced continuously and the products are removed continuously.

By volume, pure air consists of 78% nitrogen, 21% oxygen, and 1% other gases. Polluted air also contains such chemicals as carbon monoxide, hydrocarbons, sulfur dioxide, sulfuric acid, nitrogen oxide, nitrogen dioxide, and particulate matter. Some of these chemicals can react in the presence of sunlight to form dangerous photochemical smog. All these pollutants can cause health problems for people and animals and can damage plant life. However, several devices can be used to reduce emissions of harmful chemicals. They include industrial devices, such as the scrubber and the electrostatic precipitator, and emission-control devices on automobiles.

Hazardous wastes pollute the land in many areas. Plastics, pesticides, petroleum products, and the manufacture of other goods all contribute to the problem. Attempts to deal with this form of pollution have begun only recently.

The pollution of our lakes, rivers, springs, and wells is a serious problem that must be solved soon if we are to continue to have enough drinking water. Erosion, run-off, chemical spills, leaks, and the use of water for cooling electric power plants contribute to the problem. Legislation has resulted in significant increases in water quality, but there is still room for a great deal of improvement.

Key Terms

acid rain **(18.4)**
air pollution **(18.4)**
alpha (α) particle **(18.1)**
beta (β) particle **(18.1)**
brachytherapy **(18.2)**
CAT scan **(18.2)**
chain reaction **(18.1)**
curie **(18.1)**
electrolytic cell **(18.3)**
electrostatic precipitator **(18.4)**
eutrophication **(18.6)**
fuel cell **(18.3)**
gamma (γ) ray **(18.1)**
gaseous air pollutant **(18.4)**
Geiger–Müller counter **(18.1)**
half-life **(18.1)**
hazardous waste **(18.5)**
lethal dose **(18.1)**
nuclear fission **(18.1)**
nuclear fusion **(18.1)**
nuclear transformation (transmutation) **(18.1)**
particulate matter **(18.4)**
PET scan **(18.2)**
photochemical smog **(18.4)**
rad **(18.1)**
radioactive (nuclear) decay **(18.1)**
radioactivity **(18.1)**
radiopharmaceutical therapy **(18.2)**
rem **(18.1)**
röentgen **(18.1)**
scrubber **(18.4)**
teletherapy **(18.2)**
voltaic cell **(18.3)**

Nuclear materials handling technicians at the Department of Energy's Hanford site wear protective clothing while producing weapons-grade plutonium for use in the national defense program.
(Department of Energy)

CAREER SKETCH

NUCLEAR MATERIALS HANDLING TECHNICIAN

As a nuclear materials handling technician you will work with nuclear scientists and engineers to manage nuclear materials in many different stages. You will fabricate, label, package, and transport nuclear fuel elements and deal with radioactive waste. You can work as an industrial waste inspector, atomic fuel assembler, or a radioactive waste dispatcher and decontaminator.

From the mining of uranium ore to the disposing of radioactive waste, you will take part in milling, conversion, separation, and fabrication of nuclear materials and fuels. You will ship and store spent fuel, dissolve cut-up fuel, and convert fuel into solid pellets for use in nuclear power plants.

A knowledge of nuclear chemistry is essential to become a specialist in this field. You must also be familiar with the techniques of solvent extraction, precipitation, centrifuging, diffusion, and enrichment of fuels. You will also learn to operate remote control devices that are commonly used for safety, so that materials can be manipulated without being handled by humans.

Some of the college courses that you will take include nuclear technology, radiation physics, technical mathematics, electricity and electronics, technical communications, radiation detection and measurement, blueprint reading, meteorology, metallurgy, technical writing, and nuclear chemistry.

Self-Test Exercises

Learning Goal 1: Radioactivity and Atomic Energy

1. List some advantages and some disadvantages of atomic energy for the human race.
2. What would be your concern about living near a nuclear power plant?

3. If you were a scientist working on the Manhattan Project, what questions would you ask yourself about the morality of your work?
4. What problems might be encountered in disposing of waste from conventional nuclear power plants?

▲ **5.** What are the advantages and the disadvantages of having nuclear power plants supply all our electrical needs?
▲ **6.** What is meant by the term *radioactivity?*

Learning Goal 2: Types of Nuclear Radiation

▲ **7.** Describe each of the following: *(a)* alpha particles *(b)* beta particles *(c)* gamma rays
▲ **8.** *(a)* Which has the greater penetrating ability, a beta particle or a gamma ray?
(b) What kind of shield would you need to stop alpha particles?

9. What are the charge and mass of *(a)* an alpha particle. *(b)* a beta particle?
10. Why does the atomic number of an atom *increase* by 1 when the atom *emits* a beta particle?

Learning Goal 3: Writing Balanced Nuclear Equations

11. Complete each of the following nuclear equations:
(a) ${}^{197}_{78}Pt \longrightarrow {}^{0}_{-1}e + ?$
(b) ${}^{212}_{84}Po \longrightarrow {}^{4}_{2}He + ?$
(c) ${}^{10}_{5}B + {}^{4}_{2}He \longrightarrow {}^{1}_{0}n + ?$
(d) ${}^{238}_{92}U + {}^{1}_{0}n \longrightarrow ? \longrightarrow {}^{94}_{36}Kr + ?$
12. Complete each of the following nuclear equations:
(a) ${}^{226}_{88}Ra \longrightarrow {}^{4}_{2}He + ?$
(b) ${}^{234}_{90}Th \longrightarrow {}^{0}_{-1}e + ?$
(c) ${}^{234}_{91}Pa \longrightarrow {}^{0}_{-1}e + ?$
(d) ${}^{210}_{82}Pb \longrightarrow {}^{0}_{-1}e + ?$

13. Write the equation for the formation of a beta particle from a neutron.
14. What isotope is formed when ${}^{214}_{82}Pb$ emits a beta particle from its nucleus?

▲ **15.** When an isotope emits an alpha particle, what happens to *(a)* the isotope's mass number, *(b)* its atomic number?
▲ **16.** What isotope is formed when ${}^{194}_{78}Pt$ emits an alpha particle from its nucleus?

17. When an isotope emits a beta particle, what happens to *(a)* the isotope's mass number *(b)* its atomic number?
18. What isotope is formed when ${}^{218}_{85}At$ emits an alpha particle from its nucleus?

19. Complete and balance each of the following nuclear equations:
(a) ${}^{14}_{6}C \longrightarrow {}^{0}_{-1}e + ?$
(b) $? \longrightarrow {}^{0}_{-1}e + {}^{24}_{12}Mg$
(c) $? \longrightarrow {}^{4}_{2}He + {}^{234}_{90}Th$
(d) ${}^{84}_{36}Kr \longrightarrow {}^{1}_{0}n + ?$
20. Complete and balance each of the following nuclear equations:
(a) ${}^{137}_{55}Cs \longrightarrow {}^{0}_{-1}e + ?$
(b) ${}^{27}_{13}Al + {}^{4}_{2}He \longrightarrow {}^{30}_{15}P + ?$
(c) ${}^{214}_{83}Bi \longrightarrow {}^{4}_{2}He + ?$
(d) ${}^{0}_{-1}e + ? \longrightarrow {}^{7}_{3}Li$

Learning Goal 4: Half-life

▲ **21.** Define the term *half-life* and explain how scientists use it.
▲ **22.** Explain why the decay rates of radioactive elements are given in terms of half-life.

Learning Goal 5: Calculations Using Half-life

▲ **23.** The half-life of carbon-14 is 5,770 years. Suppose you have a piece of wood that originally contained 4 g of carbon-14. How old is the piece of wood today if the amount of carbon-14 remaining is only 0.25 g?
▲ **24.** The half-life of cobalt-60 is 5.3 years. Suppose a sample of cobalt-60 originally weighed 100.0 g. How

old is the sample today if the amount of cobalt-60 remaining is only 12.5 g?

25. The isotope iodine-131 has a half-life of 8 days. Suppose a person is injected with $\overline{100}$ mg of iodine-131 today. How much iodine-131 will remain in the body of this individual after 24 days (assuming that none is lost through normal body channels)?

26. A patient is injected with the radioisotope ^{99}Tc in order to diagnose a possible malfunction of the liver or gallbladder. The radioisotope ^{99}Tc is a gamma-ray emitter. The path of the isotope is traced with a gamma-ray camera and projected on a monitor. The half-life of ^{99}Tc is 6 hours. How many hours must pass for only one-fourth of the original dosage to remain in the body?

Learning Goal 6: Nuclear Fission and Fusion

▲ **27.** Define and give an example of *(a)* nuclear fission, *(b)* nuclear fusion.

▲ **28.** Complete the following statement: The splitting of an atom is an example of nuclear ________.

29. Complete the following statements:
(a) The atomic bomb is a nuclear ________ (fusion/fission) device.
(b) The hydrogen bomb is a nuclear ________ (fusion/fission) device.

30. Describe the research being done at the Princeton Plasma Physics Laboratory.

Learning Goal 7: Emergence of Nuclear Power Plants

▲ **31.** What fuel is used to power conventional nuclear power plants?

▲ **32.** How is the heat produced by nuclear fission used to generate electricity?

33. Discuss the problems associated with the disposal of radioactive waste.

34. Why are nuclear waste products dangerous?

35. Describe the accident that took place at Three Mile Island near Harrisburg, Pennsylvania.

36. What caused the accident at the Chernobyl nuclear power plant in the Ukraine to be so dangerous?

Learning Goal 8: How Radiation Affects Human Health

▲ **37.** How might ionizing radiation induce a gene within a chromosome to undergo mutation?

▲ **38.** Explain how the similarity between the elements radium and calcium led to the development of bone cancer in many workers who painted radium on watch dials in the early 1900s.

Learning Goal 9: Methods of Detecting Radiation

▲ **39.** What causes the pulses of electricity that are detected in the tube of a Geiger–Müller counter?

▲ **40.** To which type of radiation is the Geiger–Müller tube most sensitive? Why?

41. A device that records the amount *and* the energy of radiation is the ________ counter.

42. Explain how a film badge works.

Learning Goal 10: Measurement of Radiation

▲ **43.** Distinguish among the curie, the rad, and the roentgen.

▲ **44.** Explain the meaning of the term LD^{30}_{50}.

45. What is the yearly recommended radiation dose limit for the general public, measured in rems?

46. Is one exposed to more radiation from a chest x ray or a fluoroscope of the intestinal tract?

Learning Goal 11: Radioisotopes and Medicine

▲ **47.** Name a radioactive isotope used in medicine, and discuss how it is used.

▲ **48.** One treatment for malignancies known to be responsive to gamma radiation is to irradiate the area with the radioisotope cobalt-60. Is this an example of teletherapy or brachytherapy?

49. How is iodine-131 used to monitor the thyroid gland?

50. The isotope ________ can be used in performing both brain scans and bone scans.

Learning Goal 12: Voltaic and Electrolytic Cells

▲ **51.** Define and give an example of *(a)* a voltaic cell, *(b)* an electrolytic cell.

▲ **52.** A battery is connected to a cell with two electrodes, both of which are immersed in a solution of hydrochloric acid. The battery supplies electrons to the cathode, which is the negative pole of the cell. When a switch is closed, an electric circuit is completed and the following chemical reaction takes place:

$$2HCl(aq) \longrightarrow H_2(g) + Cl_2(g)$$

Is this an electrolytic cell or a voltaic cell?

53. Balance the following equation for the discharging of the lead storage battery:

$$Pb + PbO_2 + H_2SO_4 \longrightarrow PbSO_4 + H_2O$$

54. The zinc–copper cell was used to provide energy for the first transcontinental telegraph lines. It consists of a piece of copper immersed in a copper(II) sulfate solution. The copper and zinc are connected by a wire, and a salt bridge provides contact between the two solutions. Is this an electrolytic cell or a voltaic cell?

Learning Goal 13: The Lead Storage Battery and the Fuel Cell

▲ **55.** *(a)* What is a fuel cell?
(b) Discuss the advantages and disadvantages of substituting a fuel cell for the internal combustion engine as a source of power for the automobile.

▲ **56.** The reactions that occur during the discharging of a lead storage battery are

$$Pb + SO_4^{2-} \longrightarrow PbSO_4$$

$$PbO_2 + 4H^{1+} + SO_4^{2-} \longrightarrow PbSO_4 + 2H_2O$$

(a) Rewrite both half-reactions, supplying the proper number of electrons in each.
(b) Which reaction represents oxidation, and which represents reduction? (You may want to check Section 10.6 before answering this question.)

57. How does a fuel cell differ from a voltaic cell?

58. The fuel cell shown in Figure 18.10 depends on the reaction between hydrogen gas and oxygen gas. The two half-reactions for this fuel cell are

$$2H_2 \longrightarrow 4H^{1+}$$

$$O_2 + 4H^{1+} \longrightarrow 2H_2O$$

(a) Rewrite both half-reactions, supplying the proper number of electrons in each.
(b) Which reaction represents oxidation, and which represents reduction?

Learning Goal 14: Composition of the Air and Air Pollutants

▲ **59.** Define the terms *gaseous air pollutant* and *particulate matter*.

▲ **60.** Do you think that carbon dioxide in excessively high concentration would be classifed as an air pollutant? Explain your answer.

61. List all the gaseous air pollutants you can name.

62. While you are standing outdoors, a small piece of ash enters your eye. What type of air pollutant has affected you?

63. List all the particulate air pollutants you can name.

64. When Mt. St. Helens erupted on May 18, 1980, large amounts of ash were emitted into the atmosphere. What were some of the effects of the eruption on the people living in the area?

65. What two substances account for most of the composition of air by volume?

66. Name four gaseous pollutants that are produced when coal is burned.

Learning Goal 15: Reactions That Produce Air Pollution

67. Would the introduction of electrically powered cars solve the problem of pollution of the air by cars?

68. Why is photochemical smog more dangerous than ordinary smog?

69. Complete and balance each of the following equations:
(a) $C + O_2 \longrightarrow$
(b) $S + O_2 \longrightarrow$
(c) $SO_2 + O_2 \longrightarrow$
(d) $? + ? \longrightarrow NO$

70. Write the equations for the formation of acid rain.

71. What is photochemical smog?

▲ **72.** In the presence of sunlight, nitrogen dioxide decomposes into nitric oxide and atomic oxygen. Write an equation for this reaction.

▲ **73.** How are oxides of nitrogen formed in the internal combustion engine?

74. Write the equation for the formation of ozone in the atmosphere.

Learning Goal 16: Health Problems Caused by Pollution

▲ **75.** Of the gaseous and particulate air pollutants that we have discussed, which do you think would be most harmful to people suffering from respiratory and lung diseases?

▲ **76.** Many newspapers print a daily air-quality report that is issued by environmental experts. When the air quality is poor, people with heart problems or respiratory disease are advised to stay indoors. Why?

Learning Goal 17: Ways of Reducing Air Pollution

▲ **77.** What type of control device would you place on a smokestack of a city incinerator that is spewing out black smoke?

▲ **78.** What, if anything, can be done to eliminate pollution caused by emissions from a home fireplace?

79. What control devices would you place on the smokestack of a power company that is using oil as a fuel for generating electricity?

80. How have catalytic converters reduced air pollution?

81. "Dilution is the solution to pollution." Discuss the logic of this statement.

82. What type of device can be used to control pollution that is due to an incinerator burning hazardous wastes that have a high chlorine content?

83. To stop the pollution of our air, what steps must be taken by *(a)* government, *(b)* industry, *(c)* the average citizen?

84. What is the benefit of a power company burning low-sulfur-content fuel to generate electricity? Are there any disadvantages?

Learning Goal 18: Hazardous Wastes

▲ **85.** Define the term *hazardous waste,* and give an example.

▲ **86.** How do PCBs affect people?

87. How can hazardous wastes enter the food chain?

88. Would you rather see hazardous wastes incinerated and risk air pollution, or see these wastes deposited in a landfill and risk ground-water pollution? List some advantages and disadvantages of each waste-disposal method.

Learning Goal 19: Potable Water

▲ **89.** Define the term *potable,* and list four sources of potable water.

▲ **90.** Is water from the ocean classified as potable water?

Learning Goal 20: Water Pollutants

▲ **91.** List some common water pollutants. How do they get into our lakes and rivers?

▲ **92.** Name three sources of water pollution.

93. Discuss the importance of the food chain as it pertains to water pollution.

***94.** How can a body of water clean itself?

Extra Exercises

95. The potassium–argon (^{40}K–^{40}Ar) method of dating rocks has been used to date the oldest known rocks on earth. These rocks, found in Greenland, have been analyzed as follows:

$$^{40}\text{K remaining in crystal} = 3 \text{ atoms}$$

$$^{40}\text{K originally present in crystal} = 24 \text{ atoms}$$

If ^{40}K has a half-life of 1.3 billion years, how old are the rocks?

▲ **96.** Name three types of chemical compounds that are considered air pollutants. State how each originates and what its effects are.

97. Discuss the technological problems associated with the use of fusion as a source of energy.

98. In light of the near disaster at the Three Mile Island Nuclear Power Plant near Harrisburg, Pennsylvania, in March 1979 and the Chernobyl disaster of 1986, what do you think should be the policy regarding our present-day nuclear power plants?

99. Discuss which kind of pollution—air or water pollution—is a bigger problem in your area.

▲ **100.** A certain material is producing nuclear radiation. Assuming that you have the proper equipment, explain how you could determine what type of nuclear radiation the material is producing.

101. Contrast the beta particle, the electron, and the alpha particle.

102. The isotope ^{138}Cs has a half-life of 32.2 minutes. Is ^{138}Cs more or less stable than an isotope with a half-life of 3 days? Explain.

103. Complete the following equation by supplying the missing particle:

$$^{16}_{8}O \longrightarrow ^{12}_{6}C + ?$$

▲ **104.** Suggest a way to prepare $^{13}_{7}N$ from $^{10}_{5}B$.

105. The half-life of $^{45}_{19}K$ is $3\bar{0}$ minutes. Suppose you have a sample of 30 mg. How much time would elapse before the amount was reduced to 7.5 mg?

106. A nuclear power plant uses water from a nearby river as cooling water. The water returns to the river at an elevated temperature. What problems can this cause for fish?

19 Organic Chemistry, Part 1: Hydrocarbons

Learning Goals

After you have studied this chapter, you should be able to:

1. Distinguish between organic compounds and inorganic compounds.
2. Write the Lewis electron-dot structure for the carbon atom.
3. Describe the geometry of carbon bonds.
4. Draw structural formulas when you are given the molecular formulas or the names of alkanes, alkenes, and alkynes.
5. Define and illustrate the terms *saturated hydrocarbon, unsaturated hydrocarbon,* and *cyclic hydrocarbon*.
6. Define and illustrate the terms *isomer* and *homologous series*.
7. Name alkanes, alkenes, and alkynes using the IUPAC system.
8. Define the term *aromatic hydrocarbon,* and give a common example.

Introduction

During the 1700s and early 1800s, most chemists believed that there were two distinct classes of chemical compounds: *organic* and *inorganic*. They said that *organic compounds* were derived from living or once-living organisms and that *inorganic compounds* were part of the nonliving world. Organic compounds were substances like sugar, fats, oils, and wood. Inorganic compounds were substances like salt or iron.

Many chemists of those days noted that organic substances could be converted into inorganic substances by heating but that inorganic substances could not be converted into organic substances. These observations gave rise to the theory of **vitalism.** According to this theory, (1) life and substances associated with life were not bound by the laws of science, and (2) these life-connected substances had a "vital force" that human beings could not control. Many scientists of this period thought that chemists, even though they could do many marvelous things in their laboratories, could not duplicate this "vital force" and would never be able to synthesize organic compounds (that is, make them artificially).

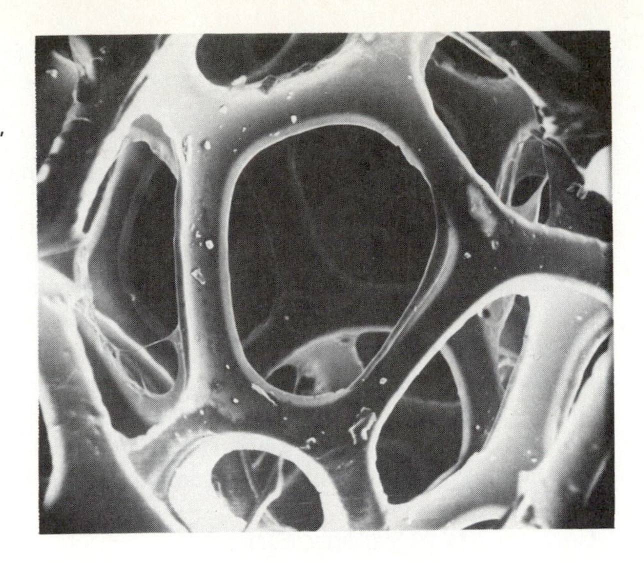

Polyurethane is an example of an organic compound.
(Courtesy John B. Vander Sande, Massachusetts Institute of Technology)

But in 1828, the German chemist Friedrich Wöhler made a startling discovery. In one of his experiments, Wöhler, who was working with cyanide compounds, heated the compound ammonium cyanate. At this time, chemists considered ammonium cyanate inorganic, so Wöhler wasn't expecting it to show any association with life. But after he had heated the ammonium cyanate, Wöhler found that this inorganic compound had undergone a chemical change into a different molecular arrangement and had formed the compound urea, an organic compound found in urine. Thus Wöhler became the first person to produce an organic compound artificially. He wrote a letter to his friend Berzelius, a Swedish chemist: "I must tell you that I can make urea, but without kidneys or even an animal being necessary, be it man or dog." This development put the theory of vitalism on shaky ground.

Now, more than 160 years after Wöhler's discovery, chemists are able to synthesize more than 10 million organic compounds. In this chapter and Chapter 20, we shall look at a small number of these compounds.

19.1 Organic Chemistry and Carbon

Organic chemistry is primarily *the study of compounds that contain carbon.* The old definition of organic substances does not apply anymore. Synthetic fibers, plastics, and pharmaceuticals (drugs and medicines) are all organic compounds because they all have numerous carbon atoms in their molecules. However, not *all* carbon-containing compounds are organic. For example, carbonate compounds (such as $CaCO_3$ and Na_2CO_3) and the oxides of carbon are considered *in*organic carbon compounds.

Learning Goal 1
Organic and inorganic compounds

It's difficult to establish an absolute definition of what an organic compound is. In most cases, **organic compounds** *are those that contain the element carbon,* along with elements such as hydrogen, oxygen, nitrogen, sulfur, and the Group VIIA elements. (Once we have examined the different types of organic compounds, a precise definition will no longer be necessary.)

Why Are There So Many Organic Compounds?

There are more than 10 million organic compounds, and there are only about 100,000 inorganic compounds, so the organic compounds far outnumber the inorganic. How is this possible? After all, organic compounds are made up primarily of carbon and a few other elements, whereas inorganic compounds are made up of combinations of all the 100 or so remaining elements.

The answer lies in *the ability of the carbon atom to bond directly to other carbon atoms in the molecule*. This unusual ability enables carbon atoms to form all kinds of chain-like and ring-shaped molecules, making the number of potential organic compounds almost infinite (Figure 19.1).

Carbon: Its Electronic Structure

Learning Goal 2
Lewis electron-dot structure of carbon

Look at the periodic table (inside the front cover) and you'll see that carbon is in Group IVA. This means that a carbon atom has four electrons in its outermost energy level—and that is an open invitation to bond: *A carbon atom can use all four of these electrons to bond with other atoms*. In organic compounds, carbon bonds covalently. In other words, it shares its four electrons with other atoms to form bonds that are very strong:

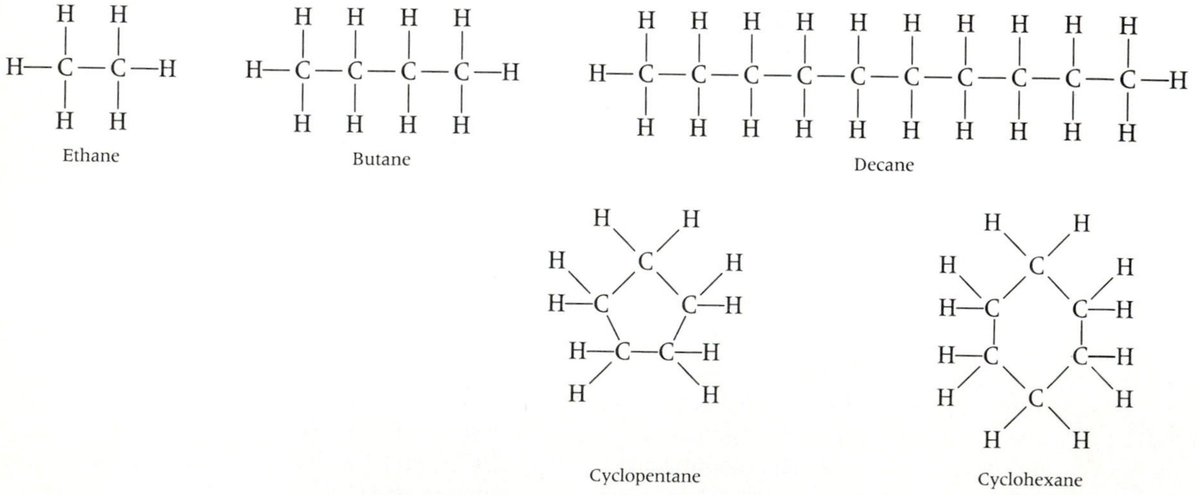

$$\cdot\dot{\underset{\cdot}{C}}\cdot$$

```
 H H
 | |
H—C—C—H
 | |
 H H
```
Ethane

```
 H H H H
 | | | |
H—C—C—C—C—H
 | | | |
 H H H H
```
Butane

```
 H H H H H H H H H H
 | | | | | | | | | |
H—C—C—C—C—C—C—C—C—C—C—H
 | | | | | | | | | |
 H H H H H H H H H H
```
Decane

Cyclopentane

Cyclohexane

Figure 19.1
Carbon atoms can form either long chains or ring-like compounds. Practically no other atoms can.

Carbon: Its Tetrahedral Bond

Learning Goal 3
Geometry of carbon bonds

When a carbon atom shares its four outer electrons with four other atoms, the result is a compound such as CH_4, the compound *methane*. In this compound, the central carbon atom is bonded to each of the four hydrogen atoms. We can write the electron-dot structure in the usual way:

$$\begin{array}{c} H \\ H:\ddot{\underset{\cdot\cdot}{C}}:H \\ H \end{array} \quad \text{or} \quad \begin{array}{c} H \\ | \\ H{-}C{-}H \\ | \\ H \end{array}$$

These diagrams give the impression that the carbon–hydrogen bonds are at 90-degree angles to each other:

$$\begin{array}{c} H \\ | \\ H{-}C{-}H \\ |\quad 90^\circ \\ H \end{array}$$

However, this is not the way the molecule actually exists. Molecules are *three*-dimensional, not two-dimensional. And the four outer electrons in the carbon move as far apart from each other as possible, because electrons repel each other. (Remember that particles with similar charges always repel one another.) The electrons can get as far apart as possible if the carbon–hydrogen bonds form a *tetrahedron* (Figure 19.2). A **tetrahedron** (*tetra* is from the Greek word for "four") is a pyramid-shaped figure with three upper sides plus a bottom side. You can make a tetrahedron this way: Take six pencils. Place three of them on a table in the form of a triangle. Hold the other three pencils upright, placing one at each corner of the triangle. Now tilt the three upright pencils inward until they touch. You have formed a tetrahedron. This tetrahedral bonding is common to many organic compounds; methane is only one example. With the carbon inside the tetrahedron and the hydrogens (in methane) at the corners, the bonds are 109.5 degrees apart (Figure 19.3).

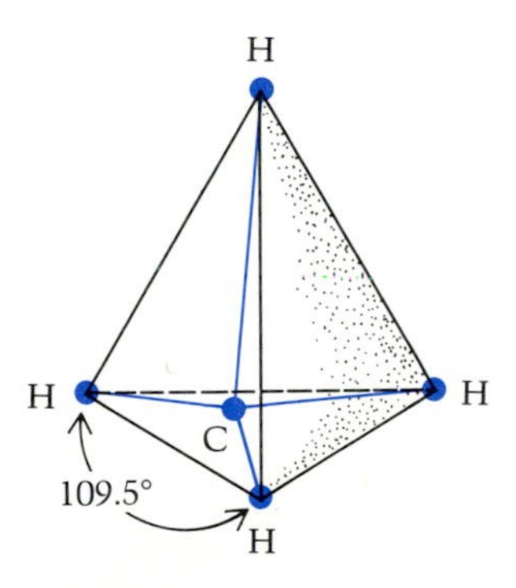

Figure 19.2
The tetrahedral structure of the CH_4 molecule

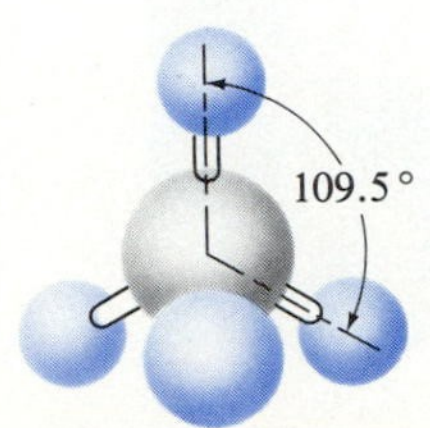

Figure 19.3
(*a*) Space-filling and (*b*) ball-and-stick models of methane. Both models are three dimensional.
(From Ebbing, Darrell D., *General Chemistry*, Third Edition. Copyright © 1990 by Houghton Mifflin Company. Used with permission.)

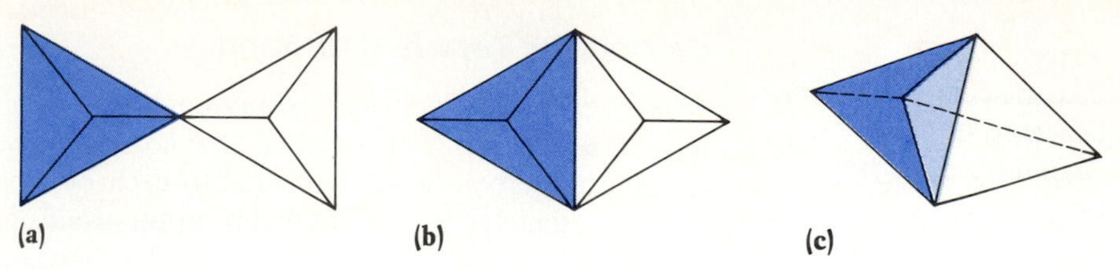

Figure 19.4
Visualization of two carbon atoms joined together by (*a*) a single bond, (*b*) a double bond, and (*c*) a triple bond

Carbon–Carbon Bonds

When two carbon atoms bond with each other, they can do so in three ways:

1. They can form a single bond: $\cdot\dot{\underset{\cdot}{C}}-\dot{\underset{\cdot}{C}}\cdot$
2. They can form a double bond: $\cdot\dot{C}=\dot{C}\cdot$
3. They can form a triple bond: $\cdot C\equiv C\cdot$

In the case of the single bond, we can visualize the two carbon atoms joining in such a way that the corners of two tetrahedrons are touching (Figure 19.4). The double and triple bonds are difficult to visualize or draw on a flat page. However, for the double bond, we can visualize two tetrahedrons with a common edge. For the triple bond, we can visualize two tetrahedrons with a common face.

Elemental Forms of Carbon

As an element, carbon exists in several forms. The two crystalline forms of carbon are diamond and graphite (Figure 19.5). Although graphite and dia-

Figure 19.5
The two crystalline forms of carbon
(*Left:* Fundamental Photographs, New York. *Right:* Courtesy of UCAR Carbon Company Inc.)

Figure 19.6
The crystalline structures of diamond and graphite

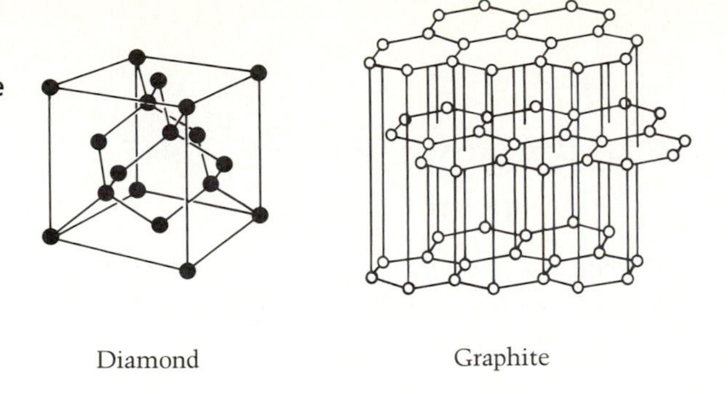

Diamond Graphite

mond are both composed of pure carbon atoms, their properties are very different. Diamond is one of the hardest materials known. Graphite, on the other hand, is one of the softest. It is found in pencils ("lead" pencils don't really contain lead, but graphite) and in lubricants. Obviously the crystalline structures of diamond and graphite must differ vastly. Figure 19.6 shows how.

In diamond, the carbon atoms exhibit a tetrahedral arrangement wherein each carbon atom is bonded to four other carbon atoms. It is this structure that gives diamond its strength and hardness. In graphite, the carbon atoms are in parallel planes or sheets. Each sheet contains carbon atoms attached to one another, in six-sided structures. Two kinds of bonding forces hold the graphite structure together. These are the bonding between carbon atoms within each sheet and the bonding between the sheets of carbon atoms. The second bonding force is much weaker than the first. In other words, the forces holding the sheets together are very weak, so the sheets can slide over each other. That is why graphite is used to lubricate sticky locks and zippers.

19.2 Writing Formulas for Organic Compounds

Learning Goal 4
Structural formulas of organic compounds

In earlier chapters, *molecular formulas* were written for various compounds. For example, water is written as H_2O. The molecular formula tells us the actual number of atoms of each element in a molecule of the compound.

For the organic chemist, however, the molecular formula is not enough. As we have seen, carbon atoms can bond together in different arrangements. For this reason, there can be many different compounds with the same molecular formula. For example, the molecular formula C_4H_{10} can represent two distinct compounds, and the formula C_8H_{18} can represent 18 distinct compounds! For this reason, when organic chemists write the formula for an organic compound, they use a **structural formula.** *The structural formula indicates how the carbon atoms are bonded to each other, as well as the number of atoms of each element*. We can write the structural formulas for the two compounds represented by C_4H_{10} in one plane like this:

$$\begin{array}{ccccccccc} & H & & H & & H & & H & \\ & | & & | & & | & & | & \\ H- & C & - & C & - & C & - & C & -H \\ & | & & | & & | & & | & \\ & H & & H & & H & & H & \end{array} \quad \text{and} \quad \begin{array}{ccccccc} & & H & H & H & & \\ & & \diagdown & | & \diagup & & \\ & H & & C & & H & \\ & | & & | & & | & \\ H- & C & - & C & - & C & -H \\ & | & & | & & | & \\ & H & & H & & H & \end{array}$$

Structural formulas are sometimes written in a more convenient form called the **condensed form.** For example, the two formulas above can also be written in this way:

$$CH_3-CH_2-CH_2-CH_3 \quad \text{and} \quad \begin{array}{c} CH_3 \\ | \\ CH_3-CH-CH_3 \end{array}$$

In the first method, all bonds are shown. In the condensed form, only bonds between carbon atoms are shown. The other bonds are understood to be present, as in a molecular formula.

Example 19.1

Write each of the following structural formulas in the condensed form:

$$(a)\ \begin{array}{ccccccccccccc} & H & & H & & H & & H & & H & & H & \\ & | & & | & & | & & | & & | & & | & \\ H- & C & - & C & - & C & - & C & - & C & - & C & -H \\ & | & & | & & | & & | & & | & & | & \\ & H & & H & & H & & H & & H & & H & \end{array} \qquad (b)\ \begin{array}{ccccccc} & & H & H & H & & \\ & & \diagdown & | & \diagup & & \\ & H & & C & & H & \\ & | & & | & & | & \\ H- & C & - & C & - & C & -H \\ & | & & | & & | & \\ & H & & C & & H & \\ & & \diagup & | & \diagdown & & \\ & & H & H & H & & \end{array}$$

$$(c)\ \begin{array}{ccccccccccc} & & & H & H & H & & & & & \\ & & & \diagdown & | & \diagup & & & & & \\ & H & & H & C & & H & & H & & \\ & | & & | & | & & | & & | & & \\ H- & C & - & C & -C- & & C & - & C & -H & \\ & | & & | & | & & | & & | & & \\ & H & & H & C & & H & & H & & \\ & & & \diagup & | & \diagdown & & & & & \\ & & & H & H & H & & & & & \end{array}$$

Solution

(a) $CH_3-CH_2-CH_2-CH_2-CH_2-CH_3$

Always check to see that you have the same number of carbon atoms and hydrogen atoms as in the original.

$$\begin{array}{c} \quad\quad CH_3 \\ \quad\quad | \\ \textbf{(b)}\ CH_3-C-CH_3 \\ \quad\quad | \\ \quad\quad CH_3 \end{array}$$

Both this formula and the more complex formula show that the four CH_3 groups are attached to a central carbon atom.

$$\begin{array}{c} \quad\quad\quad\quad CH_3 \\ \quad\quad\quad\quad | \\ \textbf{(c)}\ CH_3-CH_2-C-CH_2-CH_3 \\ \quad\quad\quad\quad | \\ \quad\quad\quad\quad CH_3 \end{array}$$

Here it is easiest to start with the central carbon atom and work outward.

Practice Exercise 19.1 Write each of the following structural formulas in the condensed form:

$$(a)\ \begin{array}{ccccccccc} & & H & & H & & H & & H & \\ & & | & & | & & | & & | & \\ H & - & C & - & C & - & C & - & C & - & H \\ & & | & & | & & | & & | & \\ & & H & & H & & H & & H & \end{array}$$

$$(b)\ \begin{array}{ccccccccccc} & & H & & H & & H & & H & & H & \\ & & | & & | & & | & & | & & | & \\ H & - & C & - & C & - & C & - & C & - & C & - & H \\ & & | & & | & & | & & | & & | & \\ & & H & & H & & C & & H & & H & \\ & & & & / & & | & \backslash & & & & \\ & & & & H & & H & & H & & & \end{array}$$

19.3 The Classification of Organic Compounds

We have said that there are more than 10 million known organic compounds. Is each of these compounds unique, or are some similar to each other? In other words, can we classify this great number of compounds into groups?

Most organic compounds can be classified into a few groups, called **series,** in such a way that compounds that react alike are placed together. We shall look at some of these series in this chapter and at others in Chapter 20.

19.4 The Alkanes

Let us begin the study of organic series with the **alkanes.** This group of compounds is one of the simplest in composition and structure, for the following reasons:

1. All alkanes are **hydrocarbons.** Hydrocarbons are compounds that contain only carbon and hydrogen.

2. The bonds between carbon atoms in alkanes are all single bonds. *When all*

Learning Goal 5
Saturated hydrocarbons

the carbon–carbon bonds in an organic compound are single bonds, we say that the compound is **saturated.**

The simplest member of this series of hydrocarbons is *methane,* CH_4, with which you are already familiar:

```
    H
    |
H—C—H
    |
    H
```

The next alkane has two carbon atoms and is called *ethane,* $CH_3—CH_3$, or

```
    H   H
    |   |
H—C—C—H
    |   |
    H   H
```

Then we come to the member of this series that has three carbon atoms. This molecule is called *propane,* $CH_3—CH_2—CH_3$, or

```
    H   H   H
    |   |   |
H—C—C—C—H
    |   |   |
    H   H   H
```

Table 19.1
The First Ten Members of the Alkane Series

n	Molecular formula	Structural formula	Name	Comments
1	CH_4	CH_4	Methane	Main ingredient of natural gas
2	C_2H_6	$CH_3—CH_3$	Ethane	Minor ingredient of natural gas
3	C_3H_8	$CH_3—CH_2—CH_3$	Propane	Minor ingredient of natural gas; usually separated from other ingredients and sold as bottled gas
4	C_4H_{10}	$CH_3—(CH_2)_2—CH_3$	Butane	Also found in natural gas; usually separated from other ingredients and used as a fuel—for example, in butane lighters
5	C_5H_{12}	$CH_3—(CH_2)_3—CH_3$	Pentane	Ingredients of gasoline
6	C_6H_{14}	$CH_3—(CH_2)_4—CH_3$	Hexane	
7	C_7H_{16}	$CH_3—(CH_2)_5—CH_3$	Heptane	
8	C_8H_{18}	$CH_3—(CH_2)_6—CH_3$	Octane	
9	C_9H_{20}	$CH_3—(CH_2)_7—CH_3$	Nonane	
10	$C_{10}H_{22}$	$CH_3—(CH_2)_8—CH_3$	Decane	An ingredient of kerosene

Note that each member of the alkane series differs from the member before it by having one more CH_2 group. You can see this more clearly by looking at the molecular formulas of these compounds:

CH_4	C_2H_6	C_3H_8
Methane	Ethane	Propane

Learning Goal 6
Isomers and homologous series

When the compounds in a series differ by the same structural group (like CH_2), we say that these compounds are members of a **homologous series.** This means that all the compounds in this series are of the same chemical type and differ only by fixed increments (an increment is an increase in quantity). In the alkane series, the fixed increment is CH_2. The general formula for this series of compounds can be written as C_nH_{2n+2}, where n represents the number of carbon atoms. If you substitute the numbers 1 through 10 for n in this formula, you will get each of the molecular formulas listed in Table 19.1, in turn. It is important that you learn the names of these alkanes, because they are used in naming many other organic compounds.

19.5 Isomers

In Section 19.2, we noted that the molecular formula for an organic compound could actually represent many different compounds. As an example, we showed that the formula C_4H_{10} can represent two compounds. This phenomenon is known as **structural isomerism.**

Isomers are *compounds with the same* ***molecular*** *formula but different* ***structural*** *formulas. In other words, the atoms of isomers are bonded together differently*. The title of Table 19.1 says that it shows the first ten members of the alkane series. However, it really shows the first ten *straight-chain* members of the alkane series. This is because there are isomers for all of the alkane series except the first three members (Table 19.2). Let us write the structures for the isomers of C_4H_{10} once again:

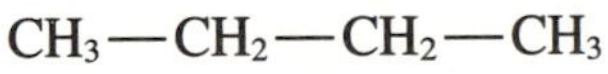

$$CH_3—CH_2—CH_2—CH_3$$

Normal butane

$$\begin{array}{c} CH_3 \\ | \\ CH_3—C—CH_3 \\ | \\ H \end{array}$$

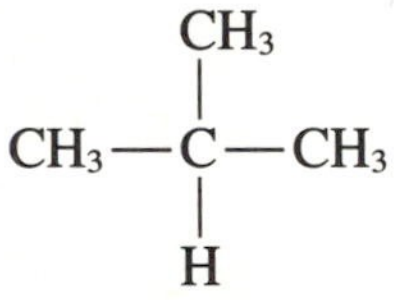

Isobutane

The structure on the left is the **straight-chain** isomer, which we call normal butane (or *n*-butane). (The word *normal* or the letter *n* before the word *butane* signifies the straight-chain isomer.) The second structure is the **branched-chain** isomer, which is commonly called *iso*butane. *Before you continue reading, be sure you understand the differences between these two structures.*

Let's see how we write the structures for the isomers of pentane (C_5H_{12}). One isomer has the five carbons bonded to each other in a straight chain.

Table 19.2
Numbers of Isomers of Some Compounds in the Alkane Series

Molecular formula	Number of isomers
C_4H_{10}	2
C_5H_{12}	3
C_6H_{14}	5
C_7H_{16}	9
C_8H_{18}	18
C_9H_{20}	35
$C_{10}H_{22}$	75

$$
\text{(1)}\quad \begin{array}{ccccccccccc}
 & & H & & H & & H & & H & & H & \\
 & & | & & | & & | & & | & & | & \\
H & - & C & - & C & - & C & - & C & - & C & - & H \\
 & & | & & | & & | & & | & & | & \\
 & & H & & H & & H & & H & & H &
\end{array}
\quad \text{or} \quad CH_3-CH_2-CH_2-CH_2-CH_3
$$

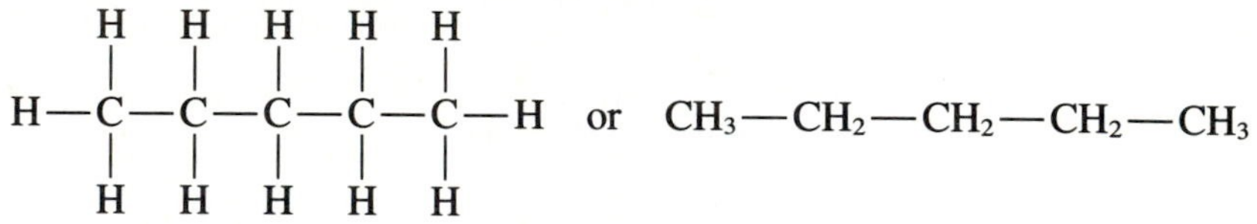

n-pentane

To determine the structures of the other isomers, consider a chain of carbon atoms one less than the maximum, which in this case would be four, and branch the remaining carbon off of an interior carbon. (For example, the second carbon from the left.)

$$
\text{(2)}\quad \begin{array}{ccccccccc}
 & & H & & H & & H & & H & \\
 & & | & & | & & | & & | & \\
H & - & C & - & C & - & C & - & C & - & H \\
 & & | & & | & & | & & | & \\
 & & H & & C & & H & & H & \\
 & & / & & | & & \backslash & & & \\
 & & H & & H & & H & & &
\end{array}
\quad \text{or} \quad
\begin{array}{l}
CH_3-CH_2-CH_2-CH_3 \\
\qquad\;\; | \\
\qquad CH_3
\end{array}
$$

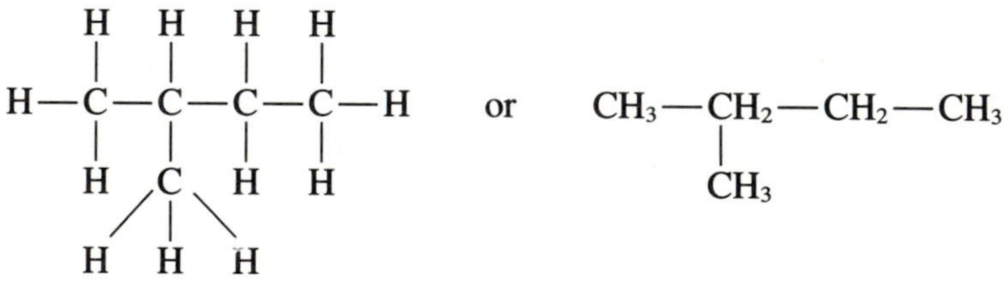

Isopentane

This structure is different from the straight-chain compound and is therefore an isomer. Try to write another structure for C_5H_{12} with a four-carbon chain by placing the fifth carbon on the third carbon from the left, i.e.,

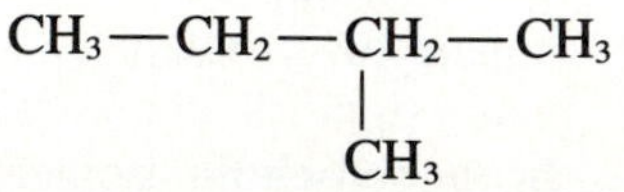

$$
\begin{array}{l}
CH_3-CH_2-CH_2-CH_3 \\
\qquad\qquad\quad\;\; | \\
\qquad\qquad\quad CH_3
\end{array}
$$

However this is not another isomer. It is the four-carbon chain structure (2) rotated 180° out of the plane of the paper. Thus there is only one isomer for C_5H_{12} contained in a four-carbon chain. Try to write another structure for C_5H_{12} by considering a three-carbon chain with the remaining two carbons branched from the chain.

(3)

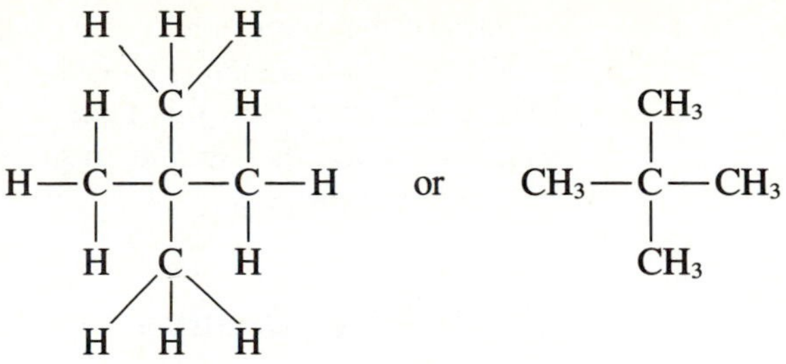

$$
\begin{array}{ccccccc}
 & H & H & H & & & \\
 & & \diagdown | \diagup & & & & \\
 & H & C & H & & & \\
 & | & | & | & & & \\
H- & C & -C- & C & -H & & \\
 & | & | & | & & & \\
 & H & C & H & & & \\
 & & \diagup | \diagdown & & & & \\
 & H & H & H & & &
\end{array}
\quad \text{or} \quad
\begin{array}{c}
CH_3 \\
| \\
CH_3-C-CH_3 \\
| \\
CH_3
\end{array}
$$

Neopentane

This structure is different from both (1) and (2) and is therefore an isomer. Any other attempts to write a different structure for C_5H_{12} will result in repeating one of the three structures already obtained.

Example 19.2

Write the structural formulas for the five isomers of C_6H_{14}.

Solution Try to work out the five different bonding sequences. Remember that *each carbon atom can be bonded to only four other atoms*.

(a) $CH_3-CH_2-CH_2-CH_2-CH_2-CH_3$

(b) 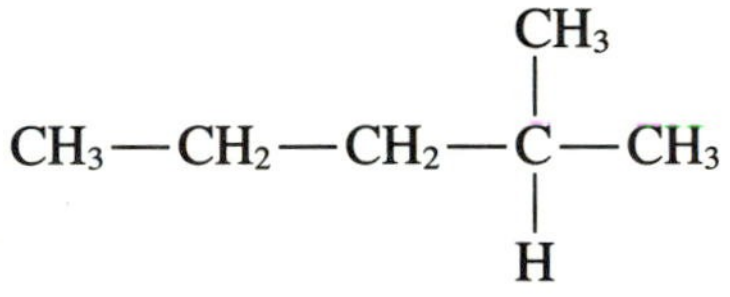
$$
\begin{array}{l}
\qquad\qquad\qquad\qquad\quad CH_3 \\
\qquad\qquad\qquad\qquad\quad\;\; | \\
CH_3-CH_2-CH_2-C-CH_3 \\
\qquad\qquad\qquad\qquad\quad\;\; | \\
\qquad\qquad\qquad\qquad\quad\;\; H
\end{array}
$$

(c)
$$
\begin{array}{l}
\qquad\qquad\quad\;\; CH_3 \\
\qquad\qquad\qquad | \\
CH_3-CH_2-C-CH_2-CH_3 \\
\qquad\qquad\qquad | \\
\qquad\qquad\qquad H
\end{array}
$$

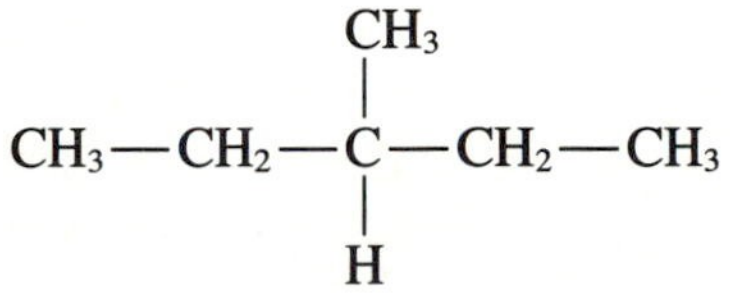

(d) 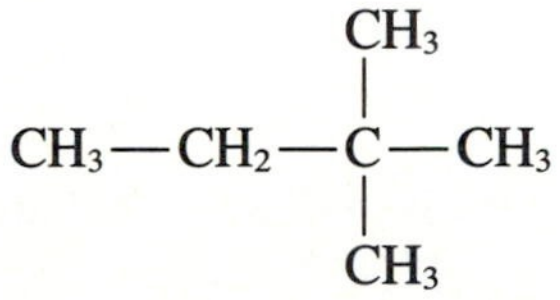
$$
\begin{array}{l}
\qquad\qquad\quad\;\; CH_3 \\
\qquad\qquad\qquad | \\
CH_3-CH_2-C-CH_3 \\
\qquad\qquad\qquad | \\
\qquad\qquad\quad\;\; CH_3
\end{array}
$$

(e) 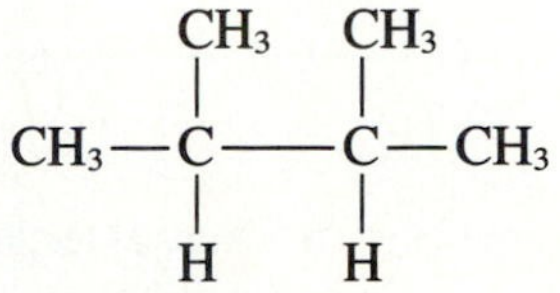
$$
\begin{array}{l}
\qquad\;\; CH_3 \quad CH_3 \\
\qquad\quad | \qquad\;\; | \\
CH_3-C---C-CH_3 \\
\qquad\quad | \qquad\;\; | \\
\qquad\quad H \qquad\; H
\end{array}
$$

Practice Exercise 19.2 Write the structural formulas for the nine isomers of C_7H_{16}.

Each of the five isomers in Example 19.2 has a distinct name. The way in which names are assigned to such compounds is called *organic nomenclature*. Before we can discuss this topic, however, we need to discuss *alkyl groups*, which are somewhat similar to groups of atoms such as SO_4^{2-} and OH^{1-} in inorganic chemistry.

19.6 Alkyl Groups

Suppose that an organic chemist begins to study a compound with the formula

$$CH_3—C≡C—CH_3$$

The chemist then extends the study to other compounds with the same type of structure:

$$CH_3—CH_2—C≡C—CH_2—CH_3$$

$$CH_3—CH_2—CH_2—C≡C—CH_2—CH_2—CH_3$$

$$CH_3—CH_2—CH_2—CH_2—C≡C—CH_2—CH_2—CH_3$$

The chemist notes that the chemical behavior of these compounds is similar. Moreover, their similar behavior is due to the triple carbon bond, which the

Table 19.3
Names of Some Alkyl Groups

Alkane name	Alkyl group name*
CH_4 Methane	$CH_3—$ Methyl
$CH_3—CH_3$ Ethane	$CH_3—CH_2—$ Ethyl
$CH_3—CH_2—CH_3$ Propane	$CH_3—CH_2—CH_2—$ *n*-propyl (or propyl)
	$CH_3—\underset{\mid}{C}H—CH_3$ Isopropyl
$CH_3—CH_2—CH_2—CH_3$ Butane	$CH_3—CH_2—CH_2—CH_2—$ *n*-butyl (or butyl)
	$CH_3—\overset{CH_3}{\overset{\mid}{C}H}—CH_2—$ Isobutyl
	$CH_3—CH_2—\overset{CH_3}{\overset{\mid}{C}H}—$ *sec*-butyl (secondary butyl)
	$CH_3—\underset{CH_3}{\underset{\mid}{\overset{CH_3}{\overset{\mid}{C}}}}—$ *tert*-butyl (tertiary butyl)

*The lowercase *n* in *n*-propyl, *n*-butyl, and so on means the hydrocarbon chains are normal. That is, the alkyl group stands for an unbranched hydrocarbon chain that has a hydrogen atom missing from carbon number 1.

chemist calls the *functional* part of the molecule, or the *functional group*. We can write the general structure for all these compounds as

$$R{-}C{\equiv}C{-}R'$$

where R and R′ (which is "R prime") may stand for $-CH_3$, $-CH_2-CH_3$, or any of the other groups of atoms attached to the triple-bonded carbons. These particular groups of carbon and hydrogen atoms are called **alkyl groups.** (The names of the alkyl groups come from the names of the individual alkanes, with the final *-ane* changed to *-yl* in each case.) Table 19.3 gives the first few alkyl groups.

We use the aklyl-group nomenclature in naming organic compounds. However, the R-group notation, in which R or R′ can stand for any structural group attached to a functional group, has much wider application. We will use it to discuss classes of organic compounds in Chapter 20.

19.7 Naming Organic Compounds

Learning Goal 7
Naming organic compounds

In the late 1800s, when many organic compounds were being discovered and put to use, organic chemists realized that they had to have some kind of logical system for naming them. Such a system—called the Geneva System—was developed in 1892 at an international meeting of chemists in Geneva, Switzerland. In 1930 this system was revised and extended at another meeting of chemists in Liège, Belgium. By this time chemists had organized, so the meeting was a convention of the International Union of Pure and Applied Chemistry (IUPAC, pronounced "EYE-you-pack"). The revised system is thus called the *IUPAC system*.

The chief feature of the IUPAC system is that it assigns numbers to carbon atoms to show where various R groups are located along the main carbon chain. In this section, we introduce some simple rules for naming alkanes by the IUPAC method.

Rule 1 Look for the longest carbon chain, and use this as the **base name** of the compound.

The base names for alkanes are those given in Table 19.1. For example, a compound whose longest chain contains ten carbon atoms has the base name *decane*.

Example 19.3

Find the longest carbon chain in each of the following, and name it. (To simplify the writing of structures, we'll show only the carbon skeletons, but keep in mind that there are always hydrogen atoms attached to the carbon atoms.)

```
(a) C—C—C—C—C—C—C        (b) C—C—C—C—C—C
          |                            |
          C                            C
          |                            |
          C                            C

        C
        |
        C
        |
(c) C—C—C
    |
    C
```

Solution We'll use color to show the longest chain. Don't be misled by the way the structure is written; just look for the longest continuous chain of carbon atoms.

```
(a) C—C—C—C—C—C—C
          |
          C
          |
          C
```

The longest chain has seven carbon atoms, so the base name of this compound is *heptane*.

```
(b) C—C—C—C—C—C
            |
            C
            |
            C
```

The longest chain in this compound has seven carbon atoms. This compound would be better written as

```
C—C—C—C—C—C—C
        |
        C
```

The base name of this compound is also *heptane*.

```
        C
        |
        C
        |
(c) C—C—C
    |
    C
```

The longest chain in this compound has five carbon atoms. This compound would be better written as

```
C—C—C—C—C
    |
    C
```

The base name of this compound is *pentane*.

Practice Exercise 19.3 Find the longest carbon chain in each of the following, and name it:

```
(a) C—C—C—C—C      (b) C—C—C
          |                |
          C                C
          |                |
          C                C

          C—C
          |
(c) C—C   C
    |  |  |
    C  C—C
```

Rule 2 Next add (as a prefix) the name of the side chain.

The side chains are alkyl groups. Their names are derived from the names of the alkanes, with *-ane* changed to *-yl* (Table 19.3).

Example 19.4

Using the compounds in Example 19.3, add the names of the side chains to the names of the compounds.

Solution

```
(a) C—C—C—C—C—C—C     (b) C—C—C—C—C—C—C
          |                         |
          C                         C
          |                  Methylheptane
          C
     Ethylheptane

(c) C—C—C—C—C
        |
        C
   Methylpentane
```

Practice Exercise 19.4 Using the compounds in Practice Exercise 19.3, add the names of the side chains to the names of the compounds.

Rule 3 Specify the location of the side chain by numbering the carbon atoms along the **longest** chain. Always number the longest chain so that the side-chain group will have the lowest possible number.

Suppose we have

$$\begin{array}{ccccccccccccc} C & - & C & - & C & - & C & - & C & - & C & - & C \\ & & & & & & & & | & & & & \\ & & & & & & & & C & & & & \end{array}$$

Do we number from the right or the left? The answer is that we number from the side that will give the methyl side-chain group the lowest possible number. So, in this compound, we number from the right:

$$\begin{array}{ccccccccccccc} 7 & & 6 & & 5 & & 4 & & 3 & & 2 & & 1 \\ C & - & C & - & C & - & C & - & C & - & C & - & C \\ & & & & & & & & | & & & & \\ & & & & & & & & C & & & & \end{array}$$

The name of this compound is *3-methylheptane*.

Example 19.5

Using the compounds in Example 19.3, give the entire name of each compound.

Solution

Understand the Problem

We are to name the compounds given in Example 19.3, using the IUPAC system.

Devise a Plan

We already know the base names of the compounds (from Example 19.3) and the names of the side chains (from Example 19.4). We can use Rule 3 to complete the name of each compound.

Carry Out the Plan

Following Rule 3, we number the carbon atoms along the longest chain so that the side chains will have the lowest possible numbers.

```
    1   2   3   4   5   6   7
(a) C—C—C—C—C—C—C
              |
              C
              |
              C
```

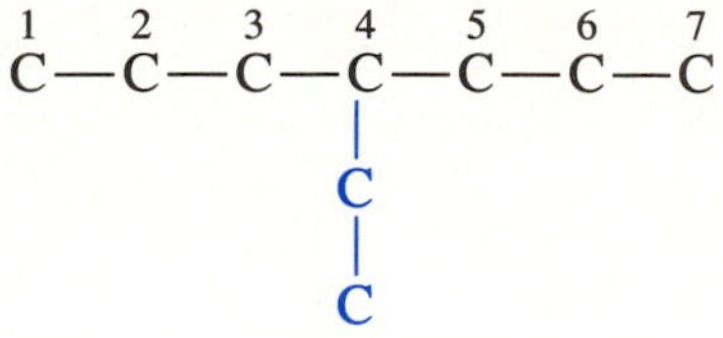

The name of this compound is *4-ethylheptane*. In this compound, we can number from left *or* right, because the side chain (the ethyl group) is attached to the middle carbon atom.

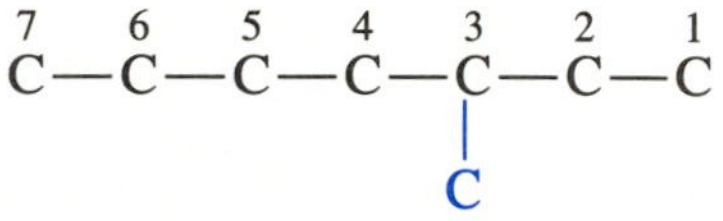

```
    7   6   5   4   3   2   1
(b) C—C—C—C—C—C—C
                  |
                  C
```

The name of this compound is *3-methylheptane*. In this compound, we must number the carbon atoms from the right so that the methyl side chain has the lowest possible number, 3.

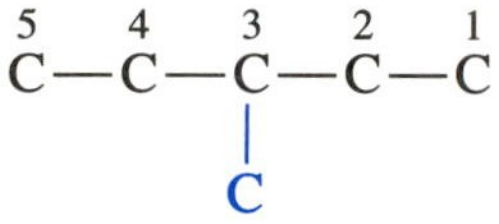

```
    5   4   3   2   1
(c) C—C—C—C—C
          |
          C
```

The name of this compound is *3-methylpentane*. In this compound, we may number from either end.

Look Back

Check that you have numbered the *longest* carbon chains in such a way that the side chains will have the lowest possible numbers: 4, 3, and 3, respectively.

Practice Exercise 19.5 Using the compounds in Practice Exercise 19.3, give the entire name of each compound.

Rule 4 If **more than one side chain** is present, use the following procedure to name the compound:

(a) Find the **longest chain,** as usual, to get the base name of the compound.

(b) Locate the **side chains.**

(c) Number the **longest chain** so that the side chains will have the lowest possible numbers.

(d) Add (as a prefix) the name and position of the **side chains,** using alphabetical order.

(e) In determining alphabetical order, ignore prefixes such as **di-, tri-,**

tetra-, and so on. The actual name of the **alkyl group** is used to determine alphabetical order.

Example 19.6

Name the compound represented by the following structure:

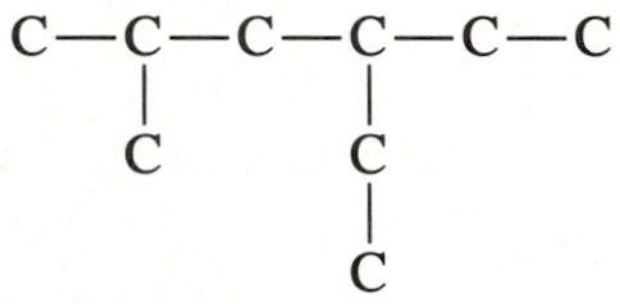

```
C—C—C—C—C—C
   |     |
   C     C
         |
         C
```

Solution We'll name this compound in steps:

(a) The longest chain is hexane.

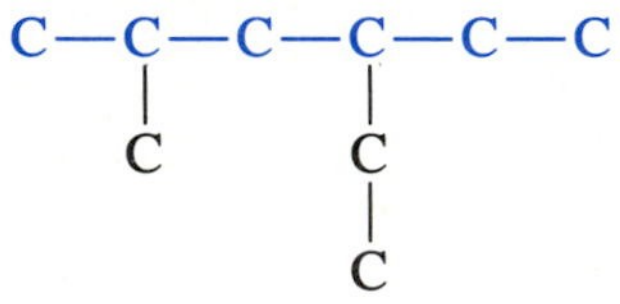

```
C—C—C—C—C—C
   |     |
   C     C
         |
         C
```

(b) The side chains are a methyl group and an ethyl group.

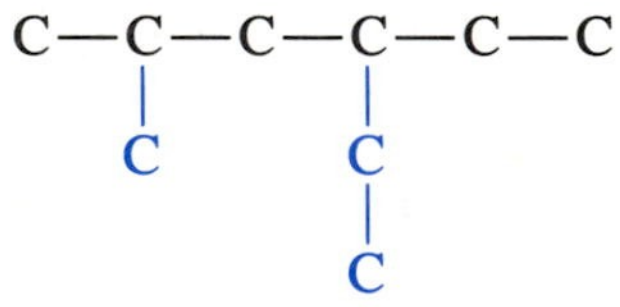

```
C—C—C—C—C—C
   |     |
   C     C
         |
         C
```

(c) We number the longest chain from the left so that the side chains have the lowest possible numbers: 2 and 4.

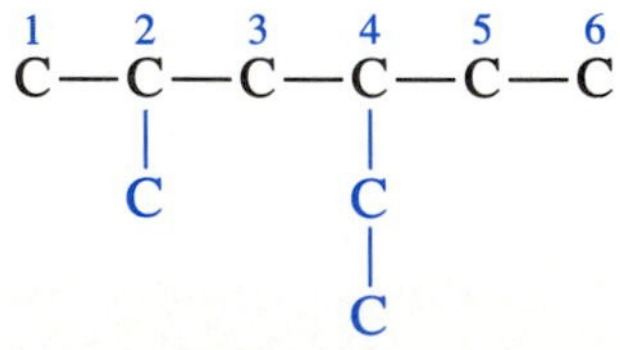

```
1  2  3  4  5  6
C—C—C—C—C—C
   |     |
   C     C
         |
         C
```

(If we had numbered from the right, the side-chain numbers would have been 3 and 5.)

(d) We can now write the full name of this compound, using the information we have obtained. (Remember, side chains are added on alphabetically, left to right.)

Longest chain	hexane
Side chain and its position	2-methyl
Side chain and its position	4-ethyl
Full name of compound	4-ethyl-2-methylhexane

Practice Exercise 19.6 Name the compound represented by the following structure:

```
C—C—C—C—C—C—C
    |   |
    C   C
        |
        C
```

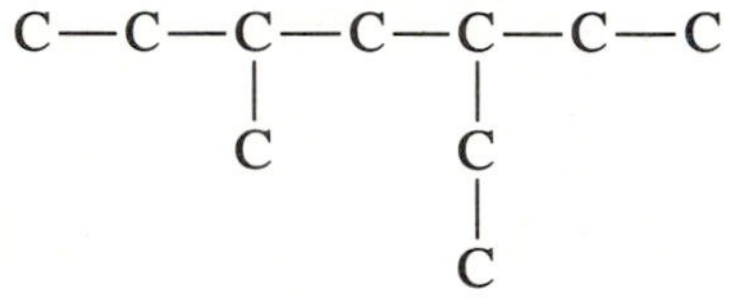

Rule 5 If two side chains in a compound are the same, use a Greek prefix instead of repeating the name of the side chains.

See Chapter 8 to review the Greek prefixes.

Example 19.7

Name the compound represented by the following structure:

```
  C     C
  |     |
C—C—C—C—C—C—C
```

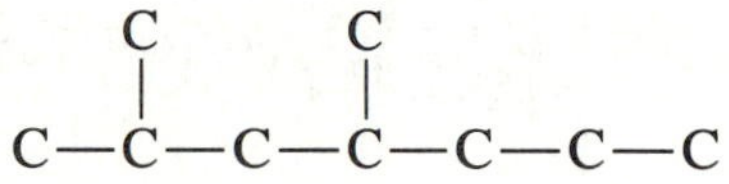

Solution We'll name this compound in steps:

(a) The longest chain is heptane.

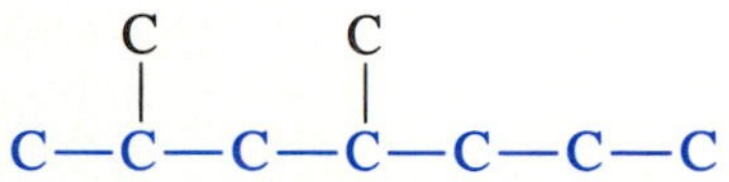

```
  C     C
  |     |
C—C—C—C—C—C—C
```

(b) The side chains are two methyl groups. We number from the left, in order to keep the numbers as low as possible.

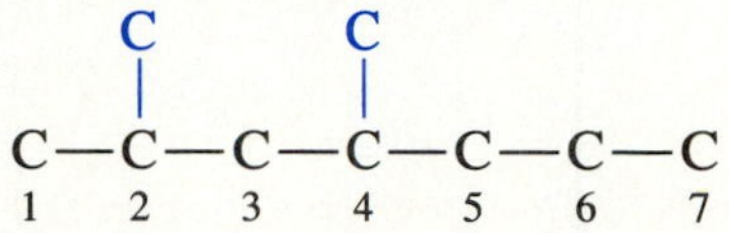

```
  C     C
  |     |
C—C—C—C—C—C—C
1 2 3 4 5 6 7
```

(c) We can now write the full name of the compound, using this information. (Instead of 2-methyl-4-methyl, we shall write 2,4-dimethyl.)

Longest chain	heptane
Side chains and their positions	2,4-dimethyl
Full name of compound	2,4-dimethylheptane

Practice Exercise 19.7 Name the compound represented by the following structure:

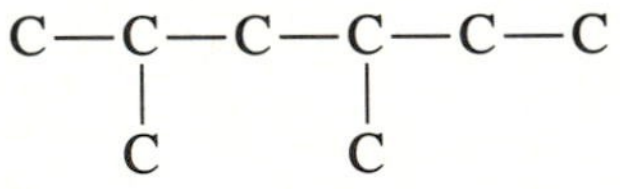

```
C—C—C—C—C—C
  |     |
  C     C
```

Example 19.8

Name each of the following compounds:

```
                C
                |
(a) C—C—C—C—C—C
                |
                C

                      C
                      |
(b) C—C—C—C—C—C—C—C
                |     |
                C     C
                      |
                      C

        C     C
        |     |
(c) C—C—C—C—C

                C
                |
(d) C—C—C—C—C
                |
                C

          C                       C
          |                       |
(e) C—C—C—C—C—C—C—C—C—C
                |     |     |
                C     C     C
                |     |
                C     C
                |
                C
```

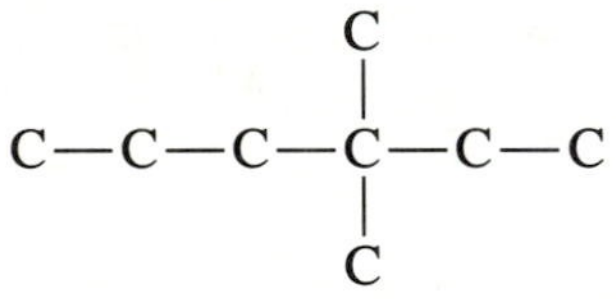

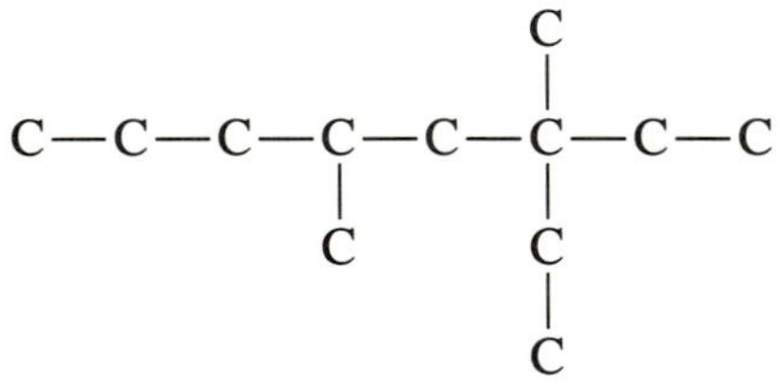

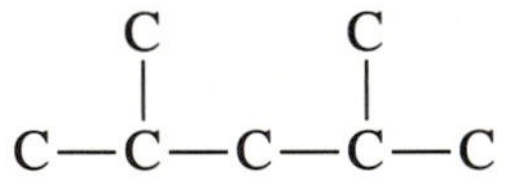

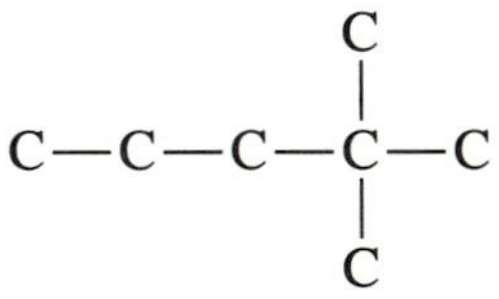

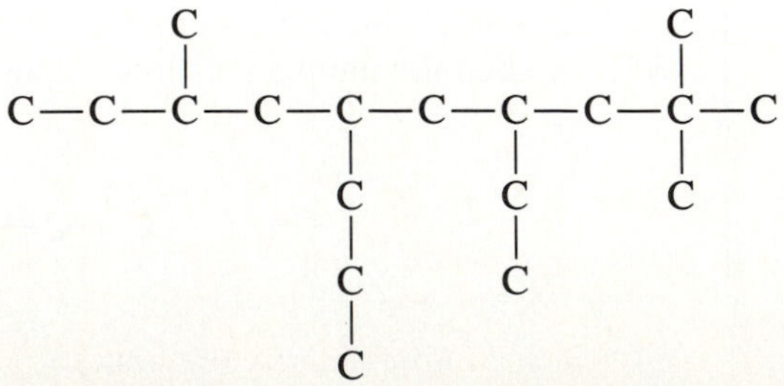

Solution

(a)

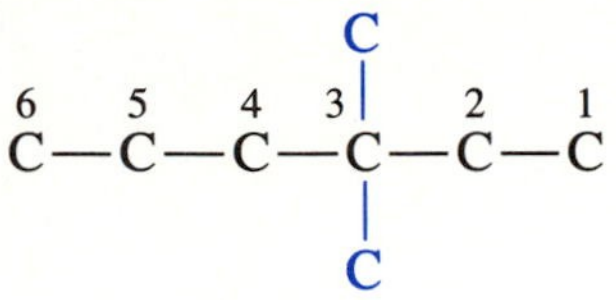

```
            C
6  5  4  3|  2  1
C—C—C—C—C—C
            |
            C
```

Longest chain	hexane
Side chains and their positions	3,3-dimethyl
Full name of compound	3,3-dimethylhexane

(b)

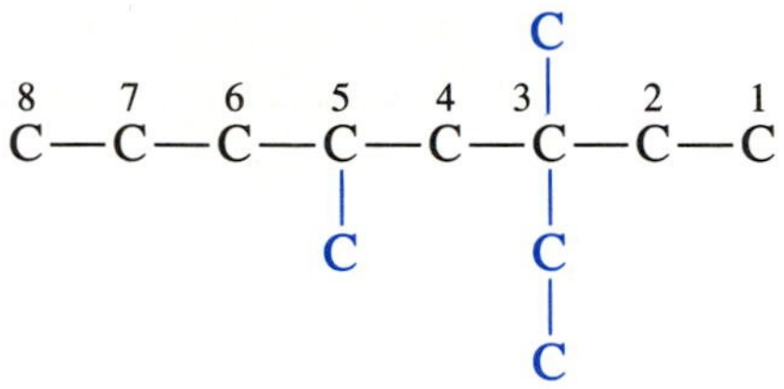

```
                    C
8  7  6  5  4  3|  2  1
C—C—C—C—C—C—C—C
         |        |
         C        C
                  |
                  C
```

Longest chain	octane
Side chains and their positions	3,5-dimethyl
Side chain and its position	3-ethyl
Full name of compound	3-ethyl-3,5-dimethyloctane

(c)

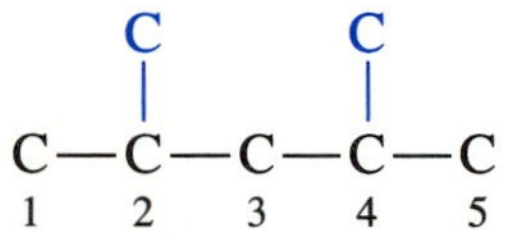

```
   C     C
   |     |
C—C—C—C—C
1  2  3  4  5
```

Longest chain	pentane
Side chains and their positions	2,4-dimethyl
Full name of compound	2,4-dimethylpentane

(d)

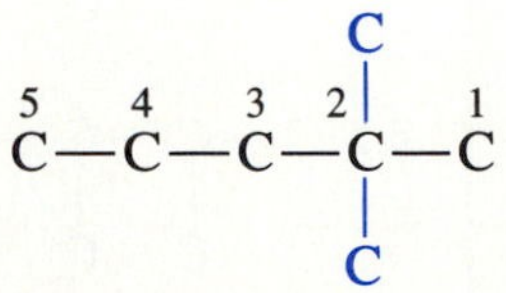

```
         C
5  4  3 2|  1
C—C—C—C—C
         |
         C
```

Longest chain	pentane
Side chains and their positions	2,2-dimethyl
Full name of compound	2,2-dimethylpentane

(e)

```
                C                             C
   10    9    8 |    7    6    5    4    3  2 |   1
   C  —  C  —  C  —  C  —  C  —  C  —  C  —  C  —  C  —  C
                          |         |         |
                          C         C         C
                          |         |
                          C         C
                          |
                          C
```

Longest chain	decane
Side chain and its position	6-propyl
Side chains and their positions	2,2,8-trimethyl
Side chain and its position	4-ethyl
Full name of compound	4-ethyl-2,2,8-trimethyl-6-propyldecane

Practice Exercise 19.8 Name each of the following compounds:

(a)
```
C—C—C—C—C—C—C—C—C
  |  |
  C  C
```

(b)
```
C—C—C—C—C—C—C—C—C
  |  |  |  |
  C  C  C  C
        |
        C
```

(c)
```
C—C—C—C—C—C
  |  |  |  |
  C  C  C  C
```

(d)
```
C—C—C—C—C—C—C
     |  |
     C  C
     |  |
     C  C
```

(e)
```
C—C—C—C—C—C—C—C
  |  |  |  |  |
  C  C  C  C  C
        |
        C
```

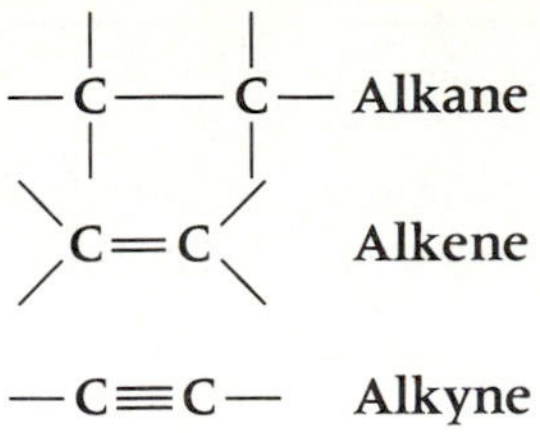

$$-\overset{|}{\underset{|}{C}}-\overset{|}{\underset{|}{C}}- \quad \text{Alkane}$$

$$>C=C< \quad \text{Alkene}$$

$$-C \equiv C- \quad \text{Alkyne}$$

Figure 19.7
The alkanes, alkenes, and alkynes

19.8 The Alkenes and Alkynes

The alk*enes* and alk*ynes,* like the alkanes, are hydrocarbons. But the alk**enes** have one carbon–carbon double bond, and the alk**ynes** have a carbon–carbon triple bond (Figure 19.7).

The simplest *alkene* has the formula C_2H_4 and the structure

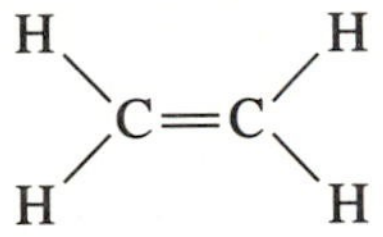

$$\begin{matrix} H & & H \\ & C=C & \\ H & & H \end{matrix}$$

(Note the double bond linking the carbons.) The common name for this compound is ethylene. It is the substance from which we make polyethylene, a plastic used for containers, electrical insulation, and packaging.

The simplest *alkyne* has the formula C_2H_2 and the structure

$$H-C \equiv C-H$$

(Note the triple bond linking the carbons.) The common name for this compound is acetylene; it is used in torches that can cut through metals.

Learning Goal 5
Unsaturated hydrocarbons

Each of these compounds is the first compound in its homologous series (Table 19.4). Both these series of hydrocarbons are said to be **unsaturated.** This means that there are double or triple carbon–carbon bonds in their molecules.

The IUPAC name for an alkene or alkyne can be found from the name of the alkane that has the same number of carbon atoms in its longest chain. We simply change the alkane ending to *ene* for alkenes or to *yne* for alkynes. For example, the IUPAC name for C_2H_4 is eth*ene,* and the IUPAC name for C_2H_2 is eth*yne.* Both names are derived from the name of the two-carbon alkane compound ethane; only the endings are changed.

Table 19.4
The Alkene and Alkyne Series

	Alkenes	**Alkynes**
Structural formula of the first member of the series	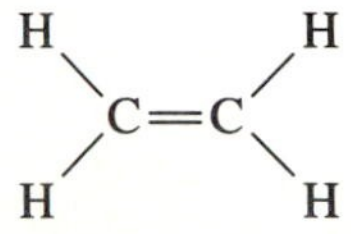 $H_2C=CH_2$ (H, H / C=C / H, H)	$H-C \equiv C-H$
Molecular formula	C_2H_4	C_2H_2
Common name	Ethylene	Acetylene
IUPAC name	Eth*ene*	Eth*yne*
General formula for the series	C_nH_{2n}	C_nH_{2n-2}

Example 19.9

Name each of the following compounds. (Again, to simplify writing the structures, we'll show only the carbon skeletons; keep in mind that there are hydrogens attached to these carbon atoms.)

(a) C—C=C *(b)* C—C—C=C

```
                       C
                       |
(c) C—C=C—C     (d) C—C—C=C
```

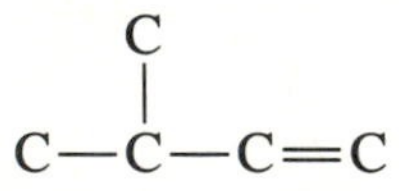

(e) C—C≡C *(f)* C—C—C≡C

```
                             C
                             |
(g) C—C≡C—C     (h) C—C—C—C—C≡C—C
                             |
                             C
                             |
                             C
```

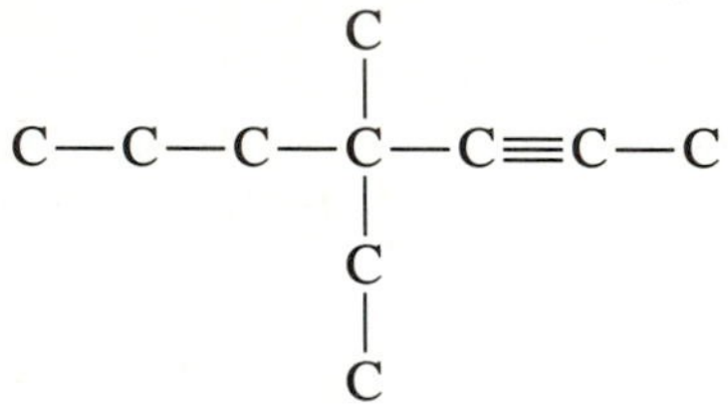

Solution

(a) C—C=C is called prop*ene*. It is a three-carbon compound that is the second member of the alkene series. It takes its name from the three-carbon alkane compound, except that it has an *-ene* ending.

```
    4  3  2  1
(b) C—C—C=C
```

is called 1-but*ene*. This third member of the alkene series takes its name from the four-carbon alkane. But *there is one additional point:* the position of the double bond must be specified in the name. *To specify the position of the double bond, we number the carbon atoms, as we did before, keeping the number of the double bond as low as possible. Then, in the compound name, we give the number of the carbon atom from which the double bond* **starts** *(the smaller of the two numbers).*

```
    4  3  2  1
(c) C—C=C—C
```

is called 2-but*ene*. This compound, similar to the compound in part (b), is also a butene. However, the position of its double bond is different.

```
       C
       |
(d) C—C—C=C
    4  3  2  1
```

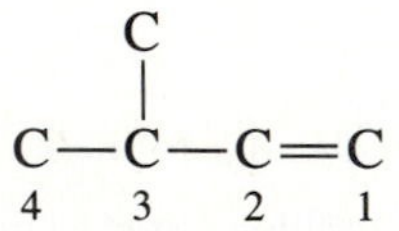

When we number the carbon chain, we give preference to the double bond and not to the side chain. In other words, we want to keep the number of the double bond as low as possible.

Longest chain and position of double bond	1-butene
Side chain and its position	3-methyl-
Full name of compound	3-methyl-1-butene

(e) C—C≡C is called prop*yne* and is the second member of the alkyne series. It takes its name from the three-carbon alkane compound, except that it has a *-yne* ending.

(f) $\overset{4}{\mathrm{C}}—\overset{3}{\mathrm{C}}—\overset{2}{\mathrm{C}}≡\overset{1}{\mathrm{C}}$ is called 1-but*yne*. It is the third member of the alkyne series and takes its name from the four-carbon alkane. Remember that we have to note the position of the triple bond when we name this compound, because there is another butyne compound.

(g) $\overset{4}{\mathrm{C}}—\overset{3}{\mathrm{C}}≡\overset{2}{\mathrm{C}}—\overset{1}{\mathrm{C}}$ is called 2-but*yne*. This compound is also a butyne, but it is different from the one in part (f), because the position of the triple bond is different.

(h)

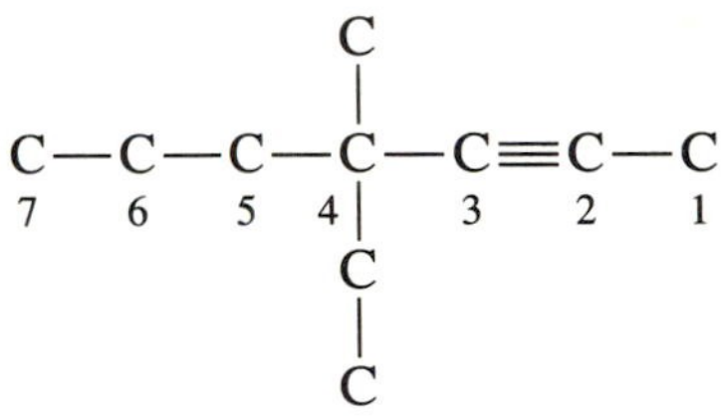

```
                  C
                  |
C — C — C — C — C≡C — C
7   6   5   4|  3  2  1
              C
              |
              C
```

Longest chain and position of triple bond	2-heptyne
Side chain and its position	4-methyl
Side chain and its position	4-ethyl
Full name of compound	4-ethyl-4-methyl-2-heptyne

Practice Exercise 19.9 Name each of the following compounds:

(a) C—C—C—C=C

(b) C—C—C—C—C=C

(c) C—C—C—C=C—C—C

(d)
```
C—C—C—C—C=C
      |
      C
```

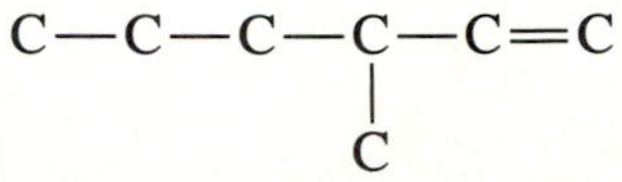

(e) C—C—C—C—C≡C

(f) C—C—C≡C—C—C

(g)
```
C—C—C—C—C≡C
  |  |
  C  C
```

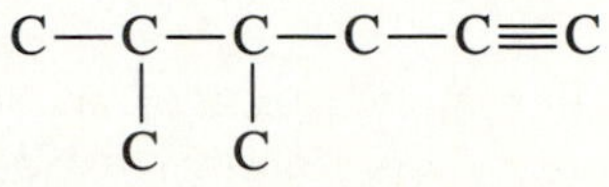

(h)
```
C—C—C—C≡C
  |  |
  C  C
```

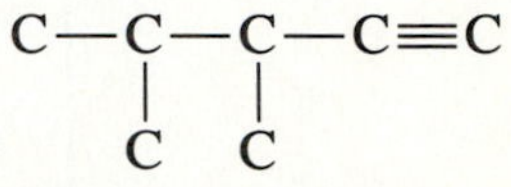

19.9 Cyclic Hydrocarbons

Learning Goal 5
Cyclic hydrocarbons

Many hydrocarbons form **cyclic structures**—that is, structures that join to form a closed chain, like a snake biting its tail. An example is

or C_3H_6

The name of this compound—cyclopropane—is the name of the three-carbon alkane (propane) with the prefix *cyclo-* added. Note that it is *not* an isomer of propane. Many cyclic hydrocarbons have useful properties. For example, cyclopropane is an anesthetic that is sometimes used in surgery. However, cyclopropane is very flammable, and great caution must be exercised in the operating room when it is used.

Another example of a cyclic hydrocarbon is

or C_5H_{10}

The name of this compound—cyclopentane—is the name of the five-carbon alkane (pentane) with the prefix *cyclo-* added. To make the writing of cyclic hydrocarbons simpler, we generally show only the shape described by the atoms in the longest chain. For example, cyclopentane is shown as

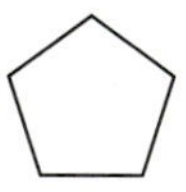

The shape of the figure tells us the name of the compound. Each line represents a carbon–carbon bond, and each corner represents a CH_2 group.

Example 19.10

Name each of the following compounds:

(a)

(b)

(c) 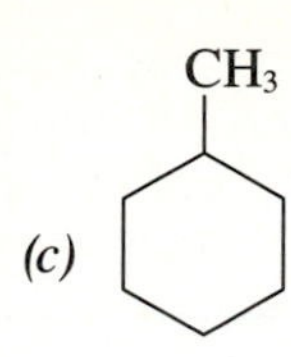

(d) 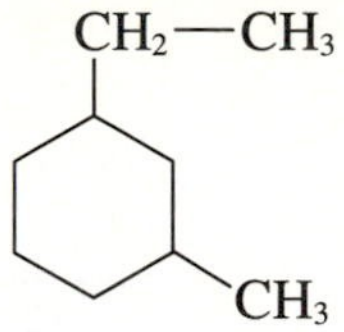 $CH_2—CH_3$ / CH_3

Solution

(a)

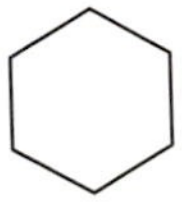

This is a cyclic hydrocarbon with six carbon atoms. It is called *cyclohexane*.

(b)

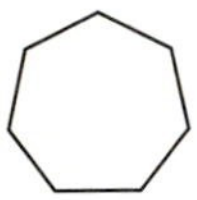

This is a cyclic hydrocarbon with seven carbon atoms. It is called *cycloheptane*.

(c) 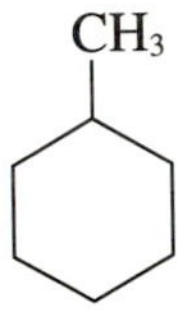 CH_3

This is a *substituted* cyclic hydrocarbon. The CH_3 side-chain group has replaced a hydrogen atom that was bonded to one of the carbon atoms. This compound is called *methylcyclohexane*.

(d) 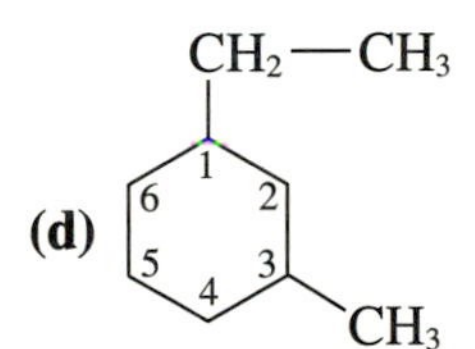

This is a *di*substituted cyclic hydrocarbon, which means that *two* hydrogens on the ring have been replaced by side-chain groups.

When more than one side chain is attached to a cyclic hydrocarbon, it is necessary to number the C atoms in the ring. Remember to number them in such a way as to keep the numbers of the side chains as low as possible.

Name of cyclic structure	cyclohexane
Side chain and its position	3-methyl
Side chain and its position	1-ethyl
Full name of compound	1-ethyl-3-methylcyclohexane

Practice Exercise 19.10 Name each of the following compounds:

(a)

(b)

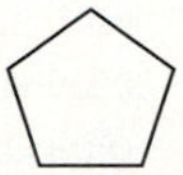

Some cyclic hydrocarbons have double and triple bonds. These hydrocarbons take their names from the corresponding alkene or alkyne, with the prefix *cyclo-* added.

Example 19.11

Name each of the following compounds:

(a) 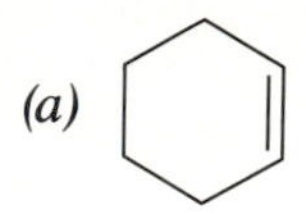

(b) 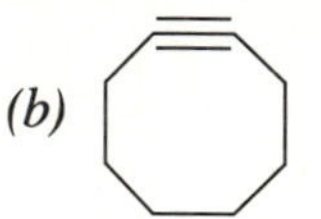

Solution

(a) 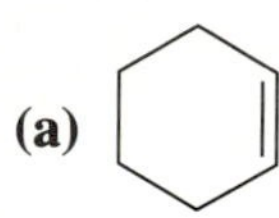

This is a six-carbon cyclic compound with a double bond. It takes its name from the corresponding *alkene*. The name of this compound is *cyclohexene*. It can be used as a stabilizer for high-octane gasolines.

(b) 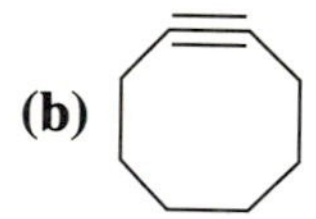

This is an eight-carbon cyclic compound with a triple bond. It takes its name from the corresponding *alkyne*. The name of this compound is *cyclooctyne*.

Practice Exercise 19.11 Name each of the following compounds:

(a) 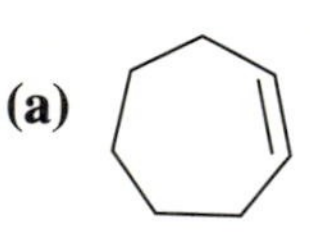

(b) 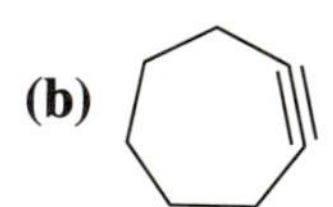

(c) 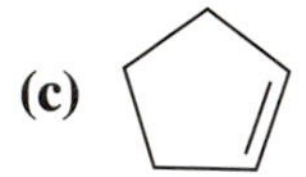

(d) 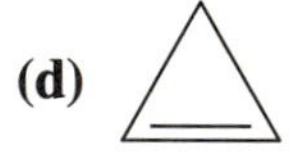

19.10 Aromatic Hydrocarbons

Learning Goal 8
Aromatic hydrocarbons

One series of cyclic hydrocarbons was once thought to have alternating single and double bonds. The members of this series were called **aromatic hydrocarbons** by early organic chemists who noticed that these compounds had pleasant odors. (Organic chemists later found that some of these compounds have unpleasant odors.) The parent compound of the aromatic hydrocarbons is benzene, C_6H_6, whose structure is

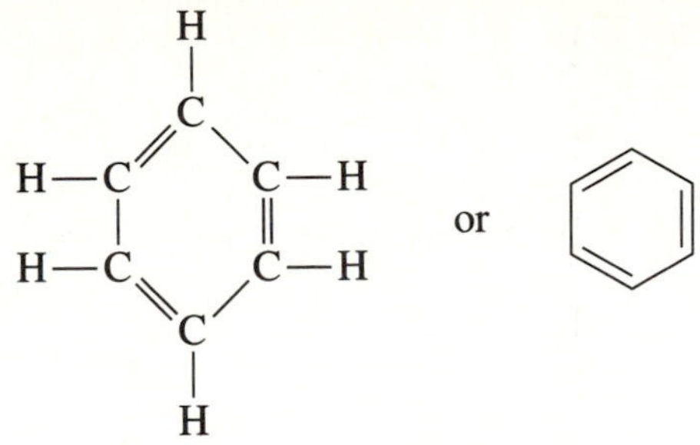

Although only three of the carbon–carbon bonds in benzene are written as double bonds, all the bonds seem to be equivalent. At one time, chemists incorrectly thought that the double bonds shifted back and forth between carbon atoms. If that were true, either of these two diagrams would give a valid picture of the benzene molecule:

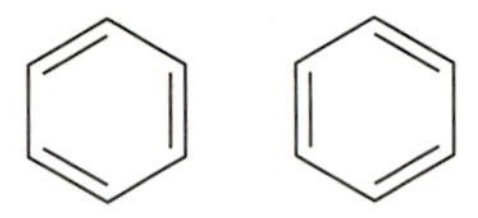

Today, chemists still don't agree on how to represent the structure of benzene. Any one of the following representations can be used:

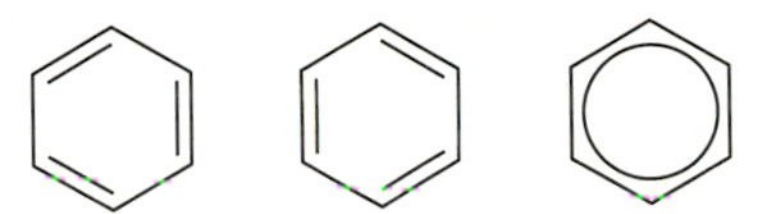

The third diagram is meant to show that all the carbon–carbon bonds in benzene are equivalent, as they are known to be.

Benzene, which can be obtained from coal tar, is one of the most important organic compounds. It is the starting material for countless products that improve the human condition, including many synthetic fibers such as nylon, perfumes, synthetic dyes, drugs, and pesticides.

Other aromatic hydrocarbons are made up of benzene rings joined together (Figure 19.8). And many compounds consist of aromatic compounds with various side chains. An example is

One name for this compound is methylbenzene, but organic chemists call it *toluene*. It is used in making explosives and dyes.

There are also rules for naming these types of compounds, but they will not be covered here.

Figure 19.8
Some aromatic hydrocarbons

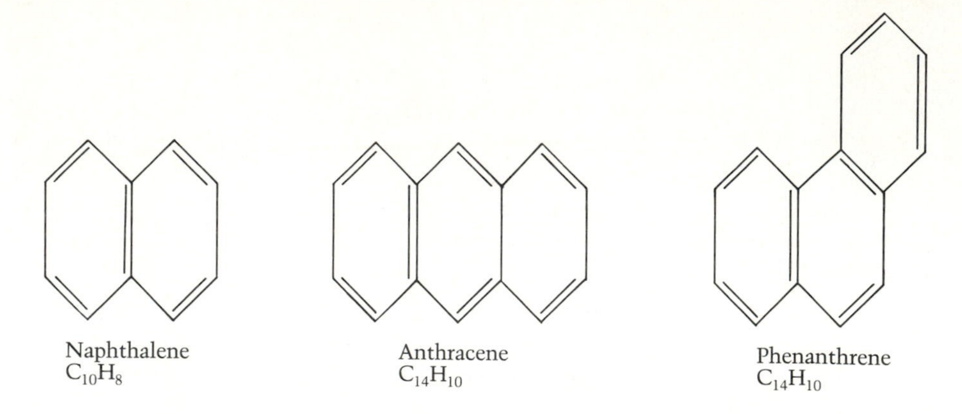

19.11 Other Types of Organic Compounds

In this chapter, we have discussed only hydrocarbons—organic compounds made up of carbon and hydrogen and nothing else. But there are many other kinds of organic compounds, and these contain other elements as well. In fact, the field of organic chemistry is enormous, and the products developed by organic chemists affect our lives in many ways every day.

In Chapter 20 we will look briefly at several of these other kinds of organic compounds. For a preview, you might want to look at Table 20.1.

Summary

Organic chemistry is essentially the study of compounds that contain carbon—along with such elements as hydrogen, oxygen, nitrogen, sulfur, and the Group VIIA elements. More than 10 million organic compounds are known.

The carbon atom has four electrons in its outermost energy level. It can use all four electrons to bond with other atoms, including other carbon atoms. Two carbon atoms can form a single, double, or triple bond. When a carbon atom bonds with four hydrogen atoms, the four carbon–hydrogen bonds define a tetrahedral shape, with the H atoms at the corners of the tetrahedron. This tetrahedral shape is common to many organic compounds.

Because carbon atoms can bond together in different ways, different organic compounds can have the same molecular formula. For this reason, structural formulas are used to specify organic compounds. The structural formula shows how the carbon atoms are bonded, as well as the elements that make up the compound.

Saturated compounds are compounds in which all bonds between carbon atoms are single bonds. The alkanes are a series of saturated hydrocarbons with the general formula C_nH_{2n+2}. The alkenes are hydrocarbons with double carbon–carbon bonds. The alkynes are hydrocarbons with triple carbon–carbon bonds. Organic compounds with double or triple bonds are said to be unsaturated. All three of these series—the alkanes, alkenes, and alkynes—are homologous series. That is, each compound in the series differs from the preceding compound by the same fixed increment.

Organic compounds are named according to the IUPAC system. In this system, the name of the compound indicates its structure. Thus isomers, or compounds with the same molecular formula but different structural formulas, are assigned different names in the IUPAC system.

Cyclic hydrocarbons are hydrocarbons in which the carbon atoms form a closed chain. An important series of cyclic hydrocarbons consists of the aromatic hydrocarbons, which include the compound benzene.

Key Terms

alkane **(19.4)**
alkene **(19.8)**
alkyl group **(19.6)**
alkyne **(19.8)**
aromatic hydrocarbon **(19.10)**
branched-chain isomer **(19.5)**
condensed form **(19.2)**
cyclic structure **(19.9)**
homologous series **(19.4)**
hydrocarbon **(19.4)**
isomer **(19.5)**
organic chemistry **(19.1)**
organic compound **(19.1)**
saturated hydrocarbon **(19.4)**
series **(19.3)**
straight-chain isomer **(19.5)**
structural formula **(19.2)**
structural isomerism **(19.5)**
tetrahedron **(19.1)**
unsaturated hydrocarbon **(19.8)**
vitalism **(Introduction)**

Career Sketch

Chemical Engineer

As a chemical engineer, you will plan, design, and construct chemical plants and equipment. You will also do research to develop and improve the processes by which large quantities of chemicals are produced for industry and commerce. You may also set up pilot plants to test new processes. You will work with materials such as plastics, detergents, textile fibers, pharmaceuticals, and synthetic rubbers.

The field of chemical engineering is so diversified that chemical engineers often specialize in a particular chemical process. You may choose to specialize in a process such as oxidation or polymerization. You also may choose to specialize in pollution control or in a particular production process such as the formulation of paints or plastics.

In the United States today, there are about 55,000 chemical engineers working in the chemical and petrochemical industry, and about 10,000 others in higher education, government, research, and consulting. If you decide to work in an area that affects the public welfare, you will need to obtain a pro-

This chemical engineer is involved in the development of specialty chemicals.
(Courtesy Pfizer Inc.)

fessional engineering license. This involves earning a baccalaureate degree, working four or five years in the field, and passing a professional engineering licensing exam.

Chemical Frontiers

Growth Factors Aid Healing

Each year, thousands of patients spend extended periods of time in the hospital recovering from wounds, burns, skin ulcers and sores, surgical incisions, and other injuries. A new class of medications can cut healing time dramatically, saving enormous medical costs and allowing patients to spend less time in the hospital. These medications, which contain epidermal growth factors (EGF), enhance the growth of tissue cells, helping the body repair itself more quickly.

The first breakthrough in speeding up the healing process came in 1974, when researchers observed that a material released from blood platelets (substances that assist in clotting) promotes the growth of cells in a variety of tissues. Further experimentation led researchers to conclude that there are five different platelet-derived growth factors that contribute to wound healing and repair. These growth factors can be genetically engineered in the laboratory and applied directly to the damaged tissues.

In clinical trials, patients with diabetes, bed sores, rheumatoid arthritis, trauma, and other conditions were treated with genetically engineered growth factors. Ninety percent of those treated achieved 100% healing in about 7 weeks, in contrast to the average usual healing time for these conditions of 96 weeks. Platelet-derived medications for wound healing are now available for medical use, and more than 30 such products have been developed by some 80 different companies.

Self-Test Exercises

Learning Goal 1: Organic and Inorganic Compounds

1. Are carbon–carbon bonds in organic compounds ionic or covalent?

▲ **2.** How is a diamond different from the graphite found in a "lead" pencil, from a structural point of view?

▲ **3.** Name two crystalline forms of carbon. What accounts for their widely differing properties?

4. What are inorganic compounds?

5. List some examples of compounds that contain carbon but are *not* organic.

6. Approximately how many organic compounds are known?

7. What accounts for the large number of organic compounds?

8. What elements besides carbon and hydrogen are commonly found in organic compounds?

Learning Goal 2: Lewis Electron-Dot Structure of Carbon

▲ **9.** Write the Lewis electron-dot structure for carbon.

▲ **10.** How many electrons does carbon possess that are available for bonding?

Learning Goal 3: Geometry of Carbon Bonds

11. The bonding pattern of the carbon atom in many organic compounds is ________.

12. What is meant by "tetrahedral carbon bonding"?

▲ **13.** Tell whether each of the following descriptions indicates a carbon–carbon single bond, a double bond, or a triple bond:
(a) two tetrahedral carbon atoms with a common edge
(b) two tetrahedral carbon atoms with a common face
(c) two tetrahedral carbon atoms with their corners touching

▲ **14.** How many degrees apart are the carbon–hydrogen bonds in methane?

Learning Goal 4: Structural Formulas of Organic Compounds

▲ **15.** Write structural formulas for the following compounds: *(a)* eth*ane* *(b)* eth*ene* *(c)* eth*yne*

▲ **16.** Why is a structural formula more important in organic chemistry than a molecular formula?

17. What is the difference between a structural formula and a molecular formula?

18. How many compounds are represented by the molecular formula C_4H_{10}?

19. Write the formulas and names for the first five straight-chain members of the alk*ene* series.

20. What are the two ways in which structural formulas can be written?

21. Write the formulas and names for the first five straight-chain members of the alk*yne* series.

22. Write the structural formula for *n*-heptane.

▲ **23.** Write the following structural formula in the condensed form:

```
     H   H   H   H   H
     |   |   |   |   |
 H—C—C—C—C—C—H
     |   |   |   |   |
     H   C   H   C   H
        /|\     /|\
       H H H   H H H
```

▲ **24.** Write the structural formula for 1-heptene.

25. Rewrite each of the following structures by adding hydrogen atoms to the carbon skeleton:

(*a*)
```
C—C—C—C—C—C
  |       |
  C       C
```

(*b*)
```
C—C—C=C—C—C
          |
          C
```

(*c*)
```
C—C—C=C—C—C≡C
  |
  C
```

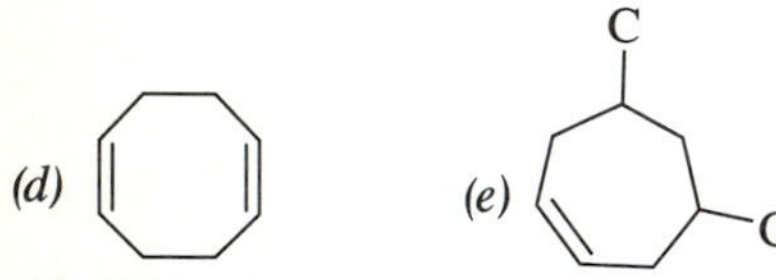

26. Write the structural formula for 1-heptyne.

27. Write the structure for each of the following hydrocarbons:

(*a*) 2-methylbutane
(*b*) 2,3,4-trimethylheptane
(*c*) 2,2-dimethylpropane
(*d*) 1-hexene
(*e*) 2-hexene
(*f*) 3-ethyl-2-methyl-1-pentene
(*g*) 3-methyl-1-butyne

28. Write the following structural formula in the condensed form:

```
     H   H   H   H   H   H   H   H
     |   |   |   |   |   |   |   |
 H—C—C—C—C—C—C—C—C—H
     |   |   |   |   |   |   |   |
     H   H   H   C   H   H   C   H
                /|\         /|\
               H H H       H H H
```

29. Write the structure for each of the following cyclic hydrocarbons:

(*a*) cyclopropane
(*b*) cycloheptane
(*c*) 3-ethyl-1-methylcyclohexane
(*d*) cyclohexyne
(*e*) cyclohexene
(*f*) 1,2-dimethylbenzene
(*g*) 1,3-dimethylbenzene

30. Write the structure for each of the following hydrocarbons:

(*a*) 2,2-dimethylbutane
(*b*) 2,3-dimethylpentane
(*c*) 3-hexene
(*d*) 3-methyl-2-butene
(*e*) 2,3,5-trimethylheptane
(*f*) 3-methyl-3-ethyl-1-hexene

31. Write the structure for each of the nine isomers of C_7H_{16}.

32. Write the structure for each of the following cyclic hydrocarbons:

(*a*) 1-methylcyclopropane
(*b*) 1-ethyl-3-methylcycloheptane
(*c*) cyclopentene
(*d*) cyclopentyne
(*e*) 1-ethyl-3-methylbenzene
(*f*) 1-methylcyclohexene

Learning Goals 5 and 6: Saturated, Unsaturated, and Cyclic Hydrocarbons; Isomers and Homologous Series

▲ **33.** Define the following terms, and give an example of each: *saturated hydrocarbon, unsaturated hydrocarbon,* and *isomer*.

▲ **34.** Distinguish between a saturated and an unsaturated hydrocarbon.

▲ **35.** Define the term *homologous series,* and give an example of such a series by listing some of its members.

▲ **36.** What is a hydrocarbon?

37. Write the formulas and names of the first ten straight-chain members of the alkane series (Don't look back in the chapter.)

38. How is a cyclic hydrocarbon different from an aromatic hydrocarbon?

39. The formula of an organic compound is C_5H_{10}. Write the structural formulas of all the possible compounds.

40. Name the first nine straight-chain members of the alkene series.

41. How many different straight-chain heptenes are there? Name each one.

42. How many different straight-chain hexenes are there? Explain.

Learning Goal 7: Naming Organic Compounds

▲ **43.** Name each of the isomers in Self-Test Exercise 31.

▲ **44.** Name each of the following alkanes by the IUPAC system:

```
(a) C—C—C—C—C        (b) C—C—C
        |                  |
        C                  C

(c) C—C—C—C—C
      |   |
      C   C
```

45. Name each of the following alkanes by the IUPAC system:

```
                            C
                            |
(a) C—C—C—C        (b) C—C—C
        |                   |
        C                   C

(c) C—C—C—C—C—C
        | | |
        C C C
```

46. Name each of the following compounds by the IUPAC system:

```
(a) C—C—C=C
(b) C—C=C—C
(c) C—C—C—C—C—C=C
(d) C—C—C—C=C
        | |
        C C
          |
          C
```

47. Name each of the following alkenes by the IUPAC system:

```
(a) C—C—C—C—C=C
(b) C—C—C—C=C—C
(c) C—C—C=C—C—C
(d) C—C—C—C=C
        | |
        C C
        |
        C
```

48. Name each of the following alkynes by the IUPAC system:

```
(a) C—C≡C
(b) C—C—C—C≡C
        |
        C
        |
        C
        |
        C
(c) C—C—C—C≡C
(d) C—C—C—C—C≡C—C
      |   |
      C   C
          |
          C
```

▲ **49.** Name each of the following alkynes by the IUPAC system:

```
(a) C—C—C—C≡C      (b) C—C—C≡C—C
(c) C—C—C≡C
      |
      C
(d) C—C—C—C≡C—C—C
        |       |
        C       C
        |
        C
```

▲ **50.** Name each of the following cyclic hydrocarbons:

(a) 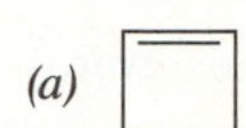

(b) 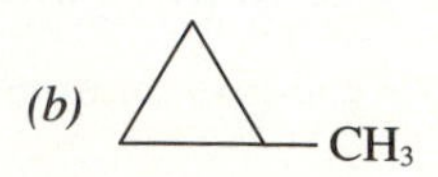

(c) 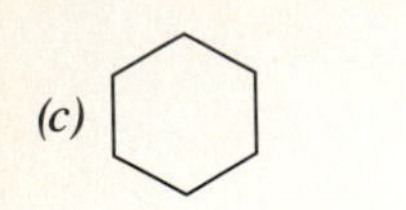

(d) 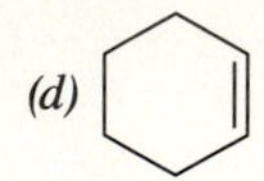

(e) 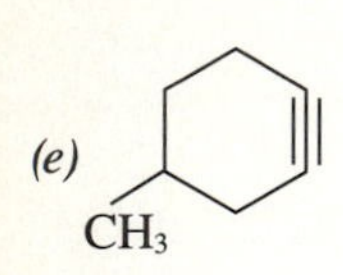

(f) 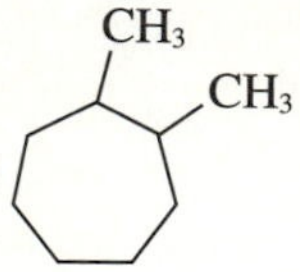 CH_3, CH_3

51. Name each of the following cyclic hydrocarbons:

(a) 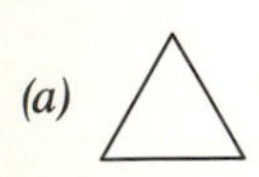

(b) 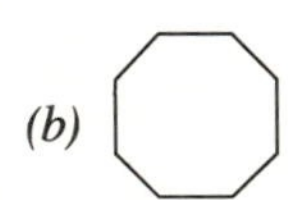

(c) 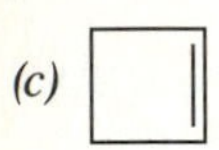

(d) 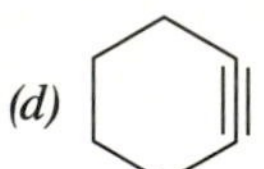

(e) 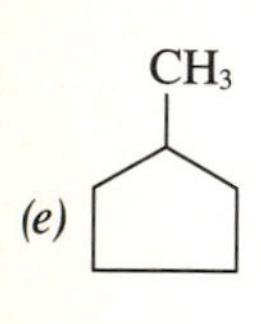

(f) 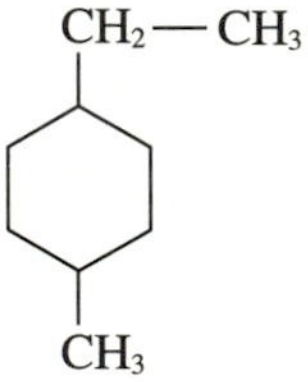 $CH_2—CH_3$, CH_3

(g) 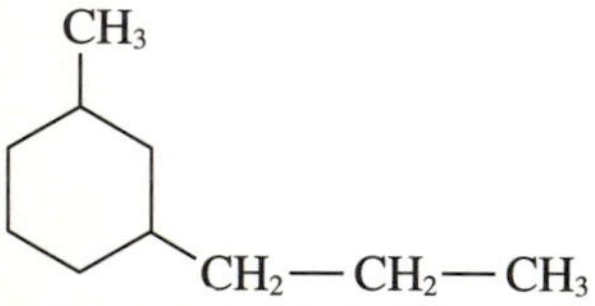 CH_3, $CH_2—CH_2—CH_3$

52. Name each of the following compounds by the IUPAC system:

(a)

(b) C—C≡C

(c) C=C—C—C—C—C—C

(d) 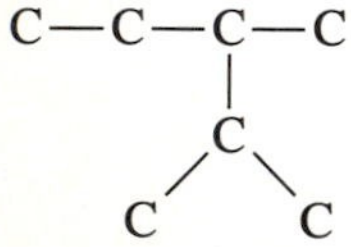 C—C—C—C (with C attached to the third carbon, bearing two C)

Learning Goal 8: Aromatic Hydrocarbons

▲ **53.** Define and give an example of an aromatic hydrocarbon.

▲ **54.** Write the structural formula for benzene.

Extra Exercises

55. Label the R group in each of the following compounds:

(a) $CH_3—CH_2—CH_2—O—CH_2—CH_2—CH_3$

(b) $CH_3—CH_2—CH_2—OH$

(c) 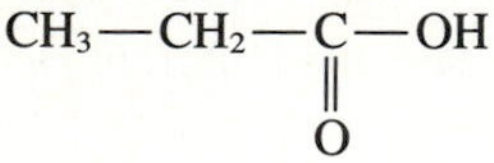 $CH_3—CH_2—C(=O)—OH$

▲ **56.** Most organic compounds have different boiling points. Could a distillation apparatus as described in Chapter 14 be used to separate some of these compounds? Explain.

***57.** Write a general formula for a dialkene, a carbon compound containing two double bonds.

58. What is the relationship between the following compounds?

$CH_3—CH=CH—CH_3$ and

$CH_2=CH—CH_2—CH_3$

59. Name the following compound:

$CH_3CH_2CH(CH_3)CH=CHCH(CH_3)CH_3$

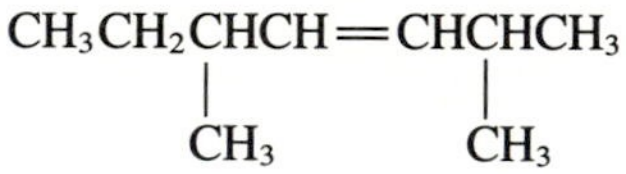

▲ **60.** What category of hydrocarbon does C_8H_{14} belong to?

61. Write a structural formula for 2-methylpropene.

62. Write a structural formula for methylcyclopentane.

63. Write a structural formula for 2-ethyl-1-butene.

▲ **64.** Name the following compound:

C—C—C(C)(C—C)—C—C=C

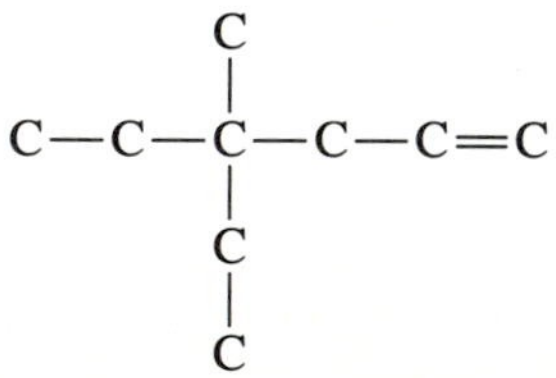

▲ **65.** Name all the possible straight-chain oct*enes*.

66. Rewrite the following structure of naphthalene, replacing the ring structure with the proper numbers of carbon atoms and hydrogen atoms:

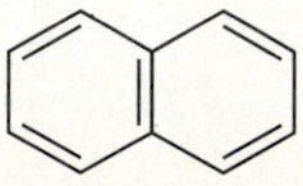

20 Organic Chemistry, Part 2: Classes of Organic Compounds

Learning Goals

After you have studied this chapter, you should be able to:

1. Explain what a functional group is.
2. Discuss or state the general formula, names, structure, and uses of alcohols.
3. Discuss or state the general formula, common names, structure, and uses of ethers.
4. Discuss or state the general formula, names, structure, and uses of aldehydes.
5. Discuss or state the general formula, names, structure, and uses of ketones.
6. Discuss or state the general formula, names, structure, and uses of carboxylic acids.
7. Discuss or state the general formula, names, structure, and uses of esters.
8. Write the product of an esterification reaction.
9. Discuss or state the general formulas, names, structure, and uses of amines.

Introduction

In Chapter 19, we discussed only hydrocarbons—organic compounds made up of carbon and hydrogen. However, we also noted that carbon can form compounds with other elements besides hydrogen. And, just as hydrocarbons are grouped in several series, the non-hydrocarbons are grouped in *classes* of compounds. In this chapter, we briefly survey seven of these classes. The basis for the classification of organic compounds is the *functional group,* which was introduced in Section 19.6.

Table 20.1
Some Classes of Organic Compounds

Class	General formula	Example	Common name	IUPAC name
Alkane	C_nH_{2n+2}	CH_4	Meth*ane* (swamp gas)	Meth*ane*
Alkene	C_nH_{2n} (n = 2 or more)	$CH_2{=}CH_2$	Ethylene	Eth*ene*
		$CH_3-CH{=}CH_2$	Propylene	Prop*ene*
Alkyne	C_nH_{2n-2} (n = 2 or more)	$H-C{\equiv}C-H$	Acetylene	Eth*yne*
		$CH_3-C{\equiv}C-H$	Methylacetylene	Prop*yne*
Alcohol	$R-OH$	CH_3-OH	Methyl alcohol (wood alcohol)	Methan*ol*
		CH_3-CH_2-OH	Ethyl alcohol (grain alcohol)	Ethan*ol*
Ether	$R-O-R$	CH_3-O-CH_3	Dimethyl ether	Methoxymethane
		$CH_3-CH_2-O-CH_2-CH_3$	Diethyl ether	Ethoxyethane
Aldehyde	$R-\underset{\underset{H}{\mid}}{C}{=}O$	$H-\underset{\underset{H}{\mid}}{C}{=}O$	Formaldehyde	Methan*al*
		$CH_3-CH_2-\underset{\underset{H}{\mid}}{C}{=}O$	Propionaldehyde	Propan*al*
Ketone	$R-\underset{\underset{O}{\Vert}}{C}-R$	$CH_3-\underset{\underset{O}{\Vert}}{C}-CH_3$	Dimethyl ketone (acetone)	Propan*one*
		$CH_3-CH_2-\underset{\underset{O}{\Vert}}{C}-CH_2-CH_3$	Diethyl ketone	3-pentan*one*
Carboxylic acid	$R-\underset{\underset{O}{\Vert}}{C}-OH$	$H-\underset{\underset{O}{\Vert}}{C}-OH$	Formic acid	Methan*oic* acid
		$CH_3-\underset{\underset{O}{\Vert}}{C}-OH$	Acetic acid	Ethan*oic* acid
Ester	$R-\underset{\underset{O}{\Vert}}{C}-O-R$	$H-\underset{\underset{O}{\Vert}}{C}-O-CH_3$	Methyl formate	Methylmethanoate
		$CH_3-\underset{\underset{O}{\Vert}}{C}-O-CH_2-CH_3$	Ethyl acetate	Ethylethanoate
Amine	$R-NH_2$	CH_3-NH_2	Methylamine	Aminomethane
		$CH_3-CH_2-NH_2$	Ethylamine	Aminoethane

Note: Remember that n = number of carbon atoms.

20.1 Functional Groups

Learning Goal 1
Functional groups

A **functional group** is *an atom or group of atoms that determines the specific chemical properties of a class of organic compounds*. This definition actually contains three statements. First, all the members of each class of organic compounds have some basic properties in common. Second, all the compounds in a specific class contain the same functional group. And finally, the functional group determines the basic properties of the class.

Other atoms are attached to the functional group. It is these attached atoms (the R groups of Section 19.6) that give a specific compound its individual properties. In other words, we can write the general formula for the compounds in a class as

$$\text{R—functional group}$$

The functional group is *the same* for every compound in the class. The R group (or groups) is *different* for each compound in the class. The R groups are usually alkyl groups.

Table 20.1 shows the classes of compounds that we will discuss in this chapter (those from alcohol down) and their general formulas. For each class, the functional group is the part that is attached to the R or R′s.

20.2 Alcohols

Learning Goal 2
Alcohols

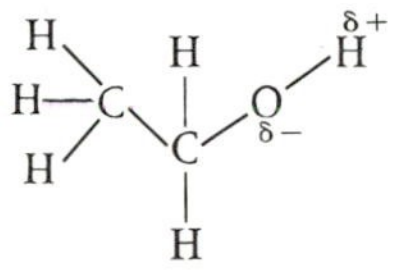

The general formula for an **alcohol** *is* R—OH. The —OH is called the **hydroxyl functional group.** The R group is usually an alkyl group. If the R group consists of only a few carbon atoms, the alcohol tends to be polar because of the polarity of the —OH group (Figure 20.1). As a result, the alcohol dissolves in polar solvents such as water. For example, the alcohol CH_3CH_2OH (ethanol) dissolves in water.

However, if the R group consists of a long chain of nonpolar carbon atoms, the polarity of the —OH group is less obvious. In this case, the alcohol appears to be nonpolar (Figure 20.2) and dissolves in nonpolar solvents like benzene and carbon tetrachloride. For example, the alcohol $CH_3CH_2CH_2CH_2CH_2CH_2OH$ (hexanol) dissolves well in benzene.

H, H, H, H, H, C, C, O, H, δ+, δ−

Figure 20.1
If the R group of an alcohol consists of only a few carbon atoms, the alcohol is polar.

Example 20.1

Determine whether each of the following alcohols is polar or nonpolar:

(a) CH_3OH (Methanol) *(b)* $CH_3—(CH_2)_9—OH$ (Decanol)

Solution

(a) Because methanol has a short R group, it is a polar alcohol.
(b) Decanol has a long R group, so it is nonpolar.

Practice Exercise 20.1 Determine whether each of the following alcohols is polar or nonpolar:

(a) $CH_3—CH_2—OH$ *(b)* $CH_3—(CH_2)_{10}—OH$

Figure 20.2
If the R group of an alcohol consists of a long chain of carbon atoms, the alcohol tends to be nonpolar (because it is mostly an alkane chain) and behaves like a nonpolar alkane.

We can name alcohols in two ways. First, we can use their common names, which we find by adding the word *alcohol* to the alkyl group. For example, CH_3CH_2OH is called *ethyl alcohol*. Second, we can use their IUPAC names, which we find by changing the alkane ending *-ane* to the alcohol ending *-anol*. For example, CH_3CH_2OH is also called *ethanol*.

Example 20.2

Give the common name and the IUPAC name for each of the following alcohols:

(a) $CH_3CH_2CH_2CH_2OH$ *(b)*

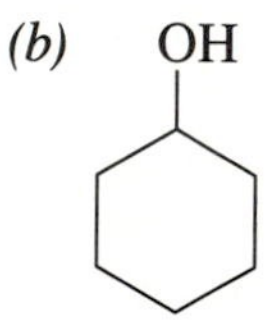

Solution

(a) $CH_3CH_2CH_2CH_2OH$ is a four-carbon alcohol. Therefore its common name is n-*butyl alcohol,* and its IUPAC name is *butanol*.

(b) OH This six-carbon cyclic alcohol has the common name *cyclohexyl alcohol*. Its IUPAC name is *cyclohexanol*.

Practice Exercise 20.2 Give the common name and the IUPAC name for each of the following alcohols:

(a) $CH_3CH_2CH_2OH$ *(b)* OH

Methanol, CH_3OH, is the simplest alcohol. This very poisonous compound is also known as *wood alcohol* because it can be obtained from the heating of wood in the absence of air (a process called *destructive distillation*). A word of warning about methanol or wood alcohol: Death can result from drinking just 30 mL of it, and permanent blindness can result from drinking lesser amounts. During the years of Prohibition in the United States, when alcoholic drinks were prohibited by law, many people drank beverages containing methanol, not realizing the dangers involved.

Ethanol, CH_3CH_2OH, is the alcohol present in wine, liquor, and beer. The ethanol found in these beverages is produced through the process of fermentation, which involves the breakdown of the sugar *glucose*.

$$\underset{\text{Glucose}}{C_6H_{12}O_6} \xrightarrow{\text{Yeast}} 2CH_3CH_2OH + 2CO_2 + \text{energy}$$

A solution that is 70% ethanol by volume acts as an antiseptic by coagulating the proteins of bacteria.

It is interesting to note that ethanol is metabolized in the human body to produce metabolic energy, carbon dioxide, and water. Other organic alcohols are not metabolized to this extent. Their short-circuited metabolism products are highly toxic to humans. The toxic effects of ethanol are related to its concentration in the blood, which is determined by the relative rate of consumption versus the rate of metabolism.

20.3 Ethers

Learning Goal 3
Ethers

The general formula for an **ether** *is* R—O—R′. If the R and R′ groups are the same, we say the ether is **symmetrical.** If the R and R′ groups are different, we say the ether is **unsymmetrical.**

Usually we refer to ethers by using their common names, which are obtained by naming the two alkyl groups in alphabetical order and adding the name *ether*. For example $CH_3CH_2OCH_3$ is called *ethyl methyl ether*.

Example 20.3

Give the common name for each of the following ethers:

(a) $CH_3CH_2OCH_2CH_3$ *(b)* $CH_3CH_2CH_2OCH_3$

Solution

(a) $CH_3CH_2OCH_2CH_3$ has an ethyl group on both sides of the oxygen. Therefore the name of this ether is *diethyl ether,* or simply *ethyl ether* (it is not imperative to use the prefix *di-* on symmetrical ethers).

(b) $CH_3CH_2CH_2OCH_3$ has a methyl group on one side of the oxygen and a propyl group on the other. We list these groups alphabetically and add the word *ether*. Therefore the name of this compound is *methyl propyl ether*.

Practice Exercise 20.3 Give the common name for each of the following ethers:

(a) $CH_3—CH_2—CH_2—O—CH_2—CH_2—CH_3$
(b) $CH_3—CH_2—O—CH_3$

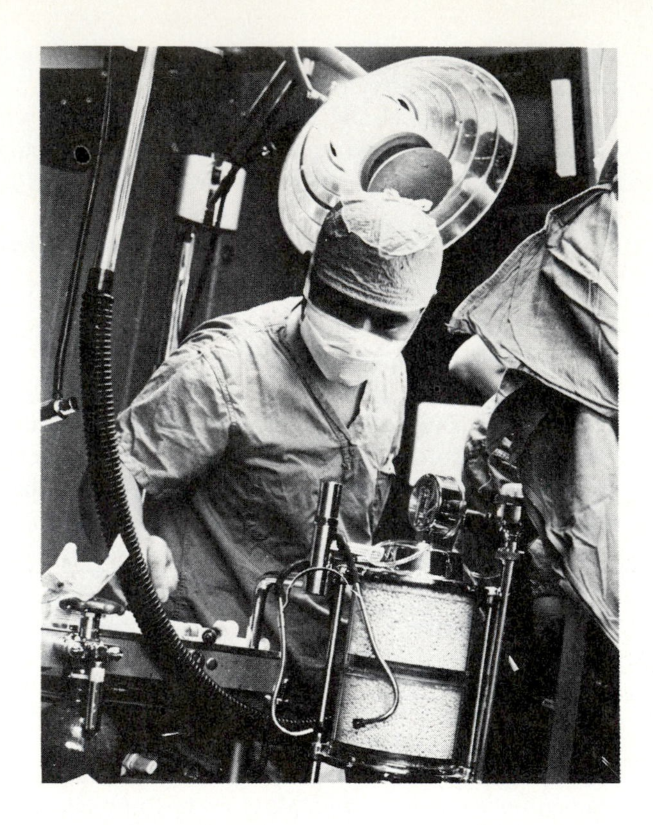

Figure 20.3
Ethyl ether was one of the earliest anesthetics used in hospitals. It is still used in certain procedures today.
(Courtesy St. Luke's–Roosevelt Hospital Center, New York)

Ethers tend to be very flammable, and many of them have some type of anesthetic properties. They also tend to vaporize easily.

Ethyl ether, $CH_3CH_2OCH_2CH_3$, is sometimes just called *ether*. It is probably the best-known member of this class of compounds. It was, and still is, used in hospitals as an anesthetic for various surgical procedures (Figure 20.3). This anesthetic is medically safe because there is a significant difference between the dosage needed to anesthetize and the dosage that will kill.

Ethylene oxide, a cyclic ether, is a gas at room temperature. It is a cyclic ether that can be used as a *fumigant* (disinfectant) for foodstuffs and as a fungicide in agriculture.

$$\overset{\quad O}{\diagup\ \ \diagdown}$$
$$CH_2 - CH_2$$

Chemical Frontiers

Dissolving Gallstones Without Surgery

Gallstones are deposits formed in the gallbladder from excess cholesterol in the body. Surgery followed by a long recuperative period is usually required for their removal. Recently, a group of researchers at the Mayo Clinic in

Rochester, Minnesota, used a chemical related to the anesthetic ethyl ether to dissolve gallstones in a group of 75 patients. This innovative technique based on an application of solution chemistry, proved to be highly effective by eliminating gallstones without requiring surgery.

To prepare for the procedure, patients were given iodine tablets with their evening meals for two days. The gallbladder, one of the organs that absorbs iodine, could now be made visible by the iodine using an x-ray device called a fluoroscope. Next, a small incision was made on the patient's right side, and a catheter (a small hollow tube) was inserted. Using the fluoroscope, the researchers guided the catheter into the gallbladder. Then the ether compound (known as methyl-tert-butyl ether) was injected into the gallbladder through the catheter and immediately suctioned. The procedure was repeated several times over a five-hour period.

The results of the study showed that in 75 patients more than 95% of the gallstones had been dissolved and flushed out of the body. Although 7 patients had enough debris left in their gallbladders to continue causing symptoms, and 6 patients eventually required surgery, the technique proved highly successful for the majority of patients. The application of solution chemistry to this problem may lead the way for researchers to develop similar procedures for other such ailments in the near future.

20.4 Aldehydes

Learning Goal 4
Aldehydes

The general formula for an **aldehyde** *is*

$$\mathrm{R{-}\underset{\underset{\displaystyle O}{\|}}{C}{-}H}$$

The functional group

$$\mathrm{{-}\underset{\underset{\displaystyle O}{\|}}{C}{-}H}$$

is the **aldehyde group.** Many aldehydes have attractive odors and are used in perfumes. Some aldehydes have pleasant tastes and are used as flavoring agents.

We can name aldehydes in two ways. First, we can use the common name, which in many cases is simply the name of the alkyl group plus the word *aldehyde*. For example,

$$\mathrm{CH_3CH_2\underset{\underset{\displaystyle O}{\|}}{C}H}$$

is called *propionaldehyde*. Second, we can use the IUPAC name, which we derive by changing the alkane ending *-e* to the aldehyde ending *-al*. For example,

$$\underset{\displaystyle \overset{\|}{O}}{CH_3CH_2CH}$$

is called *propanal*.

Example 20.4

Give the IUPAC name and the common name for each of the following aldehydes:

(a) $CH_3CH_2CH_2\underset{\overset{\|}{O}}{C}H$ *(b)* $CH_3CH_2CH_2CH_2CH_2\underset{\overset{\|}{O}}{C}H$

Solution

(a)

$$CH_3CH_2CH_2\underset{\overset{\|}{O}}{C}H$$

is a four-carbon aldehyde. Therefore the common name of this compound is *butyraldehyde;* its IUPAC name is *butanal*.

(b)

$$CH_3CH_2CH_2CH_2CH_2\underset{\overset{\|}{O}}{C}H$$

is a six-carbon aldehyde that has the common name *hexaldehyde;* its IUPAC name is *hexanal*. (This compound also has another common name, *caproaldehyde,* because it is the aldehyde of caproic acid, which is found in goat fat. *Caper* is the Latin word for "goat.")

Practice Exercise 20.4 Give the IUPAC name and the common name for each of the following aldehydes:

(a) $CH_3-CH_2-\underset{\overset{\|}{O}}{C}-H$

(b) $CH_3-CH_2-CH_2-CH_2-CH_2-CH_2-\underset{\overset{\|}{O}}{C}-H$

Methanal is the simplest and best-known aldehyde.

$$H-\underset{\overset{\|}{O}}{C}-H$$

You are more likely to recognize it as *formaldehyde*. Although it is a gas at room temperature, it is usually dissolved in water and sold in a 40%-by-mass–volume solution. In this form it is used as a germicide, disinfectant, and preservative whose common name is *formalin*. Biologists sometimes use formalin solution to preserve tissues of various organisms. Formaldehyde can be very irritating to your eyes, nose, and throat, and it has a pungent odor. It is also a suspected carcinogen.

Vanillin is used both commercially and medically as a flavoring agent.

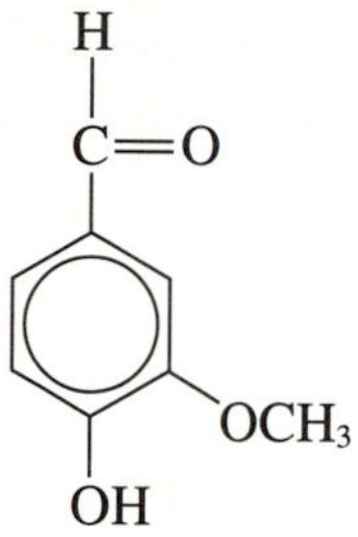

$$\text{vanillin: benzene ring bearing } -\text{CHO (H}-\text{C}{=}\text{O)},\ -\text{OCH}_3,\ -\text{OH}$$

It has a very pleasant odor and flavor that can cover up the harsh tastes of other substances. Vanillin gives us the characteristic taste of vanilla ice cream.

20.5 Ketones

Learning Goal 5
Ketones

The general formula for a **ketone** *is*

$$\mathrm{R-\overset{\displaystyle O}{\overset{\|}{C}}-R'}$$

The functional group

$$\mathrm{-\overset{\displaystyle O}{\overset{\|}{C}}-}$$

is called the **carbonyl** or **keto group.** Ketones are very similar in structure to aldehydes. However, ketones have two R groups attached to the carbonyl group, whereas aldehydes have only one R group and a hydrogen atom:

$$\underset{\text{Ketone}}{\mathrm{R-\overset{\displaystyle O}{\overset{\|}{C}}-R'}} \qquad \underset{\text{Aldehyde}}{\mathrm{R-\overset{\displaystyle O}{\overset{\|}{C}}-H}}$$

We can name ketones in two ways. First, we can use the common name, which we determine simply by naming each of the alkyl groups in alphabetical order and then adding the word *ketone*. For example,

$$\mathrm{CH_3CH_2-\overset{\displaystyle O}{\overset{\|}{C}}-CH_3}$$

is called *ethyl methyl ketone*. Second, we can use the IUPAC name, which we form by counting the number of carbon atoms in the compound (including those in the carbonyl group) and changing the *-e* of the alkane ending *-ane* to the ketone ending *-one*. For example,

$$CH_3CH_2-\underset{\underset{\displaystyle O}{\|}}{C}-CH_3$$

is called *butanone*. (*Note:* Sometimes we have to mention the number of the carbon atom that is attached to the oxygen, if isomers are possible. Example 20.5 shows what we mean.)

Example 20.5

Give the IUPAC name and the common name for each of the following ketones:

(a) $CH_3-\underset{\underset{\displaystyle O}{\|}}{C}-CH_3$ *(b)* $CH_3CH_2CH_2-\underset{\underset{\displaystyle O}{\|}}{C}-CH_2CH_3$

(c) $CH_3CH_2CH_2CH_2-\underset{\underset{\displaystyle O}{\|}}{C}-CH_3$

Solution

(a)

$$CH_3-\underset{\underset{\displaystyle O}{\|}}{C}-CH_3$$

is a compound whose IUPAC name is *propanone*. Its common name is *dimethyl ketone*. However, this compound is probably best known as *acetone*.

(b)

$$CH_3CH_2CH_2-\underset{\underset{\displaystyle O}{\|}}{C}-CH_2CH_3$$

is a six-carbon ketone; therefore its IUPAC name would be *hexanone*. However, other hexanones are possible [see part (c) of this example]. Thus we must number the carbon chain in such a way that the carbon atom attached to the oxygen atom will have the lowest possible number:

$$\overset{6}{C}H_3\overset{5}{C}H_2\overset{4}{C}H_2-\underset{\underset{\displaystyle O}{\|}}{\overset{3}{C}}-\overset{2}{C}H_2\overset{1}{C}H_3$$

In this case, the correct IUPAC name is *3-hexanone*. The common name of this compound is *ethyl propyl ketone*, which we derive from the alkyl groups attached to the carbonyl group.

(c)

$$\overset{6}{CH_3}\overset{5}{CH_2}\overset{4}{CH_2}\overset{3}{CH_2}-\underset{\underset{\displaystyle O}{\|}}{\overset{2}{C}}-\overset{1}{CH_3}$$

is a compound that is also a six-carbon ketone. The IUPAC name is *2-hexanone*. However, the common name of this compound is *butyl methyl ketone*.

Practice Exercise 20.5 Give the IUPAC name and the common name for each of the following ketones:

(a) $CH_3-CH_2-\underset{\underset{\displaystyle O}{\|}}{C}-CH_2-CH_3$

(b) $CH_3-CH_2-CH_2-CH_2-\underset{\underset{\displaystyle O}{\|}}{C}-CH_2-CH_3$

(c) $CH_3-CH_2-\underset{\underset{\displaystyle O}{\|}}{C}-CH_3$

Propanone (also known as *acetone*) is the simplest and best-known ketone. It is an excellent solvent for many organic substances. It is found in paint removers and nail-polish removers. Persons with uncontrolled diabetes mellitus produce acetone in their blood and give off an acetone smell when they exhale. A physician can easily detect this *acetone breath* in an examination of such persons. Acetone has a unique odor and sweetish taste.

Methyl phenyl ketone, also known as *acetophenone,*

$$C_6H_5-\overset{\overset{\displaystyle O}{\|}}{C}-CH_3$$

has an orange-blossom-like fragrance and therefore is used in the making of some perfumes. It has also been used in medicine as a hypnotic agent.

20.6 Carboxylic Acids

Learning Goal 6
Carboxylic acids

The general formula for a **carboxylic acid** *is*

$$R-\underset{\underset{\displaystyle O}{\|}}{C}-OH$$

The functional group

$$-\underset{\underset{\displaystyle O}{\|}}{C}-OH$$

is called the **carboxyl** group. This group acts like an acid because it releases hydrogen ions in aqueous solutions:

$$R-\underset{\underset{\displaystyle O}{\|}}{C}-OH + H_2O \rightleftharpoons R-\underset{\underset{\displaystyle O}{\|}}{C}-O^- + H_3O^+$$

This ionization occurs only to a slight degree, so carboxylic acids are weak acids.

There are two ways of naming carboxylic acids. First, we can use the common name, which (in most cases) is the common root plus the ending *-ic*, followed by the word *acid*. For example,

$$CH_3CH_2-\underset{\underset{\displaystyle O}{\|}}{C}-OH$$

is called *propionic acid*. Second, we can use the IUPAC name, which is the alkane name with the final *-e* changed to *-oic*, followed by the word *acid*. For example,

$$CH_3CH_2-\underset{\underset{\displaystyle O}{\|}}{C}-OH$$

is called *propanoic acid*.

Example 20.6

Give the IUPAC name and the common name for each of the following carboxylic acids:

(a) $CH_3CH_2CH_2-\underset{\underset{\displaystyle O}{\|}}{C}-OH$ *(b)* $CH_3-\underset{\underset{\displaystyle O}{\|}}{C}-OH$

Solution

(a)

$$CH_3CH_2CH_2-\underset{\underset{\displaystyle O}{\|}}{C}-OH$$

is a four-carbon carboxylic acid; therefore its IUPAC name is *butanoic acid*. The common name of this acid is *butyric acid*.

(b)

$$CH_3-\underset{\underset{\displaystyle O}{\|}}{C}-OH$$

The IUPAC name of this compound, which is a two-carbon carboxylic acid, is *ethanoic acid*. Its common name is *acetic acid*.

Practice Exercise 20.6 Give the IUPAC name and the common name for each of the following carboxylic acids:

(a) 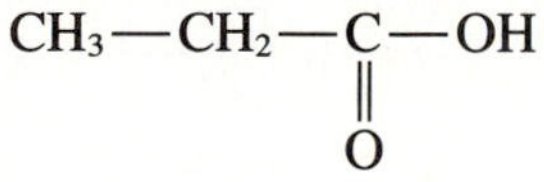 $CH_3{-}CH_2{-}\underset{\underset{O}{\|}}{C}{-}OH$ *(b)* 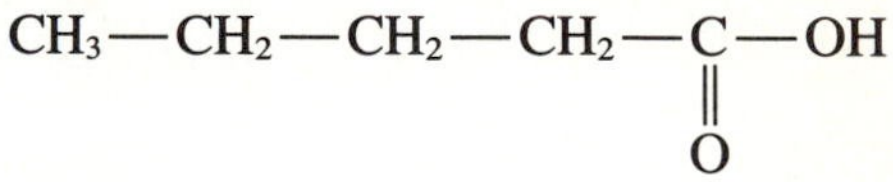 $CH_3{-}CH_2{-}CH_2{-}CH_2{-}\underset{\underset{O}{\|}}{C}{-}OH$

Methanoic acid is the simplest carboxylic acid:

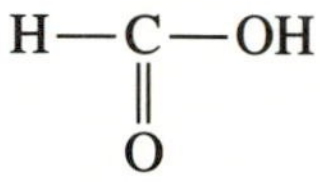

$$H{-}\underset{\underset{O}{\|}}{C}{-}OH$$

Its common name is *formic acid*. Ants and bees produce this acid and inject it into their victims when they bite or sting. Once a victim is stung by a bee, it is the formic acid that causes the skin to become inflamed.

Ethanoic acid, also known as *acetic acid,*

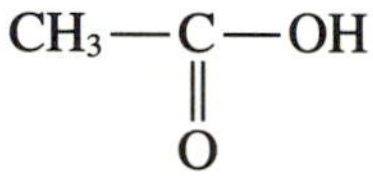

$$CH_3{-}\underset{\underset{O}{\|}}{C}{-}OH$$

is the substance found in vinegar. We discussed some of the properties of this acid in Chapter 16.

Butanoic acid, also known as *butyric acid,*

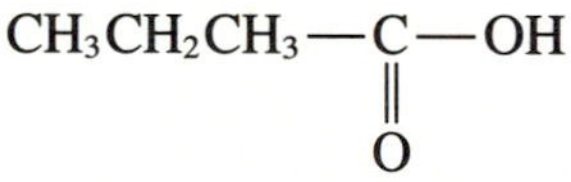

$$CH_3CH_2CH_3{-}\underset{\underset{O}{\|}}{C}{-}OH$$

is produced when butter becomes rancid. In its pure form, it has a very pungent odor. It is also one of the substances that causes body odor.

20.7 Esters

Learning Goal 7
Esters

The general formula for an **ester** *is*

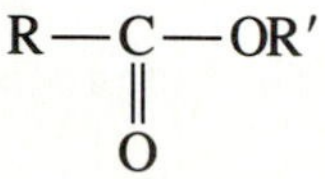

$$R{-}\underset{\underset{O}{\|}}{C}{-}OR'$$

Some of the most common flavors and fragrances are due to the properties of esters (Table 20.2).

Esters can be formed by the reaction of an alcohol with a carboxylic acid. For example,

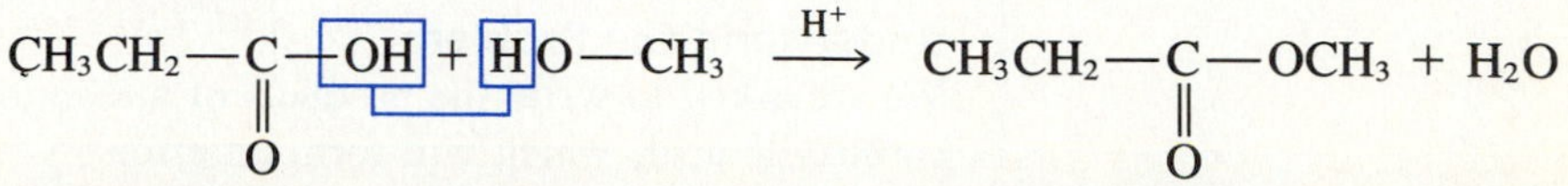

$$\underset{\text{Propanoic acid}}{CH_3CH_2{-}\underset{\underset{O}{\|}}{C}{-}\boxed{OH}} + \underset{\text{Methanol}}{\boxed{H}O{-}CH_3} \xrightarrow{H^+} \underset{\text{An ester}}{CH_3CH_2{-}\underset{\underset{O}{\|}}{C}{-}OCH_3} + H_2O$$

Table 20.2
Some Esters Used for Their Flavoring or Fragrance

Name	Formula	Flavor or fragrance
*Amyl acetate	$CH_3—(CH_2)_4—O—C(=O)—CH_3$	Banana
Amyl butyrate	$CH_3—(CH_2)_4—O—C(=O)—CH_2CH_2CH_3$	Apricot
Ethyl butyrate	$CH_3CH_2—O—C(=O)—CH_2CH_2CH_3$	Pineapple
Ethyl formate	$CH_3CH_2—O—C(=O)—H$	Rum
Octyl acetate	$CH_3—(CH_2)_7—O—C(=O)—CH_3$	Orange

*For a chain of five carbon atoms, the word *amyl* is often used as the root.

Note how the acid and alcohol parts combine to produce the ester:

$$\underbrace{R—\overset{}{C}(=O)}_{\text{Acid part}}—\underbrace{OR'}_{\text{Alcohol part}}$$

Learning Goal 8
Esterification reactions

The reaction between an alcohol and a carboxylic acid to produce an ester is called an **esterification reaction.**

Example 20.7

Write the products of the following esterification reaction:

$$CH_3CH_2CH_2—C(=O)—OH + HO—CH_2CH_3 \xrightarrow{H^+}$$

Solution

Understand the Problem

We are asked to write the products of a reaction between an alcohol and a carboxylic acid, which will form an ester.

Devise a Plan

We'll "remove" the OH and H, as was done in the preceding text, and combine the rest.

Carry Out the Plan

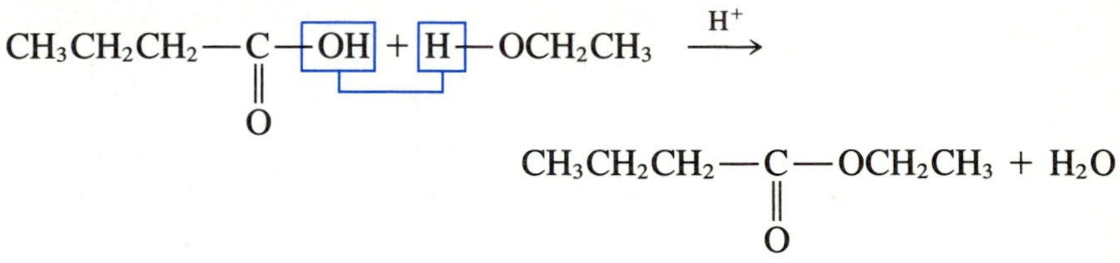

$$CH_3CH_2CH_2-\underset{\underset{O}{\|}}{C}-\boxed{OH} + \boxed{H}-OCH_2CH_3 \xrightarrow{H^+} CH_3CH_2CH_2-\underset{\underset{O}{\|}}{C}-OCH_2CH_3 + H_2O$$

Look Back

Be certain the proper product has been recorded.

Practice Exercise 20.7 Write the products of the following esterification reaction:

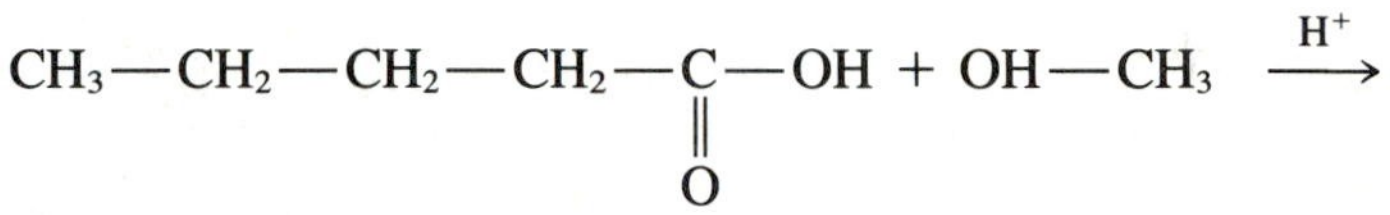

$$CH_3-CH_2-CH_2-CH_2-\underset{\underset{O}{\|}}{C}-OH + OH-CH_3 \xrightarrow{H^+}$$

There are two ways of naming esters. First, we can use the common name, which we find by adding the name of the alkane part of the alcohol group to the common name of the acid, and changing the *-ic* ending to *-ate*. For example,

$$CH_3CH_2-\underset{\underset{O}{\|}}{C}-OCH_3$$

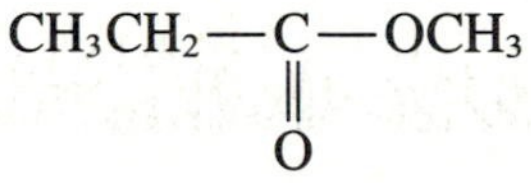

is called *methyl propionate* (written as two words). Second, we can use the IUPAC name, which we get by adding the name of the alkane part of the alcohol group to the IUPAC name of the acid, and changing the *-ic* ending to *-ate*. For example,

$$CH_3CH_2-\underset{\underset{O}{\|}}{C}-OCH_3$$

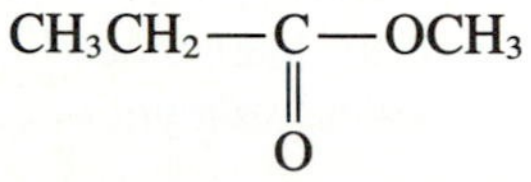

is called *methylpropanoate* (written as one word).

Example 20.8

Give the IUPAC name and the common name for each of the following esters:

(a) $CH_3CH_2-\underset{\underset{\displaystyle O}{\|}}{C}-OCH_2CH_3$ *(b)* $CH_3-\underset{\underset{\displaystyle O}{\|}}{C}-OCH_2CH_2CH_3$

Solution

(a)

$$CH_3CH_2-\underset{\underset{\displaystyle O}{\|}}{C}-OCH_2CH_3$$

is the ester of ethyl alcohol and propionic acid. Therefore its common name is *ethyl propionate*. Its IUPAC name is *ethylpropanoate*.

(b)

$$CH_3-\underset{\underset{\displaystyle O}{\|}}{C}-OCH_2CH_2CH_3$$

is the ester of propyl alcohol and acetic acid. Therefore its common name is *propyl acetate*. Its IUPAC name is *propylethanoate*. (Remember, the IUPAC name of acetic acid is ethanoic acid.)

Practice Exercise 20.8 Give the IUPAC name and the common name for each of the following esters:

(a) $CH_3-CH_2-\underset{\underset{\displaystyle O}{\|}}{C}-OCH_3$

(b) $CH_3-CH_2-CH_2-\underset{\underset{\displaystyle O}{\|}}{C}-O-CH_2-CH_2-CH_3$

Ethyl acetate is a compound that is used in artificial fruit essences:

$$CH_3-\underset{\underset{\displaystyle O}{\|}}{C}-OCH_2CH_3$$

It is also an excellent solvent for varnishes, lacquers, and airplane lubricants. In addition, this compound is used in manufacturing smokeless powder, artificial leather, and perfumes and in cleaning textiles.

Methyl salicylate, commonly known as *oil of wintergreen,*

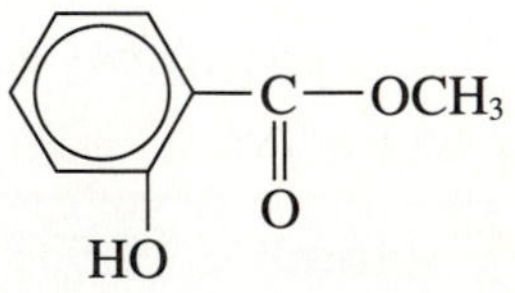

$$(HO)C_6H_4-\underset{\underset{\displaystyle O}{\|}}{C}-OCH_3$$

is the methyl ester of salicylic acid,

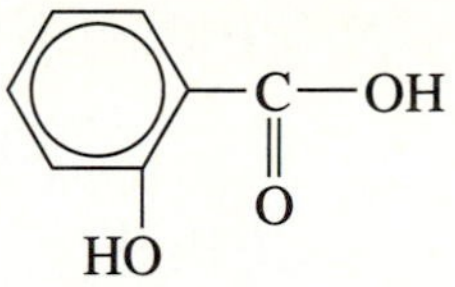

$$\text{Salicylic acid: } C_6H_4(OH)-C(=O)-OH$$

Salicylic acid

It has a pleasant spearmint-like odor. For this reason, it is used (in very low concentrations) in candies and perfumes. Because it also creates a sensation of warmth in the skin, which makes sore muscles feel better, methyl salicylate is found in many liniments. In higher concentrations it is toxic.

20.8 Amines

Learning Goal 9
Amines

There are three general formulas for **amines:**

1. $R-NH_2$ (primary amine)
2. $R-NH-R'$ (secondary amine)
3. $R-N(R')-R''$ (tertiary amine)

Amines are, essentially, ammonia (NH_3) with the hydrogen atoms replaced by R groups. If one R group has replaced a hydrogen atom, the compound is a **primary amine.** If two R groups have replaced two hydrogen atoms, the compound is a **secondary amine.** And if three R groups have replaced three hydrogen atoms, the compound is a **tertiary amine** (Figure 20.4).

Example 20.9

Determine whether each of the following compounds is a primary, secondary, or tertiary amine:

(a) $CH_3-N(CH_3)-CH_3$ *(b)* $CH_3CH_2-NH-CH_2CH_3$ *(c)*  $C_6H_5-NH_2$

Solution

(a) $CH_3-N(CH_3)-CH_3$ This compound has three R groups attached to the nitrogen atom, so it is a tertiary amine.

(b) $CH_3CH_2—NH$ (with CH_2CH_3 attached to N)

This compound has two R groups attached to the nitrogen atom, so it is a secondary amine.

(c) 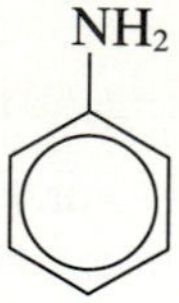 $C_6H_5NH_2$ (benzene ring with NH_2)

This compound has one R group attached to the nitrogen atom, so it is a primary amine.

Practice Exercise 20.9 Determine whether each of the following compounds is a primary, secondary, or tertiary amine:

(a) $CH_3—N—CH_3$ (with CH_3 attached to N) *(b)* $CH_3—N—H$ (with CH_3 attached to N) *(c)* $CH_2—NH_2$

We can name amines in two ways. First, we can use the common name, which is the alkane name with *-yl* substituted for the *-ane,* plus the ending *amine*. For example, CH_3NH_2 is called *methylamine*. If two R groups are attached to the nitrogen atom, we name the groups in alphabetical order. For example,

$CH_3—NH$ (with CH_3 attached to N)

is called *dimethylamine,* and

$CH_3CH_2—NH$ (with CH_3 attached to N)

is called *ethylmethylamine*. Second, we can use the IUPAC name, which for primary amines involves naming the $—NH_2$ group as an amino group along the alkane chain and numbering the carbon atoms to locate its position. For example,

$\overset{1}{CH_3}—\overset{2}{CH}—\overset{3}{CH_3}$ (with NH_2 attached to C-2)

H—N—H (with H attached to N) — Ammonia

R—N—H (with H attached to N) — Primary amine

R—N—H (with R′ attached to N) — Secondary amine

R—N—R″ (with R′ attached to N) — Tertiary amine

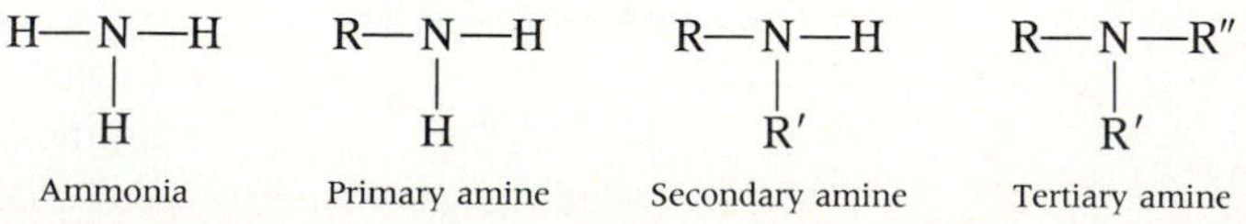

Figure 20.4
The relationship of primary, secondary, and tertiary amines to ammonia

is called *2-aminopropane*. For secondary and tertiary amines, the IUPAC rules become more complex and will not be covered here. Moreover, most people use the common names for amines.

Example 20.10

Give the common name for each of the following amines. If possible, give the IUPAC name also.

(a) $CH_3CH_2—NH_2$ *(b)* $CH_3CH_2CH_2—N(CH_3)—CH_3$

(c) $CH_3CH_2CH_2CH_2—CH(NH_2)—CH_2CH_3$

Solution

(a) $CH_3CH_2—NH_2$ is a two-carbon amine. Therefore the IUPAC name of this compound is *aminoethane*. Its common name is *ethylamine*.

(b)

$$CH_3CH_2CH_2—\underset{\displaystyle CH_3}{\underset{|}{N}}—CH_3$$

is a tertiary amine that consists of two methyl groups and a propyl group. Therefore the common name of this compound is *dimethylpropylamine*.

(c)

$$\overset{7}{C}H_3\overset{6}{C}H_2\overset{5}{C}H_2\overset{4}{C}H_2—\overset{3}{C}H(NH_2)—\overset{2}{C}H_2\overset{1}{C}H_3$$

is a primary amine that consists of seven carbon atoms, and the amino group is on the third carbon atom. Therefore the IUPAC name of this compound is *3-aminoheptane*.

Practice Exercise 20.10 Give the common name for each of the following amines. If possible, give the IUPAC name also.

(a) $CH_3CH_2—NH_2$ *(b)* $CH_3—N(CH_3)—CH_3$ *(c)* $CH_3CH_2—N(CH_3)—H$

The compound *1,5-diaminopentane*,

$$H_2N—CH_2CH_2CH_2CH_2CH_2—NH_2$$

is also called *1,5-pentanediamine*, but it is better known as *cadaverine*. This compound is produced by decaying organisms. Such amines are called *ptomaines* (pronounced "TOE-mains") and, in fact, can also be formed by the action of bacteria on meat and fish (thus the term *ptomaine poisoning*). Cadaverine has a very pungent odor; it also acts as a base.

The compound 1-phenyl-2-aminopropane,

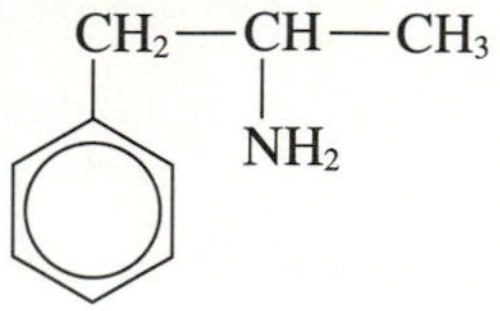

$$\begin{array}{ccccc} CH_2 & - & CH & - & CH_3 \\ | & & | & & \\ C_6H_5 & & NH_2 & & \end{array}$$

also called *(phenylisopropyl)amine* but better known as *amphetamine* (or *benzedrine*), acts as a stimulant to the central nervous system. This compound and others related to it are called "uppers" because of their ability to keep one active and awake. These substances are dangerous because they can cause drug dependence. In other words, after a person has taken them for a while, he or she must continue to take them in order to stay alert. Amphetamine has also been used in nasal inhalers to relieve nasal congestion in people who have colds.

Summary

The members of a class of organic compounds are composed of a functional group (which is the same for each compound in the class) and one or more R groups (which differ among the members of each class). The functional group consists of one or more atoms and gives the class of compounds its specific chemical properties. The R groups are usually alkyl groups, and they give the individual compounds their specific properties.

The *alcohols* all contain the —OH hydroxyl group. Their general formula is R—OH. Alcohols with short R groups, consisting of only a few carbon and hydrogen atoms, are polar. Those with long R groups are nonpolar.

Ethers have the general formula R—O—R′. They tend to be extremely flammable, and many have anesthetic properties.

Aldehydes have the general formula

$$\begin{array}{ccccc} R & - & C & - & H \\ & & \| & & \\ & & O & & \end{array}$$

Some are used in perfumes and as flavoring agents. Formaldehyde (which is not used in perfumes) is probably the best-known member of this class. It is used as a germicide, disinfectant, and tissue preservative.

Ketones have the general formula

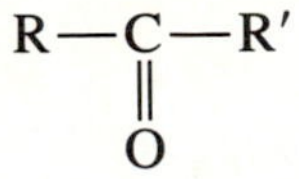

$$\begin{array}{c} R-\underset{\underset{\displaystyle O}{\|}}{C}-R' \end{array}$$

in which the functional group

$$-\underset{\underset{\displaystyle O}{\|}}{C}-$$

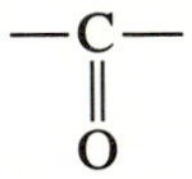

is called the carbonyl or keto group. The male hormone testosterone and the female hormone progesterone are ketones.

The general formula for the *carboxylic acids* is

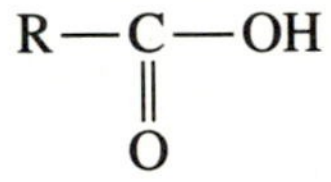

$$R-\underset{\underset{\displaystyle O}{\|}}{C}-OH$$

in which the functional group is called the carboxyl group. These compounds ionize to a slight degree in water.

Esters have the general formula

$$R-\underset{\underset{\displaystyle O}{\|}}{C}-OR'$$

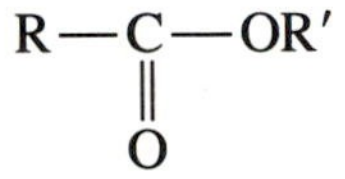

They are formed in the reaction of an alcohol and a carboxylic acid, which is called an esterification reaction. Aspirin is an ester.

Amines are, essentially, ammonia in which R groups have replaced one, two, or all three hydrogen atoms. In a primary amine, with the general formula $R-NH_2$, one hydrogen atom has been replaced by an R group. In secondary and tertiary amines, two and three hydrogen atoms, respectively, have been replaced.

The compounds in all seven of these classes have both common names and IUPAC names. The IUPAC name of a compound is derived from the name of the hydrocarbon that the compound resembles in structure.

Key Terms

alcohol **(20.2)**
aldehyde **(20.4)**
aldehyde group **(20.4)**
amine **(20.8)**
carbonyl (keto) group **(20.5)**
carboxyl group **(20.6)**
carboxylic acid **(20.6)**
ester **(20.7)**
esterification reaction **(20.7)**
ether **(20.3)**
functional group **(20.1)**
hydroxyl group **(20.2)**
ketone **(20.5)**
primary amine **(20.8)**
secondary amine **(20.8)**
symmetrical ether **(20.3)**
tertiary amine **(20.8)**
unsymmetrical ether **(20.3)**

Self-Text Exercises

Learning Goal 1: Functional Groups

▲ **1.** Define the term *functional group*.

▲ **2.** What is the difference between a functional group and an R group?

3. Circle the functional group in each of the following compounds, and state what it is called:
(a) CH_3OH *(b)* $CH_3CH_2OCH_3$
(c) $CH_3\overset{\overset{\displaystyle O}{\|}}{C}H$ *(d)* $CH_3CH_2{-}\overset{\overset{\displaystyle O}{\|}}{C}{-}OH$

4. Circle the functional group in each of the following compounds, and state what it is called:
(a) CH_3CH_2OH *(b)* CH_3OCH_3
(c) $CH_3CH_2\overset{\overset{\displaystyle O}{\|}}{C}H$ *(d)* $CH_3{-}\underset{\underset{\displaystyle CH_3}{|}}{N}{-}CH_3$

Learning Goal 2: Alcohols

▲ **5.** Write the general formula for an alcohol.

6. Complete this sentence: The —OH group is called the ________ functional group.

7. Is each of the following alcohols polar or nonpolar?
(a) $CH_3CH_2CH_2CH_2CH_2CH_2CH_2OH$
(b) CH_3CH_2OH

8. Is each of the following alcohols polar or nonpolar?
(a) $CH_3{-}(CH_2)_{12}{-}CH_2OH$ *(b)* $CH_3{-}OH$

9. Give the IUPAC name or the common name for each of the following alcohols:
(a) $CH_3CH_2CH_2OH$
(b) $CH_3CH_2CH_2CH_2CH_2CH_2OH$

10. Give the IUPAC name or the common name for each of the alcohols listed in Self-Test Exercise 8.

11. Give an example of an important alcohol, and state its uses.

▲ **12.** Why is it dangerous to drink methanol even once, but not so dangerous to drink small amounts of ethanol occasionally?

13. Use the names of the following alcohols to write their structures:
(a) 3-hexanol *(b)* butyl alcohol
(c) 2-octanol *(d)* methyl alcohol

14. Use the names of the following alcohols to write their structures:
(a) 2-hexanol *(b)* ethyl alcohol
(c) 2-pentanol *(d)* propyl alcohol

Learning Goal 3: Ethers

▲ **15.** Give the general formula for an ether.

▲ **16.** Distinguish between a symmetrical and an unsymmetrical ether.

▲ **17.** What is the common name of each of the following ethers?
(a) $CH_3CH_2{-}O{-}CH_2CH_2CH_2CH_3$
(b) $CH_3CH_2{-}O{-}CH_3$

▲ **18.** What is the common name of each of the following ethers?
(a) $CH_3{-}O{-}CH_2CH_2CH_2CH_3$
(b) $CH_3CH_2{-}O{-}CH_2CH_3$

19. Use the common name of each of the following ethers to write its structure:
(a) methyl butyl ether *(b)* dimethyl ether

20. Use the common name of each of the following ethers to write its structure.
(a) ethyl methyl ether *(b)* dibutyl ether

▲ **21.** Give an example of an important ether, and state its uses.

▲ **22.** Name the ether that is commonly used in hospitals as an anesthetic for various surgical procedures.

Learning Goal 4: Aldehydes

▲ **23.** Write the general formula for an aldehyde.

▲ **24.** Use Lewis-dot notation to show the bonding in a molecule of formaldehyde, HCHO.

25. Give the common name or the IUPAC name for each of the following aldehydes:
(a) $CH_3{-}(CH_2)_4{-}\overset{\overset{\displaystyle O}{\|}}{C}H$ *(b)* $CH_3CH_2{-}\overset{\overset{\displaystyle O}{\|}}{C}H$

26. Give the common name or the IUPAC name for each of the following aldehydes:
(a) $CH_3{-}(CH_2)_3{-}\overset{\overset{\displaystyle O}{\|}}{C}{-}H$ *(b)* $CH_3{-}\overset{\overset{\displaystyle O}{\|}}{C}{-}H$

27. Write the structure of each of the following aldehydes from its name:
(*a*) pentanal (*b*) acetaldehyde

28. Write the structure of each of the following aldehydes from its name:
(*a*) hexanal (*b*) butanal

29. Give an example of an important aldehyde, and state its uses.

30. Name the aldehyde that gives vanilla ice cream its flavor.

Learning Goal 5: Ketones

▲ 31. What is the general formula for a ketone?

▲ 32. Write the general structure of a carbonyl group.

33. Give the IUPAC name or the common name for each of the following ketones:

(*a*) $CH_3CH_2-\underset{\underset{O}{\|}}{C}-CH_2CH_3$

(*b*) $CH_3CH_2CH_2CH_2-\underset{\underset{O}{\|}}{C}-CH_3$

(*c*) 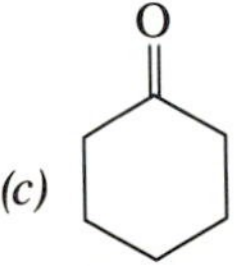

34. Give the IUPAC name or the common name for each of the following ketones:

(*a*) $CH_3-\underset{\underset{O}{\|}}{C}-CH_3$ (*b*) $CH_3CH_2-\underset{\underset{O}{\|}}{C}-CH_3$

(*c*) 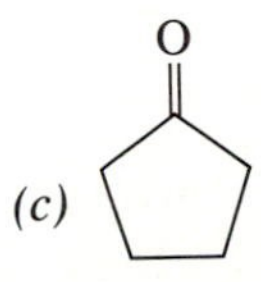

35. Write the structure of each of the following ketones from its name:
(*a*) 4-octanone (*b*) methyl propyl ketone

36. Write the structure of each of the following ketones from its name:
(*a*) 3-octanone (*b*) ethyl methyl ketone

▲ 37. Give an example of an important ketone, and state its uses.

▲ 38. Describe some of the uses of acetone.

Learning Goal 6: Carboxylic Acids

▲ 39. What is the general formula for a carboxylic acid?

▲ 40. What is the formula for the carboxyl group?

41. Give the IUPAC name or the common name for each of the following carboxylic acids:

(*a*) $CH_3-(CH_2)_5-\underset{\underset{O}{\|}}{C}-OH$ (*b*) $H-\underset{\underset{O}{\|}}{C}-OH$

(*c*) $CH_3CH_2CH_2-\underset{\underset{O}{\|}}{C}-OH$

42. Give the IUPAC name or the common name for each of the following carboxylic acids:

(*a*) $CH_3-(CH_2)_3-\underset{\underset{O}{\|}}{C}-OH$ (*b*) $CH_3-\underset{\underset{O}{\|}}{C}-OH$

(*c*) $CH_3-CH_2-\underset{\underset{O}{\|}}{C}-OH$

43. Write the structure of each of the following carboxylic acids from its name:
(*a*) heptanoic acid (*b*) propionic acid

44. Write the structure of each of the following carboxylic acids from its name:
(*a*) butanoic acid (*b*) ethanoic acid

▲ 45. Give an example of an important carboxylic acid, and state its uses.

▲ 46. Name the carboxylic acid that ants and bees produce and inject into their victims when they bite or sting.

Learning Goal 7: Esters

▲ 47. Write the general formula for an ester.

▲ 48. What two classes of organic compounds react to form an ester?

49. Give the IUPAC name or the common name for each of the following esters:

(*a*) $CH_3-(CH_2)_4-\underset{\underset{O}{\|}}{C}-OCH_2CH_3$

(*b*) $CH_3-\underset{\underset{O}{\|}}{C}-OCH_3$

50. Give the IUPAC name or the common name for each of the following esters:

(a) $CH_3-(CH_2)_3-\underset{\underset{\displaystyle O}{\|}}{C}OCH_3$

(b) $CH_3CH_2-\underset{\underset{\displaystyle O}{\|}}{C}OCH_2CH_3$

51. Write the structure of each of the following esters from its name:

(a) ethyl butyrate *(b)* butylethanoate

52. Write the structure of each of the following esters from its name:

(a) ethyl acetate *(b)* propylethanoate

▲ **53.** Give an example of an important ester, and state its uses.

▲ **54.** Which ester smells like a banana?

Learning Goal 8: Esterification Reactions

▲ **55.** Write the products of each of the following esterification reactions:

(a) $CH_3CH_2-\underset{\underset{\displaystyle O}{\|}}{C}-OH + HO-CH_2CH_2CH_3 \xrightarrow{H^+}$

(b) $CH_3-\underset{\underset{\displaystyle O}{\|}}{C}-OH + HO-CH_3 \xrightarrow{H^+}$

▲ **56.** Write the products of each of the following esterification reactions:

(a) $CH_3CH_2CH_2-\underset{\underset{\displaystyle O}{\|}}{C}-OH + OH-CH_3 \xrightarrow{H^+}$

(b) $CH_3CH_2-\underset{\underset{\displaystyle O}{\|}}{C}-OH + OH-CH_2CH_3 \xrightarrow{H^+}$

▲ **57.** Given the products of the following esterification reaction, provide the reactants:

$\_\_\_\_\_ + \_\_\_\_\_ \xrightarrow{H^+}$

$CH_3CH_2CH_2CH_2-\underset{\underset{\displaystyle O}{\|}}{C}-OCH_2CH_3 + H_2O$

▲ **58.** Given the products of the following esterification reaction, provide the reactants:

$\_\_\_\_\_ + \_\_\_\_\_ \xrightarrow{H^+}$

$CH_3CH_2-\underset{\underset{\displaystyle O}{\|}}{C}-OCH_3 + H_2O$

Learning Goal 9: Amines

▲ **59.** What are the three general formulas for amines?

▲ **60.** How are amines related to the compound ammonia?

61. Is each of the following amines primary, secondary, or tertiary?

(a) $CH_3-\underset{\underset{\displaystyle CH_3}{|}}{N}-CH_3$ *(b)* $CH_3CH_2-\underset{\underset{\displaystyle CH_3}{|}}{NH}$

(c) 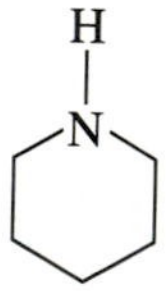 H, N *(d)* 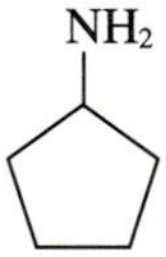 NH_2

62. Is each of the following amines primary, secondary, or tertiary?

(a) $CH_3-\underset{\underset{\displaystyle CH_2CH_3}{|}}{N}-CH_2CH_3$ *(b)* $CH_3-\underset{\underset{\displaystyle H}{|}}{N}-CH_2CH_3$

(c) 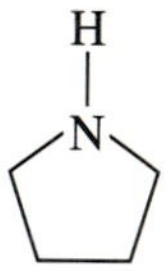 H, N *(d)* 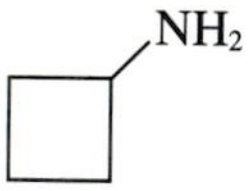 NH_2

63. Give the IUPAC name or the common name for each of the following amines:

(a) $CH_3-\underset{\underset{\displaystyle CH_3}{|}}{N}-CH_3$ *(b)* 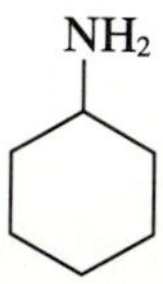 NH_2

(c) 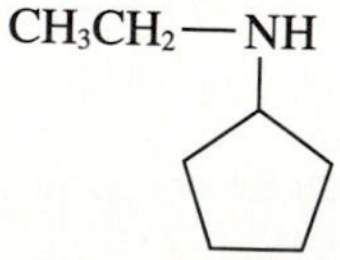 CH_3CH_2-NH

64. Give the IUPAC name or the common name for each of the following amines:

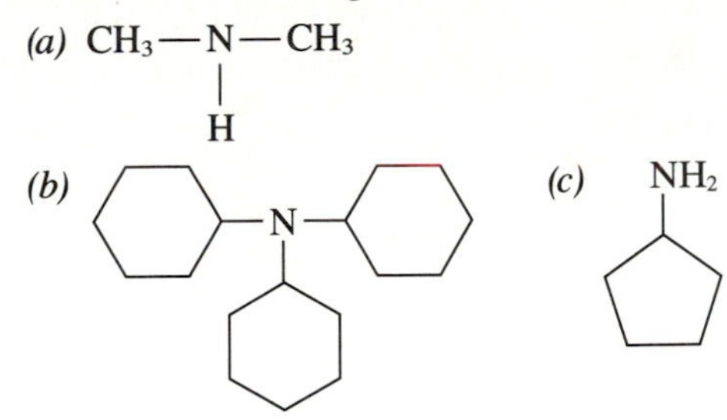

65. Write the structure of each of the following amines from its name:
(*a*) diethylamine (*b*) ethylpropylamine
(*c*) 2-aminopentane (*d*) 3-aminopentane

66. Write the structure of each of the following amines from its name:
(*a*) dimethylamine (*b*) butylethylamine
(*c*) 3-aminoheptane (*d*) 1,5-diaminopentane

▲ **67.** Give an example of an important amine, and state its uses.

▲ **68.** What is amphetamine?

Extra Exercises

▲ **69.** Explain why ethanol (C_2H_5OH) is soluble in water but ethane (C_2H_6) is not.

70. Why is it important to know whether an alcohol product to be used for human consumption contains any methanol?

▲ **71.** Distinguish between a ketone and an aldehyde.

72. Match each class on the left with the correct general formula on the right.

(*a*) amine
(*b*) acid
(*c*) alcohol
(*d*) ester
(*e*) alkane
(*f*) aromatic
(*g*) ether

1. ROH
2. $R-\underset{\underset{O}{\|}}{C}-OH$
3. $R-\underset{\underset{O}{\|}}{C}-OR'$
4. $R-O-R'$
5. $R-NH_2$
6. $R-CH_3$
7. R

73. Isomers are compounds that have the same percentage composition and molecular mass, different structures, and the same molecular formula. Draw all possible isomers for trichlorobenzene. (This is a benzene molecule in which three hydrogens have been replaced by three Cl atoms.)

74. Circle the functional group in each of the following compounds, and state what it is called:
(*a*) CH_3-NH_2 (*b*) CH_3-O-CH_3
(*c*) $CH_3-\underset{\underset{O}{\|}}{C}-OH$ (*d*) $CH_3-\underset{\underset{O}{\|}}{C}-O-CH_3$

75. Write the structure for 2-butanol.

76. Write the structure for 3-aminohexane.

77. Write the structure for dibutyl ether.

78. Name the following compound:

$$CH_3-(CH_2)_5-\underset{\underset{O}{\|}}{C}-OH$$

79. Write the formula for propionaldehyde.

80. Write the formula for 2-pentanone.

81. Write the formula for butanoic acid.

82. Name each of the following compounds:
(*a*) $H-\underset{\underset{O}{\|}}{C}-OCH_2CH_3$ (*b*) $H-\underset{\underset{O}{\|}}{C}-OCH_3$

▲ **83.** What class of organic compounds do you think is the most important? Explain your answer.

▲ **84.** Can a compound contain more than one functional group? Can you give an example of such a compound?

Cumulative Review/Chapters 18–20

1. Distinguish between alpha, beta, and gamma radiation.

2. Complete the following nuclear equation:

$$^{10}_{5}\text{B} + ^{4}_{2}\text{He} \longrightarrow ^{1}_{0}\text{n} + ?$$

3. What isotope will be formed when $^{214}_{83}\text{Bi}$ emits beta radiation from its nucleus?

4. An isotope of bismuth called bismuth-210 has a half-life of 5.00 days and emits beta particles.
(a) Write the nuclear equation for this reaction.
(b) How many grams of bismuth-210 remain after a 25.0-g sample of bismuth-210 decays over a period of 20 days?

5. If 100.0 mg of radium-226 which has a half-life of 1,620 years, begins to decay today, how many milligrams of the isotope will be left after 3,240 years?

6. At the University of California at Berkeley, Albert Ghiorso and his research team bombarded californium-249, ^{249}Cf, with oxygen-18, ^{18}O, to form element 106, whose mass number is 263. Four neutrons were also formed. Write the nuclear equation for this reaction.

7. Element 106 has a half-life of 0.9 seconds and decays to form $^{259}_{104}\text{Unq}$ and an alpha particle. Write the nuclear equation for this reaction.

8. Element 109 is formed when bismuth-209 is bombarded with iron-58. A neutron is released during the reaction. Write the nuclear equation for this reaction.

9. Element 107 is formed when element-109 decays, releasing an alpha particle. Write the nuclear equation for this reaction.

10. The process by which electrical energy is used to bring about a chemical change is called electrolysis. True or false?

11. When a lead storage battery discharges, the following reactions take place:

$$\text{Pb} + \text{SO}_4^{2-} \longrightarrow \text{PbSO}_4 \quad \text{(not balanced)}$$

$$\text{PbO}_2 + \text{SO}_4^{2-} + \text{H}^{1+} \longrightarrow \text{PbSO}_4 + \text{H}_2\text{O} \quad \text{(not balanced)}$$

Complete and balance the half-cell equations for these reactions, indicating where electrons are supplied.

12. How is an electrolytic cell different from a voltaic cell?

13. In a redox reaction, the oxidizing agent is oxidized and the reducing agent is reduced. True or false?

14. How is a gaseous air pollutant different from a particulate?

15. How is acid rain formed in the environment?

16. Why is PVC potentially hazardous?

17. Name three ways to dispose of hazardous waste safely.

18. How are PCBs dangerous to the environment?

19. Distinguish between smog and photochemical smog.

20. All gasoline that enters an automobile engine is burned completely. True or false?

21. Write the Lewis electron-dot notation for carbon, indicating how many electrons are available for bonding.

22. Two tetrahedral carbon atoms sharing a common side form a double bond. True or false?

23. Write the names and formulas for the first five straight-chain members of the alkane, alkene, and alkyne series.

24. Write the structure for each of the following:
(a) 2-hexene
(b) 4,5,5-trimethyl-2-hexene

25. Complete this sentence: ________ are compounds that have the same molecular formula but different structural formulas.

26. Write the names and draw the structural formulas for straight-chain alkyl groups that have the formula C_nH_{2n+1} and contain from six to ten carbon atoms.

27. How is a cyclic hydrocarbon different from an aromatic hydrocarbon?

28. Define the term *homologous series*. Choose one series, and list three of its members.

29. Name the cyclic hydrocarbon represented as

30. How many electron pairs are associated with a covalently bonded carbon atom?

31. Write the structural formulas for all straight-chain isomers that have the molecular formula C_4H_7I.

32. What is a functional group?

33. Write the structures for each of the following:
(a) 2-hexanol *(b)* 3-octanol

34. Which of the following alcohols is nonpolar?

$$CH_3CH_2OH \quad \text{or} \quad CH_3{-}(CH_2)_{10}{-}CH_2OH$$

35. Give the common name for

$$CH_3CH_2{-}O{-}(CH_2)_2{-}CH_3$$

36. Give the IUPAC name or the common name for

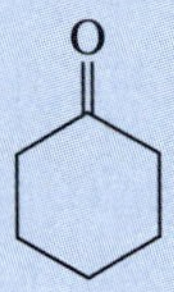

(structure: cyclohexane ring with C=O)

37. Write the structure of 2-octanone.

38. Decide whether each of the following is a primary, a secondary, or a tertiary amine:

(a) $CH_3{-}CH_2{-}\underset{\displaystyle |\atop \displaystyle H}{N}{-}CH_2{-}CH_3$

(b) (structure: N bearing CH_3 and two benzene rings)

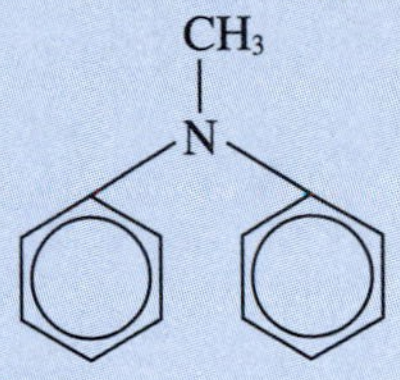

(c) (structure: six-membered ring containing N–H)

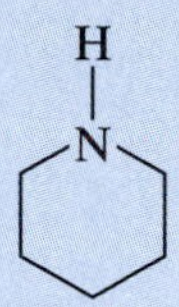

39. Given the products of the following esterification reaction, name the reactants:

$$? + ? \longrightarrow$$

$$CH_3{-}CH_2{-}\underset{\displaystyle \| \atop \displaystyle O}{C}{-}O{-}CH_2{-}CH_2{-}CH_2{-}CH_3 + H_2O$$

40. Write the structure of the ester ethylethanoate.

41. Write the structure of the aldehyde butanal.

42. Write the structure of methyl propyl ketone.

Supplement A
Basic Mathematics for Chemistry

Learning Goals

After you have studied this supplement, you should be able to:

1. Add and subtract algebraically.
2. Multiply and divide fractions.
3. Multiply and divide exponential numbers.
4. Use the factor-unit method in solving problems.
5. Add and subtract decimals.
6. Multiply and divide decimals.
7. Solve simple algebraic equations.
8. Set up and solve proportions.
9. Solve simple density problems.
10. Solve problems involving percentages.
11. Perform mathematical operations using a calculator.

Introduction

Full appreciation of many ideas in chemistry requires some basic mathematics. (There are also concepts in chemistry that require absolutely *no* mathematics.) This supplement is here so that you can review and master the mathematical skills you need to be successful in chemistry. Whenever you are in doubt about a mathematical problem as you work your way through the book, turn to this supplement for guidance.

You'll find that chemical mathematics is not very difficult, as long as you can perform some basic operations. Work your way through this supplement carefully, and try your hand at all the sample problems. Then do the exercises at the end. Remember that the knowledge you gain here will greatly increase your success in chemistry.

A.1 Adding and Subtracting Algebraically

Learning Goal 1
Adding and subtracting algebraically

In chemistry we deal with both positive and negative numbers, so it is important to know how to add and subtract both. The process is called algebraic addition and subtraction. Here are the rules to follow.

Addition

1. *When the signs are alike:* When the signs of two numbers are the same, add the numbers and keep the same sign.

$$+4 + 2 = +6$$

$$-4 - 2 = -6$$

2. *When the signs are different*: When the signs of two numbers are not the same, subtract the numbers and keep the sign of the larger number.

$$-4 + 2 = -2$$

$$+4 - 2 = +2$$

Subtraction

Change the sign of the *number being subtracted* and follow the rules for addition. (*Note:* Plus signs before positive numbers are usually left out, but minus signs before negative numbers are always written.)

$4 - (-2)$	becomes	$4 + 2 = 6$
$4 - (+2)$	becomes	$4 - 2 = 2$
$-4 - (+2)$	becomes	$-4 - 2 = -6$
$-4 - (-2)$	becomes	$-4 + 2 = -2$

A.2 Fractions

A ratio of two numbers is called a **fraction.** Some examples of fractions are $\frac{1}{4}$, $\frac{1}{2}$, $\frac{2}{5}$, $\frac{1}{11}$, and $\frac{2}{1}$. A fraction has two parts: the **numerator,** which is the top number, and the **denominator,** which is the bottom number.

$$\frac{5}{8}$$

Numerator (top, 5)
Denominator (bottom, 8)

Here are some basic rules to follow in dealing with fractions.

Multiplication

Learning Goal 2
Multiplying and dividing fractions

Multiply numerators; then multiply denominators.

$$\frac{1}{2} \times \frac{3}{4} = \frac{3}{8} \qquad \frac{5}{9} \times \frac{9}{5} = \frac{45}{45} = 1$$

Division

Invert (turn upside down) the fraction to the right of the division sign, and follow the rules for multiplication.

$$\frac{1}{2} \div \frac{1}{4} \quad \text{becomes} \quad \frac{1}{2} \times \frac{4}{1} = \frac{4}{2} = 2$$

$$\frac{3}{4} \div \frac{5}{9} \quad \text{becomes} \quad \frac{3}{4} \times \frac{9}{5} = \frac{27}{20}$$

A.3 Exponents

Exponents are numbers written to the right of and above another number (called the **base number**). A positive exponent tells you to multiply the base number by itself. The exponent tells you how many times the base number is taken. For example, $(4)^2$ means take the number 4 twice: 4×4. Here 4 is the base number and 2 is the exponent.

$(4)^2$ ← Exponent
↑ Base number

Sometimes exponents are negative. For example, we may have

$$(4)^{-2} \quad \text{which means} \quad \frac{1}{(4)^2}$$

$$\frac{1}{(4)^{-2}} \quad \text{which means} \quad \frac{(4)^2}{1}$$

Example A.1 shows how exponents are used.

Example A.1

Perform the indicated operation: *(a)* $(3)^2$ *(b)* $(4)^3$ *(c)* $(3)^1$ *(d)* $(3)^{-2}$ *(e)* $\frac{1}{(2)^{-3}}$

Solution

(a) $(3)^2 = 3 \times 3 = 9$ **(b)** $(4)^3 = 4 \times 4 \times 4 = 64$

(c) $(3)^1 = 3$ **(d)** $(3)^{-2} = \frac{1}{(3)^2} = \frac{1}{3 \times 3} = \frac{1}{9}$

(e) $\frac{1}{(2)^{-3}} = \frac{(2)^3}{1} = 2 \times 2 \times 2 = 8$

Learning Goal 3
Multiplying and dividing exponential numbers

Sometimes it is necessary to *multiply* exponential numbers that have the *same base*. This is done as shown in Example A.2.

Example A.2

Perform the indicated operation: *(a)* $(4)^8(4)^2$ *(b)* $(3)^2(3)^{-1}$ *(c)* $(8)^2(8)^1$ *(d)* $(2)^{-2}(2)^{-3}$ *(e)* $(3)^{-2}(3)^2$

Solution Keep the same base, and *add the exponents algebraically*.

(a) $(4)^8(4)^2 = (4)^{10}$ **(b)** $(3)^2(3)^{-1} = (3)^1 = 3$
(c) $(8)^2(8)^1 = (8)^3$ **(d)** $(2)^{-2}(2)^{-3} = (2)^{-5}$
(e) $(3)^{-2}(3)^2 = (3)^0 = 1$ (Any number to the zero power is equal to 1.)

Sometimes it is necessary to *divide* exponential numbers that have the *same base*. This is done as shown in Example A.3.

Example A.3

Perform the indicated operation: *(a)* $\frac{(2)^5}{(2)^2}$ *(b)* $\frac{(2)^5}{(2)^{-2}}$ *(c)* $\frac{(2)^{-5}}{(2)^2}$ *(d)* $\frac{(2)^{-5}}{(2)^{-2}}$

Solution Keep the same base, and subtract the exponent in the denominator from the exponent in the numerator *algebraically*.

(a) $\frac{(2)^5}{(2)^2} = (2)^{5-2} = (2)^3$

(b) $\frac{(2)^5}{(2)^{-2}} = (2)^{5-(-2)} = (2)^{5+2} = (2)^7$

(c) $\frac{(2)^{-5}}{(2)^2} = (2)^{-5-(+2)} = (2)^{-5-2} = (2)^{-7}$

(d) $\frac{(2)^{-5}}{(2)^{-2}} = (2)^{-5-(-2)} = (2)^{-5+2} = (2)^{-3}$

In Examples A.2 and A.3, only two numbers have been multiplied or divided. However, sometimes there are more than two numbers in the calculation, as in Example A.4.

Example A.4

Perform the indicated operation: *(a)* $(5)^3(5)^3(5)^2$ *(b)* $(a)^5(a)^3(a)^{-4}$ *(c)* $\frac{(b)^3(b)^6}{(b)^2}$ *(d)* $\frac{(2)^5(2)^{-6}}{(2)^{-9}}$ *(e)* $\frac{(a)^{-3}(a)^4(a)^6}{(a)^{-2}(a)^7}$

Solution

(a) $(5)^3(5)^3(5)^2 = (5)^8$ The exponents were added algebraically.

(b) $(a)^5(a)^3(a)^{-4} = (a)^4$ The exponents were added algebraically.

(c) $\frac{(b)^3(b)^6}{(b)^2} = \frac{(b)^9}{(b)^2} = (b)^{9-2} = (b)^7$ When you work out a problem one step at a time, it becomes easy.

(d) $\frac{(2)^5(2)^{-6}}{(2)^{-9}} = \frac{(2)^{-1}}{(2)^{-9}} = (2)^8$

(e) $\frac{(a)^{-3}(a)^4(a)^6}{(a)^{-2}(a)^7} = \frac{(a)^7}{(a)^5} = (a)^2$

A.4 Working with Units

Learning Goal 4
Solving problems using the factor-unit method

Almost all numbers, in chemistry or anywhere else, are accompanied by units. It is important to be able to work with these units, as well as with the numbers. Study the following examples carefully.

Example A.5

How many feet are there in 36 inches?

Solution We are going to use what is known as the **factor-unit method** of analysis. Here's how it works. Because there are 12 inches in 1 foot, we can express this relationship mathematically as

$$\frac{12 \text{ inches}}{1 \text{ foot}} \quad \text{(reads "12 inches per 1 foot")}$$

or as

$$\frac{1 \text{ foot}}{12 \text{ inches}} \quad \text{(reads "1 foot per 12 inches")}$$

These are called factor units. We set the problem up so that when we multiply the quantity in question (36 inches) by the appropriate factor unit, we end up with what we want (feet). Here, we have

$$? \text{ feet} = (36 \cancel{\text{inches}})\left(\frac{1 \text{ foot}}{12 \cancel{\text{inches}}}\right) = 3 \text{ feet}$$

Notice how the unit *inches* cancels out in Example A.5 when the numbers are multiplied. You may wonder how you'd know that you must use the factor unit.

$$\left(\frac{1\text{ foot}}{12\text{ inches}}\right) \quad \text{and not the factor unit} \quad \left(\frac{12\text{ inches}}{1\text{ foot}}\right)$$

That's easy. If you used the wrong factor unit, the term *inches* wouldn't cancel out. You would arrive at

$$?\text{ feet} = (36\text{ inches})\left(\frac{12\text{ inches}}{1\text{ foot}}\right) = \frac{432\text{ (inches)}^2}{1\text{ foot}}$$

which has no meaning to you. Try the problems in the following examples for practice in using the factor-unit method.

Example A.6

How many inches are there in 5 feet?

Solution Set the problem up this way:

$$?\text{ inches} = (5\text{ }\cancel{\text{feet}})\left(\frac{12\text{ inches}}{1\text{ }\cancel{\text{foot}}}\right) = 60\text{ inches}$$

Note how the unit *feet* cancels out when the numbers are multiplied with their units.

Example A.7

How many minutes are there in 4 hours? How many seconds are there in 4 hours?

Solution There are 60 minutes in 1 hour. So the factor unit is

$$\frac{60\text{ minutes}}{1\text{ hour}}$$

The problem is solved this way:

$$?\text{ minutes} = (4\text{ }\cancel{\text{hours}})\left(\frac{60\text{ minutes}}{1\text{ }\cancel{\text{hour}}}\right) = 240\text{ minutes}$$

There are 60 seconds in 1 minute. Our factor unit for this part of the problem is

$$\frac{60\text{ seconds}}{1\text{ minute}}$$

So the problem is solved as follows:

$$? \text{ seconds} = (240 \cancel{\text{minutes}})\left(\frac{60 \text{ seconds}}{1 \cancel{\text{minute}}}\right) = 14{,}400 \text{ seconds}$$

Example A.8

How many gallons are there in 60 quarts?

Solution There are 4 quarts in 1 gallon. Therefore

$$? \text{ gallons} = (60 \cancel{\text{quarts}})\left(\frac{1 \text{ gallon}}{4 \cancel{\text{quarts}}}\right) = 15 \text{ gallons}$$

Example A.9

Change 60 miles per hour to feet per second.

Solution Here you can really see the value of the factor-unit method. However, we have to know that 1 mile equals 5,280 feet and that 1 hour equals 3,600 seconds. Now, 60 miles per hour means

$$60 \frac{\text{miles}}{\text{hour}} \quad \text{which is the same as} \quad \frac{60 \text{ miles}}{1 \text{ hour}}$$

So the complete calculation reads

$$? \frac{\text{feet}}{\text{seconds}} = \left(\frac{60 \cancel{\text{miles}}}{1 \cancel{\text{hour}}}\right)\left(\frac{5{,}280 \text{ feet}}{1 \cancel{\text{mile}}}\right)\left(\frac{1 \cancel{\text{hour}}}{3{,}600 \text{ seconds}}\right)$$

$$= 88 \frac{\text{feet}}{\text{second}}$$

A.5 Decimals

Chemistry is such an exact science that the measurements we take usually consist of a string of numbers with a decimal point—for example, 2.7135 grams. When we are working with decimal numbers, we have to keep certain rules in mind.

Learning Goal 5
Adding and subtracting decimals

Addition and Subtraction

In adding or subtracting numbers, it is important to line up the columns correctly. Keep the decimal points stacked up one over the other.

Example A.10

Perform the indicated operation: *(a)* 25.8 + 107.09 + 0.011
(b) 342.78 − 14.99

Solution Line up the numbers properly.

(a)
$$\begin{array}{r} 25.8 \\ 107.09 \\ 88.004 \\ 0.011 \\ \hline 220.905 \end{array}$$

(b)
$$\begin{array}{r} 342.78 \\ -\ 14.99 \\ \hline 327.79 \end{array}$$

Multiplication

Learning Goal 6
Multiplying and dividing decimals

When we multiply 2.4 × 1.6, where does the decimal point go? First we write the numbers and perform the multiplication, ignoring the decimal point.

$$\begin{array}{r} 2.4 \\ \times\ 1.6 \\ \hline 144 \\ 24 \\ \hline 384 \end{array}$$

To find out where the decimal point goes, we count the number of digits to the right of the decimal point in each term. (There is one digit to the right of the decimal point in each term.) Then we add these numbers of digits (one plus one equals two) and count off this new number from the right-hand side of the answer. In our example, we count off two digits from the right, and we place the decimal point between the 3 and the 8. This gives us an answer of 3.84.

Example A.11

Perform the indicated operation: *(a)* (4.25)(5) *(b)* (7.12)(3.64)
(c) (5.222)(4.11)

Solution

(a)
$$\begin{array}{r} 4.25 \\ \times\ \ 5 \\ \hline 21.25 \end{array}$$

(b)
$$\begin{array}{r} 7.12 \\ \times\ 3.64 \\ \hline 2848 \\ 4272 \\ 2136 \\ \hline 25.9168 \end{array}$$

(c)
$$\begin{array}{r} 5.222 \\ \times\ 4.11 \\ \hline 5222 \\ 5222 \\ 20888 \\ \hline 21.46242 \end{array}$$

Division

Where does the decimal point go when we divide 11 by 0.56? First we write the problem in a more familiar way.

$$0.56\overline{)11}$$

(Here 0.56 is called the **divisor,** and 11 is called the **dividend.**) Our next step is to move the decimal point in the divisor so that all digits are to the left of the decimal point: 0.56 becomes 56. We must also move the decimal point in the dividend the same number of places: 11 becomes 1100.

$$0.56\overline{)11.00} \quad \text{becomes} \quad 56\overline{)1100.00}$$

We now perform the division, locating the decimal point in the answer directly above the decimal point in the dividend.

```
      19.64
  56)1100.00
     56x xx
     540
     504
      360
      336
       240
       224
        16
```

Example A.12

Perform the indicated operation: *(a)* $\dfrac{9.25}{2.5}$ *(b)* $\dfrac{24.138}{7.45}$

Solution

(a) $2.5\overline{)9.2\ 5}$ **(b)** $7.45\overline{)24.13\ 8}$

```
     3.7             3.24
 25)92.5       745)2413.80
    75 x           2235 xx
    175             1788
    175             1490
      0              2980
                     2980
                        0
```

A.6 Solving Algebraic Equations

Learning Goal 7
Solving simple algebraic equations

Very often in chemistry, we need to solve algebraic equations ("find the unknown"). So study the following examples carefully.

Example A.13

Solve each of the following equations for the unknowns a, y, and z:
(a) $3a = 9$ *(b)* $4y + 5 = 37$ *(c)* $9z - 3 = 22$

Solution To solve an algebraic equation, we try to isolate the unknown. **(a)** In $3a = 9$, we can isolate the a by dividing both sides of the equation by 3. This operation removes the 3 from the left-hand side of the equation. Remember, what we do to one side of the equation we *must* do to the other side, in order to maintain the equality.

$$\frac{\cancel{3}a}{\cancel{3}} = \frac{9}{3}$$

After simplifying both sides of the equation, we get the equality

$$a = 3$$

(b) In $4y + 5 = 37$, we want to isolate the y. Our first step is to remove the 5 from the left-hand side of the equation. We do this by subtracting 5 from both sides of the equation.

$$4y + 5 - 5 = 37 - 5 \qquad \text{so} \qquad 4y = 32$$

We now divide both sides of the equation by 4.

$$\frac{\cancel{4}y}{\cancel{4}} = \frac{32}{4}$$

After simplifying both sides of the equation, we get the equality

$$y = 8$$

(c) In $9z - 3 = 22$, the first step is to remove the 3 from the left-hand side of the equation. We do this by adding 3 to both sides of the equation.

$$9z - 3 + 3 = 22 + 3 \qquad \text{so} \qquad 9z = 25$$

Now divide both sides of the equation by 9.

$$\frac{\cancel{9}z}{\cancel{9}} = \frac{25}{9}$$

After simplifying both sides of the equation, we have the equality

$$z = \frac{25}{9}$$

Example A.14

Solve each of the following equations for the unknown: *(a)* $\frac{2a}{3} = \frac{12}{9}$
(b) $5y + 3 = 2y - 42$

Solution

(a) In the equation

$$\frac{2a}{3} = \frac{12}{9}$$

we can remove the 3 on the left-hand side by multiplying both sides of the equation by 3.

$$(3)\left(\frac{2a}{3}\right) = \left(\frac{12}{9}\right)(3) \qquad 2a = \frac{36}{9} \qquad 2a = 4$$

Now divide both sides of the equation by 2.

$$\frac{\cancel{2}a}{\cancel{2}} = \frac{4}{2}$$

After simplifying both sides of the equation, we have the equality

$$a = 2$$

(b) In the equation

$$5y + 3 = 2y - 42$$

we first gather all the terms containing the unknown on one side. Place all y terms on the left-hand side of the equation by subtracting $2y$ from each side of the equation.

$$5y + 3 - 2y = 2y - 42 - 2y \qquad \text{so} \qquad 3y + 3 = -42$$

Then subtract 3 from each side of the equation.

$$3y + 3 - 3 = -42 - 3 \qquad \text{so} \qquad 3y = -45$$

Now divide each side of the equation by 3.

$$\frac{\cancel{3}y}{\cancel{3}} = \frac{-45}{3}$$

After simplifying both sides of the equation, we have the equality

$$y = -15$$

A.7 Ratios and Proportions

Learning Goal 8
Solving proportions

There are four wheels on a car. This relationship can be expressed mathematically in the following ways:

$$4 \text{ wheels} : 1 \text{ car} \qquad \text{or} \qquad \frac{4 \text{ wheels}}{1 \text{ car}}$$

Both statements express the fact that there are four wheels on one car. Such statements are known as **ratios** because they reveal the numerical relationship between different things. A ratio is just like a factor unit, and it can be used as an alternative to the factor-unit method to solve many types of problems. But to use ratios in problem solving, you must understand the concept of proportions. A **proportion** is *an equality between two ratios,* as in

$$\frac{1}{2} = \frac{5}{10} \quad \text{or} \quad 1:2 = 5:10$$

(When the second expression is used, the symbol : is often read "is to" and the symbol = is read "as.") In a problem involving a proportion, you are usually given three of the four numbers and are asked to determine the missing number. Example A.15 shows how this works.

Example A.15

If there are 4 wheels on 1 car, how many wheels are there on 5 cars?

Solution The proportion is set up in the following way:

$$\frac{4 \text{ wheels}}{1 \text{ car}} = \frac{y \text{ wheels}}{5 \text{ cars}}$$

To solve for y, multiply each side of the equation by 5 cars.

$$(5 \cancel{\text{cars}})\left(\frac{4 \text{ wheels}}{1 \cancel{\text{car}}}\right) = \left(\frac{y \text{ wheels}}{5 \cancel{\text{cars}}}\right)(5 \cancel{\text{cars}})$$

$$20 \text{ wheels} = y \text{ wheels}$$

In other words, $y = 20$ wheels, so there are 20 wheels on 5 cars.

A.8 Solving Word Equations

Learning Goal 9
Solving simple density problems

Many of the mathematical equations we use in chemistry are word equations, such as

$$\text{Density} = \frac{\text{mass}}{\text{volume}} \quad \text{sometimes abbreviated as} \quad D = \frac{m}{V}$$

As written, the equation is set up so that we can solve for the density of a substance if we are given its mass and volume. In some problems, however, we are given the density and volume of a substance and are asked to solve for its mass. We do so as shown in Example A.16.

Example A.16

Given the following information, solve for the mass m:

$$D = 0.8\,\frac{\text{g}}{\text{mL}} \qquad V = 30\text{ mL}$$

Solution First write the formula.

$$D = \frac{m}{V}$$

Then isolate the m by multiplying both sides of the equation by V.

$$(D)(V) = \left(\frac{m}{\cancel{V}}\right)(\cancel{V}) \qquad \text{so} \qquad DV = m$$

Now substitute the numbers given in the problem, and solve for the mass in grams.

$$\left(0.8\,\frac{\text{g}}{\cancel{\text{mL}}}\right)(30\ \cancel{\text{mL}}) = m \qquad \text{so} \qquad 24\text{ g} = m$$

In some problems, we are given the density and mass of a substance and are asked to solve for its volume.

Example A.17

Given the following information, solve for the volume V:

$$D = 2\,\frac{\text{g}}{\text{mL}} \qquad m = 20\text{ g}$$

Solution Write the density formula in its original form.

$$D = \frac{m}{V}$$

We must get the V into the numerator by itself. This can be done in two steps:

1. Multiply both sides of the equation by V.

$$(D)(V) = \left(\frac{m}{\cancel{V}}\right)(\cancel{V}) \qquad \text{so} \qquad DV = m$$

2. Divide both sides of the equation by D.

$$\frac{\cancel{D}V}{\cancel{D}} = \frac{m}{D} \qquad \text{so} \qquad V = \frac{m}{D}$$

Now substitute the numbers given in the problem, and solve for the volume in milliliters.

$$V = \frac{m}{D} \quad \text{so} \quad V = \frac{20 \text{ g}}{2 \text{ g/mL}} = 10 \text{ mL}$$

Do you understand how the unit *gram* cancels out to give milliliters in the answer? We actually have

$$V = 20 \text{ g} \div \frac{2 \text{ g}}{1 \text{ mL}}$$

When we apply the *rules for division of fractions* discussed in Section A.2, this expression becomes

$$20 \text{ g} \times \frac{1 \text{ mL}}{2 \text{ g}} = 10 \text{ mL}$$

and the unit *gram* cancels out.

A.9 Calculating and Using Percentages

Learning Goal 10
Solving problems involving percentages

A **percentage** is *the number of parts of something out of 100 parts*. For example, if a chemistry class has 100 students and 40 of the students are women, the percentage of women is 40 percent (also written as 40%).

$$40 \text{ women per } 100 \text{ students} = 40\%$$

We can state this mathematically (using the factor-unit method) as follows:

$$\text{Percentage women} = \left(\frac{40 \text{ women}}{100 \text{ students}}\right)(100 \text{ students})$$
$$= 40 \text{ percent women}$$

However, we usually set the formula up as follows:

$$\textbf{Percentage} = \frac{\text{number of items of interest}}{\text{total number of items}} \times 100$$

Suppose that we have a chemistry class with 60 students and that 48 of them are women. What is the percentage of women in the class? By the formula,

$$\text{Percentage} = \frac{\text{number of items of interest}}{\text{total number of items}} \times 100$$
$$= \frac{48 \text{ women}}{60 \text{ students}} \times 100 = 80\% \text{ women}$$

Example A.18

A box of mixed vegetables has 25 carrots, 30 tomatoes, and 95 heads of lettuce. What is the percentage of each vegetable in the box?

Solution First obtain the total number of vegetables in the box.

$$25 \text{ carrots} + 30 \text{ tomatoes} + 95 \text{ lettuces} = 150 \text{ vegetables}$$

Now calculate the percentage of each vegetable, using our formula.

$$\% \text{ carrots} = \frac{25 \text{ carrots}}{150 \text{ vegetables}} \times 100 = 16.7\% \text{ carrots}$$

$$\% \text{ tomatoes} = \frac{30 \text{ tomatoes}}{150 \text{ vegetables}} \times 100 = 20\% \text{ tomatoes}$$

$$\% \text{ lettuces} = \frac{95 \text{ lettuces}}{150 \text{ vegetables}} \times 100 = 63.3\% \text{ lettuces}$$

Note that the total of the percentages equals 100%. In other words, the whole is the sum of the parts.

Sometimes we know the percentage of something and we want to solve for the particular number of items. For example, say that 30% of the people in a particular community own a car. If this community has 800 people, how many people own cars? We can solve this problem using the factor-unit method if we remember that 30% car ownership means

$$\frac{30 \text{ people own cars}}{100 \text{ people in town}}$$

Therefore

$$? \text{ people owning cars} = (800 \cancel{\text{people}})\left(\frac{30 \text{ people owning cars}}{100 \cancel{\text{people}}}\right)$$

$$= 240 \text{ people own cars}$$

However, a simpler method of solving this problem is to move the decimal point on the percent number two places to the left and multiply it by the number of people in town. In this method, 30% becomes 0.30. Therefore

$$800 \text{ people} \times 0.30 = 240 \text{ people own cars}$$

Example A.19

A box of fruit contains apples, oranges, and bananas. There are 500 pieces of fruit in the box; 20% are apples, 30% are oranges, and 50% are bananas. How many pieces of each kind of fruit are in the box?

Solution Turn each percent into its decimal equivalent and multiply it by the total number of fruit in the box.

20% becomes 0.20 (for apples)

30% becomes 0.30 (for oranges)

50% becomes 0.50 (for bananas)

Therefore

$$500 \text{ fruit} \times 0.20 = 100 \text{ apples}$$

$$500 \text{ fruit} \times 0.30 = 150 \text{ oranges}$$

$$500 \text{ fruit} \times 0.50 = 250 \text{ bananas}$$

Note that the total number of pieces of fruit is 500.

A.10 Using the Calculator

Learning Goal 11
Performing mathematical operations using a calculator

If they are used correctly, electronic calculators speed up the process of making mathematical computations. One common error is the simple mistake of pushing the buttons incorrectly (in the wrong order, for instance). Many such errors can be detected by first estimating the answer so you can tell whether the calculated answer makes sense.

You should follow the specific directions given for your particular calculator, of course, but the following rules apply to most calculators:

1. Enter the operations in the order in which they are written.

[4] [×] [5] [=] [20]

First press the "4" key, then the "×" key, then the "5" key, followed by the "=" key. The answer "20" will appear in the electronic display position.

Example A.20

Perform the following operations:

(a) [10] [×] [5] [=]
(b) [10] [+] [5] [=]
(c) [10] [÷] [5] [=]
(d) [10] [−] [5] [=]

Solution

(a) 50
(b) 15
(c) 2
(d) 5

2. Once the calculator is directed to add, subtract, multiply, or divide two numbers, the results of the mathematical operation are stored in its memory. Another operation key (+, −, ×, or ÷) can then be pressed and another calculation carried out.

Example A.21

Perform the following operations:

(a) [10] [×] [5] [=] [+] [4] [=]

(b) [20] [÷] [5] [−] [4] [=]

(c) [6] [−] [4] [×] [20] [=]

(d) [8] [+] [6] [×] [2] [=]

Solution

(a) 54
(b) 0
(c) 40
(d) 28

3. Your calculator probably has a floating decimal point. Once you reach the proper place for the decimal point, press the [·] key to position the decimal point.

4. Final zeros to the right of the decimal point are not shown in the electronic display after a calculation is complete.

Self-Test Exercises

Learning Goal 1: Adding and Subtracting Algebraically

1. *Add* the following numbers algebraically:
(a) $5 + 12$ *(b)* $-5 - 10$ *(c)* $5 + 10$
(d) $-5 + 10$

2. *Subtract* the following numbers algebraically:
(a) $3 - (+2)$ *(b)* $3 - (-2)$ *(c)* $-3 - (+2)$
(d) $-3 - (-2)$

3. *Add* the following numbers algebraically:
(a) $8 + 25$ *(b)* $8 - 25$ *(c)* $-8 + 25$
(d) $-8 - 25$

4. *Subtract* the following numbers algebraically:
(a) $55 - (+44)$ *(b)* $55 - (-44)$
(c) $-55 - (+44)$ *(d)* $-55 - (-44)$

Learning Goal 2: Multiplying and Dividing Fractions

5. Perform the following operations: *(a)* $\frac{1}{8} \times \frac{1}{2}$
(b) $\frac{1}{8} \div \frac{1}{2}$ *(c)* 8^3 *(d)* $(2)^{-5}$

6. Perform the following operations: *(a)* $\frac{3}{5} \times \frac{2}{7}$
(b) $\frac{20}{50} \times \frac{9}{5}$ *(c)* $\frac{5}{6} \div \frac{10}{3}$ *(d)* $\frac{25}{35} \div \frac{7}{5}$

Learning Goal 3: Multiplying and Dividing Exponential Numbers

7. Perform the following operations: *(a)* $(4)^{-2}(4)^{-5}$ *(b)* $(4)^2(4)^3$ *(c)* $(6)^4(6)^{-2}$ *(d)* $(5)^{-8}(5)^9$

8. Perform the following operations: *(a)* $\frac{(4)^5}{(4)^2}$ *(b)* $\frac{(4)^5}{(4)^{-3}}$ *(c)* $\frac{(a)^3(a)^2}{(a)^4}$ *(d)* $\frac{(b)^6(b)^4}{(b)^3(b)^5}$

9. Perform the following operations: *(a)* $(25)^2$ *(b)* $(10)^3$ *(c)* $(99)^1$ *(d)* $(8)^{-3}$ *(e)* $\frac{1}{(4)^{-2}}$

10. Perform the following operations: *(a)* $(5)^2(5)^3$ *(b)* $(12)^3(12)^9$ *(c)* $(10)^2(10)^{-2}$ *(d)* $(4)^{-8}(4)^{-2}$

11. Perform the following operations: *(a)* $\frac{(4)^5}{(4)^3}$ *(b)* $\frac{(12)^3}{(12)^{-2}}$ *(c)* $\frac{(20)^{-4}}{(20)^3}$ *(d)* $\frac{(20)^{-4}}{(20)^{-3}}$

12. Perform the following operations: *(a)* $(4)^4(4)^3(4)^2$ *(b)* $(x)^2(x)^3(x)^{-4}$ *(c)* $\frac{(y)^2(y)^8}{(y)^3}$ *(d)* $\frac{(p)^2(p)^{-9}}{(p)^{-8}}$ *(e)* $\frac{(10)^{-3}(10)^5(10)^7}{(10)^{-8}(10)^9}$

13. Perform the following operations: *(a)* $(2)^6$ *(b)* $(3)^{-4}$ *(c)* $\frac{1}{(4)^3}$ *(d)* $\frac{1}{(4)^{-3}}$

14. Perform the following operations: *(a)* $(10)^5(10)^{-2}$ *(b)* $(2)^{-3}(2)^{-2}$ *(c)* $(3)^{-5}(3)^2$ *(d)* $(6)^{-8}(6)^{10}$

15. Perform the following operations: *(a)* $\frac{(3)^7}{(3)^5}$ *(b)* $\frac{(2)^4(2)^{-6}}{(2)^7(2)^{-12}}$ *(c)* $\frac{(a)^{12}(a)^{-10}}{(a)^{-7}(a)^{-6}}$ *(d)* $\frac{(b)^{-20}(b)^{-10}}{(b)^{-9}(b)^{-8}}$

Learning Goal 4: Solving Problems Using the Factor-Unit Method

16. $(4 \text{ feet})\left(\frac{12 \text{ inches}}{1 \text{ foot}}\right) = ?$

17. $A = l \times w$; therefore $w = ?$ Solve the equation for w in terms of A and l.

18. How many yards are there in 18 feet?

19. How many eggs are there in 4 dozen eggs?

20. *(a)* How many hours are there in 360 minutes? *(b)* How many hours are there in 7,200 seconds?

21. How many pints are there in 8 quarts?

22. Change 1 mile per hour to feet per second.

23. $(72 \text{ inches})\left(\frac{1 \text{ foot}}{12 \text{ inches}}\right) = ?$

24. $(15 \text{ feet})\left(\frac{12 \text{ inches}}{1 \text{ foot}}\right) = ?$

25. How many feet are there in 100 yards?

26. How many inches are there in 5 yards?

27. How many dozen eggs do you have if you have 288 eggs?

28. How many minutes are there in 24 hours? How many seconds are there in 24 hours?

29. How many gallons are there in 40 quarts?

30. A rocket is traveling at 22,000 feet per second. How many miles per hour is it traveling?

Learning Goal 5: Adding and Subtracting Decimals

31. Perform the following operations:
(a) $15.4 + 117.33 + 16.909 + 0.044$
(b) $171.82 - 30.41$

32. Perform the following operations:
(a) $25.431 + 0.761 + 0.325 + 0.008$
(b) $123.25 - 19.54$
(c) $98.77 - 38.25 + 45.62$
(d) $254.37 - 68.26 - 38.33$

Learning Goal 6: Multiplying and Dividing Decimals

33. Perform the following operations: *(a)* (7.33)(4) *(b)* (9.01)(4.28) *(c)* (3.111)(8.7)

34. Perform the following operations: *(a)* $\frac{6.324}{3.1}$ *(b)* $\frac{25.30}{5.06}$

35. Perform the following operations: *(a)* (9.54)(3.27) *(b)* (6.5)(7.1) *(c)* $\frac{80.4}{2.02}$ *(d)* $\frac{640.75}{8.5}$

Learning Goal 7: Solving Simple Algebraic Equations

36. Solve each of the following equations for the unknown: *(a)* $25a = 100$ *(b)* $40y + 13 = 26$ *(c)* $12z - 12 = 24$

37. Solve each of the following equations for the unknown: *(a)* $\frac{6x}{3} = \frac{36}{12}$ *(b)* $12y + 6 = 6y + 12$

38. Solve the equation $PV = nRT$ for *(a)* P, *(b)* n, *(c)* T, *(d)* n/V.

39. Solve each of the following equations for the unknown: *(a)* $4t = 64$ *(b)* $6y + 3 = 45$ *(c)* $2x - 7 = 33 - 3x$ *(d)* $40k + 8 = 96 + 51k$

40. Solve the equation $A = bcd$ for *(a)* b, *(b)* c, *(c)* A/b.

Learning Goal 8: Solving Proportions

41. Solve each of the following equations for the unknown: *(a)* $\frac{x}{25} = \frac{4}{5}$ *(b)* $\frac{3}{y} = \frac{7}{63}$ *(c)* $\frac{4t}{12} = \frac{2}{3}$ *(d)* $\frac{2}{5k} = \frac{5}{50}$

42. Solve each of the following proportions:
(a) $\frac{k}{5} = \frac{9}{45}$ *(b)* $\frac{2}{7} = \frac{h}{28}$ *(c)* $\frac{72}{f} = \frac{9}{2}$ *(d)* $\frac{18}{6} = \frac{12}{y}$

43. Solve each equation for the missing value or values: *(a)* $2:5:9 = x:20:y$ *(b)* $7:9 = h:54$ *(c)* $2:3:5:7 = 1:a:b:c$ *(d)* $2:5 = 5:d$

Learning Goal 9: Solving Simple Density Problems

44. Given that $D = \frac{m}{V}$, solve this formula for *(a)* m and *(b)* V.

45. A substance has a mass of 25 g and a volume of 5 mL. What is its density?

Learning Goal 10: Solving Problems Involving Percentages

46. A university has the following enrollments: *(a)* liberal arts, 200 students; *(b)* laboratory technology, 100 students; and *(c)* health science technology, 300 students. This accounts for all the students at the university. What percentage of the students is enrolled in each curriculum?

47. A survey of 2,000 people shows that 40% want to travel to Europe, 30% to Asia, 10% to Alaska, and 20% to Miami. How many people want to travel to each place?

Learning Goal 11: Performing Mathematical Operations Using a Calculator

48. Perform the following operations using a calculator:
(a) $11 \times 6 = ?$
(b) $11 + 6 = ?$
(c) $30 \div 5 = ?$
(d) $30 - 5 = ?$

49. Perform the following operations using a calculator:
(a) $(12 \times 5) + 5 = ?$
(b) $(20 \div 4) - 2 = ?$
(c) $(10 + 20) \div (5 \times 6) = ?$
(d) $(100 \times 50) \div (10 \times 40) = ?$

50. Perform the following operations using a calculator:
(a) $(50 \times 6) \div 12 = ?$
(b) $(100 + 50) \div (15 \times 0.01) = ?$
(c) $(0.05 \times 0.30) \div 4 = ?$
(d) $(10{,}000 \times 5) \div 4000 = ?$

Supplement B
Important Chemical Tables

Table B.1
Prefixes and Abbreviations

nano-	= 0.000000001	$= 10^{-9}$	nanometer	= nm	centigram = cg
micro-	= 0.000001	$= 10^{-6}$	micrometer	= μm (Greek letter mu)	decimeter = dm
milli-	= 0.001	$= 10^{-3}$	millimeter	= mm	decigram = dg
centi-	= 0.01	$= 10^{-2}$	milliliter	= mL	kilometer = km
deci-	= 0.1	$= 10^{-1}$	milligram	= mg	kilogram = kg
deka-	= 10	$= 10^{1}$	centimeter	= cm	
kilo-	= 1,000	$= 10^{3}$			

Table B.2
The Metric System

Length		
1 millimeter = 0.001 meter = $\frac{1}{1000}$ meter	or	1 meter = 1,000 millimeters
1 centimeter = 0.01 meter = $\frac{1}{100}$ meter	or	1 meter = 100 centimeters
1 decimeter = 0.1 meter = $\frac{1}{10}$ meter	or	1 meter = 10 decimeters
1 kilometer = 1,000 meters	or	1 meter = 0.001 kilometer
Mass		
1 microgram = 0.000001 gram		1 gram = 1,000,000 micrograms
1 milligram = 0.001 gram		1 gram = 1,000 milligrams
1 centigram = 0.01 gram		1 gram = 100 centigrams
1 decigram = 0.1 gram		1 gram = 10 decigrams
1 kilogram = 1,000 grams		1 gram = 0.001 kilogram
Volume		
1 milliliter = 0.001 liter		
1 milliliter = 1 cubic centimeter*		

*Cubic centimeter is abbreviated cm^3 or cc.

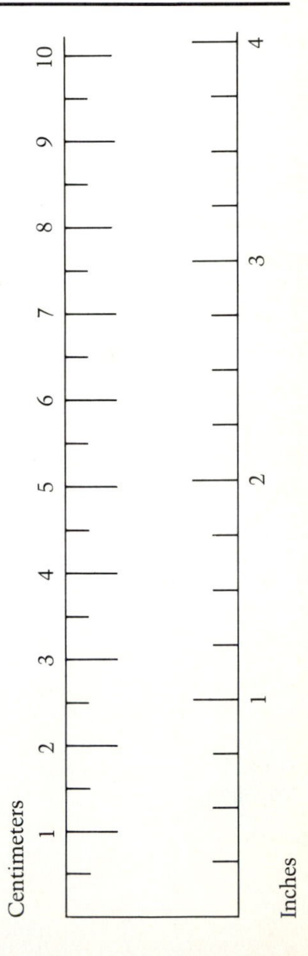

Table B.3
Conversion of Units (English–Metric)

To convert	into	multiply by
Length		
inches	centimeters	2.540 cm/in
centimeters	inches	0.3937 in/cm
feet	meters	0.30 m/ft
meters	feet	3.28 ft/m
Weight (mass)		
ounces	grams	28.35 g/oz
grams	ounces	0.035 oz/g
pounds	grams	454 g/lb
grams	pounds	0.0022 lb/g
Volume		
liters	quarts	1.057 qt/L
quarts	liters	0.9463 L/qt

Table B.4
Solubilities

	Acetate	Arsenate	Bromide	Carbonate	Chlorate	Chloride	Chromate	Hydroxide	Iodide	Nitrate	Oxide	Phosphate	Sulfate	Sulfide
Aluminum	W	a	W	—	W	W	—	A	W	W	a	A	W	d
Ammonium	W	W	W	W	W	W	W	W	W	W	—	W	W	W
Barium	W	w	W	w	W	W	A	W	W	W	W	A	a	d
Cadmium	W	A	W	A	W	W	A	A	W	W	A	A	W	A
Calcium	W	w	W	w	W	W	W	W	W	W	w	w	w	w
Chromium	W	—	W*	W	—	I	—	A	W	W	a	w	W†	d
Cobalt	W	A	W	A	W	W	A	A	W	W	A	A	W	A
Copper(II)	W	A	W	—	W	W	—	A	a	W	A	A	W	A
Hydrogen	W	W	W	—	W	W	—	—	W	W	W	W	W	W
Iron(II)	W	A	W	w	W	W	—	A	W	W	A	A	W	A
Iron(III)	W	A	W	—	W	W	A	A	W	W	A	w	w	d
Lead(II)	W	A	W	A	W	W	A	w	w	W	w	A	w	A
Magnesium	W	A	W	w	W	W	W	A	W	W	A	w	W	d
Mercury(I)	w	A	A	A	W	a	w	—	A	W	A	A	w	I
Mercury(II)	W	w	W	—	W	W	w	A	w	W	w	A	d	I
Nickel	W	A	W	w	W	W	A	w	W	W	A	A	W	A
Potassium	W	W	W	W	W	W	W	W	W	W	W	W	W	W
Silver	w	A	a	A	W	a	w	—	I	W	w	A	w	A
Sodium	W	W	W	W	W	W	W	W	W	W	d	W	W	W
Strontium	W	w	W	w	W	W	w	W	W	W	W	A	w	W
Tin(II)	d	—	W	—	W	W	A	A	W	d	A	A	W	A
Tin(IV)	W	—	W	—	—	W	W	w	d	—	A	—	W	A
Zinc	W	A	W	w	W	W	w	A	W	W	w	A	W	A

Reprinted with permission from *CRC Handbook of Chemistry and Physics,* 70th ed. Copyright © The Chemical Rubber Co., Boca Raton: CRC Press, Inc.

Abbreviations: W = soluble in water; A = insoluble in water, but soluble in acids; w = only slightly soluble in water, but soluble in acids; a = insoluble in water, and only slightly soluble in acids; I = insoluble in both water and acids; d = decomposes in water.

*$CrBr_3$

†$Cr_2(SO_4)_3$

Supplement C
Study Skills

About Learning Chemistry

Chemistry is an exciting subject. It affects our lives each day in many ways. The foods we eat and the clothes we wear, the materials used to construct our homes, the medications that cure our illnesses, and the energy we depend on are all products of the science of chemistry coupled with technology.

As you learn chemistry, you will gain knowledge that will help you understand more about our technological society. You'll be able to apply your understanding of chemistry to help solve some of the problems you will face in the real world. You'll be able to approach familiar situations with new insight and to make decisions based on the knowledge you'll acquire in this course.

It will be easier for you to learn chemistry if you start out with a positive attitude and the knowledge of how to approach the subject most effectively. Although some of the material in the course requires memorization, a large part of it involves learning how to solve problems. *Basic Concepts of Chemistry* focuses on developing your problem-solving skills. A general approach to problem solving is discussed in Chapter 2. Throughout the text are worked-out examples with solutions given step by step. Following each worked-out example is a practice exercise that offers the same type of problem whose solution has just been demonstrated. The practice exercises give you an opportunity to test your understanding of the skills presented in the chapter. At the end of each chapter, two additional sets of problems give you even more practice in problem solving.

The learning goals listed at the beginning of each chapter will help you organize the material you'll need to master. How can you be sure, as you go on with your studies, that your knowledge of the material is adequate? As you read each chapter, you will notice that the learning goals are printed in color for easy location of important material. The examples and practice exercises described above are included wherever problem-solving skills are essential to your understanding of the topic. At the end of each chapter, self-test exercises are keyed to each of the learning goals. If you are able to work through the examples, the practice exercises, and the self-test exercises correctly, you can assume you have mastered the necessary problem-solving skills. For sections that require you to memorize reactions or descriptive material, self-test exercises are designed to help you master these skills as well.

Before an exam, it is important to speak with your instructor to find out what material will be covered on the test. Instructors emphasize important concepts in class, and some will help guide your studies by offering information on the material that deserves special attention.

You can successfully complete basic chemistry if you develop good study skills. The tools to help you learn the subject are in the classroom and in the textbook. If you use your time efficiently and ask for help from your instructor when you encounter material that needs further explanation, you should be able to do well in the course.

It is also helpful to form a small study group. With three or four classmates you can practice problem solving and reasoning. Several exercises at the end of each chapter are marked for cooperative problem solving.

The sections that follow discuss some common-sense methods of achieving success in college and offer additional advice you can apply to your day-to-day study of chemistry. As you become a well-prepared and successful student, you will approach examinations with confidence. The successful habits you form will stay with you as you complete your education and will benefit you throughout your life.

Time Management for College Students

You've got an exam on Friday and a paper due on Monday. You know you've got to get to work. Just then the phone rings. A group of friends from your dorm is going out for a snack. You know you shouldn't join them, but you've got to eat anyway, so you go along. When you return home, you need a little time to unwind. You turn on the television. Two hours later you're relaxed, but you're also tired. You decide to call it a night. There's always tomorrow!

If this scenario sounds familiar, it should. It happens to all of us now and then. As a college student, however, you will find that it's very important to learn to manage your time effectively.

Setting Realistic Goals

The first step in managing your time effectively is to know where you're going. It helps to set goals for yourself. Three different types of goals that you should consider are long-range goals, medium-range goals, and short-range goals.

Long-range goals are usually personal wishes. They have to do with your career aims, your educational plans, and your social desires. Think about where you would like to be five or ten years from now. The education you are now receiving in college should be a steppingstone to your long-range goals. Depending on what your career plans are, the grades you earn in your courses will help determine whether or not you will be able to fulfill these goals.

Medium-range goals help you achieve your long-range goals. They can be set two or three times a year. For example, if you plan to enter medical school after graduation, you will need a considerable number of A's in your courses. A medium-range goal would be to get four or five A's in your courses. Another medium-range goal might be to join a club or to improve your skills in your favorite sport. Let's say that your grades last semester weren't the best. A medium-range goal for you, then, might be to improve them. If you're saving money to buy a car, watching your budget more carefully might be a reasonable goal to set.

Short-range goals can be set each week. These goals include taking care of your daily tasks and keeping up with your assignments. To read a chapter in a book, complete an assignment, or write a paper is an example of a short-range goal.

Planning to Reach Your Goals

Achieving your goals often requires the setting up of a plan. Although we'd like to think we have the will power to do all we have to do, very often we become distracted. A well-planned, flexible schedule is a useful tool to use to get things done more effectively.

The first step to managing your time is to know yourself. When do you function the best? If you're a morning person, it's best to do as much of your work as you can in the morning when you are at your best. If you function better later in the day, try to schedule most of your work for the afternoon or evening.

Setting up a monthly plan is the next step in becoming organized. Obtain a calendar and write in, on the appropriate date, the important assignments and events you need to remember for each month of the semester (Figure C.1). Term papers, exams, sports meets, and social activities can all fit in if you see the overall picture and plan in advance. For example, if you know you have a paper due the same week of two scheduled exams, you'll know you have to keep your schedule free to allow for your school work.

After you've completed your monthly plan, devise a weekly plan. Map out a schedule for each day of the week. Think over how much time you need to complete each assignment and to study for your exams, in order to fit all of your responsibilities into your schedule (Figure C.2).

It's usually best to find a quiet place to work. Accomplish your tasks in priority order, breaking the large tasks into smaller ones. Don't try to do too much in one day. It's better to do a little less than to do too much. Complete one task at a time, and avoid working on two projects at once.

You may find that after a month or so, you can manage your time effectively without a written schedule. For some students, having a written schedule is a necessity. When you are designing your plan, don't forget that you need time for enjoyment as well as for work. Giving yourself positive reinforcement is a good idea. Reward yourself when you've accomplished some-

Figure C.1
Monthly planning calendar

MONTH__________ YEAR__________

SUNDAY	MONDAY	TUESDAY	WEDNESDAY	THURSDAY	FRIDAY	SATURDAY

Figure C.2
Weekly plan

Your main goal this week: ____________________

	Monday	*Tuesday*	*Wednesday*	*Thursday*	*Friday*	*Saturday*	*Sunday*
6 A.M.							
7 A.M.							
8 A.M.							
9 A.M.							
10 A.M.							
11 A.M.							
12 P.M.							
1 P.M.							
2 P.M.							
3 P.M.							
4 P.M.							
5 P.M.							
6 P.M.							
7 P.M.							
8 P.M.							
9 P.M.							
10 P.M.							
11 P.M.							
12 A.M.							

thing worthwhile. It is important to enjoy what you're doing and to set enough time for rest and relaxation.

Study Skills for Learning Chemistry

Developing good study skills is an important goal for every student. They'll form the basis of your success in school by enabling you to keep pace with the material covered in your classes. One of the primary reasons many students are forced to drop out of school is that they lack good study skills. In this section, we'll review some of the basics of effective studying and then home in on some special tips for learning chemistry.

Study Skill Basics

There are several skills that a good student must develop. The ability to read with comprehension, to memorize when necessary, to take thorough notes, and to solve problems and analyze the material at hand are the basics of studying effectively. But there are also some negative factors that can stand in the way of good skills. Some students spend a great deal of time worrying about school work, which wastes energy and time. And some students come into chemistry class convinced that they won't be able to learn the material, which is an attitude that only invites failure.

Students are also often overburdened with activities. There are only seven days in a week, and it's impossible to fit eight days' worth of activities into seven days. Trying to do this often results in your doing things haphazardly and poorly. Effective time management is the key to overcoming the dilemma of having more to do than there are hours in a day.

In addition, it's important for you to care about your studies. If you lack interest in what you're learning and feel no enthusiasm for the subject, you'll be reluctant to put in the time required for studying chemistry. Sometimes lack of interest results from uncertainty about your long-range goals. If you know why you're going to study chemistry, your interest level will remain high.

Finally, lack of intellectual ability can be a reason for not achieving success in a subject. Your instructor is the best person to consult with if you come to believe you don't have the basic knowledge to succeed in chemistry. He or she can help you evaluate your basic skills and determine whether they are adequate.

Getting Yourself to Study

Many people find it difficult to sit down and study. They fall prey to countless distractions to avoid facing the task. If you have a problem getting started, convince yourself you'll be able to get through the material quickly by breaking it up into small, manageable units. After learning each small unit, stop and take a short break, and then continue.

Developing the ability to concentrate is an important study skill. Keep your mind on what you're doing. Don't allow yourself to be distracted. Remove the telephone and the television set, and try to eliminate interruptions. Find a quiet place where you won't be disturbed. Low-level noise such as instrumental music or a steady flow of traffic sometimes helps stimulate concentration. And don't try to study if you are hungry. That's a sure way to break your concentration.

The Importance of Attending Class

Nothing is more crucial to your study of chemistry than attending class. Regular attendance allows you to have all of the lecture notes you'll need to organize your studies, and the experiments you perform in the laboratory give you hands-on experience in chemical principles. Except for illnesses and emergencies, there is no valid reason for missing class. Punctuality is also important. Being late interrupts the class and causes you to miss some of the material being discussed.

How to Study Chemistry

The first step in your studies is to read the relevant chapter in the textbook before going to the lecture. Survey the material for the topics to be covered. Examine the types of problems to be solved. Know what the chapter is about.

Each chapter in *Basic Concepts of Chemistry* begins with a list of learning goals. Look these over and locate them in the chapter. These topics are likely to receive special emphasis in class. Also look over the figures and tables to see how they relate to the material. Glance over the summary and list of key terms at the end of each chapter.

In class you will take notes. Review them the next time you study, and reread each section of the textbook slowly for understanding. Combine what you learned in class with what you read in the textbook. Where problem solving takes place, go over the step-by-step solutions in the textbook. Then answer the practice exercises to be sure you understand how to find the solutions yourself. If you're stuck, ask the instructor for help. *Do not ignore material you do not understand.*

The next time you study, go back over your class notes and reread for a second time the sections you have learned in the textbook. *Basic Concepts of Chemistry* is known for its simple, conversational, highly readable style. You should not have a difficult time understanding the book. Spend some time being sure you can recall what you have learned. Sometimes it helps to write as you read, recording on paper what you have learned in your own words. The last step is to reread the entire chapter and go over all of your notes.

Taking Good Notes

Note taking is one of the most important skills for a student to develop. Your notes must be accurate to be useful. You will have to develop your own shorthand in order to take notes quickly. There are many symbols and abbreviations you can use in devising a shorthand. Mathematical symbols such as $>$ (greater than) and $<$ (less than), the plus sign (+), and the equal sign (=) are useful. Spelling a word phonetically by leaving out the vowels also speeds things up.

As you work through the exercises in each chapter of the textbook, clearly indicate each type of problem you are solving. Write a short description of the type of problem, and explain why you are solving this particular one. When there are several variables in a problem, be sure you can solve for each of the variables, depending on the information given.

Combine your lecture notes with your textbook notes. Be certain you understand your notes. To test your knowledge, turn to the end of the chapter and do the self-test exercises. If you can answer them, you're on your way.

Taking a Test

Being well prepared and confident is the way to approach taking a test. Cramming doesn't work in chemistry. You cannot open your book the night before the exam and learn a few weeks' worth of material in one sitting. Chemistry must be studied gradually. If you have trouble solving a particular type of problem, you may need to consult your instructor. That's difficult to do if you try to study at the last minute.

Try to study a little each day. You'll find this method more effective and less stressful. *Basic Concepts of Chemistry* contains an entire package of helpful materials. Besides the textbook, there is a study guide for students, written by James R. Braun, and a solutions manual, containing step-by-step solutions to all exercises that have specific answers. These ancillary materials can give you the extra practice you may need to achieve success in the course.

Get a good night's sleep before an exam. Be as relaxed as possible so you can think easily. You'll know whether you're prepared or not. If you've studied properly, you should do well.

Avoiding School-Related Stress

Your college years are likely to be among the most demanding and enjoyable years of your life. During this time many activities will compete for your attention. Studying, developing relationships, and handling your financial affairs are among the most important challenges that will require your energy, creativity, and brain power. You'll be able to manage successfully the different facets of your life by developing good coping skills. These are skills that will remain with you throughout your life and can be applied to just about any situation you'll encounter.

Maintaining a Good Balance

School-related stress usually results when you allow yourself to become overwhelmed and overloaded. Taking on too much responsibility often results in a high stress level, which can cause you a whole range of physical, emotional, and social problems.

When you register for your courses, be sure you're not taking on too much. Most colleges schedule a full load for their students. This means that you sign up for a set number of credits that school personnel believe a full-time student can handle successfully. The full load is calculated so that a full-time student can finish the requirements for a two-year degree in two years, or for a bachelor's degree in four years.

Students who work diligently and have good study skills and few other demands on their time can normally handle a full load. Problems often arise when a student signs up for more credits than are recommended for a full load, or when he or she has other time-consuming responsibilities. When you have too much responsibility to handle at once, stress can readily develop. Instead of completing your course work with ease, you may find yourself playing "catch-up" all semester.

Arranging a course load to fit your needs is one of your most important objectives as a college student. You can avoid overloading yourself by taking a night course during the summer at a local college or by attending summer school. Spending an extra semester or two in college is a reasonable price to pay in order to achieve success and avoid stress while you're in school.

When Money Becomes a Problem

College students very often find that they are low on money. Instead of wasting time worrying about how you'll get by, work on figuring out how you can make ends meet.

Students with a full academic load can usually work a maximum of ten hours per week. If you work more than ten hours per week, you should not carry a full load or you're likely to become overloaded. You might be able to keep up with such a schedule for a month or two, but you'll probably find yourself falling behind in your school work and putting too much stress on your body.

If you have financial problems, find out if your school has a financial aid office where you can get information about scholarships and student loans. Often a financial aid counselor can help you make up a budget. Your school may also have a job placement office to help you locate part-time work.

Being on Your Own

If you are like most young adults, your college years represent the first time you are on your own and away from your family. Loneliness is a feeling commonly shared by many students at this time. It is important to learn to meet people and make new friends when you enter college. Although it's often difficult to strike up a conversation with people you don't know, it's a skill you'll need to develop to combat loneliness.

College campuses have many clubs and organizations that provide opportunities to meet people. The college newspaper, a sports team, a political organization, or the school governing body are places to get started. Make appointments and schedule social meetings with your new acquaintances. It'll make being on your own in a new environment less threatening.

When You Socialize Too Much

Whereas some people find it difficult to socialize, others become overly involved in social activities. Attending college can be very exciting, and for some young adults the endless opportunities to work for a cause, socialize with friends, or attend frequent parties are irresistible. It is not uncommon for some college students to use study time for socialization. But around the time of midterms, panic sets in for these students and lasts until final exams. Don't get caught in this trap. Be sure you are not becoming involved in too many social activities in order to avoid your studies. If you're serious about receiving a college education, your study time will be important to you.

Maintaining Your Individuality

Peer pressure is a powerful force at all ages, and it plays a large role during the college years. As humans, we all want approval from our friends and are sometimes afraid we'll lose their respect or companionship if we disagree with them or refuse to go along with the group.

The truth is, you must set your own goals for yourself. If doing well in school is very important to you, and your friends criticize you for working so hard, then they probably don't share your goals or values. This doesn't necessarily mean you should give up your studies or your friends. It means that it's time to assert yourself and do what you think is right. If you feel you must compromise too much to remain a member of your circle of friends, then perhaps it's time to develop new relationships.

If College Isn't Right for You

Sometimes we find ourselves in situations that aren't right for us at a particular time in our lives. Some students may have to work a full-time job, leaving little time for course work or studies. Problems with relationships, drugs, alcohol, or one's family may be other reasons that college isn't working out. A lack of good study skills or simply a general lack of interest in school may also suggest that it's not the right time for an individual to be in college. If you find yourself in one of these situations, consider taking a leave of absence from college, or even dropping out. You can return when the time is right for you.

If a major problem causes you to consider leaving college, the first task at hand is to work on solving that problem. If you can't solve the problem yourself, don't hesitate to seek professional help. Your college counseling office may offer some useful solutions.

Sometimes leaving college works out for the best. It might give you the opportunity to develop a skill or talent that you truly enjoy. It might encourage you to sit down and think about your goals for the future. Colleges today enroll a large number of adults who wish to begin or complete their studies. They tend to be serious students who are interested in learning, and so they are an asset to their classes.

Words of Encouragement

As you begin your study of chemistry, we hope you do so with an open mind and a positive attitude. You are probably enrolled in a chemistry course because it is part of the path you have chosen in order to reach your career goals. We hope you will enjoy the course and find the textbook a useful tool to help you learn the subject matter.

This supplement has been added to the text to provide guidance for those who will benefit from it. Our goal is not to offer advice, but rather to offer suggestions. Chemistry may be new to you, and we hope that some of these tips for studying more effectively will be useful to you. If you have taken a chemistry course before, or if you have a different method of studying that works for you, we invite you to write to us and share your suggestions.

As you enter the fascinating world of chemistry, we hope your experience is a good one and that you come away from the course with an understanding of why we remain so enthusiastic about this exciting subject.

Supplement D Glossary

In each entry, the number in parentheses is the number and section of the chapter in which the term is first discussed.

absolute zero (13.4) The coldest possible temperature that matter can reach. It is equal to 0 K (−273°C).

accuracy (2.4) The closeness of a measurement to its true value

acid (10.4) A substance that releases hydrogen ions in solution, counteracts bases, and donates protons

acid rain (18.4) Rain containing acids that can be harmful to plant and animal life

activation energy (17.1) The energy required to begin a chemical reaction

activity series (10.5) The elements listed in decreasing order of their reactivity, or ability to react chemically

air pollution (18.4) The presence of a contaminant in the outdoor atmosphere in a concentration large enough to injure human, plant, or animal life or to interfere with the enjoyment of life or property

alcohol (20.2) Any of a class of organic compounds with the general formula R—OH

aldehyde (20.4) Any of a class of organic compounds with the general formula

$$\begin{array}{c} R—C—H \\ \| \\ O \end{array}$$

alkali (16.2) Any substance that dissolves to give a basic solution

alkane (19.4) Any of a series of organic hydrocarbons whose carbon–carbon bonds are all single bonds

alkene (19.8) Any of a series of organic hydrocarbons that contain a carbon–carbon double bond

alkyl group (19.6) An alkane less one hydrogen atom. For example, removing a hydrogen atom from methane (CH_4) yields the alkyl group called methyl (CH_3).

alkyne (19.8) Any of a series of organic hydrocarbons that contain a carbon–carbon triple bond

alpha (α) particle (4.4) Particle with a mass of 4 atomic mass units and a charge of +2 radiated from an unstable nucleus. This particle is like a helium atom with its electrons removed.

amine (20.14) Any of a class of organic compounds with the general formula R—NH_2, where one, two, or three R groups can replace the hydrogen atoms.

amorphous solid (3.5) A solid that does not have a well-defined crystalline structure

anhydrous salt (14.7) A salt molecule that has no water of hydration bonded to it

anion (6.7) A negatively charged ion

anode (4.3) A positive electrode

area (2.6) A measure of the extent of a surface, equal to an object's length times its width

aromatic hydrocarbon (19.10) Any of a series of cyclic, or closed-chain, hydrocarbons based on the molecule benzene and containing alternating single and double bonds

Arrhenius acid (16.1) A substance that releases hydrogen ions when it dissolves in water

Arrhenius base (16.2) A substance that releases hydroxide ions when it dissolves in water

atmosphere (atm) (13.2) A unit of pressure equal to 760 torr or 14.7 pounds per square inch

atom (3.10) The smallest part of an element that can enter into chemical combinations

atomic mass (3.15) The mass of an element in relation to the mass of an atom of carbon-12

atomic mass unit (amu) (4.5) The unit used to compare the relative masses of atoms. One atomic mass unit is one-twelfth the mass of a carbon-12 atom.

atomic number (4.6) The number of electrons or protons in a neutral atom

atomic radius (6.6) The distance from the center of the nucleus to the outermost electron

Avogadro's number (9.1) The number of atoms whose mass is the gram-atomic mass of any element. It is equal to 6.02×10^{23}.

Avogadro's principle (13.9) Equal volumes of all gases, under the same conditions of temperature and pressure, contain the same number of molecules.

balanced equation (10.2) A chemical equation that has the same number of atoms of each element on the reactant side as on the product side

barometer (13.2) A device used to measure air pressure

base (10.5) A substance that releases hydroxide ions in solution, counteracts acids, and accepts protons

base number (2.5) In 10^3 the base number is 10 and 3 is the exponent.

beta (β) particle (18.1) A particle with a negligible mass and a charge of −1 radiated from an unstable nucleus. This particle is electron-like in nature, but it originates from the nucleus of a radioactive atom.

binary compound (8.2) A compound composed of two elements

boiling point (14.3) The temperature at which the equilibrium vapor pressure of a liquid equals the atmospheric pressure

boiling-point elevation (15.9) The temperature at which a solution boils, which is higher than the boiling point of the pure solvent alone

boiling-point-elevation constant (15.10) The boiling point of a solution minus the boiling point of the pure solvent alone. This constant is unique for each solvent.

Boyle's Law (13.3) At constant temperature, the volume of a given mass of gas varies inversely with the pressure.

brachytherapy (18.2) Radiation therapy in which an enclosed radioactive isotope is placed under the skin to radiate cancerous tissue in a particular area of the body

branched-chain isomer (19.5) An organic compound in which the carbon atoms branch off the main chain. A branched-chain isomer has the same molecular formula as the straight-chain isomer; only its structure is different

Brønsted–Lowry acid (16.1) A substance that donates protons to another substance

Brønsted–Lowry base (16.2) A substance that accepts protons from another substance

buffer solution (17.8) A solution prepared by mixing a weak acid and its salt or a weak base and its salt. It tends to maintain its pH when an acid or base is added to it.

calorie (cal) (12.2) A unit of heat energy representing the amount of heat needed to raise the temperature of 1 gram of water by 1 Celsius degree

carbonyl (keto) group (20.8) An organic functional group with the formula

$$-\underset{\underset{\displaystyle O}{\|}}{C}-$$

carboxyl group (20.10) An organic functional group with the formula

$$-\underset{\underset{\displaystyle O}{\|}}{C}-OH$$

carboxylic acid (20.10) Any of a class of organic compounds with the general formula

$$R-\underset{\underset{\displaystyle O}{\|}}{C}-OH$$

catalyst (17.1) A substance that accelerates a chemical reaction without being used up (or undergoing any permanent change) in the process

cathode (4.3) A negative electrode

cation (6.7) A positively charged ion

CAT scan (18.2) A three-dimensional x-ray scanning technique. CAT stands for "computerized axial tomography."

Celsius (centigrade) temperature scale (2.11) A temperature scale on which the freezing point of water is 0 degrees and the normal boiling point of water is 100 degrees

chain reaction (18.1) A self-sustaining nuclear reaction

Charles's Law (13.4) At constant pressure, the volume of a given amount of gas varies directly with the temperature in kelvins.

chemical bond (Chapter 7, Introduction) The binding force due to electron loss or gain that holds atoms or ions together in molecules or formula units

chemical change (3.6) A change in the chemical composition of a substance

chemical equilibrium (17.2) A dynamic state in which two or more opposing processes take place simultaneously and at the same rate

chemical formula (3.14) The combination of the symbols of the particular elements that form a chemical compound, showing the number of atoms of each element

chemical kinetics (17.1) The study of how chemical reactions occur and the rates at which they occur

chemical nomenclature (Chapter 8, Introduction) A system for naming chemical compounds

chemical property (3.6) Those properties that show how one substance reacts with another substance

coefficient (10.3) A number placed before an element or compound in a chemical equation to balance the equation

colligative property (15.9) Any of the collective properties of solutions, such as boiling-point elevation and freezing-point depression, that depend mainly on the number of solute particles present in a given quantity of solvent

collision theory (17.1) A theory stating that a chemical reaction can occur only if reactant molecules collide with each other

combination reaction (10.4) A reaction in which two or more substances combine to form a more complex substance

Combined Gas Law (13.5) The gas law formula that relates the initial pressure, volume, and temperature of a gas to its final pressure, volume, and temperature. For a given quantity of gas, $P_i V_i/T_i = P_f V_f/T_f$.

common ion effect (17.6) The reduction in concentration of one ion as the concentration of another ion is increased at chemical equilibrium

common name (Chapter 8, Introduction) A name for a chemical compound that is derived from common usage or has been handed down through chemical history

compound (3.11) A chemical combination of two or more elements

concentration (15.3) The amount of solute in a solution. It can be expressed in terms of percentage, molarity, normality, or molality.

condense (14.5) To change from the gaseous state to the liquid state

condensed form (19.2) A shorthand notation of the formulas of organic compounds

continuous spectrum (5.1) A series of colors in which one color merges into the next, like a rainbow

coordinate covalent bond (7.3) A chemical bond in which one atom donates all the electrons used to form the bond

covalent bond (7.2) A chemical bond formed by the sharing of electrons between two atoms

crystal (14.6) A solid that has geometrically arranged planar surfaces and a symmetrical structure on the inside

crystal lattice (14.6) The symmetrical structure formed by the particles in a crystal

crystalline solid (3.5) A solid that has a fixed, regularly repeating, symmetrical internal structure

curie (18.1) A unit of radiation which is equal to 3.7×10^{10} disintegrations per second

cyclic structure (19.9) A closed-chain or ring-like structure characteristic of many hydrocarbons

Dalton's Law of Partial Pressures (13.7) The total pressure of a mixture of gases is equal to the sum of the partial pressures of all the gases in the mixture.

decomposition reaction (10.4) A reaction in which a complex substance is broken down into simpler substances is a decomposition reaction.

deliquescence (14.7) The ability to absorb enough water from the atmosphere to form a solution

density (2.10) The mass per unit volume of a substance

desiccant (14.7) A hygroscopic substance that is used as a drying agent

desiccator (14.7) A container used to hold chemicals and keep them dry

diatomic element (7.2) An element that is found in molecules made up of two like atoms. For example, hydrogen, oxygen, and nitrogen exist in the diatomic forms H_2, O_2, and N_2, respectively.

diffusion (15.11) The process by which a solute moves from an area of high solute concentration to an area of low solute concentration

dissociate (10.4) To separate an ionic substance into ions by the action of a solvent

dissociation constant (17.6) The equilibrium constant for a weak acid or weak base. This constant expresses the degree of dissociation of the acid or base

double covalent bond (7.2) A chemical bond in which two pairs of electrons are shared between two atoms

double-replacement reaction (10.4) A reaction in which two compounds exchange ions with each other

efflorescence (14.7) The spontaneous loss of water of hydration by a hydrated salt when exposed to the atmosphere

electrolysis (14.11) The process of decomposing a compound into simpler substances by passing electricity through it

electrolytic cell (18.3) A device that uses electrical energy to produce a chemical reaction

electrolytic solution (15.7) A solution that conducts electric current

electromagnetic spectrum (5.1) The full range of electromagnetic radiation, including radio waves and x rays

electron (4.3) A particle with a relative negative charge of one unit and a mass of 0.0005486 atomic mass units

electron affinity (6.8) The energy released when an additional electron is added to a neutral atom

electron configuration (5.8) The positions of the electrons in the various energy levels of an atom

electron-dot structure (Lewis electron-dot structure) (7.1) A notation that shows the symbol for an element and, by the use of dots, the number of outer electrons in an atom of the element

electronegativity (7.6) The attraction that an atom has for the electrons it is sharing with another atom

electrostatic precipitator (18.4) A pollution-control device used to eliminate particulate matter from the gases leaving a smokestack

element (3.9) Any of the basic building blocks of matter that cannot be broken down physically or by chemical means into simpler substances

empirical formula (9.2) A chemical formula showing the simplest whole-number ratio of the atoms that make up a molecule of a compound

endothermic reaction (12.1) A reaction that absorbs energy from the surroundings

end point (16.10) The point in a titration at which the indicator changes color

energy (3.2) The ability to do work. Energy appears in many forms—for example, as heat and as chemical, electrical, mechanical, and radiant (light) energy

energy level (5.3) Any of the various regions outside the nucleus of an atom in which electrons move

energy sublevel (subshell) (5.6) Any of the more specific regions within an energy level in which electrons move

enthalpy (12.2) The heat content of a chemical substance, represented by the symbol H

equilibrium constant (17.3) The product of the concentrations of a chemical reaction's products (in moles per liter), each raised to the power of its coefficient in the balanced equation, divided by the product of the concentrations of the reactants (in moles per liter), each raised to the power of its coefficient in the balanced equation.

equilibrium expression (17.3) The mathematical expression that relates the concentrations of the reactants and products to the equilibrium constant

equilibrium vapor pressure (14.2) The pressure exerted by a vapor when it is in equilibrium with its liquid at any given temperature

equivalence point (16.9) The point in an acid–base titration at which the equivalents of acid are equal to the equivalents of base

equivalent (of an acid or base) (15.5) The mass of the acid or base that contains Avogadro's number of hydrogen ions or hydroxide ions

ester (20.12) Any of a class of organic compounds with the general formula

$$\mathrm{R{-}\underset{\underset{\displaystyle O}{\|}}{C}{-}O{-}R'}$$

esterification reaction (20.12) A reaction between an alcohol and a carboxylic acid that produces an ester

ether (20.4) Any of a class of organic compounds with the general formula

$$\mathrm{R{-}O{-}R'}$$

eutrophication (18.6) The overproduction of algae that occurs in a body of water when plant nutrients are introduced through erosion or run-off

evaporate (14.2) To change from the liquid state to the gaseous (or vapor) state

excited state (5.3) The state of an electron when it is at an energy level higher than its ground-state level. An electron is in an excited state after it has absorbed energy.

exothermic reaction (12.1) A reaction that releases energy to the surroundings

exponent (2.5) A number written to the right of and above another number (called the base number)

Fahrenheit temperature scale (2.11) A temperature scale on which the freezing point of water is 32 degrees and the normal boiling point of water is 212 degrees

family (6.2) A vertical column of elements in the periodic table

formula mass (3.16) The sum of the atomic masses of all the atoms that make up a formula unit of a compound

formula unit (3.13) For an ionic compound, the smallest part of the compound that retains the properties of that compound

freeze (14.4) To change from the liquid state to the solid state

freezing point (14.4) The temperature at which the solid and liquid states of a substance exist together

freezing-point depression (15.9) The temperature at which a solution freezes, which is lower than the freezing point of the pure solvent alone

freezing-point-depression constant (15.10) The freezing point of a solution minus the freezing point of the pure solvent alone, divided by the molality. This constant is unique for each solvent.

fuel cell (18.3) A high-efficiency battery in which reactants are introduced continuously and products are removed continuously

fumigant (20.5) A substance that has disinfecting properties

functional group (20.1) An atom or group of atoms that determines the specific chemical properties of a class of organic compounds

gamma (γ) ray (18.1) A high-energy form of electromagnetic radiation emitted from the nuclei of radioactive atoms

gas (3.5) The least compact of the three physical states of matter. Gases have no definite shape or volume; they are easily compressible and will spread to fill the container in which they are placed.

gaseous air pollutant (18.4) A gaseous substance in the air, such as carbon monoxide or sulfur dioxide, that is not part of the normal composition of air and that, in sufficient concentration, may injure plant and animal life

Gay-Lussac's Law of Combining Volumes (13.8) When gases at the same temperature and pressure combine, the volumes of the reactants and products are in the same ratio as the coefficients of the balanced equation for the reaction.

Geiger-Müller Counter (18.1) A device used for detecting ionizing radiation

gram-atomic mass (9.1) The atomic mass of an element expressed in grams

gram-formula mass (9.3) The formula mass of a substance expressed in grams

ground state (5.3) The most stable state of an atom. In this state, the electrons are in their lowest possible energy levels.

group (of elements) (6.2) A vertical column of elements in the periodic table

Haber process (Chapter 11, Introduction) A method used to synthesize ammonia from its elements, nitrogen and hydrogen, for commercial purposes

half-life (18.1) The time required for the decay of half the atoms originally present in a sample of a radioactive substance

hazardous waste (18.5) Any waste that may cause death or irreversible or incapacitating illness and that threatens human health and the environment because of its composition

heat of crystallization (solidification) (14.4) The quantity of heat released to the environment when a quantity of liquid changes to a solid at the freezing point. It is usually expressed in calories per gram.

heat of formation (12.7) The heat released or absorbed when 1 mole of a compound is formed from its elements

heat of reaction (12.2) The change in heat content of a substance during a chemical reaction

heat of vaporization (14.3) The quantity of heat required to change a quantity of liquid to a vapor at the normal boiling point. It is usually expressed in calories per gram.

heterogeneous matter (3.7) Matter made up of parts with different properties; nonuniform matter

heterogeneous mixture (3.7) A mixture that consists of two or more substances that retain their own characteristic properties

homogeneous matter (3.7) Matter made up of parts with similar properties; uniform matter

homogeneous mixture (3.7) A mixture that consists of two or more substances but is uniform in composition—that is, every part of the mixture is exactly like every other part.

homologous series (19.4) A series of compounds of the same chemical type that differ only by fixed increments of the constituent elements

hydrate (14.7) A compound that contains chemically combined water in definite proportions

hydrocarbon (19.4) Any of the organic compounds that contain only carbon and hydrogen

hydrogen bond (as applied to water molecules) (14.7) A chemical bond formed between the hydrogen atom of one water molecule and the oxygen atom of another water molecule

hydronium ion (16.1) The ion formed by the addition of a hydrogen ion to a water molecule; written as H_3O^+

hydroxyl functional group (20.2) The —OH group in an organic compound such as an alcohol

hygroscopic substance (14.7) A substance that can absorb water from the atmosphere

ideal gas (13.1) A gas that obeys the Ideal Gas Law. For a gas to be "ideal," the molecules of the gas should have no attraction for each other under all conditions of temperature and pressure.

Ideal Gas Law (13.10) The relationship among the pressure, volume, temperature, and number of moles of a gas expressed by the mathematical formula $PV = nRT$

immiscible (15.1) Incapable of forming a solution when mixed

indicator (16.10) A chemical substance that has the ability to change color depending on the pH of the solution in which it is placed

ion (7.4) An atom or group of atoms that has gained or lost one or more electrons and therefore has a positive or negative charge

ionic bond (7.4) A bond formed by the transfer of electrons from one atom to another. The atoms involved are always of different elements.

ionization (6.7) The process by which ions are formed from atoms or molecules by the transfer of electrons

ionization (dissociation) constant (17.6) The equilibrium constant for a weak acid or weak base that expresses the degree of dissociation of the acid or base

ionization potential (ionization energy) (6.7) The energy needed to remove an electron from an isolated atom

irreversible reaction (17.2) A chemical reaction in which one or more reactants form one or more products, but the products cannot then react to form the reactants

isomer (19.4) A compound with the same molecular formula as another compound but with a different structural formula

isotope (4.7) One of two or more atoms that have the same number of electrons and protons but different numbers of neutrons

Kelvin scale (2.11) An absolute temperature scale in which kelvins equal degrees Celsius + 273

ketone (20.8) A class of organic compounds with the general formula

$$\begin{array}{c} R-\underset{\Large \overset{\|}{O}}{C}-R' \end{array}$$

kinetic energy (3.4) Energy that an object possesses by virtue of its motion

kinetic theory of gases (13.1) A theory that attempts to explain the properties of gases in terms of the forces between the gas molecules and the energy of these molecules

Law of Conservation of Energy (3.3) Energy can neither be created nor destroyed.

Law of Conservation of Mass (3.3) Matter can neither be created nor destroyed.

Law of Conservation of Mass and Energy (3.3) Matter and energy can neither be created nor destroyed, but they can change from one form to another, and the sum of all the matter and energy in the universe always remains the same.

Law of Definite Composition (3.11) A law that states every compound is composed of elements in a certain fixed proportion

Law of Mass Action (17.3) For a system at equilibrium, the ratio of the product of the product concentrations to the product of the reactant concentrations is a constant.

LeChatelier's Principle (17.5) The principle that states when stress is applied to a system at equilibrium, the system adjusts to a new equilibrium position, if possible, in such a way as to reduce the effect of the stressful condition

lethal dose (18.1) The amount of radiation that will kill 50% of all exposed organisms within 30 days; written as LD^{30}_{50}

line spectrum (5.1) A series of colors that shows bright lines separated by dark bands

liquid (3.5) The physical state of matter in which particles are held together but are free to move about. Liquids have a definite volume but take the shape of the container in which they are placed.

mass (2.9) A measure of the quantity of matter in an object

mass number (4.7) The mass of a particular atom in atomic mass units. It is essentially the total number of protons and neutrons in the nucleus of the atom.

matter (3.2) Anything that occupies space and has mass

melt (14.4) To change from the solid state to the liquid state

melting point (14.4) The temperature at which the solid and liquid states of a substance exist together

metal (3.9) An element that conducts electricity and heat, has luster, and takes on a positive oxidation number when it bonds with another element

metalloid (3.9) An element that has some of the properties of metals and some of the properties of nonmetals
metric system (2.8) A system of measurement based on multiples of ten
miscible (14.5) Capable of forming a solution when mixed
mixture (3.7) A combination of two or more substances that can be separated by physical means
molality (*m*) (15.8) A concentration unit for solutions: moles of solute per kilogram of solvent
molarity (15.4) A concentration unit for solutions: moles of solute per liter of solution
molar mass (9.3) A general term used to describe the gram-formula mass or gram-atomic mass of a substance
molar volume of a gas (13.9) The volume of 1 mole of a gas at 0°C and 1 atm, which is 22.4 liters for an ideal gas
mole (9.1) 6.02×10^{23} items
molecular formula (9.4) A chemical formula showing the number of atoms of each element in a molecule of a compound
molecular mass (3.16) The sum of the atomic masses of all of the atoms that make up a molecule
molecule (3.12) The smallest particle of a compound that can enter into chemical reactions and retain the properties of the compound
monatomic ion (7.4) An ion consisting of a single atom that has taken on a positive or negative charge
negative ion (anion) (6.7) An atom that has gained one or more electrons and thereby taken on a negative charge
neutralization reaction (10.4) A reaction in which an acid and a base react to form a salt and water
neutron (4.5) A particle with no electric charge and a mass of 1.0086650 atomic mass units
nonelectrolytic solution (15.7) A solution that does not conduct electric current
nonmetal (3.9) An element that is not a good conductor of heat or electricity and that usually takes on a negative oxidation number when it bonds with a metal
nonpolar covalent bond (7.8) A covalent bond in which the electrons are shared equally by the atoms forming the bond
normality (*N*) (15.5) A concentration unit for solutions: equivalents of solute per liter of solution
nuclear fission (18.1) The splitting of an atom into two or more different atoms when it is bombarded by neutrons
nuclear fusion (18.1) The combining of two atoms to form a new, heavier atom. This process releases a huge amount of energy.
nuclear transformation (transmutation) (18.1) The changing of one element to another by the process of radioactive decay
nucleus (4.5) The center of an atom, containing most of the mass and one or more units of positive charge
octet rule (5.6) A general rule stating that atoms with eight valence electrons tend to be nonreactive
orbital (5.4) A region of space, near the atomic nucleus, in which there is a 95% probability of finding an electron. The orbital defines the outer boundaries of the atom.
organic chemistry (19.1) The branch of chemistry that deals with organic (carbon-containing) compounds and their properties
organic compound (19.1) Any of the compounds containing the element carbon, along with the elements hydrogen, oxygen, nitrogen, sulfur, and the Group VIIA elements
osmometry (15.11) A method for determining molecular masses that relates osmotic pressure to the molar concentration of a solute in a solution
osmosis (15.11) The passage of a solvent through a semipermeable membrane
osmotic pressure (15.11) The amount of pressure that must be applied to prevent the flow of a solvent through a semipermeable membrane
outermost shell (5.9) The energy level farthest away from the nucleus of an atom
oxidation (10.6) The loss of electrons by a substance undergoing a chemical reaction
oxidation number (8.3) A number that expresses the combining capacity of an element or a polyatomic ion in a compound
oxidation-reduction (redox) reaction (10.6) A reaction in which one chemical substance is oxidized and another chemical substance is reduced is a redox reaction.
oxidizing agent (10.6) A substance that causes something else to be oxidized, or to lose electrons
partial pressure (13.7) The pressure of a particular gas in a mixture of gases
particulate matter (18.4) Particles of smoke, dust, fumes, and aerosols that are found in the atmosphere and may pollute the environment

percentage composition (9.5) The percentage by mass, volume, or mass–volume of each element in a compound

period (6.2) A horizontal row of elements in the periodic table

PET scan (18.2) A computerized analysis of radiation emitted by radioisotopes administered to a patient for medical diagnosis or therapy. PET stands for "positron emission tomography."

photochemical smog (18.4) Smog that results from atmospheric chemical reactions in which sunlight is a catalyst

pH scale (16.6) A scale that expresses the acidity of a solution

physical property (3.6) A characteristic property such as color, odor, taste, boiling point, or melting point that can be measured by nonchemical means

plasma (3.5) A form of matter composed of electrically charged atomic particles

polar covalent bond (7.8) A covalent bond in which there is an unequal sharing of electrons by the atoms forming the bond

polar molecule (dipole) (7.8) A molecule that is positive at one point and negative at another point

polyatomic ion (8.4) A charged group of covalently bonded atoms

positive ion (cation) (6.7) An atom that has lost one or more electrons and thereby taken on a positive charge

potential energy (3.4) Energy that is stored in an object by virtue of its position

precipitate (10.4) A solid substance that separates out of a solution in the course of a chemical reaction

precision (2.4) The closeness of repeated measurements to each other

primary amine (20.14) An amine in which one alkyl group replaces one hydrogen atom

product (10.1) A substance produced in a chemical reaction

proton (4.4) A particle with a relative positive charge of one unit and a mass of 1.0072766 atomic mass units

pure substance (3.7) Matter that has a definite and fixed composition is a pure substance. Elements and compounds are pure substances.

quantum (plural, *quanta*) (5.3) A specific bundle of energy emitted by an electron as it moves from one energy level to another

quantum mechanics (5.4) A mathematical model of the atom based on the probability of finding electrons in a particular region of space surrounding the nucleus of an atom

rad (18.1) A unit used in measuring the amount of radiation absorbed by living tissue. Rad stands for "radiation absorbed dose."

radioactive (nuclear) decay (18.1) Emission of radiation by a radioactive atom

radioactivity (18.1) Emissions from the nucleus of an atom as it undergoes nuclear decay, usually in the form of alpha particles, beta particles, or gamma rays

radiopharmaceutical therapy (18.2) Administration of a radioactive isotope either orally or intravenously for medical diagnosis or therapy

reactant (10.1) Any of the starting materials in a chemical reaction

reaction mechanism (17.1) The path of a chemical reaction taken by atoms or molecules to arrive at a product

real gas (13.1) Any gas that actually exists, such as oxygen, nitrogen, hydrogen, or helium gas

redox (10.6) An abbreviation for the term *oxidation–reduction*

reducing agent (10.6) A substance that causes something else to be reduced, or to gain electrons, while it itself is oxidized

reduction (10.6) The gain of electrons by a substance undergoing a chemical reaction

rem (18.1) A unit used in measuring the biological effect of radiation regardless of the source. Rem stands for "röentgen equivalent man."

representative element (6.2) An A-group element. For the A-group elements, the group number indicates how many outer electrons there are.

reversible reaction (17.2) A chemical reaction in which the products, once they are formed, can react to yield the original reactants

röentgen (18.1) A unit for measuring radiation, representing the amount of radioactivity that produces 2×10^9 ion pairs in 1 cubic centimeter of air

rounding off (2.4) A process in which one or more digits at the right end of a number are dropped in order to attain the correct number of significant figures

salt (10.4) A compound composed of the positive ion of a base and the negative ion of an acid

saturated hydrocarbon (19.4) Any of the organic compounds that contain only carbon and hydrogen and whose carbon–carbon bonds are all single bonds

saturated solution (15.2) A solution in which no more solute can be dissolved
saturation point (15.3) The level of concentration at which no more solute can dissolve in a given amount of solvent at a particular temperature
scientific method (3.1) A series of logical steps used by scientists to approach a problem and solve it effectively
scientific notation (2.5) A number expressed in exponential notation. The number 0.00625 can be 6.25×10^{-3} in scientific notation.
scrubber (18.4) A pollution-control device used to remove gaseous pollutants from exhaust gases
secondary amine (20.14) An amine in which two alkyl groups replace two hydrogen atoms
series (19.3) In organic chemistry, a collection of organic compounds that share a common structure or have similar properties
shells (5.3) See energy levels.
significant figures (2.4) Digits that express information that is reasonably reliable
single covalent bond (7.2) A chemical bond in which a single pair of electrons is shared by two atoms
single replacement reaction (10.4) A reaction in which an uncombined element replaces another element that is in a compound is a single replacement reaction.
solid (3.5) The physical state of matter in which particles are held in a definite arrangement. Solids have a definite shape and definite volume.
solubility product (17.7) The equilibrium expression for a slightly soluble salt, which is the product of the concentrations of its ions in a saturated solution at a particular temperature
solute (15.1) In a solution, the substance that is being dissolved
solution (3.8) A solution is a homogeneous mixture.
solvent (15.1) In a solution, the substance that is doing the dissolving
specific heat (12.3) The number of calories needed to raise the temperature of 1 gram of a substance by 1 Celsius degree
standard temperature and pressure (STP) (13.6) The conditions 273K (0°C) and 1 atm pressure
stoichiometry (11.1) The calculation of the quantities of substances involved in chemical reactions is stoichiometry.
straight-chain isomer (19.5) An organic compound in which the carbon atoms are linked in a straight chain; in other words, there are no branched chains
strong acid (16.3) An acid that completely dissociates into ions
strong base (16.3) A base that completely dissociates into ions
structural formula (19.2) A representation of the bonding of the carbon atoms in an organic compound
structural isomerism (19.5) The existence of two or more compounds that have the same molecular formula but different structural formulas
sublimation (14.4) The process in which a substance changes directly from the solid state to the gaseous state
supersaturated (15.2) A solution that contains more solute than is needed for saturation for a given quantity of solvent at a particular temperature
symmetrical ether (20.4) An ether that has the same two alkyl groups (R groups)
synthesize (Chapter 11, Introduction) Combining reactants to make a product.
systematic chemical name (Chapter 8, Introduction) Any of the names for chemical compounds derived from the naming system developed by the International Union of Pure and Applied Chemistry
teletherapy (18.2) Radiation therapy in which a high-energy beam of nuclear radiation is aimed at cancerous tissue
temperature (2.11) A measure of the intensity of heat
ternary compound (8.1) A compound composed of three elements
tertiary amine (20.14) An amine in which three alkyl groups replace the three hydrogen atoms in ammonia
tetrahedron (19.1) A solid figure having four faces. The most common tetrahedron is a pyramid whose base and three sides are equilateral triangles. The carbon–hydrogen bonds in a methane molecule, CH_4, form the shape of a tetrahedron.
thermochemical equation (12.4) A chemical equation that includes information about the physical state of the reactants and products and the amount of heat released or absorbed during the chemical reaction
titration (16.9) The process of determining the amount of a substance present in a solution by measuring the volume of a different solution of known strength that must be added to complete a chemical change
torr (13.2) A unit of pressure equal to 1/760 of an atmosphere
transition metal (6.2) A B-group element.

triple covalent bond (7.2) A chemical bond in which three pairs of electrons are shared by two atoms
unit cell (14.6) A repeated unit in a crystal lattice
unsaturated hydrocarbon (19.8) Any of the organic compounds that contain only carbon and hydrogen and have some carbon–carbon double bonds or carbon–carbon triple bonds in their molecules
unsaturated solution (15.2) A solution that contains less solute than can be dissolved in it at a particular temperature
unsymmetrical ether (20.4) An ether that has two different alkyl groups (R groups)
vacuum (4.3) An enclosed space from which all matter has been removed
vitalism (Chapter 19, Introduction) A now-defunct theory prevalent in the 1700s and early 1800s stating that organic substances have a vital force associated with them that human beings cannot control
volatile (14.2) Easily vaporized
voltaic cell (18.3) A device that produces electrical energy from a chemical reaction
volume (2.6) A measure of the capacity of a three-dimensional object
volumetric flask (15.4) A type of flask used by chemists to prepare a solution with a particular concentration
weak acid (16.3) An acid that partially ionizes in aqueous solution
weak base (16.3) A base that partially ionizes in aqueous solution
weight (2.9) The gravitational attraction of an object to the earth or any other body

Answers to Selected Exercises

This part of the text contains answers to self-test exercises and extra exercises whose numbers are in color at the end of each chapter. In addition, answers to all practice exercises, cumulative review exercises, and exercises in Supplement A are included.

Answers to Practice Exercises

2.1 27 **2.2** 53.44 **2.3** 6.3 cm^2 **2.4** 10 **2.5** 82 **2.6** *(a)* 2 *(b)* 4 *(c)* 6 *(d)* 4 *(e)* 2 *(f)* 5 **2.7** *(a)* 4.2×10^3 *(b)* 5.60×10^4 *(c)* 6.023×10^6 *(d)* 1.23×10^{-3} **2.8** 301 in^3 **2.9** 3,500 cm **2.10** 0.350 g **2.11** 1,800 cm **2.12** 66 cm **2.13** *(a)* 5.5 cm *(b)* 0.055 m *(c)* 0.000055 km or 5.5×10^{-5} km **2.14** *(a)* 4,400 g *(b)* 44,000 dg *(c)* 440,000 cg *(d)* 4,400,000 mg or 4.4×10^6 mg **2.15** 13.8 ft by 4.10 ft **2.16** 143 L **2.17** 2.7 g/cm^3 **2.18** 8.0 g/cm^3 **2.19** 5.50 g/cm^3 liquid A, 2.00 g/cm^3 liquid B. Liquid A is more dense. **2.20** $V = 1{,}300$ cm^3 **2.21** 118°C **2.22** 302°F **3.1** *(a)* 2 carbon atoms, 7 hydrogen atoms, 1 nitrogen atom *(b)* 2 nitrogen atoms, 8 hydrogen atoms, 1 sulfur atom, 4 oxygen atoms **3.2** *(a)* 138.9 *(b)* 55.8 *(c)* 39.9 *(d)* 118.7 **3.3** *(a)* 106.0 *(b)* 129.9 *(c)* 87.0 *(d)* 92.0 **4.1** *(a)* 14 p, 14 e, 14 n *(b)* 15 p, 15 e, 16 n *(c)* 3 p, 3 e, 3 n **4.2** Average mass = 6.94 **4.3** $^{100}Y = 20.0\%$, $^{110}Y = 80.0\%$ **4.4** 909,200 **5.1** *(a)* 32 electrons *(b)* 98 electrons **5.2** *(a)* 32 electrons *(b)* 4 sublevels: *s*, *p*, *d*, and *f* *(c)* *s* can hold 2, *p* can hold 6, *d* can hold 10, and *f* can hold 14. *(d)* *s* sublevel has 1 orbital, *p* sublevel has 3 orbitals, *d* sublevel has 5 orbitals, and *f* sublevel has 7 orbitals. **5.3** *(a)* $1s^2\,2s^2\,2p^6\,3s^2$ *(b)* $1s^2\,2s^2\,2p^6\,3s^2\,3p^6\,4s^2\,3d^{10}\,4p^4$ **6.1** The order is S, B, N, and O from longest to shortest radius. **6.2** The order is O, N, B, and S from highest to lowest ionization potential.

7.1 *(a)* H:N̤̈:H (with H below N) *(b)* H:C̤̈l: *(c)* :Ö:S̤̈ (with :Ö̤ below S)
7.2 H—O←S→O—H (with ‖O below S)

7.3 *(a)* covalent *(b)* ionic *(c)* covalent *(d)* covalent **7.4** *(a)* 84% ionic, 16% covalent *(b)* 63% ionic, 37% covalent *(c)* 55% ionic, 45% covalent *(d)* 59% ionic, 41% covalent **7.5** CS_2 is most covalent, H_2S is next, then $AsCl_3$ and CO_2 tie for least covalent. **7.6** *(a)* non-polar *(b)* polar *(c)* polar **7.7** *(a)* BF_3 bonds are polar; molecule is nonpolar. *(b)* Cl_2O bonds are polar; molecule is polar. *(c)* CO_2 bonds are polar; molecule is nonpolar. **8.1** *(a)* SO_2 *(b)* N_2O_5 *(c)* N_2O_4 *(d)* PBr_5 **8.2** KF **8.3** MgO **8.4** SrS **8.5** K_2O **8.6** $AlBr_3$ **8.7** *(a)* HgO *(b)* $FeBr_3$ *(c)* CoI_2 *(d)* MnO_2 *(e)* Cu_2S *(f)* Fe_2S_3 **8.8** *(a)* AgCl *(b)* ZnO **8.9** *(a)* $Fe(NO_2)_3$ *(b)* $Ba_3(PO_4)_2$ *(c)* $CuSO_4$ *(d)* K_2CrO_4 **8.10** *(a)* diphosphorus pentoxide *(b)* oxygen difluoride *(c)* sulfur dioxide *(d)* carbon monoxide **8.11** 2+ **8.12** *(a)* 4+ *(b)* 5+ **8.13** *(a)* rubidium chloride *(b)* gallium sulfide *(c)* strontium oxide *(d)* zinc iodide *(e)* nickel(II) chloride *(f)* iron(III) iodide *(g)* mercury(II) sulfide *(h)* copper(I) oxide **8.14** *(a)* ammonium oxide *(b)* magnesium cyanide *(c)* aluminum nitrite *(d)* zinc phosphate *(e)* calcium oxalate *(f)* nickel(II) borate **8.15** *(a)* hydrochloric acid *(b)* hydroselenic acid *(c)* hydroiodic acid **8.16** *(a)* phosphoric acid *(b)* chloric acid **8.17** *(a)* chlorous acid *(b)* nitrous acid **8.18** *(a)* perchloric acid *(b)* chloric acid *(c)* chlorous acid *(d)* hypochlorous acid **9.1** *(a)* 6.00 ft *(b)* $12\overline{0}$ in. **9.2** *(a)* 10.0 moles *(b)* 32.0 g **9.3** *(a)* 0.500 mole *(b)* 0.100 mole *(c)* 20.0 moles *(d)* 1.00×10^{-4} mole **9.4** *(a)* 981 g *(b)* 16.0 g *(c)* $12\overline{0}$ g **9.5** *(a)* 9.03×10^{24} atoms *(b)* 1.20×10^{23} atoms *(c)* 1.81×10^{24} atoms **9.6** H_2O **9.7** Cr_2S_3 **9.8** ZnN_2O_4

9.9 *(a)* 5.0 moles *(b)* 0.0010 mole **9.10** *(a)* 184 g *(b)* 5.0 g *(c)* 395 g *(d)* 7.9 g **9.11** *(a)* 2.41×10^{24} *(b)* 3.0×10^{22} *(c)* 1.51×10^{24} *(d)* 3.6×10^{22}
9.12 $C_9H_8O_4$ **9.13** 32.9% K, 67.1% Br
9.14 *(a)* 85.7% C, 14.3% H *(b)* 39.3% Na, 60.7% Cl *(c)* 92.3% C, 7.7% H **9.15** 75.0 g
10.1 $2CuO \longrightarrow 2Cu + O_2$
10.2 $Zn + H_2SO_4 \longrightarrow ZnSO_4 + H_2$
10.3 *(a)* $2Al + 3Cl_2 \longrightarrow 2AlCl_3$
(b) $2Na + 2H_2O \longrightarrow 2NaOH + H_2$
(c) $2KNO_3 \longrightarrow 2KNO_2 + O_2$
(d) $2HNO_3 + Ba(OH)_2 \longrightarrow Ba(NO_3)_2 + 2H_2O$
10.4 *(a)* $4K + O_2 \longrightarrow 2K_2O$
(b) $2Ca + O_2 \longrightarrow 2CaO$
(c) $H_2 + Br_2 \longrightarrow 2HBr$
10.5 *(a)* $Sr(OH)_2 \longrightarrow SrO + H_2O$
(b) $SrCO_3 \longrightarrow SrO + CO_2$
(c) $2KClO_3 \longrightarrow 2KCl + 3O_2$
(d) $2KCl \longrightarrow 2K + Cl_2$
10.6 *(a)* $2Li + 2H_2O \longrightarrow 2LiOH + H_2$
(b) $2Al + 3H_2SO_4 \longrightarrow Al_2(SO_4)_3 + 3H_2$
(c) $3Mg + 2Al(NO_3)_3 \longrightarrow 3Mg(NO_3)_2 + 2Al$
10.7 *(a)* $2BiCl_3 + 3H_2S \longrightarrow Bi_2S_3 + 6HCl$
(b) $2Fe(OH)_3 + 3H_2SO_4 \longrightarrow Fe_2(SO_4)_3 + 6H_2O$
10.8 *(a)* No reaction occurs. *(b)* Reaction does occur.
10.9 *(a)* Copper is oxidized; silver ion is reduced. *(b)* Sodium is oxidized; chlorine is reduced. *(c)* Bromide ion is oxidized; chlorine is reduced. *(d)* Zinc is oxidized; sulfur is reduced. **10.10** *(a)* Copper is the reducing agent; silver nitrate is the oxidizing agent. *(b)* Sodium is the reducing agent; chlorine is the oxidizing agent. *(c)* Sodium bromide is the reducing agent; chlorine is the oxidizing agent. *(d)* H_2SO_4 is both the oxidizing and reducing agent. **11.1** 2.43 g Mg, 1.60 g O **11.2** 0.900 g Al, 7.10 g $Al(NO_3)_3$, 3.27 g Zn **11.3** 49.0 g **12.1** $4\overline{0}{,}000$ cal or $4\overline{0}$ kcal **12.2** 12,500 cal or 12.5 kcal **12.3** 5380 cal or 5.38 kcal **12.4** 2680 cal or 2.68 kcal
12.5 $C_2H_6(g) + 3\frac{1}{2}O_2(g) \longrightarrow 2CO_2(g) + 3H_2O(l) + 372 \text{ kcal}$
12.6 $2C_6H_{14}(l) + 19O_2(g) \longrightarrow 12CO_2(g) + 14H_2O(l) + 198\overline{0} \text{ kcal}$
12.7 204.9 kcal heat are produced.
12.8 $PCl_5(g) \longrightarrow PCl_3(g) + Cl_2(g) - 22.2 \text{ kcal}$
12.9 $\Delta H = -75$ kcal **12.10** $\Delta H = -135.4$ kcal
12.11 $\Delta H = -216.6$ kcal **12.12** $\Delta H = -6$ kcal
12.13 $\Delta H = -74.4$ kcal **13.1** *(a)* 57 torr *(b)* 7370 torr **13.2** *(a)* 0.010 atm *(b)* 12.0 atm **13.3** 250 mL
13.4 $5\overline{0}0$ mL **13.5** 6080 torr **13.6** 1875 mL or 1880
13.7 $t_i = 17.0°C$ **13.8** $V_f = 5.15$ L

13.9 $t_f = 12.0°C$ **13.10** $V_f = 2.50$ L **13.11** $V_f =$ 235 mL **13.12** $V_f = 187$ mL **13.13** $V_f = 4.70$ L
13.14 40.0 L of HF gas **13.15** 0.0200 mole
13.16 4.48 L **13.17** 36 L **13.18** 0.0217 mole
13.19 M.M. = 410 **14.1** 40.50% **15.1** *(a)* Salt is the solute and water is the solvent. *(b)* Sulfur trioxide is the solute and water is the solvent. *(c)* Water is the solute and isopropyl alcohol is the solvent. *(d)* Hydrogen gas is the solute and oxygen gas is the solvent. *(e)* Carbon dioxide is the solute and water is the solvent. **15.2** *(a)* No *(b)* Yes *(c)* Yes **15.3** *(a)* supersaturated *(b)* unsaturated *(c)* saturated **15.4** *(a)* 44.4% *(b)* 30.0% **15.5** *(a)* 20.0% *(b)* 55.6% **15.6** *(a)* 4.5% *(b)* 20.00% **15.7** *(a)* 0.200 *M* *(b)* 0.0164 *M* *(c)* 1.07 *M* **15.8** 3.71×10^{-2} g **15.9** 49.9 mL
15.10 2.00 *N* and 2.00 *M* **15.11** 0.440 *N* and 0.440 *M* **15.12** 12.0 g **15.13** 2.00 L of the 6.00 *N* HCl solution are needed. **15.14** 16.7 mL of the 12.0 *N* H_3PO_4 solution are needed.
15.15 *(a)* 1.0 *(b)* 0.0400 *(c)* 0.160 **15.16** *(a)* 100.83°C *(b)* 100.02°C **15.17** *(a)* −2.98°C *(b)* −0.0744°C
15.18 Water will move out of the cell.
16.1 *(a)* pH = 5 *(b)* pH = 10 *(c)* pH = 1
16.2 *(a)* pH = 4 *(b)* pH = 3 *(c)* pH = 9 *(d)* pH = 3 *(e)* pH = 13 **16.3** 0.33 *N* **16.4** 0.086 *N*
16.5 0.409 *N* **16.6** 0.102 *N* **17.1** $K = \dfrac{[COCl_2]}{[CO][Cl_2]}$
17.2 $K = 2\overline{0}$ **17.3** [glucose–6–phosphate] = 2.0
17.4 1.20 moles of AB, 0.400 mole of A_2, and 0.400 mole of B_2 **17.5** *(a)* left *(b)* right *(c)* left *(d)* right
17.6 *(a)* right *(b)* left **17.7** *(a)* no change *(b)* left
17.8 *(a)* left *(b)* left *(c)* right *(d)* left
17.9 $K_a \doteq 4.5 \times 10^{-4}$ **17.10** $[H^{1+}]$ = $[H_2citrate$ ion$] = 2.9 \times 10^{-2}$ *M* [citric acid] = 0.97 *M*
17.11 $[NH_4^{1+}] = [OH^{1-}] = 4.2 \times 10^{-3}$ *M* $[NH_3] = 1.0$ *M* **17.12** *(a)* $K_{sp} = [Pb^{2+}][SO_4^{2-}]$ *(b)* $K_{sp} = [Ag^{1+}]^2[CrO_4^{2-}]$ *(c)* $K_{sp} = [Ag^{1+}][I^{1-}]$
17.13 0.034 g/L **17.14** The equilibrium will shift to the left, causing a decrease in the H^{1+} concentration, with a resultant increase in pH. A condition known as respiratory alkalosis results. **18.1** *(a)* $^{86}_{36}Kr$ *(b)* $^{208}_{82}Pb$ *(c)* $^{32}_{16}S$ **18.2** 0.391 g
19.1 *(a)* $CH_3—CH_2—CH_2—CH_3$

(b)
```
CH3—CH2—CH—CH2—CH3
         |
        CH3
```

19.2
```
(1) C—C—C—C—C—C—C
(2) C—C—C—C—C—C
                |
                C
```

```
(3) C—C—C—C—C—C
             |
             C

                  C
                  |
(4) C—C—C—C—C
                  |
                  C

          C
          |
(5) C—C—C—C—C
          |
          C

(6) C—C—C—C—C
          |  |
          C  C

(7) C—C—C—C—C
       |     |
       C     C

(8) C—C—C—C—C
          |
          C
          |
          C

          C
          |
(9) C—C—C
       |  |
       C  C
```

19.3 *(a)* hexane *(b)* butane *(c)* octane
19.4 *(a)* methylhexane *(b)* methylbutane *(c)* *n*-octane
19.5 *(a)* 3-methylhexane *(b)* 2-methylbutane *(c)* *n*-octane **19.6** 3-ethyl-5-methylheptane
19.7 2,4-dimethylhexane **19.8** *(a)* 2,3-dimethylnonane *(b)* 4-ethyl-2,3,5-trimethylnonane *(c)* 2,3,4,5-tetramethylhexane *(d)* 3,4-diethylheptane *(e)* 4-ethyl-2,3,5,6-tetramethyloctane **19.9** *(a)* 1-pentene *(b)* 1-hexene *(c)* 3-heptene *(d)* 3-methyl-1-hexene *(e)* 1-hexyne *(f)* 3-hexyne *(g)* 4,5-dimethyl-1-hexyne *(h)* 3,4-dimethyl-1-pentyne **19.10** *(a)* cyclobutane *(b)* cyclopentane **19.11** *(a)* cycloheptene *(b)* cycloheptyne *(c)* cyclopentene *(d)* cyclopropene
20.1 *(a)* polar *(b)* nonpolar **20.2** *(a)* propyl alcohol or propanol *(b)* cyclopentyl alcohol or cyclopentanol
20.3 *(a)* dipropyl ether *(b)* ethyl methyl ether
20.4 *(a)* propionaldehyde or propanal *(b)* heptylaldehyde or heptanal **20.5** *(a)* diethyl ketone or 3-pentanone *(b)* butyl ethyl ketone or 3-heptanone *(c)* ethyl methyl ketone or butanone
20.6 *(a)* propionic acid or propanoic acid *(b)* valeric acid or pentanoic acid

20.7
```
CH3CH2CH2CH2—C—OCH3 + H2O
             ‖
             O
```

20.8 *(a)* methyl propionate or methylpropanoate *(b)* propyl butyrate or propylbutanoate
20.9 *(a)* tertiary *(b)* secondary *(c)* primary
20.10 *(a)* ethyl amine or aminoethane *(b)* trimethyl amine *(c)* ethyl methyl amine

Chapter 2

1. *(a)* five *(b)* one *(c)* three *(d)* five *(e)* seven *(f)* one *(g)* four *(h)* three *(i)* three *(j)* four **3.** *(a)* seven *(b)* two *(c)* one *(d)* three *(e)* seven *(f)* three *(g)* six *(h)* four *(i)* six *(j)* four **5.** 12 cm; two significant figures **7.** 1.7 cm^2; two significant figures
9. 20.60 cm **11.** 214 cm^2 **13.** *(a)* two *(b)* three *(c)* four *(d)* four *(e)* one *(f)* two *(g)* three *(h)* four *(i)* four *(j)* three **15.** *(a)* 8×10^6 *(b)* 2×10^{-3} *(c)* 4×10^{-5} *(d)* 9.05×10^5 *(e)* 2.07×10^{-3} *(f)* 3.05×10^8 **17.** *(a)* 1.0581×10^4 *(b)* 2.05×10^{-3} *(c)* 1×10^6 *(d)* 8.02×10^2
19. *(a)* 8.5×10^8 *(b)* 6.07×10^{-6} *(c)* 6.308×10^6 *(d)* 6.005×10^{-2} *(e)* 5×10^2 *(f)* 5.0×10^2 *(g)* 5.00×10^2 *(h)* 5.000×10^2 *(i)* 2.300×10^7 *(j)* 9.30×10^{-8} **21.** 18 m^2 **23.** 426 cm^2 or 66.1 in^2 (depending on how you measure the page)
25. $6\bar{0}0$ m^2 **27.** $45\bar{0}$ cm^3 **29.** 42,400 cm^3
31. $4,5\bar{0}0$ cm^3 **33.** $1,7\bar{0}0$ cm^3 **35.** 1,728 in^3
37. *(a)* 150 g *(b)* 1,500 dg *(c)* 150,000 mg **39.** *(a)* 31 dm *(b)* 310 cm *(c)* 3,100 mm **41.** *(a)* 14.9 cm *(b)* 0.149 m *(c)* 0.000149 km **43.** *(a)* $7,85\bar{0}$ mm *(b)* 785.0 cm *(c)* 78.50 dm *(d)* 7.850 m *(e)* 0.007850 km **45.** *(a)* 2,500 m *(b)* 25,000 dm *(c)* 250,000 cm
47. *(a)* 5,678.0 cg *(b)* 567.80 dg *(c)* 56.780 g *(d)* 0.056780 kg **49.** *(a)* 3.500 L *(b)* 35.00 dL
51. 1.00×10^6 cm^3 **53.** *(a)* 82.0 ft *(b)* 985 in.
55. *(a)* 0.32 gal *(b)* 1.3 qt **57.** *(a)* 0.0110 lb *(b)* 0.176 oz **59.** 35.40 ft^3/m^3 **63.** 10.8 ft^2/m^2
69. *(a)* 3.0 m *(b)* 3.0×10^2 cm *(c)* 3.0×10^3 mm
71. *(a)* 1.83 m *(b)* 183 cm **73.** *(a)* 91.4 m *(b)* 9,140 cm **75.** *(a)* 18.9 L *(b)* 18,900 mL
79. 4.80 g/cm^3 **81.** 203 cm^3 or $2\bar{0}0$ cm^3 (for the proper number of significant figures) **83.** 53 g
85. 0.019 g **87.** 6.83 g/cm^3 **89.** 1.25 g/cm^3
91. 110 g **93.** 163 g **95.** 9.32 g/cm^3 **97.** 28,950 g or $29,\bar{0}00$ (for the proper number of significant figures)
99. 22.1 cm^3 **101.** 3.67 g/cm^3 **103.** 19,500 g
105. 50.0 cm^3 **107.** −2°C **109.** −40°C. At this temperature, degrees F equal degrees C. **111.** 68°F
113. *(a)* 10°C *(b)* −70°C *(c)* 215°C *(d)* −90.0°C

115. *(a)* 203°F *(b)* −112°F *(c)* 176°F *(d)* 410°F
118. No; the person would have to be about 13.1 feet tall and weigh $44\bar{0}$ pounds. **119.** 48 cm^3
120. 218°C **123.** 820 **124.** −8.89°C
125. *(a)* 61°F *(b)* 392°F **126.** *(a)* 149°C *(b)* −101°C
127. 22.6 miles/hour

Chapter 3

5. *(a)* 2 *(b)* 5 *(c)* 3 *(d)* 4 *(e)* 1 **7.** *(a)* chemical *(b)* physical *(c)* chemical *(d)* chemical *(e)* physical *(f)* chemical **11.** *(a)* metal *(b)* metalloid *(c)* nonmetal *(d)* metal *(e)* metalloid *(f)* metal *(g)* metal **13.** Co is the element cobalt, and CO is the compound carbon monoxide. **17.** *(a)* 12 carbon atoms, 22 hydrogen atoms, 11 oxygen atoms *(b)* 2 potassium atoms, 1 chromium atom, 4 oxygen atoms *(c)* 8 hydrogen atoms, 2 nitrogen atoms, 3 oxygen atoms, 2 sulfur atoms *(d)* 1 zinc atom, 2 nitrogen atoms, 6 oxygen atoms **21.** The atomic mass of sulfur would be 2 amu. **23.** *(a)* 85.5 *(b)* 52.0 *(c)* 238.0 *(d)* 79.0 *(e)* 74.9 **25.** *(a)* 102.0 *(b)* 223.3 *(c)* 98.0 *(d)* 42.4 *(e)* 74.1 *(f)* 399.9 *(g)* 68.1 *(h)* 129.9 **27.** *(a)* 60.1 *(b)* 82.1 *(c)* 121.6 *(d)* 104.5 *(e)* 187.5 *(f)* 218.7 *(g)* 149.0 *(h)* 60.0 **35.** *(a)* magnesium and chlorine *(b)* nitrogen and oxygen *(c)* nitrogen, hydrogen, sulfur, and oxygen *(d)* hydrogen, phosphorus, and oxygen **37.** *(a)* element *(b)* mixture *(c)* compound *(d)* mixture *(e)* compound **39.** *(a)* 254.2 *(b)* 63.0 *(c)* 89.8 *(d)* 601.9

Cumulative Review/Chapters 1–3

1. True **2.** False **3.** False **4.** True **5.** False **6.** True **7.** False **8.** True **9.** True **10.** True **11.** 128 m^2 **12.** 4,548 cm^2 **13.** 16,540 cm^2
14. *(a)* 280 g *(b)* 2,800 dg *(c)* 280,000 mg
15. *(a)* 68 dm *(b)* 680 cm *(c)* 6,800 mm
16. *(a)* 12.5 cm *(b)* 0.125 m *(c)* 0.000125 km
17. *(a)* 25.595 L *(b)* 255.95 dL
18. *(a)* 164 ft *(b)* 1970 in **19.** *(a)* 0.0562 lb *(b)* 0.899 oz **20.** 16.4 cm^3/in^3 **21.** 0.363 lb
22. *(a)* 45.3 L *(b)* 1,540 fl oz **23.** 0.289 g/cm^3
24. 0.177 g/mL. The density of air is 1.18×10^{-3} g/mL, so this gas is heavier than air and will not float in air. **25.** 6.00 g/mL **26.** 4.00 cm **27.** *(a)* 31.8 *(b)* 9.29 *(c)* 79.3 *(d)* 98.3 **28.** *(a)* 5×10^3 *(b)* 5×10^{-4} *(c)* 6.023×10^8 *(d)* 3.5000×10^7
29. *(a)* 5 *(b)* 4 *(c)* 5 *(d)* 4 **30.** *(a)* 7.22°C *(b)* −23.3°C *(c)* 230°C *(d)* −73.33°C **31.** *(a)* $19\bar{0}$°F *(b)* 9.50°F *(c)* $15\bar{0}$°F *(d)* 752°F **32.** Amorphous solids have no definite internal structure or form. Crystalline solids have a fixed, regularly repeating, symmetrical inner structure. **33.** *(a)* chemical *(b)* physical *(c)* chemical *(d)* physical **34.** Heterogeneous matter is made up of different parts with different properties. Homogeneous matter is made up of parts with the same properties throughout. **35.** *(a)* 1 zinc atom, 4 carbon atoms, 6 hydrogen atoms, 4 oxygen atoms *(b)* 2 nitrogen atoms, 8 hydrogen atoms, 1 chromium atom, 4 oxygen atoms **36.** Molecular mass is used for compounds composed of molecules. Formula mass is used for compounds composed of ions. **37.** Mercury would have a mass of 5 amu. **38.** *(a)* 173.0 *(b)* 210.0 *(c)* 31.0 *(d)* 107.9 **39.** *(a)* 232.8 *(b)* 132.1 *(c)* 180.0 *(d)* 251.1 **40.** The formula of $Al_2(SO_4)_3$ means that this formula unit contains 2 aluminum atoms, 3 sulfur atoms, and 12 oxygen atoms. **41.** *(a)* mixture *(b)* element *(c)* compound *(d)* compound *(e)* element *(f)* mixture *(g)* mixture **42.** *(a)* heterogeneous *(b)* homogeneous *(c)* heterogeneous *(d)* homogeneous **43.** Actinium (Ac), aluminum (Al), americium (Am), argon (Ar), arsenic (As), astatine (At), gold (Au), and silver (Ag) **44.** Helium (He), carbon (C), nitrogen (N), oxygen (O), fluorine (F), neon (Ne), phosphorus (P), sulfur (S), chlorine (Cl), argon (Ar), selenium (Se), bromine (Br), krypton (Kr), iodine (I), xenon (Xe), and radon (Rn) **45.** *(a)* potassium and sulfur *(b)* silver, chromium, and oxygen *(c)* potassium, manganese, and oxygen *(d)* mercury, phosphorus, and oxygen **46.** *(a)* N_2O *(b)* K_2CrO_4 *(c)* NH_3 *(d)* $SrSO_4$ **47.** *(a)* chemical *(b)* physical *(c)* chemical *(d)* physical **48.** No is the element nobelium, and NO is the compound nitrogen monoxide. **49.** No! During the burning process, some of the atoms that compose paper react with oxygen in the air and form gaseous compounds. **50.** Add water to the mixture and shake. The salt dissolves and the sand settles to the bottom of the container. Filter the sand and collect the salt water (filtrate). To retrieve the pure salt, evaporate the water.

Chapter 4

13. *(a)* 11 p, 11 e, and 12 n *(b)* 87 p, 87 e, and 136 n *(c)* 92 p, 92 e, and 146 n **15.** *(a)* 8 p, 8 e, and 8 n *(b)* 8 p, 8 e, and 9 n *(c)* 8 p, 8 e, and 10 n *(d)* 10 p, 10 e, and 10 n *(e)* 10 p, 10 e, and 11 n *(f)* 10 p, 10 e, and 12 n **17.** *(b)* is not an isotope of the others **19.** *(a)* 25 p, 25 e, and 32 n *(b)* 27 p, 27 e, and 33 n *(c)* 36 p, 36 e, and 44 n *(d)* 52 p, 52 e, and 76 n **21.** 3 p, 3 e, and 8 n

23.

Symbol	Protons	Electrons	Neutrons	Mass number	Atomic number
$^{174}_{70}Yb$	70	70	104	174	70
$^{141}_{59}Pr$	59	59	82	141	59
$^{104}_{44}Ru$	44	44	60	104	44
$^{45}_{21}Sc$	21	21	24	45	21
$^{50}_{22}Ti$	22	22	28	50	22
$^{25}_{12}Mg$	12	12	13	25	12

27. Protons, electrons, neutrons **29.** *(a)* $^{103}_{48}Cd$ *(b)* $^{123}_{55}Cs$ *(c)* $^{223}_{87}Fr$ *(d)* $^{32}_{17}Cl$ **31.** Atomic mass of Ga = 69.72. **33.** 35.46 amu **35.** 52.00 amu **37.** The percentage abundance of ^{79}Br is 50.87%, and the percentage abundance of ^{81}Br is 49.13%. **39.** ^{121}Sb = 57.696%, ^{123}Sb = 42.304 **41.** 1,500 atoms ^{2}H **43.** 2.0×10^7 oxygen atoms **45.** 7.42×10^6 atoms ^{6}Li **47.** 5.59×10^5 atoms ^{26}Mg **49.** 222 atoms ^{13}C **51.** 3.7×10^{12} atoms ^{15}N **53.** 1.0×10^9 hydrogen atoms **56.** *(a)* C *(b)* D *(c)* 222 *(d)* D *(e)* Po **57.** They are the same. **59.** The Lenard experiment using ultraviolet light **60.** *(a)* 6 p, 6 e, 7 n *(b)* 5 p, 5 e, 5 n *(c)* 7 p, 7 e, 7 n **61.** Isotopes exist. **62.** 2,570 Ne-21 atoms **63.** 1,836 electrons

Chapter 5

3. By examining the line spectrum of an extraterrestrial body with a spectroscope, scientists can determine which elements are present. **9.** Energy levels in atoms **11.** Electrons **13.** Seven energy levels: *K, L, M, N, O, P, Q* or 1, 2, 3, 4, 5, 6, 7, respectively **23.** One sublevel **25.** The number of electrons in an energy level is the sum of the electrons in the sublevels. **27.** *(a)* 5 *(b)* 50 **31.** Total of four orbitals **35.** c and d **37.** *(a)* $1s^22s^22p^2$ *(b)* $1s^22s^22p^63s^23p^64s^23d^2$ *(c)* $1s^22s^22p^63s^23p^64s^23d^{10}4p^6$ *(d)* $1s^22s^22p^63s^23p^64s^23d^{10}4p^65s^24d^{10}5p^5$ **39.** *(a)* $_{56}Ba$ *(b)* $_{20}Ca$ *(c)* $_{38}Sr$ *(d)* $_{4}Be$ **41.** *(a)* 31 *(b)* 8 *(c)* 13 *(d)* 10 *(e)* 31 **43.** Elements 6 and 14, elements 3 and 19, elements 17 and 35 **45.** 98 electrons **47.** *(a)* $1s^22s^22p^63s^23p^64s^23d^{10}4p^1$ *(b)* $1s^22s^22p^63s^23p^64s^23d^{10}4p^65s^24d^{10}5p^3$ *(c)* $1s^22s^22p^63s^23p^64s^23d^{10}4p^65s^24d^{10}5p^66s^24f^{14}5d^{10}6p^2$ *(d)* $1s^22s^22p^63s^23p^64s^23d^{10}4p^65s^24d^{10}5p^66s^24f^{14}5d^{10}6p^67s^2$

49. *(a)* $_{12}Mg$ *(b)* $_{38}Sr$ *(c)* $_{70}Yb$ *(d)* $_{81}Tl$ **51.** *(a)* 60 *(b)* 12 *(c)* 24 *(d)* 20 *(e)* 4 *(f)* 60 **55.** The Group VIIIA elements tend to be chemically unreactive. **65.** The atomic number would be 280.

Chapter 6

7. Period, group or family **9.** The same number of electrons in the outermost energy level **11.** The number of electrons in the outermost energy level **13.** A-group elements, B-group elements **15.** Rb, Sr, Ca, S, Cl **17.** P, As, I, Sb, In **25.** *(a)* decreases *(b)* increases *(c)* decreases **27.** Cl, S, Ca, Sr, Rb **29.** Sb, I, As, P **31.** *(d)* **33.** *(c)* **35.** *(d)* **39.** *(a)* $_{11}Na$, $1s^22s^22p^63s^1$ *(b)* $_{13}Al$, $1s^22s^22p^63s^23p^1$ *(c)* $_{8}O$, $1s^22s^22p^4$ *(d)* $_{16}S$, $1s^22s^22p^63s^23p^4$ *(e)* $_{5}B$, $1s^22s^22p^1$ *(f)* $_{3}Li$, $1s^22s^1$ **41.** Atomic number 118 **42.** *(a)* Group IA *(b)* Group VIIA **44.** *(a)* K *(b)* K *(c)* K **48.** *(a)* K *(b)* Se **49.** *(a)* K *(b)* Se **50.** Group IVA

Cumulative Review/Chapters 4–6

1. Protons and neutrons in nucleus, electrons surrounding nucleus **2.** Cathode rays came from the cathode. **3.** A magnet would deflect particles, not waves. An electric field would easily deflect particles. **4.** UV light is directed onto a metal and makes it emit electrons. **5.** An atom is a sphere of positive electric charge in which electrons are embedded. The positive particles balance the negative electrons. **6.** Rutherford found that one alpha particle in 20,000 ricocheted. This led him to believe that an atom has a small but dense center of positive charge. **7.** New evidence disproved Thomson's theory. **8.** Neutrons **9.** Isotopes are atoms of the same element with different atomic masses. **10.** *(a)* 27 protons *(b)* 40 protons *(c)* 34 protons **11.** 20.17 **12.** 370 N atoms **13.** *(a)* $^{35}_{17}Cl$ *(b)* $^{28}_{14}Si$ **14.** 75% ^{50}Z and 25% ^{52}Z **15.** 32 p, 32 e, 41 n **16.** 26 electrons **17.** ^{16}O is most abundant. **18.** 1900 ^{7}Li atoms **19.** 9.258×10^5 atoms **20.** 1.00×10^7 atoms **21.** Electrons **22.** *(a)* **23.** Six electrons **24.** *(a)* *K* has 1. *(b)* *K* has 2 and *L* has 1. *(c)* *K* has 2, *L* has 8, and *M* has 1. *(d)* *K* has 2, *L* has 8, *M* has 8, and *N* has 1. **25.** Two electrons **26.** Sulfur **27.** *(b)* **28.** It has five electrons in its outermost energy level. **29.** *(a)* *K* has 1. *(b)* *K* has 2 and *L* has 1. *(c)* *K* has 2, *L* has 8, and *M* has 1. *(d)* *K* has 2, *L* has 8, *M* has 8, *N* has 1. **30.** $_{60}Nd$: *K* has 2, *L* has 8, *M* has 18, and *N* has 32. **31.** Raindrops act as prisms. **32.** Line spectra are used to analyze the sun and other extraterrestrial bodies. **33.** Line spectra of his hair indicated the presence of the element arsenic.

34. Our major source of radiant energy is the sun. **35.** Electromagnetic waves travel through the vacuum of space. **36.** *(c)* **37.** By producing electromagnetic waves that were longer than visible light waves **38.** Luminous watch dials absorb light energy and re-emit it when they glow. **39.** A sample of an element is heated and it begins to glow. **40.** Quanta **41.** The periodic law states that the chemical properties are periodic functions of their atomic numbers. This means that elements with similar chemical properties recur at regular intervals and are placed accordingly on the periodic table. **42.** Moseley was able to determine the nuclear charge or the atomic numbers of the atoms of known elements. **43.** $1s^22s^22p^63s^23p^63d^{10}4s^24p^64d^{10}4f^{14}5s^25p^65d^{10}5f^{14}6s^26p^66d^57s^2$ Element 107 is a Group VIIB element. **44.** $1s^22s^22p^63s^23p^63d^{10}4s^24p^64d^{10}4f^{14}5s^25p^65d^{10}5f^{14}6s^26p^66d^47s^2$. Element 106 is a Group VIB element. **45.** Elements 18–19, 27–28, 52–53, and 92–93 **46.** False **47.** True **48.** False **49.** Fe, Ru, Os **50.** $_{11}Na$ **51.** True **52.** False **53.** Fluorine **54.** $1s^22s^22p^63s^23p^2$ **55.** True **56.** *(a)* Group IIA *(b)* Group VIA **57.** *(a)* Na *(b)* Ca **58.** Mg **59.** Cs **60.** Li

Chapter 7

3. *(a)* ·C̈l: *(b)* :C̈l:$^{1-}$ **5.** *(a)* Ḃe· *(b)* Be^{2+}
9. *(a)* Mg^{2+} *(b)* Na^{1+} *(c)* Al^{3+} **11.** *(a)* :B̈r:$^{1-}$ *(b)* :S̈e:$^{2-}$
(c) :P̈:$^{3-}$ **15.** Double

19. *(a)* H—C—Br: (with :Br: bonded above and :Br: bonded below the C) *(b)* H—C≡C—H
(c) H—P̈—H (with H bonded below the P) *(d)* H—C≡N̈

21. *(a)* :F̈—C—F̈: (with :F: bonded above and :F: bonded below the C) *(b)* H—C=C—H (with H bonded below each C)
(c) H—Äs—H (with H bonded below the As) *(d)* H—S̈· (with H bonded below the S)

23. P bonded to five :Cl: atoms. Ten electrons surround the P in this compound.

25. *(a)* :Ö=N—Ö—H (with :Ö: bonded to N by an arrow) *(b)* H—Ö—S—Ö—H (with :Ö: bonded to S by an arrow)

29. *(a)* ionic *(b)* covalent *(c)* ionic *(d)* covalent *(e)* covalent **31.** *(a)* 4% ionic, 96% covalent *(b)* 0% ionic, 100% covalent *(c)* 51% ionic, 49% covalent *(d)* 63% ionic, 37% covalent
33. HI, HBr, HCl, HF (HI: Most covalent; HF: Least covalent)
35. *(a)* 43% ionic, 57% covalent *(b)* 0.5% ionic, 99.5% covalent *(c)* 74% ionic, 26% covalent *(d)* 47% ionic, 53% covalent **37.** H_2O, H_2S, H_2Se, H_2Te (H_2O: Least covalent; H_2Te: Most covalent)
39. CsCl is the most ionic. **41.** *(a)* covalent *(b)* covalent *(c)* ionic *(d)* covalent **45.** *(a)* Bonds are polar; molecule is polar. *(b)* Bonds are polar; molecule is polar. *(c)* Bonds are polar; molecule is polar. *(d)* Bond is nonpolar; molecule is nonpolar. **47.** *(a)* Bonds are polar; molecule is nonpolar. *(b)* Bonds are polar; molecule is polar. *(c)* Bonds are polar; molecule is nonpolar. *(d)* Bond is nonpolar; molecule is nonpolar.

50. *(a)* [H—P—H, with H bonded to P by an arrow above and H bonded below]$^{1+}$ *(b)* O=C=O
(c) O bonded to two F atoms (F—O—F, bent)

52. O=O N—O C—O
Nonpolar bond; Most polar bond of the three
53. C—S bond

54. *(a)* $\left[\begin{array}{c} \text{H} \\ \uparrow \\ \text{H—N—H} \\ | \\ \text{H} \end{array}\right]^{1+}$ *(b)* Cl—Br **55.** *(a)* MgO *(b)* Al_2O_3 *(c)* Cs_2S **59.** There is no way that the duet rule can be satisfied for a molecule containing three hydrogen atoms.

Chapter 8

3. Hydrogen oxide **5.** *(a)* 8 *(b)* 3 *(c)* 4 *(d)* 9 *(e)* 5 *(f)* 7 **7.** *(a)* P_2O_5 *(b)* ClO_2 *(c)* N_2O_4 *(d)* Cl_2O_7
9. *(a)* Al_2O_3 *(b)* Li_2S *(c)* Na_2S *(d)* Ca_3N_2 *(e)* AgI *(f)* $ZnCl_2$ **11.** *(a)* III *(b)* IV *(c)* V *(d)* X
13. *(a)* metal ion with the higher charge *(b)* metal ion with the lower charge *(c)* the suffix for the atom or ion that has a negative charge in a chemical compound
15. *(a)* Cu_2S *(b)* $HgCl_2$ *(c)* FeO *(d)* SnI_2 *(e)* $CoCl_3$ *(f)* Hg_3N_2 **17.** *(a)* $(NH_4)^{1+}$ *(b)* $(OH)^{1-}$ *(c)* $(BO_3)^{3-}$ *(d)* $(HCO_3)^{1-}$ *(e)* $(C_2O_4)^{2-}$ *(f)* $(SO_3)^{2-}$
19. *(a)* arsenate *(b)* permanganate *(c)* sulfite *(d)* ammonium *(e)* hydrogen sulfite *(f)* acetate
21. *(a)* $Hg_3(PO_4)_2$ *(b)* $Sn_3(AsO_4)_2$ *(c)* $Fe(C_2H_3O_2)_3$ *(d)* Li_3PO_4 *(e)* $Al_2(SO_3)_3$ *(f)* $Zn(NO_2)_2$ **23.** *(a)* $PbSO_4$ *(b)* $Co_3(PO_4)_2$ *(c)* $(NH_4)_2Cr_2O_7$ *(d)* CaC_2O_4 *(e)* $Sn(NO_3)_2$ *(f)* $Mg(HSO_4)_2$ **25.** *(a)* diphosphorus pentasulfide *(b)* carbon monoxide *(c)* silicon dioxide *(d)* chlorine dioxide **27.** *(a)* aluminum oxide *(b)* sodium iodide *(c)* zinc chloride *(d)* magnesium nitride *(e)* silver sulfide *(f)* lithium iodide **29.** *(a)* 1− *(b)* 2.67+ *(c)* 4+ *(d)* 5+ *(e)* 1− **31.** *(a)* 5+ *(b)* 1+ *(c)* 1+ *(d)* 15+ *(e)* 3− *(f)* 1− **33.** *(a)* osmium(VIII) oxide *(b)* mercury(I) phosphide or mercurous phosphide *(c)* iron(II) sulfide or ferrous sulfide *(d)* cobalt(II) chloride or cobaltous chloride *(e)* copper(I) nitride or cuprous nitride *(f)* copper(I) oxide or cuprous oxide
35. *(a)* silver carbonate *(b)* mercury(II) phosphate or mercuric phosphate *(c)* iron(III) sulfate or ferric sulfate *(d)* sodium nitrate *(e)* copper(II) chromate or cupric chromate *(f)* zinc hydroxide **37.** *(a)* calcium sulfate *(b)* potassium cyanide *(c)* aluminum phosphate *(d)* copper(I) oxalate or cuprous oxalate *(e)* iron(III) chromate or ferric chromate *(f)* copper(II) nitrite or cupric nitrite **39.** *(a)* HClO *(b)* HBr *(c)* HNO_3 *(d)* $HClO_2$ *(e)* $HBrO_4$ *(f)* H_2SO_4 **41.** *(a)* hydrocyanic acid *(b)* hydrosulfuric acid *(c)* bromic acid *(d)* sulfuric acid *(e)* hypochlorous acid *(f)* bromous acid
43. *(a)* $Mg(OH)_2$ *(b)* H_2SO_4 *(c)* $NaNO_3$ *(d)* N_2O
45. *(a)* quicklime *(b)* marble *(c)* gypsum *(d)* oil of vitriol **47.** *(a)* SrO *(b)* AlN *(c)* Rb_2S **49.** *(a)* aluminum nitride *(b)* vanadium(V) oxide *(c)* iron(III) hydroxide or ferric hydroxide *(d)* ammonium sulfide
53. *(a)* GaF_3 *(b)* $Pd(NO_3)_2$ *(c)* AuP *(d)* $La(C_2H_3O_2)_3$ *(e)* PuO_2 *(f)* RuO_4

Cumulative Review/Chapters 7–8

1. *(a)* $\dot{\text{C}}\text{a}\cdot$ *(b)* $[\text{Ca}]^{2+}$ **2.** *(a)* $\cdot\ddot{\underset{\cdot\cdot}{\text{Cl}}}\text{:}$ *(b)* $[\text{:}\ddot{\underset{\cdot\cdot}{\text{Cl}}}\text{:}]^{1-}$
3. An ionic bond is formed by the transfer of electrons from one atom to another. The atoms are always of different elements. A covalent bond is formed by the sharing of electrons between two atoms. A coordinate covalent bond is a bond in which one element donates the electrons to form the bond.

4.
```
   H  H  H
   x· x· x·
 H:C··C··C:H
   x· x· x·
   H  H  H
```

5. HCl = 0.9, $CaCl_2$ = 2.0, $BaCl_2$ = 2.1, CaF_2 = 3.0 **6.** CsF **7.** *(a)* Cu_2SO_3 *(b)* OsO_4 *(c)* $(NH_4)_2CO_3$ *(d)* N_2O
8. *(a)* W^{5+} *(b)* Sn^{4+} *(c)* V^{5+} *(d)* In^{3+}
9. *(a)* ammonium iodide *(b)* calcium carbide *(c)* copper(I) chromate **10.** X_2Y_3 **11.** False **12.** True **13.** True **14.** False **15.** True **16.** MgO **17.** C—H **18.** Na_2SO_4 **19.** 2−

20. *(a)*
```
[  xx   xx  ]1−
[ xO:N:O:x  ]
[    ··  xx ]
[    xx     ]
[   xOx     ]
```
(b)
```
[   H   ]1+
[   ··  ]
[ H:N:H ]
[   ·x  ]
[   H   ]
```

21. *(a)* 6+ *(b)* 4+ *(c)* 7+ *(d)* 3+ **22.** *(a)* Hydrogens have partial positive charges, and oxygen has a partial negative charge. *(b)* Hydrogen has a partial positive charge, and iodine has a partial negative charge. *(c)* Chlorines have partial positive charges, and oxygen has a partial negative charge. *(d)* Phosphorus has a partial positive charge, and chlorines have a partial negative charge.

23. *(a)* :F:F: *(b)* :Cl:C::C:Cl: *(c)*
```
 :Cl:      Cl:
     \    /
:Cl — P — Cl:
      |
     :Cl:
```
(d)
```
:Cl:C::C:Cl:
   :Cl::Cl:
```

24. *(a)* 1+ *(b)* 1− *(c)* 2+ *(d)* 3+ **25.** *(a)* $Fe_3(PO_4)_2$ *(b)* $Hg(CN)_2$ *(c)* $Zn(HCO_3)_2$ *(d)* $Fe_2(Cr_2O_7)_3$ **26.** *(a)* In_2S_3 *(b)* $Mg_3(AsO_4)_2$ *(c)* Ga_2O_3 *(d)* Rb_2Se *(e)* FeO *(f)* $CoCl_2$ *(g)* Cu_3N *(h)* Cu_2O **27.** *(a)* HIO_4 *(b)* $HClO_3$ *(c)* HBr *(d)* $HClO_2$ **28.** A coordinate covalent bond is a bond in which one atom donates both electrons to form the bond. **29.** Yes. A molecule like carbon tetrachloride has four polar C—Cl bonds, but due to the shape of the molecule, the dipoles cancel out and the center of positive charge in the molecule coincides with the center of negative charge. **30.** *(a)* Group VIA *(b)* 6 electrons *(c)* 2−

Chapter 9

1. 6.02×10^{23} atoms **3.** 2.103 moles **5.** 2.53×10^8 dollars or 3×10^8 dollars (for one significant figure) **7.** *(a)* 0.600 mole *(b)* 5.00 moles *(c)* 0.00250 mole *(d)* 15.00 moles **9.** 81.25 g **11.** 44.2 g **13.** *(a)* $27\bar{0}$g *(b)* 0.479 g *(c)* 8,020 g *(d)* 1.28 g **15.** 1.81×10^{24} atoms **17.** 9×10^{-20} dollars/atom **19.** 1×10^{10}miles **21.** NaCl, sodium chloride (table salt) **23.** $C_{27}H_{46}O$ **25.** Empirical formula is CH_3; molecular formula is C_2H_6. **27.** Empirical formula is C_2H_4O; molecular formula is $C_4H_8O_2$. **29.** 1.998 moles **31.** 2.74 moles **33.** 0.0500 mole **35.** *(a)* 10.8 g water *(b)* 1,280 g SO_2 *(c)* 0.022 g CO_2 *(d)* 2.05 g $Al_2(SO_4)_3$ **37.** 13 g of SO_2 **39.** 2,270 g of propane or 5.00 pounds of propane **41.** 2.71×10^{24} molecules **43.** 1.50×10^{23} formula units **45.** 3.25×10^{24} formula units **47.** $C_6H_8O_6$ **49.** C_6H_6 **51.** $C_3H_6N_6$ **53.** $C_2H_4O_2$, acetic acid; $C_6H_{12}O_6$, dextrose **55.** H = 2.0%, S = 32.7%, O = 65.2% **57.** *(a)* B = 15.9%, F = 84.1% *(b)* U = 67.6%, F = 32.4% *(c)* C = 39.1%, H = 8.7%, O = 52.2% **59.** 34 g of sugar present **61.** *(a)* S = 50.1%, O = 49.9% *(b)* C = 75.0%, H = 25.0% *(c)* C = 42.1%, H = 6.4%, O = 51.5% *(d)* Ca = 40.1%, C = 12.0%, O = 48.0% **63.** 1,278 g of copper **65.** 244.7 g of iron **67.** 199 g of oxygen **69.** 304 g of Na_2S and 196 g of Fe_2O_3 **71.** 1.9×10^{16} years **72.** CH_2Cl **73.** $C_2H_4Cl_2$ **74.** 362 g **75.** 1.76×10^{22} atoms **76.** Hematite, 69.9% Fe, and magnetite, 72.3% Fe **77.** Cu_5FeS_4 and $CuFeS_2$ **78.** S = 50.1%, O = 49.9% **79.** 1.79×10^{-21} g **80.** 4.00 moles **81.** 124 g **82.** 1.66×10^{-24} g **83.** *(a)* 0.20 mole *(b)* 1.2×10^{23} molecules *(c)* 0.40 mole C atoms *(d)* 1.2 moles H atoms *(e)* 0.20 mole O atoms *(f)* 2.4×10^{23} C atoms *(g)* 7.2×10^{23} H atoms *(h)* 1.2×10^{23} O atoms **85.** *(a)* 0.400 mole H_2O *(b)* 3.00 moles MnO_2 *(c)* 2.5×10^{-3} mole N_2 *(d)* 15.00 moles $(NH_4)_2SO_4$ **87.** *(a)* 4.00 moles *(b)* 2.41×10^{24} molecules *(c)* 8.00 moles C *(d)* 20.0 moles H *(e)* 8.00 moles O *(f)* 4.00 moles N *(g)* 96.0 g C *(h)* 20.2 g H *(i)* 128 g O *(j)* 56.7 g N *(k)* They should be the same.

Chapter 10

1. *(a)* $2Li + 2H_2O \longrightarrow 2LiOH + H_2$
(b) $2HNO_3 + Ca(OH)_2 \longrightarrow Ca(NO_3)_2 + 2H_2O$
(c) $Mg + Zn(NO_3)_2 \longrightarrow Mg(NO_3)_2 + Zn$
(d) $K_2O + H_2O \longrightarrow 2KOH$
(e) $2ZnS + 3O_2 \longrightarrow 2ZnO + 2SO_2$
(f) $2Al + Fe_2O_3 \longrightarrow Al_2O_3 + 2Fe$
3. *(a)* $4Fe + 3O_2 \longrightarrow 2Fe_2O_3$
(b) $3H_2SO_4 + 2Al(OH)_3 \longrightarrow Al_2(SO_4)_3 + 6H_2O$
(c) $2AgNO_3 + BaCl_2 \longrightarrow 2AgCl + Ba(NO_3)_2$
(d) $Cu_2S + O_2 \longrightarrow 2Cu + SO_2$
5. *(a)* decomposition *(b)* double replacement *(c)* single replacement *(d)* combination *(e)* combination *(f)* double replacement **7.** *(a)* decomposition *(b)* decomposition *(c)* decomposition *(d)* combination *(e)* combination *(f)* single replacement *(g)* no reaction *(h)* double replacement **9.** *(a)* combination *(b)* double replacement *(c)* double replacement *(d)* single replacement
11. *(a)* $H_2 + Br_2 \longrightarrow 2HBr$
(b) $BaO + H_2O \longrightarrow Ba(OH)_2$
(c) $2Na + Cl_2 \longrightarrow 2NaCl$
(d) $N_2 + 2O_2 \longrightarrow 2NO_2$
(e) $CaCO_3 \longrightarrow CaO + CO_2$
(f) $2KOH \longrightarrow K_2O + H_2O$
(g) $Hg(ClO_3)_2 \longrightarrow HgCl_2 + 3O_2$
(h) $PbCl_2 \longrightarrow Pb + Cl_2$
(i) $2Li + 2H_2O \longrightarrow 2LiOH + H_2$
(j) $Zn + H_2SO_4 \longrightarrow ZnSO_4 + H_2$
(k) $Ni + Al(NO_3)_3 \longrightarrow$ no reaction (Aluminum is above nickel in the activity series.)
(l) $2Al + 3Hg(C_2H_3O_2)_2 \longrightarrow 2Al(C_2H_3O_2)_3 + 3Hg$
(m) $H_2SO_4 + 2NH_4OH \longrightarrow (NH_4)_2SO_4 + 2H_2O$
(n) $2AgNO_3(aq) + BaCl_2(aq) \longrightarrow Ba(NO_3)_2(aq) + 2AgCl(s)$
(o) $3H_2SO_3 + 2Al(OH)_3 \longrightarrow Al_2(SO_3)_3 + 6H_2O$
(p) $NaNO_3(aq) + KCl(aq) \longrightarrow$ no reaction
13. *(a)* $2N_2O \longrightarrow 2N_2 + O_2$
(b) $H_2CO_3 \longrightarrow H_2O + CO_2$
(c) $2NaNO_3 \longrightarrow 2NaNO_2 + O_2$
(d) $H_2 + F_2 \longrightarrow 2HF$
(e) $N_2 + 3H_2 \longrightarrow 2NH_3$
(f) $Ca + 2HCl \longrightarrow CaCl_2 + H_2$

(g) $Cu + NiCl_2 \longrightarrow$ no reaction
(h) $3AgNO_3(aq) + K_3AsO_4(aq) \longrightarrow 3KNO_3(aq) + Ag_3AsO_4(s)$
15. (a) $2K + Cl_2 \longrightarrow 2KCl$
(b) $K_2O + H_2O \longrightarrow 2KOH$
(c) $MgO + H_2O \longrightarrow Mg(OH)_2$
(d) $CaO + CO_2 \longrightarrow CaCO_3$
(e) $2NH_3 \longrightarrow N_2 + 3H_2$
(f) $SrCO_3 \longrightarrow SrO + CO_2$
(g) $2NaClO_3 \longrightarrow 2NaCl + 3O_2$
(h) $Mg(OH)_2 \longrightarrow MgO + H_2O$
17. (a) NaOH and HCl (b) KOH and H_2SO_4
(c) $Al(OH)_3$ and HNO_3
19. (a) $Zn + 2HNO_2 \longrightarrow Zn(NO_2)_2 + H_2$
(b) $Ag + NiCl_2 \longrightarrow$ no reaction
(c) $Zn + 2AgNO_3 \longrightarrow Zn(NO_3)_2 + 2Ag$
(d) $2Cs + 2H_2O \longrightarrow 2CsOH + H_2$
(e) $3HCl + Al(OH)_3 \longrightarrow AlCl_3 + 3H_2O$
(f) $KNO_3 + ZnCl_2 \longrightarrow$ no reaction
(g) $Al(NO_3)_3 + 3NaOH \longrightarrow Al(OH)_3 + 3NaNO_3$
(h) $K_2CrO_4 + Pb(NO_3)_2 \longrightarrow PbCrO_4 + 2KNO_3$
23. (a) $2K + Cl_2 \longrightarrow 2KCl$ (The K is oxidized; the Cl is reduced.)
(b) $2NH_3 \longrightarrow N_2 + 3H_2$ (The N is oxidized; the H is reduced.)
(c) $CuO + H_2 \longrightarrow Cu + H_2O$ (The H is oxidized; the Cu is reduced.)
(d) $Sn + 2Cl_2 \longrightarrow SnCl_4$ (The Sn is oxidized; the Cl is reduced.)
25. (a) K is the reducing agent; Cl_2 is the oxidizing agent. (b) The NH_3 is both oxidizing and reducing agent. (c) H_2 is the reducing agent; CuO is the oxidizing agent. (d) Sn is the reducing agent; Cl_2 is the oxidizing agent. **27.** (a) Zn is the reducing agent; HNO_2 is the oxidizing agent. (b) No reaction took place. (c) Zn is the reducing agent; $AgNO_3$ is the oxidizing agent. (d) Cs is the reducing agent; H_2O is the oxidizing agent. **29.** (a) vanadium pentoxide (b) zinc (c) zinc (d) vanadium **30.** $MgCO_3 \longrightarrow MgO + CO_2$
$CaCO_3 \longrightarrow CaO + CO_2$
31. (a) $N_2 + 3H_2 \longrightarrow 2NH_3$
(b) $3Fe + 4H_2O \longrightarrow Fe_3O_4 + 4H_2$ (c) balanced
33. (a) $2C + O_2 \longrightarrow 2CO$ (b) $H_2 + Cl_2 \longrightarrow 2HCl$
(c) $2Na + 2H_2O \longrightarrow 2NaOH + H_2$ (d) no reaction
34. (a) no reaction
(b) $Mg + Zn(NO_3)_2 \longrightarrow Mg(NO_3)_2 + Zn$
(c) no reaction
(d) $Zn + Cu(NO_3)_2 \longrightarrow Zn(NO_3)_2 + Cu$
36. (a) $Ca(OH)_2 + 2HCl \longrightarrow CaCl_2 + 2H_2O$ (double replacement) (b) $Zn + H_2SO_4 \longrightarrow ZnSO_4 + H_2$ (single replacement) (c) $2Ba + O_2 \longrightarrow 2BaO$ (combination) (d) $Cs_2O + H_2O \longrightarrow 2CsOH$ (combination)
37. $4NH_3 + 7O_2 \longrightarrow 2N_2O_4 + 6H_2O$ **38.** (a) Fe^{2+} (b) chromium in $Cr_2O_7^{2-}$ (c) $Cr_2O_7^{2-}$ (d) Fe^{2+}
39. $\underset{\text{Sodium}}{12Na} + \underset{\text{Phosphorus}}{P_4} \longrightarrow \underset{\text{Sodium phosphide}}{4Na_3P}$

Chapter 11

1. (a) $H_2SO_4 + 2KCN \longrightarrow K_2SO_4 + 2HCN$
(b) 9.81 g sulfuric acid (c) 17.4 g potassium sulfate, 5.40 g hydrogen cyanide
3. (a) $NH_4NO_3 \xrightarrow{\text{Heat}} N_2O + 2H_2O$ (b) 4.0 g ammonium nitrate (c) 1.8 g water
5. (a) $C_7H_8 + 3HNO_3 \longrightarrow C_7H_5N_3O_6 + 3H_2O$
(b) 405 g of toluene, 832 g nitric acid (c) 238 g water
7. (a) $2NH_4Cl + H_2SO_4 \longrightarrow (NH_4)_2SO_4 + 2HCl$
(b) 14.6 g sulfuric acid (c) 19.7 g ammonium sulfate, 10.8 g hydrogen chloride **9.** Percent $KClO_3$ = 51.2%
11. (a) $3BaO_2 + 2H_3PO_4 \longrightarrow 3H_2O_2 + Ba_3(PO_4)_2$
(b) 68.0 g hydrogen peroxide (c) $13\bar{0}$ g phosphoric acid **13.** 11.9 g hydrogen **15.** 1460 g air
17. (a) $2C_4H_{10} + 13O_2 \longrightarrow 8CO_2 + 10H_2O$
(b) 83.2 g oxygen (c) 70.4 g CO_2 and 36.0 g H_2O
19. 64.0 g ozone
21. (a) $2Al(OH)_3 + 3H_2SO_4 \longrightarrow Al_2(SO_4)_3 + 6H_2O$
(b) 228 g $Al(OH)_3$ and 430 g H_2SO_4 (c) 158 g H_2O
23. (a) $Fe_2O_3 + 3CO \longrightarrow 2Fe + 3CO_2$ (b) $65\bar{0}$ g Fe_2O_3 (c) 342 g CO and 536 g CO_2
25. (a) $4NH_3 + 5O_2 \longrightarrow 4NO + 6H_2O$
(b) $1\bar{0}00$ g O_2 (c) $75\bar{0}$ g NO and 675 g H_2O
27. 502 g $Zn(NO_3)_2$
29. (a) $CaC_2 + 2H_2O \longrightarrow Ca(OH)_2 + C_2H_2$
(b) 148 g $Ca(OH)_2$ and 52.0 g C_2H_2
31. (a) $CO + 2H_2 \longrightarrow CH_3OH$ (b) 228 g of methyl alcohol **33.** (a) $2H_2 + O_2 \longrightarrow 2H_2O$ (b) 72 g water
35. (a) $Ca_3P_2 + 6H_2O \longrightarrow 2PH_3 + 3Ca(OH)_2$
(b) 136 g PH_3 (c) 445 g $Ca(OH)_2$ **37.** $2\bar{0}$ moles $CaSiO_3$, 33 moles CO and 6.7 moles P_2
39. (a) $3Fe + 4H_2O \longrightarrow Fe_3O_4 + 4H_2$
(b) 10.8 g H_2 (c) $31\bar{0}$ g Fe_3O_4 **41.** (a) 0.128 mole (b) 19.1 g
42. (a) $2Cu(NO_3)_2 \longrightarrow 2CuO + 4NO_2 + O_2$
(b) 8.54 g (c) $11\bar{0}$ g **43.** (a) oxygen (b) 206 g
44. (a) $2H_2 + O_2 \longrightarrow 2H_2O$ (b) 0.8 g H_2 left and 72.0 g H_2O produced **45.** $39\bar{0}$ g
46. (a) $Zn + H_2SO_4 \longrightarrow ZnSO_4 + H_2$ (b) 19,600 g

(or $2\bar{0},000$ g for two significant figures) **47.** *(a)* The second experiment, $Pb + Cu(NO_3)_2 \longrightarrow Pb(NO_3)_2 +$ Cu *(b)* 79.8 g $Pb(NO_3)_2$ and 15.3 g Cu **48.** 6.7 g **49.** 58.5 g NaCl produced and 40.0 g NaOH left **50.** 112 g

Cumulative Review/Chapters 9–11

1. 0.0100 mole **2.** 0.40 mole **3.** *(a)* 73.6 g H_2SO_4 *(b)* 90.0 g H_2O *(c)* 0.434 g H_2CO_3 *(d)* 3.000×10^3 moles $HC_2H_3O_2$ **4.** 1.2×10^{23} atoms **5.** SnO_2 **6.** $C_{40}H_{64}O_{12}$ **7.** $C_6H_{12}O_6$ **8.** *(a)* 0.200 mole *(b)* 1.20×10^{23} molecules *(c)* 1.80 moles C atoms *(d)* 1.60 moles H atoms *(e)* 0.800 moles O atoms **9.** N_2O_4 **10.** *(a)* 310.3 *(b)* 16.0 *(c)* 46.0 *(d)* 255.3 **11.** *(a)* 5.9% H, 94.1% S *(b)* 39.3% Na, 60.7% Cl *(c)* 24.0% Fe, 30.9% C, 3.9% H, 41.2% O *(d)* 44.9% K, 18.4% S, 36.7% O **12.** C_6H_{14}
13. 74.1% C, 8.6% H, 17.3% N **14.** 2.41×10^{22} molecules **15.** 2.50 g $CaCO_3$
16. $CaBr_2 + H_2SO_4 \longrightarrow 2HBr + CaSO_4$
17. $2Mg + O_2 \longrightarrow 2MgO$
18. $NH_3(g) + HCl(g) \longrightarrow NH_4Cl(s)$
19. *(a)* decomposition *(b)* combination *(c)* single replacement *(d)* double replacement
20. *(a)* $H_2 + Cl_2 \longrightarrow 2\ HCl$
(b) $2NaCl + Pb(NO_3)_2 \longrightarrow 2NaNO_3 + PbCl_2$
(c) $2Al + 3H_2SO_4 \longrightarrow Al_2(SO_4)_3 + 3H_2$
(d) $SrCO_3 \longrightarrow SrO + CO_2$
21. $CaCO_3 \longrightarrow CaO + CO_2$
22. *(a)* $TiCl_4 + 2H_2O \longrightarrow TiO_2 + 4HCl$
(b) $P_4O_{10} + 6H_2O \longrightarrow 4H_3PO_4$
23. *(a)* $Fe(OH)_3 + H_3PO_4 \longrightarrow FePO_4 + 3H_2O$
(b) $Pb(OH)_2 + 2HNO_3 \longrightarrow Pb(NO_3)_2 + 2H_2O$
24. $SO_3 + H_2O \longrightarrow H_2SO_4$
25. $2C_8H_{18} + 25O_2 \longrightarrow 16CO_2 + 18H_2O$
26. $2C_3H_7OH + 9O_2 \longrightarrow 6CO_2 + 8H_2O$
27. $CaSO_4 + Na_2CO_3 \longrightarrow CaCO_3(s) + Na_2SO_4$
28. $C_{12}H_{22}O_{11} \longrightarrow 12C + 11H_2O$ **29.** True
30. False **31.** True **32.** True **33.** True **34.** True
35. $C + 2H_2SO_4 \longrightarrow CO_2 + 2SO_2 + 2H_2O$
36. 5 moles CuO **37.** 0.20 mole NaOH
38. 88 grams **39.** 726 grams HCl
40. 12.6 grams CO_2 **41.** 1860 grams H_3PO_4
42. 77.6 grams $Ca_3(PO_4)_2$ 55.6 grams $Ca(OH)_2$
43. 15.0 moles, 1810 grams **44.** 119 grams H_2O_2
45. 13.9 grams $PbCl_2$ **46.** 96.1 grams LiOH
47. 72.9 grams Mg **48.** 171.3 grams $Ba(OH)_2$, 2.0 grams H_2, 36.0 grams excess H_2O

Chapter 12

7. 20.0 kcal **9.** 0.345 kcal **11.** 2.09 kcal **13.** 30.0°C **15.** $\Delta H_R = -68.5$ kcal/mole **21.** -14 kcal **23.** 3210 kcal **25.** *(a)* $\Delta H_R = -621.4$ kcal *(b)* -62100 kcal **27.** *(a)* $\Delta H_R = -55.7$ kcal *(b)* 1520 g **29.** $\Delta H_f = -183.0$ kcal **31.** ΔH_f (calculated) = -94.8 kcal **33.** $\Delta H_f = -107$ kcal **35.** -14.2 kcal **37.** 2.5 kcal **39.** -47.0 kcal **41.** *(a)* -47.0 kcal *(b)* 141 kcal **43.** -21.36 kcal
45. *(a)* $PCl_5 + 4H_2O \longrightarrow H_3PO_4 + 5HCl$ *(b)* -139.8 kcal *(c)* -27.96 kcal **47.** $\Delta H_f = -37.2$ kcal/mole **49.** 6.30 kcal **50.** -288.5 kcal **51.** *(a)* and *(d)*, energy is absorbed; *(b)* and *(c)*, energy is released **52.** Acetylene releases more energy; $\Delta H = -300.2$ kcal/mole **53.** Endothermic
54. 5500 kcal **55.** H_2 produces 29 kcal/g, CH_4 produces 13.3 kcal/g, C produces 7.8 kcal/g
57. 10,600 kcal **58.** 10.5 kcal/mole

Chapter 13

3. *(a)* 0.329 atm *(b)* 2.04 atm *(c)* $1\bar{0}$ atm **5.** *(a)* 2280 torr *(b)* 228 torr *(c)* 114 torr **7.** *(a)* 0.10 atm *(b)* 3.50 atm *(c)* 0.00500 atm **9.** *(a)* 7.60 torr *(b)* 9120 torr *(c)* 4180 torr **11.** $V_f = 180$ L **13.** $V_f = 3\bar{0}\bar{0}0$ mL or 3.00 L **15.** $P_f = 95\bar{0}0$ torr **17.** $t_f = 427°C$
19. $t_f = 527°C$ **21.** $V_f = 33.3$ L **23.** *(a)* 227°C *(b)* $-269°C$ **25.** *(a)* 283 K *(b)* 263 K *(c)* $2\bar{0}\bar{0}$ K
27. *(a)* $-273°C$ *(b)* $-173°C$ *(c)* $-73°C$ *(d)* 27°C
29. *(a)* 293 K *(b)* 273 K *(c)* 173 K *(d)* 373 K
31. $P_f = 2$ atm **33.** $P_f = 1590$ torr **35.** $V_f = 421$ mL **37.** $V_f = 709$ mL **39.** $V_f = 394$ mL **41.** It would increase to 2.7 times its original volume.
43. $P_f = 59.8$ atm **45.** $V_f = 819$ mL **47.** $V_f = 276$ mL **49.** *(a)* increases *(b)* increases *(c)* increases
51. $P_{total} = 7\bar{0}\bar{0}$ torr **53.** 608 torr or 610 torr for two sig. figs. **55.** 24 L H_2O vapor **57.** 89.6 L
59. 20.5 g Na **61.** O_2 uncombined = 40 mL, $H_2O(g)$ formed = 80 mL
63. *(a)* $2SO_2 + O_2 \longrightarrow 2SO_3$ *(b)* 75.0 L *(c)* $15\bar{0}$ L
65. *(a)* $2C_4H_{10} + 13O_2 \longrightarrow 8CO_2 + 10H_2O$ *(b)* 97.5 L *(c)* 60.0 L CO_2 and 75.0 L H_2O
67. *(a)* $N_2 + 2O_2 \longrightarrow 2NO_2$ *(b)* 37.5 mL *(c)* 75.0 mL **69.** 6.83 g N_2
71. $R = 0.0821 \dfrac{\text{liter-atm}}{\text{K-mole}}$ **73.** 11.2 L **75.** 2.24 g
77. $1\bar{0}\bar{0}$ atm **79.** 1.52 K or $-271.5°C$ **81.** *(a)* C_2H_5

(b) MM = 59 *(c)* C_4H_{10} **83.** MM = 41 **85.** D_{O_2} = 1.43 g/L **87.** MM = 24.0 **89.** *(a)* NO_2 *(b)* MM = 92.3 *(c)* N_2O_4 **92.** 50.0 L **93.** 124 L **94.** 1797 L or $18\bar{0}0$ L for three sig. figs. **95.** *(a)* 277 K *(b)* 355 K *(c)* 373 K **96.** 22.4 L **97.** 0.714 g/L **99.** 429 mL **100.** 31.5 g/mole; O_2 is a likely candidate

Chapter 14

1. *(a)* The boiling point of a liquid increases with increasing atmospheric pressure. This is because a greater vapor pressure is needed for molecules to escape from the liquid. This can happen only at a higher temperature. *(b)* No effect. The melting phenomenon is not affected by changes in atmospheric pressure. **5.** *(a)* liquid *(b)* gas *(c)* solid **9.** Once the water is boiling, it doesn't get any hotter. Turning up the heat on the stove doesn't change the temperature of the water and doesn't speed up the cooking of the potatoes. **11.** If the mountain is high enough, the water boils at a significantly lower temperature because of the decrease in atmospheric pressure. Because the temperature of the water is lower when it begins to boil, it takes longer to cook a hardboiled egg on a mountain than at sea level. **15.** It is the high-energy molecules of liquid that escape from the container. The low-energy ones get left behind. The liquid therefore has less energy—in other words, less heat. The temperature of the liquid drops. **37.** Amorphous **39.** Well-defined (lattice) structure **43.** A molecular solid **45.** An ionic solid **47.** *(a)* metallic *(b)* ionic **49.** *(a)* atoms *(b)* molecules *(c)* ions **61.** 59 g **65.** 1.00 L ($1\bar{0}00$ mL) **67.** *(a)* $Mg(OH)_2$ *(b)* H_2SO_4 *(c)* NaOH + H_2 **69.** *(a)* $SrO + H_2O \longrightarrow Sr(OH)_2$ *(b)* $SO_2 + H_2O \longrightarrow H_2SO_3$ *(c)* $2Li + 2H_2O \longrightarrow 2LiOH + H_2$ **71.** 36.1% **75.** *(a)* 51.1% *(b)* 62.9% *(c)* 45.5% **77.** *(a)* $CaSO_4 \cdot 2H_2O \longrightarrow CaSO_4 + 2H_2O$ *(b)* $KAl(SO_4)_2 \cdot 12H_2O \longrightarrow KAl(SO_4)_2 + 12H_2O$ **79.** $CoSO_4 \cdot 6H_2O$ **84.** It is a repeating unit. **85.** Yes, some substances are already gases at *normal* atmospheric pressure (1.00 atm), for example, CO_2. The *normal boiling point* is defined as the boiling point at one atmosphere pressure. Therefore many substances do not have normal boiling points. **88.** Glass is an amorphous solid. **92.** *(a)* solid and liquid *(b)* solid *(c)* gas **94.** Higher

Cumulative Review/Chapters 12–14

1. Endothermic is heat taken in from the surroundings; exothermic is heat given off to the environment **2.** $10,\bar{0}00$ calories **3.** 138 calories **4.** 291 calories **5.** We use standard state conditions so we can measure heat given up or absorbed during thermochemical reactions, as opposed to measuring heat involved in temperature or pressure changes. **6.** $25,\bar{0}00$ calories **7.** $H_2(g) + \frac{1}{2}O_2(g) \longrightarrow H_2O(l) + 68.3$ kcal **8.** −54 calories **9.** 281 calories **10.** 7.05 kcal **11.** 24 mL **12.** 18.8 mL **13.** 616 torr **14.** $25\bar{0}0$ mL **15.** 551°C **16.** 5.94 L **17.** 7.31 L **18.** 0.371 L **19.** 4.16 L **20.** 16.0 moles **21.** 336 L **22.** 274 L **23.** 0.0412 mole **24.** MM = 286 **25.** 10.0 L **26.** 3.12 g/L **27.** Because $P_f < P_i$, therefore $V_f > V_i$ **28.** 2.62 L **29.** MM = 93.7 **30.** Forces of attraction between gas molecules vary. **31.** It will decrease. **32.** CO_2 = 342 torr, H_2 = 326 torr, N_2 = 138 torr, CH_4 = 0.0245 torr **33.** 3*a*, 2*b*, 1*c* **34.** $CaCO_3 \cdot 6H_2O$ **35.** 49.3% H_2O **36.** 20.9% **37.** $BaCl_2 \cdot 2H_2O$ **38.** $Na_2CO_3 \cdot 10H_2O$ **39.** 28.1 grams **40.** 3*a*, 1*b*, 2*c* **41.** 2*a*, 3*b*, 1*c* **42.** *b* **43.** Sublimation occurred. **44.** Molecular **45.** Ionic **46.** Amorphous **47.** Molecules gain energy; kinetic energy increases; attractive forces between molecules decrease. **48.** Benzene **49.** In some crystals, electrons are free to wander throughout the crystal. **50.** Attraction of opposite charges makes the bonds so strong.

Chapter 15

3. *(a)* Sodium hydroxide is the solute and water is the solvent. *(b)* Ether is the solute and carbon tetrachloride is the solvent. *(c)* Iodine is the solute and alcohol is the solvent. **7.** True **9.** Point of saturation **11.** *(a)* unsaturated *(b)* supersaturated **15.** *(a)* $2\bar{0}$% mass *(b)* 1.5% by mass *(c)* 15% by mass **17.** $2\bar{0}$ g **19.** *(a)* $2\bar{0}$% by volume *(b)* 6% by volume **21.** 12.5 mL **23.** $2\bar{0}$ mL **25.** *(a)* 3.3% by mass–volume *(b)* 5.0% by mass–volume **27.** 15 g **29.** 0.05% by mass **31.** *(a)* 12.5% *(b)* 5.00% **33.** $3\bar{0}$ g **35.** *(a)* 5.0% *(b)* 0.833% **37.** 28 mL **39.** $2\bar{0}0$ mL **41.** *(a)* 3.13% *(b)* 5.00% **43.** 12.5 g **45.** *(a)* 0.30 *M* *(b)* 0.5 *M* **47.** 2.74 *M* **49.** 8.4 g **51.** 0.109 *M* **53.** 0.15 *M* **55.** *(a)* 0.167 *M* *(b)* 0.200 *M* **57.** 0.525 *M* **59.** 16.4 g **61.** *(a)* 0.10 equivalent *(b)* 0.20 equivalent *(c)* 0.0902 equivalent

65. *(a)* 0.0299 equivalent *(b)* 0.750 equivalent **67.** *(a)* 1120 g *(b)* 9.27 g *(c)* 0.990 **69.** *(a)* 2.97 *N* *(b)* 0.13 *N* *(c)* 0.99 *M*, 0.063 *M* **71.** 59 g **73.** 250 mL **75.** *(a)* 1.00 *N* *(b)* 0.500 *N* **77.** 3.25 g **79.** 599 mL **81.** Take 125 mL of 16 *N* H_2SO_4; dilute to $50\bar{0}$ mL with water. **83.** Take 50 mL of 6 *N* NaOH; dilute to 3L with water. **85.** Take 104 mL of 12.0 *M* HCl; dilute to $25\bar{0}$ mL with water. **87.** Take 42 mL of 3.00 *M* H_2SO_4; dilute to 5.00 L with water. **97.** 0.15 molal **99.** *(a)* 0.80 molal *(b)* 0.333 molal **105.** −33.1°C **107.** 109.26°C **109.** f.p. = −2.79°C, b.p. = 100.78°C **111.** *(a)* osmosis *(b)* diffusion **116.** 1×10^{-2} *M* or 0.01 *M* **117.** 0.01 *N* and 0.005 *M* **118.** $50\bar{0}$ g **119.** 0.510 *M* and 1.02 *N* **122.** Ethanol **123.** 41.3 mL **124.** $15\bar{0}$ g/mole **125.** 2.00 *molal* solution **126.** 3.71 g/mole **127.** A 0.104°C increase in boiling point; therefore, if the boiling point of the pure water is 100.000°C, the boiling point of the solution is 100.104°C **128.** *(a)* 2.00 equivalents *(b)* 2.00 equivalents **129.** *(a)* 1.00 equivalent *(b)* 2.00 equivalents

Chapter 16

3. Acetic acid **7.** $NH_3 + H_2O \rightleftharpoons NH_4^{1+} + OH^{1-}$
15. Organic **17.** *(a)* blue, red *(b)* red, blue
19. $HBr + H_2O \longrightarrow H_3O^{1+} + Br^{1-}$
23. *(a)* $K_2O + H_2O \longrightarrow 2KOH$
(b) $MgO + H_2O \longrightarrow Mg(OH)_2$
25. *(a)* $SO_2 + H_2O \longrightarrow H_2SO_3$
(b) $CO_2 + H_2O \longrightarrow H_2CO_3$
27. *(a)* $2H_3PO_4 + 3Mg(OH)_2 \longrightarrow Mg_3(PO_4)_2 + 6H_2O$
(b) $P_2O_5 + 3H_2O \longrightarrow 2H_3PO_4$
(c) $K_2O + H_2O \longrightarrow 2KOH$
(d) $Mg(OH)_2 + CO_2 \longrightarrow MgCO_3 + H_2O$
(e) $3Zn + 2H_3PO_4 \longrightarrow Zn_3(PO_4)_2 + 3H_2$ **29.** Use a *weak* acid; for example, dilute acetic acid. Under no circumstances should you use a strong mineral acid (even if it is diluted).
35. *(a)* $Zn + H_2SO_4 \longrightarrow ZnSO_4 + H_2$
(b) $HCl + LiOH \longrightarrow LiCl + H_2O$
(c) $Ca + H_2SO_4 \longrightarrow CaSO_4 + H_2$
(d) $3Ca(OH)_2 + 2H_3PO_4 \longrightarrow Ca_3(PO_4)_2 + 6H_2O$
37. *(a)* $Zn + 2HCl \longrightarrow ZnCl_2 + H_2$
(b) $2HCl + Ca(OH)_2 \longrightarrow CaCl_2 + 2H_2O$
(c) $Mg(OH)_2 + CO_2 \longrightarrow MgCO_3 + H_2O$
41. *(a)* HNO_3 *(b)* HF *(c)* HI **43.** *(a)* sodium hydroxide *(b)* calcium hydroxide *(c)* iron(II) hydroxide *(d)* aluminum hydroxide **45.** *(a)* nitric acid *(b)* phosphoric acid *(c)* hydroiodic acid *(d)* sulfuric acid **47.** *(a)* LiOH *(b)* $Ba(OH)_2$ *(c)* $Fe(OH)_2$ *(d)* $Fe(OH)_3$ **59.** *(a)* 100 times more acidic *(b)* 100,000 times more acidic **63.** *(a)* 3 *(b)* 4 *(c)* 2 *(d)* 11 *(e)* 10 *(f)* 5 **65.** *(a)* 10^{-5} *M* *(b)* 10^{-9} *M* *(c)* 10^{-2} *M* *(d)* 10^{-12} *M* (All answers for $[H^+]$ concentration)
67. *(a)* $[H^{1+}] = 10^{-9}$ *M* *(b)* $[H^{1+}] = 10^{-6}$ *M* *(c)* $[H^{1+}] = 10^{-4}$ *M* *(d)* $[H^{1+}] = 10^{-12}$ *M*
69. *(a)* 10^{-11} *M* *(b)* 10^{-10} *M* *(c)* 10^{-6} *M* *(d)* 10^{-5} *M* *(e)* 10^{-8} *M* *(f)* 10^{-12} *M* **71.** *(a)* sodium sulfide *(b)* potassium bromide *(c)* sodium sulfate *(d)* potassium nitrate *(e)* calcium carbonate *(f)* barium iodide
73. *(a)* $AgNO_3$ *(b)* $CoCl_2$ *(c)* Cu_2SO_3 *(d)* $Hg_3(PO_4)_2$
79. $2LiOH + H_2SO_4 \longrightarrow Li_2SO_4 + 2H_2O$
81. $Zn + 2HBr \longrightarrow ZnBr_2 + H_2$
83. *(a)* $Mg + 2HCl \longrightarrow MgCl_2 + H_2$
(b) $Zn + H_2SO_4 \longrightarrow ZnSO_4 + H_2$
85. *(a)* $3KOH + H_3PO_4 \longrightarrow K_3PO_4 + 3H_2O$
(b) $Ca(OH)_2 + 2HNO_3 \longrightarrow Ca(NO_3)_2 + 2H_2O$
87. *(a)* $2K + Cl_2 \longrightarrow 2KCl$
(b) $Ca + Br_2 \longrightarrow CaBr_2$ **89.** 0.63 *N* **91.** 0.2778 *N* **93.** 0.180 *N* **95.** 0.667 *N* **97.** *(a)* acid *(b)* base **98.** *(a)* base *(b)* acid **99.** *(a)* base *(b)* acid
100. pH = 11, $[OH^{1-}] = 10^{-3}$
101. *(a)* $HCl + NaOH \longrightarrow NaCl + H_2O$
(b) $BaO + 2HCl \longrightarrow BaCl_2 + H_2O$
(c) $Zn + H_2SO_4 \longrightarrow ZnSO_4 + H_2$
102. *(a)* $OH^{1-} + H_2O \longrightarrow H{-}OH + OH^{1-}$
(b) $NH_3 + H_2O \longrightarrow NH_4^{1+} + OH^{1-}$
(c) $HSO_4^{1-} + H_2O \longrightarrow H_2SO_4 + OH^{1-}$
103. *(a)* $H_2O + H{-}OH \longrightarrow H_3O^{1+} + OH^{1-}$
(b) $HCO_3^{1-} + H_2O \longrightarrow H_3O^{1+} + CO_3^{2-}$
(c) $HNO_3 + H_2O \longrightarrow H_3O^{1+} + NO_3^{1-}$

105. $$H^{1+} + :PCl_3 \longrightarrow \left[\begin{array}{c} \quad\quad Cl \\ \quad\quad | \\ H \leftarrow P{-}Cl \\ \quad\quad | \\ \quad\quad Cl \end{array}\right]^{1+}$$

Chapter 17

1. The matching pairs are as follows: *(a)* 2 *(b)* 3 *(c)* 1 **9.** The catalyst lowers the activation energy of a reaction. **11.** The four factors affecting the rate of a chemical reaction are: a change in the concentration of the reactants; a change in the temperature; a change in the pressure on gaseous products and reactants; the presence of a catalyst. **15.** Reaction rate approximately doubles for each 10°C rise in the temperature. **17.** 80 torr per minute

21. $HC_2H_3O_2 \xrightarrow{H_2O} H^{1+} + C_2H_3O_2^{1-}$ (forward)
$H^{1+} + C_2H_3O_2^{1-} \xrightarrow{H_2O} HC_2H_3O_2$ (reverse)
23. $NaCl(s) \xrightarrow{H_2O} Na^{1+} + Cl^{1-}$ (forward)
$Na^{1+} + Cl^{1-} \xrightarrow{H_2O} NaCl(s)$
27. *(a)* False *(b)* True *(c)* False *(d)* True
29. *(a)* $K_{eq} = \frac{[C]^5[D]^4}{[A]^2[B]^3}$ *(b)* $K_{eq} = \frac{[HI]^2}{[H_2][I_2]}$
(c) $K_{eq} = \frac{[HCl]^2}{[H_2][Cl_2]}$
(d) $K_{eq} = \frac{[N_2O_4]}{[NO_2]^2}$ **31.** *(a)* $K_{eq} = \frac{[NH_3]^2}{[N_2][H_2]^3}$
(b) $K_{eq} = \frac{[NO_2]^2}{[NO]^2[O_2]}$ *(c)* $K_{eq} = \frac{[PCl_3][Cl_2]}{[PCl_5]}$
(d) $K_{eq} = \frac{[CO]^2[O_2]}{[CO_2]^2}$ **33.** $K_{eq} = 67.2$
35. $K_{eq} = 51.5$ **37.** $K_a = 6.4 \times 10^{-5}$
39. $K_a = 2.1 \times 10^{-4}$
41. $[H^{1+}] = [C_2H_3O_2^{1-}] = 1.34 \times 10^{-3}\ M$
43. $[H^{1+}] = [CN^{1-}] = 1.41 \times 10^{-5}\ M$
45. $[CuOH^{1+}] = [OH^{1-}] = 7.07 \times 10^{-5}\ M$
47. $[NH_4^{1+}] = [OH^{1-}] = 1.34 \times 10^{-3}\ M$
49. *(a)* $K_{sp} = [Ag^{1+}][I^{1-}]$ *(b)* $K_{sp} = [Fe^{3+}][OH^{1-}]^3$
(c) $K_{sp} = [Ag^{1+}]^2[CrO_4^{2-}]$
51. *(a)* $K_{sp} = [Ba^{2+}][CrO_4^{2-}]$ *(b)* $K_{sp} = [Ca^{2+}][F^{1+}]^2$
(c) $K_{sp} = [Cu^{2+}][S^{2-}]$ **53.** $K_{sp} = 1.5 \times 10^{-9}$
55. $K_{sp} = 1.90 \times 10^{-12}$
57. Solubility of $CuS = 9.1 \times 10^{-21}$ g/L
59. Solubility of $Ag_2CrO_4 = 2.6 \times 10^{-2}$ g/L
63. $H_2O + CO_2 \rightleftarrows H_2CO_3 \rightleftarrows H^{1+} + HCO_3^{1-}$
65. *(a)* Equilibrium shifts to the left *(b)* Equilibrium shifts to the right *(c)* Equilibrium shifts to the right
67. $HCO_3^{1-} + H^{1+} \longrightarrow H_2CO_3 \longrightarrow H_2O + CO_2$
69. *(a)* No shift *(b)* Equilibrium shifts to the right *(c)* Equilibrium shifts to the right **70.** *(a)* Equilibrium shifts to the left *(b)* Equilibrium shifts to the right *(c)* Equilibrium shifts to the right
71. *(a)* $K_{eq} = \frac{[NO]^2}{[N_2][O_2]}$ *(b)* $K_{eq} = \frac{[CO]^2[O_2]}{[CO_2]^2}$
(c) $K_{eq} = \frac{[SO_3]^2}{[SO_2]^2[O_2]}$ **72.** $K_{eq} = 3.5 \times 10^{-3}$
73. $K_a = 1.8 \times 10^{-4}$
74. $[H^{1+}] = [NO_2^{1-}] = 6.7 \times 10^{-3}\ M$
75. $K = 100$ **76.** *(a)* $K_{sp} = [Mn^{2+}][OH^{1-}]^2$
(b) $K_{sp} = [Ag^{1+}][Cl^{1-}]$ *(c)* $K_{sp} = [A^{3+}]^2[X^2]^3$
77. Solubility of $CaF_2 = 1.6 \times 10^{-2}$ g/L
78. Ag_2CrO_4

Cumulative Review/Chapters 15–17

1. *(a)* Oxygen is the solute and nitrogen is the solvent. *(b)* Sulfur dioxide is the solute and water is the solvent. *(c)* Lemon juice and sugar are solutes and water is the solvent. **2.** *(a)* H—F is polar; H_2O is polar. Therefore a solution forms. *(b)* Benzene is nonpolar; *trans*-dichloroethane is nonpolar. Therefore a solution forms. **3.** *(a)* 66.7% by mass salt *(b)* 11.7% by mass sugar **4.** *(a)* 5.49% by volume alcohol in water *(b)* 36.8% by volume benzene in CCl_4 **5.** 6.7% mass–volume benzene in CCl_4 **6.** 0.027 *M* **7.** 13.3 g NaOH **8.** 234 mL **9.** Normality is the number of equivalents of solute per liter of solution, and molarity is the number of moles of solute per liter of solution. **10.** 65.4 g H_3PO_4 **11.** 1.00 *N* or 0.500 *M* **12.** Dilute 0.203 L of 36.0 *N* with water to 5.50 L. **13.** 51.0 g $AgNO_3$ **14.** False **15.** 177 g H_2SO_4 **16.** *(a)* 545 g of sugar in 2180 g solution *(b)* 0.795 M solution **17.** 0.90 g solute **18.** Dilute the solution to a total volume of 3.82 L. **19.** 101.7°C **20.** 0.1 *m* **21.** Arrhenius acid is a substance that releases hydrogen ions when dissolved in water. Brønsted–Lowry acid is a substance that is a proton donor. **22.** False **23.** *(a)* $2HC_2H_3O_2 + Ca(OH)_2 \longrightarrow Ca(C_2H_3O_2)_2 + 2H_2O$
(b) $N_2O_5 + H_2O \longrightarrow 2HNO_3$
24. $H_2O + SO_3 \longrightarrow H_2SO_4$ or $H_2O + SO_2 \longrightarrow H_2SO_3$ **25.** pH = 2 **26.** pH = 6 **27.** 18,$\bar{0}$00 mL **28.** 1.00 *M* Na^{1+} and 1.00 *M* Cl^{1-} **29.** 1.20 *M* HCl **30.** $M_{NaOH} = 0.0450$
31. Ionization is the process by which ions are formed from atoms or molecules by the transfer of electrons. Dissociation is the process by which an ionic substance separates into ions by the action of a solvent.
32. 1 *M* sulfuric acid is more acidic than 1 *N* sulfuric acid. **33.** 1 *M* hydrochloric acid is more acidic than 1 *M* acetic acid. **34.** 0.04 *M* ions **35.** If the Cl^{1-} concentration is 0.500 *M*, the Sr^{2+} concentration must be half of that (or 0.250 *M*). **36.** 0.125 *M* **37.** The term refers to ionization. **38.** Twice as many Cl^{1-} as Sr^{2+} **39.** pH = 4 **40.** $[OH^{1-}] = 10^{-6}$ **41.** Concentration, temperature, presence of a catalyst and the reactants themselves affect the rate of a chemical reaction. **42.** As equilibrium approaches, the rate of the forward reaction slows and the rate of the reverse reaction increases. At equilibrium the forward and reverse reactions occur at the same rate. **43.** $H_2O(l) \rightleftharpoons H_2O(g)$ **44.** Exothermic means heat is released as a result of the reaction. **45.** False **46.** The equilibrium shifts

to the left to remove H^{1+} ions. **47.** $K_{eq} = [CO_2]$ **48.** $K_{sp} = 2.6 \times 10^{-13}$ **49.** Solubility is 2.5×10^{-17} moles/L. **50.** Solubility is 0.073 g/L. **51.** 0.75 mole/Liter **54.** AgCl has the greatest molar solubility. **55.** It will shift to the left to relieve the stress.

Chapter 18

11. *(a)* ${}^{197}_{78}\text{Pt} \longrightarrow {}^{0}_{-1}\text{e} + {}^{197}_{79}\text{Au}$
(b) ${}^{212}_{84}\text{Po} \longrightarrow {}^{4}_{2}\text{He} + {}^{208}_{82}\text{Pb}$
(c) ${}^{10}_{5}\text{B} + {}^{4}_{2}\text{He} \longrightarrow {}^{1}_{0}\text{n} + {}^{13}_{7}\text{N}$
(d) ${}^{238}_{92}\text{U} + {}^{1}_{0}\text{n} \longrightarrow {}^{239}_{92}\text{U} \longrightarrow {}^{94}_{36}\text{Kr} + {}^{145}_{56}\text{Ba}$
19. *(a)* ${}^{14}_{6}\text{C} \longrightarrow {}^{0}_{-1}\text{e} + {}^{14}_{7}\text{N}$
(b) ${}^{24}_{11}\text{Na} \longrightarrow {}^{0}_{-1}\text{e} + {}^{24}_{12}\text{Mg}$ *(c)* ${}^{238}_{92}\text{U} \longrightarrow {}^{4}_{2}\text{He} + {}^{234}_{90}\text{Th}$
(d) ${}^{84}_{36}\text{Kr} \longrightarrow {}^{1}_{0}\text{n} + {}^{83}_{36}\text{Kr}$ **23.** 23,080 years
25. 12.5 mg **29.** *(a)* fission *(b)* fusion **31.** ${}^{235}\text{U}$
40. beta **41.** scintillation counter
45. 0.5 rem/year
53. $Pb + PbO_2 + 2H_2SO_4 \longrightarrow 2PbSO_4 + 2H_2O$
65. nitrogen and oxygen **69.** *(a)* $C + O_2 \longrightarrow CO_2$
(b) $S + O_2 \longrightarrow SO_2$ *(c)* $2SO_2 + O_2 \longrightarrow 2SO_3$
(d) $N_2 + O_2 \longrightarrow 2NO$ **77.** An electrostatic precipitator to remove particulates **79.** An electrostatic precipitator for particulates and a scrubber for sulfur oxides **95.** 3.9 billion years **102.** less stable
103. ${}^{16}_{8}\text{O} \longrightarrow {}^{12}_{6}\text{C} + {}^{4}_{2}\text{He}$
104. ${}^{10}_{5}\text{B} + {}^{4}_{2}\text{He} \longrightarrow {}^{13}_{7}\text{N} + {}^{1}_{0}\text{n}$ **105.** 60 minutes
106. The warmer the water, the less the amount of dissolved oxygen. Fish could die from lack of oxygen.

Chapter 19

1. Covalent **3.** Diamond and graphite **9.** ·Ċ· (with dots above and below)
11. Tetrahedral **13.** *(a)* double bond *(b)* triple bond *(c)* single bond **15.** *(a)* $CH_3—CH_3$ *(b)* $CH_2═CH_2$ *(c)* $CH≡CH$

19. $CH_2═CH_2$ (Ethene) $CH_3CH═CH_2$ (Propene) $CH_3CH_2CH═CH_2$ (1-butene)
$CH_3CH_2CH_2CH═CH_2$ (1-pentene)
$CH_3CH_2CH_2CH_2CH═CH_2$ (1-hexene)

23.
```
CH3—CH—CH2—CH—CH3
    |       |
    CH3     CH3
```

27. *(a)*
```
CH3CH2—CH—CH3
       |
       CH3
```

(b)
```
CH3—CH—CH—CH—CH2CH2CH3
    |   |   |
    CH3 CH3 CH3
```

(c)
```
       CH3
       |
CH3—C—CH3
       |
       CH3
```

(d) $CH_3CH_2CH_2CH_2CH═CH_2$
(e) $CH_3CH_2CH_2CH═CHCH_3$
(f) and *(g)*
```
(f) CH2═C—CHCH2CH3   (g) CH≡CCHCH3
       |  |                    |
       CH3 CH2CH3              CH3
```

31 and 43.
```
C—C—C—C—C—C—C      n-heptane

C—C—C—C—C—C        2-methylhexane
        |
        C

C—C—C—C—C—C        3-methylhexane
      |
      C

      C
      |
C—C—C—C—C          2,2-dimethylpentane
      |
      C

    C
    |
C—C—C—C—C          3,3-dimethylpentane
    |
    C

C—C—C—C—C          2,3-dimethylpentane
    | |
    C C

C—C—C—C—C          2,4-dimethylpentane
  |   |
  C   C

C—C—C—C—C          3-ethylpentane
    |
    C
    |
    C

    C
    |
C—C—C—C            2,2,3-trimethylbutane
  | |
  C C
```

41. Three straight-chain heptenes, 1-heptene, 2-heptene, and 3-heptene. **45.** *(a)* 2-methylbutane *(b)* 2,2-dimethylpropane *(c)* 2,3,4-trimethylhexane
47. *(a)* 1-hexene *(b)* 2-hexene *(c)* 3-hexene

(d) 3-ethyl-2-methyl-1-pentene **49.** *(a)* 1-pentyne *(b)* 2-pentyne *(c)* 3-methyl-1-butyne *(d)* 5-ethyl-2-methyl-3-heptyne **51.** *(a)* Cyclopropane *(b)* Cyclooctane *(c)* Cyclobutene *(d)* Cyclohexyne *(e)* Methylcyclopentane *(f)* 1-ethyl-4-methylcycloheptane *(g)* 3-methyl-1-propylcyclohexane **56.** Yes

57. 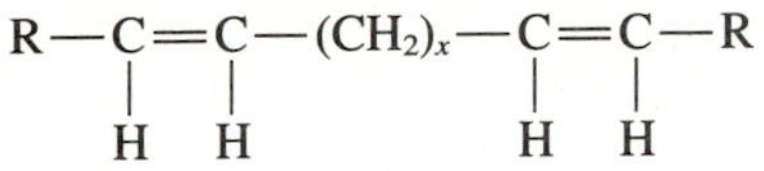 $R-\underset{H}{\underset{|}{C}}=\underset{H}{\underset{|}{C}}-(CH_2)_x-\underset{H}{\underset{|}{C}}=\underset{H}{\underset{|}{C}}-R$

58. Structural isomers **59.** 2,5-dimethyl-3-heptene

60. alkyne, C_nH_{2n-2}

61. 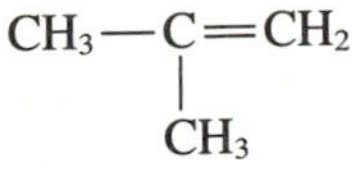 $CH_3-\underset{CH_3}{\underset{|}{C}}=CH_2$

62. 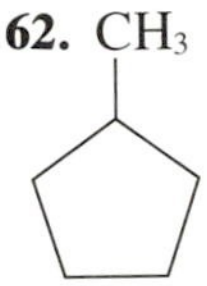 CH_3 (on a cyclopentane ring)

63. 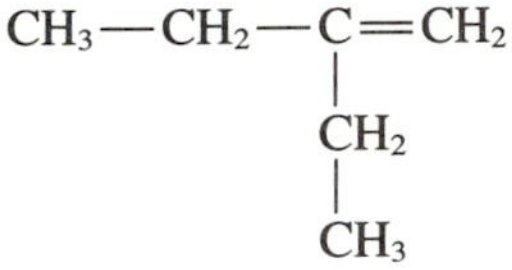 $CH_3-CH_2-\underset{\underset{CH_3}{\underset{|}{CH_2}}}{\underset{|}{C}}=CH_2$

64. 4-ethyl-4-methyl-1-hexene **65.** 1-octene, 2-octene, 3-octene, 4-octene

66. 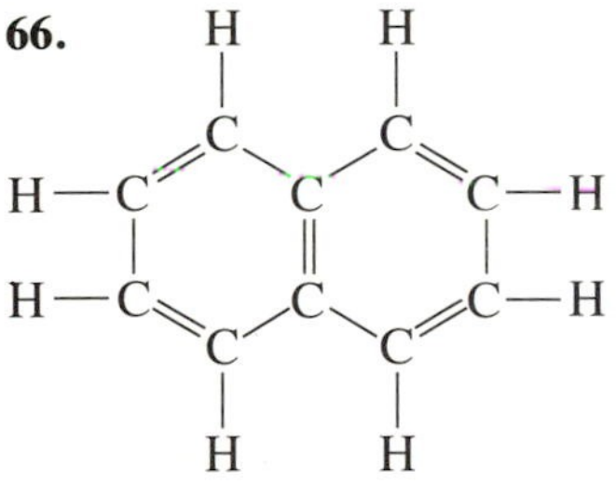 (naphthalene, $C_{10}H_8$: two fused six-membered carbon rings with alternating double bonds, each outer C bearing an H)

```
          H       H
          |       |
          C       C
        //  \   /  \\
  H — C       C       C — H
      |       ||      |
  H — C       C       C — H
        \\  /   \   //
          C       C
          |       |
          H       H
```

Chapter 20

3. *(a)* alcohol *(b)* ether *(c)* aldehyde *(d)* carboxylic acid **7.** *(a)* nonpolar *(b)* polar **9.** *(a)* propanol *(b)* 1-hexanol

13. *(a)* 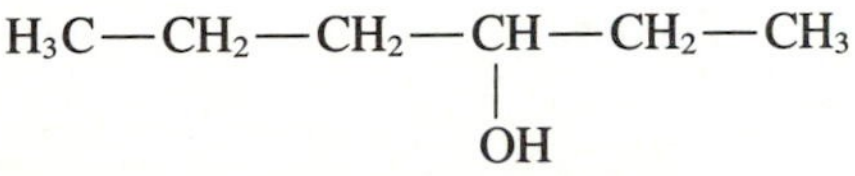 $H_3C-CH_2-CH_2-\underset{OH}{\underset{|}{CH}}-CH_2-CH_3$

(b) $H_3C-CH_2-CH_2-CH_2-OH$

(c) 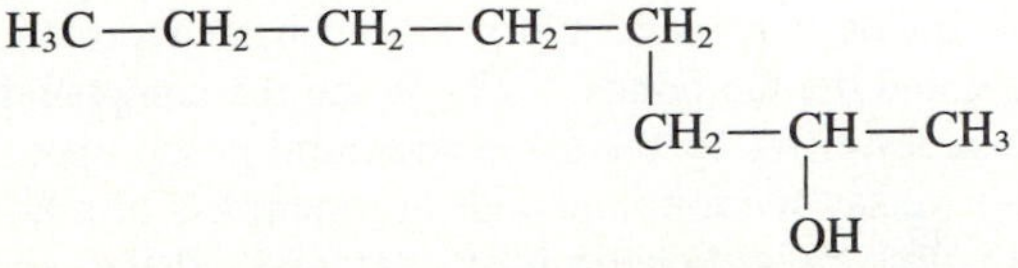 $H_3C-CH_2-CH_2-CH_2-\underset{\underset{CH_2-\underset{OH}{\underset{|}{CH}}-CH_3}{|}}{CH_2}$

(d) CH_3OH **17.** *(a)* butyl ethyl ether *(b)* ethyl methyl ether **19.** *(a)* $CH_3-O-CH_2CH_2CH_2CH_3$ *(b)* CH_3-O-CH_3 **25.** *(a)* hexanal *(b)* propanal

27. *(a)* 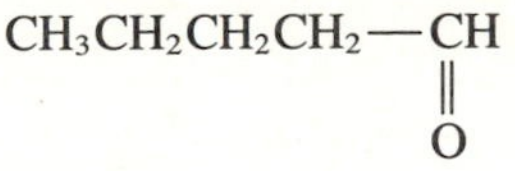 $CH_3CH_2CH_2CH_2-\underset{O}{\underset{\|}{CH}}$ *(b)* 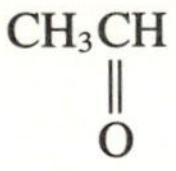 $CH_3\underset{O}{\underset{\|}{CH}}$

33. *(a)* 3-pentanone *(b)* 2-hexanone *(c)* cyclohexanone

35. *(a)* 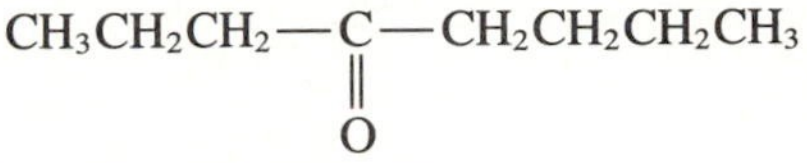 $CH_3CH_2CH_2-\underset{O}{\underset{\|}{C}}-CH_2CH_2CH_2CH_3$

(b) 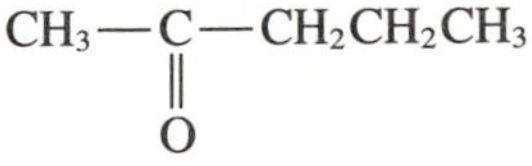 $CH_3-\underset{O}{\underset{\|}{C}}-CH_2CH_2CH_3$

41. *(a)* heptanoic acid *(b)* formic acid (or methanoic acid) *(c)* butanoic acid

43. *(a)* 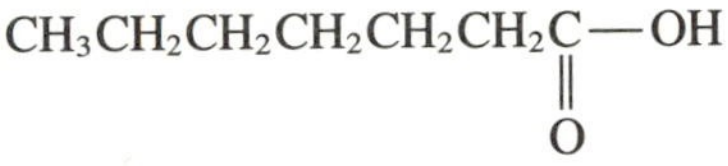 $CH_3CH_2CH_2CH_2CH_2CH_2\underset{O}{\underset{\|}{C}}-OH$

(b) 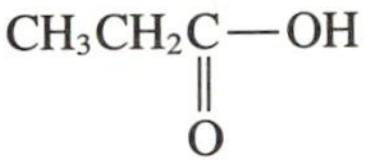 $CH_3CH_2\underset{O}{\underset{\|}{C}}-OH$

49. *(a)* ethylhexanoate *(b)* methylethanoate

51. *(a)* 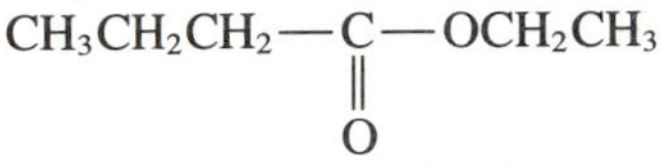 $CH_3CH_2CH_2-\underset{O}{\underset{\|}{C}}-OCH_2CH_3$

(b) 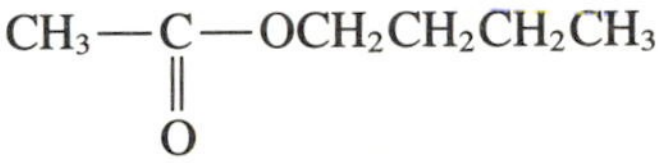 $CH_3-\underset{O}{\underset{\|}{C}}-OCH_2CH_2CH_2CH_3$

55. *(a)* 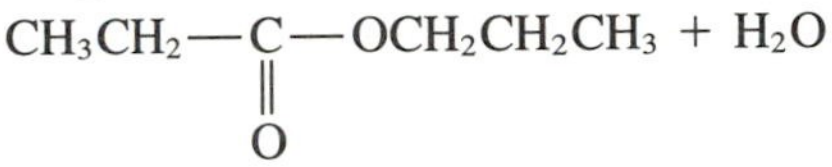 $CH_3CH_2-\underset{O}{\underset{\|}{C}}-OCH_2CH_2CH_3 + H_2O$

(b)  $CH_3-\underset{O}{\underset{\|}{C}}-OCH_3 + H_2O$

57. 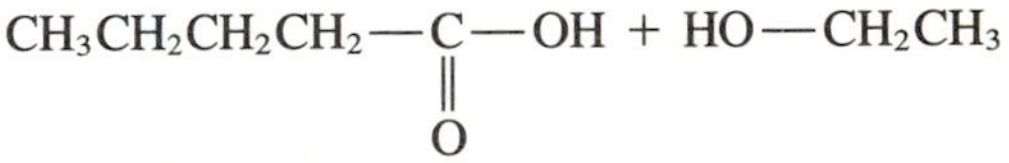 $CH_3CH_2CH_2CH_2-\underset{O}{\underset{\|}{C}}-OH + HO-CH_2CH_3$

61. *(a)* tertiary *(b)* secondary *(c)* secondary *(d)* primary

63. *(a)* trimethylamine *(b)* cyclohexylamine *(c)* cyclopentylethylamine

65. *(a)* 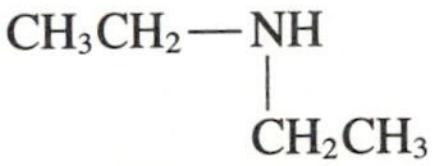 $CH_3CH_2-\underset{CH_2CH_3}{\underset{|}{NH}}$

(b) 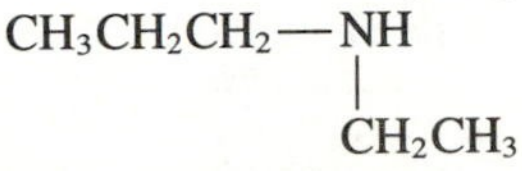 $CH_3CH_2CH_2-\underset{CH_2CH_3}{\underset{|}{NH}}$

(c) 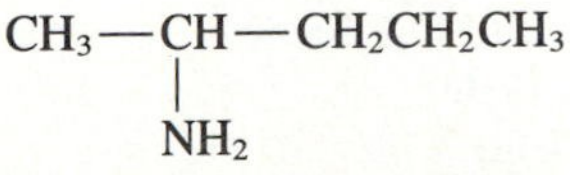 $CH_3-\underset{NH_2}{\underset{|}{CH}}-CH_2CH_2CH_3$

(d) 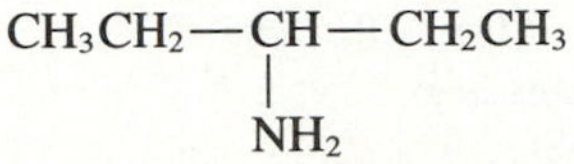 $CH_3CH_2-\underset{NH_2}{\underset{|}{CH}}-CH_2CH_3$

72. 1(C), 2(B), 3(D), 4(G), 5(A), 6(E), 7(F)

73. Cl Cl Cl Cl Cl Cl Cl Cl Cl

74. *(a)* amine *(b)* ether *(c)* carboxylic acid *(d)* ester
75. $CH_3—CH_2—CH—CH_3$ (with OH on the CH)
76. $CH_3—CH_2—CH—CH_2—CH_2—CH_3$ (with NH_2 on the CH)
77. $CH_3—CH_2—CH_2—CH_2—O—CH_2—CH_2$ (with $CH_2—CH_3$ on the last CH_2)
78. Heptanoic acid
79. $CH_3—CH_2—C—H$ (with =O on the C)
80. $CH_3—CH_2—CH_2—C—CH_3$ (with =O on the C)
81. $CH_3—CH_2—CH_2—C—OH$ (with =O on the C)
82. *(a)* ethyl formate *(b)* methyl formate

Cumulative Review/Chapters 18–20

1. Alpha particles are helium nuclei. They carry a double positive charge. Beta particles are identical to an electron in mass and size. They have a negative electrical charge. Gamma rays are made up of energy. They have no mass and carry no electrical charge.
2. ${}^{13}_{7}N$ **3.** ${}^{214}_{84}Po$ **4.** *(a)* ${}^{210}_{83}Bi \longrightarrow {}^{210}_{84}Po + {}^{0}_{-1}e$ *(b)* 1.56 g **5.** 25 mg
6. ${}^{249}_{98}Cf + {}^{18}_{8}O \longrightarrow {}^{263}_{106}Unh + 4\,{}^{1}_{0}n$
7. ${}^{263}_{106}Unh \longrightarrow {}^{259}_{104}Unq + {}^{4}_{2}He$
8. ${}^{209}_{83}Bi + {}^{58}_{26}Fe \longrightarrow {}^{266}_{109}Une + {}^{1}_{0}n$
9. ${}^{266}_{109}Une \longrightarrow {}^{262}_{107}Uns + {}^{4}_{2}He$ **10.** True
11. The balanced equations are:
(a) $Pb + SO_4^{2-} \longrightarrow PbSO_4 + 2e^{1-}$
(b) $PbO_2 + 4H^{1+} + SO_4^{2-} + 2e^{1-} \longrightarrow PbSO_4 + 2H_2O$

12. An electrolytic cell uses electricity to produce a chemical reaction. A voltaic cell produces electrical energy from a chemical reaction. **13.** False **14.** A gaseous air pollutant is composed of gas molecules, whereas a particulate is a tiny particle of dust, smoke, fume, or aerosol. **15.** Sulfur in coal is oxidized to sulfur dioxide, which eventually becomes sulfuric acid. The sulfuric acid remains as a mist in the atmosphere and can mix with moisture in clouds to produce acid rain. **16.** If PVC burns it can produce hydrogen chloride gas, which is poisonous. **17.** Three methods of hazardous waste disposal are incineration, landfilling, and reacting the hazardous waste with chemicals to form harmless products. **18.** PCBs can leach out of plastics and end up free in the environment where they can enter the food chain. **19.** Smog is a mixture of smoke and fog. Photochemical smog is smog that results from atmospheric reactions in which sunlight is present. **20.** False **21.** $\cdot\dot{\underset{\cdot}{C}}\cdot$ **22.** False
23. *Alkane:* (1) methane, CH_4; (2) ethane, C_2H_6; (3) propane, C_3H_8; (4) butane, C_4H_{10}; (5) pentane, C_5H_{12}. *Alkene:* (1) ethene, C_2H_4; (2) propene, C_3H_6; (3) butene, C_4H_8; (4) pentene, C_5H_{10}; (5) hexene, C_6H_{12}. *Alkyne:* (1) ethyne, C_2H_2; (2) propyne, C_3H_4; (3) butyne, C_4H_6; (4) pentyne, C_5H_8; (5) hexyne, C_6H_{10}.
24. *(a)* $—C—C{=}C—C—C—C—$
(b) $—C—C{=}C—C—C—C—$ (with CH_3 above the fifth C, and CH_3 CH_3 below the fourth and fifth C)
25. Isomers
26.

$—C_6H_{13}$	hexyl
$—C_7H_{15}$	heptyl
$—C_8H_{17}$	octyl
$—C_9H_{19}$	nonyl
$—C_{10}H_{21}$	decyl

27. Cyclic hydrocarbons can have single, double, or triple bonds. Aromatic hydrocarbons have alternating single and double bonds. **28.** When the compounds in a series differ by the same structural group (like CH_2) we say these compounds are members of a homologous series. For example, methane, ethane, and propane are all members of the alkane series.
29. Cyclobutane **30.** One, two or three pairs of shared electrons are possible for covalently bonded compounds.

31. C_4H_7I

```
        |    |    |    |
(1)  —C=C—C—C—
        |
        I

        |         |    |
(2)  —C=C—C—C—I
             |    |    |

        |         |    |
(3)  —C=C—C—C—
             |    |    |
             I

        |    |    |    |
(4)  —C=C—C—C—
                  |    |
                  I

        |    |    |    |
(5)  —C—C=C—C—
        |              |
        I

        |         |    |
(6)  —C—C=C—C—
             |         |
             I

             C
             |
(7)  C—C=C
             |
             I

             C
             |
(8)  I—C—C=C

        C
        |
(9)  C—C=C—I

(10) C—C—I
     |    |
     C—C
```

32. A functional group is an atom or group of atoms that determines the specific chemical properties of a class of organic compounds.

33. *(a)*

```
    |    |    |    |    |    |
  —C—C—C—C—C—C—      2-hexanol
    |    |    |    |    |    |
         OH
```

(b)

```
    |    |    |    |    |    |    |    |
  —C—C—C—C—C—C—C—C—
    |    |    |    |    |    |    |    |
              OH
                                      3-octanol
```

34. CH_3CH_2OH: short R group is polar
$CH_3(CH_2)_{10}CH_2OH$: long R group is nonpolar
35. $CH_3CH_2—O—(CH_2)_2—CH_3$: ethyl propyl ether
36. cyclohexanone

37.

```
          O
          ‖
CH3—C—(CH2)5CH3
```

38. *(a)* secondary *(b)* tertiary *(c)* secondary

39.

```
CH3CH2—C—OH + H—O—(CH2)3—CH3
            ‖
            O
```

40.

```
CH3—C—O—CH2CH3   ethylethanoate
       ‖
       O
```

41.

```
CH3CH2CH2—C—H   butanal
                ‖
                O
```

42.

```
       O
       ‖
CH3—C—(CH2)2CH3   methyl propyl ketone
```

Supplement A

1. *(a)* 17 *(b)* -15 *(c)* 15 *(d)* 5 **2.** *(a)* 1 *(b)* 5 *(c)* -5 *(d)* -1 **3.** *(a)* 33 *(b)* -17 *(c)* 17 *(d)* -33 **4.** *(a)* 11 *(b)* 99 *(c)* -99 *(d)* -11 **5.** *(a)* $\frac{1}{16}$ *(b)* $\frac{1}{4}$ *(c)* 512 *(d)* $\frac{1}{32}$ **6.** *(a)* $\frac{6}{35}$ *(b)* $\frac{18}{25}$ *(c)* $\frac{1}{4}$ *(d)* $\frac{25}{49}$ **7.** *(a)* $(4)^{-7}$ *(b)* $(4)^5$ *(c)* $(6)^2$ *(d)* $(5)^1$ **8.** *(a)* $(4)^3$ *(b)* $(4)^8$ *(c)* $(a)^1$ *(d)* $(b)^2$ **9.** *(a)* 625 *(b)* 1000 *(c)* 99 *(d)* $\frac{1}{512}$ *(e)* 16 **10.** *(a)* $(5)^5$ *(b)* $(12)^{12}$ *(c)* $(10)^0 = 1$ *(d)* $(4)^{-10}$ **11.** *(a)* $(4)^2$ *(b)* $(12)^5$ *(c)* $(20)^{-7}$ *(d)* $(20)^{-1}$ **12.** *(a)* $(4)^9$ *(b)* x *(c)* $(y)^7$ *(d)* p *(e)* $(10)^8$ **13.** *(a)* 64 *(b)* $\frac{1}{81}$ *(c)* $\frac{1}{64}$ *(d)* 64 **14.** *(a)* $(10)^3$ *(b)* $(2)^{-5}$ *(c)* $(3)^{-3}$ *(d)* $(6)^2$ **15.** *(a)* $(3)^2$ *(b)* $(2)^3$ *(c)* $(a)^{15}$ *(d)* $(b)^{-13}$ **16.** 48 inches **17.** $w = A/l$ **18.** 6 yards **19.** 48 eggs **20.** *(a)* 6 hours *(b)* 2 hours **21.** 16 pints **22.** 1.47 ft/sec **23.** 6 feet **24.** 180 inches **25.** 300 feet **26.** 180 inches **27.** 24 dozen **28.** 1440 minutes, 86,400 seconds **29.** 10 gallons **30.** 15,000 mi/hr **31.** *(a)* 149.683 *(b)* 141.41 **32.** *(a)* 26.525 *(b)* 103.71 *(c)* 106.14 *(d)* 147.78 **33.** *(a)* 29.32 *(b)* 38.5628 *(c)* 27.0657 **34.** *(a)* 2.04 *(b)* 5 **35.** *(a)* 31.20 *(b)* 46.15 *(c)* 39.80 *(d)* 75.38 **36.** *(a)* $a = 4$ *(b)* $y = \frac{13}{40} = 0.325$ *(c)* $z = 3$ **37.** *(a)* $x = \frac{3}{2}$ *(b)* $y = 1$ **38.** *(a)* $P = \frac{nRT}{V}$ *(b)* $n = \frac{PV}{RT}$ *(c)* $T = \frac{PV}{nR}$ *(d)* $\frac{n}{V} = \frac{P}{RT}$ **39.** *(a)* $t = 16$ *(b)* $y = 7$ *(c)* $x = 8$ *(d)* $k = -8$ **40.** *(a)* $b = \frac{A}{cd}$ *(b)* $c = \frac{A}{bd}$ *(c)* $\frac{A}{b} = cd$ **41.** *(a)* $x = 20$ *(b)* $y = 27$ *(c)* $t = 2$ *(d)* $k = 4$ **42.** *(a)* $k = 1$ *(b)* $h = 8$ *(c)* $f = 16$ *(d)* $y = 4$ **43.** *(a)* $x = 8$, $y = 36$ *(b)* $h = 42$ *(c)* $a = 1.5$, $b = 2.5$, $c = 3.5$ *(d)* $d = 12.5$ **44.** *(a)* $m = DV$ *(b)* $V = \frac{m}{D}$ **45.** 5 g/mL **46.** LA = 33.3%, LT = 16.7%, HT = 50.0% **47.** 800 to Europe, 600 to Asia, 200 to Alaska, 400 to Miami **48.** *(a)* 66 *(b)* 17 *(c)* 6 *(d)* 25 **49.** *(a)* 65 *(b)* 3 *(c)* 1 *(d)* 12.5 **50.** *(a)* 25 *(b)* 1000 *(c)* 0.00375 *(d)* 12.5

Index